SOIL FERTILITY EVALUATION and CONTROL

CHARLES A. BLACK

Professor Emeritus
Department of Agronomy
Iowa State University
Ames, Iowa

LEWIS PUBLISHERS
Boca Raton   Ann Arbor   London   Tokyo

Library of Congress Cataloging-in-Publication Data

Black, C. A. (Charles Allen)
    Soil fertility evaluation and control / Charles A. Black.
        p.   cm.
    Includes bibliographical references (p.   ) and index.
    ISBN 0-87371-834-8
    1. Soil fertility.   2. Fertilizers.   I. Title.
S596.7.B545   1992
631.4'22--dc20   92-25089
                                                                                      CIP

PRINTED IN THE UNITED STATES OF AMERICA
    2 3 4 5 6 7 8 9 0

Printed on acid-free paper

Charles A. Black is Distinguished Professor Emeritus in the Department of Agronomy at Iowa State University. He received B.S. degrees in chemistry and soil science at Colorado State University in 1937, and the M.S. and Ph.D. degrees in soil fertility at Iowa State University in 1938 and 1942.

Dr. Black's professional experience includes soil survey; field and laboratory experimental work in soil fertility and chemistry; teaching undergraduate and graduate courses in soil science, soil fertility, and soil chemistry laboratory; organizing multidisciplinary task forces of scientists to prepare reports on current controversial issues in food and agriculture for members of Congress and the media; and editing and publishing the documents prepared by these scientists. He has been author or coauthor of more than 100 scientific publications.

He has published two editions of a graduate level textbook, *Soil-Plant Relationships;* he has been the American editor of a monograph on *Soils Derived From Volcanic Ash in Japan;* and he has served as editor-in-chief of a two-volume monograph on *Methods of Soil Analysis.*

He has been the president of the Soil Science Society of America and the American Society of Agronomy. He was an organizer of the Council for Agricultural Science and Technology, a consortium of 29 scientific societies in food and agriculture. He has served that organization as president, executive vice president, and executive chairman of the board.

Preface

THIS BOOK DEALS WITH THE SOIL FERTILITY ASPECTS of the most important humanitarian problem of the next few decades — feeding all the people due to arrive. To judge from the past, it is doubtful that new arrivals in many countries will be adequately fed. The problem of producing enough food lies principally in the developing countries. To meet population needs, soil fertility must be greatly increased, and fertilizers must play a key role in the process.

Developed countries generally produce more than enough food for their own populations, and they could produce still more. Their soil fertility problem is one of efficient management, which requires improved precision in evaluating the factors that influence plant responses and in applying the correct treatments in order to obtain the desired results.

In all countries, the problem that must be faced eventually is dwindling resources of most plant nutrients. Efficient use of available resources thus is important not only for the current generation, but also for those to follow.

This book is an outgrowth of a course in advanced soil fertility taught previously by the author to graduate students at Iowa State University, where it follows a course in soil-plant relationships taken by graduate students in soil science and other disciplines. The first course covers important factors influencing soil productivity, emphasizing plant nutrition. In the advanced course, the subject matter was limited to evaluation and control of the supplies of plant nutrients in soils, with emphasis on the theoretical background for the principal practical applications with which soil fertility specialists are concerned. This book follows the same pattern. In contrast to the traditional coverage of the detailed behavior of individual nutrients in soils, emphasis here is on general principles of nutrient-plant behavior, with secondary attention to individual nutrients.

In a subject such as soil fertility evaluation and control, in which the theoretical background at the level of treatment attempted here is still in an evolutionary stage, one cannot confidently stray far from the scientific literature. As a consequence, numerous references are made to the scientific literature, with emphasis on recent publications. The book, however, is not intended to be a literature review. Rather, the literature citations are to give credit for concepts, to provide concrete examples, to provide a broad perspective not always found in the brief publications of the day, and to provide convenient sources of pertinent information for those desiring to read more extensively.

The subject matter of soil fertility evaluation and control does not lend itself to pictorial representation. Graphs are used as a substitute to aid in conveying a mental picture of important concepts. A few tables are included where they seemed more effective than graphs. Although the graphs and tables are credited to the authors who published the data, all graphs have been redrawn; many have been relabeled to fit the context of their use in the book, which is not always the context in which they were published by the authors; some have been constructed from data not shown graphically by the authors; and some have been plotted in ways different than those used by the authors. Apologies are extended to authors who may be offended by these liberties taken in use of their data. This explanation is provided in the preface in preference to extending the text to give the specific details in each instance. Errors are inevitable in a work as extensive as this book, but hopefully the errors in use of the experimental data included in the text are few.

Throughout the book, crop responses are emphasized as the ultimate basis for soil fertility evaluation and control. The first chapter introduces response curves as an object of scientific study. Subsequent chapters make use of response curves for evaluating the supplies of nutrients in soils, the degree of sufficiency of nutrients for crops, the quantities of nutrients needed to meet specified objectives, the value of soil pH alterations by liming, the value of different fertilizer placements, the residual effects of fertilization, the relative value of different sources of a common nutrient, and the economics of fertilization and liming. Environmental aspects of plant nutrients are discussed in general in the chapter on economics of fertilization, and specific aspects are discussed in connection with other topics, but the principal emphasis of the book is on soil nutrient supplies in relation to crop production. Experimental design, which is an important adjunct of soil fertility research, is discussed only tangentially. Researchers generally will need to consult with a statistician in designing experiments.

An understanding of the theoretical background is valuable to every soil fertility specialist, whether a researcher who conducts the basic experiments, an extension worker or farm advisor who interprets research data for growers, or a teacher who instructs the next generation. This book summarizes the theoretical understanding, as developed up to the present, and attempts to explain it in terms appropriate for graduate students in soil fertility.

Special thanks are due the many scientists who supplied information and reviewed portions of the text dealing with their work. Critical reviews by graduate students at Iowa State University who used preliminary drafts of chapters as a text in the current course in advanced soil fertility were most helpful. I am grateful to the Department of Agronomy and the Parks Library at Iowa State University for providing offices and to Dr. Alfred M. Blackmer of the Department of Agronomy, who allowed me to use his computer equipment in preparing the manuscript.

The generosity of the following copyright owners in permitting free use of data, graphs, or text from their publications is gratefully acknowledged: Academic Press, Académie d'Agriculture de France, Académie des Sciences (Paris), *Acta Botanica, Agronomie,* American Agricultural Economics Association, American Society for Horticultural Science, American Society of Agronomy,

American Society of Plant Physiologists, *Annals of Botany,* Association Française pour la Production Fourragère, Association Internationale des Fabricants de Superphosphate, Association of Official Analytical Chemists, *Australian Forest Research,* Australian Institute of Agricultural Science, *Australian Journal of Agricultural Research, Australian Journal of Experimental Agriculture, Australian Journal of Soil Research,* Bureau for Scientific Publications (South Africa), Cambridge University Press, Centre for Agricultural Publishing and Documentation, CSIRO (Australia), Department of Agricultural Technical Services (South Africa), Department of Scientific and Industrial Research (New Zealand), Elsevier Applied Science Publishers B.V., Fertiliser Association of India, The Fertiliser Society (London), Fertilizer Society of South Africa, Fluid Fertilizer Foundation, Food and Agriculture Organization of the United Nations, Gordon and Breach Science Publishers, Harvard Black Rock Forest, Hawaiian Sugar Planters Association, The Controller of Her Britannic Majesty's Stationery Office, Indian Society of Agronomy, Indian Society of Soil Science, Institut für Pflanzenernährung, Institut National de la Recherche Agronomique, International Fertilizer Development Center, International Grassland Society, International Institute for Sugar Beet Research, International Potash Institute, International Society of Soil Science, Iowa State University Press, *Journal of Soil Science* (Blackwell Scientific Publications), Kluwer Academic Publishers, *Madras Agricultural Journal,* Marcel Dekker, Martinus Nijhoff Publishers, Meister Publishing Co., National Fertilizer Development Centre (Pakistan), New Zealand Grassland Association, *Oléagineux,* Oxford University Press, *Physiologia Plantarum,* Punjab Agricultural University, *Revista Brasileira de Ciência do Solo,* Rothamsted Experimental Station, Society of Chemical Industry, *Soil Science* (Williams and Wilkins), Soil Science Society of America, *Soils and Fertilizers* (C.A.B. International), Springer Verlag, Stam Tijdschriften B.V., University of Tennessee, University Press of Virginia, VCH Verlagsgesellschaft mbH, Verlagsbuchhandlung Paul Parey, and Western Co-Operative Fertilizers Limited (Calgary).

Charles A. Black
Iowa State University
August 1992

Contents

CHAPTER 1

Nutrient Supplies and Crop Yields: Response Curves

C HARACTERISTICALLY, A PLOT OF THE YIELD of a crop against the quantity of an essential nutrient added to the soil may be represented by a curve. At first the yield increases. Eventually it reaches a maximum value. In some instances, there is a broad plateau region in which the yield remains about the same over a wide range of quantities of the nutrient. In other instances, the plateau region is very narrow. Beyond the plateau, the yield decreases with increasing additions of the nutrient. The complete curve or any segment, generally beginning with the yield of the control that does not receive the nutrient, is called a response curve because it represents the response of a crop to a treatment applied to the soil.

Response curves are a consequence of the interaction of plants with their environment as one or more nutrients or other factors are varied systematically. Although plants respond to their total environment, the total environment cannot be measured, and sometimes the only measurements made are the crop yields obtained with different quantities of a nutrient applied in a fertilizer. Special measurements on the crop, such as the concentrations of individual nutrients, can provide useful information on the contributions of these specific aspects of environment to crop production, but no analysis or combination of analyses provides the information on the many environmental factors that would permit a comprehensive theoretical integration of the causal factors into a mathematical system.

Experimentally, response curves are discontinuous, being sampled at intervals, with experimental error. Thus, in addition to the inadequacy of information about causal factors and the way they interact in crop production, the precise courses of response curves are in some doubt. The usual reaction to these problems is to represent the trends by smooth curves for purposes of interpolation and interpretation. The curves may be "eye fitted," or they may be mathematical functions. Mathematical functions generally are fitted to the data by the method of least squares. Mathematical functions may be based upon some theoretical concept, which is never comprehensive, or they may be looked upon merely as a convenient fiction for representing the data in summary form. Both types of functions are used and are useful.

This chapter introduces response curves as a basis for evaluating the effects of soil treatments. Response curves play an important role in much of the subject matter to be covered in later chapters. Crop yield responses are emphasized because yields are the most common response measurements, and they are the most widely used criterion of crop production in the commercial world. The chief applications are to fertilizers and fertilization as means of increasing production.

As the subject is addressed here, emphasis is placed upon developing a basis for understanding the relationships between nutrient supplies and crop yields. The mathematical and statistical aspects of fitting functions or models to experimental data are a complex and important part of translating observations and understanding into practical applications, but they are not considered in detail.

The first two sections discuss the Liebig and Mitscherlich concepts of plant response to limiting factors. These two concepts are historically of greatest importance. They provide approximations that are useful in further study of the subject. The third section reviews factors that affect the shape of response curves. The complexity of the relationships explained with examples in this section makes evident the fact that plant responses to limiting factors cannot be represented in a fundamental sense by any simple mathematical device that provides no way to take into account most of the factors involved. Succeeding sections then summarize the limitations of response functions and discuss their use.

1-1. Liebig's "Law of the Minimum"

Several early scientists commented on the general nature of plant responses to nutrient supplies, but Liebig's "law of the minimum" is the most widely quoted. Liebig (1855, pages 23 to 25) stated this "law" in three parts. (1) "By the deficiency or absence of one necessary constituent, all the others being present, the soil is rendered barren for all those crops to the life of which that one constituent is indispensable." (2) "With equal supplies of the atmospheric conditions of the growth of plants, the yields are directly proportional to the mineral nutrients supplied in the manure." (3) "In a soil rich in mineral nutrients, the yield of a field cannot be increased by adding more of the same substances."

For a single deficient nutrient, the law of the minimum usually is interpreted to say that with increasing additions of a fertilizer, the crop yield may be represented as a straight line sloping upward from the control yield to a higher yield at which another factor becomes limiting. At this point, the linear increase ceases, and with further additions of the nutrient, the yield remains substantially constant.

The foregoing is the interpretation history has made of the statements quoted from Liebig's work. Whether he would have agreed with this interpretation seems uncertain, for on page 26 of his book, a little beyond the passage containing the three parts of the law of the minimum, he stated the following as the general proposition from which all the others were derived: ". . . the nutrition, the growth, and the development of the plant depend upon the uptake of certain substances that act by virtue of their composition. This action is within certain limits directly proportional to their mass and inversely proportional to the resistances that impede their action." This statement of relationships is similar to that in the Balmukand "resistance" equation in Section 1-5.2.

Although Liebig's personal interpretation must remain in doubt, the historical interpretation has been useful. The Liebig law was invoked by Blackman (1905) in his analysis of the relationship between photosynthesis and various limiting factors. Blackman's statement of the law was as follows: "When a process is conditioned as to its rapidity by a number of separate factors, the rate of the process is limited by the pace of the slowest factor." In his paper, he referred to the results of an earlier experiment by Reinke ". . . in which with increasing light the rate of assimilation (as measured by the rate of bubbling of *Elodea*) suddenly ceased its proportional increase and remained stationary while the light increased yet another tenfold." According to Blackman's interpretation, carbon dioxide probably became a limiting factor and prevented the rate of assimilation from increasing further when the light was increased beyond a certain intensity.

Blackman's concept, which is in accord with the usual interpretation of the Liebig law, is illustrated in Fig. 1-1. In this figure derived from Blackman's paper, assimilation of carbon dioxide by leaves is represented as a function of carbon dioxide concentration. Blackman evidently considered the rate of carbon dioxide assimilation to be proportional to the concentration of carbon dioxide as long as light was not limiting. When a critical concentration of carbon dioxide was reached, light became a limiting factor, and no further increase in carbon dioxide assimilation took place. When the light intensity was increased to a new value, the carbon dioxide assimilation increased at the same rate as before in proportion to the atmospheric carbon dioxide up to the concentration at which light again acted as a limiting factor and prevented a further increase in the assimilation.

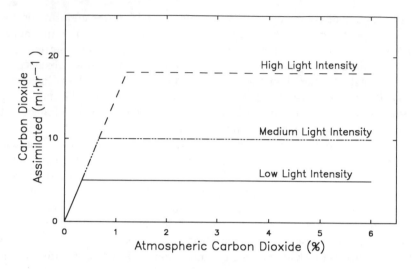

Fig. 1-1. A concept of plant response to limiting factors, illustrated by the assimilation of carbon dioxide by leaves with different concentrations of atmospheric carbon dioxide at low, medium, and high light intensities. (Blackman, 1905)

At present, the behavior illustrated in Fig. 1-1 is sometimes called the "constant-returns" type of response because the increase in yield per unit of the nutrient is constant throughout the portion of the response function represented by the upward-sloping straight line. Some experimental data correspond closely to this behavior. See, for example, a paper by Pinkerton (1991), in which the yield responses of oilseed rape and Indian mustard to phosphorus at three levels of nitrogen in nutrient solutions applied to sand cultures were fitted well by the Blackman stepwise model. The existence of such data has sparked renewed interest in the law of the minimum, as will be noted in Section 1-5.3. Most data are of the "diminishing-returns" type, in which the response to successive equal increments of a nutrient declines. One formulation of this concept is discussed in Section 1-2. Others are found in Sections 1-5.1 and 1-5.2. Portions of some response curves are of the "increasing returns" type. That is, the magnitude of the response to succeeding equal increments of the nutrient increases. Possible causes of increasing-returns responses are discussed in Section 1-3.2.

1-2. Mitscherlich's "Law of Physiological Relationships" or "Effect Law of the Growth Factors"

1-2.1. Development of the Mitscherlich Equation

Four years after the publication of Blackman's paper, Mitscherlich published the first in a long series of papers presenting a different concept of the effect

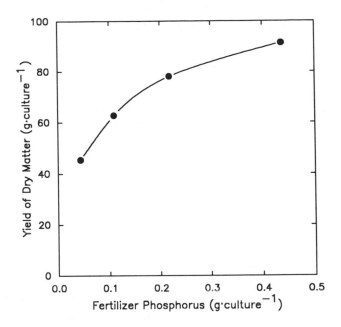

Fig. 1-2. Response of oat plants to different quantities of phosphorus supplied as dibasic calcium phosphate in sand cultures. (Mitscherlich, 1909)

of limiting factors (Mitscherlich, 1909).[1] In his first work, he grew oat plants in sand cultures treated with different quantities of phosphorus as dibasic calcium phosphate. (See Appendix A for scientific names of plants referred to in the text by their common names.) The results are shown in Fig. 1-2.

The data in the figure indicate that the oat yield was not a simple straight-line function of the quantity of fertilizer phosphorus supplied, as would be inferred from Blackman's concept and Fig. 1-1. Rather, the rate of increase in yield, which is given by the tangent of the yield curve, becomes smaller with each succeeding increment of fertilizer phosphorus. Similar results were obtained with monobasic calcium phosphate as the source of phosphorus. The rate of increase of yield is relatively small at the point corresponding to the greatest application of fertilizer phosphorus, so that the curve seemingly is approaching a maximum beyond which additional quantities of phosphorus would produce no further increase in yield. This state of affairs suggested to Mitscherlich the concept that the rate of increase in yield produced by adding fertilizer phosphorus is proportional to the decrement from the maximum yield.

[1]Spillman (1933) published a classic treatise on an equation he had developed independently of Mitscherlich, but which turned out to be a different mathematical form of the Mitscherlich equation. Spillman emphasized the mechanics of fitting the response function to experimental data by the method of least squares.

Mitscherlich's concept is expressed mathematically as

$$\frac{dy}{dx_1} = c_1(A - y).$$

Integrating, evaluating the constant of integration, changing to logarithms to the base 10, and separating the total quantity x_1 of the nutrient into its component parts x and b,

$$\mathrm{Log}(A - y) = \mathrm{Log}A - c(x + b)$$

or

$$y = A(1 - 10^{-c(x+b)}).$$

In these equations, y is the yield with total nutrient quantity x_1 or with nutrient quantity x added plus b, the effective quantity in soil and seed expressed in the same units as x (see Fig. 4-1 for a graphic representation of b); A is the maximum yield attainable as x increases indefinitely; and c is the proportionality factor.[2]

---

[2]Mitscherlich used the equation in the logarithmic form and designated the parameters with the letters A, b, and c as shown here. His conventions will be used in this book, and the equation will be called the Mitscherlich equation. Many authors call it the exponential equation, and they represent it in a variety of mathematical forms and use a variety of symbols for the parameters. Some exponential equations differ from the Mitscherlich equation, and authors usually call them modified exponential equations or functions.

For the possible benefit of those who may wish to follow the mathematics involved in proceeding from the differential equation to the integrated form, the differential equation may be rewritten as

$$\int \frac{dy}{(A - y)} = \int c_1 dx_1, \tag{1}$$

and also as

$$-\int \frac{d(A - y)}{(A - y)} = c_1 \int dx_1, \tag{2}$$

which may be integrated to

$$-Ln(A - y) = c_1 x_1 + K, \tag{3}$$

where K is the constant of integration. When $x_1 = 0$, $y = 0$, and

$$-LnA = K. \tag{4}$$

Thus,

$$-Ln(A - y) = c_1 x_1 - LnA \tag{5}$$

The fundamental point Mitscherlich (1909) made in his original paper was that his response data were well fitted mathematically by curves. He did not comment on or refer to the angular representation of response used by Blackman (Fig. 1-1). The ''law of physiological relationships'' was a descriptive term Mitscherlich used later, as his ideas developed about the constancy of the parameter c in the equation he used to represent plant responses.

1-2.2. The Proportionality Constant c

1-2.2.1. Significance of c

Mitscherlich called the proportionality constant c the *Wirkungsfaktor*, which may be translated as *effect factor* or *efficiency factor*. The significance of c may be inferred from Fig. 1-3, which illustrates the relationship between the shape of the response curve and the value of c, where the y intercept is 20 and A is 100. With high values of c, the yield approaches A rapidly, and with low values, the yield approaches A slowly. Practical use of this property will be made later in connection with fertilizer evaluation and prediction of yield response to fertilization.

---

or

$$Ln(A - y) = LnA - c_1x_1. \tag{6}$$

Taking the antilogarithm of equation (6) yields

$$y = A(1 - e^{-c_1x_1}). \tag{7}$$

Changing equation (6) to logarithms to the base 10,

$$Log(A - y) = LogA - cx_1, \tag{8}$$

where $c = 0.4343c_1$. Taking the antilogarithm of equation (8) yields

$$y = A(1 - 10^{-cx_1}). \tag{9}$$

The total quantity x_1 of the nutrient available to the plants may now be divided into its component parts, x and b, where x is the quantity of the nutrient added in the fertilizer, and b is the effective quantity of the nutrient supplied by the soil and seed. The effective quantity b is measured in the same units as x. Equations (8) and (9) thus become

$$Log(A - y) = LogA - c(x + b) \tag{10}$$

and

$$y = A(1 - 10^{-c(x+b)}). \tag{11}$$

Equations (6) and (7) may be modified in like manner.

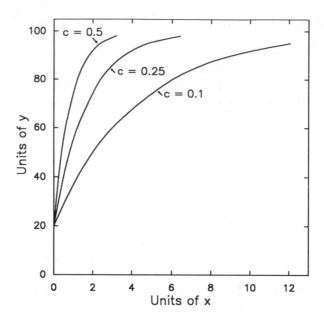

Fig. 1-3. Hypothetical response curves showing the effect of the magnitude of the *c* value in the Mitscherlich equation under circumstances in which the yield without fertilization is 20 and the maximum yield *A* is 100.

1-2.2.2. Implications of the Constancy of c

In his original paper, Mitscherlich was not particularly concerned about the behavior of the effect-factor *c*. As he conducted more experimental work and developed his ideas further, however, he came to the conclusion that with the exception of a few situations (see Mitscherlich et al., 1923), there is a characteristic and constant effect-factor *c* for each growth factor that is independent of the nature of the crop and the experimental conditions.

With the development of the concept that each growth factor has a characteristic and constant effect-factor *c*, the point became clearer, especially to more recent writers, that Mitscherlich's concept was fundamentally different from Blackman's representation of the Liebig law in Fig. 1-1. Whereas according to Blackman the factors other than the one being supplied in variable quantity limit the yield only at the maximum, the Mitscherlich equation with constancy of *c* implies that these factors limit the yield throughout the entire yield range produced by adding different quantities of the variable factor. In the words of Sumner and Farina (1986), Mitscherlich's law of physiological relationships states that "yield can be increased by each single growth factor even when it is not present in the minimum so long as it is not present in the optimum."

As will be explained in Section 1-2.2.5, the evidence indicates that values of *c* for the various nutrients are not constants, as supposed by Mitscherlich. At this point, however, it is important to develop an understanding of the implications of the constancy of c. The approximation that c values are constant is widely used in soil fertility work, in some instances without full appreciation for the origin and meaning of the usage.

Constancy of the values of c for the plant nutrients leads to two important inferences that may be explained in connection with Fig. 1-4. The first is that if nutrient 1 is applied in increasing quantities at each of two levels of nutrient 2, resulting in two response curves, the ratio of the yields (y values) on the two response curves is the same for all quantities of nutrient 1. Fig. 1-4 is drawn in accordance with this situation. For example, the y values on the next to lowest curve produced by a constant addition of nutrient 2 are twice those on the lowest curve. Thus in Fig. 1-4,

$$\frac{y'_1}{y_1} = \frac{y'_2}{y_2} = 2.$$

If a range of quantities of nutrient 2 is applied at each quantity of nutrient 1, the result is a series of response curves with successively higher yields. With equal increments of nutrient 2, the curves are located closer together as the maximum yield obtainable with nutrient 2 is approached, but there is always a

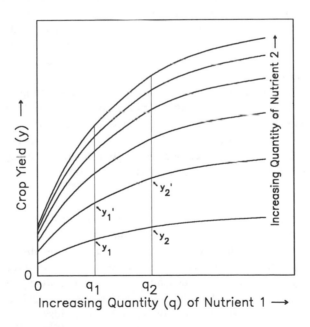

Fig. 1-4. Hypothetical response curves showing crop yields versus the quantity of nutrient 1 at increasing quantities of nutrient 2 where the value of c in the Mitscherlich equation is constant for each nutrient.

constant yield ratio between any two curves at equal quantities of nutrient 1, whatever these quantities may be. All the curves can be made to coincide if the yields on each curve are multiplied by a constant factor that is appropriate for that curve.

The second inference is that the ratio of yields obtained with any two specific quantities of nutrient 1 is the same on all the response curves obtained with different levels of nutrient 2. Thus in Fig. 1-4

$$\frac{y_2'}{y_1'} = \frac{y_2}{y_1} \quad .$$

The numerical value of the ratio will depend upon the magnitude of the difference between q_2 and q_1 as well as the value of q_1 (that is, the segment of the response curves under consideration). The absolute increase in yield obtained from the application of the nutrient increment $q_2 - q_1$, however, increases with the level of nutrient 2 and reaches a maximum at the level of nutrient 2 that leads to the maximum yield.

1-2.2.3. Relative Yields

If the values of c in the Mitscherlich equation are constant for the individual plant growth factors, the curves representing crop response to one factor at different levels of other factors can be made to coincide by expressing the yields in the various experiments or experimental series as decimal fractions or percentages of the respective maximum yields or A values, that is, as relative yields. This behavior was mentioned in the preceding section in connection with the first implication of the constancy of c.

Relative yields commonly are used as a means of representing the results of field experiments with different quantities of a given fertilizer. The combined results often can be expressed as a single function with little or no obvious segregation of results for the individual experiments or experimental series.

Colwell et al. (1988), however, were strongly critical of the use of relative yields. They summarized their concerns in five objections, which are given here because of their significance:

Objection 1. Relative yields do not provide a basis for estimating economic fertilizer rates. Farmers derive their income from absolute crop yields and must justify their expenditures for fertilizers on the basis of economic returns from absolute increases in yields. Relative yields of course can be converted to absolute yields, but the absolute yields corresponding to a given relative yield may vary widely. Thus, the problem of dealing with differences in yield levels among experiments is only postponed by calculating relative yields.

Objection 2. The maximum yield A, which is used for calculating relative yields, is often poorly defined by experimental data and often must be estimated subjectively.

Objection 3.   The functional relationship between yield and quantity of fertilizer or soil analyses differs from one site to another. The yields associated with a given value on the X axis are not simply proportional to the corresponding maximum yields or A values.

Objection 4.   Calculating relative yields can produce statistical bias. This problem arises because dividing the yields in the various experiments or experimental series by their respective A values in effect weights the error deviations by the reciprocals of the maximum yields. The bias can affect the overall regression line calculated for the relative yields.

Objection 5.   Tests of significance of regressions are invalid where the results of different experiments are combined by the relative yield device. The reason is that the deviations from the overall regression line contain different types of error — some of it within experiments and some of it among experiments.

These points about relative yields to which Colwell et al. (1988) call attention are valid and worth bearing in mind. The first three objections deal with the kind of subject matter treated in this chapter. The last two are more strictly statistical concerns.

The matter of postponing consideration of the issue of absolute yields mentioned in the first objection is certainly true in the many instances in which researchers do not proceed beyond the relative yield stage to make use of the information. On the other hand, representing the values as relative yields can be an effective aid in developing a general relationship for use in dealing with absolute yields in specific situations. Such a situation is described in Section 3-4.2.7.

Not mentioned by the authors in their fourth objection is the point that if the absolute yields vary widely in the various experiments or experimental series represented by relative yields, the original variances may be inhomogeneous, and the homogeneity may be improved by expressing the results as relative yields. Thus, if an overall analysis is to be made of the results of several experiments with inhomogeneous variance, statistical bias exists if no allowance is made for the inhomogeneity of the variances of the untransformed data, and bias of another sort exists if the values are transformed to cause the variances to become homogeneous.

Transforming data by calculating relative yields is frowned upon by statisticians for reasons given by Colwell et al. (1988). On the other hand, statisticians probably are less concerned about the bias introduced by data transformations they consider acceptable than are soil scientists.

The logarithmic transformation is a common one. A few soil scientists have transformed their crop yield data or fertilizer quantities by use of a logarithmic scale, but most seem reluctant to do so. In part, the reluctance to use a logarithmic transformation may result from concern for the fact that logarithmic transformations can affect the shapes of response curves. For example, the sigmoid response curves that are of both theoretical and practical concern to soil scientists (see Section 1-3.2) tend to lose their sigmoid shape if the yield data are plotted on a logarithmic scale.

In part, however, the reluctance to use a logarithmic transformation may be born of habit, because soil scientists are not offended by the logarithmic transformation of hydrogen-ion activities to pH values, nor are they offended by the practice of averaging pH values instead of averaging hydrogen-ion activities. If pressed, soil scientists justify these conventions on the basis that the actual measurement made by the glass electrode pH meter is a voltage, and the hydrogen-ion activity is theoretically logarithmically related to this voltage. Experimentally, the pH value of a composite soil sample is estimated more closely by the mean of the pH values of the individual samples than by the mean of the hydrogen-ion activities that has been converted back to pH values (Baker et al., 1981). Moreover, the relationships between pH values and biological effects are generally more nearly linear than those between hydrogen-ion activities and biological effects.

The responses to fertilizers under different environmental conditions are often more multiplicative than additive (expressed mathematically by the Baule generalization of the Mitscherlich equation in Section 1-2.2.4), a property made use of when relative yields are calculated. This behavior reinforces an argument of some statisticians that part of the heterogeneous variance in fertilizer response data could be eliminated by the use of a logarithmic scale for crop yields instead of the usual linear scale.

The logarithmic transformation does not have the visually pleasing result of the relative yield transformation in bringing data from different experimental series together to approximate a single curve. Rather, for data in which the Baule generalization of the Mitscherlich equation applies, it results in a series of response curves in which the vertical distance between two successive curves is independent of the quantity of fertilizer.

Because of the diversity of situations that must be dealt with in soil fertility work, researchers are always seeking ways to correlate data into a unified framework. The relative yield device has been useful in this regard. Part of the seeming usefulness, however, is a consequence of the usual inadequacy of field experiments. Most field experiments with fertilizers do not include enough different quantities of fertilizer or enough replications of the treatments to define response curves with the desired or expected degree of precision. Additionally, too little is generally known about the effects of other factors, such as temperature and rainfall distribution, pest damage, and cultivar effects, to take these factors into account in a quantitative way in interpreting the results of a given experiment for specific situations other than those in the experiment even if precise response curves were available.

Under these circumstances, an assembly of data by the relative yield device may be viewed as the result of a single experiment for a general situation. A statistical test of significance may be invalid, as pointed out by Colwell et al. (1988), but a test of significance can readily be dispensed with. The basis for action can always be improved by further research, but the data at hand are the best available at the moment.

Section 1-2.2.5 includes examples of experimental data in which the relative yield transformation worked well and others in which it did not. Succeeding chapters include examples in which the relative yield concept was used to advantage. Bolland et al. (1989) published data from phosphorus fertilization experiments in southwestern Australia in which the transformation generally made some improvement in bringing groups of response curves together, but still was unsatisfactory.

1-2.2.4. Baule's Concepts

Baule (1918) introduced the concept of the *effect quantity*, which is the quantity of the nutrient or other growth factor necessary to produce half the maximum yield. This quantity is known now as a *Baule unit*. When one Baule unit of a nutrient is present, $x + b = 1$ and $y = 0.5A$, so that

$$\text{Log}(A - 0.5A) = \text{Log}A - c,$$

and

$$\text{Log}\frac{A}{0.5A} = \text{Log}2 = 0.301 = c.$$

According to this generalization, the value of c for all growth factors may be represented by 0.301 if the supplies of the various factors are expressed in Baule units. The Baule-unit concept has the advantage that if the responses are given by the Mitscherlich equation, the yields form a geometric series of $0.5A$, $0.75A$, $0.875A$, . . ., A with 1, 2, 3, . . ., ∞ Baule units of the growth factor (see Fig. 1-5).

Similarly, if a soil has the inherent supply b of a nutrient that produces the control yield y_0, addition of one Baule unit of a nutrient will increase the yield by 50% of the difference between y_0 and the maximum yield A. The next Baule unit will produce an additional increase in yield equal to half of that produced by the first Baule unit, and so on.

The concept of a regular progression of diminishing responses with succeeding equal increments of fertilizer appeared explicitly in the Spillman (1933) form of the Mitscherlich equation,

$$Y = M - AR^x,$$

where Y is the yield obtained when x units of the growth factor are applied in the fertilizer, M is the maximum yield approached as x increases indefinitely, A is the theoretical maximum increase in yield approached as x increases indefinitely, and R is the ratio of the increase in yield produced by a given increment

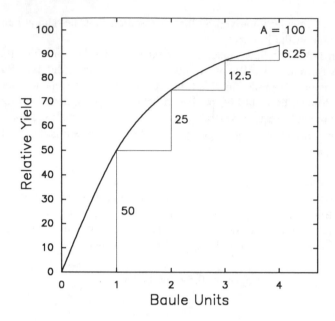

Fig. 1-5. A hypothetical response curve with construction illustrating the concept that if additions of a nutrient are made in multiples of the quantity that produces 50% of the maximum yield (one Baule unit), the increase in yield produced by each successive increment is half of that produced by the preceding increment. (Baule, 1918)

of x to the increase in yield produced by the preceding equal increment. That is, R is the ratio of a decreasing geometric series. The value of R depends upon the size of the increment of x. For the special case in which the increments of x are Baule units, $R = 0.5$. (Note that the Spillman equation deals with increases in yield from fertilization and does not include a term for the effective quantity of the nutrient in the soil.)

Baule (1918) also generalized Mitscherlich's equation to represent the response of plants to all growth factors. Baule's equation represents the state of affairs if plants respond to each growth factor in the manner specified by the Mitscherlich equation, and if the effect factor or efficiency factor c is constant for each growth factor. Baule's general equation is

$$y = A_{max}(1 - e^{-c_1 x_1})(1 - e^{-c_2 x_2}) \ldots (1 - e^{-c_n x_n}),$$

where $x_1, x_2, \ldots, x_n$ are the available quantities of the various growth factors, where $c_1, c_2, \ldots, c_n$ are the efficiency factors of the respective growth factors, and where e is the base of natural logarithms. If each of the values of x in the equation is infinite, then each of the values of e^{-cx} is equal to zero, and the observed yield y is equal to A_{max}, the absolute maximum yield, which is determined by the genetic nature of the plant. But if all the growth factors are at a level sufficient to produce the maximum yield except, for example, phosphorus, which is at 0.9 of that level, the yield y will be $0.9A_{max}$. If phosphorus is 0.9 sufficient and potassium is 0.5 sufficient, the yield y will be $0.9 \cdot 0.5A_{max} = 0.45A_{max}$.

Baule's concepts often have been referenced and sometimes criticized, but they have not been given much explicit attention as a basis for practice. Middleton (1984) incorporated the Baule unit concept in a theoretical approach he developed for extending advice on economic applications of fertilizers to practical conditions. As empirical justification, he presented a graph of crop response data from various published sources showing close correspondence of the experimental data to the line representing the Baule equation.

Baule's multiplicative concept was used more recently by Johnson (1991) in a different framework to represent the joint effects of phosphorus and potassium. For the response to individual factors, however, he used a quadratic form rather than the exponential response curve employed by Mitscherlich and Baule. Johnson's formulation made provision for use of soil test values to estimate the degree of nutrient sufficiency and the fertilizer requirement. For example, extractable soil phosphorus of 10 milligrams per kilogram by the method used corresponded to 80% sufficiency of phosphorus and a requirement of 20 kilograms of fertilizer phosphorus per hectare irrespective of the yield level. To identify the nitrogen requirement, the yield goal or target yield concept described in Sections 4-4.6.4 and 4-4.7 was used.

Arthur and Garn Wallace discussed the nature of crop responses to limiting factors in a series of papers constituting Numbers 3 and 4 of Volume 13 of the *Journal of Plant Nutrition*. The concepts are described in an introduction (Wallace and Wallace, 1990) and the first two papers (Wallace, 1990a,b). The other papers are examples, some representing original data and others based upon data from the scientific literature.

According to the Wallace concept, there are two basic types of responses to limiting factors: the Liebig type and the Mitscherlich type. In the Liebig type (corresponding to the first part of the Liebig law in Section 1-1), one or more nutrients or other growth factors are so severely deficient that little response to additions of other nutrients or to improvement in other growth factors will occur until the most severe limitations have been removed.

The Wallaces denoted the responses obtained under conditions of more mild deficiencies as Mitscherlich-type responses, particularly for those limiting factors that when present individually in quantities producing less than the maximum yield result in relative yields that can be multiplied to provide an estimate of the relative yield obtained when the factors are supplied together. This concept is an application of the multiplicative principle in the equation representing Baule's generalization of the Mitscherlich equation based upon the constancy of the c values for the various growth factors.

Perhaps less obviously, the Baule multiplicative principle provides a way to calculate the response to a joint application of given quantities of each of two or more plant growth factors when the responses to the factors supplied singly are known. From Fig. 1-4,

$$\frac{y_2'}{y_1'} = \frac{y_2}{y_1} \quad .$$

Thus y_2', which is the yield obtained from joint application of the quantity $q_2 - q_1$ of nutrient 1 and the quantity of nutrient 2 that is responsible for raising the yield level from the lowest response curve to the next to lowest curve, is given by

$$y_2' = \frac{y_1' \cdot y_2}{y_1} \quad .$$

For example, if $y_1 = 30$, $y_1' = 60$, and $y_2 = 40$, the foregoing relation gives a value of 80 for y_2'.

Wallace (1990b) multiplied the relative responses to the nutrients applied individually to obtain the relative response to joint application. Using the same numerical values as before, $(y_1'/y_1)(y_2/y_1) = (60/30)(40/30) = 2.667$, which is equal to the ratio of 80/30 where absolute yields are used. Wallace (1990b) gave various numerical instances of this behavior from published experimental data. As one example, with relative yields of 1.00 for the control and 1.20 and 1.41 for phosphorus and nitrogen applied individually, the estimated relative yield when they were applied jointly would be $1.20 \times 1.41 = 1.69$. The observed relative yield from joint application was 1.75, which is close to the estimated value.

Wallace (1990a) gave other examples of joint responses that exceeded the product of the individual responses, and he interpreted these as being a manifestation of Liebig-type limiting factors. He emphasized the practical importance of first eliminating Liebig-type limiting factors because so little benefit could be derived from other efforts to increase yields as long as any such marked deficiencies remain.

In the discussion thus far, the multiplicative concept in the Baule equation has been considered in terms of the interaction of only two nutrients. In theory, however, it applies to all nutrients and to other factors as well. The theory does not seem to have been tested explicitly to determine the effects of joint deficiencies of a number of nutrients. It probably would apply less well to nutrient solutions than to soils because plants are capable of absorbing virtually all of all nutrients from a nutrient solution. Thus, if a nutrient solution contains enough of each of six nutrients to produce a relative yield of 0.875 or 87.5% (three Baule units), why should the relative yield be only $0.875 \times 0.875 \times 0.875 \times 0.875 \times 0.875 \times 0.875 = 0.45$? Fig. 1-6 shows that according to the multiplicative principle, 3 Baule units of each of six nutrients would produce a relative

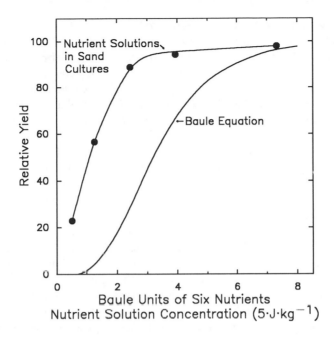

Fig. 1-6. The percentage of the maximum yield as calculated from the Baule generalization of the Mitscherlich equation for the situation in which a mixture of six nutrients is added in different quantities, together with the relative yields of barley plants grown in sand cultures supplied with a solution containing six nutrients (nitrogen, phosphorus, potassium, calcium, magnesium, and sulfur) in different concentrations. The nutrient solution data were obtained experimentally by Hoagland. (Baule, 1918; Hoagland, 1919)

yield of 0.45 or 45%, and 8 Baule units would produce a relative yield of approximately 0.97 or 97%. In an experiment with plants grown in nutrient solutions at different concentrations, in which the maximum concentration used resulted in a relative yield of about 0.97, reducing the concentration of six nutrients to 3/8 of that producing a relative yield of about 0.97 did not have the drastic effect on yield predicted by the multiplicative concept expressed in the Baule equation. The relative yield was about twice that predicted by the Baule equation. In this experiment, the number of Baule units of nutrients corresponding to the relative yield of about 0.97 was unknown. Better evidence from experimental work designed specifically for the purpose is needed.

The multiplicative effect of the nutrients that would be a consequence of the constancy of the c values for various nutrients in the Mitscherlich equation is mathematically simple, and it often provides a useful approximation. Nonetheless, the biological behavior is complex. As explained in the next section, c values are not constant, and one cannot be confident that effects will be multiplicative.

1-2.2.5. Evidence on the Constancy of c

Most of the controversy over the Mitscherlich equation as a means of representing plant responses has revolved around the alleged constancy of c. Whether or not the value of c for each growth factor may be regarded as constant is important, not only because it involves the validity of the theoretical concepts proposed by Mitscherlich and Baule, but also because the inferences that can be made from experimental data are greater if c is constant than if it varies from one set of conditions to the next.

Verification of Mitscherlich's hypothesized constancy of c is provided by a greenhouse experiment conducted by Lange (1938), in which the yield of oat plants was determined with different additions of dibasic calcium phosphate and water to sand cultures. The results, shown in Fig. 1-7, provide a particularly striking verification of the constancy of c because the range in absolute yields is so great. Actually, Lange (1938) used slightly different c values for calculating the curves in Fig. 1-7A, but the relative yields for the different water levels in Fig. 1-7B could not have been represented to such a good approximation by a single curve if the c values had been widely different.

The most extensive verification brought forward by Mitscherlich for the hypothesized constancy of c is found in a summary of the data from some 27,000 field experiments in Germany in which different crops were grown with graded applications of fertilizer nitrogen, phosphorus, and potassium. Mitscherlich averaged the yields obtained in the various experiments with each crop and evaluated the constants in the Mitscherlich equation from the resulting response curves. He found that, for all crops, the observed and calculated yields agreed closely if the effect factor c for each nutrient was assumed to be constant and equal to the value he had been using for a number of years. Fig. 1-8 illustrates the results obtained with different quantities of fertilizer phosphorus.

The data on responses to fertilizer phosphorus in the same summary of German field experiments published by Mitscherlich (1947) as evidence for the constancy of c were cited by Van der Paauw (1952) as evidence for the inconstancy of c. He found that whereas Mitscherlich's value for c was 0.6, the best-fitting values of c calculated by the method of least squares ranged from less than 0.74 for clover to 3.36 for barley (in the same units used by Mitscherlich). With the best-fitting value of c, the coefficient of variation of yields of potato was much reduced, but with the other crops there was no great reduction. The relative insensitivity of the coefficient of variation to arbitrary changes in the value of c results from the fact that three parameters (A, b, and c) are involved in fitting the equation. If one parameter is fixed, the other two change in such a way as to give the best fit for the particular value chosen for the fixed parameter. The adaptability permitted by the uncontrolled variation of A and b, together with the generally large experimental errors in yield data, account in large part for Mitscherlich's success in obtaining a reasonably good fit with diverse sets of data when c for a particular nutrient was assigned a constant value.

For the data shown in Fig. 1-8 and the many other results from the same summary of more than 27,000 field experiments, another reason that can be

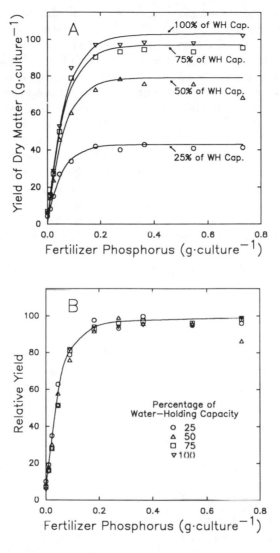

Fig. 1-7. A. Yields of oat plants in sand cultures with different additions of dibasic calcium phosphate at four levels of water supply. The curves have been plotted from Mitscherlich equations given in the original article. B. Relative yields calculated from data in A. (Lange, 1938)

given for the good fit of observed responses to the values calculated on the basis of a constant value of c for each nutrient is the limited average response to fertilization. With limited responses, the curvature of the response curves is small. Under these circumstances, the form of the response function becomes less critical. For example, straight lines would have given almost as good a fit as the Mitscherlich equation for many of the data.

Further evidence for the inconstancy of c in the Mitscherlich equation was obtained by Rippel et al. (1926/1927). They grew oat in sand cultures with different additions of potassium sulfate at two levels of ammonium nitrate and obtained the data shown in Fig. 1-9. In contrast to Fig. 1-7B, in which all the individual response curves could be represented to a good approximation by a single curve when the yields were expressed as a percentage of the maximum, the two response curves in Fig 1-9B were quite distinct, a behavior that indicates different values of c. Similar results were obtained by Meyer and Storck (1927/1928) in an experiment in which sand cultures were treated with increasing quantities of fertilizer nitrogen at two levels of phosphorus.

The marked difference in responses between Figs. 1-7 and 1-9 may be explained rather simply. In the experiment in Fig. 1-7, ample supplies of nutrients other than phosphorus were supplied to all cultures, so that these nutrients did not become deficient when the yield was increased by adding phosphorus and increasing the water level. Water deficiencies were not created by adding phosphorus because the cultures were watered continually to return the water supply to the predetermined level. Phosphorus deficiencies were not created by increas-

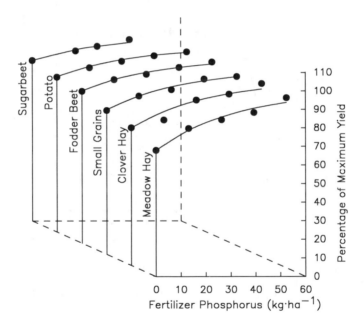

Fig. 1-8. The percentages of the maximum yields of crops in field experiments in Germany with different additions of fertilizer phosphorus (data points), and the yields as calculated from the Mitscherlich equation (solid lines) with the c value of 1.37 hectares per 100 kilograms or per 0.1 megagram, which was Mitscherlich's standard c value for phosphorus. The results shown represent 1,535 experiments with meadow hay, 105 experiments with clover hay, 3,045 experiments with small grains (rye, oat, wheat, and barley), 600 experiments with fodder beet, 1,642 experiments with potato, and 238 experiments with sugarbeet. (Mitscherlich, 1947)

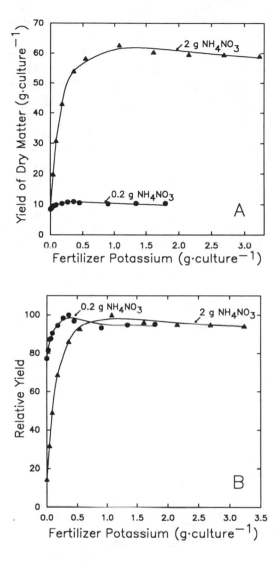

Fig. 1-9. A. Yields of oat plants in sand cultures with different additions of potassium sulfate and two quantities of ammonium nitrate. B. Relative yields calculated from data in A. (Rippel et al., 1926/1927)

ing the water level because the phosphorus was added as dibasic calcium phosphate, which is a sparingly soluble substance that was distributed throughout the cultures as solid particles. Thus, as the water supply was increased and more phosphorus was absorbed by the plants, more of the dibasic calcium phosphate dissolved, replacing that used.

In the experiment in Fig. 1-9, nutrients other than nitrogen and potassium were added in ample amounts to all cultures. The potassium supply in the sand

and seed was enough to produce a low yield with no added potassium. In the series receiving 0.2 gram of ammonium nitrate per culture, the nitrogen supply was exhausted with addition of only a little fertilizer potassium, and so the maximum yield attainable by potassium fertilization was low. In the series receiving 2 grams of ammonium nitrate per culture, much more fertilizer potassium was required to exhaust the nitrogen supply. The maximum yield was much higher because of the extra nitrogen.

The results of a series of cooperative experiments reported by Mitscherlich (1939) supply additional evidence on the behavior of c. In this work, oat plants were grown with uniform nutrient mixtures in sand cultures by scientists at various locations in Europe. Mitscherlich assembled the data and calculated the values of the parameters in the Mitscherlich equation. The data in Fig. 1-10 indicate that the c values were not constant, but that they decreased with an increase in the maximum yield attainable with fertilizer phosphorus. Similar results were obtained with nitrogen. The c values for potassium were scattered, but there was no significant trend.

Kletschkowsky and Shelesnow (1931) earlier had summarized data from the literature and reported original findings on the behavior of c values for fertilizer nitrogen as a function of differences in supply of fertilizer phosphorus and vice versa. They found that as the A values increased in both the nitrogen series and

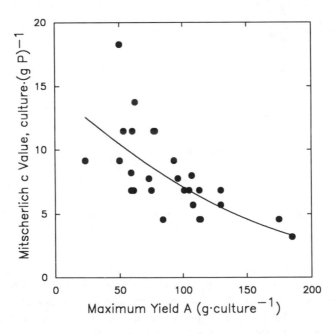

Fig. 1-10. A plot of the Mitscherlich c values against the A values obtained in cooperative experiments on phosphorus fertilization of oat plants in sand cultures at various locations in Europe. Each point represents a separate experiment. (Mitscherlich, 1939)

the phosphorus series, the c values decreased. In a more recent verification of the same principle, Dahnke and Olson (1990) found that the quantities of fertilizer phosphorus and nitrogen required to reach the maximum yield increased with the maximum yield in a group of field experiments with wheat in North Dakota.

These findings regarding the downward trends of c values with increased growth of the crop are logical expectations from the fact that plants must contain at least a minimum concentration of the nutrients required to produce them. If plant growth increases fivefold as a result of improvement in conditions, more nitrogen and more phosphorus will be required to produce the additional growth. The "efficiency" of a given amount of the nutrient in the soil in advancing the yield toward the maximum thus will decrease.

A constant c value implies that the efficiency of a given quantity of a nutrient in increasing the yield of dry matter remains the same, no matter how great the increase in yield may be. The fact that absorption of nutrients from a soil by plants depletes the supply remaining in the soil is overlooked.

One must conclude that the c value in the Mitscherlich equation is not constant for a particular nutrient, as supposed by Mitscherlich, but that it is variable. His vision of a theoretical equation of universal application has faded, and the applications that could be made if the c value were constant cannot be made at all or at best are approximations.

Nonetheless, the Mitscherlich equation is a good one for representing plant responses to nutrients and other limiting factors in many situations. The parameters in the Mitscherlich equation have the special merit of denoting important biological concepts that are readily understood: A is the maximum yield attainable by supplying the nutrient, b is the supply of the nutrient in the soil, and c is the efficiency of the added nutrient in increasing the yield. This advantage plus the large body of published experimental data that relates to the Mitscherlich equation make this equation the best one available as an aid to soil fertility specialists in developing an understanding of response curves and their use.

The quadratic equation also has been widely used. Like the Mitscherlich equation, it has three parameters, a, b, and c, but the only one with biological meaning is a, the yield of the control. The efficiency of the nutrient is expressed by a combination of parameters b and c with the quantity x of the nutrient. The maximum yield does not appear as one of the parameters in the equation, although it too can be calculated from b, c, and x.

1-3. Factors Affecting the Shape of Response Curves

This section reviews and illustrates some of the important factors that contribute to responses observed experimentally and explains in brief how they may be interpreted and in some instances taken into account in soil fertility evaluation and control. More details are given in succeeding chapters.

The influence of some factors upon the shape of response curves may be perceived from a single response curve. Others cannot be inferred from a single curve, but require two curves to make the effects evident. The combination of graphic examples with theoretical explanations in this section is intended to provide an appreciation for the complexity of the factors that dictate the shapes of response curves and a basis for interpreting the curves in terms of causal factors. The illustrations will help to explain why mathematical response functions that use only the quantity of fertilizer as an independent variable cannot be expected to provide a general representation of plant response to limiting factors.

By making appropriate measurements before a crop is grown and by using suitable theoretical or empirical relationships, the effects of some of the factors that affect crop responses can be taken into account to make site-specific predictions of responses based upon the controlling causal factors as opposed to predictions based upon average responses measured in the past that provide an expression of the relationship of average outputs to average inputs.

1-3.1. Soil Nutrient Supply

An obvious effect on the shape of response curves that may be related to a factor measurable before a crop is produced is differences in the ratio of the control yield to the maximum yield obtainable by adding increasing quantities of a nutrient in the form of fertilizer. As the supply of a particular nutrient in the soil increases, the ratio of the control yield to the maximum yield obtained by adding more of the nutrient increases. When the supply of the nutrient in the soil is adequate to produce the maximum yield, the response curve is horizontal for at least small additions of the same nutrient as fertilizer. This relationship between the soil supply and the shape of response curves is made use of in the development and calibration of chemical soil tests for nutrient availability, discussed in Chapter 4.

1-3.2. Effective Nutrient Supply Versus Added Nutrient Quantity

The usual convention in production research is to express the output as a function of the input. This may be done graphically or mathematically. However it is done, the implication is that the differences in measured units of input are responsible for the differences in output. In a sense, of course, they are. Where the inputs are fertilizers and the output is crop yield or some other measure of response, however, the crop may be responding to quantities or concentrations of the nutrient that differ from the sum of the initial supply in the soil and the quantities added because of interactions of the added fertilizer with the soil and loss of some of the fertilizer from the soil. Other interactions occur, some involving the crop, that affect the supplies of nutrients and toxic factors other than the one added, and these also may affect the response curve. All these interactions are related to the added quantity of the nutrient producing the response curve, so that the influence they may have on the response curve is not obvious. With choice of a suitable equation, the crop yield thus may be expressed as a function of the quantity of nutrient added without appreciation for the devious ways in which the results were produced.

The existence of differences between nominal and effective additions of nutrients is well known, but the differences are not always thought of in relation to their effects on response curves. Discrepancies between measured and effective quantities may have important effects in terms of both a single response curve obtained at a given location and the response curve for one soil or condition versus another.

If an added nutrient reacts with the soil, the degree of reaction characteristically is greatest with the smallest additions and decreases with increasing additions. The scale of the increase in effective quantities thus is not proportional to the scale of added quantities. That is, the scale of the increase in effective quantities cannot be obtained by multiplying the scale of added quantities by a constant factor. An interaction with the soil that decreases the solubility of an added nutrient depresses the lower portion of the response curve, but has no direct effect upon the upper part of the curve where the supply of the nutrient is not limiting.

The results of an experiment by Borden (1945) provide a verified example of an interaction of an added nutrient with the soil that was evident in the lower portion of the response curve. Borden added different quantities of ammonium nitrate to soil with and without a supplemental addition of sugarcane leaves. He allowed the samples of soil to incubate in a moist condition for 8 weeks; then he analyzed them for mineral nitrogen (ammonium and nitrate) and grew panicumgrass as a test crop.

The data in Fig. 1-11A show that a plot of yield of grass against nitrogen added produced two separate response curves. The data in Fig. 1-11B show that in both the control cultures and the cultures with sugarcane leaves added, the mineral nitrogen present in the soil at planting increased with the quantity of ammonium nitrate added. But the amounts of mineral nitrogen present at planting were not proportional to and generally were not equal to the sum of the mineral nitrogen initially present and that added as fertilizer.

The mineral nitrogen present at planting was reduced by addition of sugarcane leaves, as would be supposed from Fig. 1-11A. Almost all of the mineral nitrogen initially present in the soil was immobilized in the absence of fertilizer nitrogen, but addition of ammonium nitrate hastened the mineralization of organic nitrogen, so that with the greatest addition the amount of mineral nitrogen present at the time the crop was planted was almost the same as the sum of the amount added and the amount present initially. In the control soil without sugarcane leaves, the mineral nitrogen present at the time of planting was almost equal to that present initially where no ammonium nitrate was added. But application of ammonium nitrate increased the mineralization of soil nitrogen, so that the increase over the control due to addition of the fertilizer exceeded the amount added in the fertilizer. The scale representing the quantities of nitrogen added

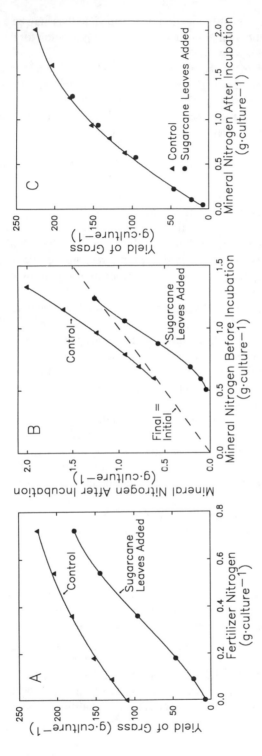

Fig. 1-11. Transformations of fertilizer nitrogen in soil and their effect on response curves. A. Yields of panicumgrass versus quantity of nitrogen added as ammonium nitrate to soil in the presence and absence of 36 grams of sugarcane leaves per culture. The cultures were incubated with the additions of ammonium nitrate and sugarcane leaves for 8 weeks before the test crop was grown. B. Mineral nitrogen (ammonium plus nitrate) in soil before and after incubation of the soil with additions of sugarcane leaves and different quantities of ammonium nitrate. C. Yields of panicumgrass versus mineral nitrogen found at the time of planting by analysis of soil cultures that had been incubated with and without sugarcane leaves and different quantities of ammonium nitrate. (Borden, 1945)

was evidently not the correct one for use as an index of the increase in nitrogen availability in the soil due to fertilization in either the presence or absence of the sugarcane leaves.

In Fig. 1-11C, in which the yield of panicumgrass is plotted against the quantities of mineral nitrogen found in the soil at the time of planting, all the results can be represented to a good approximation by a single response curve that has the usual downward concavity. The scale representing the quantities of ammonium and nitrate nitrogen found by analysis at the time of planting thus appears to be a more accurate representation of the differences in supply of nitrogen to which the plants were responding than was the scale representing the quantities of nitrogen added. The convergence of the data for the different treatments into a single response curve when plotted against the mineral nitrogen in the soil as opposed to the fertilizer nitrogen added is not an invariable occurrence with soil testing methods. Probably more often than not, the curves do not coincide. See data published by Bolland and Baker (1987) for an example, which very likely reflects a deficiency in the soil testing method.

The behavior illustrated in Fig. 1-11 is a manifestation of well-known biological processes that occur in soils. Chemical processes also have received attention. Sigmoid response curves sometimes are obtained with phosphorus. The stronger retention of fertilizer phosphorus by soil where the quantities added are small than where they are large provides a partial explanation. Fig. 1-12 shows the characteristic upward concavity of plots of phosphorus in solution versus phosphorus added to soils. The five soils illustrated have been selected from a much larger number to illustrate the kinds of differences that may be observed. Other nutrient ions that react strongly with soils behave similarly.

Fig. 1-13 shows sigmoid response curves that were produced experimentally by adding freshly prepared synthetic hydrous ferric oxide in different quantities to soil to react with added fertilizer phosphorus and reduce its availability to the test crop of subterranean clover. The sigmoid tendency of the curves increased with the quantity of hydrous ferric oxide added.

The reaction of fertilizer phosphorus with hydrous iron oxides in soils appears to be responsible for the finding by Brandon and Mikkelsen (1979) that wheat and barley grown after flooded rice responded markedly to fertilizer phosphorus and produced sigmoid response curves, whereas in two of the soils studied no response would be expected if the soils had not been flooded, and only slight response would have been expected in the third soil.

While soil is flooded for rice production, the solubility of some of the iron is greatly increased by reduction to the ferrous form. When the soil is drained and molecular oxygen returns, the ferrous iron is oxidized to amorphous hydrous ferric oxides, which occur as coatings on the soil solids and have a high capacity to react with inorganic orthophosphate. As the oxides age, they presumably become more crystalline and less reactive, and the phosphorus availability increases, gradually approaching the availability in soil that has not been flooded (Sah and Mikkelsen, 1986).

Although the biological and chemical reactions that reduce the availability of nutrients added to soils probably occur rather generally, the reason most response curves are not sigmoid appears to be that the net curvature is a resultant of opposing forces. In most instances, the tendency for an increase in slope of response curves with an increase in the effective quantity of an added nutrient per unit of the nutrient added can be seen only if the curve becomes concave upward. To make a response curve concave upward requires overcoming the effects of the factors that cause the curves to be concave downward. The net result is that in most instances an increase in effectiveness of unit quantity of a nutrient with an increase in the quantity added merely causes the initial portion of the response curve to be more nearly a straight line than it otherwise would have been.

The existence of differences in response curves due to differences from one soil to another in the extent of changes a nutrient may undergo upon addition to soils is indicated by the results of an experiment by Holford (1982). To eliminate differences in environmental conditions associated with location, he collected samples of 15 soils and brought them to a greenhouse. To each soil, he added five different quantities of monobasic potassium phosphate along with uniform quantities of sulfur, magnesium, and micronutrients, and potassium in

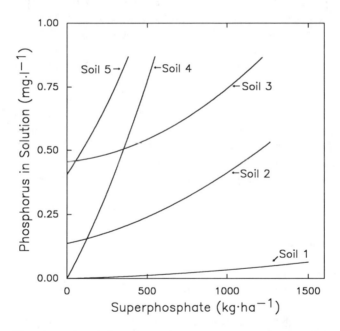

Fig. 1-12. Phosphorus in solution as orthophosphate after 6 hours of shaking two parts of water with one part of soil to which different quantities of superphosphate had been added. The addition of phosphorus corresponding to 500 kilograms of superphosphate per hectare was equivalent to about 8.7 milligrams of phosphorus per kilogram of soil. If all this phosphorus had remained in solution, the concentration would have been 4.4 milligrams per liter. (Demolon and Boischot, 1951)

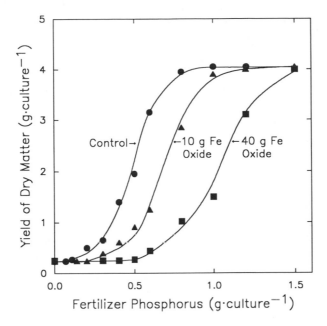

Fig. 1-13. Yields of subterranean clover tops versus fertilizer phosphorus added to 3-kilogram quantities of a soil that had been treated with different quantities of synthetically prepared hydrous iron oxide. (Bolan et al., 1983)

quantities that presumably compensated for the original differential applications of phosphorus. He then grew ladino clover as a test crop. Fig. 1-14 shows that the effectiveness of the fertilizer phosphorus in increasing the crop yield decreased with an increase in sorptivity of the soils for phosphorus.[3] In terms of response curves with equal control yields and equal maximum yields, the findings in Fig. 1-14 correspond to curves that ascend steeply and bend sharply to the

---

[3]The axis labels on the figure are equivalent to those used by Holford (1982). He expressed the yields of the ladino clover test crop as percentages of the maximum yields on the respective soils. His Y axis units of percent per milligram were obtained as a result of expressing the yields as a percentage of the maximum on each soil and fitting the data with the linear response and plateau model (see Section 1-5.3), which uses a straight line between the control yield and the maximum yield to represent the ascending segment of the response curve. Thus, the greater the percentage increase in yield per milligram of fertilizer phosphorus, the steeper is the slope of the response curve, and the greater is the effectiveness of unit quantity of fertilizer phosphorus.

For the measurements on the X axis, Holford used the term sorptivity because, although he employed the Langmuir adsorption equation to calculate the results of the soil measurements (see Olsen and Watanabe, 1957), the reaction of phosphorus with soils is more complex than adsorption. The Langmuir equation is responsible also for the strange units of $l \cdot kg^{-1}$ for sorptivity. According to Holford's use of this equation, the sorptivity is the increase in number of milligrams of phosphorus held in sorbed form by a kilogram of soil per milligram increase in concentration of phosphorus per liter of solution in equilibrium with the soil at zero concentration. With the quantities of phosphorus sorbed expressed as milligrams per kilogram of soil and the concentration in solution expressed as milligrams per liter, the milligram units cancel out, and the resulting units are liters per kilogram.

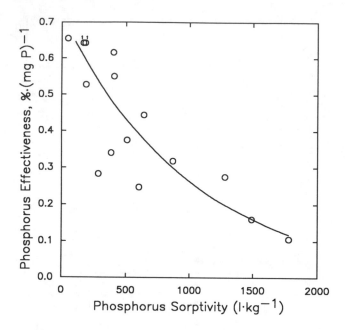

Fig. 1-14. Effectiveness of fertilizer phosphorus versus the sorptivity of 15 soils of New South Wales for phosphorus in a greenhouse experiment with ladino clover as test crop. See the text and footnote 3 for an explanation of the nature of the measurements. (Holford, 1982)

right for soils with low phosphorus sorptivity, changing gradually to curves that ascend slowly and bend only gradually for soils with high phosphorus sorptivity. Holford and Cullis (1985) published the results of 29 field experiments on phosphate fertilization of wheat in New South Wales that showed a similar trend of shapes of the response curves with an increase in phosphorus sorptivity of the soils.

Interactions of an added nutrient with the soil can be measured before a crop is grown and can be taken into account in the process of predicting the effect an addition of fertilizer may have on the response curve in a particular situation. Ways in which the effects of such interactions have been taken into account in field experiments for soil test calibration purposes and in chemical soil tests will be described in Chapter 4.

Complications exist, so that the application is not necessarily as simple as might be inferred from the preceding paragraphs. For example, sigmoid response curves may result from increasing quantities of a granular form of soluble phosphate fertilizer, where the increased supply of phosphorus is in discontinuous locations (Burns et al., 1963; Kafkafi and Putter, 1965). In acid soils with low phosphorus availability and toxic concentrations of aluminum, a sigmoid tendency of curves showing response to phosphorus may be promoted by a reduction in aluminum toxicity as well as an increase in phosphorus availability when the sorption of added phosphorus by the soil results in part in precipitation of soluble and exchangeable aluminum in the form of aluminum phosphates of low solubility. Where deficiencies of a nutrient are extreme, sigmoid curves may also result from supplies of other nutrients that are excessive for the lowest yields.

1-3.3. Other Nutrient Deficiencies

When the crop yield is increased by applying a nutrient under test, greater quantities of other nutrients are required. If they cannot be supplied by the soil without creating a deficiency or enhancing an existing deficiency, the maximum yield obtainable by applying the nutrient under test will be lower than it otherwise could have been.

An experimental example showing the mutual effects of two nutrients is found in Fig. 1-15, which reproduces some of the results of an experiment on a phosphorus-sorptive soil in South Africa involving fertilization with both phosphorus and nitrogen. The response curve for each nutrient ascended more steeply and produced a considerably higher yield where the other nutrient was present in ample supply than where it was deficient. Thus, both nutrients were deficient in the control soil, and the response curve produced by additions of each nutrient was affected by the addition of the other nutrient. The responses to both nitrogen and phosphorus, of course, would differ among years, but on average the pre-

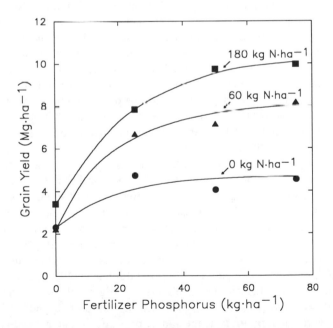

Fig. 1-15. Yields of corn grain versus additions of fertilizer phosphorus at three levels of fertilizer nitrogen in a field experiment in South Africa. The yields are for the sixth year in a long-term experiment in which the same quantities of fertilizers were applied annually. (Sumner and Farina, 1986)

cision with which the response to each nutrient could be predicted could be improved by prior knowledge of the supply of the other nutrient.

As another example, Munk (1985) found in an analysis of the results of 82 experiments on phosphate fertilization of soils of Germany that the fertilizer phosphorus requirement associated with a given value of extractable soil phosphorus increased with the yield of the crop supplied with nitrogen and potassium in the absence of fertilizer phosphorus. Increased soil productivity, due to whatever cause or causes, thus was related to a need for more fertilizer phosphorus.

The principle that the effects of individual nutrients on crop responses are not independent of each other is generally recognized, and conceivably an allowance for the effect might be made. In practice, however, it is more difficult to take into account the mutual effects of nutrients on response curves than it is to deal with a single nutrient. There are many "other" nutrients on which measurements might be needed to include the few most critical nutrients, and understanding generally is not good enough to provide the desired confidence in the interpretations if the measurements were made.

For example, there is some special interaction between nitrogen and phosphorus such that applying fertilizer nitrogen tends to increase the phosphorus percentage in plant tissue, making it appear that plants have received an application of phosphorus as well as nitrogen. Such an interaction was noted in the experiment from which Fig. 1-15 was derived.

The situation is complicated by the different kinds of mutual effects that may occur. Ionic competition in one way or another in the soil and in the plant may result in a low concentration of a nutrient in plants without actual exhaustion of the supply in the soil. An example is potassium-induced magnesium deficiency in crops on soils low in magnesium. This phenomenon is important also in "grass tetany," an abnormally low concentration of magnesium in the blood of grazing animals that may occur even though the magnesium concentration in the forage is not low enough to represent a deficiency for plants (Grunes and Rendig, 1979). Another example is the partial exclusion of molybdate from plants by sulfate (Stout et al., 1951).

A physiological interaction of sodium and potassium is important in some crops. Response to sodium at low levels of potassium but not at the maximum yield with potassium is interpreted as partial substitution of sodium for potassium (Shepherd et al., 1959). An increase in the maximum yield obtainable with potassium fertilization in the presence of sodium suggests an independent effect of sodium (Draycott et al., 1976; McCree and Richardson, 1987).

Some nutrient interaction effects reflected in response curves are related to differences in soil pH. For example, Lorenz and Johnson (1953) noted that where potato plants were fertilized with phosphorus on a slightly alkaline, noncalcareous sandy loam soil, the response curves indicated marked phosphorus deficiency in the soil with calcium nitrate as the source of nitrogen, but an adequate supply of phosphorus with ammonium sulfate. Differential effects of the two fertilizers on soil pH and soil phosphorus availability apparently were responsible.

Liming sometimes significantly increases soil nitrogen availability, as indicated by a higher control yield and less response on limed soil than on unlimed soil (Mulder, 1950). Micronutrient availability often is affected. Harry and Graham (1981) found that the response of triticale to raising the pH of a copper-deficient soil from 4.0 to 8.4 was a small increase in grain production in the presence of added copper, but a precipitous drop to zero grain production without added copper. A considerable difference among crops was observed in the same experiment. Wheat produced no grain at any soil pH without added copper, and the yield of rye was not greatly affected by either soil pH or addition of copper. In some instances, the response of crops to molybdenum on acid, unlimed soils and the lack of a response with addition of sufficient limestone indicates that the response to liming is largely a response to the associated increase in soil molybdenum availability (Anderson and Moye, 1952; Parker and Harris, 1962; Anderson, 1970).

Instead of making measurements to aid in predicting the effects of other nutrients on the response curve for a nutrient under test, an alternative often used is to add ample quantities of nutrients that might possibly be deficient at one or more levels of the nutrient under test without measuring their effects. This practice tends to increase the response to the nutrient under test.

1-3.4. Water Supply

Water is a plant nutrient, although it is not ordinarily classified as such. Fig. 1-16 illustrates the response of wheat and barley to nitrogen fertilization observed in 66 experiments in North Dakota classified according to the amount of available water in the soil at planting. The response evidently increased with the stored water supply, a measurement that could be made before planting and fertilization. The experiments were conducted at various locations and on various kinds of soils over a 3-year period.

With enough experiments such as those in Fig. 1-16, predictions could be refined to apply specifically to individual groups of similar soils. Alternatively, the influence of stored water within seasons could be determined experimentally by applying different amounts of irrigation water before planting. The influence of seasons could be evaluated, largely independent of locations, by repeating the experimental work in successive years at adjacent sites. Ramig (1960) used this approach on winter wheat in Nebraska in an experiment repeated in 3 years. His work produced results similar to those in Fig. 1-16, but the numerical values were different.

1-3.5. Toxicities

The practice of adding an ample supply of nutrients other than the one under test may lead to unappreciated toxicities, particularly in soil and sand culture experiments. Because of the small volumes usually available for roots, adding at planting the total amount of nutrients needed to produce the plants may depress the yields due to osmotic effects until the growth of the test crop has been sufficient to absorb part of the early excess and reduce the external concentration.

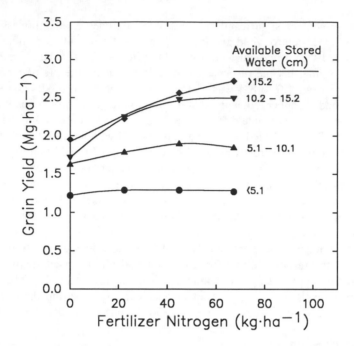

Fig. 1-16. Response of cereals to nitrogen fertilization on various soils during a 3-year period in North Dakota with results classified according to the available water stored in the soils at the time of planting. The data summarize results from 64 experiments on spring wheat and two experiments on spring barley. (Bauer et al., 1965)

The effects of the soluble salts are dictated mainly by those supplying nitrogen and potassium, the two nutrients required in greatest quantities. Most of the research has been done with nitrogen. Fig. 1-17 includes two illustrations. Fig. 1-17A shows the response of oat plants to increasing quantities of nitrogen when the ammonium sulfate source was added at planting and the plant response measurements were made at different stages of growth. Note particularly the responses of the plants receiving 0.42 and 1.70 grams of nitrogen per culture. At 53 days after planting, 0.42 gram of nitrogen per culture was enough to produce the maximum yield, and the 1.70-gram addition was far too much. At 120 days after planting, the 0.42-gram addition had been exhausted; it produced an intermediate yield. But the plants receiving the 1.70-gram addition, which was initially excessive, had surpassed the others and then produced the highest yield. The estimated concentrations of salts in solution with the greater additions of ammonium sulfate were in the range in which growth depressions due to osmotic effects occur in sensitive plants (Wadleigh and Ayers, 1945).

The experimental data in Fig. 1-17B are consistent with the interpretation of the data in Fig. 1-17A in terms of osmotic effects. When the osmotic effects of increasing concentrations of calcium nitrate were eliminated by supplying the nitrate in combination with an anion exchange resin, the yield was not depressed by large nitrate additions.

Toxicities due to osmotic effects of nutrients other than the one under test, along with their undesirable effects on the response curve or curves being produced, can be avoided in experimental work in various ways, one being adding the nutrients in smaller quantities at intervals. Janssen (1990) described a double-pot technique that would be useful. In this technique, nutrients other than the one under test are contained in a nutrient solution in a pot directly below the one containing the soil.

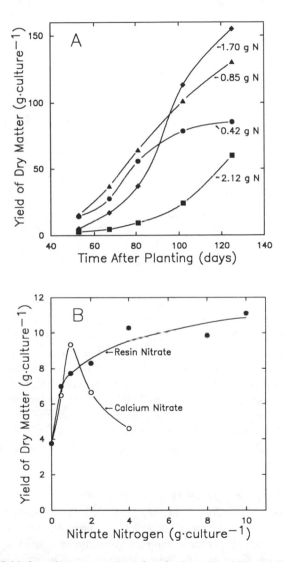

Fig. 1-17. A. Yield of oat plants versus time after planting, with addition of different quantities of nitrogen as ammonium sulfate to sand cultures at the time of planting. (Meyer and Storck, 1927/1928) B. Yields of yellow mustard plants with different quantities of nitrate nitrogen as calcium nitrate and resin nitrate in cultures containing 18 kilograms of soil. (Poulsen, 1959)

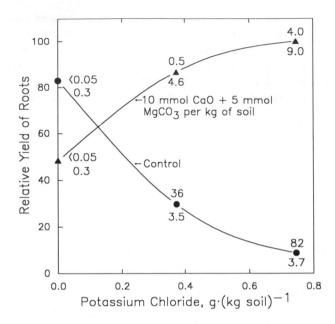

Fig. 1-18. Relative yields of roots of sorghum in an acid subsoil with different additions of potassium chloride, calcium oxide, and magnesium carbonate. The numbers above and below each experimental observation are the concentrations of aluminum in milligrams per liter and of calcium plus magnesium in millimoles per liter of saturation extract. The soil treated with calcium oxide and magnesium carbonate was only partly neutralized. (Ragland and Coleman, 1959)

Another type of toxicity is a consequence of chemical interactions of an added nutrient in soils. Fig. 1-18 is an example. The negative response to potassium chloride in the unlimed acid soil changed to a positive response in the limed soil. Aluminum toxicity induced by the increasing concentration of aluminum in solution in the soil as a result of replacement of some of the exchangeable aluminum by the added potassium was probably a major cause of the downward trend of the response curve in the unlimed soil. An enhanced deficiency of calcium, magnesium, or both may have been a contributing factor because addition of potassium tends to decrease the concentrations of calcium and magnesium in plants. Effects of this kind on response curves can be predicted from appropriate preliminary measurements.

1-3.6. Secondary Components of Fertilizers

It is conventional and usually satisfactory to attribute the response curve obtained with increasing quantities of a fertilizer to the nutrient supplied intentionally. Most fertilizers, however, contain more than one nutrient, and they may contain other substances that can modify the response curve. The occasional instances in which the usual inference is incorrect could be avoided by preliminary analyses and experiments, but usually such efforts are bypassed because checking out all the possibilities would be inefficient use of time.

When the unrecognized but significant substance is toxic, the yield response curves to the main nutrient rise less rapidly, and the maximum yield is decreased. In extreme instances, the response curve may be negative, as illustrated for a different situation in Fig. 1-18.

Response curves with negative slopes were obtained in the eastern United States many years ago with some batches of fertilizer. Eventually the problem was traced to relatively high concentrations of boron in potassium chloride derived from salt deposits in Searles Lake, California (Schreiner et al., 1920). When the cause was identified, it was possible to modify the processing of the fertilizer to reduce the content of boron and eliminate the toxicity.

In Florida, the phenomenon of "yellow tipping" of citrus leaves had been observed for more than 20 years before the cause was discovered. Suspecting the presence of a toxic substance because the symptom was associated with the use of potassium nitrate from Chile, Stewart and Leonard (1952) eventually identified the toxin as perchlorate. This problem similarly was eliminated by a change in processing of the fertilizer.

Correcting an unrecognized deficiency by applying increasing quantities of a deficient nutrient in a fertilizer conventionally applied for its content of another nutrient would make the response curves rise more steeply and reach a higher maximum than they would if the deficiency did not exist. This condition would create a source of error in instances in which response curves are being related to soil test values for calibration purposes.

Thus, to calibrate soil test values for phosphorus, experiments with concentrated superphosphate (which is primarily monobasic calcium phosphate) would be safer than those with ordinary superphosphate (which is about half monobasic calcium phosphate and half calcium sulfate dihydrate). Similarly, to calibrate soil test values for sulfur, experiments with calcium sulfate would be more suitable than those with ordinary superphosphate. Calcium, of course, is a secondary nutrient in both of these fertilizers. Calcium deficiencies are more rare than those of sulfur and phosphorus, but for instances such as soils that are very low in exchangeable calcium and for plants such as peanut with a relatively high requirement for calcium, the experiments could be done in such a way as to avoid confounding the sulfur and phosphorus effects with those of calcium.

Instances have been reported in which responses to one nutrient have been thought to be due to another. In New Zealand, ordinary superphosphate had been the principal fertilizer for many years. Although sulfur was known to be an essential nutrient and some sulfur deficiencies had been found in that country, it was conventional to attribute the widespread responses to superphosphate primarily to the phosphorus it contained and to assume that whatever sulfur deficiencies might exist were eliminated by the use of this fertilizer. According to Walker (1955), the discovery that responses to sulfur might constitute an important part of the response to superphosphate resulted from an observation by Sears (1953) that in a field experiment on a soil testing high in extractable phosphorus, the pasture vegetation responded to application of superphosphate.

Another effect, discovered in the same way as the sulfur-phosphorus effect that may occur with ordinary superphosphate, has been a chloride effect with potassium chloride. In the Northern Great Plains of the United States, wheat and barley were found to be giving small but fairly consistent responses to potassium chloride even though the soils contained so much exchangeable potassium that no responses would be expected on the basis of the potassium supply. Experiments in which the chloride was replaced by sulfate (Timm et al., 1986) or by nitrate and in which the potassium was replaced by calcium (Fixen et al., 1986a,b) indicated that the effect was indeed due to chloride and not potassium. The effects appear to be complex, including enhanced resistance to some diseases (Timm et al., 1986; Fixen et al., 1986a; McCree and Richardson, 1987; Grybauskas et al., 1988; Mason et al., 1991).

In recent work in New Zealand, another example of importance of a secondary nutrient was encountered in a different setting. In a group of experiments in which Sechura phosphate rock and concentrated superphosphate were compared as sources of phosphorus, it was discovered that the Sechura phosphate rock contained enough molybdenum (43 milligrams per kilogram) to cause the pasture vegetation on molybdenum-deficient soils to yield more after 4 to 6 years of treatment with this phosphate rock than with equal quantities of phosphorus as concentrated superphosphate that contained no molybdenum (Sinclair et al., 1990). The molybdenum effect meant that the yield response curves were not suitable for comparing the phosphorus availability effects. In this investigation, the experimental areas on which molybdenum deficiency was expected had been fertilized with molybdenum, and the response to molybdenum appeared in some of the experiments on soils on which response to molybdenum was considered unlikely. It was only after the responses to molybdenum occurred that the trace of molybdenum was discovered in the phosphate rock.

1-3.7. Biological Effects

A number of biological factors affect the shape of response curves and are of great importance in agriculture. The effects can be measured by suitable experiments while a crop is growing, but predictions generally cannot be site-specific and must be based upon experiments from which average effects may be assessed.

1-3.7.1. Plant Character

The shape of response curves may differ considerably among plants. The data by Mulder (1952) in Fig. 1-19 show differential responses of alfalfa and grasses to molybdenum. In this instance, the shapes of the curves suggest that the soil provided enough molybdenum for almost the maximum yield of grasses, but that it was deficient in molybdenum for alfalfa. Molybdenum is an essential component of the nitrate reductase enzyme and others in plants in general, and in legumes it is an essential component of the nitrogenase enzyme that reduces molecular nitrogen to ammonia in nitrogen fixation. The nitrogenase function may have been responsible for the special sensitivity of alfalfa to molybdenum shown in Fig. 1-19.

In some instances, experimental work has provided evidence of causes of differences among plants. Greenwood and Draycott (1988) obtained evidence of differences in nitrogen requirement that would affect the shape of response curves to nitrogen. They found that for radish, winter cabbage, and sugarbeet, the respective quantities of fertilizer nitrogen required to produce the maximum nitrogen uptake were 157, 280, and 729 kilograms per hectare, and the corresponding uptakes of nitrogen were 40, 190, and 303 kilograms per hectare. The maximum yields of dry matter presumably were in the same relative order. Other examples were published by Foy et al. (1980, 1981) and Godbold et al. (1984).

Differences in response curves among plants have been exploited by plant breeders in developing cultivars that will produce high yields at high levels of soil fertility. In small grain crops, the maximum yield and the quantity of fertilizer nitrogen associated with the maximum yield have been increased by production of cultivars with short, stiff straw resistant to the lodging that previously limited the supply of nitrogen that could be used. For response curves, see Mulder (1954) and Biswas and Singh (1982). Older corn cultivars were prone to barrenness when planted in thick stands, and new cultivars were needed to take advantage of the potential for increased yields with high stand densities and increased supplies of plant nutrients. Effects of this kind on response curves are important in determining the appropriate economic quantities of fertilizers.

The lower end of response curves for phosphorus may differ greatly among plant species. In New Zealand, Davis (1991) found that two lupins grew well

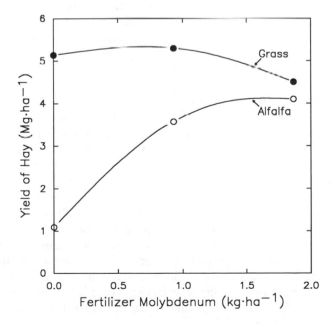

Fig. 1-19. Yields of alfalfa and grass in The Netherlands with different additions of molybdenum as sodium molybdate. The grass was a mixture of four species. (Mulder, 1952)

and did not respond to phosphorus fertilization on an acid sandy loam soil. On the same soil, seven different pasture legumes failed without phosphorus fertilization although they produced well with added phosphorus. In Germany, Föhse et al. (1988) found that ryegrass and wheat produced 80% of the maximum yield in soil with a phosphorus concentration of about 1.3 micromoles per liter of soil solution, but onion required 6.9 micromoles per liter. The concentration of phosphorus in the plant dry matter at 80% of the maximum yield was about 0.3% in ryegrass and wheat, but only 0.14% in onion. Onion thus had a low internal phosphorus requirement but a high external requirement, which the authors related to the lower ratio of roots to tops in onion than in ryegrass and wheat.

Further potential exists to develop cultivars that will make more efficient use of limited soil supplies of certain nutrients, thus raising the lower end of response curves and reducing dependency upon fertilizers. This avenue has been pursued most vigorously in developing cultivars for countries in which the cost of fertilizers is relatively high in relation to crop values. The International Rice Research Institute in the Philippines, for example, screens selections of rice to determine their capability to produce relatively well where phosphorus and zinc are deficient (Soil Chemistry and Plant Breeding Departments, 1976; Soil Chemistry/Physics, Plant Breeding, and Plant Physiology Departments, and International Rice Testing Program, 1984).

Differences in plant character that affect response curves in the ways described in preceding paragraphs can be taken into account by using the species or cultivar of interest in the experiments used to develop calibrations relating soil nutrient supplies to plant responses. As important examples, forage legumes are generally more sensitive to deficiencies of phosphorus, potassium, and molybdenum, and less sensitive to deficiencies of nitrogen, than are the grasses commonly grown with them. These are strictly empirical relationships.

1-3.7.2. Plant Population Density

Plant population density is a factor that can be taken into account in predicting the shape of response curves because the density can be controlled to a considerable extent by the operator. Increasing densities of genetically similar plants result in increasing competition of like entities for the environmental necessities — nutrients, light, and water.

Two general classes of relationships appear to exist where yield is plotted against plant population density (Willey and Heath, 1969). In one, the yield increases to a maximum value and then remains substantially constant with further increases in the plant population. In the other, the yield passes through a maximum and then decreases with further increases in population. The total yield of above-ground dry matter of corn is an example of the first type, and the yield of corn grain is an example of the second type. These effects are superimposed on the effects of fertilizers in experiments that include different levels of plant populations and fertilizers.

Interactions of plant population density with nutrient supplies are probably most marked with nitrogen. For example, a high population density raises the total dry matter produced by corn plants at all levels of nitrogen fertilization, but it depresses the grain yield at low nitrogen levels and raises the maximum grain yield at high nitrogen levels, as found by Knapp and Reid (1981) in New York.

The amount of nitrogen required to reach the maximum yield increases with the population density. From experiments in Florida, Rhoads et al. (1988) concluded that irrigated corn should be fertilized according to the population density. In 2 years with different growing conditions and a range of populations and quantities of fertilizer nitrogen, they found that the grain yield per hectare with 3.0 kilograms of fertilizer nitrogen per 1000 plants was greater than that with 0, 1.0, or 2.0 kilograms and about the same as that with 5.0 kilograms.

The general behavior described for corn applies in principle to wheat and other small grains. The capacity of the small grains for tillering, however, partly obscures the effects of differences in plant population (Wollring, 1990).

1-3.7.3. Weed Infestations

Weed infestations act for the most part like crop plant population densities in that they involve competition. The effects are far more complex than plant population densities because they involve different types of weeds, population densities that are uncontrolled, interactions of densities with level of nutrient supply, and other factors. Papers including data illustrating different types of response curve effects include those by Singh et al. (1984), Agenbag and De Villiers (1989), Moody (1981), Pons and Van Wieringen (1985), Carlson and Hill (1985), and Kim and Moody (1980). The best predictions of crop response curves are those made for crops with very low weed populations.

1-3.7.4. Plant Diseases and Insect Infestations

As with weed effects, the best predictions of crop response curves are those made in the absence of the pest as a result of treatment with appropriate pesticides or use of resistant cultivars where available. The most common effect of pest infestations on response of crops to nitrogen is a depression of the upper part of the response curve. An infestation with a disease or insect often depresses the lower portion of curves depicting the response of crops to potassium. This behavior appears to be in part a consequence of the practice of supplying a constant amount of nitrogen that is excessive at low levels of potassium and more nearly in balance at high levels of potassium.

Huber (1980) published an extensive tabulation of plant responses to pathogens observed in many investigations. Singh and Agarwal (1983) published a useful summary table showing the effects of fertilization on the change in population of insects or plant injury from insects.

1-3.7.5. Mycorrhizas

Mycorrhizas are in effect extensions of the root system. They are fungi that are attached to the roots of most plant species and affect the shape of response curves for plant nutrients by absorbing organic nutrients from the roots and by transferring to the roots a portion of the mineral nutrients they absorb from the soil.

The importance of mycorrhizas in plant nutrition is greatest with cultivars of a given species having small root systems (Azcon and Ocampo, 1981), with plant species having thick roots and few or no root hairs (Baylis, 1975; St. John, 1980), when plants are deficient in nutrients (Sanders and Tinker, 1973), and for nutrients such as phosphorus that require extensive contact between roots and soil for adequate uptake. Cassava, for example, has fleshy roots, but grows well on impoverished soils if the roots are mycorrhizal.

Most of the research has been done on phosphorus. Howeler et al. (1982), for example, found that to support near maximum growth, the phosphorus soil test value for nonmycorrhizal cassava had to be about 13 times higher than that for mycorrhizal cassava. Viebrock (1988) found that nonmycorrhizal *Capsicum annuum* plants measurably depleted the soil of phosphorus to a distance of about 1 millimeter from the root surfaces, but that mycorrhizal plants measurably depleted the phosphorus from soil 20 to 40 millimeters distant from the root surfaces.

The principal effect of mycorrhizas on response curves is to raise the lower portion of the curves by supplying extra amounts of the nutrients under conditions of deficiency. There are other effects as well, of which only two will be mentioned here.

In an experiment on a soil that was deficient in both zinc and phosphorus for wheat, Singh et al. (1986) obtained the results in Fig. 1-20, which show that fertilizer phosphorus increased the yield of wheat plants, but decreased the percentage of the roots infected with mycorrhizal fungi and decreased the total content of zinc in the plants. In terms of response curve effects due to mycorrhizas, the findings in Fig. 1-20 suggest that the upper end of the response curve was depressed because while supplying phosphorus needed by the plants, the fertilizer was also decreasing the percentage of the roots infected with mycorrhizal fungi, and loss of these mycorrhizas deprived the plants of part of their zinc supply. See Tinker and Gildon (1983) for a review of such interactions.

Cumming and Weinstein (1990) observed that mycorrhizal seedlings of pitch pine were relatively insensitive to concentrations of aluminum that were severely inhibitory to nonmycorrhizal seedlings. This behavior would have the effect of raising the level of the entire response curve for some nutrients and, for nutrients such as calcium, magnesium, and phosphorus that show interactions with aluminum, of reducing the addition required to reach the maximum yield.

From a comprehensive review of the literature, Harley and Smith (1983) were moved to comment that ''Mycorrhizas are indeed the chief organs involved in nutrient uptake of most land plants.'' Their view may be controversial for fertilized crop plants, especially the young plants commonly used in experimental work, because mycorrhizas usually do not become well developed for the first

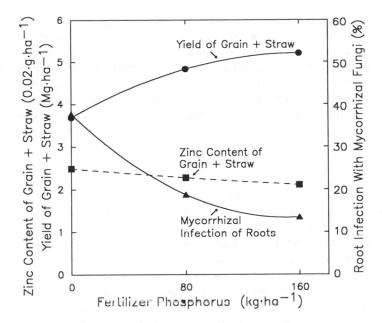

Fig. 1-20. Mycorrhizal infection of wheat roots, and yield and zinc content of the grain and straw in a field experiment with different additions of fertilizer phosphorus 5 years previously in Saskatchewan. Root infection with mycorrhizal fungi is the percentage of microscopic fields of roots in which infection was observed. (Singh et al., 1986)

few weeks. Root systems can be measured after plants are grown, but no comparable methods have been developed to measure the mycorrhizal extension. Plants with root systems free of mycorrhizas can be grown and studied experimentally, but the treatments to eliminate the mycorrhizas may have other effects, so that the contribution of mycorrhizas is not necessarily inferred accurately from the difference between plant responses with and without the treatment. Some predictions about root systems can be made that have important implications for plant responses to limiting factors, as will be brought out in Chapter 8 on liming, but in most work, roots and their associated mycorrhizas must be looked upon as an interacting complex that is related to roots.

1-3.8. Quality Characteristic Selected

Crop yield response is important with all crops. But high yields are not invariably the most valuable. An obvious example is Christmas trees, which increase in value up to a desirable size, but then diminish in value with greater size or yield.

Multiplying the yield by the price per unit produces a response curve expressed in terms of crop value. If the perceived quality of the crop is independent of the nutrient supply, so that the price per crop unit is constant, the shapes of the yield curve and the crop value curve are the same.

If the perceived value of the crop varies with the nutrient supply, so that the price per crop unit is variable, the shape of the crop value response curve obtained by multiplying the yield by the price per crop unit will differ from that of the yield response curve. The portion of the curve with the lower value per unit will be depressed relative to the yield response curve. Depending upon the nature of the crop, the quality factor, and the market, the crop value curve may be smooth or discontinuous. Holbrook et al. (1983) illustrated a situation with wheat in the United Kingdom in which the crop value curve was discontinuous, rising by 15% when the protein content of the grain reached 11%. The 15% premium was a consequence of improved quality of the high-protein flour (produced with higher levels of nitrogen fertilization) for bread making. Prediction of the quantity of fertilizer nitrogen required to move far enough up the response curve to produce the premium grain thus would be a matter of concern. Halvorson et al. (1986) found that the premium for high-protein wheat in the northern Great Plains of the United States can pay for 50 to 75% of the nitrogen fertilizer added to produce the high protein content.

With malting barley, the desired product is plump, high-starch grain with medium protein content. The plump kernels can be separated from the thin kernels by sieving, so that the premium connected with plump kernels is continuous and not abrupt, as with the premium for high-protein wheat. See Fig. 1-21 for a summary of results of nitrogen fertilization experiments in Idaho showing the yield of total grain, the yield of plump kernels, and the protein concentration in the grain versus the sum of fertilizer nitrogen and residual nitrate nitrogen in the soil. Note that the yield of plump kernels was well below that for total grain in the higher reaches of the response curve and that the quantity of nitrogen required to produce the maximum yield of total grain exceeded that required to produce the maximum yield of plump kernels. Experimental data such as those in Fig. 1-21 are a valuable aid in making site-specific predictions of nitrogen fertilizer needs. Consistent with the data in Fig. 1-21, Withers and Dyer (1990) observed in 39 experiments on nitrogen fertilization of barley in England and Wales that the median application of fertilizer nitrogen to obtain the maximum economic return over the cost of the fertilizer was 36 kilograms per hectare less than the median optimum application for yield.

1-3.9. Other Factors

Other factors also affect the shape of response curves. Weather variations and hazards cannot be predicted, but their effects can be taken into account in a probability sense on the basis of weather records. The time at which a given crop is fertilized is an important factor that can be taken into account. See Fig. 5-14B, which shows a major effect of delayed application in increasing the steepness of the curve (increasing the efficiency) representing the response of corn to fertilizer nitrogen applied to a sandy soil in Minnesota. The same graph shows that adding a nitrification inhibitor with nitrogen fertilizer applied before planting also greatly increased the slope of the response curve.

The effect of the length of time since fertilization is also of importance. One aspect of this subject, the residual effect on succeeding crops, is considered in Chapter 6.

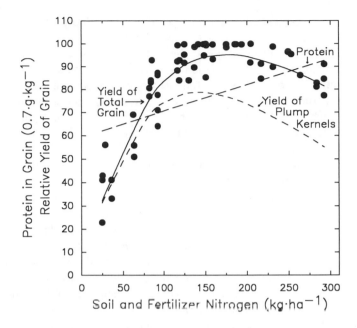

Fig. 1-21. Relative yields of barley grain and protein concentration in the grain versus the sum of the nitrate nitrogen found in the soil to a depth of 60 centimeters and the fertilizer nitrogen, which was added as urea or ammonium nitrate. The results shown are from nine field experiments conducted over an 8-year period at four locations in southern Idaho. The data points apply to the line for total yield of grain. In each experiment, the maximum observed yield was given a relative value of 100. The curve for yield of plump grain is derived from the regressions of total yield of grain and of the percentage of plump grain by weight on the sum of the soil and fertilizer nitrogen. In Canada and the United States, plump grain is defined as grain retained on a screen with 2.38-millimeter openings. (Stark and Brown, 1987)

The effects of the quantity of a nutrient required are of extreme importance. For example, a crop may take up 200 kilograms of nitrogen but less than 1 kilogram of molybdenum per hectare. The shapes of the response curves would be similar if the quantities applied were expressed as Baule units, but far different if the quantities were expressed as kilograms per hectare. Such differences among nutrients are reflected in the *c* value or efficiency factor in the Mitscherlich equation, and normally are taken into account automatically.

Also reflected in the *c* value of the Mitscherlich equation is the relative efficiency of different sources of a given nutrient. Chapter 5 is devoted to elaboration of this subject.

1-4. Limitations of Response Functions

According to Bliss (1970, p. 55), "A mathematical model developed from an intimate understanding of the biological mechanism should be utilized whenever feasible. The agreement of a series of observations with a postulated model is a necessary but not a sufficient test of its validity.''

Although Bliss's advice is good, the fact is that where plant responses to fertilizers are concerned, the situation is exceedingly complex, as was emphasized in Section 1-3 on factors affecting the shape of response curves. No simple mathematical function will describe the responses of field crops to fertilizers under all conditions. Moreover, the only sense in which simple functions are likely to be found valid is that they may represent certain experimental data to a good approximation.

Four significant limitations of mathematical representations of plant responses to fertilizers must be recognized. First, and most important, all the equations must be classed as empirical. None are really theoretical. The standard procedure of minimizing the sums of squares of deviations in curve fitting thus gives the best fit of data to a chosen response function, which must be selected arbitrarily. Different minimum sums of squares are obtained with different functions.

Second, the degree of correspondence to reality differs among equations. Although there is no doubt that a maximum yield is attainable with a finite supply of each growth factor, and that with greater supplies of the growth factor the yield decreases, this situation is not recognized by a number of equations. In some — the Mitscherlich equation for example — a maximum value of y is approached as x approaches infinity. In some other equations, y increases without limit as x increases. Still other equations pass through a maximum value of y and decrease with finite values of x, but describe curvatures that do not correspond to the shape of many yield-response curves. The commonly used quadratic equation is an example.

The deviations from reality that characterize the application of mathematical equations to heavy applications of fertilizers are of greater concern from the theoretical standpoint than they are in practice. The reason is that fertilizers are used to increase crop yields, not decrease them, so that from the practical standpoint one is concerned with the portion of the response curve that lies to the left of the maximum yield.

Third, although the availability of computers has facilitated the once tedious calculations required to evaluate the parameters in response functions by the method of least squares, some functions still are troublesome because of the computer time involved. Fitting a Mitscherlich-type equation, for example, requires a series of successive approximations through which the best-fitting values of the parameters are found by trial and error, a time-consuming approach with most programs now available. With suitable programming, however, a computer can make these long computations and can fit a number of successive data sets without continual input by the operator. With some programs, the time requirement is no longer a problem.

The complexity of the problem of curve fitting is increased when an equation is made more flexible by adding more parameters. At one time, Mitscherlich

(1928) published a paper describing a "Zweite Annäherung," or "second approximation," to cover situations in which decreases in yield indicate the association of injurious as well as beneficial effects with increasing supplies of a growth factor. In this equation,

$$y = A(1 - 10^{-cx})(10^{-kx^2}),$$

the greater the "injury factor" k, the lower is the observed maximum yield, and the more rapid is the downturn of the response curve with increasing values of x. (The parameter b for the supply of the nutrient in the soil does not appear in this equation, but it would need to be added or taken into account in some other way if the equation were used in practice.)

The regular Mitscherlich equation is difficult to fit by the method of least squares because of the successive approximations required, but the second approximation is even more difficult to fit. Mitscherlich did not use it in practice, but was steadfast in his dedication to his original equation, which may be considered the first approximation. He apparently used a graphic method in place of the method of least squares for fitting even the first approximation equation. Neeteson and Wadman (1987) allowed for the possible decrease in yield with heavy applications of fertilizer by subtracting from the Mitscherlich equation a linear term that is proportional to the quantity of fertilizer applied. This modification involves fitting a fourth parameter, but it is easier to deal with mathematically than Mitscherlich's second approximation. Neeteson and Wadman used a simplified method of curve fitting.

Fourth, in general, the parameters in the equations have no biological significance, but are mere mathematical devices. The Mitscherlich equation is an exception. All three of the parameters, A, b, and c, represent clear biological concepts — respectively, the maximum yield attainable by addition of the fertilizer, the supply of the nutrient in the soil when no fertilizer is added, and the effectiveness of unit quantity of the fertilizer in increasing the yield.

1-5. Other Response Functions

In preceding pages, principal emphasis has been on the Mitscherlich equation. The biological relevance of the constants A, b, and c and the many examples available because of the extensive use of the equation make the Mitscherlich equation an excellent basis for developing understanding as long as its empirical nature is kept in mind.

Reviews of certain response functions have been published by Mason (1957), Heady et al. (1961), Mead and Pike (1975), and others. The publications by Mason and Heady et al. are particularly relevant to crop response to fertilizers. Especially notable for clarity and philosophy are papers by Wood (1980) and Colwell (1983).

1-5.1. Polynomial Functions

As noted by Mead and Pike (1975), the most frequently used type of response function is the polynomial of degree p, where p is the exponent of x with the highest degree. The value of p is 1 in the linear equation,

$$y = a + bx,$$

2 in the quadratic equation,

$$y = a + bx + cx^2,$$

3 in the cubic equation,

$$y = a + bx + cx^2 + dx^3,$$

and so on. In these equations, y is the yield, x is the quantity of the nutrient added in the fertilizer, and a, b, c, and d are constants that must be evaluated for the specific set of data at hand.

Mead and Pike (1975) pointed out that for biological processes in general, aside from the straight line, the most commonly used form is the quadratic equation. The reasons for the popularity of this equation are probably: (1) it involves merely the addition of an extra term to the linear equation, which for most people makes it the simplest curvilinear relationship; (2) it has a simply defined maximum equal to $-b/2c$; and (3) the method of least squares commonly used in fitting mathematical functions to experimental data produces estimates of the constant terms with relatively simple calculations.

The quadratic equation once used so extensively to represent plant response to increasing supplies of nutrients has been criticized repeatedly in recent years as being inappropriate for many response curves observed experimentally. The quadratic equation gives a relatively poor representation of response curves that rise rapidly with small applications of a nutrient and then have a broad maximum range. In such instances, the quadratic equation peaks more sharply than the data, and the peak may represent unattainably high yields (Cerrato and Blackmer, 1990). Moreover, the slope of the curve often exceeds the slope of the data in the range just below the maximum. As has been pointed out by a number of investigators, this behavior has the important practical consequence of overestimating the quantity of fertilizer required to yield the maximum net profit, a subject that will be discussed further in Section 2-8.3.

The relatively sharp peaking of the quadratic curve was responsible for a finding by Heady et al. (1955) in an experiment used as an example in Chapter 2 that a "square root" function provided a better fit than the traditional quadratic function. In the square root equation, $x^{1/2}$ replaces x in the quadratic equation:

$$y = a + bx^{1/2} + cx.$$

The square-root equation is more realistic than the quadratic equation in many cases because it has a broader and flatter maximum than the quadratic equation. On the other hand, the slope of the curve approaches infinity as x approaches zero. This aspect of the square-root function is less realistic than the quadratic function.

1-5.2. Inverse Polynomial Functions

Inverse polynomials are equations that have a polynomial in the denominator. The one most widely referenced is the Balmukand (1928) equation. This equation represents a curve that approaches a maximum asymptotically, like the Mitscherlich equation. For a single nutrient — nitrogen, for example — the equation may be written

$$\frac{1}{y} = \frac{a_n}{N + n} + C,$$

where y is the yield, N is the quantity of fertilizer nitrogen, a_n and C are constants appropriate for nitrogen, and n is the supply of nitrogen in the soil in the same units as N. The parameter n thus is analogous to the Mitscherlich b value. In this equation, y approaches a maximum value $1/C$ as N approaches infinity.

The more general form for nitrogen, phosphorus, potassium, and other nutrients is

$$\frac{1}{y} = \frac{a_n}{N + n} + \frac{a_p}{P + p} + \frac{a_k}{K + k} + \ldots + C.$$

With P and K constant at suboptimum levels, the value of y in this equation approaches a maximum value as N approaches infinity. With N at infinity, the value of y approaches a new and higher maximum value as P approaches infinity, and so on.

A significant point about the Balmukand equation is the way it treats interactions. As mentioned in Section 1-2.2.2, one of the consequences of the hypothesized constancy of c in the Mitscherlich equation is that if a crop receives increasing quantities of one nutrient at each of two quantities of another, the ratios of the yields on the two response curves at corresponding quantities of the first nutrient will be independent of the quantities of the first nutrient. This postulated interaction was shown to represent the facts to a good approximation under some circumstances (Fig. 1-7), but not others (Fig. 1-9).

According to the Balmukand equation, the differences in reciprocals of the yields with the two levels of the second nutrient should be constant at different levels of the first nutrient. Balmukand illustrated this point by some data on yields of barley with 15 and 1215 milligrams of fertilizer nitrogen per culture at five levels of fertilizer phosphorus. The ratio of the yields with high nitrogen to low nitrogen increased from 4.9 with the smallest quantity of phosphorus up to 24 at the greatest. The difference between the reciprocals of the yields, however, ranged from 0.24 to 0.30, and there was no significant trend with the addition of phosphorus.

The results of a study by Greenwood (1970) support the finding by Balmukand. Greenwood analyzed the response data obtained in 67 field experiments in which effects of the fertilizer treatments were large and in which there were nutrient interactions in the statistical sense. His purpose was to find whether there was a transformation that would cause the treatment effects to be additive, thus eliminating the need for an interaction term in the response function. His findings showed that the maximum likelihood of additivity occurred with the reciprocals of the yield in 51 experiments, with the logarithm of yield in 8 experiments, and with the untransformed yields in 8 experiments.

The Balmukand equation is an example of a simple form of inverse polynomial. Enlarging on work by Nelder (1966), Wood (1980) illustrated three types of inverse polynomials. (1) The linear-over-linear type. In this equation, y (the yield) approaches a maximum as x (the quantity of fertilizer) approaches infinity. As represented by Wood, the form of the equation was $y = (a + bx)/(1 + cx)$. (2) The linear-over-quadratic type. In this equation, y reaches a maximum and then declines at a slower rate with values of x exceeding the maximum y than with values of x below the maximum y. Wood represented this equation as $y = (a + bx)/(1 + cx + dx^2)$. (3) The quadratic-over-linear type. According to Wood, the shape of the curve described by this form is similar to (2) except that y declines at a slower rate at values of x exceeding the maximum y. He represented the equation as $y = (a + bx + cx^2)/(1 + dx)$.

Greenwood et al. (1971) proposed an inverse polynomial equation that was modeled after the general form of the Balmukand (1928) equation. The Greenwood equation incorporates a modification to account for the yield-depressing effect of high levels of nitrogen, and it makes provision for substituting soil test values for nitrogen, phosphorus, and potassium for the n, p, and k parameters in the general Balmukand equation. The equation was used later by Greenwood et al. (1974, 1980a,b,c) as a basis for estimating the relative responsiveness of different crops to nitrogen, phosphorus, and potassium fertilization on different soils.

1-5.3. Segmented Functions

Moving further in the same direction as the square root equation, Swanson (1963) wrote a theoretical paper supportive of the Liebig concept. Boyd (1972) wrote a descriptive paper introducing, as an alternative to curvilinear response functions, the concept of a segmented function. He showed that the responses of crops to fertilizer nitrogen in some field experiments could be represented to a good approximation by two intersecting straight lines — an ascending line to represent the response, joining to a (usually) horizontal line to represent the yield plateau.

Waugh et al. (1973) published the first procedure for fitting the segmented linear function, and there have been improvements by Anderson and Nelson

(1975, 1987). The statistical procedure is described by Spector et al. (1985). Boyd et al. (1976), Cate (1979), and others have made further contributions, and segmented functions are now in common use. Segmented functions are sometimes called discontinuous functions because of the mathematical discontinuity that occurs at the junction of the lines that constitute the segments. Compared with nonsegmented or continuous functions that have essentially the same shape, the advantage of the segmented functions is that they are simpler and easier to fit.

In effect, the new trend represents an almost full-circle return to the Liebig law of the minimum, as represented by Blackman (1905) in Fig. 1-1. In fact, Fig. 1 in the bulletin by Waugh et al. (1973) is analogous to Fig. 1-1 from Blackman in this chapter. One difference is that although modern authors usually represent the plateau region beyond the linear response region as having zero slope, this is probably more a matter of convention than belief.

Waugh et al. (1973) used the term *linear response-and-plateau* to describe the response function consisting of intersecting straight lines. They represented the plateau as horizontal. Anderson and Nelson (1975) used the term *linear-plateau* model, implying the existence of "a region of linear response (with possibly more than one slope) and a plateau." Colwell et al. (1988) used the term *broken-stick*. The term *linear response and plateau* will be used here.

In the general case, the plateau may be level, or it may slope upward or downward. Whether the plateau is level, downward sloping, or level followed by a downward sloping segment, however, is not of much practical concern for the current crop because quantities of fertilizer great enough to advance the response into the region in which the yield increase is zero or negative are uneconomic and may have undesirable environmental effects.

The linear response and plateau function has received the greatest attention. Others are possible, however, and a quadratic response curve joined to a linear plateau has received some attention, as will be noted at the end of this section.

Fig. 1-22 shows three examples of crop responses to fertilization that have been represented as intersecting straight lines. The linear response and plateau function gives a good representation of the data in these examples, at least in the portion of the data of practical interest. The data from the experiment with corn in Fig. 1-22A provide an especially good illustration because of the wide range of quantities of fertilizer used in small increments and the small experimental error. A review of the response curves shown previously in this chapter will verify that although the data have been represented as curves, the linear response and plateau function would give a good representation of the data in a number of instances. In fact, the data in Fig. 1-22B, shown here as an example of intersecting straight-line responses, in accordance with the way the data were fitted by the authors, were represented in Fig. 1-21 by a curve.

Limited biological justification can be advanced for both a linear increase and a level plateau. Under some circumstances, and within the range of marked deficiency of a nutrient, there can be an essentially constant concentration of a nutrient in plants as the supply of the nutrient is increased. That is to say, a

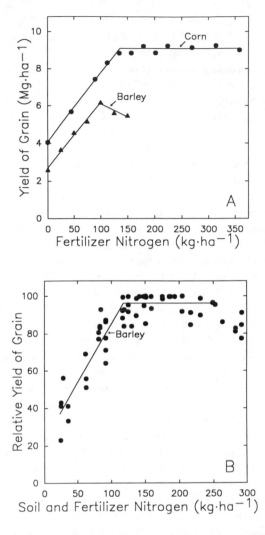

Fig. 1-22. Illustrations of the use of intersecting straight lines to represent the response of crops to fertilizer nitrogen in field experiments. A. Data from one experiment on corn in Oregon (Hunter and Yungen, 1955) and one experiment on barley in England (Boyd et al., 1976). B. Data on barley in Idaho (Stark and Brown, 1987). B uses the relative yield method to correlate the results of nine experiments over an 8-year period at four locations. The highest-yielding treatment in each experiment was given a relative yield of 100.

given quantity of a nutrient supports the production of a given quantity of plant dry matter, and the quantity of dry matter produced is proportional to the quantity of the nutrient in the plants.

A yield plateau is the expected biological consequence when two conditions are satisfied. First, the supply of a nutrient is great enough to produce the maximum yield attainable under the existing conditions. When this condition exists, increasing the supply of the nutrient beyond the minimum needed to produce the maximum yield of course will not increase the yield further. And second, there is a further range, beyond the minimum needed to produce the maximum yield, in which the nutrient is not present in great enough quantity to depress the yield as a result of the extra supply of the nutrient in the soil or in the plants (for example, as a result of toxicity of the nutrient itself or susceptibility of the plants to insects or diseases). The second condition obviously does not exist in the range where yields are depressed, and in some instances one or more factors may restrict the range of the plateau so much as to make it virtually nonexistent (see, for example, the results for barley in Fig. 1-22A).

The linear-response or constant-returns concept perhaps can be visualized most clearly as a logical outcome of experiments in which plants are grown in nonrenewed nutrient solutions or sand cultures with adequate supplies of all nutrients except one, which is supplied in increasing quantities within the range of marked deficiency. Under these circumstances, the plants will virtually exhaust the deficient nutrient from the culture medium, and if enough time is provided, they will produce as much dry matter as can be produced with the amount of the nutrient absorbed.

The exhaustion of the limiting nutrient from the medium, as described for nonrenewed nutrient solutions or sand cultures, may occur with chloride in soils. Other nutrients do not behave in this way. Nitrogen comes closest because nitrate is the principal form of nitrogen available to plants, and plants can virtually exhaust the nitrate from soils. After the pre-existing nitrate has been absorbed, however, nitrate production in the soil and absorption by plants continue, as may be inferred from experimental data reported by Bishop (1930). The situation in which plants either have or do not have a source of nitrate thus does not exist in soils that are producing nitrate.

With most nutrients, which are released from the soil solids as the supplies in the soil solution are depleted by plants, the soil supplies are not exhausted by plants as plants are caused to make greater growth with additional supplies of other nutrients or in other ways. Rather, the supplies are merely reduced. Moreover, although the forms supplied in fertilizers may be water-soluble when added, they generally do not remain so in the soil, so that once the added nutrients are present in the soil, they may take on some of the character of the indigenous supplies in the soil. This nutrient-buffering effect of soils would be expected to increase the range of fertilizer additions over which the transition occurs between the upward-sloping portion of response curves and the plateau portion. The transition should occur most rapidly with fertilizer sources of nutrients that can be exhausted from the soil by plants and least rapidly with fertilizer sources of nutrients that cannot be exhausted.

The most biologically unrealistic aspect of the linear response and plateau function is the implied abrupt change of slope at the intersection of the two lines. In many instances, the seeming resemblance of response curves to two inter-secting straight lines is enhanced by low experimental precision and few data points in the critical range. Few experiments provide good enough data to establish precise curvatures. Reid (1970) reported data from a field experiment on perennial ryegrass with 21 rates of fertilizer nitrogen replicated three times in increments of 28 or 56 kilograms per hectare in which the R^2 values for the response curve he fitted exceeded 0.99. Although the curvature seems unmistakable, even these data could be fitted well by two intersecting straight lines if the second linear segment is assigned an upward slope less than that of the first steeply ascending segment. In any event, the linear response and plateau function often performs well, as will be noted in Section 1-7, and it has now become popular.

The conceptual application of the linear response and plateau function to use of fertilizer nitrogen and phosphorus in different quantities is illustrated in Fig. 1-23. The data are from an experiment on corn reported by Heady et al. (1955).

In this experiment, the control yield with no fertilizer averaged 0.96 megagram per hectare. What Waugh et al. (1973) called the "threshold" yields for nitrogen and phosphorus (the yields with no addition of the nutrient named but with an adequate supply of the other or others, as the case might be) were 1.76 and 1.02 megagrams, respectively. Phosphorus thus was the first limiting factor, and nitrogen was the second. The plateau yield attainable with joint application of both nutrients in adequate quantities was estimated as 7.98 megagrams per hectare.[4]

The slope of the ascending segment of the function for fertilizer nitrogen was estimated by linear regression, with the result that $y = 1.87 + 0.0531N$. The minimum quantity of fertilizer nitrogen required to reach the plateau yield in the presence of an adequate supply of fertilizer phosphorus was calculated by setting y equal to the plateau yield and solving for N to obtain N = 115.3.

For fertilizer phosphorus, the results were represented as two linear segments plus a level plateau. The procedure was to calculate first the linear regression for the second linear segment, which corresponded to yields of 5.77, 6.68, 6.93, and 7.86 megagrams per hectare. The intersection of this segment with the plateau was calculated by setting the yield of the linear regression equal to the plateau yield (7.98 megagrams per hectare) and solving for the quantity of fertilizer phosphorus to obtain 84.3 kilograms. The first ascending linear segment then was represented as a straight line joining the threshold yield of 1.02 megagrams

---

[4]The data shown in the graphs were derived from an experiment with an incomplete 2×9 factorial design. Except for the yield of the control, which is the average yield from only two plots, the points are average yields of from 6 to 18 plots in which the indicated quantities of each nutrient were accompanied by from three to eight different quantities of the other nutrient. The groupings of treatments used are those employed by Anderson and Nelson (1987). The plateau yield is the average yield obtained on the 44 plots with the five highest levels of fertilizer phosphorus and the six highest levels of fertilizer nitrogen. The data points lying along the plateau in the figures were derived from these same yields, but they represent different numbers of plots and would need to be properly weighted to obtain the plateau yield. The original publication by Anderson and Nelson (1987) should be consulted to obtain a full appreciation for the data manipulations involved.

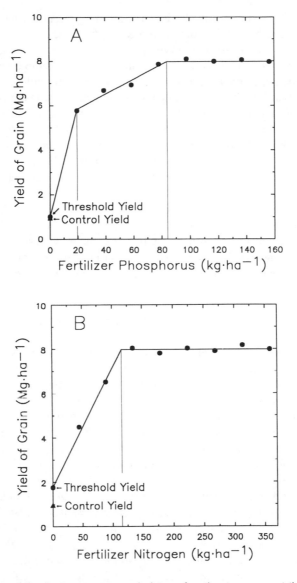

Fig. 1-23. Use of the linear response and plateau function to represent the response of corn to fertilizer phosphorus (A) and nitrogen (B) in an experiment in Iowa. (Heady et al., 1955)

with the point at which the regression for the second linear ascending segment intersected the quantity of fertilizer phosphorus (19.57 kg per hectare) used in the first increment of fertilizer; this yield was 5.83 megagrams per hectare, a little higher than the measured yield of 5.77 megagrams.

The existence of curvature in response data below the maximum yield may be allowed for by the ''quadratic-plateau'' function (Ihnen and Goodnight, 1985).

This function consists of a quadratic equation to represent the increase of yield with dose and a horizontal line to represent the plateau. In addition to the advantage of providing for the curvature observed in many circumstances, the quadratic-plateau function reduces or eliminates the undesirable tendency of the quadratic equation to overestimate the rate of increase of yield in the range just below the maximum yield. An application of the quadratic response and plateau model to experimental data will be described in Chapter 2. The data for phosphorus in Fig. 1-23 could have been fitted by a quadratic response and plateau function. Bock and Sikora (1990) described a modified quadratic response and plateau function with four or five parameters that is more flexible than the original version and provides improved estimates of economically optimum quantities of fertilizers.

If desired, the quadratic response and plateau function could have a plateau that slopes upward or downward. A continuous function with this feature that would be expected to perform about like the quadratic response and plateau function is the modified Mitscherlich equation used by Neeteson and Wadman (1987). This equation includes a linear term with an adjustable slope.

1-6. Selecting a Response Function

With a set of response data at hand, which of the many response functions that have been suggested should be used to represent the data for viewing or for further calculations? Other than "a function suitable for the purpose and available resources," no categorical answer can be given to this question.

One of the problems in selection is that the data may be fitted about equally well by diverse functions, as illustrated by Fig. 1-24 for four of the most popular equations. The commonly used R^2 test for goodness of fit yielded values ranging from a low of 0.663 with the quadratic equation to a high of 0.684 with the Mitscherlich equation. The square root equation and the linear response and plateau functions had R^2 values of 0.683. Colwell (1983) extended the picture in Fig. 1-24 by using each of nine equations to represent two sets of experimental data yielding response curves differing in shape. His paper includes the original numerical data, the numerical values of the parameters in the fitted equations, the R^2 values, and the response curves assembled in sequential order with the equations for easy comparison. Cerrato and Blackmer (1990) published a similar graphic comparison of the fit of five different response equations to a given set of data.

Campbell and Keay (1970) suggested as the basis for a series of functions representing the law of diminishing returns the hypothesis that the rate of increase of yield with respect to the supply of the nutrient (dy/dx) is proportional to the decrement from the maximum yield $(A - y)$ raised to an adjustable power:

$$\frac{dy}{dx} = \alpha(A - y)^{m+1},$$

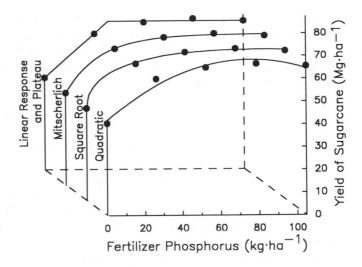

Fig. 1-24. Yields of sugarcane with different quantities of fertilizer phosphorus in an experiment in Brazil. The points show the experimental data, and the lines show the way four different response functions fit the data. (Colwell et al., 1988)

where α is a constant and m is the adjustable element in the exponent. They noted that when $m = 0$, the differential equation is the same as the one from which the Mitscherlich equation was developed.

Mombiela and Nelson (1981) independently used this concept and developed it into a way to decide among mathematical functions for representing a particular set of response data. With use of A for the theoretical maximum yield and y as the yield obtained with a particular quantity x of a nutrient, the objective of the procedure they described is to find which power of $(A - y)$ is directly proportional to the rate of increase of yield with respect to the supply of the nutrient (dy/dx). This is done by finding a transformation of the decrements from the maximum $(A - y)$ that yields a linear plot of the transformed decrements against x. When the chosen function eliminates the curvature from the plot, and the points are scattered around a straight line, the function fits the data. The special advantage of the linearization test is that it leads relatively easily to an appropriate response function for the data to be fitted. The transformations are made by estimating the theoretical maximum yield (A_{est}), calculating the decrements from the estimated maximum for all observed values of y, applying a trial exponent c to all the decrements, and calculating the resulting values of $(A - y)^c$. The transformed values then are plotted against the corresponding values of x to see if linearity has been achieved.

Mombiela and Nelson (1981) showed by some mathematical manipulations that the equations used most commonly to represent plant responses to nutrient supplies are different members of the "power series," in which the decrements from the maximum yield are raised to some arbitrary power. The "resistance" formula of Balmukand (1928) and the linear inverse polynomial equation of

Nelder (1966) may be derived from the assumption that the rate of increase of yield with respect to the supply of the nutrient is proportional to the square of the decrement from the maximum yield, that is, $dy/dx = c(A - y)^2$, where c is the proportionality factor. Testing conformance to this equation may be done by plotting values of $(A_{est} - y)^{-1}$ against x.

The Mitscherlich equation may be derived from the assumption that the rate of increase of crop yield with respect to the supply of the nutrient is proportional to the decrement from the maximum yield, that is, $dy/dx = c(A - y)$. To test conformance of data to this function, values of $\log(A_{est} - y)$ are plotted against x.

The quadratic equation may be derived from the assumption that the rate of increase of the yield with respect to the supply of the nutrient is proportional to the square root of the decrement from the maximum yield, that is, $dy/dx = c(A - y)^{1/2}$. Conformance to this equation may be tested by plotting values of $(A_{est} - y)^{1/2}$ against x.

The square root equation, $y = a + bx^{1/2} + cx$, may be derived from the hypothesis that $dy/d\sqrt{x} = c(A - y)^{1/2}$. According to Mombiela and Nelson, a plot of $(A - y)^{1/2}$ against $x^{1/2}$ will be linear for data that fit the square root function.

The exponents of $(A - y)$ that correspond to the well-known equations are all arbitrary. Other exponents thus may be used if they provide a better fit.

Examples of transformations based upon precise response data from the National Fertilizer Development Centre (1983) are found in Fig. 1-25. Fig. 1-25A shows the basic yield data and a plot of untransformed $(A_{est} - y)$ values against x. The National Fertilizer Development Centre employed the quadratic equation to fit the data, and the value 4.76 used for A_{est} in the figure is the maximum yield calculated from their equation. The usual upward concavity of the plot of untransformed values of $(A - y)$ against x is evident.

Fig. 1-25B includes plots of two sets of transformed values of $(A_{est} - y)$. The values for $\log (5.4 - y)$ correspond to the Mitscherlich equation, and the values for $(6.9 - y)^{-1}$ correspond to the Balmukand resistance formula and the Nelder linear inverse polynomial. Both plots are approximately linear, but the A values required to obtain the best linearity are not the same, and both exceed the calculated maximum value 4.76 for the quadratic function.

Fig. 1-25C includes plots of three sets of transformed values of $(A_{est} - y)$. The plot corresponding to the transformation $(4.76 - y)^{2/3}$ is slightly concave upward, but the curvature is less than that of the untransformed values of $(4.76 - y)$ in Fig. 1-25A, indicating that the transformation $(4.76 - y)^{2/3}$ gives a better fit than no transformation, for which the exponent of $(4.76 - y)$ is 1. The plot corresponding to the transformation $(4.76 - y)^{1/3}$ is slightly concave downward. The change of curvature from upward, where the exponent is 2/3, to downward, where the exponent is 1/3, indicates that the best linearity will be obtained with an intermediate exponent. The central line, corresponding to the quadratic equation [transformation $(4.76 - y)^{1/2}$], is a good linear plot, verifying that the quadratic equation is appropriate for use with the data.

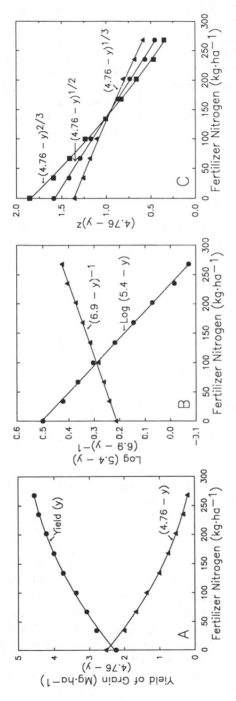

Fig. 1-25. Curve-fitting procedures employing the average yields of wheat grain in 165 experiments with different quantities of fertilizer nitrogen in Pakistan. A. Plots of the average yields (*y*) and the differences between the estimated maximum yield 4.76 and the observed yield *y* against the quantity of fertilizer nitrogen. The maximum yield was estimated by the quadratic equation used to represent the data. B. Plots showing that with suitably selected values for the estimated maximum yield, the Mitscherlich equation [(plot of log (5.4 − *y*)] and the Nelder inverse linear polynomial and Balmukand equations [plot of (6.9 − *y*)⁻¹] would represent the average yields to a good approximation. C. Plots of three exponential values of the difference between the estimated maximum yield 4.76 and the observed yield *y* against the quantity of fertilizer nitrogen. Note that with the exponent *z* = 2/3, the plot is slightly concave upward; with the exponent *z* = 1/3, the plot is slightly concave downward; and with the exponent *z* = 1/2 (corresponding to the quadratic equation), the plot can be represented by a straight line. These observations indicate that the response curve corresponding to the exponent *z* = 1/2 fits the data better than response curves corresponding to the exponents *z* = 1/3 and 2/3. (Data from the National Fertilizer Development Centre, 1983)

Additional calculations on the transformations made with the 2/3 and 1/3 exponents yielded other results that provide insight into the linearization process. All the plots in Fig. 1-25C were made using 4.76 as the value for A_{est}. This value was chosen because it was the maximum yield in the quadratic equation calculated by the National Fertilizer Development Centre (1983) to represent the results. The upward concavity in the plot of $(A_{est} - y)^{2/3}$ against x suggested that the value of A was too high. The choice of 4.6 instead of 4.76 yielded a linear plot with only the point for no fertilizer slightly off the line. The downward concavity of the plot of $(A_{est} - y)^{1/3}$ against x could be eliminated by using A_{est} = 4.9. This transformation resulted in a linear plot with only a slight waviness in the line. Similarly, plotting $\log (A_{est} - y)$ against x (a test of applicability of the Mitscherlich equation — Mombiela and Nelson noted that the log transformation is equivalent to an exponent of 0) yielded a curve that was concave downward when A_{est} was taken as 4.76, but as shown in Fig. 1-25B, use of A_{est} = 5.4 yielded a good approximation of a linear plot.

As described, the Mombiela and Nelson (1981) procedure does not apply to logarithmic and negative exponent transformations of data sets in which one or more experimental yield values are equal to or greater than the estimated maximum yield. To circumvent the mathematical impasse presented by these situations, Mombiela and Nelson added 1 to the value of $(A_{est} - y)$ to all the values in the experimental series they used as an example before making the transformation. In general terms, this procedure implies that a number equal to $1 + (y_t - A_{est})$ should be added before applying the transformation to all the values in the experimental series, where y_t is the yield value that exceeds what seems to be the maximum for the response curve as a whole.

Several comments seem appropriate regarding the data from the National Fertilizer Development Centre (1983) employed in the examples in Fig. 1-25. First, nine different quantities of nitrogen fertilizer were employed under conditions of marked nitrogen deficiency in the control plots; this gave a good coverage of the portion of the response curve that is most important from the practical standpoint. Second, the figures are averages of the results obtained in 165 experiments. The deviations of experimental values from a smooth curve thus are relatively small, and this makes the effects of the various transformations more evident than they would be if the transformations were applied to the data from a single experiment with only a few replicates and much scattering of points. And finally, the shape of a response curve derived from the means of responses measured in many individual experiments is not necessarily the same as the shapes of the response curves for the individual experiments that contribute to the mean responses. This point has been emphasized particularly for means of responses that can be represented by intersecting straight lines. Means of the experimental observations in several such response functions with intersections at different additions of a nutrient would yield a curve.

If desired, a least-squares linear regression could be calculated for a linearized transformation to obtain calculated values for the parameters. These parameters would be more precise than the approximations that might be derived from the graphs. Although the values would represent least-squares estimates based upon the transformed data rather than the original untransformed data, this may be considered an advantage if the range of yields is so great that the variance is not homogeneous (Pairunan et al., 1980). Campbell and Keay (1970) took the logarithm of yields of forage crops before using the method of least squares to fit the data on responses to fertilizer phosphorus additions. Then they transformed the data back to yields for presentation. They noted that their procedure stabilized the variance and also eliminated a sigmoid tendency that appeared in some of the data and would have required a special equation to represent the sigmoid tendency if the logarithmic transformation had not been used.

None of the functions mentioned in preceding paragraphs apply specifically to sigmoidal response curves. Cochrane (1988) proposed the Mitscherlich equation for use in representing the portion of sigmoidal response curves to the right of the inflection. The best-known function for fitting sigmoidal curves beginning at $x = 0$ is the logistic equation (Nair, 1954). Murray and Nunn (1987) listed five equations for sigmoid-type curves and tested two more.

Barrow and Mendoza (1990) described a versatile sequence of three equations for fitting crop responses to fertilizers. They pointed out that a function that would fit sigmoid responses could be produced by adding an exponent n to the quantity of fertilizer nutrient x in a modified exponential form of the Mitscherlich equation to obtain

$$y = A - Be^{-cx^n},$$

where y is the yield with the quantity x of the nutrient applied in the fertilizer, A is the maximum yield, e is the base of natural logarithms, c is the Mitscherlich efficiency factor, and B and n are parameters. B is equal to Ae^{-cb}, where b is the effective quantity of the nutrient in soil and seed expressed in the same units as the quantity x of nutrient added. This function leads asymptotically to a maximum yield like the unmodified Mitscherlich equation, but it can also fit data with an increasing slope near the origin. When $n = 1$, the modified version becomes the Mitscherlich equation. As n increases, the sigmoid tendency becomes more pronounced, and the maximum yield is approached more rapidly.

Barrow and Mendoza (1990) continued by noting that the Mitscherlich equation is a special case of the law of diminishing returns in which the rate of increase of yield y with respect to x is proportional to the decrement from the maximum yield. That is, $dy/dx = c(A - y)$, where A is the maximum yield, y is the yield with a particular quantity x of the nutrient, and c is the proportionality constant. In a more general case, $dy/dx = c(A - y)^{1+m}$. When $m = 0$, the equation reduces to the basic assumption from which the Mitscherlich equation is derived. When m is not equal to 0, the integrated version becomes an equation given by Campbell and Keay (1970), which Barrow and Mendoza described as a generalized hyperbola:

$$y = A - \frac{B}{(1 + mcx)^{1/m}} \quad .$$

This equation does not yield sigmoid curves, but it is more flexible than the Mitscherlich equation. Combining this equation with the modified version of the Mitscherlich equation gives

$$y = A - \frac{B}{(1 + mcx^n)^{1/m}} \quad .$$

This equation combines the capability to represent sigmoid responses introduced by adding the exponent n with flexibility in rate at which the maximum yield is approached.

Selecting a suitable response function for the situation at hand is an important consideration in research. But the functions are not the ultimate objective. Rather, they are to be regarded as tools. The practical purpose of the functions is to fit experimental data and to express the physical and biological relationships in a way that will be useful to the researcher and appropriate for the use to be made of the information. The response functions are all empirical, even though some might embody an aspect of fundamental behavior, such as reaction of an added nutrient with the soil, resulting in reduced availability.

In the usual situation in which several different response functions may fit a set of data about equally well, one must make an arbitrary decision as to how far to go in attempts to reduce the sum of squares of deviations. In a larger sense, if the research objective is, for example, to estimate the response to applications of particular quantities of fertilizers to farmers' fields on the basis of the responses obtained in a group of experiments conducted for calibration purposes, together with the values of soil tests on samples of soils from the farmers' fields, the appropriate procedure would not be to find the response functions and the values of the parameters in these functions that lead to the minimum sums of squares of deviations in the individual experiments. The objective of putting together the data from the individual experiments into a useful overall framework cannot be accomplished in any convenient way when the findings in the individual experiments are represented by a heterogeneous group of response functions. In experimental work such as that indicated, the group of experiments is to be looked upon as an experiment, and a response function suitable for the group of experiments is to be sought.

1-7. Response Functions Compared

The results described in connection with the linearization process in Section 1-6 show that several different response models represented the experimental data to a good approximation. These findings are analogous to the situation discussed in Section 1-2.2 for the Mitscherlich equation, which provides a good approximation for a variety of response curves with a constant value of the proportionality factor c if the values for A and b are allowed to vary at will.

The experimental series chosen to serve as an example in Section 1-6 did not include data in the plateau region. As a consequence, different A_{est} values could be used without causing points near the maximum to deviate markedly from the curve. Had the range of additions of fertilizer nitrogen been increased to include yields in the plateau region, fewer transformations would have given a good fit.

Numerous comparisons have been made of the goodness of fit of experimental data to two or more response functions. The most significant comparisons involve many experiments, each using a number of quantities of fertilizer. Findings made in comparisons meeting these criteria more or less well have been published by several investigators, including Waugh et al. (1973), Jónsson (1974), Anderson and Nelson (1975), Boyd et al. (1976), Sparrow (1979a,b), Neeteson and Wadman (1987), Colwell et al. (1988), and Cerrato and Blackmer (1990).

Boyd et al. (1976) and Needham and Boyd (1976) noted that the linear response and plateau function was generally superior to curvilinear functions for representing the response of barley to nitrogen fertilization in experiments in southwestern England, but that curvilinear responses were associated with sites at which there was considerable leaf disease. Aside from this specific observation, the results of the various comparisons of response functions will not be reviewed in detail because of the lack of information about causal factors. Suffice it to say here that the findings do not agree, the linear response and plateau function often showed up well, and the quadratic equation was never favored. The performance of certain functions will be discussed further in connection with specific applications.

Some comments by Colwell (1983) on differences among response functions are relevant from the mathematical standpoint. He pointed out that "The ease with which actual functions can be estimated from data, their suitability for standard analyses of variance, their suitability for generalization for two or more nutrients, and the ease of calculation of optimal rates, all vary enormously between models." The polynomial equations are the most convenient. (The two on the list of equations he discussed were the square root equation and the now often maligned quadratic equation.) He regarded most of the others as "near impossibly complex, with highly unstable parameters . . . when extended to two or more nutrients."

Colwell (1983) described instability as "large compensatory variations in the parameter values" that may result from small variations in data. For example, a high value for yield with the maximum application of fertilizer in an experimental series would increase the estimated A value for maximum yield in the Mitscherlich equation, at the same time decreasing the c value for curvature (which reflects the efficiency of the nutrient in increasing the yield) and increasing the b value (which estimates the soil supply of the nutrient in the absence of the fertilizer). As an example of a function with unstable parameters, Colwell referred to the Baule generalization of the Mitscherlich equation (see Section 1-2.2.3).

Moreover, increasing the number of parameters in a function to increase its flexibility requires correspondingly elaborate experiments in terms of number of treatment levels. The expense of obtaining the basic data is thus increased. Computations then become more time consuming.

The commonly used R^2 criterion for goodness of fit has the advantage of allowing comparisons between functions having different numbers of parameters. Along with this advantage come two limitations. (1) Comparisons are best made only among different functions applied to a common set of data. A reason for this is that the R^2 value increases with the range of the data. (2) R^2 values can be affected by the number and distribution of data points. Thus, for example, a concentration of many points in the zone representing small quantities of fertilizer would tend to decrease the R^2 value for the square root equation in comparison with the quadratic model because the square root equation fits many sets of data poorly in that region. Conversely, a concentration of many points in the region of maximum response would tend to decrease the R^2 value for the quadratic model relative to the square root model because the quadratic model fits many sets of data poorly in the region of maximum response.

In the usual situation, where several response functions may provide a good fit to a specific set of data, ease of curve-fitting is a rational basis for choice among functions. In most instances, no other check is made. Holford and Cullis (1985), however, described a situation in which reanalysis of a given set of response data with use of a different response function led to choice of a different soil test method and to much improved estimates of the phosphorus fertilizer requirement in the experiments involved. The original soil test method, which had been widely adopted, was relatively ineffective according to the new analysis. Situations in which the choice of a response curve has such important conscquences may well be rare, but they are not evident until the proof is at hand. In the instance encountered by Holford and Cullis, the authors attributed the superiority of the reanalysis to the use of a response function in which the two important factors — attainable increase in yield and efficiency of the fertilizer in increasing the yield — were represented explicitly by parameters in the function. Although relatively easy to fit, the response function employed originally did not recognize the two important factors explicitly, but distributed their effects between other parameters.

1-8. Statistical Computations

Directions for using the least-squares method to fit mathematical models commonly employed to represent crop responses to fertilizers have been given by Spillman (1933), Pimentel Gomes (1953), Statistics Division (1966), Guinard (1982), Waggoner and Norvell (1979), and others. The calculations, however, are arduous where more than three levels of fertilization with a single nutrient are used. Most calculations now are made with the aid of computer programs. Colwell (1978) published programs with explanation and worked examples specifically for soil fertility and fertilizer studies. A more recent report on the same subject has been published by the same author with colleagues (Colwell et al., 1988). Widely used generalized computer programs for statistical analyses are available through the SAS Institute (1985).

For the purpose of fitting mathematical functions to plant response data, advances in computer software by Jandel Scientific have made it possible to circumvent most of the need for mathematical and statistical knowledge and almost all of the drudgery connected with the actual fitting. These problems have constituted a significant hurdle to many soil fertility specialists in the past. With the software, it is possible to make selections from several thousand equations, to rank the equations in order of goodness of fit, to show the data and a selected fitted curve on the screen, and to export the information to another program that prints a high quality graph. The software makes the process easy and fast. The large number of equations that will fit a given set of plant response data with R^2 values or F values almost equal, however, emphasizes the empirical nature of the process and the absence of a theoretical response curve. The most important need of all, that of formulating the basic theoretical relationships and developing means to make the appropriate measurements for experimental use, is unfortunately not a matter of computer programming.

1-9. Practical Applications

The results of the comparisons of various mathematical response functions are consistent with the fact that response curves may have different shapes, depending upon the circumstances. When one looks at experimental data with hindsight, a function can be selected by inspection and trial and error that fits the data as well as or better than various other functions that may be of interest. The parameters in the function may be estimated by the method of least squares, and estimates of goodness of fit may be obtained. Although supplying some scientific enlightenment, the results of such exercises in mathematics and statistics are of no practical value for the experiments involved because the experiments have become history.

The ultimate goal of scientific work in soil fertility evaluation and control is to make predictions about what will happen if certain practices are followed in the future. Analyses of what happened in the past are useful primarily in developing understanding and in guiding predictions. Precise predictions require knowledge of the shapes of the response curves as well as the locations on the response curves where the results are expected to fall.

Predicting the shapes of the response curves requires information on the conditions that modify the shape. Some of the most important needed information, that related to weather conditions, cannot be predicted. As indicated in

Section 1-3, however, estimates of the effects of a number of other factors not often taken into account can be made before a crop is grown. With the aid of suitable calibration experiments, this information can be used to develop modifications of response functions that will lead to improved predictions. Thus, although comparisons of the goodness of fit of various response functions found in the literature and in this chapter to data from experiments that have been conducted in the past may be of no practical value for use with completed experiments, they may be useful in providing a place to start in developing more complex prediction functions that incorporate additional measurable factors. Predictions may be expected to improve with more research.

An important goal for the future is an experimentally based system for integrating soil and plant analyses into a scheme of multinutrient response curves for advisory work. Progress in this direction seems most likely to occur if a program of experiments designed specifically for the purpose can be developed.

Well along toward completion in many countries is the computerization of data processing leading to routine advisory recommendations. With the new system, desired recommendations can be worked out and printed automatically, as opposed to the step by step approach used in the past, in which some of the input information may not have been formalized. Computerization has largely eliminated the limitations due to data processing.

Along with the computerization of existing data and data-manipulation processes have been continuing efforts to develop new and improved ways to estimate crop responses to environmental conditions, including aspects of soil fertility. Efforts have proceeded along two lines. In one approach, researchers have gone as far as they considered worthwhile in incorporating measurements of environmental conditions to develop and operate their models. Research of this kind is valuable for finding what can be done with present knowledge and for pointing the way to simpler modifications, but the models generally require so much input information that they are impractical. In the other approach, efforts are made to develop improved systems of prediction based upon limited information that is easily obtained, so that the models may have practical use. Several models developed for specific purposes will be discussed in succeeding chapters, and a few of the simpler ones will be discussed in some detail as examples. As the science of soil fertility evaluation and control advances, the collection of needed input information will become increasingly the primary limiting factor in precision of predictions.

1-10. Yield Versus Nutrient-Uptake Plots

The preceding sections in this chapter have dealt with the matter of direct practical interest in fertilizer application, namely, "what is the relation of the output, in terms of crop quantity, quality, or both, to the fertilizer input?" Input-output curves unfortunately are situation-specific. Differences in response curves from one situation to another may be a consequence of differences in uptake of the nutrient by the crop, differences in the effectiveness of the absorbed nutrient in promoting the yield of the crop, or both.

De Wit (1953) developed a useful way of expressing data that provides information on both the uptake and utilization aspects. Fig. 1-26 shows, for an experiment on nitrogen fertilization of permanent pasture, the way the data are represented graphically and illustrates some of the kinds of results that may be obtained. Plotted in the first quadrant is the yield of dry matter versus the yield of nitrogen in the crop. The fourth quadrant contains a plot of the yield of nitrogen in the crop versus the quantity of nitrogen applied in the fertilizer. The second quadrant shows a plot of the yield of dry matter against the quantity of nitrogen applied in the fertilizer; this is the conventional representation of the input-output results of a fertilizer experiment except that the direction of plotting the fertilizer application is reversed because the first quadrant has already been used.

In Fig. 1-26B, the yield of dry matter associated with a given yield of nitrogen in the crop, shown in the first quadrant, appears to be the same for the early and late applications of fertilizer nitrogen. Consequently, a single line has been used to express the relationship. A single line means that whether the nitrogen was absorbed from the early application or the late application, equal amounts of absorbed nitrogen were equally effective in promoting the yield of dry matter.

The plots of yield of nitrogen against quantity of fertilizer nitrogen applied, shown in the fourth quadrant, do not coincide. The slope of the line representing the late application exceeds the slope of the line representing the early application. Thus, the recovery of the fertilizer nitrogen in the crop was greater when the fertilizer was applied late than early. The ratio of the slopes is equal to the ratio of the recoveries in the crop of the fertilizer nitrogen applied at the two different times.

In Fig. 1-26A, the yield of dry matter versus yield of nitrogen plots for early and late applications coincide at low yields, but diverge as the yields increase. The relative locations of the curves signify that at the higher yields a given amount of nitrogen absorbed from the early application was more effective in producing dry matter than an equal amount absorbed from the late application. The yield-of-nitrogen versus nitrogen-applied relationships in the fourth quadrant were again represented by diverging straight lines, but in this instance, the relative recoveries of fertilizer nitrogen in the vegetation are the reverse of those in Fig. 1-26B. That is, the percentage recovery of nitrogen by the crop from the early application exceeded the recovery from the late application.

The plots in Fig. 1-26 do not give the causes of the observations, but the results are much more enlightening than the standard response curves that show only input and output. They provide a basis for hypotheses about causes that could be tested by other appropriate experiments.

In the same publication from which the data for Fig. 1-26 were taken were analogous experiments on peat and clay soils. In 2 years on the peat soil and in

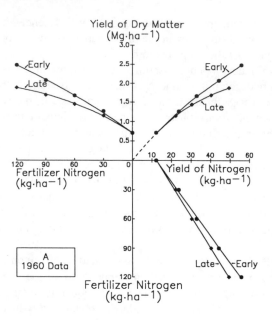

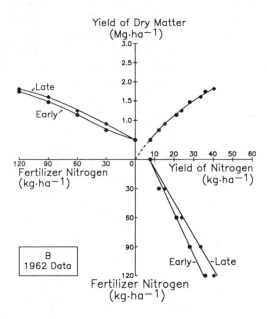

Fig. 1-26. A representation of the yield of dry matter and yield of nitrogen data in a nitrogen fertilization experiment on permanent pasture on a sandy soil in The Netherlands in 1960 and 1962. The experiment compared the effects of early and late applications of ammonium nitrate limestone. (Oostendorp, 1964)

1 year on the clay soil, the plots of yields of nitrogen in the crop versus the quantities of fertilizer nitrogen applied could be represented by a single line, signifying that the percentage recoveries of early and late applications were approximately equal. Although the recoveries were approximately equal, the plots of yield of dry matter against yield of nitrogen showed that in each instance the effectiveness of a given amount of absorbed nitrogen in producing dry matter was greater at high yields with the early application than with the late application — another item of information that conceivably could be explained by appropriate experiments.

In Fig. 1-26, the plots of nitrogen uptake versus fertilizer nitrogen supplied are essentially straight lines within the range of applications in the experiments. This behavior is commonly observed with nitrogen, but less commonly with phosphorus, for example.

A publication by Van Keulen and Van Heemst (1982) contains many examples from the literature, some involving phosphorus and potassium. As is well illustrated in this publication, the plots of yield of dry matter against the yield of a nutrient under different conditions tend to coincide at low levels of the nutrient and to diverge at higher levels. The reason for this behavior is that at low levels of a nutrient the yield is controlled mainly by the supply of the nutrient, and at high levels the yield is controlled mainly by other factors. In this connection, Van Keulen and Van Heemst (1982) made the significant observation that the initial slopes of the plots of yield of dry matter versus yield of nutrient are fairly crop-specific. For example, the initial slope has a value of about 70 kilograms of rice and wheat grain per kilogram of nitrogen absorbed by the crop and about 100 kilograms of potato tuber dry matter (450 kilograms of fresh tubers) per kilogram of nitrogen absorbed by the crop. The values for yield of nitrogen are for the above-ground portion of the crop or, in the case of potato, the above-ground portion plus the tubers.

The behavior of nutrients in crops will be discussed further in Chapter 3. Other uses of plots of the type shown in Fig. 1-26 will be made in Chapters 5 and 7 on fertilizer evaluation and placement.

1-11. Literature Cited

Agenbag, G. A., and O. T. de Villiers. 1989. The effect of nitrogen fertilizers on the germination and seedling emergence of wild oat (*A. fatua* L.) seed in different soil types. *Weed Research* 29:239-245.

Anderson, A. J. 1970. Trace elements for sheep pastures and fodder crops in Australia. *Journal of the Australian Institute of Agricultural Science* 36:15-29.

Anderson, A. J., and D. V. Moye. 1952. Lime and molybdenum in clover development on acid soils. *Australian Journal of Agricultural Research* 3:95-110.

Anderson, R. L., and L. A. Nelson. 1975. A family of models involving intersecting straight lines and concomitant experimental designs useful in evaluating response to fertilizer nutrients. *Biometrics* 31:303-318.

Anderson, R. L., and L. A. Nelson. 1987. *Linear-plateau and plateau-linear-plateau models useful in evaluating nutrient responses.* North Carolina Research Service Technical Bulletin 283. Raleigh.

Azcon, R., and J. A. Ocampo. 1981. Factors affecting the vesicular-arbuscular infection and mycorrhizal dependency of thirteen wheat cultivars. *New Phytologist* 87:677-685.

Baker, A. S., S. Kuo, and Y. M. Chae. 1981. Comparisons of arithmetic average soil pH values with the pH values of composite samples. *Soil Science Society of America Journal* 45:828-830.

Balmukand, B. 1928. Studies in crop variation. V. The relation between yield and soil nutrients. *Journal of Agricultural Science (Cambridge)* 18:602-627.

Barrow, N. J., and R. E. Mendoza. 1990. Equations for describing sigmoid yield responses and their application to some phosphate responses by lupins and subterranean clover. *Fertilizer Research* 22:181-188.

Bauer, A., R. A. Young, and J. L. Ozbun. 1965. Effects of moisture and fertilizer on yields of spring wheat and barley. *Agronomy Journal* 57:354-356.

Baule, B. 1918. Zu Mitscherlich's Gesetz der physiologischen Beziehungen. *Landwirtschaftliche Jahrbücher* 51:363-385.

Baylis, G. T. S. 1975. The magnolioid mycorrhiza and mycotrophy in root systems derived from it. Pp. 373-389. In F. E. Sanders, B. Mosse, and P. B. Tinker (Eds.), *Endomycorrhizas.* Academic Press, London.

Bishop, L. R. 1930. The nitrogen content and "quality" of barley. *Journal of the Institute of Brewing (London)* 36:352-364.

Biswas, S. K., and K. N. Singh. 1982. Nitrogen management in barley and wheat I: Yield, yield attributes and economics of production. *Indian Journal of Agronomy* 27:241-247.

Blackman, F. F. 1905. Optima and limiting factors. *Annals of Botany* 19:281-295.

Bliss, C. I. 1970. *Statistics in Biology. Statistical Methods for Research in the Natural Sciences.* Volume II. McGraw-Hill, NY.

Bock, B. R., and F. J. Sikora. 1990. Modified-quadratic plateau model for describing plant responses to fertilizer. *Soil Science Society of America Journal* 54:1784-1789.

Bolan, N. S., A. D. Robson, and N. J. Barrow. 1983. Plant and soil factors including mycorrhizal infection causing sigmoidal response of plants to applied phosphorus. *Plant and Soil* 73:187-201.

Bolland, M. D. A., D. G. Allen, and R. J. Gilkes. 1989. The influence of seasonal conditions, plant species and fertilizer type on the prediction of plant yield using the Colwell bicarbonate test for phosphate. *Fertilizer Research* 19:143-158.

Bolland, M. D. A., and M. J. Baker. 1987. Increases in soil water content decrease the residual value of superphosphate. *Australian Journal of Experimental Agriculture* 27:571-578.

Borden, R. J. 1945. Nitrogen depletion by soil organisms. *Hawaiian Planters Record* 49:251-257.

Boyd, D. A. 1972. Some recent ideas on fertilizer response curves. Pp. 461-473. In *Role of Fertilization in the Intensification of Agricultural Production.* Proceedings, 9th Congress, International Potash Institute, Berne, Switzerland.

Boyd, D. A., L. T. K. Yuen, and P. Needham. 1976. Nitrogen requirement of cereals. I. Response curves. *Journal of Agricultural Science (Cambridge)* 87:149-162.

Brandon, D. M., and D. S. Mikkelsen. 1979. Phosphorus transformations in alternately flooded California soils: I. Cause of plant phosphorus deficiency in rice rotation crops and correctional methods. *Soil Science Society of America Journal* 43:989-994.

Burns, G. R., D. R. Bouldin, and C. A. Black. 1963. A possible cause of sigmoid yield-of-phosphorus curves. *Soil Science Society of America Proceedings* 27:715-716.

Campbell, N. A., and J. Keay. 1970. Flexible techniques in describing mathematically a range of response curves of pasture species. Pp. 332-334. In M. J. T. Norman (Ed.), *Proceedings of the XI International Grassland Congress, Queensland, Australia*. University of Queensland Press, St. Lucia.

Carlson, H. L., and J. E. Hill. 1985. Wild oat (*Avena fatua*) competition with spring wheat: Effects of nitrogen fertilization. *Weed Science* 34:29-33.

Cate, R. B. 1979. Law of the minimum. Pp. 258-260. In R. W. Fairbridge and C. W. Finkl, Jr. (Eds.), *The Encyclopedia of Soil Science Part 1. Physics, Chemistry, Biology, Fertility, and Technology*. Dowden, Hutchinson & Ross, Stroudsburg, PA.

Cerrato, M. E., and A. M. Blackmer. 1990. Comparison of models for describing corn yield response to nitrogen fertilizer. *Agronomy Journal* 82:138-143.

Cochrane, T. T. 1988. A differential equation to estimate fertilizer response curves. *Soil Science Society of America Journal* 52:525-529.

Colwell, J. D. 1978. *Computations for Studies of Soil Fertility and Fertilizer Requirements*. Commonwealth Agricultural Bureaux, Slough, England.

Colwell, J. D. 1983. Fertilizer requirements. Pp. 795-815. In *Soils: An Australian Viewpoint*. CSIRO, Division of Soils, Melbourne. Academic Press, London.

Colwell, J. D., A. R. Suhet, and B. van Raij. 1988. *Statistical procedures for developing general soil fertility models for variable regions*. CSIRO (Australia) Division of Soils, Divisional Report No. 93.

Cumming, J. R., and L. H. Weinstein. 1990. Aluminum-mycorrhizal interactions in the physiology of pitch pine seedlings. *Plant and Soil* 125:7-18.

Dahnke, W. C., and R. A. Olson. 1990. Soil test correlation, calibration, and recommendation. Pp. 45-71. In R. L. Westerman (Ed.), *Soil Testing and Plant Analysis*. Third Edition. Soil Science Society of America, Madison, WI.

Davis, M. R. 1991. The comparative phosphorus requirements of some temperate perennial legumes. *Plant and Soil* 133:17-30.

Demolon, A., and P. Boischot. 1951. Réaction des sols à l'apport de phosphates solubles. Doses isodynames. *C. R. Acad. Sci., Paris* 233:509-512.

De Wit, C. T. 1953. *A physical theory on placement of fertilizers*. Verslagen van Landbouwkundige Onderzoekingen No. 59.4. Centrum voor Landbouwpublicaties en Landbouwdocumentatie, Wageningen, The Netherlands.

Draycott, A. P., M. J. Durrant, D. B. Davies, and L. V. Vaidyanathan. 1976. Sodium and potassium fertilizer in relation to soil physical properties and sugar-beet yield. *Journal of Agricultural Science (Cambridge)* 87:633-642.

Fixen, P. E., G. W. Buchenau, R. H. Gelderman, T. E. Schumacher, J. R. Gerwing, F. A. Cholick, and B. G. Farber. 1986a. Influence of soil and applied chloride on several wheat parameters. *Agronomy Journal* 78:736-740.

Fixen, P. E., R. H. Gelderman, J. Gerwing, and F. A. Cholick. 1986b. Response of spring wheat, barley, and oats to chloride in potassium chloride fertilizers. *Agronomy Journal* 78:664-668.

Föhse, D., N. Claassen, and A. Jungk. 1988. Phosphorus efficiency of plants. I. External and internal P requirement and P uptake efficiency of different plant species. *Plant and Soil* 110:101-109.

Foy, C. D., A. L. Fleming, and J. W. Schwartz. 1981. Differential resistance of weeping lovegrass genotypes to iron-related chlorosis. *Journal of Plant Nutrition* 3:537-550.

Foy, C. D., P. W. Voigt, and J. W. Schwartz. 1980. Differential tolerance of weeping lovegrass genotypes to acid coal mine spoils. *Agronomy Journal* 72:859-862.

Godbold, D. L., W. J. Horst, J. C. Collins, D. A. Thurman, and H. Marschner. 1984. Accumulation of zinc and organic acids in roots of zinc tolerant and non-tolerant ecotypes of *Deschampsia caespitosa*. *Journal of Plant Physiology* 116:59-69.

Greenwood, D. J. 1970. Interactions between the beneficial effects of nitrogen, phosphate and potassium fertiliser on yield. P. 49. In *National Vegetable Research Station, Twentieth Annual Report 1969*. Wellesbourne, Warwick.

Greenwood, D. J., T. J. Cleaver, and M. K. Turner. 1974. Fertiliser requirements of vegetable crops. *Fertiliser Society (London) Proceedings* 145:5-31.

Greenwood, D. J., T. J. Cleaver, M. K. Turner, J. Hunt, K. B. Niendorf, and S. M. H. Loquens. 1980a. Comparison of the effects of potassium fertilizer on the yield, potassium content and quality of 22 different vegetable and agricultural crops. *Journal of Agricultural Science (Cambridge)* 95:441-456.

Greenwood, D. J., T. J. Cleaver, M. K. Turner, J. Hunt, K. B. Niendorf, and S. M. H. Loquens. 1980b. Comparison of the effects of phosphate fertilizer on the yield, phosphate content and quality of 22 different vegetable and agricultural crops. *Journal of Agricultural Science (Cambridge)* 95:457-469.

Greenwood, D. J., T. J. Cleaver, M. K. Turner, J. Hunt, K. B. Niendorf, and S. M. H. Loquens. 1980c. Comparison of the effects of nitrogen fertilizer on the yield, nitrogen content and quality of 21 different vegetable and agricultural crops. *Journal of Agricultural Science (Cambridge)* 95:471-485.

Greenwood, D. J., and A. Draycott. 1988. Recovery of fertilizer-N by diverse vegetable crops: Processes and models. Pp. 46-61. In D. S. Jenkinson and K. A. Smith (Eds.), *Nitrogen Efficiency in Agricultural Soils*. Elsevier Applied Science, London.

Greenwood, D. J., J. T. Wood, T. J. Cleaver, and J. Hunt. 1971. A theory for fertilizer response. *Journal of Agricultural Science (Cambridge)* 77:511-523.

Grunes, D. L., and V. V. Rendig (Eds.). 1979. *Grass Tetany*. ASA Special Publication No. 35. American Society of Agronomy, Crop Science Society of America, and Soil Science Society of America, Madison, WI.

Grybauskas, A. P., A. L. Their, and D. J. Sammons. 1988. Effect of chloride fertilizers on development of powdery mildew of winter wheat. *Plant Disease* 72:605-608.

Guinard, A. 1982. Economic optimization of fertilizer applications: a method for field staff based on response curves and surfaces. *Tropical Agriculture* 59:257-264.

Halvorson, A. D., A. L. Black, D. L. Watt, and A. G. Leholm. 1986. Economics of a one-time phosphorus application in the northern Great Plains. *Applied Agricultural Research* 1:137-144.

Harley, J. L., and S. E. Smith. 1983. *Mycorrhizal Symbiosis*. Academic Press, London.

Harry, S. P., and R. D. Graham. 1981. Tolerance of triticale, wheat and rye to copper deficiency and low and high soil pH. *Journal of Plant Nutrition* 3:721-730.

Heady, E. O., L. G. Albaugh, J. C. Engibous, B. L. French, E. W. Kehrberg, D. D. Mason, and J. T. Pesek. 1961. *Status and Methods of Research in Economic and Agronomic Aspects of Fertilizer Response and Use*. National Academy of Sciences — National Research Council Publication 918. Washington.

Heady, E. O., J. T. Pesek, and W. G. Brown. 1955. *Crop response surfaces and economic optima in fertilizer use.* Iowa Agricultural Experiment Station Research Bulletin 424. Ames.

Hoagland, D. R. 1919. Relation of the concentration and reaction of the nutrient medium to the growth and absorption of the plant. *Journal of Agricultural Research* 18:73-117.

Holbrook, J. R., W. R. Byrne, and W. J. Ridgman. 1983. Assessment of results from wheat trials testing varieties and application of nitrogenous fertilizer. *Journal of Agricultural Science (Cambridge)* 101:447-452.

Holford, I. C. R. 1982. Effects of phosphate sorptivity on long-term plant recovery and effectiveness of fertilizer phosphate in soils. *Plant and Soil* 64:225-236.

Holford, I. C. R., and B. R. Cullis. 1985. Effects of phosphate buffer capacity on yield response curvature and fertilizer requirements of wheat in relation to soil phosphate tests. *Australian Journal of Soil Research* 23:417-427.

Howeler, R. H., L. F. Cadavid, and E. Burckhardt. 1982. Response of cassava to VA mycorrhizal inoculation and phosphorus application in greenhouse and field experiments. *Plant and Soil* 59:327-339.

Huber, D. M. 1980. The role of mineral nutrition in defense. Pp. 381-406. In J. G. Horsfall and E. B. Cowling (Eds.), *Plant Disease. An Advanced Treatise. Vol. V. How Plants Defend Themselves.* Academic Press, NY.

Hunter, A. S., and J. A. Yungen. 1955. The influence of variations in fertility levels upon the yield and protein content of field corn in eastern Oregon. *Soil Science Society of America Proceedings* 19:214-218.

Ihnen, L. A., and J. H. Goodnight. 1985. The NLIN procedure. Pp. 575-606. In *SAS User's Guide: Statistics.* Version 5 Edition. SAS Institute, Cary, NC.

Janssen, B. H. 1990. A double-pot technique as a tool in plant nutrition studies. Pp. 759-763. In M. L. van Beusichem (Ed.), *Plant Nutrition — Physiology and Applications.* Kluwer, Dordrecht, The Netherlands.

Johnson, G. V. 1991. General model for predicting crop response to fertilizer. *Agronomy Journal* 83:367-373.

Jónsson, L. 1974. On the choice of a production function model for nitrogen fertilization on small grains in Sweden. *Swedish Journal of Agricultural Research* 4:87-97.

Kafkafi, U., and J. Putter. 1965. Some aspects of sigmoidal yield-response curves related to the geometry of granule fertilizer availability. *Israel Journal of Agricultural Research* 15:169-178.

Keulen, H. van, and H. D. J. van Heemst. 1982. *Crop response to the supply of macronutrients.* Verslagen van Landbouwkundige Onderzoekingen (Agricultural Research Reports) No. 916. Centre for Agricultural Publishing and Documentation, Wageningen, The Netherlands.

Kim, S. C., and K. Moody. 1980. Effect of plant spacing on the competitive ability of rice growing in association with various weed communities at different nitrogen levels. *Journal of the Korean Society of Crop Science* 25, No. 4:17-27.

Kletschkowsky, W. M., and P. A. Shelesnow. 1931. Ueber Verschiebungen der Wirkungsfaktoren von Stickstoff und Phosphorsäure. *Landwirtschaftliche Jahrbücher* 74:353-404.

Knapp, W. R., and W. S. Reid. 1981. *Interactions of hybrid maturity class, planting date, plant population, and nitrogen fertilization on corn performance in New York.* Search: Agriculture, No. 21. Cornell University Agricultural Experiment Station, Ithaca, NY.

Lange, A. 1938. Untersuchungen über den Wachstumsfaktor Wasser. *Landwirtschaftliche Jahrbücher* 85:465-499.

Liebig, J. von. 1855. *Die Grundsätze der Agrikultur-Chemie mit Rücksicht auf die im England angestellten Untersuchungen.* Friedrich Vieweg und Sohn, Braunschweig. (This book was translated to English and appeared as Liebig, J. von. 1855. *Principles of Agricultural Chemistry, With Special Reference to the Late Researches Made in England.* Walton & Maberly, London.)

Lorenz, O. A., and C. M. Johnson. 1953. Nitrogen fertilization as related to the availability of phosphorus in certain California soils. *Soil Science* 75:119-129.

Mason, D. D. 1957. Functional models and experimental designs for characterizing response curves and surfaces. Pp. 76-98. In Baum, E. L., E. O. Heady, and J. Blackmore (Eds.), *Methodological Procedures in the Economic Analysis of Fertilizer Use Data.* Iowa State College Press, Ames.

Mason, R., T. L. Jackson, and L. D. Calvin. 1991. Supplementing experimental results with survey data. *Journal of Production Agriculture* 4:272-277.

McCree, K. J., and S. G. Richardson. 1987. Salt increases the water use efficiency in water stressed plants. *Crop Science* 27:543-547.

Mead, R., and D. J. Pike. 1975. A review of response surface methodology from a biometric viewpoint. *Biometrics* 31:803-851.

Meyer, R., and A. Storck. 1927/1928. Ueber den Pflanzenertrag als Funktion der Stickstoffgabe und der Wachstumszeit bei Hafer. *Zeitschrift für Pflanzenernährung Düngung Bodenkunde* 10A:329-347.

Middleton, K. R. 1984. Economic control of fertiliser in highly productive pastoral systems. V. Extending advice through the Mitscherlich-Liebig model. *Fertilizer Research* 5:77-93.

Mitscherlich, E. A. 1909. Das Gesetz des Minimums und das Gesetz des abnehmenden Bodenertrages. *Landwirtschaftliche Jahrbücher* 38:537-552.

Mitscherlich, E. A. 1928. Die zweite Annäherung des Wirkungsgesetzes der Wachstumsfaktoren. *Zeitschrift für Pflanzenernährung Düngung Bodenkunde* 12A:273-282.

Mitscherlich, E. A. 1939. *Klimatische Einflüsse auf die ertragssteigernde Wirkung der Pflanzennährstoffe.* Schriften der Königsberger Gelehrten Gesellschaft, Naturwissenschaftliche Klasse, 15 Jahr, Heft 5.

Mitscherlich, E. A. 1947. Das Ergebnis von über 27000 Feld-Düngungsversuchen. *Zeitschrift für Pflanzenernährung Düngung Bodenkunde* 38:22-35.

Mitscherlich, E. A., F. Dühring, S. v. Saucken, and C. Böhm. 1923. Die pflanzenphysiologische Lösung der chemischen Bodenanalyse. *Landwirtschaftliche Jahrbücher* 58:601-617.

Mombiela, F. A., and L. A. Nelson. 1981. Relationships among some biological and empirical fertilizer response models and use of the power family of transformations to identify an appropriate model. *Agronomy Journal* 73:353-356.

Moody, K. 1981. *Weed-fertilizer interactions in rice.* IRRI [International Rice Research Institute] Research Paper Series No. 68.

Mulder, E. G. 1950. Effect of liming of an acid peat soil on microbial activity. *Trans. Fourth International Congress of Soil Science* 2:117-121.

Mulder, E. G. 1952. Fertilizer vs. legume nitrogen for grasslands. *Proceedings of the Sixth International Grassland Congress* 1:740-748.

Mulder, E. G. 1954. Effect of mineral nutrition on lodging of cereals. *Plant and Soil* 5:246-306.

Munk, H. 1985. Ermittlung wirtschaftlich optimaler Phosphatgaben auf Löss- und Geschiebelehmböden auf Basis der CAL-Methode. *Zeitschrift für Pflanzenernährung Bodenkunde* 148:193-213.

Murray, A. W. A., and P. A. Nunn. 1987. A non-linear function to describe the response of % nitrogen in grain to applied nitrogen fertiliser. *Aspects of Applied Biology* 15:219-225.

Nair, K. R. 1954. The fitting of growth curves. Pp. 119-132. In O. Kempthorne, T. A. Bancroft, J. W. Gowen, and J. L. Lush (Eds.), *Statistics and Mathematics in Biology*. Iowa State College Press, Ames.

National Fertilizer Development Centre. 1983. *The physical response of crops to fertilizer applications: 1, Wheat*. Study Paper Series-1. National Fertilizer Development Centre, Government of Pakistan, Islamabad.

Needham, P., and D. A. Boyd. 1976. Nitrogen requirement of cereals. 2. Multi-level nitrogen tests with spring barley in south-western England. *Journal of Agricultural Science (Cambridge)* 87:163-170.

Neeteson, J. J., and W. P. Wadman. 1987. Assessment of economically optimum application rates of fertilizer N on the basis of response curves. *Fertilizer Research* 12:37-52.

Nelder, J. A. 1966. Inverse polynomials, a useful group of multi-factor response functions. *Biometrics* 22:128-141.

Olsen, S. R., and F. S. Watanabe. 1957. A method to determine a phosphorus adsorption maximum of soils as measured by the Langmuir isotherm. *Soil Science Society of America Proceedings* 21:144-149.

Oostendorp, D. 1964. Stikstofbemesting en grasgroei in het voorjaar. *Landbouwkundige Tijdschrift* 76:101-110.

Paauw, F. van der. 1952. Critical remarks concerning the validity of the Mitscherlich effect law. *Plant and Soil* 4:97-106.

Pairunan, A. K., A. D. Robson, and L. K. Abbott. 1980. The effectiveness of vesicular-arbuscular mycorrhizas in increasing growth and phosphorus uptake of subterranean clover from phosphorus sources of different solubilities. *New Phytologist* 84:327-338.

Parker, M. B., and H. B. Harris. 1962. Soybean response to molybdenum and lime and the relationship between yield and chemical composition. *Agronomy Journal* 54:480-483.

Pimentel Gomes, F. 1953. The use of Mitscherlich's regression law in the analysis of experiments with fertilizers. *Biometrics* 9:498-516.

Pinkerton, A. 1991. Critical phosphorus concentrations in oilseed rape (*Brassica napus*) and Indian mustard (*Brassica juncea*) as affected by nitrogen and plant age. *Australian Journal of Experimental Agriculture* 31:107-115.

Pons, T. L., and H. G. W. van Wieringen. 1985. Effect of nitrogen fertilizer on competition of *Marsilea crenata* Presl. with rice. *Biotrop Bulletin in Tropical Biology* 23:59-65. [*Soils and Fertilizers* 49, Abstract 9874, 1986.]

Poulsen, E. 1959. Nitrate fertilization by means of ion exchange. *Physiologia Plantarum* 12:826-833.

Ragland, J. L., and N. T. Coleman. 1959. The effect of soil solution aluminum and calcium on root growth. *Soil Science Society of America Proceedings* 23:355-357.

Ramig, R. E. 1960. *Relationships of Soil Moisture at Seeding Time and Nitrogen Fertilization to Winter Wheat Production*. Ph.D. Thesis, University of Nebraska, Lincoln.

Reid, D. 1970. The effects of a wide range of nitrogen application rates on the yields from a perennial ryegrass sward with and without white clover. *Journal of Agricultural Science (Cambridge)* 74:227-240.

Rhoads, F. M., F. G. Martin, and R. L. Stanley, Jr. 1988. Plant population as a guide to N fertilization of irrigated corn. *Journal of Fertilizer Issues* 5:67-71.

Rippel, A., W. Estor, and R. Meyer. 1926/1927. Zur experimentellen Widerlegung des Mitscherlich-Baule'schen Wirkungsgesetzes der Wachstumsfaktoren. *Zeitschrift für Pflanzenernährung Düngung Bodenkunde* 8A:65-80.

Sah, R. N., and D. S. Mikkelsen. 1986. Sorption and bioavailability of phosphorus during the drainage period of flooded-drained soils. *Plant and Soil* 92:265-278.

Sanders, F. E., and P. B. Tinker. 1973. Phosphate flow into mycorrhizal roots. *Pesticide Science* 4:385-395.

SAS Institute. 1985. *SAS User's Guide: Basics.* Version 5 Edition. *SAS User's Guide: Statistics.* Version 5 Edition. (These are two separate volumes.) SAS Institute, Cary, NC.

Schreiner, O., B. E. Brown, J. J. Skinner, and M. Shapovalov. 1920. *Crop injury by borax in fertilizers.* U.S. Department of Agriculture, Department Circular 84.

Sears, P. D. 1953. Pasture growth and soil fertility. I. The influence of red and white clovers, superphosphate, lime, and sheep grazing, on pasture yields and botanical composition. *New Zealand Journal of Science and Technology* 35, Sec. A, Supplement 1:1-29.

Shepherd, L. N., J. C. Shickluna, and J. F. Davis. 1959. The sodium-potassium nutrition of sugar beets produced on organic soil. *Journal of the American Society of Sugar Beet Technologists* 10:603-608.

Sinclair, A. G., P. W. Shannon, and W. H. Risk. 1990. Sechura phosphate rock supplies plant-available molybdenum for pastures. *New Zealand Journal of Agricultural Research* 33:499-502.

Singh, J. P., R. E. Karamanos, and J. W. B. Stewart. 1986. Phosphorus-induced zinc deficiency in wheat on residual phosphorus plots. *Agronomy Journal* 78:668-675.

Singh, R., and R. A. Agarwal. 1983. *Fertilizers and pest incidence in India.* Potash Review, Subject 23, Plant Protection and Plant Diseases, 64th Suite, No. 11.

Singh, S., G. Singh, and R. Vasisht. 1984. Chemical weed control to economise fertilizer use in wheat. II. Yield attributes, response and economics of production. *Indian Journal of Agronomy* 29:424-428.

Soil Chemistry and Plant Breeding Departments. 1976. Tolerance to phosphorus deficiency. Pp. 168-170. In *International Rice Research Institute Annual Report for 1975.* International Rice Research Institute, Los Baños, Laguna, Philippines.

Soil Chemistry/Physics, Plant Breeding, and Plant Physiology Departments, and International Rice Testing Program. 1984. Screening. Pp. 94-97. In *International Rice Research Institute Annual Report for 1983.* International Rice Research Institute, Los Baños, Laguna, Philippines.

Sparrow, P. E. 1979a. Nitrogen response curves of spring barley. *Journal of Agricultural Science (Cambridge)* 92:307-317.

Sparrow, P. E. 1979b. The comparison of five response curves for representing the relationship between the annual dry-matter yield of grass herbage and fertilizer nitrogen. *Journal of Agricultural Science (Cambridge)* 93:513-520.

Spector, P. C., J. H. Goodnight, J. P. Sall, and W. S. Sarle. 1985. The GLM procedure. Pp. 433-506. In *SAS User's Guide: Statistics.* Version 5 Edition. SAS Institute, Cary, NC.

Spillman, W. J. 1933. *Use of the exponential yield curve in fertilizer experiments.* U.S. Department of Agriculture Technical Bulletin 348.

Stark, J. C., and B. D. Brown. 1987. Estimating nitrogen requirements for irrigated malting barley. *Communications in Soil Science and Plant Analysis* 18:433-444.

Statistics Division. 1966. *Statistics of Crop Response to Fertilizers.* Food and Agriculture Organization, Rome.

Stewart, I., and C. D. Leonard. 1952. The cause of yellow tipping in citrus leaves. *Proceedings of the Florida State Horticultural Society* 65:25-27.

St. John, T. V. 1980. Root size, root hairs and mycorrhizal infection: A re-examination of Baylis's hypothesis with tropical trees. *New Phytologist* 84:483-487.

Stout, P. R., W. R. Meagher, G. A. Pearson, and C. M. Johnson. 1951. Molybdenum nutrition of crop plants. I. The influence of phosphate and sulfate on the absorption of molybdenum from soils and solution cultures. *Plant and Soil* 3:51-87.

Sumner, M. E., and M. P. W. Farina. 1986. Phosphorus interactions with other nutrients and lime in field cropping systems. *Advances in Soil Science* 5:201-236.

Swanson, E. R. 1963. The static theory of the firm and three laws of plant growth. *Soil Science* 95:338-343.

Timm, C. A., R. J. Goos, B. E. Johnson, F. J. Sobolik, and R. W. Stack. 1986. Effect of potassium fertilizers on malting barley infected with common root rot. *Agronomy Journal* 78:197-200.

Tinker, P. B., and A. Gildon. 1983. Mycorrhizal fungi and ion uptake. Pp. 21-32. In D. A. Robb and W. S. Pierpoint (Eds.), *Metals and Micronutrients: Uptake and Utilization by Plants.* Academic Press, London.

Van der Paauw, F. 1952. Critical remarks concerning the validity of the Mitscherlich effect law. *Plant and Soil* 4:97-106.

Van Keulen, H., and H. D. J. van Heemst. 1982. *Crop response to the supply of macronutrients.* Verslagen van Landbouwkundige Onderzoekingen (Agricultural Research Reports) No. 916. Centre for Agricultural Publishing and Documentation, Wageningen, The Netherlands.

Viebrock, H. 1988. *Ursachen der Erhöhung des Phosphat-Aneignungsvermögens von Pflanzen durch VA-Mykorrhiza.* Ph.D. Thesis, Göttingen. (Quoted by Jungk, A., and N. Claassen. 1989. Availability in soil and acquisition by plants as the basis for phosphorus and potassium supply to plants. *Zeitschrift für Pflanzenernährung Bodenkunde* 152:151-157.)

Von Liebig, J. 1855. *Die Grundsätze der Agrikultur-Chemie mit Rücksicht auf die im England angestellten Untersuchungen.* Friedrich Vieweg und Sohn, Braunschweig. (This book was translated to English and appeared as Liebig, J. von. 1855. *Principles of Agricultural Chemistry, With Special Reference to the Late Researches Made in England.* Walton & Maberly, London.)

Wadleigh, C. H., and A. D. Ayers. 1945. Growth and biochemical composition of bean plants as conditioned by soil moisture tension and salt concentration. *Plant Physiology* 20:106-132.

Waggoner, P. E., and W. A. Norvell. 1979. Fitting the law of the minimum to fertilizer applications and crop yields. *Agronomy Journal* 71:352-354.

Walker, T. W. 1955. Sulphur responses on pastures in Australia and New Zealand. *Soils and Fertilizers* 18:185-187.

Wallace, A. 1990a. Crop improvement through multidisciplinary approaches to different types of stresses — law of the maximum. *Journal of Plant Nutrition* 13:313-325.

Wallace, A. 1990b. Interactions of two parameters in crop production and in general biology: Sequential additivity, synergism, antagonism. *Journal of Plant Nutrition* 13:327-342.

Wallace, A., and G. Wallace. 1990. The meaning of multiple fraction yield plot (MFYP) in conventional and high technological agriculture and also in environmentally safe agriculture. *Journal of Plant Nutrition* 13:309-312.

Waugh, D. L., R. B. Cate, Jr., and L. A. Nelson. 1973. *Discontinuous models for rapid correlation, interpretation, and utilization of soil analysis and fertilizer response data.* International Soil Fertility Evaluation & Improvement Program, North Carolina State University, Technical Bulletin No. 7. Raleigh.

Willey, R. W., and S. B. Heath. 1969. The quantitative relationships between plant population and crop yield. *Advances in Agronomy* 21:281-321.

Wit, C. T. de. 1953. *A physical theory on placement of fertilizers.* Verslagen van Land-bouwkundige Onderzoekingen, No. 59.4. Centrum voor Landbouwpublikaties en Landbouwdocumentatie, Wageningen, The Netherlands.

Withers, P. J. A., and C. Dyer. 1990. The effect of applied nitrogen on the acceptability of spring barley for malting. *Aspects of Applied Biology* 25:329-337.

Wollring, J. 1990. Das optimale N-Angebot im Frühjahr bei unterschiedlicher Keimdichte von Winterweizen. *Journal of Agronomy and Crop Science* 164:137-143.

Wood, J. 1980. The mathematical expression of crop responses to inputs. Pp. 263-271. In *Physiological Aspects of Crop Productivity.* Proceedings of the 15th Colloquium of the International Potash Institute. International Potash Institute, Worblaufen-Bern, Switzerland.

CHAPTER 2

Economics of Fertilization

*If we wanted to, we could grow crops on billiard balls.
It's just a matter of the economics.*

THIS SPUR-OF-THE-MOMENT ANSWER by the late Charles E. Kellogg to a question at the end of a seminar he gave at Iowa State University many years ago was his way of emphasizing the importance of economics in practical crop production. Because it involved supplying crops with nutrients, the quotation dramatizes also the importance of economics in use of fertilizers on soils, which is the subject of this chapter.

The primary objective of commercial producers in applying fertilizers is to make a profit. The extent to which their use of fertilizers contributes to this objective depends not only upon the kinds and amounts of fertilizers they apply, the way they apply them, and the crop responses that result, but also upon the costs of fertilization and the prices received for the crops. Both the physical and economic realities must be recognized.

The first part of this chapter reviews the concepts and procedures involved in interpreting crop responses in economic terms. A discussion follows on the limitations of the basic crop response data, the problems of interpretation created by attempts to apply mathematical models to these data, and the consequent poor precision with which economic theory can be applied to make predictions of practical value. Ways to deal with the limitations are indicated. The chapter ends with a discussion of certain environmental aspects of the economics of fertilization. The economics of residual effects will be considered in Chapter 6.

2-1. Optimum Quantity of Fertilizer

The term *optimum* means best. No single quantity of fertilizer, however, can be designated as the optimum without qualifying the frame of reference. Agronomists like to think in terms of the maximum yield they can produce by whatever means. Economists prefer to think in terms of the magnitude of economic returns.

The economic optimum for one person may not be the same as the economic optimum for another. For example, if the tenant buys the fertilizer but the landlord gets half of the crop, the optimum application for the landlord is the quantity that produces the maximum crop yield obtainable by applying the fertilizer. The optimum application for the tenant is lower, and in some instances might be zero.

If the farm is owner-operated, the optimum fertilizer application depends upon the supply of capital. Making the best economic decision about the amount of money to invest in fertilizer requires that the operator know not only the costs of fertilization and the returns to be obtained, but also the costs of possible competing uses of available funds and the returns to be obtained from them. Such knowledge probably is never available. If capital is limited, a decision must be made as to how much to divert to other enterprises so as to obtain the maximum overall net profit. If the farmer borrows money to buy the fertilizer, the optimum application will depend upon the rate of interest charged on the loan.

Another economic imponderable that influences the definition of optimum is the uncertainty about the benefits to be obtained. The benefits are projected from existing information based upon history to a time in an uncertain future. Who knows when projections may be too low or too high because the economic conditions lead to greater or lesser returns from fertilizer than were predicted? And who knows when projections may be vitiated and money invested in fertilizer lost because a crop is damaged or destroyed by hail, drought, frost, pests, or some other force that is beyond the control of the operator?

From the standpoint of consumers of agricultural products, the optimum economic applications of fertilizers would be those that would result in the lowest prices of the products. Such applications would be expected to exceed the economic optima for producers. The complexity of the situation is increased if the government enters the economic arena by buying surplus crops at prices above those on the world market in which it sells the crops. Taxpayers must subsidize this operation. The societal issue is clouded by speculations about the unevaluated costs of nitrate in groundwater, loss of atmospheric ozone as a result of nitrous oxide derived from nitrogen fertilizers, and degradation of the quality of surface waters as a result of the growth of unwanted algae and other vegetation due to the presence of nutrients derived from fertilizers.

All in all, the economics of fertilization is complex. For situations in which fertilizer use is profitable enough to compete with other enterprises, the rational range of use by farmers having other profitable investment opportunities but limited capital for investment will fall somewhere between the break-even application and the application yielding the maximum net profit from fertilization. Two intermediate optima that have been suggested are the maximum profit per dollar invested in fertilization and the maximum profit per dollar invested in crop production. Each of these economic concepts and others are discussed in this chapter.

2-2. Economic Principles Applied to Response Functions

The principles involved in the economics of fertilization are independent of the mathematical functions used to represent crop responses to fertilizers. That is to say, the same economic principles apply to all response functions. When fitted to actual data, however, different response functions do not yield the same economic answers because the functions impose their individual peculiarities on the data. See Fig. 1-24 for an example. Some of the practical problems will be discussed in Section 2-8.

The various curvilinear response functions may be said to represent different versions of the "law of diminishing returns," according to which succeeding equal increments of a nutrient or other input result in succeedingly smaller increases in crop yield. Economic interpretations are made from these functions in the same general way, but the mathematical manipulations differ in complexity. To illustrate the principles, the quadratic response function is used here. The relatively simple manipulations of the quadratic function help to avoid complicating the illustrations. Procedures for fitting different response functions to experimental data are available (see Section 1-8), but procedures for applying economic principles to experimental data via different response functions generally must be worked out by the individuals who desire to use the functions. Many of these calculations are laborious and are best done by computers. Computer programs must be scrutinized carefully for assumptions and omissions that may reduce their utility for the specific purpose the user has in mind.

The linear response and plateau function implies constant returns, which means that the application of economic principles to this function is different from and simpler than the applications to curvilinear diminishing returns equations. The linear response and plateau function is given special consideration in Section 2-7.9.

2-3. Maximum Profit From Fertilization

The application of economic principles to fertilizer use is usually illustrated by the relatively uncomplicated situation in which the supply of capital is ample and no other investment possibilities would yield a more profitable return. The optimum application of fertilizer is then the quantity that will return the maximum net profit.

As an example, Fig. 2-1 shows the yield of barley in an experiment with different quantities of fertilizer nitrogen. The yield is expressed in dollars, and the fixed and variable costs of fertilization are expressed in the same way.

The *maximum net profit* from fertilization is returned by the fertilizer application at which the value of the crop most exceeds the total cost of fertilization. In Fig. 2-1A, this is the fertilizer application corresponding to the point on the X axis at which the length of a vertical line between the crop value function and the cost of fertilization is greatest. This application may be found approximately by plotting the data and measuring the length of the line, or it may be found more precisely by mathematical manipulations involving the slope of the mathematical function used to represent the relationship between crop value and quantity of fertilizer applied.

The tangent to the crop value function represents the rate of change of the crop value with respect to the quantity of fertilizer. The greater the slope of the crop value function in relation to the slope of the cost of fertilization function, the greater is the profit from an infinitesimal increase in the total application of fertilizer. Fig. 2-1A indicates that the profit derived from an infinitesimal addition of fertilizer decreases with an increase in the total amount of fertilizer.

The maximum net profit is obtained with the application of fertilizer at which the tangent to the crop value curve has the same slope as the cost of fertilization function. In Fig. 2-1A, this occurs at a slope of 0.40. For each infinitesimal increase in amount of fertilizer applied below the quantity that corresponds to the maximum net profit from fertilization, the increase in crop value exceeds the increase in cost of fertilization. Hence, the net profit is increased by applying the extra fertilizer. Above the quantity of fertilizer corresponding to the maximum net profit, each infinitesimal increase in the amount of fertilizer applied returns less in crop value than the increase in cost of fertilization. Hence, the net profit is reduced. An infinite number of situations is possible, depending upon the shape of the response curve, the price of the crop, and the cost of fertilization.

Where the response curve is represented by a mathematical function, the quantity of fertilizer that returns the maximum net profit may be estimated mathematically. The process perhaps can be understood most easily if the yield response function is first multiplied by the price per unit of crop to convert the function to monetary units. Then the procedure can be visualized in terms of Fig. 2-1A and the preceding text discussion about it.

The derivative of the crop value function is taken to find the rate of change of crop value with respect to the quantity of fertilizer. The derivative of the crop

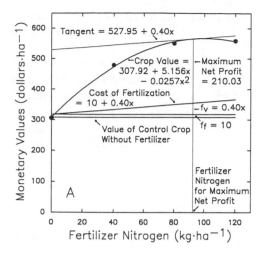

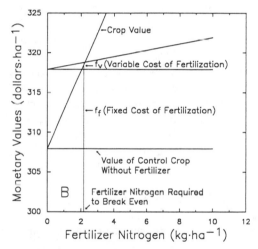

Fig. 2-1. Yield of barley in monetary terms with different quantities of fertilizer nitrogen. The construction in *A* shows the quantity of fertilizer nitrogen required to obtain the maximum net profit from fertilization. The fixed and variable costs of fertilization are represented by f_f and f_v, respectively. *B* is an enlargement of the portion of *A* showing the effects of small quantities of fertilizer nitrogen and illustrating the quantity required to break even. The experimental data are from Biswas and Singh (1982).

value function is next equated to the derivative of the fertilizer cost function, which usually is equal to the cost per unit of fertilizer. The quantity of fertilizer found by solving the resulting equation is the theoretical quantity that would have produced the maximum net profit according to the response function chosen and the assumptions involved in curve fitting. (If the crop response function is left in terms of yield, the same result is obtained by equating the derivative of the crop yield function to the ratio of the price per unit of fertilizer to the price per unit of crop and then solving the resulting equation for the quantity of fertilizer.)

To go through the process numerically, the yields of 3.07, 4.80, 5.52, and 5.60 megagrams of barley grain per hectare with fertilizer nitrogen applications of 0, 40, 80, and 120 kilograms per hectare were obtained in the experiment from which Fig. 2-1A was developed. The yield values were multiplied by a unit crop value of $99.70 per megagram to obtain the crop values in dollars per hectare. These values were fitted with the quadratic function

$$y = a + bx + cx^2$$

to obtain the crop value function

$$Crop\ value = V = 307.92 + 5.156x - 0.02570x^2,$$

where x is the quantity of fertilizer nitrogen in kilograms per hectare.

As indicated in Fig. 2-1A, fertilization is assumed to have a fixed cost f_f (taken here as $10 per hectare to show up well in the figure) that is the same for each unit area of land fertilized, plus a variable cost f_v that is proportional to the quantity of fertilizer applied per unit of land. The fixed cost of fertilization generally is considered to include the cost of applying the fertilizer. The variable cost generally is represented as the product of the number of units of fertilizer applied (x) and the price per unit of fertilizer (taken here as $0.40 per kilogram of fertilizer nitrogen), although it could include the small part of the application cost that would increase with the quantity of fertilizer. The costs of fertilization are then

$$Costs\ of\ fertilization = C = 10 + 0.40x.$$

The derivatives of the crop value and cost functions are

$$\frac{dV}{dx} = 5.156 - (2)(0.02570x)$$

and

$$\frac{dC}{dx} = 0.40.$$

Setting the two derivatives equal and solving for x,

$$5.156 - (2)(0.02570x) = 0.40,$$

and $x = 92.5$ kilograms of fertilizer nitrogen per hectare to yield the maximum net profit from fertilization, according to the data as interpreted through the assumptions inherent in the selection and fitting of the response function.

Note that in taking the derivative of the crop value function, the value of the control disappears because it is a constant. This is a significant point because the value of the control has no direct connection to the increase in crop value produced by the fertilizer, but the control embodies costs of production other than those due to fertilization. (The significance of these costs in relation to the economics of fertilization will be considered in subsequent sections.) When calculating the profit P_f from fertilization, the value of the control thus must be subtracted from the crop value function, so that the value of the control will not appear as part of the fertilizer effect:

$$P_f = V - 307.92 - C$$
$$= 307.92 + 5.156x - 0.02570x^2 - 307.92 - (10 + 0.40x).$$

If $x = 92.5$ kilograms of fertilizer nitrogen per hectare, $P_f = \$210.03$. To verify that \$210.03 is indeed the maximum net profit, the foregoing equation may be used to calculate the profit with other values of x.

The model developed here has some deficiencies that usually are ignored. The fixed costs of fertilization may not be constant, as represented. The cost of application will increase with the quantity applied. This cost could be considered a part of the cost per unit of fertilizer, although that is not done here. The cost of handling the crop will increase with the yield and, hence, with the quantity of fertilizer. This cost could be estimated by multiplying the increase in crop value by an appropriate constant factor and subtracting the cost in the profit equation.

Colwell (1983) emphasized that the quantity of fertilizer required to yield the maximum net profit is to be regarded as an academic concept that lacks validity in the real world. This is because the rate of increase in net profit becomes very small as the quantity of fertilizer required to yield the maximum net profit is approached. The producer thus has other options that will produce a greater return than can be obtained by investing in the last increments of fertilizer required to obtain the maximum net profit. The maximum overall profit will be realized where the fertilizer application yields a rate of return equal to that obtainable with the best alternative option. Colwell noted that deciding upon the appropriate alternative net return may not be easy, but that, for example, if there are safe investment opportunities (such as government securities) yielding about 10% interest, a return of that magnitude would be an absolute minimum. He considered 20% more appropriate as the minimum return that should be expected from the last increment of fertilizer in farm operations.

Still another complication is introduced in estimating the quantity of fertilizer required to produce the maximum net return when a price differential associated with quality is added to crop yield as a basis for economic evaluation. In some instances, a crop must achieve a certain quality to be acceptable for a particular purpose. Products that meet or exceed the standard are purchased at a premium price, and those that do not meet the standard must be sold at a lower price or perhaps are not salable. Melons, for example, are commonly selected in the field for salability, and the unacceptable melons may be left in the field.

In instances in which the crop units can be segegrated like melons, the proportion of the crop that meets the standard may vary with the level of fertilization. In such a situation, the shape of the response curve for the salable product would differ from that for the total product, but the same price would apply to all salable units.

A different situation prevails for a crop such as wheat, in which an entire lot either does or does not meet the standard. As noted by Holbrook et al. (1982, 1983), a premium is paid in the United Kingdom for wheat to be used for bread making if it contains 11% protein or more. The consequence of this practice is a vertical increase in the curve for crop value versus fertilizer nitrogen when the quantity of fertilizer nitrogen is just sufficient to increase the protein content to the critical value of 11%.

2-4. Minimum Applications of Fertilizer

The minimum application of fertilizer is of course zero. And this is the appropriate economic application under circumstances in which a fertilizer decreases the yield of the crop or in which the cost of fertilization exceeds the value of the increase in the crop at all levels of fertilization.

Three minimum finite economic applications of fertilizer have been suggested for the usual response curves that are concave downward. The first is called here the break-even application — the application that must be exceeded to make a profit. The second is the application required to obtain the greatest profit per dollar spent for fertilization, and the third is the application required to obtain the greatest profit per dollar spent in producing the fertilized crop. In this section, these concepts are considered under the circumstances in which the production of the crop is profitable in the absence of fertilizer.

2-4.1. Break-Even Application
In Fig. 2-1, the effect of fertilization on crop value is indicated by the increase in crop value above the control. To obtain the increase in crop value, the control value of $307.92 is subtracted from the equation for total crop value. Because the increase in crop value from fertilization is obtained at the expense of the costs of fertilization, the horizontal line representing zero cost of fertilization is located at the value of the crop without fertilization, and the costs are indicated by lines lying above the control value. These conventions are consistent with the fact that the value of the control crop and the costs of producing it are not a part of the effects and costs of fertilization.

In Fig. 2-1, fertilizer costs for a given application are represented as the sum of fixed and variable costs, f_f and f_v, and the profit from fertilization is represented by the length of a vertical line between the curve for crop value and the upward-sloping line representing the costs of fertilization. Fig. 2-1B shows more clearly than Fig. 2-1A that for very small applications of fertilizer, the profit is negative because the increase in crop value from fertilization is smaller than the costs of fertilization. The profit increases to zero with the break-even application, at which the curve for crop value intersects the line representing the total cost of fertilization, $f_f + f_v$. Then it becomes positive with greater applications of fertilizer. (If the crop value curve remains below the fertilizer costs at all levels, the profit is negative.)

To find the break-even application mathematically, the function representing the cost of fertilization is set equal to the increase in value of the crop as a function of the quantity of fertilizer, and the resulting equation is solved for the quantity of fertilizer. Continuing with the same numerical example used to calculate the application corresponding to the maximum profit from fertilization in Section 2-3, this equation is

$$5.156x - 0.02570x^2 = 10 + 0.40x$$

or

$$-10 + 4.756x - 0.02570x^2 = 0.$$

Solving for x with the aid of the quadratic formula,

$$x = \frac{-4.756 \pm \sqrt{4.756^2 - (4)(-0.02570)(-10)}}{(2)(-0.02570)}.$$

The numerical values yield $x = 2.1$ kilograms of fertilizer nitrogen per hectare for the break-even application.

Note that where the quadratic equation is used to represent response curves, the terms ordinarily are listed in the reverse order from that used in algebra. The parameters a, b, and c for the equation as it is usually used for response curves thus correspond to c, b, and a, respectively, in the quadratic formula.

The break-even application is of only theoretical interest because the purpose of crop production is not to break even, but to make a profit. A crop is not intentionally produced just to break even.

2-4.2. Maximum Profit per Dollar Invested in Fertilization

The application that corresponds to the maximum profit per dollar invested in fertilization is of greater practical relevance than the break-even application. When the quantity of fertilizer exceeds the break-even application, the profit at first increases, but eventually decreases. The profit per dollar invested in fertilization P_{df}, given by the ratio of the profit from fertilization P_f to the sum of f_f and f_v, also increases at first with the quantity of fertilizer, but then decreases, beginning at an application of fertilizer below that leading to the maximum net profit.

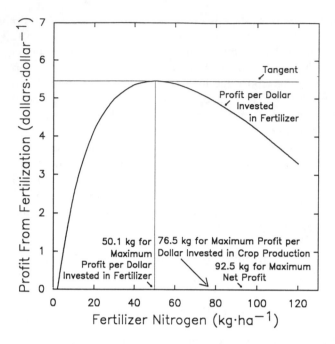

Fig. 2-2. Profit per dollar invested in fertilizer nitrogen versus the quantity of fertilizer nitrogen added per hectare to barley. The figure, based upon data in Fig. 2-1, shows that the maximum profit per dollar invested in fertilizer is obtained when the slope of the tangent to the profit curve is zero. The experimental data are from Biswas and Singh (1982).

Mathematically, the quantity of fertilizer producing the maximum profit per dollar invested in fertilization P_{df} is found by taking the derivative of the function for profit per dollar invested in fertilization, setting the derivative equal to zero, and solving the resulting equation for the quantity of fertilizer. This procedure is equivalent to finding the application of fertilizer corresponding to the point on the profit-per-dollar curve at which the tangent has a slope of zero (see Fig. 2-2). The peak of the curve for increase in crop value per dollar spent on fertilization coincides with the point of tangency of a line with zero slope.

As a numerical example, the function for profit per dollar invested in fertilization is assumed to be derived from the crop value and fertilizer cost functions used in Section 2-3. The derivative of P_{df} with respect to x is

$$\frac{dP_{df}}{dx} = \frac{[10 + 0.40x][5.156 - (2)(0.02570x)]}{[10 + 0.40x]^2}$$

$$- \frac{[(5.156x - 0.02570x^2)(0.40)]}{[10 + 0.40x]^2} .$$

Setting the derivative equal to zero and collecting terms,

$$-51.56 + 0.514x + 0.01028x^2 = 0.$$

Solving this equation for x by use of the quadratic formula,

$$x = \frac{-0.514 \pm \sqrt{0.514^2 - (4)(0.01028)(-51.56)}}{(2)(0.01028)}.$$

The numerical values yield $x = 50.1$ kilograms of fertilizer nitrogen per hectare to obtain the maximum profit per dollar invested in fertilization. Note that in these calculations the value of the control crop without fertilization has been subtracted from the equation for total crop value to obtain the increase due to fertilization. The reasoning here is the same as that for calculating the break-even application.

The value 50.1 kilograms of fertilizer nitrogen obtained for the maximum profit per dollar invested in fertilization lies between the 2.1 kilograms for breaking even and the 92.5 kilograms needed to obtain the maximum net profit. If the amount of fertilizer is insufficient to apply 50.1 kilograms of fertilizer nitrogen per hectare to all the land to be fertilized, this quantity should be applied as long as the fertilizer lasts, and the remaining area should not be fertilized. Reducing the application so as to cover all the land would reduce the return per dollar invested in fertilization.

In practice, an entrepreneur often will be growing more than one crop, and may have other enterprises as well. Where such is the case, the returns from the alternative enterprises may need to be taken into account in deciding how much fertilizer to apply. The question of competing enterprises is discussed in Section 2-6.

2-4.3. Maximum Profit per Dollar Invested in Crop Production

The profit per dollar invested in fertilization, considered in the previous section, has to do with the economics of fertilization independently of the economics of all other aspects of crop production. In a sense, separating the investment in fertilization from the investments in other aspects of crop production is artificial. Although obtaining the most profit from applying a fertilizer is a valid and important objective, the basic objective in producing a crop commercially is to make a profit. Fertilizers contribute to this overall objective. From this perspective, one may prefer to look at the rationale of fertilization in the broader sense and calculate the quantity of fertilizer required to yield the maximum profit per dollar invested in crop production.

The procedure for estimating the quantity of fertilizer required to yield the maximum profit per dollar invested in crop production is analogous to that used to estimate the maximum profit per dollar invested in fertilization. The differences are that one takes into account (a) the total crop yield instead of the increase in crop yield from fertilization and (b) the total costs of production instead of the increase in cost of production due to fertilization. With crop and cost values in monetary terms, one takes the derivative of the profit function, sets the derivative equal to zero, and solves the resulting expression for the quantity of fertilizer.

As a numerical example, the equation for profit per dollar invested in crop production for the experimental data on nitrogen fertilization of barley used in preceding sections is

$$P_{dc} = \frac{307.92 + 5.156x - 0.02570x^2 - (10 + 0.40x + 140)}{10 + 0.40x + 140}$$

$$= \frac{307.92 + 5.156x - 0.02570x^2}{150 + 0.40x} - 1$$

where 140 is an assumed value for the production costs per hectare other than those involved in fertilization. The derivative of the profit equation with respect to x is

$$\frac{dP_{dc}}{dx} = \frac{(150 + 0.40x)(5.156 - 0.0514x)}{(150 + 0.40x)^2}$$

$$- \frac{(307.92 + 5.156x - 0.02570x^2)(0.40)}{(150 + 0.40x)^2} \quad .$$

Setting the derivative equal to zero and solving the resulting expression for x with the aid of the quadratic formula,

$$x = \frac{(-2)(-0.02570)(150) \pm \sqrt{z}}{(2)(-0.02570)(0.40)} \quad ,$$

where $z = [(2)(-0.02570)(150)]^2 - [(4)(-0.02570)(0.40)][(5.156)(150) - (307.92)(0.40)]$. The numerical values work out to $x = 76.5$ kilograms of fertilizer nitrogen per hectare to produce the maximum profit per dollar invested in crop production. As for the maximum profit per dollar invested in fertilizer, all areas to be fertilized should receive the quantity to yield the maximum profit per dollar invested in crop production, and the area for which this much fertilizer cannot be applied should receive no fertilizer.

Although the quantity 76.5 kilograms calculated here differs from the 50.1 kilograms for the maximum profit per dollar invested in fertilizer, both are valid. The difference is in the objective. Voss (1975) showed that an increase in costs of handling the extra crop produced with fertilizer could be included in the equation for P_{dc} without effect on the final result if the handling costs are proportional to the yield. The quantities of fertilizer yielding the maximum profit per dollar invested in fertilization and per dollar invested in crop production depend upon the values for all the terms involved in the calculations, and these of course vary with the circumstances.

2-5. Fixed Costs

Emphasis upon the economics of fertilization in preceding sections has tended to obscure the significance of costs other than those related to fertilization. All costs, including those of fertilization, must be taken into account in appraising the overall profit from producing a crop.

As a documented example, Pesek and Heady (1958) made some economic calculations based upon a response function derived from the average yields of corn with different quantities of nitrogen fertilizer in 54 experiments in North Carolina. In these experiments, the nitrogen fertilizer was sidedressed alone in different quantities, and other fertilizers were applied at or before planting.

The response function Pesek and Heady used for the 54 experiments was

$$y = 1.64 + 0.0437N - 0.000118N^2$$

where y is the yield of corn grain in megagrams per hectare and N is the application of fertilizer nitrogen in kilograms per hectare. The value of corn, as given by Pesek and Heady, was equivalent to \$59.10 per megagram. Multiplying the yield equation by \$59.10 gives an equation for crop value V in dollars per hectare:

$$V = 96.92 + 2.58N - 0.00697N^2.$$

The derivative of this equation is

$$\frac{dV}{dN} = 2.58 - 0.01394N.$$

Fixed costs of production, including fixed costs of fertilization, were estimated at \$112.53 per hectare. With fertilizer nitrogen at \$0.33 per kilogram, the total costs of crop production C in dollars per hectare thus were

$$C = 112.53 + 0.33N.$$

The derivative of this equation is 0.33.

The quantity of fertilizer nitrogen to produce the maximum net profit from fertilization is then estimated as described in Section 2-3 by setting the derivative of the crop value equation equal to the derivative of the cost equation and solving the resulting equation for the quantity of nitrogen:

$$2.58 - 0.01349N = 0.33,$$

with $N = 166.7$ kilograms per hectare.

The net profit from crop production is given by the difference between the crop value and the costs:

Net profit $= 96.92 + 2.58N - 0.00697N^2 - (112.53 + 0.33N).$

The net profit calculated from this equation thus is $-\$15.61$ per hectare with no fertilizer nitrogen and $\$165.78$ with the quantity of fertilizer nitrogen estimated to yield the maximum net profit.

According to a private communication from D. F. Neuman at North Carolina State University, the critical economic values for 1988 were $\$116.90$ per megagram for corn, $\$0.45$ per kilogram for fertilizer nitrogen, and $\$620.79$ per hectare for fixed costs. Applying the new crop value to the same yield equation employed by Pesek and Heady (1958) for the 54 experiments in North Carolina resulted in the equation

$$V = 191.72 + 5.11N - 0.0138N^2,$$

where V is the crop value in dollars per hectare and N is the quantity of fertilizer nitrogen applied in kilograms per hectare. The derivative of this equation is

$$\frac{dV}{dN} = 5.11 - 0.0276N.$$

The updated cost equation is

$$C = 620.79 + 0.45N,$$

and the derivative of this equation is 0.45.

The quantity of fertilizer nitrogen producing the maximum net profit then is estimated by equating the derivatives of the crop value and cost equations and solving for N as before to obtain

$$5.11 - 0.0276N = 0.45$$

and $N = 168.8$ kilograms per hectare. The equation for the net profit then is

Net profit $= 191.72 + 5.11N - 0.0138N^2 - (620.79 + 0.45N),$

and from this equation the net profit per hectare may be estimated at $-\$429.07$ with no fertilizer nitrogen and $-\$35.67$ with 168.8 kilograms of fertilizer nitrogen per hectare.

To summarize these calculations, the changes in prices of nitrogen and corn between the first evaluation and the second had only a small effect on the quantity of nitrogen to produce the maximum net profit. In the first evaluation, the crop was produced at a loss without nitrogen fertilization, but yielded a fair profit with the quantity of nitrogen estimated to yield the maximum net profit. Although the value of the crop had increased by the time of the second evaluation, the fixed costs had increased so much more that the crop was produced at a loss even with nitrogen fertilization. The loss was decreased from a maximum of $429.07 per hectare to a minimum of $35.67 per hectare by applying nitrogen fertilizer, but the enterprise was still unprofitable. In this instance, the quantity of nitrogen that yielded the maximum net profit from fertilization corresponded to the quantity that resulted in the least loss in crop production.

These calculations have not taken into account the government subsidy for corn that would have made the enterprise profitable in 1988, the time for which the economic data employed were appropriate. The changes with the circumstances emphasize the joint importance of the physical and economic consequences of fertilization in guiding the rational use of fertilizers in practice.

2-6. Allocating Fertilizers Among Crops

The general economic principle involved in allocating funds among competing enterprises is that the greatest overall profit is obtained when the marginal productivity of all enterprises is equal. That is to say, the dollars invested in various enterprises or inputs that contribute to output are used in the most profitable way when the last dollar invested in each enterprise or input returns the same number of dollars in profit. In fertilizer use, the two principal allocation decisions involving this basic concept are allocating fertilizers among crops and allocating funds among the different nutrients that make up a fertilizer mixture. The first of these decisions is discussed in this section. The second is discussed in the following section.

The most profitable allocation of fertilizer among crops is that at which the rate of increase of crop value with respect to fertilizer cost is equal for all crops. Under these conditions,

$$\frac{(dy_1)(Price\ of\ 1y_1)}{(dx)(Price\ of\ 1x)} = \frac{(dy_2)(Price\ of\ 1y_2)}{(dx)(Price\ of\ 1x)}$$

$$= \ldots = \frac{(dy_n)(Price\ of\ 1y_n)}{(dx)(Price\ of\ 1x)} \quad ,$$

where y_1, y_2, ..., y_n denote the monetary yields of crops 1, 2,..., n on a per-unit-area basis, and x denotes the quantity of fertilizer on the same basis.

If capital is ample, the most profitable allocation of fertilizer is attained if each crop is fertilized with the application that yields the maximum net profit. Under these circumstances, each of the terms in the foregoing equation is equated to unity, which is in essence the same procedure employed with a single crop considered independently of all the others.

If some other amount of capital is invested in fertilizer, the optimum application is found by a series of simultaneous equations obtained by equating successive terms of the foregoing equation (that is, the first with the second, the second with the third, and so on; this leads to $n - 1$ equations with n unknown x values), together with the equation for total fertilizer cost, which is

Total fertilizer cost =
(No. of units of x applied for crop 1)(Price of 1x) +
(No. of units of x applied for crop 2)(Price of 1x) + . . . +
(No. of units of x applied for crop n)(Price of 1x)

The resulting system of equations contains n unknown x values (one for each crop). Solution of the equations yields the desired x values.

The first equation in this section applies to the condition in which equal areas are devoted to all crops. When the areas differ, the principle still applies that the net profit is maximized when the last dollar invested in fertilizer for each crop returns the same number of dollars in profit, but the equation must be modified to take the areas into account. To incorporate the area factor, the equation would become

$$\frac{(dy_1)(Price\ of\ 1y_1)(a_1)}{(dx)(Price\ of\ 1x)} = \frac{(dy_2)(Price\ of\ 1y_2)(a_2)}{(dx)(Price\ of\ 1x)}$$

$$= \ . \ . \ . \ = \frac{(dy_n)(Price\ of\ 1y_n)(a_n)}{(dx)(Price\ of\ 1x)} \ ,$$

where a_1, a_2, ..., a_n are the areas planted to crops 1, 2, ..., n. Because the areas are known, incorporating this additional factor in the equations merely adds constants and does not complicate the solution.

The question of allocating fertilizers among crops is an important one to farmers operating on limited capital, and it becomes an important national problem when fertilizers are in short supply and must be rationed. The classic paper on allocation of fertilizers among crops was published by Crowther and Yates (1941). This paper represents a summary of their work in devising a fertilizer policy for the United Kingdom during World War II.

In conducting the work, the following steps were taken:

1. The results of all field experiments with nitrogen, phosphorus, and potassium fertilizers conducted subsequent to 1900 in the United Kingdom were summarized. (Results of experiments in other northern European countries were summarized also for comparative purposes.)
2. The Mitscherlich equation was fitted to data from experiments including several quantities of fertilizer nitrogen, phosphorus, or potassium, and the c values describing the curvature were determined (see Section 1-2 for information on the Mitscherlich equation and c values). The c values for individual sets of data were averaged to obtain a single c value for use with nitrogen, a second for use with phosphorus, and a third for use with potassium.
3. With the aid of the c values, the data from all the experiments were recalculated to estimate the responses to a standard addition of the nutrient being tested. The equation

$$R = \frac{1 - 10^{-cx}}{1 - 10^{-cs}}$$

expresses the relationship, where R is the ratio of the increase in yield from a given application x of a nutrient to the increase in yield from a standard application s of the nutrient, and c is the average Mitscherlich c value for the nutrient. This procedure made it possible to estimate the response to a standard application of the nutrient irrespective of the applications actually used in the individual experiments. The estimated responses to the standard applications then were averaged by nutrients and crops to obtain the average response of each crop to the standard application of each nutrient. The average response to the standard application, together with the corresponding c values (a single c value was used for a given nutrient for all crops), formed the basis for calculating the average response curve for each crop.
4. The quantity of each fertilizer yielding the maximum net profit for each crop was calculated on the basis of the average response curves and the current prices.

Crowther and Yates illustrated in a novel way the principle involved in allocating fertilizers among crops. See Fig. 2-3 for their findings for nitrogen for four crops.

They plotted the value of the average response in monetary terms against the quantity of the nutrient added in the fertilizer, using the same vertical and horizontal scales for all crops. The plots obtained for the various crops then were superimposed. Because the same c value for a given nutrient was used for all crops, a single response curve fitted all crops, but the crops were located at different places on the curve according to the value of their responses. For each crop, the point 0,0 was located horizontally and vertically so that the values of the increases in yield would correspond to the curvature. The increases in yield were most valuable for potato. Crops with succeedingly less valuable responses

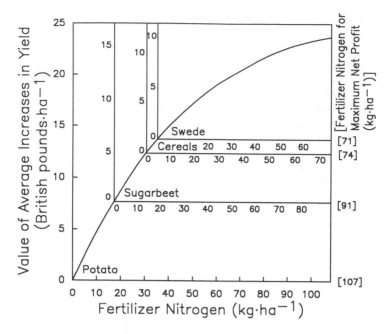

Fig. 2-3. Monetary value of the average increases in yield of different crops receiving increasing quantities of fertilizer nitrogen. The potato, sugarbeet, and swede (rutabaga) crops were grown with manure, and cereals were grown without manure. The figure summarizes the results of all experiments published in the United Kingdom between 1900 and World War II. (Crowther and Yates, 1941)

were located higher on the curve and farther to the right, where the rate of increase of value with respect to the nutrient addition was smaller. With each crop, the quantity of the nutrient required to produce the maximum net profit was located at the right edge of the graph. The lengths of the X axes for the various crops thus were proportional to the quantities of the nutrient required to produce the maximum net profit.

If equal areas of each crop were to receive the quantity of a particular nutrient estimated to yield the maximum net profit, the model employed would be expected to produce the maximum financial return. The quantities to apply per hectare under these circumstances would correspond to the values at the right edge of the graph: 107 kilograms of fertilizer nitrogen for potato, 91 kilograms for sugarbeet, 74 kilograms for cereals, and 71 kilograms for swede (rutabaga).

The total amount of fertilizer available under war-time conditions was less than that estimated to supply all crops with the quantities yielding the maximum net profit. The most profitable allocation of fertilizer among crops under these circumstances would be obtained by decreasing the allocation of fertilizer to all crops by the same number of kilograms per hectare until the sum of the products of the number of hectares of each crop and the quantities of fertilizer per hectare is equal to the total amount of fertilizer available. The rate of increase in crop value with respect to the fertilizer application then would be the same for all crops. Thus, with a small enough total supply, the less responsive crops would receive no fertilizer, and only the more responsive crops would be fertilized. To start from the other direction, if 1 hectare were to be planted to each of the crops shown in Fig. 2-3, all the fertilizer should be applied to potato up to the intersection with sugarbeet. From that point onward, the fertilizer should be divided equally between potato and sugarbeet up to the intersection with cereals. From that point onward, the fertilizer should be divided equally among potato, sugarbeet, and cereals up to the intersection with swede, and so on.

The Crowther and Yates technique was used later by Carey and Robinson (1953) in an extensive economic analysis of sugarcane response to fertilizers. This paper gives more of the mathematical details than does the paper by Crowther and Yates.

Although Crowther and Yates (1941) and Carey and Robinson (1953) used the Mitscherlich equation in their work, other response curves could be used, and the form used might differ among crops. Isfan (1986) adapted the allocation principle to two crops for which the responses could be represented by quadratic equations. In a subsequent paper, Isfan (1989) illustrated the principle with data from response curves for corn and wheat treated with different quantities of fertilizer nitrogen, fitting the data to quadratic equations and representing in graphic form the effects of crop price ratios, crop area ratios, and total amount of fertilizer available for allocation.

2-7. Allocating Funds Among Fertilizers

Of concern in this section are the economic concepts involved in the most efficient use of two or more plant nutrients as applied in a fertilizer mixture. From the economic standpoint, the different fertilizer ingredients are considered separate enterprises, analogous to producing two or more different crops.

Different crops usually are grown on separate areas of land, and the effect of fertilizer on the crop grown on one area is independent of the effect on other crops on other areas. The situation with different nutrients applied to a given crop is more complex because the various nutrients contribute to the same product, and the presence of one nutrient may influence the response of the crop to other nutrients.[1]

---

[1] More complex is the situation in which two or more crop species are grown together for hay or pasture. The individual species in the mixed vegetation normally respond differentially to the different fertilizer components. For example, grass crops respond most strongly to nitrogen, and legumes respond most strongly to phosphorus and potassium. These differences among species are accompanied by indirect effects of competition among the components of the mixed vegetation. The net effect is a change in the shape of the response curves, as mentioned in Chapter 1; however, the mixture of species is harvested and treated as a single crop. Still more complex is the situation in which different crops are grown together on the same land in intercropping systems, are fertilized with different combinations of plant nutrients, and are harvested separately, usually at different times. The economics of fertilizing such systems is not considered here.

The basic principle involved in allocating funds among two or more nutrients applied to a given crop is analogous to the one involved in allocating a given fertilizer among different crops. That is, the most efficient allocation of a given amount of money to purchases of different nutrients is achieved when the last dollar invested in each nutrient produces the same increase in crop value. The complication of interactions among the effects of different nutrients is reflected in the mathematical formulations representing the responses as functions of the nutrients supplied.

The economic concepts and procedures for making the calculations are outlined and illustrated in the following sections. Further discussion of the concepts and their use may be found in publications by Baum et al. (1956, 1957), Heady and Dillon (1961), Heady et al. (1955, 1961), Munson and Doll (1959), Dillon (1977), Colwell (1983), and others.

2-7.1. Graphic Representation

The basic concepts involved in interpreting curvilinear multinutrient response functions in terms of economic optima may be visualized by reference to Fig. 2-4, which is a simplified adaptation of one devised by Brown (1956). This figure facilitates an understanding of the following sections that deal with the mathematical relationships.

The data from which Fig. 2-4 was derived were obtained in an experiment reported by Heady et al. (1955), in which corn received different quantities of fertilizer nitrogen and phosphorus. The response function calculated to represent the data was

$$y = -0.357 - 0.0177N - 0.0535P + 0.376\sqrt{N}$$

$$+ 0.764\sqrt{P} + 0.0289\sqrt{(NP)} \quad ,$$

where y = yield of corn in megagrams per hectare, N = kilograms of nitrogen applied as ammonium nitrate per hectare, and P = kilograms of phosphorus applied as superphosphate per hectare.

The lines leading from upper left toward the lower right in Fig. 2-4 represent specific yields of corn (called *isoquants* because they represent equal quantities) that have been used as examples. Although each of the specific yields theoretically could have been produced with many different combinations of fertilizer nitrogen and phosphorus, as implied by the range of X and Y coordinates corresponding to the lines, the combination that would have produced a given yield at minimum cost depends upon the relative prices of the two nutrients. If nitrogen is more expensive than phosphorus, the least-cost combination would employ less nitrogen and more phosphorus. If phosphorus is more expensive than nitrogen, the least-cost combination would employ less phosphorus and more nitrogen.

Each of the yield lines has a range of hypothetical slopes or tangents, which have a theoretical function analogous to that of the tangent to the yield curve in

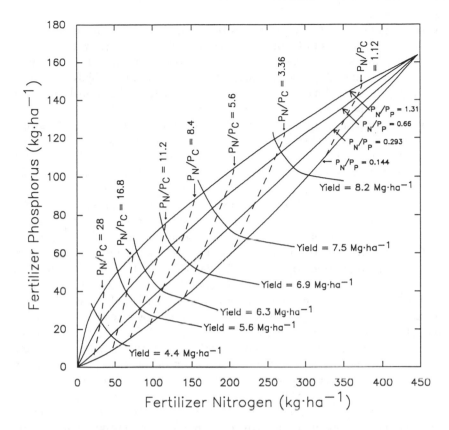

Fig. 2-4. Principal theoretical economic concepts involved in allocating funds among fertilizers applied to a single crop, as illustrated by the results of an experiment on the response of corn to the application of fertilizer nitrogen and phosphorus in various ratios and quantities. The lines leading from upper left to lower right representing certain crop yields are called isoquants. The lines leading from lower left to upper right representing different P_N/P_P ratios are called isoclines. The broken lines representing numerical values of P_N/P_C are different values of the ratio of the price per kilogram of fertilizer nitrogen to the price per kilogram of corn grain. See the text for further details. The figure is a simplified adaptation of one devised by Brown (1956).

Fig. 2-1. When the tangent in Fig. 2-1 has the same slope as the fertilizer cost function, the increase in crop value produced by an infinitesimal increase in quantity of fertilizer at the point of tangency is equal to the cost of the infinitesimal increment of fertilizer. The net profit is then at a maximum.[2] In Fig. 2-4, the

---

[2]Colwell's (1983) criticism of selecting the quantity of fertilizer required to yield the maximum net profit as the economic optimum quantity applies here as well as in Section 2-3, where it was introduced. To take into account the additional financial return that could be obtained by investing in a higher paying alternative enterprise the money that otherwise would have been used to purchase the last relatively unprofitable increments of fertilizer, Colwell would in effect multiply the investment in fertilizer by the quantity $(1 + R)$, where R is the return as a decimal fraction from the alternative investment.

hypothetical tangents to the yield lines or isoquants represent the ratios of the infinitesimal quantities of fertilizer nitrogen and phosphorus that would produce the same infinitesimal increase in yield at the points of tangency. The minimum-cost combination of the total amounts of fertilizer nitrogen and phosphorus that will produce the yield in question is that at which the last infinitesimal increment of each of the two nutrients would have equal cost.

In discussing his equivalent of Fig. 2-4, Brown used $0.33 and $0.50 as the prices per kilogram of fertilizer nitrogen and phosphorus, respectively, leading to an N/P price ratio of 0.66. This ratio would apply to all yields. Therefore, on each of the specific yield lines, the ratio of the infinitesimal quantities of the two nutrients (the tangents to the yield lines) leading to equal costs would be 1/0.66 = 1.52 kilograms of fertilizer nitrogen per kilogram of phosphorus. All these tangents would have the same slope, and the line (called an *isocline* because it represents equal inclines or slopes of the tangents) labeled $P_N/P_P = 0.66$ and leading from lower left to upper right in Fig. 2-4 would join these points of tangency on the different yield lines. The line $P_N/P_P = 0.66$ then may be used to estimate the total amounts of fertilizer phosphorus and nitrogen required for the least-cost combination that would have produced any yield within the range of the data. The total amount of phosphorus is found by extending a horizontal line leftward to the phosphorus axis and reading off the amount of phosphorus; the total amount of nitrogen is found similarly by dropping a vertical line to the nitrogen axis and reading off the amount of nitrogen. (Since the time when Brown constructed the original graph, the P_N/P_P price ratio has declined to approximately half of the value he used.)

The fact that the P_N/P_P lines are curved implies that the relative amounts of nitrogen and phosphorus in the least-cost combination depend upon the crop yield. And the fact that all the lines converge to a point at the upper right implies that according to the response function employed, the maximum yield is attainable only with a specific combination of nitrogen and phosphorus additions irrespective of the price ratio. (Not all functions behave in this way. Thus, although the basic theoretical rationale is the same for different functions, the numerical values obtained with a given data set will depend upon the function fitted to the data.)

The next step in using Fig. 2-4 is to move up the line for the least-cost combination of nitrogen and phosphorus to find the amounts of these nutrients that yielded the maximum net profit. These amounts depend upon the ratio of the price of one unit of fertilizer to the price of one unit of crop. To apply the concept discussed previously for a single nutrient, the maximum profitability would have been achieved in the experiment by adding the nitrogen and phosphorus fertilizers in the least-cost ratio up to the amount at which the rate of increase of fertilizer cost just equals the rate of increase of crop value. The values used by Brown for illustration in the figure were equivalent to $0.33 per kilogram of nitrogen and $0.0394 per kilogram for corn; that is, the ratio P_N/P_C (price of nitrogen/price of corn) was 8.4. Moving up the line for $P_N = 0.66P_P$ to the point at which it intersects the broken line labeled 8.4 for the ratio of the price of nitrogen to the price of corn, one finds by reading off the amounts of phosphorus and nitrogen on the vertical and horizontal axes of the figure that the desired fertilizer mixture would contain about 69 kilograms of phosphorus and 140 kilograms of nitrogen per hectare. The same result could have been obtained on the basis of the ratio P_P/P_C (the ratio of the price of phosphorus to the price of corn) because the broken line for $P_N/P_C = 8.4$ is also the line for $P_P/P_C = 12.7$.

The yield of corn corresponding to application of the least-cost combination of nitrogen and phosphorus fertilizers in the amount that produced the maximum net profit from fertilization may be estimated at about 7.1 megagrams per hectare by interpolating between the curves representing the 6.9 and 7.5 megagram yields. As may be inferred from the figure, the most profitable application of fertilizer and the corresponding yield both increase with decreasing price of fertilizer relative to the price of corn.

2-7.2. Marginal Product

The *marginal product* of a production input variable is the rate of change of the product with respect to that variable with other inputs constant. If the product is a crop and the input variable is a fertilizer bearing a particular nutrient, the marginal product is the partial derivative of the yield function or equation with respect to that fertilizer or nutrient. Thus, where the yield y is represented as a function of the variables $x_1, x_2, \ldots, x_n$, the equations for the marginal products are given by $\partial y/\partial x_1, \partial y/\partial x_2, \ldots, \partial y/\partial x_n$.

If the response function has a maximum, setting the marginal products equal to zero and solving the respective equations for $x_1, x_2, \ldots, x_n$ gives the values of the individual variables that correspond to the maximum yield with any specified values of the remaining variables. If the marginal products are set equal to zero, and if the resulting equations are solved simultaneously, the resulting values of $x_1, x_2, \ldots, x_n$ represent the single combination of all n variables that corresponds to the maximum yield described by the equation.

A response function calculated by Pizarro Villanueva (1982) for fertilization of wheat with nitrogen and phosphorus serves as a numerical example. This response function is used throughout most of the remainder of this section on allocating funds among fertilizers because it illustrates multinutrient procedures with only two nutrients and because the mathematical form does not require the use of successive approximations to obtain the desired numerical values for the variables. See a paper by Guinard (1982) for additional worked examples.

The basic equation is

$$y = 2.91 + 0.0143N + 0.0514P - 0.0000774N^2$$

$$- 0.000705P^2 + 0.0000755NP, \tag{1}$$

where y = wheat yield in megagrams per hectare, N = kilograms of fertilizer nitrogen per hectare, and P = kilograms of fertilizer phosphorus per hectare (the original equation has been modified to change the units of P_2O_5 to units of P).

The partial derivatives of equation (1) with respect to nitrogen and phosphorus are

$$\frac{\partial y}{\partial N} = 0.0143 - 0.000155N + 0.0000755P \tag{2}$$

and

$$\frac{\partial y}{\partial P} = 0.0514 - 0.00141P + 0.0000755N. \tag{3}$$

Setting the partial derivatives in (2) and (3) equal to zero and solving for N and P, respectively, one obtains $N = 92.4 + 0.488P$ and $P = 36.5 + 0.0535N$. Therefore, at $P = 30$, the estimated maximum yield would have been obtained with $N = 107$ kilograms per hectare. With $N = 30$, the estimated maximum yield would have obtained with $P = 38$ kilograms per hectare.

Setting the partial derivatives in (2) and (3) equal to zero and solving the two equations simultaneously, one obtains $N = 113$ and $P = 43$ kilograms. These are the estimated applications per hectare that would have led to the maximum yield attainable by applying both nutrients according to the experimental data as fitted by the quadratic response function employed.

2-7.3. Isoquants

Isoquants (sometimes called isoproduct or equal-product lines) are contours of equal quantity or yield that may be obtained by different combinations of the *x* variables. In Fig. 2-4, the relatively short curved lines leading from upper left toward lower right are isoquants. In that figure, the lines represent specific yields of corn that theoretically could have been obtained with different combinations of fertilizer nitrogen and phosphorus, according to the response function employed.

The isoquant equation is obtained by solving the original response function

$$y = f(x_1, x_2, x_3, \ldots, x_n)$$

for one of the *x* variables, for example,

$$x_1 = f_1(y, x_2, x_3, \ldots, x_n).$$

A series of points on any selected isoquant then may be obtained by inserting the desired value of *y* and a series of different values for the *x* variables on the right-hand side of the equation and by solving for the value of *x* on the left-hand side of the equation with each set of *x* values on the right.

The isoquant equation for the two-nutrient response function calculated for wheat by Pizarro Villanueva (1982) is derived by rewriting equation (1) as follows:

$$0.000705P^2 - 0.0514P - 0.0000755NP - 2.91$$
$$-0.0143N + 0.0000774N^2 + y = 0 \qquad (4)$$

An isoquant equation solved for phosphorus then is obtained by use of the quadratic formula:

$$P = \frac{0.0514 + 0.0000755N \pm \sqrt{z}}{(2)(0.000705)}, \qquad (5)$$

where $z = (0.0514 + 0.0000755N)^2 - (4)(0.000705)(-2.91 - 0.0143N + 0.0000774N^2 + y)$. (An analogous isoquant equation could be derived by solving equation (1) for N.) If $y = 4$ megagrams per hectare and N $= 100$ kilograms per hectare are substituted in equation (5) as a specific example, the quadratic formula yields P $= 8$ and 75 kilograms per hectare. P $= 8$ is the desired value. P $= 75$ applies to the descending portion of the response function, with applications exceeding those corresponding to the estimated maximum yield. Substituting other values of N in equation (5) and solving each time for P, while keeping the value of y at 4, would provide a series of points that could be used in a plot of P against N to trace the isoquant for $y = 4$.

Equation (5) is the basic isoquant equation for Pizarro Villanueva's response function. Selecting a different value of y, say 4.2, and solving each time for P, upon substituting a series of different values for N, would result in a series of points that could be used to plot the isoquant for $y = 4.2$.

As indicated by the fact that two values for phosphorus were obtained when $y = 4$ and N $= 100$ were substituted into equation (5), the isoquants extend to the descending portion of the response function. The quantities of nitrogen and phosphorus of practical interest, however, are those corresponding to the ascending portion of the function, where the slopes of the isoquants are between minus infinity and zero.

2-7.4. Rate of Substitution
The range of slopes of an isoquant represents the range in *rates of substitution*[3] of the x variables concerned, that is, the rate of change of one x variable with

---

[3]Use of the term "substitution" by economists in dealing with responses of crops to different nutrients has irritated some scientists, for example Barber (1973) and Middleton (1983). The problem scientists with a physiological outlook have with this term is that one nutrient does not substitute for an essential function of another, although there may be some partial overlapping of functions, as with sodium and potassium and with the cations that balance the soluble anions in plants. The issue, however, is not one of error, but of definition. Economists look at the effects and use substitution to indicate that within limits the use of one production input may be able to produce the same increase in yield of product as use of some quantity of a second input.

respect to another at a particular yield and with any desired combination of the remaining x variables. In economic context, rate of substitution, sometimes called marginal rate of substitution, means the number of infinitesimal units of x_2 that would produce the same infinitesimal increase in yield as one infinitesimal unit of x_1 if added at a given yield.

The rate of substitution is used to determine the most economical combination of nutrients to produce a given yield. For example, with nitrogen and phosphorus, as in Fig. 2-4 and the Pizzarro Villanueva (1982) data, the most economical combination is that at which the rate of substitution (given by $\partial P / \partial N$) is equal to the inverse price ratio of the two nutrients,

$$\frac{\partial P}{\partial N} = \frac{Price\ of\ 1N}{Price\ of\ 1P},$$

which means that the values of the infinitesimal quantities of fertilizer phosphorus and nitrogen that produce the last infinitesimal increase in yield are equal:

$$(\partial P)(Price\ of\ 1P) = (\partial N)(Price\ of\ 1N).$$

The total cost of the two nutrients with which the yield is produced is then at a minimum. With any other combination, the increase in cost of one nutrient exceeds the decrease in cost of the other.

The rate of substitution of one x variable for another x variable may be found by taking the partial derivative of the first x variable with respect to the second in the function represented by the yield equation

$$y = f(x_1, x_2, \ldots, x_n).$$

where y is considered to be constant. This partial derivative is the equation for the slope of the isoquant, which is the rate of substitution of the x variables.

As a numerical example of calculating rates of substitution, equations (2) and (3) give the partial derivatives of wheat yield with respect to nitrogen and phosphorus, respectively, in the Pizarro Villanueva (1982) response function used in previous sections. Dividing equation (2) by equation (3), one obtains

$$\frac{\partial y / \partial N}{\partial y / \partial P} = \frac{\partial P}{\partial N} = \frac{0.0143 - 0.000155N + 0.0000755P}{0.0514 - 0.00141P + 0.0000755N}, \tag{6}$$

which is the rate of substitution of phosphorus for nitrogen along any isoquant.

To take into account the fact that the slopes of yield isoquants are negative in the range of practical interest, Heady and Dillon (1961, p. 38) added a negative sign before their equivalent of equation (6). Then they eliminated the negative sign by setting the ratio of the partial derivatives equal to $-k$, which represented the rate of substitution they wished to use for developing an isocline.

The isocline concept discussed in Section 2-7.5 extends the concept of the least-cost combination for a given yield to the full range of yields within a given response function. Section 2-7.6 brings together the rate of substitution and isocline concepts and shows how to calculate the minimum-cost mixture of nutrients for any given yield.

2-7.5. Isoclines

An *isocline* is a line that intersects all isoquants at points having a given slope or rate of substitution. For example, an isocline representing a rate of substitution of 2 would join points on successive yield isoquants at which addition of nutrients x_1 and x_2 individually in the ratio of 2 to 1 would produce equal infinitesimal increases in yield at each given yield level. A different isocline exists for each possible rate of substitution. In Fig. 2-4, the solid lines leading from lower left to upper right are isoclines.

The isocline of special practical interest is the one that represents the least-cost combination of any two x variables. Along the least-cost isocline, the costs of the infinitesimal quantities of the two variables required to produce equal infinitesimal increases in yield are equal. The precise location of the least-cost isocline depends upon the price ratio of the nutrients, as illustrated in Fig. 2-4.

The general equation for the least-cost isocline is obtained by equating the rate of substitution to the inverse price ratio of the nutrients. In Pizarro Villanueva's (1982) response function for fertilization of wheat with nitrogen and phosphorus, the rate of substitution, equation (6), is equated to α, which is the ratio of the price of nitrogen to the price of phosphorus:

$$\frac{\partial P}{\partial N} = \frac{0.0143 - 0.000155N + 0.0000755P}{0.0514 - 0.00141P + 0.0000755N} = \alpha. \tag{7}$$

Solving the equation for P,

$$P = \frac{0.0514\alpha - 0.0143 + 0.000155N + 0.0000755\alpha N}{0.0000755 + 0.00141\alpha}. \tag{8}$$

The points needed for plotting the least-cost isocline are those at which this isocline intersects the various isoquants. The general equation for the N values is obtained by equating equations (5) and (8), thus eliminating P:

$$\frac{0.0514 + 0.0000755N \pm \sqrt{z}}{(2)(0.000705)}$$
$$= \frac{0.0514\alpha - 0.0143 + 0.000155N + 0.0000755\alpha N}{0.0000755 + 0.00141\alpha}, \tag{9}$$

where $z = (0.0514 + 0.0000755N)^2 - (4)(0.000705)(-2.91 - 0.0143N + 0.0000774N^2 + y)$. By substituting in equation (9) the appropriate value of α along with a series of y values for the different isoquants and solving each time for N, one obtains a series of N values. These and the associated y values may be substituted in equation (5) to obtain the corresponding values of P. The least-cost isocline then can be plotted, as in Fig. 2-4.

For example, if fertilizer nitrogen costs \$0.40 per kilogram and phosphorus costs \$1.26 per kilogram, the price ratio α is $0.40/1.26 = 0.317$. If the $y = 4$ isoquant is selected for a first calculation, equation (9) may be solved to obtain N = 35.3 and 191.7. When $y = 4$ and N = 35.3 are substituted into equation (5), the values P = 15.9 and 60.8 are obtained. The lower values for both nitrogen and phosphorus are the ones sought because the higher values are located on the descending portion of the response surface, with applications of nitrogen and phosphorus exceeding those estimated to produce the maximum yield.

2-7.6. Minimum Cost for a Specified Yield
To find the least-cost combination of any two x variables that will produce a specified yield, the equation for the least-cost isocline is solved simultaneously with the yield isoquant equation after substituting the desired yield value in the latter. Where there are only two variables, this procedure will lead to a unique numerical solution. Where there are more than two x variables, the result will be a function of the remaining x variables.

In the Pizarro Villanueva (1982) response function, the applications of nitrogen and phosphorus fertilizers leading to the minimum fertilizer cost of producing a given yield were calculated in the preceding section as an illustration of the procedure for calculating isoclines. For a yield of 4 megagrams of wheat per hectare, the least-cost fertilizer would have supplied 35.3 kilograms of nitrogen and 15.9 kilograms of phosphorus per hectare with a nitrogen/phosphorus price ratio of 0.317.

2-7.7. Most Profitable Combination for a Given Expenditure
The most profitable combination of n x-variables at a specified total expenditure is found by solving simultaneously the equation representing the total expenditure T for fertilizers,

$$T = (x_1)(Price\ of\ 1x_1) + (x_2)(Price\ of\ 1x_2)$$
$$+ \ldots, + (x_n)(Price\ of\ 1x_n),$$

and the $(n - 1)$ equations specifying the least-cost isocline. For the Pizarro Villanueva (1982) response function for wheat receiving different quantities of fertilizer nitrogen and phosphorus, the total expenditure T for fertilizer is

$$T = (N)(Price\ per\ kilogram\ of\ nitrogen)$$
$$+ (P)(Price\ per\ kilogram\ of\ phosphorus).$$

With $0.40 as the price of a kilogram of fertilizer nitrogen and $1.26 as the price of a kilogram of fertilizer phosphorus, and with an assumed expenditure of $20 per hectare in applying the least-cost combination of nitrogen and phosphorus fertilizers,

$$\$20 = 0.40N + 1.26P. \tag{10}$$

The least-cost isocline is obtained by substituting the N/P price ratio (0.317) for α in equation (7). Collecting and rearranging the terms leads to

$$0.0199 = -0.00179N + 0.00522P. \tag{11}$$

Solving equations (10) and (11) simultaneously yields N = 18.3 and P = 10.1.

2-7.8. Combination for Maximum Profit

The maximum profit from fertilization with two or more nutrients occurs when the nutrients are applied in the least-cost combination and when the amount of fertilizer applied leads to the crop value that most exceeds the cost of the fertilizer. If fertilization does not produce a profit, the numerical value obtained in the calculations is the amount and combination of nutrients at which the loss is a minimum.

The maximum profit or minimum loss is achieved when the rate of increase in crop value from fertilization is equal to the rate of increase in cost of fertilization with the least-cost combination of nutrients. The application of nutrients that corresponds to this situation is found by solving simultaneously the set of equations resulting when the partial derivatives of the crop value equation with respect to each of the nutrients are equated individually to the derivatives of the corresponding fertilizer cost functions (or when the partial derivatives of the yield function with respect to each of the nutrients are equated individually to the ratio of the price per unit of each nutrient to the price per unit of crop).

Pizarro Villanueva's (1982) response function representing the effect of nitrogen and phosphorus fertilizers on the yield of wheat is used once more to provide a numerical example of the calculations. The partial derivatives of the crop yield equation with respect to nitrogen and phosphorus (equations 2 and 3) are multiplied by the price of wheat (taken here as $128 per megagram) and then equated to the derivatives of the nitrogen and phosphorus cost functions (0.40 and 1.26 per kilogram, respectively):

$$\frac{\partial y}{\partial N} = 1.83 - 0.0198N + 0.00966P = 0.40$$

and

$$\frac{\partial y}{\partial P} = 6.58 - 0.180P + 0.00966N = 1.26.$$

Simultaneous solution of the two equations yields 89.0 kilograms of nitrogen and 34.3 kilograms of phosphorus as the quantities that would have been needed to yield the maximum net profit in the situation represented by the response function.

When N = 89.0 and P = 34.3 are substituted in equation (1), the crop yield is found to be 4.73 megagrams per hectare. At $128 per megagram for wheat, the crop value is 4.73 × $128 = $605.44 per hectare. The fertilizer cost is (89.0 × $0.40) + (34.3 × $1.26) = $78.82 per hectare. The difference between the crop value and the fertilizer cost is $605.44 − $78.82 = $526.62 per hectare. Subtracting the value of the control (29.1 × $12.80 = $372.48) yields $154.14 as the maximum net profit that could have been obtained from fertilization under the circumstances represented by the experimental data, together with the assumptions regarding the response function and the prices. These calculations do not include a fixed cost of fertilization. As noted previously, the maximum net profit from fertilization has to do only with the change in crop value due to fertilization. It may be positive, but it may also be negative if the fixed costs are high enough. If negative, it is obtained with the quantity of fertilizer at which the net loss is smallest.

Economists have given further consideration to development of techniques for identifying the least-cost fertilizer sources of a given combination of nutrients, blending costs, and application costs. Babcock et al. (1984) addressed this problem in a computer program based on linear programming, and they referred to other less comprehensive computer programs that had been written for the purpose.

2-7.9. Technique for Segmented Functions

The economic concepts in preceding sections have been presented in the framework of crop responses that are represented by continuous curvilinear functions. Segmented functions are different. These functions consist of generally two separate functions that are joined at some point. The linear response and plateau function (see Section 1-5.3) is the most common example. It consists of two straight lines, one ascending and one generally with zero slope, that join at the left edge of the plateau. The quadratic response and plateau function consists of a curvilinear ascending segment that is fitted by a quadratic function and is joined to a straight line, usually with zero slope, that constitutes the plateau.

If the ratio of fertilizer cost to crop price is high enough so that the maximum net profit falls on the curvilinear segment of a quadratic response and plateau model, the economic concepts discussed in previous sections may be applied. Otherwise, there are some differences.

Basic to an understanding of the interpretation of segmented functions according to the concept put forward by Waugh et al. (1973) and studied further by Anderson and Nelson (1975, 1987) is an appreciation for the procedure used to obtain the data. Waugh et al. were the pioneers in adapting segmented functions for use with several nutrients.

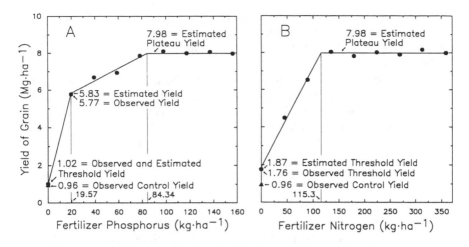

Fig. 2-5. Use of the linear response and plateau model for economic interpretations of the effects of *A* phosphorus and *B* nitrogen fertilizers on the yields of corn in an experiment in Iowa reported by Heady et al. (1955).

These investigators noted that their approach for more than one nutrient was based upon Liebig's Law of the Minimum, and they used an illustrative graph that was analogous to the one Blackman published in 1905, an adaptation of which is shown as Fig. 1-1. But they made one significant modification. The Liebig concept as interpreted by Blackman implied that crop yields below the maximum obtainable by increasing the supply of the factors under consideration would reflect a limitation imposed by a single factor, and that when this factor was supplied in adequate quantity, a second limiting factor might be found, and so on. In their work, however, Waugh et al. (1973) studied the effect of each nutrient on field plots to which other nutrients had been added. They defined the *threshold yield* as the yield limited by a single nutrient, and they used the crop yield on plots to which adequate but not excessive quantities of nutrients other than the one under investigation had been added as a measure of the threshold yield for each nutrient. If three nutrients were being studied, for example, there would be three threshold yields, one for each nutrient.

In the work by Waugh et al. (1973), the threshold yield was equivalent to what would be called the control yield in many soil fertility experiments. It was not the yield without fertilizer, however, which is the control yield in the absolute sense that applies for the producer. If the quantities of nutrients added to avoid deficiencies of nutrients other than the one under investigation are not excessive, the threshold yield will equal or exceed the yield of the crop without fertilizer, which is labeled as the control yield in Fig. 2-5.

The procedure used by Waugh et al. (1973) to evaluate the threshold yield should improve the linearity of the responses and the fit of data to the Blackman concept in Fig. 1-1. The procedure should also increase the response and improve the correlations between the crop yields and the concentration of the nutrient in question in the plant tissues (Bouma et al., 1969). At the same time, however, the procedure affects the economic interpretation of the results, as will be explained in subsequent paragraphs.

In the linear response and plateau function, the ascending segment of the response function is generally represented as a single straight line terminating at the plateau or maximum yield. According to the procedure by Waugh et al. described in a preceding paragraph, the line begins at the threshold yield and not at the yield of the control. If the fixed cost of fertilizer application is neglected, the economic implication of the linear model is that the return in crop value per dollar spent for fertilizer is the same throughout the range of the upward-sloping line. Applying the fertilizer either is or is not profitable, depending upon whether the slope of the crop value function is greater or less than the slope of the fertilizer cost function. If applying the fertilizer is profitable, the financial return in excess of the cost of the fertilizer is proportional to the quantity applied up to the point of intersection of the ascending line with the line representing the plateau. For additional quantities of fertilizer, there is no further increase in yield or crop value if the plateau is level or downward sloping, and the cost of the additional fertilizer then subtracts from the overall net profit. According to this way of representing the data, therefore, the quantity of fertilizer to apply is either zero or that corresponding to the intersection of the ascending line with the plateau.

The linear response and plateau function supplies a simplified procedure for estimating the quantities of individual fertilizers to apply when two or more fertilizers are applied together. The data from Pizarro Villanueva's experiment on fertilizing wheat with nitrogen and phosphorus fertilizers used in previous sections in this chapter do not provide a good fit to the linear response and plateau model. Accordingly, a better fitting set of data from an experiment on fertilizing corn with nitrogen and phosphorus fertilizers is used as a numerical example in Fig. 2-5. The nature of the data in the figure and the mechanics of fitting the lines were discussed in Section 1-5.3. In the following calculations, the costs per kilogram of fertilizer phosphorus and nitrogen are assumed to be $1.26 and $0.40, respectively, as used in other examples in this chapter.

The estimated yield increment of $5.83 - 1.02 = 4.81$ megagrams per hectare for the first segment of the response function for phosphorus in Fig. 2-5A was obtained with 19.57 kilograms of fertilizer phosphorus costing $(19.57)(\$1.26) = \$24.66$. To make this application profitable in the presence of an adequate supply of fertilizer nitrogen, the crop would have to be worth more than $24.66/4.81 = \$5.13$ per megagram. Similarly, the second yield increment of $7.98 - 5.83 = 2.15$ megagrams was obtained with an additional quantity of $84.34 - 19.57 = 64.77$ kilograms of fertilizer phosphorus costing $(64.77)(\$1.26) = \$81.61$. To make this application profitable in the presence of an adequate supply of fertilizer nitrogen, the crop would have to be worth more than $81.61/2.15 = \$37.96$ per megagram. In Fig. 2-5B, the yield increment of $7.98 - 1.87 = 6.11$ megagrams was obtained with 115.3 kilograms of fertilizer nitrogen costing

(115.3)($0.40) = $46.12. To make this application profitable in the presence of an adequate supply of fertilizer phosphorus would require that the crop value exceed $46.12/6.11 = $7.55 per megagram.

The stepwise approach to economic evaluation used in the preceding paragraph for data fitted by the linear response and plateau function is much simpler than the procedure for multinutrient response curves. No difficult-to-follow mathematical concepts or complex computations are involved. From the practical standpoint, however, the economic interpretations are not entirely straightforward because of the practice of measuring the response to one nutrient when adequate supplies of other nutrients have been provided. When the effects of individual nutrients are not independent, the same is true for the economics of their use.

According to the groupings of treatments used by Anderson and Nelson (1987) for the experiment in Fig. 2-5, the yield response of 5.83 − 1.02 = 4.81 megagrams estimated for addition of 19.57 kilograms of fertilizer phosphorus probably could have been obtained with 90 kilograms of fertilizer nitrogen. Thus, the total cost of the fertilizer would have been (19.57)($1.26) + (90)($0.40) = $24.66 + $36.00 = $60.66 instead of the $24.66 calculated in a preceding paragraph for the cost of the fertilizer phosphorus alone. To obtain a profit thus would have required corn valued at more than $60.66/4.81 = $12.61 per megagram as opposed to $5.13 per megagram when calculated on the basis of the cost of only the fertilizer phosphorus. Similarly, the second upward-sloping segment of the phosphorus function probably could have been obtained with an additional 45 kilograms of fertilizer nitrogen, so that the fertilizer cost for this segment would have been (64.77)($1.26) + (45)($0.40) = $81.61 + $18.00 = $99.61 instead of $81.61 when calculated on the basis of only the fertilizer phosphorus.

In the experiment in Fig. 2-5, large quantities of both phosphorus and nitrogen were needed to produce the plateau yield, but only small responses were obtained when the nutrients were applied singly. Under such circumstances, it is important to take into account the costs of both nutrients when calculating the returns from fertilization. In many experiments, some nutrients are not needed at all, and others are needed in only small quantities, and the need may become apparent only when other nutrients have been added. Because the predictability of the results is generally rather poor, being prepared to make appropriate estimates of total costs of fertilization requires that the experiments include enough different quantities of each nutrient applied alone and in combinations with others to provide a basis for calculation. This complicates the situation and is a compelling reason for experimental designs such as the incomplete factorial design used to obtain the data in Fig. 2-5 that will include enough different quantities of each of the nutrients under test so the investigator will be prepared to make a reasonable interpretation of the data that may emerge from the experiments. Complete factorial experiments would be better, but they quickly become too large to handle.

The problem just discussed for economic interpretation of the results of multinutrient experiments applies in a different way to many experiments in which response to a single nutrient is being determined for soil test calibrations for example. In such experiments, applications of nutrients other than the one or ones under test are commonly made to all plots. This practice is designed to eliminate deficiencies of the other nutrients, and it usually increases the response obtained to the nutrient or nutrients under test. Economic interpretations of the results of such experiments without some allowance for the cost of the nutrients not directly tested, however, are not directly applicable to practical conditions in which a farmer starts with the soil as it is and must pay for all the nutrients added in fertilizers.

An especially significant point about the use of segmented functions for representing crop response to nutrients is the effect of the discontinuity or discontinuities on the economic interpretations. From considerations of the economics of fertilization in preceding sections, it is evident that applying enough fertilizer to attain the maximum yield results in less than the maximum net profit in curvilinear response functions unless the cost of fertilization is zero. The same is not necessarily true for the linear response and plateau function. The reason is that the linear response and plateau function overlooks the curvature that occurs between the ascending and plateau segments of the response function (see examples in Fig. 1-24), so that the quantity of fertilizer corresponding to the point of intersection of these two lines is generally well below the quantity corresponding to the maximum yield estimated by a curvilinear function.

The quantity of fertilizer corresponding to the maximum yield is generally poorly defined with curvilinear response functions that permit a decrease in yield with large quantities of fertilizer because small experimental errors in crop yields in the region of the maximum can have a large effect on the location of the maximum found by the curve-fitting process. For the same reason, the quantity of fertilizer that produces the maximum net profit is poorly defined when the corresponding yields are only a little below the maximum. The bias of the linear response and plateau function, however, provides a stabilizing influence the curvilinear functions lack. The point of intersection of the two straight lines usually is a more conservative estimate of the economic optimum quantity of fertilizer than the estimates provided by curvilinear functions.

Too few empirical tests are available at this point to evaluate the significance of the difference between the location of the discontinuity in the linear response and plateau function and the quadratic response and plateau function. Because of the comparative behavior of the functions, the point of intersection of the quadratic curve and the plateau should occur at a larger quantity of the nutrient or nutrients in question, and the quantities estimated to provide the maximum net profit should generally exceed those derived from the point of intersection of a single linear response segment with the plateau.

Because the quadratic segment of the function could fit data ranging from a straight line to data with considerable curvature, the quadratic response and plateau function should represent a "second approximation" that is closer to biological reality than the linear response and plateau function. In comparison with the linear response and plateau representation of the response behavior, the quadratic response and plateau function is advantageous if there is a long range of yield increases with curvature before the plateau. If a steep response is followed by an early plateau, and particularly if the number of data points is small, the linear response and plateau function is adequate.

2-8. Theory Versus Practice

The theoretical aspects of the mathematics of fertilization, including the design of the experiments, the statistical treatment of data, and the economic principles, have been developed at length. Reviews of these subjects will be found in publications by Baum et al. (1956, 1957), Heady et al. (1955, 1961), Munson and Doll (1959), Nelson et al. (1985), Colwell et al. (1988), and others.

The rigor of the mathematical elaboration of the numerous considerations involved has served to point out the complexity of the system under consideration and also some of the problems that exist in attempts to apply the theory to practice. Agriculture is beset by uncertainty, especially as regards weather and prices. In the absence of good prior information on the productivity of resources to be expended in production and on the prices of the final products, the precise application of economic principles to yield an economically optimum plan for an individual field, farm, state, or country is impossible.

The practical applications of economic calculations based upon individual experiments are exceedingly limited. The results of individual experiments apply to particular situations encountered in the past, but not necessarily to any specific situation that may be encountered in the future. The problem of predicting the response to be obtained is agronomic, not economic. The agronomic limitations are legion.

2-8.1. Assumptions

Although specific mention should be unnecessary in view of the emphasis in Chapter 1 on the empirical nature of mathematical response functions, the first important assumption in applying economic theory to practice via a response function is that the function selected to represent the data trends is applicable. The assumption that the function is applicable leads to the next assumption that the economic interpretations are valid. "Valid" and especially "applicable" do not have precise meanings in the present context, so that various functional forms could be used. Different functions may lead to different interpretations because each function imposes its own peculiarities on the fit. There is no definitive answer to this problem.

Economic analyses of the results of fertilizer experiments can show the significance of the findings under different conditions of costs of inputs and prices of outputs by using different assumptions about the numerical values of these variables. The problem about such assumptions is that although the principal utility of the analyses is to provide a guide to the future, the future is unknown. The economics thus must always be presented in the sense that for an assumed set of physical facts and economic circumstances, the economic results are expected to be thus and so.

Other variables may be considered important, but not included in the analysis for lack of suitable economic values. Two such factors of current concern are possible health consequences of introducing fertilizers into the environment and of fertilizer-induced changes in nutritional values of crops. Economists could bring such factors into play by making assumptions about their economic value, but this usually is not done for purposes of providing guidance to producers until there has been opportunity for members of society to express their views on values in the market or through legislation and regulation.

As an example of the economic indecision resulting from possible health consequences, Sylvester-Bradley and George (1987) made an economic analysis of the results of 58 experiments on spring topdressing of winter wheat with nitrogen fertilizers in England and Wales. They concluded that considering the average increases in yield, the value of the crop, the premium for high protein, and the cost of the fertilizers, it would be profitable to apply 50 kilograms of fertilizer nitrogen per hectare in excess of the quantities recommended by the Ministry of Agriculture, Fisheries, and Food.

They noted, however, that wheat crops produced with an extra 50 kilograms of fertilizer nitrogen would contain only 9 kilograms of additional nitrogen in the grain per hectare in the experiments they analyzed. Reflecting public health concerns for the loss of nitrate to the groundwater, they suggested that if the extra 50 kilograms of fertilizer nitrogen were applied repeatedly, the fate of the 41 kilograms of nitrogen not recovered in the grain could have significant effects on nitrogen fertilizer policy in the long term, meaning presumably that governmental action might be taken to place a price on the nitrate lost. The price, which might be reflected in a tax on nitrogen fertilizer, or in some other way, then would have an economic value that could be taken into account in economic analyses designed to provide a guide to producers.

2-8.2. Seasonal Effects

Seasonal effects are a major limiting factor in applying theory to practice. Because of unpredictable effects of differences in weather and pest infestations, what applies this year in terms of productivity and economics of resources may not apply next year.

On the basis of field experiments in Suriname, Boxman et al. (1985) found that the optimum economic application of fertilizer nitrogen for corn decreased with an increase in the number of days in which the corn suffered moisture stress during the drought-sensitive period from 17 days before to 32 days after silking (including the period when pollination occurs). The range was from 220 kilograms of fertilizer nitrogen per hectare with no moisture stress days to 64 kilograms per hectare with 35 stress days. The corn yields were much more variable with the optimum economic application of fertilizer nitrogen (coefficient of variation 11%) than with no fertilizer nitrogen (coefficient of variation 4%). In effect, greater rainfall and fewer stress days permitted the corn to take advantage of the greater supply of nitrogen.

In South Africa, Korentajer et al. (1987) found that the quantity of nitrogen fertilizer required to maximize the profit from fertilizing wheat was affected far more by differences in physiological stress of the wheat due to water deficiency from one year to another than by the change in the ratio of the price of nitrogen to the price of wheat over a 25-year period. Eck (1988) simulated seasonal effects in experimental work on wheat in Texas by varying the irrigation regime. From his data, one may calculate that the economic optimum quantities of fertilizer nitrogen in kilograms per hectare were 184 where the wheat had an ample supply of water, 114 where it was stressed during jointing and grain filling, and 4 where it was stressed throughout the spring. The numerical values used for the prices of fertilizer nitrogen and wheat in making these calculations were $0.40 per kilogram for fertilizer nitrogen and $128.00 per megagram for wheat, as in Fig. 2-6B. In both of these investigations on wheat, the stress of water deficiency had a far greater effect on the economic optimum quantity of fertilizer nitrogen than on the quantity of nitrogen required to produce the maximum yield.

Figs. 2-6A and B show the results of two experiments in which the response of the crop to annually applied nitrogen fertilizer was determined in successive years on the same land. Predictions about the most profitable fertilizer applications for individual years based upon data such as these may be rather different than the values that would be found after the fact from calculations based upon the responses actually obtained. The range in quantities of fertilizer nitrogen in kilograms per hectare estimated to produce the maximum net profit in individual years is from 119 to 238 for the data in Fig. 2-6A and from 2 to 121 for Fig. 2-6B.

For purposes of prediction, the general practical approach currently available to deal with the effects of weather variations among years is replication of experimental measurements over a period of years such that the results will delineate a central tendency for the productivity of the resources. When a large enough body of response data has been obtained to represent the central tendency with an acceptable degree of precision, various economic problems connected with fertilizer application can be worked out and used to make acceptable predictions for those conditions.

If the experiments are of 1-year duration and are conducted on different sites, the effect of years will be confounded with the effect of sites. And if the experiments are conducted on the same experimental plots year after year, the effect of years will be confounded with whatever buildup may exist of residual effects from fertilizers applied in prior years. This dilemma could be avoided by laying out an experiment with many extra plots at a given site and by using some of these plots each year. This alternative is rarely if ever adopted.

According to Middleton (1973, 1980), the maximum profit in the long run is obtained from predictions of fertilizer applications based upon an average year. Data from a 5-year experiment on irrigated corn in Nebraska (Bock and Hergert, 1991) were in agreement with this generalization. Brown and Oveson (1958)

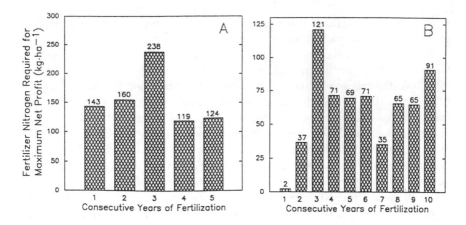

Fig. 2-6. Quantity of fertilizer nitrogen required to produce the maximum net profit in identical experiments on two crops in successive years on the same land. A. Paddy rice in Punjab (@ 3.91 rupees per kilogram for fertilizer nitrogen and 1300 rupees per megagram for rice). (Dev et al., 1984) B. Dryland wheat in Oregon (@ $0.40 per kilogram for fertilizer nitrogen and $128.00 per megagram for wheat). (Brown and Oveson, 1958)

found that the estimated net profit from fertilizing wheat annually over a 10-year period was greater with applications based upon the average response than upon the modal response.

How long it takes to approximate an average year will depend upon the variability among years in the chosen area, the acceptable departure from the average, and the nature of the measurements made to characterize a year. Whether the objective is to approximate the mean or to make adjustments according to weather cycles (see Van Rooyen and Dannhauser, 1988), the magnitude of the confidence limits in Fig. 2-7 implies that repetition of experiments over a number of years is likely to be a major requirement of a research program to develop useful advisory predictions. For example, if one desires a degree of precision such that the true mean nitrogen requirement to produce the maximum net profit will lie within the range bounded by the observed mean requirement plus or minus 30% of the observed mean with a probability of 95%, 5 to 6 years of experiments would be needed to achieve that degree of precision for the irrigated corn in Nebraska. For the dryland wheat in Oregon, 15 to 16 years of experiments would be needed. These estimates apply to the specific locations at which the experiments were conducted. Making predictions for a broader area involving differences among locations is another problem that would require more experiments and a different approach. This matter is addressed in Chapter 4.

2-8.3. Response Function Effects
Another important limitation in applying economic theory to practice is the sensitivity of the economic calculations to the mathematical function selected to

represent the response (Sutherland et al., 1986). For example, Cerrato and Black-mer (1990) determined the yields of corn obtained in each of 12 similarly designed field experiments with replicated treatments of a control plus nine different quantities of nitrogen fertilizer. They fitted five different response functions to the yield data from each experiment, and then calculated the quantities of fertilizer nitrogen required to yield the maximum net profit under a constant set of assumptions about the price of corn and the cost of fertilizer. The results in Table 2-1 show that the estimated economic optimum quantities derived from the various response functions do not agree. In fact, there is a three-fold range in the values. In an analogous analysis of data from experiments on phosphate fertilization of wheat in Australia, Colwell (1983) noted a three-fold range among response functions in the estimated quantities of fertilizer required to yield the maximum net profit in one experiment and a two-fold range in another.

The last column in Table 2-1 shows R^2 values, which provide a measure of goodness of fit of the various response functions to the experimental yield values. R^2 is the fraction of the mean sum of squares of deviations of observed values from their mean that is accounted for by the function employed. According to this criterion, the functions fit the data about equally well. The relatively high R^2 values indicate that the differences in estimated quantities of fertilizer nitrogen

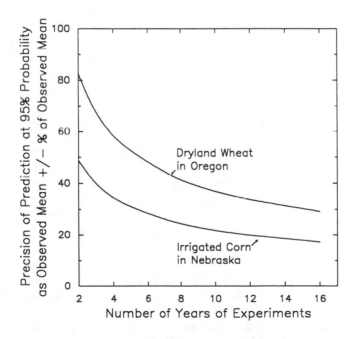

Fig. 2-7. Range of crop responses to nitrogen fertilization expressed as the observed mean plus or minus a percentage of the observed mean within which the true mean is expected to fall with 95% probability, based upon experiments in successive years under two circumstances. The results are for dryland wheat in Oregon (Brown and Oveson, 1958) and irrigated corn in Nebraska (Bock and Hergert, 1991).

Table 2–1. Quantities of Fertilizer Nitrogen Estimated by Five Different Response Functions to Yield the Maximum Net Profit in 12 Experiments on Corn in Iowa (Cerrato and Blackmer, 1990)

Response Function[a]	Estimated Quantity of Nitrogen (kg · ha⁻¹)	Mean R^2
Linear response and plateau	128	0.83
Quadratic response and plateau	184	0.84
Quadratic	225	0.82
Exponential (Mitscherlich)	252	0.82
Square root	379	0.79

[a] See Chapter 1 for a discussion of the various equations.

to obtain the maximum net return are not simply a consequence of poorly defined response curves.

The R^2 values indicate that by a small margin the quadratic response and plateau function was the best of the ones compared. Although this model is biologically unrealistic (as are all the others), it lacks the principal disadvantages of the linear response and plateau function and the quadratic function, namely, the lack of curvature in the ascending portion of the first function and the excessive slope of the second with fertilizer applications a little below those producing the maximum yield.

As pointed out previously, fertilizer applications sufficient to cause yield depressions are beyond the rational range for practical use of fertilizers, and thus should not be given much weight in the selection of functions designed for practical use. Nonetheless, experiments designed to estimate the quantities of fertilizers needed to obtain the maximum net profit require some relatively heavy applications to assure that the response data will cover the critical range, which usually is close to the maximum yield. Yield depressions thus may appear inadvertently even though the experimenter would prefer to avoid the range beyond the maximum yield. When yield depressions do occur, something must be done about them. If they are included in the analysis of the data, they will affect the shape of response curves calculated to represent applications below the maximum yield. If the reduced yields obtained with excessive amounts of fertilizer are not included in the calculations, the number of observations omitted will also affect the shape of the curve calculated to represent applications below the maximum yield, and there may be some statistical quibbling about the arbitrariness used in deciding which values to omit.

In a particularly lucid account of the problems of estimating the quantity of fertilizer to yield the maximum net profit, Wimble (1980) addressed the yield depression phenomenon by recommending that the response function employed be capable of showing a downturn of yield with fertilizer applications exceeding those required to produce the maximum yield. In his view, response functions such as the quadratic and Mitscherlich equations with only three parameters do not have sufficient flexibility to meet the needs. He was of the opinion that functions with four parameters are needed, and he listed four functions he considered appropriate.

Colwell et al. (1988) preferred the square-root equation (see Section 1-5.1), which they called the "square root quadratic equation," for general use in soil fertility work. They gave four reasons for their choice. The square-root equation is a three-parameter equation related to the quadratic function, but it lacks the unrealistically high slope of the quadratic equation with quantities of fertilizer a little below those resulting in the maximum yield. It also has a relatively flat plateau, and it permits decreases in yield with excessive applications of fertilizer. It has the disadvantage of an unrealistically high slope with small quantities of fertilizer, but Colwell and coworkers considered this property of minor practical importance because the economic optimum is generally in the range of fertilizer applications where the equation performs well, and not in the low range where it deviates from reality. For application to biological assay of effective quantities of nutrients in soils, however, the square-root equation would be unsuitable, as is noted in Chapter 4, because this application depends upon realistic performance of the response function in the range down to zero application of fertilizer.

The failure of economic calculations based upon different response functions to agree has been known for many years (Anderson, 1957). Mathematics tells what is true about a hypothetical situation, but the "real-world" responses to fertilizers rarely correspond precisely to the hypothetical situations supposed by the various response functions. The discrepancies appear to be an unavoidable consequence of a combination of experimental error with the fact that different mathematical functions used to describe crop responses to fertilizers have inherently different shapes and the fact that the shapes of response curves may vary with the circumstances. Many factors affect the shapes of response curves, as indicated in Chapter 1.

More specifically, in estimating the economic optimum quantities of fertilizers, it is of prime importance that the experiments have a substantial number of points in the critical range around the optimum, and that the mathematical function fit the points in this range. Where enough different quantities of fertilizer have been used, the existence of biases can be visualized by plotting the deviations from the estimated yields against the estimated yields. If the mathematical function used does not bias the representation of the data, the deviations should be distributed at random above and below the estimated yields throughout the range of the data. If the function fails to fit the data well in a particular range, the majority of the deviations will lie above or below the estimated yields in that range.

The mathematical nature of the response function may be expected to remain a perpetual problem. The best solution is probably to base practical applications upon a response function that provides a good approximation to response data derived from many experiments over a period of years. With enough experimental data, it may be possible to divide the experiments into logical groups for which different response functions are appropriate. The more homogeneous the circumstances included in the average, the more precise the estimates will become.

The first step in the direction of adapting the response function to individual circumstances is to use a single response function and to add variables other than the nutrients supplied in the fertilizers tested. The first variables that come to mind from the soil fertility standpoint are the inherent soil supplies of the nutrients added in the fertilizers. This matter is discussed in Chapter 4.

But there are numerous other variables. These include the cultural practices, crop, crop variety, crop stand, pest infestations, soil, and weather. Of these variables, the first three are under the control of the farmer, and the fourth and fifth are to a considerable extent under the farmer's control. An allowance can be made for the soil, either as an overall variable or through measurements of important soil properties. Weather factors may be of overriding importance, as may be determined in retrospect, but little can be done about predictions beyond using the average. An intermediate situation is the use of the available water content of the soil in the spring as a basis for predicting the success of the current crop and the response to fertilization in the Great Plains, where water availability is such a predominant factor in crop production (Read et al., 1982; Ulmer et al., 1988; see also Fig. 1-16).

Additional factors may be represented mathematically in any desired way, but factors such as the clay content or organic matter content of the soil are often introduced as linear variables. The "general linear hypothesis" is often useful as a first approximation. When introducing additional factors as linear variables in response functions, however, it may be useful to bear in mind that the added factor is assumed to have a linear effect with constant slope over all levels of the other variables. See Fig. 2-8 for a plot of an equation containing two linear terms.

In some instances, a different formulation may be justified. For example, Probert (1987) was confronted with a highly variable stand of soybean in an experiment in which the intended seeding rate was uniform. The usual procedure in such instances is to use covariance to adjust the yields to a uniform stand, but the author pointed out that this procedure assumes the relationship between yield and stand to be linear and to have the same slope over all treatments. In his experiment, the slope differed among treatments (see Fig. 2-9), and Probert combined a hyperbola-type response curve for plant density with the Mitscherlich equation for fertilizer phosphorus to produce a response function that would represent each factor in a realistic way.

Hildreth (1957) studied the results of experiments on nitrogen fertilization of corn in North Carolina in which the weather was classed as good, fair, or dry. In this instance, the general linear hypothesis provided a less precise representation of the data than did the assumption that the yields on high-yielding plots with heavy applications of fertilizer nitrogen were reduced by dry weather to the same percentage as were lower yields with light applications of nitrogen. To take this observation into account, Hildreth modified the response function by transforming the yields to the logarithms of the yields.

Addiscott et al. (1991) suggested for nitrogen fertilization the radical departure of substituting what they called the "surplus nitrate curve" (residual nitrate nitrogen in the soil at harvest versus fertilizer nitrogen applied) for the usual yield response curve. They suggested using the quantity of fertilizer nitrogen at

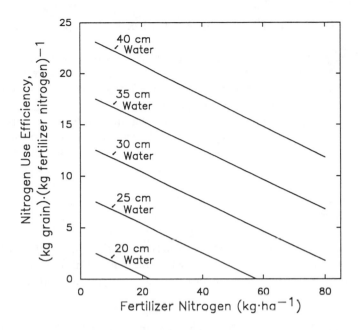

Fig. 2-8. Nitrogen use efficiency in experiments on wheat in Punjab as a function of quantity of fertilizer nitrogen and the sum of (a) the available water storage in the soil at planting and (b) the rainfall during the growing season. The plot illustrates the hypothetical way in which a dependent variable is affected by two independent linear variables. The equation was $y = -0.1680 - 0.00142N + 0.010W$, where y = kilograms of wheat grain per kilogram of fertilizer nitrogen, N = kilograms of fertilizer nitrogen per hectare, and W = centimeters of available water. (Prihar and Singh, 1983)

which the residual nitrate starts to increase sharply as "a fairly clear indication of when the crop is satisfied, and therefore of the optimum." They noted that this indication of when the crop has enough nitrogen offers four advantages. (1) It is often fairly close to the economic optimum. (2) It is "uncontaminated by economics." (3) It has "far greater environmental relevance than the economic optimum." (4) It probably can be assessed better by statistical methods than the economic optimum.

2-8.4. Experimental Precision and Price Ratios

The value of response data for economic interpretations is conditioned by their precision. An important associated factor is the ratio of the price of the fertilizer to the price of the crop. At high price ratios, the economic optimum quantity of fertilizer is small. The corresponding yield is on the steeply ascending portion of the response curve, where small differences in quantity of fertilizer have a large effect on crop yield. A given difference in crop yield then has only a small effect on the estimated optimum quantity of fertilizer. As the price ratio decreases, it becomes profitable to fertilize the crop to yields closer to the maximum. At the same time, a given difference in quantity of fertilizer has a progressively smaller effect on the yield, and a given difference in yield has a progressively greater effect on the estimated economic optimum quantity of fertilizer.

In developed countries, the price ratio is so low in most instances that the crop yields corresponding to the economic optimum quantities are on the relatively flat portion of the response curve. The consequence is poorly defined estimates of the economic optimum quantities.

Several examples will be given. In reviewing the results of 146 experiments on nitrogen fertilization of winter wheat in the United Kingdom, Sutherland et al. (1986) noted that an estimated economic optimum could be found within the range of quantities of nitrogen applied in only 76 experiments. In these 76 experiments, the standard error of the estimated optimum application in kilograms per hectare ranged from 0 to 10 in 28 trials, from 10 to 20 in 22 trials, from 20 to 30 in 11 trials, and from 30 to 50 in 7 trials, and it exceeded 50 kilograms per hectare in 8 trials. Results for winter wheat that appeared to be more concordant were reported by Dilz et al. (1982) in Belgium and Wehrmann et al. (1988) in Germany in experiments in which the mineral nitrogen in the soil at the end of the winter was related to the economic optimum application of fertilizer nitrogen.

In The Netherlands, Neeteson and Wadman (1987) analyzed the results of 133 experiments on sugarbeet and 76 experiments on potato in which the esti-

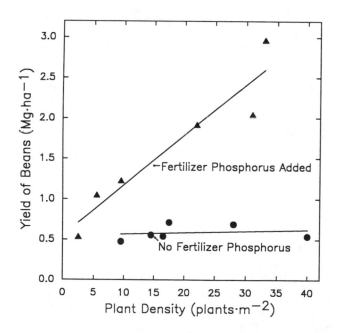

Fig. 2-9. Yield of soybean versus plant density in the absence of fertilizer phosphorus and with 100 kilograms of fertilizer phosphorus per hectare in an experiment in Northern Territory, Australia. (Probert, 1987)

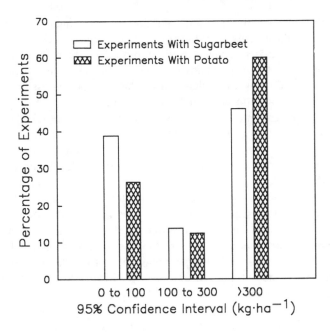

Fig. 2-10. Frequency distribution of the 95% confidence interval for the estimated quantities of fertilizer nitrogen to produce the maximum net profit in 133 experiments on sugarbeet and 76 experiments on potato in The Netherlands. (Neeteson and Wadman, 1987)

mated economic optimum rate of application fell within the range of fertilizer nitrogen applied. The results in Fig. 2-10 show that in 46% of the sugarbeet experiments and 60% of the potato experiments, the 95% confidence intervals for the estimated economic optimum quantities of fertilizer nitrogen exceeded 300 kilograms per hectare. The sugarbeet experiments involved six different quantities of fertilizer nitrogen with four replicates, and the potato experiments involved seven different quantities of fertilizer nitrogen with three replicates. Similar results from research on potassium fertilization of potato in The Netherlands were reported by Neeteson et al. (1987).

The foregoing examples illustrate the difficulty that has been encountered in practice in estimating the economic optimum quantity of fertilizer from field experimental data. But there is another side to the issue, namely, the profitability side.

Withagen (1983) demonstrated the profitability side in an analysis of the results obtained in experiments on nitrogen fertilization of sugarbeet conducted over a 3-year period in The Netherlands. As shown in Fig. 2-11A, he found considerable scatter in a plot of the estimated quantities of fertilizer nitrogen required to produce the maximum net profit against the content of mineral nitrogen in the surface 60 centimeters of soil in early spring. In most of the experiments, however, the results could be made to fall on the average regression by arbitrarily changing the maximum net profit by amounts up to ±1% (Fig. 2-11B). These adjustments increased or decreased the associated estimates of the quantities of fertilizer nitrogen required to produce the maximum net profit, and the changes in these quantities were sufficient to move most of the data points upward or downward so they fell on the average regression line. Thus, although the deviations in estimated optimum quantities of fertilizer nitrogen from the average regression line were substantial, they were of relatively little economic significance because of the flatness of the response curve and the low ratio of the price of fertilizer nitrogen to the price of the crop.

In further work on potato, Neeteson et al. (1988) developed a model for estimating the optimum quantity of fertilizer nitrogen on the basis of information available at planting, including the nitrate content of the soil to the expected depth of rooting, the subsequent mineralization of nitrogen, the soil type, average weather conditions, the dates and quantities of fertilizer nitrogen applied, the dates of planting, and other factors. Using the model, they calculated the expected response curves and optimum nitrogen applications for individual experiments. They found that if the optimum quantities of fertilizer nitrogen calculated from their model had been applied, the yields of tubers would have been within 2% of those obtained with the experimentally measured optimum applications in 84% of the experiments. The explanation for the close agreement of the calculated and observed optimum yields was apparently a satisfactory model and a low enough price ratio of fertilizer to tubers to place the optimum applications on the relatively flat response plateau, where large differences in quantity of nitrogen would have little effect on yields.

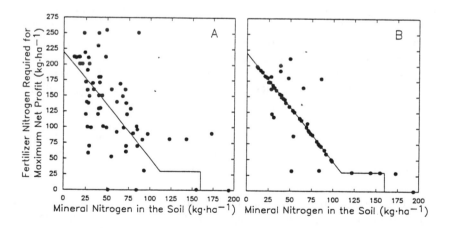

Fig. 2-11. Estimated quantities of fertilizer nitrogen required to produce the maximum net profit versus mineral nitrogen in the surface 60 centimeters of soil before April 1 in experiments on sugarbeet in The Netherlands. Each point represents an experiment. The original data are plotted in *A* on the left. *B* shows the effect of adjusting the maximum net profit by up to ±1% and moving the associated estimated quantities of fertilizer nitrogen upward or downward, as the case might be, toward the average regression line. (Withagen, 1983)

2-8.5. Isoquantal Theory Effects

Middleton (1980) was sharply critical of the "isoquantal theory," and he listed five reasons for his view. His principal point appears to be that to apply over a period of years the physiologically unbalanced mixture of nutrients that could be selected on the basis of the isoquantal theory would result in gradual accumulation of excesses of certain nutrients in the soil and unfavorable effects on crop production.

Middleton's criticism applies to the inappropriate use of the results of short-term experiments as a basis for recommendations or decisions about fertilizer practices to be followed on a long-term basis. For long-term decision-making, experiments in which an opportunity has been provided for the development of cumulative effects are appropriate as a basis for inferring the performance of treatments for which cumulative effects may be significant. If conducted over a long enough time, "long-term" experiments would provide an automatic evaluation of the average response to seasonal differences.

An alternative philosophy has the short-term objective of predicting the most appropriate investments in fertilizer resources for the immediate future, with the idea that the situation may change from year to year and should be reevaluated as needed. As a basis for such predictions, long-term experiments are inappropriate because the gradual build-up of residual effects with increasing quantities of fertilizer alters the shape of the response curves. If the residual effect is positive, a long-term response curve would overestimate short-term responses. And if the residual effect is negative, a long-term response curve would underestimate short-term responses.

Applying the long-term philosophy requires conducting long-term experiments over the necessary number of seasons. Applying the short-term philosophy also requires the conduct of short-term experiments over the necessary number of seasons, but the experiments must be on different sites.

As a substitute for the isoquant concept, with which economic optimization was connected in Section 2-7, Middleton (1980) proposed the use of preliminary trials to find the composition of the fertilizer mixture that produces the greatest increase in yield and then testing that mixture for as many years as needed, so that the average results could be said to represent an average season. The trials would be scattered over the area for which information is needed. Leaf analysis might be used in succeeding years to monitor the results and to serve as a basis for adjusting the composition of the fertilizer applied in the individual experiments. A similar technique presumably would be employed when the fertilizer mixture tested in experiments is used on farmers' fields on which no experiments have been conducted.

Middleton's procedure would substitute a maximum-yield criterion for the minimum-fertilizer-cost criterion associated with the isoquant concept. The justification for such a substitution could be the presumed deterioration of results that would occur if a fertilizer mixture based upon the isoquant-isocline concept were used without change for a number of years.

2-8.6. Psychological Factors

The quantity of fertilizer required to yield a specified economic goal for the producer increases as the ratio of fertilizer cost to crop value decreases. Superimposed upon the effect of this generally recognized principle on decision making is a subjective tendency of many producers to limit their risk of economic loss by applying less than the estimated economic optimum quantity of fertilizer if the fertilizer is expensive and more if it is inexpensive.

The subjective tendency is soundly based. With response curves of the diminishing-returns type, the loss of profit from underfertilizing by a given number of units exceeds the loss of profit from overfertilizing by the same number of units. This principle was illustrated graphically by Bock and Hergert (1991, page 162) using data from experiments on nitrogen fertilization of corn in different years. This form of insurance, however, has been criticized on environmental grounds when the fertilizer is a source of nitrogen and the ratio of fertilizer price to the value of the crop produced is low enough so that the economic optimum quantity of fertilizer results in yields close to the maximum. Under these conditions, extra nitrogen added for insurance will increase the crop yield only slightly. Much of the nitrogen added for insurance will not be absorbed by the crop and will be subject to loss by leaching.

2-9. Environmental Aspects

In this chapter, coverage of the economics of fertilization has emphasized the use of fertilizers as a production input that can lead to increased yields of quality products and to financial gain for the producer if properly used. There are additional off-site or environmental effects, and in recent years concerns about these effects have come to the fore. See a paper by Bateman (1991) for an overview of the big picture within which the environmental aspects of fertilization are a microcosm.

Off-site effects of fertilization include influences of agricultural products on people as well as indirect influences through the environment. Some effects are positive and some are negative, but support for research generally emphasizes the negatives. Off-site effects are not necessarily economic, although economics is involved, and it is primarily economists who attempt to integrate the various facets in analyses of costs versus benefits. See papers by Conrad (1988) and Follett and Walker (1989) for lucid discussions. The significance attached to these effects by society is reflected in the numerous recent scientific publications dealing at least in part with environmental aspects of nitrogen. These include a number of books: Prins and Arnold (1980), Brogan (1981), Stevenson (1982), International Institute for Sugar Beet Research (1983), Hauck (1984), Kang and Van der Heide (1985), Neeteson and Dilz (1985), Lambers et al. (1986), Van der Meer et al. (1986), Hargrove (1988), Jenkinson and Smith (1988), Follett (1989), Germon (1989), Welte and Szabolcs (1989), Follett et al. (1991), Addiscott et al. (1991), and Groot et al. (1991). Most were published in Europe.

The symposium edited by Follett et al. (1991) is notable for its applied perspective. The book by Addiscott et al. (1991) is notable for the interesting way it is written and for the sociological perception displayed in presenting the agricultural aspects of nitrogen and nitrate to nonagriculturists.

Only two environmental aspects of fertilization — the nitrate and animal manure issues — will receive emphasis in this section, but there are other aspects. For example, much research is being done to determine in effect how much of a given kind of sewage sludge can be applied to soils with various properties without immediate or eventual undesirable effects on crops, animals, and people from the heavy metals in the sludges. The heavy metal content of sludges varies greatly with the kind and proportion of industrial discharges to sewage systems.

As an example of an indirect environmental relation to agriculture that is outside the scope of soil fertility research, questions have been raised about the cancer hazard to occupants of homes built on land reclaimed from phosphate rock mining operations because of the radioactivity of the uranium that is present as a trace constituent in the phosphate rock. As applied to agricultural soils with fertilizer phosphorus, the radioactivity of the uranium seems to be accepted as negligible. More concern has been evinced for the contribution of fertilizer phosphorus to the undesired growth of algae in surface waters.

Societal attention has been focused on nitrogen fertilizers. The principal environmental concern about nitrogen fertilizers and about fertilizer use in general is the potential unfavorable impacts on human health of fertilizer-derived nitrate present in groundwater used for drinking. Of secondary concern is the enhanced growth of algae in surface waters containing extra nitrogen (and phosphorus). Algal "blooms" on surface waters are obvious and unsightly, and decomposition of the organic matter reduces the oxygen supply needed by fish and other aquatic animals. Of additional currently minor concern is the evolution of small quantities of gaseous nitrous oxide from soils and the possible contribution of the evolved nitrous oxide to the breakdown of stratospheric ozone and to global warming.

Because nitrogen fertilizers are of first importance in agriculture, much of the research in soil fertility has been redirected to nitrate-related topics. The following paragraphs emphasize the various facets of the drinking water issue because of their crucial importance in terms of future decisions that will affect both agriculture and the general public.

2-9.1. Loss of Nitrate by Leaching and Denitrification

Of major public concern is the appearance of nitrate in groundwater as a result of agricultural practices. The management and estimation of nitrogen losses from soils by leaching was the principal theme of a symposium edited by Follett et al. (1991).

Under some conditions, much nitrate is lost from soils by leaching. The amount depends upon the volume of drainage and the concentration of nitrate in the soil solution. From the agricultural standpoint, the amount is of principal importance, but from the environmental standpoint, the concentration is emphasized.

In a survey of 83 sites in valleys in California, Rible et al. (1979) reported concentrations of nitrate nitrogen in the drainage water ranging from 6 to 220 milligrams per liter. In another survey, also in California, Letey et al. (1977) measured concentrations ranging from a minimum of 1 milligram of nitrate nitrogen per liter at two locations under cotton to a maximum of 196 milligrams per liter at one location cropped to melons. Ritter (1989) reviewed the experimental work on loss of nitrate by leaching under irrigation in the United States.

Maidl and Fischbeck (1987) sampled fields in Germany to depths up to 10 meters and found that the nitrate concentrations below the root zone increased with the quantity of fertilizer nitrogen that had been used. The highest concentrations were found under soils that had been used for vegetable crops, which characteristically are heavily fertilized and recover only a minor portion of the fertilizer nitrogen added. With heavy fertilization, the nitrate tended to move downward in bands that appeared to coincide with the years in which the fields had been in vegetables. Wehrmann and Scharpf (1989) demonstrated the value of timely measurements of mineral nitrogen in soils for adjusting the application of fertilizer nitrogen to vegetable crops and for reducing the loss of mineral nitrogen by leaching during the winter.

Research by Simon et al. (1988) and Gölz-Huwe et al. (1989) showed that leaching of nitrate from the soil in the state of Baden-Württemburg, Germany, occurs principally during the winter months. Although fertilization and mineralization may produce a relatively high nitrate content of the soil in the early summer, this is not of direct importance to drainage losses because of the uptake of nitrate and water by the crop and the lack of drainage during the summer months. See Fig. 2-12 for data obtained in measurements where the crop grown during the summer was corn.

The rapid increase in content of nitrate nitrogen in the soil in the spring in Fig. 2-12 was due to mineralization of organic nitrogen, but part of the peak during the month of June was a consequence of application of fertilizer nitrogen. The smaller peak of soil nitrate that appeared in late autumn was due to mineralization of organic nitrogen because no fertilizer was applied at that time. Nitrate accumulated during the fall is at risk of loss during the winter. Similar data were published by Machet and Mary (1989), showing a continuous record for mineral nitrogen in a soil in France over a period of 1.5 years beginning before sugarbeet harvest in late autumn and continuing through harvest of winter wheat in the following year and beyond for another 200 days. Another graph in the same paper shows much more marked accumulation and loss of mineral nitrogen over a 2-year period in a fallow soil after plowing of grassland.

In 13 experiments on winter wheat in the United Kingdom, MacDonald et al. (1989) found that where ^{15}N-tagged fertilizer nitrogen was applied in quantities up to 234 kilograms per hectare, the ammonium plus nitrate nitrogen remaining in the surface 23 centimeters of soil at harvest was low and about the same as that in the control plots that received no fertilizer nitrogen. From 79 to 98% of the ammonium plus nitrate nitrogen in the soil at harvest was derived from

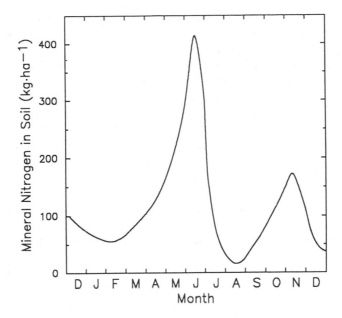

Fig. 2-12. Nitrogen present as nitrate in the surface 90 centimeters of a soil in Germany throughout a period of 1 year. Nitrogen-fertilized corn was grown during the summer. (Simon et al., 1988)

mineralization of organic forms of nitrogen and not from unused fertilizer nitrogen. These findings also emphasize the role of organic nitrogen mineralization late in the season as the major source of nitrate subject to leaching over the winter under the conditions of the experiments.

Fertilizer nitrogen that has been immobilized as organic nitrogen of course is subject to subsequent mineralization. Hart et al. (1982) found that when ^{15}N-tagged fertilizer was added to winter wheat in one year, from 6 to 11% of the ^{15}N remaining in the stubble and topsoil at the end of that year was recovered in the crop of wheat in the following year.

Consistent with these findings, Prins et al. (1988) stated in a summary of research done in the European Economic Community that direct losses of fertilizer nitrogen by leaching from crops such as cereals and sugarbeet are negligible if the application of fertilizer nitrogen does not exceed the economic optimum. Virtually all the nitrate leached comes from organic sources, including soil organic matter, crop residues, and manure. Shallow-rooting crops, such as potato and many vegetables, however, are less effective in recovering soil nitrate. When fertilized with economically optimum quantities of nitrogen, these crops, as well as crops that leave high-nitrogen residues in the soil, can lead to large losses of nitrate by leaching.

Research by Meisinger (1976), Broadbent and Rauschkolb (1977), Singh et al. (1978), Kamprath (1986), Groot and Van Keulen (1990), and others indicates that little nitrate is lost by leaching under a variety of conditions if fertilizer nitrogen is applied in timely fashion in quantities up to those producing crop yields approaching the maximum.

In the Platte River Valley in central Nebraska, where a large-scale nitrogen management project with farmers had been in progress for a number of years, Schepers et al. (1991) found that the concentration of nitrate nitrogen in the shallow irrigation wells employed in that area was closely related to deviations of quantities of fertilizer nitrogen applied for corn in 1988 from the recommended applications. The range of concentrations in the wells was from about 7 to 25 milligrams of nitrate nitrogen per liter, and the concentration associated with the recommended application was about 17 milligrams per liter. Because the nitrate losses by leaching in a single year would be insufficient to produce the observed effects, the relationship found in this work may be regarded as an estimate of the long-term consequences of a tendency of individual producers to apply consistently greater or lesser quantities of fertilizer nitrogen than those recommended.

Prins et al. (1988) and Juergens-Gschwind (1989) pointed out that from both the agronomic standpoint and the environmental standpoint it is important to apply no more fertilizer nitrogen than is needed and to maintain a plant cover on the soil as long as possible. Particularly for short-season crops, this means planting a cover crop that will be effective in absorbing nitrate after the main crop has been harvested. The cover crop may serve only as a green manure.

Grasses are particularly effective in absorbing nitrate from soil, and crops of the grass family are often used as cover crops. If grassland is mowed for hay, much of the nitrogen absorbed from the fertilizer is removed, and there is little loss by leaching. Grazed grassland is another matter. Little nitrogen is removed in the animals, and too much nitrogen per unit area is added to the small spots receiving animal excreta (particularly urine) for effective uptake by the grass. When, in addition, heavy applications of fertilizer nitrogen are made to promote grass growth because most of the grazed area has not received the concentrated doses of excretal nitrogen, large losses of nitrate by leaching may occur (Ball and Ryden, 1984).

With a given application of nitrogen and a given water supply, the potential for nitrate loss by leaching decreases as soil conditions for crop growth improve. With greater crop growth, nitrogen uptake increases, and if drainage occurs it generally decreases. Benbi et al. (1991) found in 16 cycles of a corn-wheat-cowpea rotation in Punjab that the increases in crop yield from adding phosphorus and phosphorus plus potassium were paralleled by increases in recovery of fertilizer nitrogen in the crops. The cumulative nitrogen recovery increased from 25% with annual applications of nitrogen alone to 42% with nitrogen and phosphorus and 56% with nitrogen, phosphorus, and potassium. All nitrogen treatments increased the residual soil nitrate, but the plots receiving nitrogen, phosphorus, and potassium, which increased the crop yield and the nitrogen recovery the most, had the least residual nitrate in the surface 2.1 meters of soil.

Although the absolute values vary with the circumstances, the general principle is that where the nitrate supply is low the crop is effective in removing the nitrate from the soil solution. Where the nitrate supply is high, the crop cannot absorb all of it, and more is susceptible to loss in drainage water. As illustrated graphically by Pratt (1984), a plot of the amount of nitrogen leached against the quantity of fertilizer nitrogen applied is concave upward (the same would be true of a plot of concentration against the quantity of fertilizer nitrogen applied). With increasing volumes of water lost by drainage, the general shape of the curves remains the same, but the nitrate losses increase (and the concentration in the drainage water decreases). Thus, while reducing applications of nitrogen-bearing fertilizers through the range covered by the relatively flat portion of the response curve near the maximum may result in a significant reduction in residual nitrate available for leaching, the additional reduction that occurs as a result of extending the limitation to the steeper portion of the response curve is generally much smaller.

From the standpoint of regulatory attempts to reduce the loss of nitrate from soils by leaching, a reasonable inference from the foregoing discussion of research findings is that establishing a fixed limit on the amount of fertilizer nitrogen that can be applied may not be the best approach because it would not take into account the differences in nitrogen supply and crop requirement that exist from one situation to another. A suitable soil test made before fertilization could provide a basis for adjusting the current application to take into account the residual nitrate in the soil.

To test the value of a scientific approach to controlling loss of nitrate to groundwater, a special demonstration project was conducted on more than 16,000 hectares of predominantly furrow-irrigated corn in the Platte River Valley in Nebraska, where concern had developed regarding accumulation of nitrate in groundwater (Power and Schepers, 1989). The Nebraska project was located in a subhumid area where the annual precipitation is about 65 centimeters and there is little normal leaching. Under irrigation, however, leaching docs occur. Producers typically end the season with the soil at approximately the field capacity, and the rainfall of more than 20 centimeters received between April 1 and June 15 exceeds evapotranspiration during that time.

The cycling of nitrate from the irrigation water back to the shallow groundwater from which it is pumped is fairly rapid. The groundwater is as shallow as 1 meter in some of the loamy sand soils near the Platte River, and the depth increases with distance from the River. In a personal communication, Dr. Schepers said that the transit time for nitrate from the surface of loam soils to groundwater at a depth of 3 meters appears to be about 8 to 12 months.

In the demonstration project, fertilizer nitrogen was applied in accordance with soil tests for residual nitrate to a depth of 1 meter or more, and irrigation water was applied through water meters in accordance with the need estimated from climatologic data. Over a period of 5 years, the corn yields were maintained with average annual reductions of 4.7 centimeters of irrigation water and 88 kilograms of fertilizer nitrogen per hectare.

Before the project, the nitrate concentration in the groundwater had been increasing (Schepers et al., 1984). The average concentration of nitrate nitrogen in irrigation well water was 15.6 milligrams per liter in the first year of the project and 15.1 milligrams in the fifth year, which indicates that the project efforts inhibited an increase in groundwater nitrate concentration, but did not decrease it appreciably.

A different situation was encountered by Pratt et al. (1972) in an experiment on a nitrogen-fertilized and irrigated citrus orchard under dry conditions in southern California (the annual precipitation at Riverside averages 29 centimeters). Here the water table was far below the surface of the soil. The authors analyzed cores of unsaturated soil to a depth of 30 meters and calculated that with the amount of drainage occurring as a result of the excess water added to prevent salt accumulation, up to 23 years would be required before nitrate added at the surface would reach that depth. The time required to reach the groundwater would be even greater.

A still different situation was described by Meisinger (1976) for an experiment on Long Island, New York. Here the annual precipitation averages 110 centimeters, and about 55 centimeters of water are added to the groundwater by natural drainage each year. In this situation, high nitrate concentrations in the upper groundwater from prior additions could be reduced promptly if the nitrate concentration is low in current drainage. Meisinger found that when potato was given 160 kilograms of fertilizer nitrogen per hectare (enough to produce the maximum yield), the estimated annual leaching loss was 18 kilograms per hectare, and the estimated concentration of nitrate nitrogen in the drainage water was 3.3 milligrams per liter.

On the basis of old and currently inadequate data (Walton, 1951), the U.S. public health standard for nitrate in drinking water was set at 45 milligrams of nitrate per liter or 10 milligrams of nitrate nitrogen per liter (Public Health Service, 1962). Although this standard often is applied to nitrate in drainage from soils, the practice is of dubious value because drainage water is seldom drunk directly, and important changes in nitrate concentration may occur between drainage and drinking.

One change is that drainage water from soils commonly is diluted with precipitation, deeper groundwater, or both before it is used for drinking. Precipitation and deep groundwater are normally low in nitrate. The concentration of nitrate in deep groundwater may of course increase in time with continuing recharge by drainage water high in nitrate. If low-nitrate drainage water were diluted with high-nitrate groundwater, the change in concentration would be in the positive direction.

A second change that may occur is loss of nitrate. Denitrification has been suggested as a cause. Many papers have described losses of nitrate under conditions suggestive of denitrification while the nitrate is still in transit through unsaturated soil to groundwater, and some loss may continue in groundwater. See a paper by Strebel et al. (1989) for a brief review.

By complex calculations based upon unrecovered nitrogen in field experiments involving spring applications of fertilizer tagged with nitrogen-15 to winter wheat in England, Addiscott and Powlson (1992) were enabled to partition the loss between leaching and denitrification. In 13 experiments conducted over a period of 4 years, the estimated total losses from a standard application of 150 kilograms of fertilizer nitrogen per hectare averaged 15.7%, of which 10.0% was from denitrification and 5.7% was from leaching.

Direct evidence for denitrification was obtained by Gillham et al. (1990) in measurements made in a shallow sandy aquifer beneath agricultural land in Ontario. The measurements were made at a depth where the profiles of nitrate and dissolved oxygen indicated that denitrification was occurring. They found that in a period of 10 days the concentration of added nitrate nitrogen decreased from 30 to 15 milligrams per liter, and at the same time the concentration of nitrous oxide nitrogen increased from 0 to 15 milligrams per liter. These measurements verify the occurrence of denitrification because of the equivalence of the quantities of the two forms and the fact that nitrous oxide is a product of denitrification. Normally nitrous oxide is reduced further to molecular nitrogen. In this investigation, however, acetylene was added to block the reduction of nitrous oxide so that it could be used as a measure of nitrate reduction. Other papers providing indirect evidence of loss of nitrate by denitrification include those by Shaw (1962), Lind and Pedersen (1976), Letey et al. (1977), and Van Boheemen (1987).

2-9.2. Health Hazards

The two principal health hazards attributed to nitrate in drinking water are methemoglobinemia and stomach cancer. The health hazards of nitrate are not a part of either soil science or economics, but they are reviewed here in brief as important background.

The term methemoglobinemia refers to the occurrence of methemoglobin in the blood. Methemoglobin is a form of hemoglobin in which the iron, normally present in the ferrous form, has been oxidized to the ferric form. Hemoglobin combines loosely with oxygen (forming oxyhemoglobin) and carries the oxygen to body cells. Methemoglobin is not an oxygen carrier.

The presence of 0.5 to 2% of the total hemoglobin as methemoglobin is normal (Bøckman and Bryson, 1989), but when the concentration exceeds about 10%, the skin has a bluish tinge, a condition called cyanosis and often methemoglobinemia. (Strictly speaking, cyanosis is a consequence of inadequate aeration of the blood, and it may be caused by heart defects and other conditions not related to the presence of excess methemoglobin.) Concentrations of methemoglobin exceeding 40% of the total hemoglobin can be fatal if not treated. Cyanosis due to methemoglobin can be alleviated, usually in minutes, by injection of methylene blue or ascorbic acid (vitamin C), which change methemoglobin back to hemoglobin.

Methemoglobinemia has a variety of possible causes, but the one of concern here is nitrite. Both nitrite and nitrate are absorbed from the digestive tract into the blood. Nitrite oxidizes hemoglobin to methemoglobin, producing some nitrate in the process. Nitrate is removed from the blood by the kidneys.

Methemoglobinemia is a hazard for infants in their first few months of life, when the activity of the enzyme that changes methemoglobin back to hemoglobin is low and when the stomach acidity is still inadequate to prevent development of bacteria and reduction of nitrate to nitrite in the stomach and small intestine. After the digestive system has developed to the stage at which its acidity prevents bacterial activity, nitrate is absorbed into the blood from the stomach and small intestine and is excreted in the urine. Only small quantities reach the large intestine, where the pH is high enough to permit growth of bacteria and reduction of nitrate to nitrite.

Adults tolerate large doses of nitrate without developing methemoglobinemia (Cornblath and Hartmann, 1948). In fact, Addiscott et al. (1991) pointed out that doses of sodium nitrate and ammonium nitrate up to 9 grams per day are used without adverse effects to treat phosphatic kidney stones in humans. This dosage of ammonium nitrate would be equivalent to that in 158 liters of water containing 10 milligrams of nitrate nitrogen per liter, the Public Health Service standard in the United States.

In the first report linking infant methemoglobinemia to nitrate in drinking water, Comly (1945) noted that the condition was associated with consumption of water from undesirable, poorly covered wells located near barnyards and pit privies. The bacterial counts in the water were usually high, the nitrate-nitrogen concentrations were high (90 and 140 milligrams per liter in the two cases he dealt with personally), and the affected babies suffered from gastroenteritis. In the rural settings investigated by Comly, the nitrate concentration in the well water served as an indicator of contamination with human and animal excreta and their bacterial populations.

There is no question but that nitrite, which can be derived from ingested nitrate, is a cause of infantile methemoglobinemia. Because the concentration of nitrate in water can be measured easily, the emphasis on nitrate as a causal factor is understandable. According to Hegesh and Shiloah (1982), however, the role of bacterial contamination of the water supply, which leads to the almost invariable association of gastroenteritis (evidenced by acute diarrhea) with infantile methemoglobinemia, has not been adequately appreciated. Their evidence led them to conclude that diarrhea "results in endogenous, de novo synthesis of nitrites, and this is the principal cause of infantile methemoglobinemia."

It has long been known that during acute diarrhea the pH of the contents of the stomach increases in infants, leading to conditions more favorable for growth of bacteria and reduction of nitrate to nitrite. But Hegesh and Shiloah (1982) found that infants on a low-nitrate diet excreted more nitrate than they ingested — an average of nine times more when they were affected by acute diarrhea.

The new evidence helps to explain why the nitrate concentration in drinking water has been a poor indicator of risk of methemoglobinemia. Bøckman and Bryson (1989) found in a review of worldwide clinical evidence that 90 cases of infant methemoglobinemia had been reported beginning with 1964 in which the nitrate intake was small or negligible. In an investigation by Hegesh and Shiloah (1982) involving 58 infants on low-nitrate diets who were hospitalized for acute diarrhea, the concentration of methemoglobin in the blood constituted more than 8% of the total hemoglobin in 12 cases.

Gruener and Shuval (1969) stated that there were no reports of methemoglobinemia in infants fed water from public water supplies in the United States even though the concentration of nitrate in some may be characteristically in excess of the public health standard of 10 milligrams of nitrate nitrogen per liter. In California, McKee and Wolf (1963) reported that many well waters with more than 500 milligrams of nitrate nitrogen per liter had never been linked to reported cases of methemoglobinemia. The new evidence suggests further that the indicator value the nitrate concentration once had for contamination of drinking water with human and animal excreta has now been compromised by the use of chemical fertilizers, the nitrogen of which does not connote such contamination.

Methemoglobinemia is too uncommon to be reported among causes of death by the National Center for Health Statistics in the United States. Reviewing the statistics in the United Kingdom, Addiscott et al. (1991) noted that the last death from infant methemoglobinemia occurred in 1950 and the last confirmed case was in 1972. Methemoglobinemia has become rare since discovery of the cause, partly because of improved water supplies and partly because of steps taken to avoid the hazard. A standard preventive measure is supplying newborn infants with bottled water for their first year if the home source of water is contaminated.

Dietary nitrate presents to ruminant livestock a hazard that probably exceeds the hazard for human infants. In ruminant livestock, the food enters the rumen, an anaerobic digestive compartment in which bacterial digestion takes place and reduction of nitrate to nitrite occurs. Part of the nitrite is absorbed into the blood, where it converts part of the hemoglobin to methemoglobin, as in humans. Symptoms of nitrate poisoning (methemoglobinemia) occur when 50% or more of the hemoglobin has been converted to methemoglobin (Kemp, 1982). Tests on cattle grazing grass vegetation on pastures heavily fertilized with nitrogen (Den Boer, 1980) showed that the methemoglobin content of the blood was about normal (2 to 3% of the sum of hemoglobin and methemoglobin) even when the vegetation contained 30 to 40 grams of nitrate or 6.8 to 9.0 grams of nitrate nitrogen per kilogram on the dry matter basis. The highest methemoglobin level measured was 6%. The hazard is greater when cattle are fed hay because nitrate then is taken in more rapidly and released more rapidly from the plant material. According to a graph in Kemp's (1982) publication, the nitrate intake that led to conversion of 50% of the hemoglobin to methemoglobin when the source of the nitrate was hay resulted in less than 10% conversion to methemoglobin when the source was freshly mown herbage.

The possible contribution of nitrate to human gastric cancer has also been of concern. A small proportion of the nitrate ingested in food and drink is reduced to nitrite by bacteria in the mouth, and is changed to nitrous acid upon reaction with the hydrochloric acid in the stomach. Nitrous acid combines with a variety of nitrogenous compounds, including some present in human diets and others produced during digestion, to form N-nitroso compounds. A number of N-nitroso compounds are highly toxic and have been found to be animal carcinogens. They are mutagenic and are presumed to be human mutagens and carcinogens as well, but this has never been verified. Epidemiologic evidence has not verified the significance of the theoretical mechanism as a cause of human cancer (Owen and Jürgens-Gschwind, 1986; Addiscott et al., 1991) although the possibility remains.

Van Broekhoven (1980) observed the development of up to five parts of dimethyl nitrosamine per billion in silage made from grass originally containing nitrate. Formation of N-nitroso compounds in the rumen of cattle paralleling the reduction of nitrate to nitrite has been reported by Van Broekhoven and Stephany (1978). As with humans, the question is whether the concentrations encountered have significant physiological effects.

2-9.3. Sources of Dietary Nitrate

Although major political and media attention has been focused on the nitrate content of the water supply and the influence agricultural practices may have on it, drinking water is generally only a minor source of nitrate in the human diet. The average daily dietary intake of nitrate-nitrogen by a U.S. adult is estimated at 17 milligrams, to which vegetables contribute 14.7 milligrams and water only 0.45 milligram (Assembly of Life Sciences, 1981). On average, therefore, little reduction in nitrate intake could be accomplished even if nitrate could be eliminated completely from the water supply.[4]

Several European countries have passed legislation establishing permissible concentrations of nitrate in vegetables, with an upper limit of around 4,000 milligrams of nitrate (904 milligrams of nitrate nitrogen) per kilogram of fresh produce (Van Diest, 1986b). In The Netherlands, some attention has been given to breeding cultivars of lettuce and spinach that have a relatively low content of nitrate (Groenwold, 1988; Groenwold and Reinink, 1988). Van Diest (1986b) reviewed means for reducing nitrate accumulation in vegetables and pasture

---

[4]In some circumstances, however, water is the major dietary source of nitrate. The Assembly of Life Sciences (1981) noted that a nitrate-nitrogen concentration of 22.6 milligrams per liter (comparable to the average concentration in the Sangamon River in Illinois) would correspond to a daily intake of 36 milligrams of nitrate-nitrogen per person consuming 1.6 liters of water per day.

In the era when infant methemoglobinemia was still being encountered with some frequency, Bosch et al. (1950) surveyed 389 wells used as sources of drinking water in Minnesota and found 51 with nitrate-nitrogen concentrations exceeding 99 milligrams per liter. With intake of water at 1.6 liters per person per day, the daily intake of nitrate nitrogen would exceed 159 milligrams — more than ten times the average intake from vegetables. The wells studied by Bosch et al. (1950) were generally shallow, and none were constructed and located satisfactorily so as to limit contamination from livestock and human sources. Most wells tested were contaminated with coliform bacteria. The average nitrate-nitrogen concentration in the water of wells suspected of causing methemoglobinemia was 102 milligrams per liter.

plants, looking forward to the possible need for such efforts, should the risk of penalties for producing high-nitrate vegetables and for contributing to nitrate in the groundwater become more important in production decision-making than the risk of losing a small amount of product due to nitrogen deficiency.

In the United States, an official advisory committee on drinking water standards (Hopkins and Burke, 1962) proposed some "limits and ranges related to nitrate water standards," including (a) no nitrate in squash and tomato and 188 milligrams of nitrate nitrogen per kilogram in spinach in strained baby foods (wet weight basis), (b) 11 and 813 milligrams per kilogram in green asparagus and spinach (dry weight basis), and (c) 45 milligrams per kilogram of preserved meat and fish. The drinking water standards were promulgated as regulations in the *Federal Register*, and the limit on nitrate and nitrite in meat was already in effect. As yet, however, no official standards exist for nitrate in vegetables.

2-9.4. Economics of Limiting Nitrate Loss to Groundwater

The amounts of nitrogen fertilizers applied by producers in a free market depend primarily upon economic factors: cost of fertilization, expected crop responses, and expected crop prices. Environmental concerns are not reflected in this framework. Through legislation and regulation, however, members of society may develop ways to achieve desired environmental goals, such as limiting the concentration of nitrate in water used for drinking.

Swanson (1982) classified the societal approaches as direct and indirect. The direct approaches he outlined included (1) restrictions on quantities of fertilizer nitrogen that can be applied per hectare, (2) restrictions on the concentration of nitrate in the drainage water, (3) restrictions on the nitrogen balance, that is, the difference between inputs of nitrogen and removal in harvested crops, and (4) treatment of the water to reduce the nitrate content before it is distributed for household use. The indirect procedures included (1) an excise tax on nitrogen fertilizers, (2) a charge to producers based upon the nitrate in the drainage water from their land, (3) a market for rights to use nitrogen fertilizers, and (4) an education program to improve the basis for decisions on use of nitrogen fertilizers. The first three indirect procedures are based upon economic incentives. An education program now becoming standard is that of encouraging an appropriate allowance for differences in the amount of nitrogen that can be provided to the current crop by the soil. Testing the soil for mineral nitrogen before fertilization and making an allowance for nitrogen mineralization from the soil, legumes, and manures are at the top of the educational list.

Follett and Walker (1989) classified the public policy alternatives somewhat differently, and they introduced another alternative. According to them, the public policy alternatives are regulation (emission standards, required devices, bans against polluting practices), charges (a tax on the amount of nitrate lost by leaching), and subsidies (payments to producers for emissions below a specified level). These authors considered subsidies the most politically feasible way to achieve a desired reduction in nitrate leaching.

As may be inferred from the list of alternative policies in the preceding paragraphs, economists recognize several possible sources of payments to compensate for pollution. These include the polluter, the user of the polluted product, and society. The arguments involved in deciding who should pay, to what degree, and for what reasons are complex and may differ from one situation to another.

In the state of Baden-Württemburg in Germany, the issue of nitrate loss in drainage water from agricultural fields to groundwater pumped for household use is being addressed by a prototype political arrangement. In this instance, it is the users of the polluted product who pay. In Baden-Württemburg, the consumers, who desire water low in nitrate, indirectly pay farmers for the privilege of having such water. The private companies that pump the drinking water pay a tax to the state on the amount of water they pump. The tax is paid indirectly by the consumers who purchase the water. With the tax money, the state then offers a bonus to farmers in the areas that contribute to the groundwater used for drinking if they limit the nitrate content of the drainage water from their fields. To provide an incentive, the bonus is intended to be more than enough to reimburse farmers for the loss of crop yield they experience by withholding nitrogen. For farmers in the critical areas to qualify for the bonus, analyses of the surface 90 centimeters of the soil of their fields in late autumn must indicate less than 45 kilograms of ammonium plus nitrate nitrogen per hectare.

In the United Kingdom, farmers in designated nitrate-sensitive areas are offered a subsidy if they reduce their use of nitrogen fertilizer to an amount less than the economic optimum and if they use a green cover crop during the winter. A higher subsidy is paid to farmers who will convert nitrate-sensitive cropland to low-intensity grassland or woodland (Addiscott et al., 1991).

Swanson (1982) considered at length the consequences of localized versus national controls because that issue had been of much concern in Illinois, where hearings were being held on the contribution of nitrogen fertilizers to nitrate in drinking water supplies. According to his analysis, a decrease in production would generally occur in areas affected by controls on nitrogen fertilizer usage. If controls were applied to the localized areas where the nitrate concentration in ground and surface waters is greatest, or on a state basis, producers in the affected areas would be penalized economically, and producers in unaffected areas would benefit.

If controls on nitrogen fertilizer usage were applied, losses of nitrate by leaching would decrease from fields that had been receiving excessive applications of fertilizer nitrogen. Although the amounts applied to some vegetable crops might not be excessive from the economic standpoint, they may exceed the amounts absorbed and can result in considerable loss of nitrate by leaching. Because of the high value of the crops, the maximum net profit is obtained on the relatively flat portion of the response curve where the yield is relatively insensitive to the quantity of fertilizer nitrogen. Thus, it should be possible to reduce the application of fertilizer nitrogen and hence the loss of nitrate by leaching without much effect on the yield. For potato in The Netherlands, it was estimated that reducing the application of fertilizer nitrogen by 25% below the quantity to yield the maximum net profit would reduce the yield by 2% (Neeteson, 1990).

Establishing controls on nitrogen fertilizer usage would cause agriculture to move in the direction of legumes as sources of nitrogen. For the majority of situations in which there is probably little loss of nitrate from cultivated fields to which fertilizer nitrogen is now applied, substituting legume nitrogen for fertilizer nitrogen would be expected to be of little value in decreasing leaching and could at times increase it. See the results of a long-term experiment reported by Neeteson (1989) in which the effects of legume nitrogen and fertilizer nitrogen were investigated in The Netherlands.

If controls were applied on a national basis, a variety of consequences might develop, depending upon the nature of the controls. If nitrogen fertilizer use were banned or severely restricted, farm output would decrease, food prices would rise, income to agricultural producers would increase, more and poorer land would be brought into production, and import restrictions to protect agricultural producers against lower priced foreign products probably would be put into effect. Feed grain production would decrease because of lack of nitrogen and because some of the land previously used for feed grains would be cropped to legumes. The cost of leguminous nitrogen in this context was emphasized by Mengel's (1990) observation that 44% of the cropland would need to be in a legume supplying 250 kilograms of nitrogen per hectare for a farm to be self-sufficient in nitrogen if the nitrogen requirement of the nonleguminous crop following the legume was 200 kilograms per hectare.

With current hay-baling technology, shipping hay is relatively expensive compared with grain because of the low bulk density of the hay and its limited feeding value per unit volume. Thus, if the large feedlots that have developed in the United States in response to the abundance of low-priced grain for finishing cattle were required to substitute legume hay for all or most of the grain they now use, the less favorable economics probably would cause the system of large feedlots to give way to a more dispersed cattle industry, in which the legume hay grown to supply nitrogen is used for cattle feeding on or near the site of production.

With new technology now available, the quantity of hay shipped per unit volume can be increased about three-fold over that in the average baled hay at present, making hay more competitive with feed grains. The nutritional value of compacted alfalfa hay per unit volume relative to that of corn is about 1.0 for crude protein, 0.4 for total digestible nutrients, and 0.2 for net energy for growth. Adoption of the new technology would help keep the feedlots competitive, while leading to carcasses with a lower fat content than those produced when the cattle are finished with high-energy rations based on feed grains.

Various economic analyses of the consequences of possible regulation of nitrogen fertilizers have been made since Swanson's review. One of special interest because it illustrates both the economics and the foibles of the political process of environmental regulation was published by Brink (1987), who considered options for reducing the loss of nitrate by leaching from soils cropped to wheat. In Sweden, wheat producers receive a bonus for high protein in the grain, and thus for applying nitrogen fertilizer to produce high protein and, incidentally, to increase the potential for loss of nitrate by leaching. According to Brink's analysis, eliminating the bonus for high protein in the wheat would not have enough of an effect on the optimum economic application of fertilizer nitrogen to change substantially the loss of nitrate by leaching. He reached the same conclusion regarding the 25% tax farmers were already paying on nitrogen fertilizers.

On the one hand, the Swedish government is under pressure to reduce the loss of nitrate by leaching. On the other hand, according to Brink (1987), the government supports the price of domestically produced wheat at a level three times the price of wheat on the world market. As a consequence of this policy, the government is in effect subsidizing the use of nitrogen fertilizers, which encourages farmers to use more fertilizer, thereby exacerbating the nitrate-loss effect the government would like to avoid. The situation found in Sweden parallels to some extent that in some other countries, including the United States. According to one estimate (see Black, 1989), taxpayers bought 28% of all the fertilizer used in the United States in 1986.

A continuing problem with government policies is that they may have effects that were not intended and were not foreseen at the time they were established. An attempt to solve one problem may create another. For example, Addiscott et al. (1991) pointed out that although the introduction of milk quotas in the European Community had the desired effect of reducing total milk production, the larger farms bought the quotas of the smaller farms. The consequences, per unit area in dairy farms, then included greater manure production and an increase in nitrate available for leaching.

In the United Kingdom, Sylvester-Bradley et al. (1987) estimated that the cost of fertilizer nitrogen relative to the value of winter wheat grain would have to increase threefold to reduce the optimum application of nitrogen by a third. This reduction in quantity of fertilizer nitrogen would decrease grain production by only about 5%, and the reduction in loss of nitrogen from the soil by leaching would be very small. Also in the United Kingdom, McCorriston and Sheldon (1989) investigated the potential consequences of reducing the amount of nitrogen fertilizers used by levying a tax on nitrogen fertilizers to increase the price and by establishing a quota on nitrogen fertilizer production. According to their findings, farmers would lose from both policies. The main fertilizer manufacturers might gain in the short run from a quota policy, but profits would be reduced with a tax policy. Government revenue would increase from taxes, but there would be a net loss of social welfare with both policies. In Germany, Finck (1986) concluded that if a given reduction in economic optimum application of fertilizer nitrogen is to be realized, farm income would be reduced more by decreasing the price of the product than by taxing the fertilizer.

If farmers' use of nitrogen fertilizers were reduced in a nation with free markets, and if imports of lower cost food were prevented, farmers could gain economically because of the tendency for the percentage increase in price of agricultural products to exceed the percentage decrease in supply. Consumers would pay the higher prices. The fertilizer manufacturers would lose sales and profits.

The indirect societal effect may be negligible where only a small portion of the total production is involved, as in the Baden-Württemberg example. The effect would be more significant if a policy that decreases production efficiency is nationwide. Moreover, the effect on prices of the processed products used by consumers would be greater than the effect on the prices farmers receive for the raw products for two reasons. First, raw product prices are generally not the major part of the cost of the processed products to consumers. And second, when prices of the raw products are increased, processors tend to increase their prices to maintain or increase their profit margin, so that the economic effect of differences in production efficiency is multiplied. These are some of the economic consequences borne by society from introduction of modifications in agricultural practices that lead to reduced economic efficiency.

Follett and Walker (1989) illustrated diagrammatically the economic theory of costs and benefits to producers and society, and the concept of the maximum social benefit as applied to nitrogen fertilization. Although their diagram was in terms of dollars of costs and benefits per hectare versus quantity of nitrogen added per hectare, quantification of the social costs and benefits in terms of dollars has not been achieved.

No one knows the formula for maximizing the social benefits of environmental issues in which the costs and benefits cannot be expressed in common terms, such as dollars. Members of society are not objectively appraised of the scientific basis for either proposed environmental policies or their consequences. The propensity of the media to emphasize matters that can be sensationalized tends to bias perceptions in the direction of exaggerated concerns for human health as opposed to sober appraisals of proposed policies in terms of potential effects on health hazards and economics. In this environment, the political process leads to answers that tend to satisfy some of the more vocal members of the public and to perpetuate the tenure of the responsible politicians. Doubts may be expressed regarding the extent to which the political answers maximize the social benefit, but doubters face the same hurdles as politicians in defining social benefit.

The groundwater nitrate issue is joined, and agricultural scientists are producing and publishing a great deal of information about it. The eventual outcome should not lack for effort from the agricultural side.

2-9.5. Animal Manures
The principles of economics discussed in Sections 2-1 through 2-8 apply to the use of animal manures as fertilizers as well as to chemical fertilizers, except that

the concepts related to fertilizer composition are not relevant. Animal manures supply some quantities of all plant nutrients, and each manure comes as a "package" with certain composition.

The nutrients supplied by manures are in chemical forms that differ from those in chemical fertilizers, and this influences their efficiency. Evaluation of the efficiency of individual nutrients in organic sources relative to those in chemical sources is considered in Chapter 5, and residual effects are considered in Chapter 6.

The nitrogenous components in animal manures degrade to produce nitrate as an end product under aerobic conditions, as is true also of the various forms of nitrogen added in commercial fertilizers. As a consequence, the concern about nitrate movement to groundwater extends to the nitrate supplied by animal manures.

The accumulation of plant nutrients in the land surrounding domestic animal quarters as a result of animal excrement derived from the feed produced in outlying areas has been a normal part of agriculture since time immemorial. The development of commercial fertilizers in the last century created the opportunity to counter the depletion of nutrients in outlying production areas by applying fertilizers.

In recent years, the economics of livestock production in developed countries has resulted in increasing the transport of feedstuffs to centers of consumption to such an extent that there is concern about the environmental consequences of the practice. The reason is that although the value of feedstuffs is great enough to support their transport over long distances, the value of animal manure per unit weight as a fertilizer is relatively low, especially in animal confinement systems in which the manure is accumulated and applied as a slurry with a high water content. The economic distance to which manure can be transported away from the site of concentrated animal production units thus is relatively small.

In some instances, the concentration of livestock per hectare is so great and the land available for manure application so limited that the concern is not that of supplying nutrients for crop production but of disposing of the excess manure. According to estimates by Van Boheemen (1987), 111 megagrams of manure were produced per hectare in 1984 in the Peel region in The Netherlands, which represents an extreme case. Only a small part of the manure was disposed of elsewhere. In that year, an estimated 49 megagrams of manure were produced per hectare of agricultural land in The Netherlands as a whole.

Herlihy (1980) calculated that in The Netherlands, Belgium, and Denmark, the nutrients in the animal manures exceeded the amounts that could be applied to the agricultural land without creating an excess of nitrogen, phosphorus, or potassium over removals by crops. The Netherlands was at the top of the list. Other western European countries (Germany, Ireland, the United Kingdom, France, and Italy) were in a deficit mode.

According to Van Diest (1986a), the ratios of nutrients imported in livestock feedstuffs to nutrients exported from the soils in agricultural products in The

Netherlands in 1983 were 3.9 for nitrogen, 4.8 for phosphorus, and 6.6 for potassium. The ratios of the total imports of nutrients in feedstuffs and fertilizers to the exports from the soils in agricultural products were 9.2 for nitrogen, 6.6 for phosphorus, and 11.3 for potassium.

In recognition of the potential environmental impacts of the great accumulation of nutrients, legislative action was taken in The Netherlands to limit the annual applications of manure to the land (Brussaard and Grossman, 1990). Beginning in 1987, the annual applications of nutrients were not to exceed those associated with 55 kilograms of phosphorus per hectare for arable land (except for fodder production), 109 kilograms for permanent pasture, and 153 kilograms for fodder production (essentially corn grown for fodder or silage). The maximum permissible additions are to decrease with time. Final limits have not been set, but are scheduled to be decided in about the year 2000.

The two simpler possibilities for applying legal limitations on excessive applications of nutrients in manure are (a) to restrict the application of the nutrient in greatest supply relative to the needs and (b) to restrict the application of the nutrient in greatest need. Phosphorus is usually in greatest supply relative to the need, and this was the basis selected in The Netherlands, as mentioned in the preceding paragraph. Other nutrients then will be deficient relative to phosphorus, and they may or may not be added in commercial fertilizers to suit the circumstances.

Nitrogen is usually the nutrient in greatest need. If the limit established for nitrogen is commensurate with the need of the crop (for example, about 400 kilograms per hectare annually for grassland in The Netherlands), phosphorus and potassium usually will be in excess. The limit on nitrogen application of course does not have to be equivalent to the need. According to Conrad (1988), nitrogen has been selected as the legal basis in the North Rhine-Westphalia area in Germany, and the ordinance adopted there restricts the application of liquid animal manures (slurries) to amounts supplying no more than 240 kilograms of nitrogen per hectare annually.

In Denmark (Hansen, 1989), regulations require that 65% of the field area per farm be covered by a green crop during the autumn. Additional regulations designed to increase the efficiency of nitrogen utilization and to reduce leaching require that liquid manure be incorporated in the soil within 12 hours after application and that no liquid manure be applied to uncropped soil between harvest and November 1.

As noted by Gasser (1987) and others, the ratio of potassium to phosphorus is much higher in cattle manure slurries than in swine or poultry manure slurries, but the concentration of phosphorus in pig slurry and especially in poultry slurry tends to exceed that in cattle slurry. These differences are primarily a consequence of differences in composition of the feedstuffs consumed by the several classes of animals. Thus, if phosphorus is selected as a standard for the maximum legal application, the amount of potassium accompanying the phosphorus will be far greater in the slurries derived from cattle than from swine or poultry. Some concern has been expressed about the potential for this nutrient imbalance to promote magnesium deficiency (grass tetany or hypomagnesemia) in cattle grazing pastures heavily fertilized with cattle slurry, the reason being that the extra potassium decreases the magnesium concentration in the forage. According to Kemp (1982), the magnesium supply in the forage is insufficient to meet the magnesium requirement of dairy cows in many intensive high-producing grassland systems.

2-10. Literature Cited

Addiscott, T. M., and D. S. Powlson. 1992. Partioning losses of nitrogen fertilizer between leaching and denitrification. *Journal of Agricultural Science (Cambridge)* 118:101-107.

Addiscott, T. M., A. P. Whitmore, and D. S. Powlson. 1991. *Farming, Fertilizers and the Nitrate Problem*. CAB International, Wallingford, Oxon, United Kingdom.

Anderson, R. L. 1957. Some statistical problems in the analysis of fertilizer response data. Pp. 187-206. In E. L. Baum, E. O. Heady, J. T. Pesek, and C. G. Hildreth (Eds.), *Economic and Technical Analysis of Fertilizer Innovations and Resource Use*. Iowa State College Press, Ames.

Anderson, R. L., and L. A. Nelson. 1975. A family of models involving intersecting straight lines and concomitant experimental designs useful in evaluating response to fertilizer nutrients. *Biometrics* 31:303-318.

Anderson, R. L., and L. A. Nelson. 1987. *Linear-Plateau and Plateau-Linear-Plateau Models Useful in Evaluating Nutrient Responses*. North Carolina Research Service Technical Bulletin 283. Raleigh.

Assembly of Life Sciences. 1981. *The Health Effects of Nitrate, Nitrite, and N-Nitroso Compounds*. National Academy Press, Washington.

Babcock, B., M. E. Rister, R. D. Kay, and J. A. Wallers. 1984. Identifying least-cost sources of required fertilizer nutrients. *American Journal of Agricultural Economics* 66:385-391.

Ball, P. R., and J. C. Ryden. 1984. Nitrogen relationships in intensively managed temperate grasslands. *Plant and Soil* 76:23-33.

Barber, S. A. 1973. The changing philosophy of soil test interpretations. Pp. 201-211. In L. M. Walsh and J. D. Beaton (Eds.), *Soil Testing and Plant Analysis*. Revised Edition. Soil Science Society of America, Madison, WI.

Bateman, D. 1991. Paradigms of American agriculture. Pp. 19-24. In G. E. Gaull and R. A. Goldberg (Eds.), *New Technologies and the Future of Food and Nutrition*. John Wiley & Sons, NY.

Baum, E. L., E. O. Heady, and J. Blackmore (Eds.). 1956. *Methodological Procedures in the Economic Analysis of Fertilizer Use Data*. Iowa State College Press, Ames.

Baum, E. L., E. O. Heady, J. T. Pesek, and C. G. Hildreth (Eds.). 1957. *Economic and Technical Analysis of Fertilizer Innovations and Resource Use*. Iowa State College Press, Ames.

Benbi, D. K., C. R. Biswas, and J. S. Kalkat. 1991. Nitrate distribution and accumulation in an Ustochrept soil profile in a long term fertilizer experiment. *Fertilizer Research* 28:173-177.

Biswas, S. K., and K. N. Singh. 1982. Nitrogen management in barley and wheat I: Yield, yield attributes and economics of production. *Indian Journal of Agronomy* 27:241-247.

Black, C. A. 1989. *Reducing American Exposure to Nitrate, Nitrite, and Nitroso Compounds: The National Network to Prevent Birth Defects Proposal.* Council for Agricultural Science and Technology, Comments from CAST 1989-1.

Bock, B. R., and G. W. Hergert. 1991. Fertilizer nitrogen management. Pp. 139-164. In R. F. Follett, D. R. Keeney, and R. M. Cruse (Eds.), *Managing Nitrogen for Groundwater Quality and Farm Profitability.* Soil Science Society of America, Madison, WI.

Bøckman, O. C., and D. D. Bryson. 1989. Well-water methaemoglobinaemia: The bacterial factor. Pp. 239-244. In D. Wheeler, M. L. Richardson, and J. Bridges (Eds.), *Watershed 89: The Future for Water Quality in Europe, Volume II.* Pergamon Press, Oxford.

Boer, D. J. den. 1980. Effect of extreme rates of nitrogen application on herbage nitrate content and health of grazing dairy cattle. P. 166. In W. H. Prins and G. H. Arnold (Eds.), *The Role of Nitrogen in Intensive Grassland Production.* European Grassland Federation, Proceedings of an International Symposium. Centre for Agricultural Publishing and Documentation, Wageningen, The Netherlands.

Boheemen, P. J. M. van. 1987. Extent, effects and tackling of a regional manure surplus; a case-study for a Dutch region. Pp. 175-193. In H. G. van der Meer, R. J. Unwin, T. A. van Dijk, and G. C. Ennik (Eds.), *Animal Manure on Grassland and Fodder Crops. Fertilizer or Waste?* Martinus Nijhoff Publishers, Dordrecht, The Netherlands.

Bosch, H. M., A. B. Rosenfield, R. Huston, H. R. Shipman, and F. L. Woodward. 1950. Methemoglobinemia and Minnesota well supplies. *Journal of the American Water Works Association* 42:161-170.

Bouma, D., K. Spencer, and E. J. Dowling. 1969. Assessment of the phosphorus and sulphur status of subterranean clover pastures. 3. Plant tests. *Australian Journal of Experimental Agriculture and Animal Husbandry* 9:329-340.

Boxman, O., D. Goense, B. H. Janssen, J. J. Neeteson, and J. F. Wienk. 1985. The effect of moisture stress on the response to nitrogen by maize in the humid tropics of Suriname. Pp. 199-214. In B. T. Kang and J. van der Heide (Eds.), *Nitrogen Management in Farming Systems in Humid and Subhumid Tropics.* Institute for Soil Fertility, Haren, The Netherlands, and International Institute of Tropical Agriculture, Ibadan, Nigeria.

Brink, N. 1987. Economic measures against leaching losses of nitrogen. Pp. 555-562. In W. van Duivenbooden and H. G. van Waegeningh (Eds.), *Vulnerability of Soil and Groundwater to Pollutants.* National Institute of Public Health and Environmental Hygiene, Verslagen en Mededelingen der Commissie Voor Hydrologisch Onderzoek No. 38. The Hague.

Broadbent, F. E., and R. S. Rauschkolb. 1977. Nitrogen fertilization and water pollution. *California Agriculture* 31, No. 5:24-25.

Broekhoven, L. W. van. 1980. Formation of volatile N-nitrosamines during the fermentation of grass silages. Pp. 113-117. In W. H. Prins and G. H. Arnold (Eds.), *The Role of Nitrogen in Intensive Grassland Production.* European Grassland Federation, Proceedings of an International Symposium. Centre for Agricultural Publishing and Documentation, Wageningen, The Netherlands.

Broekhoven, L. W. van, and R. W. Stephany. 1978. *Environmental Aspects of N-nitroso Compounds. International Agency for Research on Cancer, Scientific Publications* 19. (Cited by Kemp, A. 1982. The importance of the chemical composition of forage for optimizing animal production. Pp. 95-116. In *Optimizing Yields — The Role of Fertilizers.* Proceedings of the 12th IPI-Congress. International Potash Institute, Worblaufen-Bern, Switzerland.)

Brogan, J. C. (Ed.). 1981. *Nitrogen Losses and Surface Run-Off From Landspreading of Manures.* Martinus Nijhoff/Dr. W. Junk Publishers, The Hague.

Brown, W. G. 1956. Practical applications of fertilizer production functions. Pp. 151-157. In E. L. Baum, E. O. Heady, and J. Blackmore (Eds.), *Methodological Procedures in the Economic Analysis of Fertilizer Use Data.* Iowa State College Press, Ames.

Brown, W. G., and M. M. Oveson. 1958. Production functions from data over a series of years. *Journal of Farm Economics* 40:451-457.

Brussaard, W., and M. R. Grossman. 1990. Legislation to abate pollution from manure: The Dutch approach. *North Carolina Journal of International Law & Commercial Regulation* 15:85-114.

Carey, T. M., and P. Robinson. 1953. The manuring of sugar-cane. *Empire Journal of Experimental Agriculture* 21:99-115.

Cerrato, M. E., and A. M. Blackmer. 1990. Comparison of models for describing corn yield response to nitrogen fertilizer. *Agronomy Journal* 82:138-143.

Colwell, J. D. 1983. Fertilizer requirements. Pp. 795-815. In *Soils: An Australian Viewpoint.* Division of Soils, CSIRO. CSIRO, Melbourne. Academic Press, London.

Colwell, J. D., A. R. Suhet, and B. van Raij. 1988. *Statistical Procedures for Developing General Soil Fertility Models for Variable Regions.* CSIRO (Australia) Division of Soils, Divisional Report No. 93.

Comly, H. H. 1945. Cyanosis in infants caused by nitrates in well water. *Journal of the American Medical Association* 129:112-116.

Conrad, J. 1988. Nitrate debate and nitrate policy in FR Germany. *Land Use Policy* 5:207-218.

Cornblath, M., and A. F. Hartmann. 1948. Methemoglobinemia in young infants. *Journal of Pediatrics* 33:421-425.

Crowther, E. M., and F. Yates. 1941. Fertilizer policy in war-time: The fertilizer requirements of arable crops. *Empire Journal of Experimental Agriculture* 9:77-97.

Den Boer, D. J. 1980. Effect of extreme rates of nitrogen application on herbage nitrate content and health of grazing dairy cattle. P. 166. In W. H. Prins and G. H. Arnold (Eds.), *The Role of Nitrogen in Intensive Grassland Production.* European Grassland Federation, Proceedings of an International Symposium. Centre for Agricultural Publishing and Documentation, Wageningen, The Netherlands.

Dev, G., N. S. Dhillon, and J. S. Brar. 1984. Determination of fertiliser nitrogen dose for rice on basis of marginal returns. *Fertiliser Marketing News* 15, No. 7:1-3.

Diest, A. van. 1986a. The social and environmental implications of large-scale translocations of plant nutrients. Pp. 289-299. In *Nutrient Balances and the Need for Potassium.* Proceedings of the 13th IPI-Congress. International Potash Institute, Worblaufen-Bern, Switzerland.

Diest, A. van. 1986b. Means of preventing nitrate accumulation in vegetable and pasture plants. Pp. 455-471. In H. Lambers, J. J. Neeteson, and I. Stulen (Eds.), *Fundamental, Ecological and Agricultural Aspects of Nitrogen Metabolism in Higher Plants.* Martinus Nijhoff Publishers, Dordrecht, The Netherlands.

Dillon, J. L. 1977. *The Analysis of Response in Crop and Livestock Production.* Second Edition. Pergamon Press, NY.

Dilz, K., A. Darwinkel, R. Boon, and L. M. J. Verstraeten. 1982. Intensive wheat production as related to nitrogen fertilisation, crop protection and soil nitrogen: Experience in the Benelux. *Fertiliser Society (London) Proceedings* No. 211:93-124.

Eck, H. V. 1988. Winter wheat response to nitrogen and irrigation. *Agronomy Journal* 80:902-908.

Finck, H.-F. 1986. Ansatzpunkte zur Vermeidung der Nitratbelastung des Grundwassers unter besonderer Berücksichtigung einer Besteuerung von Stickstoff. *Agrarwirtschaft* 35:211-222.

Follett, R. F. (Ed.). 1989. *Nitrogen Management and Ground Water Protection.* Elsevier Science Publishers B.V., Amsterdam, The Netherlands.

Follett, R. F., D. R. Keeney, and R. M. Cruse (Eds.). 1991. *Managing Nitrogen for Groundwater Quality and Farm Profitability.* Soil Science Society of America, Madison, WI.

Follett, R. F., and D. J. Walker. 1989. Ground water quality concerns about nitrogen. Pp. 1-22. In R. F. Follett (Ed.), *Nitrogen Management and Ground Water Protection.* Elsevier Science Publishers B.V., Amsterdam, The Netherlands.

Gasser, J. K. R. 1987. The future of animal manures as fertilizer or waste. Pp. 259-278. In H. G. van der Meer, R. J. Unwin, T. A. van Dijk, and G. C. Ennik (Eds.), *Animal Manure on Grassland and Fodder Crops. Fertilizer or Waste?* Martinus Nijhoff Publishers, Dordrecht, The Netherlands.

Germon, J. C. (Ed.). 1989. *Management Systems to Reduce Impact of Nitrates.* Elsevier Applied Science, London.

Gillham, R. W., R. C. Starr, and D. J. Miller. 1990. A device for in situ determination of geochemical transport parameters 2. Biochemical reactions. *Ground Water* 28:858-862.

Gölz-Huwe, H., W. Simon, B. Huwe, and R. R. van der Ploeg. 1989. Zum jahreszeitlichen Nitratgehalt und zur Nitratauswaschung von landwirtschaftlich genutzten Böden in Baden-Württemberg. *Zeitschrift für Pflanzenernährung und Bodenkunde* 152:273-280.

Groenwold, R. 1988. Nitraat in oude en gangbare spinazierassen. *Prophyta* No. 6:163-165.

Groenwold, R., and K. Reinink. 1988. De overerving van het nitraatgehalte in sla. *Prophyta* No. 3:92-95.

Groot, J. J. R., and H. van Keulen. 1990. Prospects for improvement of nitrogen fertilizer recommendations for cereals: A simulation study. Pp. 685-692. In M. L. van Beusichem (Ed.), *Plant Nutrition — Physiology and Applications.* Kluwer Academic Publishers, Dordrecht, The Netherlands.

Groot, J. J. R., P. de Willigen, and E. L. J. Verberne. 1991. *Nitrogen Turnover in the Soil-Crop System.* Kluwer Academic Publishers, Dordrecht, The Netherlands.

Gruener, N., and H. I. Shuval. 1969. Health aspects of nitrates in drinking water. Pp. 89-106. In *Developments in Water Quality Control Research.* Proceedings of the Jerusalem International Conference of Water Quality and Pollution Research. Humphrey Science Publishers, Ann Arbor. (Quoted by Hegesh and Shiloah, 1982)

Guinard, A. 1982. Economic optimization of fertilizer applications: a method for field staff based on response curves and surfaces. *Tropical Agriculture* 59:257-264.

Hansen, J. F. 1989. Nitrogen balance in agriculture in Denmark and ways of reducing the loss of nitrogen. Pp. 1-15. In J. C. Germon (Ed.), *Management Systems to Reduce Impact of Nitrates*. Elsevier Applied Science, London.

Hargrove, W. L. (Ed.). 1988. *Cropping Strategies for Efficient Use of Water and Nitrogen*. ASA Special Publication 51. American Society of Agronomy, Madison, WI.

Hart, P. B. S., D. S. Jenkinson, A. E. Johnston, D. S. Powlson, and G. Pruden. 1982. The uptake by wheat of fertiliser N applied to the preceding crop. *Rothamsted Experimental Station Report for 1981*. Part 1:254.

Hauck, R. D. (Ed.). 1984. *Nitrogen in Crop Production*. American Society of Agronomy, Crop Science Society of America, and Soil Science Society of America, Madison, WI.

Heady, E. O., L. G. Albaugh, J. C. Engibous, B. L. French, E. W. Kehrberg, D. D. Mason, and J. T. Pesek. 1961. *Status and Methods of Research in Economic and Agronomic Aspects of Fertilizer Response and Use*. National Academy of Sciences-National Research Council Publication 918. National Academy of Sciences-National Research Council, Washington.

Heady, E. O., and J. Dillon. 1961. *Agricultural Production Functions*. Iowa State University Press, Ames.

Heady, E. O., J. T. Pesek, and W. G. Brown. 1955. *Crop Response Surfaces and Economic Optima in Fertilizer Use*. Iowa Agricultural Experiment Station Research Bulletin 424. Ames.

Hegesh, E., and J. Shiloah. 1982. Blood nitrates and infantile methemoglobinemia. *Clinica Chimica Acta* 125:107-115.

Herlihy, P. D. 1980. The use of mathematical models in studying the landspreading of animal manures. Pp. 446-458. In J. K. R. Gasser (Ed.), *Effluents From Livestock*. Applied Science Publishers, London.

Hildreth, C. G. 1957. Discrete models with qualitative restrictions. Pp. 62-75. In E. L. Baum, E. O. Heady, and J. Blackmore (Eds.), *Methodological Procedures in the Economic Analysis of Fertilizer Use Data*. Iowa State College Press, Ames.

Holbrook, J. R., and W. R. Byrne. 1983. Assessment of results from wheat trials testing varieties and application of nitrogenous fertilizer. *Journal of Agricultural Science (Cambridge)* 101:447-452.

Holbrook, J. R., J. D. Osborne, and W. J. Ridgman. 1982. An attempt to improve the yield and quality of direct-drilled winter wheat grown continuously. *Journal of Agricultural Science (Cambridge)* 99:163-172.

Hopkins, O. C. (chairman), and G. W. Burke, Jr. (secretary). 1962. *Public Health Service Drinking Water Standards 1962*. Public Health Service Publication No. 956. U.S. Government Printing Office, Washington.

International Institute for Sugar Beet Research. 1983. *Symposium "Nitrogen and Sugar Beet."* International Institute for Sugar Beet Research, Brussels, Belgium. (*Soils and Fertilizers* 47, Abstract 5204. 1984)

Isfan, D. 1986. Sharing the available fertilizer rate between two crops in a limited-capital situation. *Agronomy Journal* 78:346-347.

Isfan, D. 1989. A graphical method for fertilizer sharing in a limited-capital situation. *Communications in Soil Science and Plant Analysis* 20:1045-1051.

Jenkinson, D. S., and K. A. Smith (Eds.). 1988. *Nitrogen Efficiency in Agricultural Soils*. Elsevier Applied Science Publishers, London.

Juergens-Gschwind, S. 1989. Ground water nitrates in other developed countries (Europe) — Relationships to land use patterns. Pp. 75-138. In R. F. Follett (Ed.), *Nitrogen Management and Ground Water Protection*. Elsevier Science Publishers B.V., Amsterdam, The Netherlands.

Kamprath, E. J. 1986. *Nitrogen Studies With Corn on Coastal Plain Soils*. North Carolina Agricultural Research Service Technical Bulletin 282. Raleigh.

Kang, B. T., and J. van der Heide (Eds.). 1985. *Nitrogen Management in Farming Systems in Humid and Subhumid Tropics*. Institute for Soil Fertility, Haren, The Netherlands, and International Institute of Tropical Agriculture, Ibadan, Nigeria.

Kemp, A. 1982. The importance of the chemical composition of forage for optimizing animal production. Pp. 95-116. In *Optimizing Yields — The Role of Fertilizers*. Proceedings of the 12th IPI-Congress. International Potash Institute, Worblaufen-Bern, Switzerland.

Korentajer, L., P. R. Berliner, and J. Van Zyl. 1987. The effect of drought on economically optimal nitrogen fertilization rates of dryland wheat in the summer rainfall area. *Agrekon* 26, No. 2:20-26.

Lambers, H., J. J. Neeteson, and I. Stulen (Eds.). 1986. *Fundamental, Ecological and Agricultural Aspects of Nitrogen Metabolism in Higher Plants*. Martinus Nijhoff Publishers, Dordrecht, The Netherlands.

Letey, J., J. W. Blair, D. Devitt, L. J. Lund, and P. Nash. 1977. Nitrate-nitrogen in effluent from agricultural tile drains in California. *Hilgardia* 45:289-319.

Lind, A.-M., and M. B. Pedersen. 1976. Nitrate reduction in the subsoil II. General description of boring profiles, and chemical investigations on the profile cores. *Tidsskrift for Planteavl* 80:82-99.

MacDonald, A. J., D. S. Powlson, P. R. Poulton, and D. S. Jenkinson. 1989. Unused fertiliser nitrogen in arable soils — Its contribution to nitrate leaching. *Journal of the Science of Food and Agriculture* 46:407-419.

Machet, J. M., and B. Mary. 1989. Impact of agricultural practices on the residual nitrogen in soil and nitrate losses. Pp. 126-146. In J. C. Germon (Ed.), *Management Systems to Reduce Impact of Nitrates*. Elsevier Applied Science, London.

Maidl, F. X., and G. Fischbeck. 1987. Nitratgehalte tieferer Bodenschichten bei unterschiedlichen Fruchtfolgen auf intensiv genutzten Ackerbaustandorten. *Zeitschrift für Pflanzenernährung und Bodenkunde* 150:213-219.

McCorriston, S., and I. Sheldon. 1989. The welfare implications of nitrogen limitation policies. *Journal of Agricultural Economics* 40:143-151.

McKee, J. E., and H. W. Wolf. 1963. Water quality criteria. *State Water Agency of California, State Water Quality Control Board Publication* 3-A:224-225. (Quoted by Hegesh and Shiloah, 1982)

Meer, H. G. van der, J. C. Ryden, and G. C. Ennik (Eds.). 1986. *Nitrogen Fluxes in Intensive Grassland Systems*. Martinus Nijhoff Publishers, Dordrecht, The Netherlands.

Meisinger, J. J. 1976. *Nitrogen Application Rates Consistent With Environmental Constraints for Potatoes on Long Island*. Search Agriculture 6, No. 7. Cornell University Agricultural Experiment Station, Ithaca, NY.

Mengel, K. 1990. Impacts of intensive plant nutrient management on crop production and environment. *Transactions 14th International Congress of Soil Science, Plenary Papers, Contents, Author Index*:42-52. International Society of Soil Science, Kyoto, Japan.

Middleton, K. R. 1973. Design and analysis of superphosphate trials on high-producing permanent pasture. *New Zealand Journal of Agricultural Research* 16:497-502.

Middleton, K. R. 1980. The fertiliser economy of high producing pastoral systems. *Fertilizer Research* 1:5-27.

Middleton, K. R. 1983. Economic control of fertilizer in highly productive pastoral systems. I. A theoretical framework for the fertilization problem. *Fertilizer Research* 4:301-313.

Munson, R. D., and J. P. Doll. 1959. The economics of fertilizer use in crop production. *Advances in Agronomy* 11:133-169.

Neeteson, J. J. 1989. Effect of legumes on soil mineral nitrogen and response of potatoes to nitrogen fertilizer. Pp. 89-93. In J. Vos, C. D. van Loon, and G. J. Bollen (Eds.), *Effects of Crop Rotation on Potato Production in the Temperate Zones.* Kluwer Academic Publishers, Dordrecht, The Netherlands.

Neeteson, J. J. 1990. Development of nitrogen fertilizer recommendations for arable crops in the Netherlands in relation to nitrate leaching. *Fertilizer Research* 26:291-298.

Neeteson, J. J., and K. Dilz (Eds.). 1985. *Assessment of Nitrogen Fertilizer Requirement.* Institute for Soil Fertility and Netherlands Fertilizer Institute, Haren, The Netherlands.

Neeteson, J. J., D. J. Greenwood, and A. Draycott. 1988. A dynamic model to predict the optimum nitrogen fertiliser application rate for potato. Pp. 384-393. In D. S. Jenkinson and K. A. Smith (Eds.), *Nitrogen Efficiency in Agricultural Soils.* Elsevier Applied Science, London.

Neeteson, J. J., and W. P. Wadman. 1987. Assessment of economically optimum application rates of fertilizer N on the basis of response curves. *Fertilizer Research* 12:37-52.

Neeteson, J. J., W. P. Wadman, and P. A. I. Ehlert. 1987. Assessment of optimum application rates of fertilizer K on the basis of response curves. Pp. 395-402. In *Methodology in Soil-K Research.* Proceedings of the 20th Colloquium of the International Potash Institute held in Baden bei Wien/Austria 1987. International Potash Institute, Worblaufen-Bern, Switzerland.

Nelson, L. A., R. D. Voss, and J. Pesek. 1985. Agronomic and statistical evaluation of fertilizer response. Pp. 53-90. In O. P. Engelstad (Ed.), *Fertilizer Technology and Use.* Third Edition. Soil Science Society of America, Madison, WI.

Owen, T. R., and S. Jürgens-Gschwind. 1986. Nitrates in drinking water: a review. *Fertilizer Research* 10:3-25.

Pesek, J., and E. O. Heady. 1958. Derivation and application of a method of determining minimum recommended rates of fertilization. *Soil Science Society of America Proceedings* 22:419-423.

Pizarro Villanueva, J. B. 1982. Metodos para evaluar economicamente resultados de fertilizacion en trigo. *Instituto Nacional de Tecnologia Agropecuaria, Estacion Experimental Regional Agropecuaria Pergamino, Informe Tecnico* No. 179.

Power, J. F., and J. S. Schepers. 1989. Nitrate contamination of groundwater in North America. *Agriculture, Ecosystems and Environment* 26:165-187.

Pratt, P. F. 1984. Nitrogen use and nitrate leaching in irrigated agriculture. Pp. 319-333. In R. D. Hauck (Ed.), *Nitrogen in Crop Production.* American Society of Agronomy, Crop Science Society of America, and Soil Science Society of America, Madison, WI.

Pratt, P. F., W. W. Jones, and V. E. Hunsaker. 1972. Nitrate in deep soil profiles in relation to fertilizer rates and leaching volume. *Journal of Environmental Quality* 1:97-102.

Prihar, S. S., and R. Singh. 1983. Advances in fertiliser management for rainfed wheat. *Fertiliser News* 28, No. 9:49-56.

Prins, W. H., and G. H. Arnold (Eds.). 1985. *The Role of Nitrogen in Intensive Grassland Production*. Centre for Agricultural Publishing and Documentation, Wageningen, The Netherlands.

Prins, W. H., K. Dilz, and J. J. Neeteson. 1988. Current recommendations for nitrogen fertilisation within the E.E.C. in relation to nitrate leaching. *Fertiliser Society (London) Proceedings* No. 276.

Probert, M. E. 1987. Incorporating the effects of plant density into fertilizer response models. *Fertilizer Research* 11:143-148.

Public Health Service. 1962. *Public Health Service Drinking Water Standards 1962*. U.S. Department of Health, Education and Welfare, Public Health Service Publication No. 956.

Read, D. W. L., F. G. Warder, and D. R. Cameron. 1982. Factors affecting fertilizer response of wheat in southwestern Saskatchewan. *Canadian Journal of Soil Science* 62:577-586.

Rible, J. M., P. F. Pratt, L. J. Lund, and K. M. Holtzclaw. 1979. Nitrates in the saturated zone of freely drained fields. Pp. 297-320. In P. F. Pratt (Ed.), *Nitrate in Effluents From Irrigated Lands*. National Technical Information Service, Springfield, Virginia. (Cited by Pratt, P. F. 1984. Nitrogen use and nitrate leaching in irrigated agriculture. Pp. 319-333. In Hauck, R. D. (Ed.), *Nitrogen in Crop Production*. American Society of Agronomy, Crop Science Society of America, and Soil Science Society of America, Madison, WI.)

Ritter, W. F. 1989. Nitrate leaching under irrigation in the United States — a review. *Journal of Environmental Science and Health. Part A. Environmental Science and Engineering* A24:349-378.

Rooyen, P. J. van, and C. S. Dannhauser. 1988. The optimization of nitrogen and phosphorus application to cultivated *Digitaria eriantha* ssp. *eriantha* pasture. *South African Journal of Plant and Soil* 5:11-14.

Schepers, J. S., K. D. Frank, and D. G. Watts. 1984. Influence of irrigation and nitrogen fertilization on groundwater quality. *Proceedings of the International Union of Geodesy and Geophysics, Hamburg, Germany, 1983*. (Cited by Schepers, J. S., K. D. Frank, and C. Bourg. 1986. Effect of yield goal and residual soil nitrogen considerations on nitrogen fertilizer recommendations for irrigated maize in Nebraska. *Journal of Fertilizer Issues* 3:133-139.)

Schepers, J. S., M. G. Moravek, E. E. Alberts, and K. D. Frank. 1991. Maize production impacts on groundwater quality. *Journal of Environmental Quality* 20:12-16.

Shaw, K. 1962. Loss of mineral nitrogen from soil. *Journal of Agricultural Science (Cambridge)* 58:145-151.

Simon, W., B. Huwe, and R. R. van der Ploeg. 1988. Die Abschätzung von Nitratausträgen aus landwirtschaftlichen Nutzflächen mit Hilfe von Nmin-Daten. *Zeitschrift für Pflanzenernährung und Bodenkunde* 151:289-294.

Singh, B., C. R. Biswas, and G. S. Sekhon. 1978. A rational approach for optimizing application rates of fertilizer nitrogen to reduce potential nitrate pollution of natural waters. *Agriculture and Environment* 4:57-64.

Stevenson, F. J. (Ed.). 1982. *Nitrogen in Agricultural Soils.* Agronomy 22. American Society of Agronomy, Crop Science Society of America, and Soil Science Society of America, Madison, WI.

Strebel, O., W. H. M. Duynisveld, and J. Böttcher. 1989. Nitrate pollution of groundwater in western Europe. *Agriculture, Ecosystems and Environment* 26:189-214.

Sutherland, R. A., C. C. Wright, L. M. J. Verstraeten, and D. J. Greenwood. 1986. The deficiency of the "economic optimum" application for evaluating models which predict crop yield response to nitrogen fertiliser. *Fertilizer Research* 10:251-262.

Swanson, E. R. 1982. Economic implications of controls on nitrogen fertilizer use. Pp. 773-790. In F. J. Stevenson (Ed.), *Nitrogen in Agricultural Soils.* Agronomy 22. American Society of Agronomy, Crop Science Society of America, and Soil Science Society of America, Madison, WI.

Sylvester-Bradley, R., T. M. Addiscott, L. V. Vaidyanathan, A. W. A. Murray, and A. P. Whitmore. 1987. Nitrogen advice for cereals: Present realities and future possibilities. *Fertiliser Society (London) Proceedings* No. 263.

Sylvester-Bradley, R., and B. J. George. 1987. Effects of quality payments on the economics of applying nitrogen to winter wheat. *Aspects of Applied Biology* No. 15:303-318.

Ulmer, M. G., J. W. Enz, D. D. Patterson, R. J. Goos, and E. J. Deibert. 1988. Influence of plant available water at seeding on spring wheat and sunflower production in North Dakota. *North Dakota Farm Research* 45, No. 5:7-12, 31.

Van Boheemen, P. J. M. 1987. Extent, effects and tackling of a regional manure surplus; a case-study for a Dutch region. Pp. 175-193. In H. G. van der Meer, R. J. Unwin, T. A. van Dijk, and G. C. Ennik (Eds.), *Animal Manure on Grassland and Fodder Crops. Fertilizer or Waste?* Martinus Nijhoff Publishers, Dordrecht, The Netherlands.

Van Broekhoven, L. W. 1980. Formation of volatile N-nitrosamines during the fermentation of grass silages. Pp. 113-117. In W. H. Prins and G. H. Arnold (Eds.), *The Role of Nitrogen in Intensive Grassland Production.* European Grassland Federation, Proceedings of an International Symposium. Centre for Agricultural Publishing and Documentation, Wageningen, The Netherlands.

Van Broekhoven, L. W., and R. W. Stephany. 1978. Environmental aspects of N-nitroso compounds. *International Agency for Research on Cancer, Scientific Publications* 19:461-463. (Cited by Kemp, A. 1982. The importance of the chemical composition of forage for optimizing animal production. Pp. 95-116. In *Optimizing Yields — The Role of Fertilizers.* Proceedings of the 12th IPI-Congress. International Potash Institute, Worblaufen-Bern, Switzerland.)

Van der Meer, H. G., J. C. Ryden, and G. C. Ennik (Eds.). 1986. *Nitrogen Fluxes in Intensive Grassland Systems.* Martinus Nijhoff Publishers, Dordrecht, The Netherlands.

Van Diest, A. 1986a. The social and environmental implications of large-scale translocations of plant nutrients. Pp. 289-299. In *Nutrient Balances and the Need for Potassium.* Proceedings of the 13th IPI-Congress. International Potash Institute, Worblaufen-Bern, Switzerland.

Van Diest, A. 1986b. Means of preventing nitrate accumulation in vegetable and pasture plants. Pp. 455-471. In H. Lambers, J. J. Neeteson, and I. Stulen (Eds.), *Fundamental, Ecological and Agricultural Aspects of Nitrogen Metabolism in Higher Plants.* Martinus Nijhoff Publishers, Dordrecht, The Netherlands.

Van Rooyen, P. J., and C. S. Dannhauser. 1988. The optimization of nitrogen and phosphorus application to cultivated *Digitaria eriantha* ssp. *eriantha* pasture. *South African Journal of Plant and Soil* 5:11-14.

Voss, R. D. 1975. Fertilizer N: The key to profitable corn production with changing prices and production costs. Pp. 215-229. In *1975 Proceedings of the Thirtieth Annual Corn and Sorghum Research Conference*. American Seed Trade Association, Washington.

Walton, G. 1951. Survey of literature relating to infant methemoglobinemia due to nitrate-contaminated water. *American Journal of Public Health* 41:986-996.

Waugh, D. L., R. B. Cate, Jr., and L. A. Nelson. 1973. *Discontinuous Models for Rapid Correlation, Interpretation, and Utilization of Soil Analysis and Fertilizer Response Data*. International Soil Fertility Evaluation & Improvement Program, Technical Bulletin No. 7. North Carolina State University, Raleigh.

Wehrmann, J., and H. C. Scharpf. 1989. Reduction of nitrate leaching in a vegetable farm — fertilization, crop rotation, plant residues. Pp. 247-253. In E. Welte and I. Szabolcs (Eds.), *Protection of Water Quality From Harmful Emissions With Special Regard to Nitrate and Heavy Metals*. Proceedings of the 5th International Symposium of CIEC. International Scientific Centre of Fertilizers, Budapest, Hungary.

Wehrmann, J., H.-C. Scharpf, and H. Kuhlmann. 1988. The N_{min}-method — An aid to improve nitrogen efficiency in plant production. Pp. 38-45. In D. S. Jenkinson and K. A. Smith (Eds.), *Nitrogen Efficiency in Agricultural Soils*. Elsevier Applied Science Publishing Co., London.

Welte, E., and I. Szabolcs (Eds.). 1989. *Protection of Water Quality From Harmful Emissions With Special Regard to Nitrate and Heavy Metals*. Proceedings of the 5th International Symposium of CIEC. International Scientific Centre of Fertilizers, Budapest, Hungary.

Wimble, R. 1980. Theoretical basis of fertiliser recommendations. *Chemistry and Industry* 1980:680-683.

Withagen, L. M. 1983. Possible deviations from the attainable maximum financial yield, when applying N-fertilizer on base of N_{min} in the soil. Pp. 545-548. In *Nitrogen and Sugar Beet*. International Institute for Sugar Beet Research, Brussels, Belgium.

CHAPTER **3**

Plant Testing and Fertilizer Requirement

T HE CHEMICAL COMPOSITION OF CROPS changes with the nutrient supplies, although the change is by no means commensurate with the variation in external supplies. For example, Asher and Loneragan (1967) found that a 625-fold range of concentrations of phosphorus in flowing nutrient solutions elicited only a 4- to 16-fold range of phosphorus concentrations in the tops of eight species of plants grown in these solutions. In similar work with potassium, Spear et al. (1978) found that a 16,000-fold range in potassium concentrations external to the roots resulted in only a 7- to 13-fold range in potassium concentrations in the tops of three species of plants grown in the solutions.

Measurements of the chemical composition of crops may be made to evaluate the crops for commercial purposes, as sometimes is true for the nitrogen concentration in wheat grain as an index of protein concentration and bread-making quality, and almost always is true for the concentration of sucrose in sugarbeet and sugarcane. Much research has been done on the concentration of nutrient elements and nutrient-related entities in crops as indexes of the sufficiency of nutrient supplies for crop growth and as an aid in estimating fertilizer needs. This chapter reviews the findings and describes some applications that have been made of the relationships developed. More extensive reviews of plant analysis

155

and research on many crops may be found in books edited by Martin-Prével et al. (1987) and Westerman (1990).

As indicators of the nutrient status of crops, the final arbiters are the crops themselves. A crop combines its inherent requirements with the nutrient supplies and the environmental conditions, and reaches an integrated outcome. Plant composition, when properly measured and interpreted, thus should provide a better index of the sufficiency of individual nutrients for plant growth and functioning than should soil composition because of closer connection with the objective of the measurements. The mineral composition of crops, however, yields no primary information about nutrient deficiencies. Rather, the significance of the mineral composition must be inferred from the results of experiments to determine what happens to the crop yield, quality, and mineral composition when the supplies of the nutrients are varied experimentally or in other ways.

After appropriate standards have been developed, the most the plants can be expected to disclose when analyses are made in a nonexperimental situation is the relative sufficiencies of the various nutrients. Additional empirical information is required to quantify the treatments needed to achieve the desired objectives in terms of yield and quality of the crop.

Experiments conducted in the field are recognized as the "primary standard" for assessing mineral deficiencies in plants grown under field conditions and for determining the fertilizer requirement. If predictions based upon any other method cannot be confirmed by the results of field experiments, the method must be regarded as unsatisfactory.

Field experiments are valuable both for qualitative determination of which nutrients may be deficient and for quantitative evaluations of the degree of the deficiency. But field experiments are time consuming and expensive, and the results apply strictly to only the circumstances under which the experiments are conducted. Much effort therefore has been devoted to developing ways to make indirect use of field experiments to evaluate nutrient supplies for crops grown on farmers' fields where no experiments have been conducted and to estimate the quantities of various nutrients that may be needed to obtain desired improvements in crop yield and quality.

This chapter explains the scientific aspects of questioning plants about their supplies of individual nutrients and of interpreting the answers received. Both successes and problems will be illustrated. The practice of calling attention to problems and exceptions should not be inferred to mean that nothing useful has been achieved. Rather, the purposes of this approach are to display the range of observed behaviors and to assure that an otherwise one-sided account will not prove misleading.

The chapter begins with a discussion of qualitative diagnosis of nutrient deficiencies. Following this is background information on the seasonal course of nutrient absorption and dry matter production and the relation of nutrients to photosynthesis. The two major sections deal with concentrations and ratios of nutrients. The use of response curves for purposes of evaluation is continued.

As is true in other chapters dealing primarily with plant nutrients, the subject matter is presented mostly in general terms, and the examples involve mostly macronutrients because the literature dealing with these nutrients provides the most complete selection. The peculiarities of individual nutrients are discussed to some extent where it seems appropriate, but a nutrient by nutrient approach is not used because it would extend the text considerably, while adding little to the principles.

3-1. Qualitative Diagnosis of Deficiencies

3-1.1. Deficiency Symptoms

For qualitative diagnosis of mineral deficiencies in plants exhibiting deficiency symptoms, atlases such as *The Diagnosis of Mineral Deficiencies in Plants by Visual Symptoms* (Wallace, 1961) and *Hunger Signs in Crops* (Sprague, 1964) that illustrate with photographs in color the symptoms of deficiencies of various nutrients in different kinds of plants are very useful. A more definitive method is *plant injection*, by which generally is meant introducing a dissolved compound containing a nutrient into a plant through an artificial opening. The injection may involve a portion of a leaf, a leaf, a stem, a branch, or an entire plant, which may be even a tree. A monograph on the injection method was published by Roach (1938).

Injection can be done in the field, and in principle it is analogous to a qualitative field experiment in which arbitrary quantities of different nutrients are added as fertilizers to individual plots to determine whether perceived symptoms are alleviated by any of the fertilizers. Injections into a small segment of a leaf can be more precise than field plot experiments for determining the nature of a deficiency that has produced visual symptoms because a comparison can be made between immediately adjacent segments of a given leaf in which there is a sharp boundary between injected and noninjected tissue. The diagnosis can be made in a week or less, which is sooner than the diagnosis can be made from an experiment in which nutrients are applied to the soil in the field. Validation by comparison with the results of field plot experiments is not required.

The injection method is strictly qualitative. For diagnostic purposes, its application has been limited to situations in which a nutrient deficiency is great enough to affect the appearance of a leaf or a plant. Other methods must be used for situations in which the deficiency is less pronounced. The same may be said of the practice of applying nutrients externally by spraying or painting plants or leaves with solutions containing nutrients to be tested. Externally applied nutrients must penetrate the tissues to be effective, and assuring suitable penetration may require preliminary experiments.

3-1.2. Atterberg's Principles

A qualitative chemical basis for diagnosing nutrient deficiencies that is independent of deficiency symptoms was developed many years ago by Atterberg

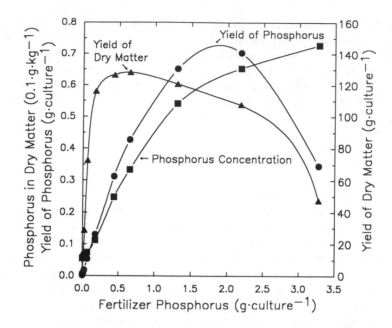

Fig. 3-1. Yield of dry matter, yield of phosphorus, and phosphorus concentration in the dry matter of oat plants grown in sand cultures with increasing additions of phosphorus as monobasic sodium phosphate. (Mitscherlich, 1935)

(1901). As a basis for his concepts, he reviewed the results of many chemical analyses that had been made on plants in fertilizer experiments by other investigators, and he reported the results of numerous original experiments. He described his understanding of the relationships of mineral nutrient supplies to the concentrations of nutrients in plants in the form of four general principles, which represent adaptations of the law of mass action employed in chemistry.

Atterberg recognized that there were exceptions to his generalizations. Nonetheless, his principles are still useful and are still used on a qualitative basis for interpreting plant composition data.

3-1.2.1. Principle 1
If the quantity of a nutrient available for oat plants increases, the nutrient is taken up and assimilated in greater quantity, and the concentration thereof in the plants increases also.

This principle may be illustrated by Mitscherlich's (1935) data in Fig. 3-1. Mitscherlich's experiment shows that as increasing quantities of monobasic sodium phosphate were added to sand cultures supplied with a minus-phosphorus nutrient solution, the phosphorus concentration increased in the oat plants grown in the cultures. The phosphorus concentration continued to increase beyond the phosphate application with which the maximum yields of dry matter and phosphorus were obtained.

The qualitative nature of principle 1 and an exception to it are illustrated in Fig. 3-2. Fig. 3-2A shows that in oat plants harvested 60 days after planting, the potassium concentration in the plants increased with the quantity of fertilizer potassium applied. At plant maturity 130 days after planting, the potassium concentration in the plants decreased with small applications of fertilizer potassium and then increased with larger additions. The decrease represents an exception to Atterberg's first principle. Fig. 3-2B shows the trend with time in the potassium concentrations in the plants with the two smallest additions of fertilizer potassium. The two curves cross. The plants with the greater supply of potassium had the greater potassium concentration at the early harvests but the lower potassium concentration at the final harvest when the plants were mature. The potassium concentration in the plants with the lower supply of potassium passed through a minimum at 80 to 90 days after planting and then increased.

Analogous results in field experimental work have been published for nitrogen in sugarcane by Baver (1960), nitrogen in cotton by Braud (1972), and phosphorus in potato by Jacob et al. (1949). Dean and Fried (1953) represented what is termed here an exception to Atterberg's first principle in a schematic diagram for phosphorus, implying that it is the usual behavior for phosphorus. Bates (1971) listed a number of investigations in which the behavior had been observed.

The exceptional behavior [sometimes called the Piper-Steenbjerg effect, after Piper (1942) and Steenbjerg (1945), who called attention to it] occurs most commonly when the nutrient in question is strongly deficient in the control plants. The explanation for the behavior has not been established, although several

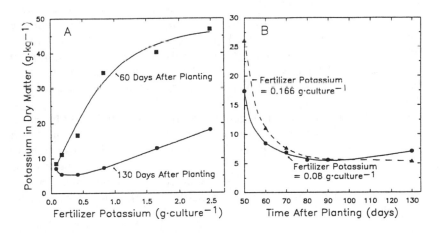

Fig. 3-2. Potassium concentrations in oat plants grown in sand cultures containing 1.5% peat and supplied with different quantities of potassium as potassium sulfate. *A* shows the potassium concentrations in the plants 60 and 130 days after planting. At 130 days, the plants were mature. *B* shows the potassium concentrations in the plants at different times after planting with the two smallest applications of fertilizer potassium. (Alten et al., 1940)

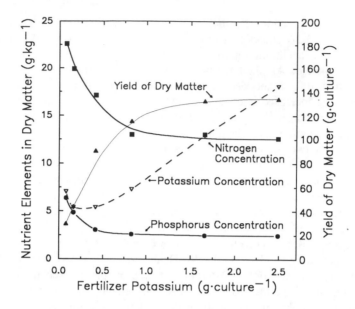

Fig. 3-3. Yield of oat plants, and nitrogen, phosphorus, and potassium concentrations in the dry matter versus quantity of fertilizer potassium added in sand cultures containing 1.5% peat. The data are from the same experiment as the one from which the preceding figure was derived. (Alten et al., 1940)

hypotheses have been suggested (see especially Hiatt and Massey, 1958; Bates, 1971; Loneragan, 1978).

The suggestion that the effect is principally a consequence of sampling plants of different physiological stages at the same time seems to provide as comprehensive an explanation as any for the Piper-Steenbjerg effect. Thus, in terms of foliar analysis procedures and interpretation, Walworth and Sumner (1988) considered that the behavior is to be regarded as a manifestation of the larger problems often associated with time of sampling, which will be discussed in Section 3-4.2.

Bates (1971) discussed the ambiguity in interpretations that might be caused by the occurrence of two yields corresponding to one value for nutrient concentration as a consequence of the Piper-Steenbjerg effect. Whether this ambiguity exists with analyses of crops that have not received prior applications of the nutrient in a fertilizer is not certain, but it could confuse the interpretation if the plant analysis is made some time after a fertilizer containing the nutrient has been applied.

3-1.2.2. Principle 2

If with increasing supply of a nutrient an increase in yield results, the supply of the other nutrients becomes smaller relative to the greater yield, and the concentration of these nutrients in the plants must then decrease.

This principle may be illustrated by the data in Fig. 3-3 on the composition of oat plants grown in a sand-peat mixture receiving different quantities of fertilizer potassium. The nitrogen and phosphorus concentrations in the oat plants decreased, in accordance with principle 2. The potassium concentration increased with potassium additions, in accordance with principle 1 (although there was a small decrease with the two smallest additions — an exception described and illustrated by data from the same experiment in Section 3-1.2.1).

Another version of Atterberg's second principle has received attention in plant nutrition literature without reference to Atterberg. This version, which has been developed for nitrogen, states that as the yield of a crop increases, the nitrogen concentration in the crop decreases. In this version, an increase in the yielding capability of the crop with a given supply of nutrients rather than an increase in yield due to addition of a particular nutrient is responsible for the dilution of the nitrogen in the crop.

In a corn-breeding study in Nebraska, Kiesselbach (1948) found that a high yield of dry matter was associated with a low nitrogen concentration in the dry matter. Calling attention to this behavior and citing other examples from the literature, Willcox (1948, 1949) used it as the basis for what he called the *inverse yield-nitrogen law*: "The yields of all agrotypes, without any clearly proved exceptions to date, are inversely proportional to the percentage of nitrogen contained in their whole dry, above-ground substance" (Willcox, 1948). That is,

$$Yield \ of \ Dry \ Matter = \frac{k}{Nitrogen \ Concentration},$$

where k is a constant. The quantity k is equal to the product of yield of dry matter and the nitrogen concentration. The novelty in Willcox's inverse yield-nitrogen law is the postulate that the value of k is constant for all plants, with the qualifications that they be grown to maturity on the same normal soil. The inverse yield-nitrogen law implies that the nitrogen concentration controls the yield. If the nitrogen concentration is low, the yield is high, and vice versa.

Greenwood (1982) published a paper on nitrogen supply and crop yield, which included data verifying in a qualitative way the inverse yield-nitrogen law. In a graph summarizing data on 22 crop species, he found that the nitrogen concentration in the plant dry matter at harvest decreased from about 40 to 11 grams per kilogram as the yield increased from about 2 to 20 megagrams per hectare. Although the relationship between yield of dry matter and nitrogen concentration was unmistakable, the products of yield and nitrogen concentration were not constant. The individual crops, however, were not grown on the same soil.

The approximate constancy of the product of yield and nitrogen concentration where different cultivars are grown side by side on the same soil appears to result from the fact that plants have the capability of virtually exhausting the supply of mineral nitrogen and especially the nitrate from the soil. For example, in an experiment by Borden (1946), in which three sugarcane cultivars were grown separately in containers of the same soil with an initial content of 9.1 grams of ammonium plus nitrate nitrogen per culture, the ammonium plus nitrate nitrogen remaining in the soil at harvest was about 0.1 gram or less per culture for all cultivars.

Thus, for plants that exhaust the mineral nitrogen from the soil and have similar proportions of their total nitrogen in the tops, the yields of nitrogen in the tops should be similar. White and Black (1954) found experimentally, however, that differences in yield of nitrogen among cultivars and crops appeared when the supply of nitrogen in the soil was great enough to avoid exhaustion. Moreover, from data in the literature, they found no evidence of constancy of yield of other nutrients among cultivars grown on the same soil, as would be expected from the fact that plants do not exhaust the supply of these nutrients as they may for nitrogen.

To address in brief the plant-breeding aspects of the inverse yield-nitrogen law, the nitrogen concentration in different cultivars of a given nonleguminous crop grown side by side on soil with low to moderate supplies of available nitrogen may be viewed as an indirect indication of the yield of dry matter. Although it may be impossible to introduce genetic alterations that will sub-stantially increase the nitrogen uptake from the soil under such circumstances, it is possible to modify the distribution of nitrogen within the plants so that, for example, more of the nitrogen will be translocated to the grain and will raise the protein content (Johnson et al., 1969).

Numerous exceptions to Atterberg's second principle have been recorded in the literature. For example, in research on oat plants grown in nutrient solutions with different concentrations of potassium, Lundegårdh (1941) found that the yield of the plants and the potassium concentration in the leaves increased with the potassium supply and that the concentrations of nitrogen and phosphorus decreased, in accordance with Atterberg's second principle. With increasing phosphorus concentrations in the solution at a constant nitrogen concentration, however, the nitrogen concentration in the leaves continued to increase, even with phosphorus concentrations exceeding those required to produce the maximum yield. These results imply a special connection between nitrogen and phosphorus.

Kamprath (1987) and others have found increases in phosphorus concentrations in vegetative tissues of plants when the nitrogen concentration was increased by fertilization. Prevot and Ollagnier (1961) noted a general correlation between nitrogen and phosphorus concentrations in tissues of a number of plant species. Part of the cause of the nitrogen-phosphorus connection in plant composition lies in the internal behavior of plants, but part is also in the soil (Grunes, 1959).

App et al. (1956) found that nitrogen fertilization increased the yield of bromegrass in the early spring and also increased the concentrations of nitrogen, phosphorus, and potassium in the dry matter. Leigh and Johnston (1983a) found that nitrogen fertilization increased the yield of barley grain and also increased the concentration of potassium in the dry matter of the plants in the field during most of the season. Their findings will be referred to again in Section 3-5.2, and an explanation will be given that has possible application to many of the exceptions to Atterberg's second principle.

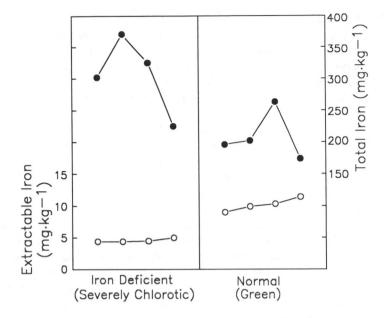

Fig. 3-4. Total and extractable iron in iron-deficient and normal leaves of peanut plants from a field experiment on an alkaline soil in India. The extractable iron was removed from the fresh leaves with an acid solution (pH 3) containing 15 grams of orthophenanthroline per liter. Values for total iron are on the dry weight basis, and those for extractable iron are on the fresh weight basis. The samples are arranged in order of increasing concentrations of extractable iron. (Rao et al., 1987)

3-1.2.3. Principle 3

Where the yield of oat plants is low, a low concentration in the plants is to be expected only for the particular nutrient present in smallest quantity in relation to the requirement of the plants. Other nutrients present in only small quantity may show a more or less high concentration.

The data in Fig. 3-3 illustrate principle 3 as well as principle 2. The low-yielding plants obtained with the smallest additions of potassium had a relatively low concentration of potassium, but relatively high concentrations of nitrogen and phosphorus.

An exception to principle 3 sometimes is found with iron, a nutrient that is relatively immobile in plants. As indicated in Fig. 3-4, the iron concentration may be greater in leaves with iron deficiency chlorosis than in green leaves. The cause of this behavior appears to be immobilization of iron within the plant in forms that are not of value in chlorophyll formation. Fig. 3-4 shows that the ferrous iron extractable from the fresh leaf tissue with an acid solution of or-thophenanthroline was greater in green leaves than in severely chlorotic leaves. The values for extractable iron thus are in agreement with Atterberg's third principle.

3-1.2.4. Principle 4

Which nutrient is present in the minimum may be determined in the following way. One compares the concentration of the nutrient found by analysis with the corresponding average and minimum concentrations of the nutrient in oat plants. The nutrient with concentration furthest below the average, or least exceeding and most closely approaching the minimum concentration, is present in the minimum.

This principle logically follows from the first three, and it is useful to the extent that the first three principles reflect the behavior of plants. The procedure described by Atterberg is useful in diagnosing particular cases in which a mineral deficiency is suspected to be the cause of poor growth or foliage symptoms.

Complete data on average and minimum concentrations of the various nutrients in all plants have not been obtained. Major summaries of available information, however, have been published by Beeson (1941), Goodall and Gregory (1947), Chapman (1966), Walsh and Beaton (1973), and Reuter and Robinson (1986).

Bergmann (1983) summarized data from the literature on the ''satisfactory'' ranges of concentrations of individual nutrients in various plants. Bergmann and Bergmann (1985) then developed a computer program that would print for each kind of plant a chart that shows the lower and higher boundaries of the satisfactory range on a relative basis such that despite variations in absolute concentrations and differences in the ratio of the higher limit to the lower limit, all the satisfactory ranges appear in the center of the chart and have equal width. The nutrient status of a given sample relative to the satisfactory ranges for the individual nutrients is displayed by superimposing on the printout a bar for each nutrient extending from the left of the chart and ending at a location dependent upon the concentration relative to the satisfactory range. The finished chart, all produced by computer, provides a quick overview of the nutrient status of the sample in relation to the standards chosen.

Because nutrient problems vary with location, sometimes within small areas, it is often feasible to analyze samples from good and poor areas within a single field. Such comparisons are generally more critical than a comparison of analyses from a poor area with published data because conditions other than the supply of the nutrient suspected as the cause are more nearly comparable. Table 3-1 shows the results of use of this procedure to diagnose the cause of poor growth

Table 3-1. Nutrient Concentrations in Petioles From Vigorous and Poorly Growing Sugarbeet Plants Within a Field in California (Ulrich et al., 1959)

Beet Growth	Nutrient Concentration in Petiole Dry Matter (g·kg⁻¹)			
	Nitrate Nitrogen	Phosphate Phosphorus	Potassium	Sulfate Sulfur[a]
Vigorous	2.5	2.1	31.5	0.530
Poor	9.0	3.0	38.0	0.085

[a] The leaf blades provided a more sensitive indication of sulfur deficiency than did the petioles. The sulfate sulfur concentration in the leaf blades was 1.88 grams per kilogram in samples from the vigorously growing plants and 0.155 gram per kilogram in samples from the poorly growing plants.

of sugarbeet in an area within a field in California. The leaves appeared nitrogen deficient, but analysis showed that the nitrate concentration was higher in the samples from the poor area than from the good area. Sulfate was much lower in samples from the poor area than from the good area, and application of sulfate to plots within the poor area confirmed that the plants were deficient in sulfur.

3-1.3. Krauss Foliar Vector Diagnosis

Krauss (1965) devised a way of diagramming changes in plant composition that has been found useful by several subsequent investigators concerned with fertilization of conifers in which the number of needles developing from a bud is determined in the bud year, which precedes the year in which the needles develop on the annual growth increment. The response of the needles in the buds in the year in which a fertilizer is applied provides a basis for judging the increase in volume of wood that will be produced in succeeding years, and thus as a basis for judging whether fertilizing a given area with a given nutrient is likely to be useful.

The manner of presenting the data has been modified from the original version by Krauss. Fig. 3-5 is an assembly from several sources. The general concept is diagrammed in Fig. 3-5A, and an example from an experiment on fertilization of jack pine is shown in Fig. 3-5B.

For a single nutrient, the nutrient concentration in the needle dry matter is plotted on the Y axis, and the quantity of the nutrient per needle is plotted on the X axis. The diagonal lines represent weights per needle, increasing from left to right. The data for the needles from unfertilized and fertilized trees are plotted in the same graph, and an arrow is drawn from the point representing the control to the point representing the fertilized trees. Fig. 3-5A shows arrows in various directions, indicating different possible outcomes. The most useful way of plotting the data is to use relative values, representing the control as having a nutrient concentration, nutrient quantity per needle, and weight per needle as 100. Then all nutrients can be plotted on a single graph, and the effect of fertilization on all nutrients can be displayed in a form convenient for comparison.

If the arrow leading from the control to the fertilized treatment is in direction A, the concentration has decreased, but the quantity of the nutrient per needle and the weight per needle have increased. Thus, the extra nutrient absorbed due to fertilization has been diluted by extra growth. Sometimes this effect is observed with a nutrient added in the fertilizer if the nutrient is strongly limiting in the control. More often it is noted when the nutrient in question is not added in the fertilizer and is not limiting, and the first limiting nutrient is added in the fertilizer.

If the arrow leading from the control to the fertilized treatment is a horizontal shift in direction B, the concentration in the needle remains the same, and the weight per needle and the quantity of the nutrient per needle increase to the same degree. This situation may result where the nutrient is at the minimum concentration (see Section 3-4.2.1) in both the control and fertilized trees or where the nutrient concentration would have decreased with a smaller addition and then increased to the level shown with the nutrient addition employed. In

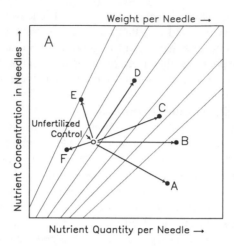

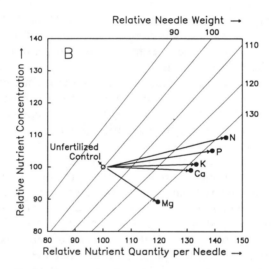

Fig. 3-5. The modified Krauss foliar vector diagnosis. A. Schematic representation of possible effects of fertilization on conifer needle weight, nutrient concentrations in the needles, and quantities of nutrients per needle. (Derived mostly from Timmer and Stone, 1978, with one modification from Weetman and Wells, 1990.) B. Response of jack pine needles to nitrogen fertilization. (Timmer and Morrow, 1984)

either case, the nutrient added is limiting. If the horizontal displacement is observed with a nutrient not added, the indications are that this nutrient is not limiting.

A displacement upward and to the right, as in *C*, indicates an increase in needle weight, an increase in content of the nutrient per needle, and an increase in nutrient concentration. For a nutrient added in the fertilizer, this shift indicates that the nutrient is deficient in the control. For a nutrient not added in the fertilizer, movement in the *C* direction would indicate an increase in availability resulting from some interaction between the fertilizer and the other nutrient.

A displacement upward and to the right, as in *D*, would signify increases in nutrient concentration and in nutrient quantity per needle, but no increase in needle weight. Such a result with the nutrient added in the fertilizer would indicate that the nutrient is not limiting in the control, and that the addition in the fertilizer has merely increased the uptake, but not to the extent at which toxicity occurs. Displacement in the *D* direction is unlikely for a nutrient not added in the fertilizer.

A shift upward and to the left, as in direction *E*, indicates an increase in nutrient concentration, a decrease in nutrient quantity per needle, and a decrease in weight per needle. Movement in direction *E* for a nutrient added in the fertilizer would indicate that the nutrient is toxic in the quantity applied. Results of this type are unlikely for nutrients not added in the fertilizer.

An arrow downward and to the left, as in direction *F*, indicates a decrease in nutrient concentration, a decrease in nutrient quantity per needle, and a decrease in weight per needle. A result of this type is to be expected only for a nutrient not added in the fertilizer that has been reduced in availability by some antagonistic effect. Induction of magnesium deficiency in a soil low in magnesium by a heavy addition of fertilizer potassium would be an example.

In Fig. 3-5B, nitrogen fertilization increased the weight per needle and the quantities of all the elements tested per needle. Nitrogen was deficient in the unfertilized control. The concentration of phosphorus was increased, illustrating the commonly observed nitrogen-phosphorus effect, and the concentration of magnesium was reduced. The additional magnesium absorbed was diluted by the extra growth. Concentrations of potassium and calcium were essentially unaffected by fertilization. The relative increases in quantities of these nutrients taken up almost matched the relative increase in dry weight per needle, indicating that they were not limiting.

The descriptions of the various possible outcomes may be perceived to have much in common with Atterberg's principles. The Krauss foliar vector approach would be unsuitable for annual crop plants, but it could be applied to the total above-ground dry matter.

3-2. Course of Nutrient Absorption and Dry Matter Production

As background for sections that follow on attempts to develop semiquantitative relationships between nutrient concentrations in plants and nutrient sufficiency, this section reviews in brief the overall picture of the uptake of nutrients by plants and the changes in concentration that may occur during a growing season. Fig. 3-6 shows the course of nutrient absorption and dry matter production for two crops, corn and soybean. The uptake data in Figs. 3-6A and B are expressed as percentages of the maximum amounts accumulated in the above-ground tissues during the growing season to permit use of the same scale throughout. The

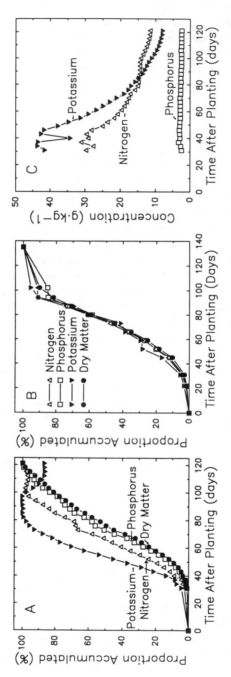

Fig. 3-6. A. Percentage of the maximum amounts of nitrogen, phosphorus, potassium, and dry matter present in the above-ground tissues of corn plants different lengths of time after planting in Ohio. (Sayre, 1948) B. Percentage of the maximum amounts of nitrogen, phosphorus, potassium, and dry matter present in the above-ground tissues of soybean plants different lengths of time after planting in Iowa. (Hammond et al., 1951) C. Concentrations of nitrogen, phosphorus, and potassium in the dry above-ground tissues of corn plants different lengths of time after planting in Ohio. The data are from the same analyses from which A was derived. (Sayre, 1948)

characteristic sigmoid shape of the nutrient-absorption and dry-matter-production curves is evident with both crops.

Plots of the type in Figs. 3-6A and B usually show that the nutrient-absorption curves lead the dry-matter-production curves. That is, a given percentage of the seasonal total of nutrients is absorbed by an earlier date than the same percentage of the total dry matter has been accumulated. The figures show that the extent to which this occurs is not necessarily the same with different nutrients or different crops, and the same might be said for different conditions. For example, Fig. 3-6A shows some net loss of potassium from corn plants at the end of the season, but no comparable loss of nitrogen and phosphorus. Schjørring et al. (1989) observed a decrease in fertilizer-derived nitrogen in the above-ground portion of barley plants beginning about 40 days after emergence in a field experiment in Denmark, and on the basis of leaf loss and observations in the literature, they attributed it to volatilization. The soil-derived nitrogen in the plants did not show an analogous decrease, presumably because mineralization and absorption of soil nitrogen continued during the season.

The tendency of the nutrient-absorption curves to lead the dry-matter-accumulation curves may be visualized in another way by plotting the concentrations of nutrients in the plants against the plant age. Fig. 3-6C shows such a plot for the same corn crop from which Fig. 3-6A was derived. The change in concentration of the nutrients in the dry matter with time was marked with potassium and nitrogen but slight with phosphorus. Hanway's (1962) data showed changes in the same order, that is, potassium > nitrogen > phosphorus, but the drop in phosphorus concentration with time was more pronounced than that in Fig. 3-6C. Analogous data for the soybean crop in Fig. 3-6B are not shown, but as may be inferred from the figure, the changes in concentration with time were considerably smaller than those for corn.

The behavior exhibited by corn has been reported more commonly than that shown by soybean, a possible reason being that fertilizers usually are applied at the beginning of the season in experiments in which the uptake of nutrients is determined throughout the season. No fertilizers were applied for soybean in the experimental work from which Fig. 3-6B was derived, and only a small quantity of fertilizer was applied to the corn crop analyzed to obtain the results in Figs. 3-6A and 3-6C. A second possible reason is the fact that soybean is a legume, and its nitrogen concentration is controlled by an internal mechanism. The nitrogen concentration may exert an indirect controlling influence on the concentration of phosphorus and probably other nutrients, as will be noted later. The effect of fertilizers will be considered in more detail in subsequent sections.

Although the concentrations of nitrogen, phosphorus, and potassium in vegetative tissues of plants generally decrease as the season advances, the decline can be reduced by late applications of fertilizers as topdressings. Fertilizers supplying nitrate nitrogen are the most effective in this regard because nitrate does not react chemically with most soils and is carried down into the root zone with water. Most other nutrients move downward less readily. Fig. 3-7 illustrates a pronounced although delayed effect of potassium fertilization in increasing the

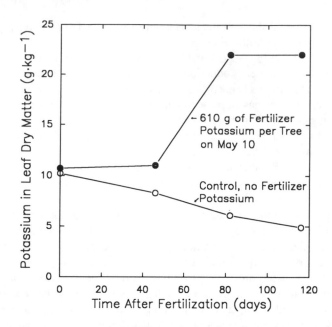

Fig. 3-7. Potassium concentration in leaves of peach trees at different times on a gravelly loam soil in Maryland. The fertilized trees received 610 grams of potassium as potassium sulfate per tree on May 10, which was the first date of sampling. Subsequent sampling dates were June 25, July 31, and September 3. (Waugh and Cullinan, 1940)

concentration of potassium in peach leaves as the season progressed. Retention of the potassium in exchangeable form with resultant slowness in movement down into the root zone is a probable explanation for the delay.

Fig. 3-8 shows the behavior of five nutrients in the leaves of peach trees in Washington. Concentrations of nitrogen, phosphorus, and potassium decreased with increasing age of the leaves, in accordance with the usual behavior. Concentrations of calcium and magnesium increased, a common behavior of these nutrients in leaves of fruit trees. Shear and Faust (1971) found similarly that the calcium content of apple leaves increased with age. In the period from 10 weeks before harvest to harvest, the calcium concentration in the leaves of field grown trees in Pennsylvania increased from about 12 to 18 grams per kilogram, but the calcium concentration in the fruit decreased from about 0.6 to 0.2 gram per kilogram.

De Villiers and Beyers (1961) followed the concentrations of nine nutrients in leaves from fruiting terminals of orange trees in monthly samples from 2 to 16 months and found that the concentrations of phosphorus and potassium decreased with increasing leaf age; nitrogen, magnesium, boron, and copper concentrations remained about the same; and calcium, manganese, and iron concentrations increased. There are evidently differences among species and apparently also some differences among nutrients. Nutrients that tend to be immobile in plants (iron is probably the best example) seem to be the most likely to exhibit increases in concentration in vegetative parts with age of the tissues. Smith (1962) published a table summarizing findings on changes in nutrient concentrations in tissues of various plant species with aging.

3-3. Nutrients and Photosynthesis

The concept of photosynthesis as a function of the nutrient composition of plants is introduced at this point to call attention to another aspect of plant nutrition that is related to the use of plant composition data as an index of nutrient sufficiency. Although nutrients have a variety of effects, they all have a direct or indirect influence on crop yield through photosynthesis.

Greenwood et al. (1991) pointed out that the ultimate carbon dioxide fixing enzyme in photosynthesis is ribulose-1,5-bisphosphate carboxylate/oxygenase, which constitutes a large portion of the protein in leaves. The nitrogen concentration in leaves is highly correlated with the concentration of this enzyme. Hence, they argued that the potential maximum rate of photosynthesis would be expected to be linearly related to the nitrogen concentration in the leaf. With this relationship as a central theoretical concept, they developed several models connecting relative growth rates to relative nitrogen concentrations. Because

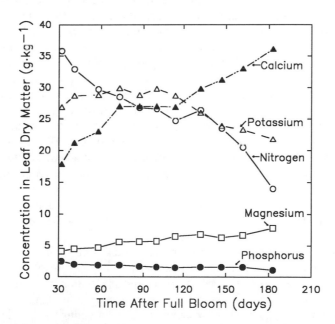

Fig. 3-8. Concentrations of nitrogen, phosphorus, potassium, calcium, and magnesium in peach leaves at 11 sampling dates from 32 to 183 days after full bloom in Washington. (Batjer and Westwood, 1958)

their models have to do with critical concentrations of nutrients, their work is discussed following the development of the critical concentration concept. See Section 3-4.2.5.

A leaf low in chlorophyll has a low capacity for photosynthesis per unit of leaf area (Terry, 1980). Iron is an essential constituent of chlorophyll, and this is no doubt a reason for the stunting of plants with chlorosis due to iron deficiency. In addition to its quantitative importance in the carbon dioxide fixing enzyme ribulose-1,5-bisphosphate carboxylate/oxygenase, nitrogen is of qualitative importance as an essential constituent of chlorophyll. Lawlor et al. (1989) found that supplying nitrogen-deficient wheat plants with fertilizer nitrogen increased photosynthesis by increasing both the leaf area and the photosynthesis rate per unit of leaf area. With some nutrients, as found for phosphorus in white clover by Hart and Greer (1988), the greater leaf area that results when the nutrient is supplied in increasing quantities is responsible for most of the difference in photosynthesis.

The organs that provide the most unambiguous indication of the relationship between the rate of photosynthesis and the sufficiency of the current nutrient supply for the tissues in question are probably the newly formed leaves. Having just been formed, these leaves must contain nutrients that have been moved into them only recently. Older tissues gained their original supply of nutrients at earlier times, and they may have lost or gained quantities of the various nutrients since those times. Thus, the current rate of photosynthesis in older tissues, as well as the nutrients and dry matter accumulated, are affected by what has happened in the past, when the supply may have been different.

With potassium, which is readily translocated in plants, the concentration in new leaf tissue may be suboptimum for photosynthesis, but older leaves that once were active may be dying because much of their potassium has been transferred to the new tissue. With iron, which is relatively immobile in plants, older leaves may remain active in photosynthesis while a deficiency develops in the new tissue because of a low rate of current uptake and limited translocation from older tissues. Moreover, some portion of the iron may have been inactivated in older tissues, so that it no longer contributes to photosynthesis. Inclusion of the inactive fraction in an analysis of the tissue for iron will reduce the rate of photosynthesis associated with a given concentration of iron and will cause a deterioration of the correlation between the iron concentration and photosynthesis because of differences in the amount of the inactive fraction under different circumstances.

The use of photosynthesis measurements is illustrated by the work of Nable et al. (1984) on manganese nutrition of subterranean clover seedlings in nutrient solutions. Fig. 3-9 shows that when 20-day-old plants that had been grown in nutrient solutions containing manganese were transferred to manganese-free nutrient solutions, the manganese concentrations in the youngest open leaf blades decreased rapidly. These leaves developed after the plants were transferred to the manganese-free solutions. The rate at which these same leaf blades carried

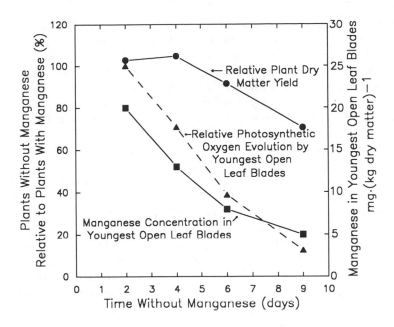

Fig. 3-9. Dry matter yield of young subterranean clover plants, photosynthetic oxygen evolution from the youngest open leaf blades, and the manganese concentration in these leaf blades with time after plants initially supplied with manganese were transferred to a nutrient solution lacking manganese. Values for dry matter yield and photosynthetic oxygen evolution are percentages of those obtained with control plants that were transferred at time 0 to a fresh nutrient solution with the same manganese concentration as before. (Nable et al., 1984)

out photosynthesis also decreased rapidly relative to the photosynthesis rate of corresponding leaves from control plants that had been transferred to fresh nutrient solutions containing the same manganese concentrations as before. These results indicate that the current supply of manganese to the young leaves was decreasing rapidly.

As indicated in Fig. 3-9, the relative dry weight of the subterranean clover plants decreased to a much smaller extent than the manganese concentration and relative photosynthesis rate in the young leaves. This distinction in behavior between the whole plant dry weight and the manganese concentration and relative photosynthesis rate of the young leaves may be attributed to a combination of (a) the limited mobility of manganese in the above-ground tissues of subterranean clover and (b) the mass of dry matter already accumulated. The limited mobility, verified by Nable and Loneragan (1984), would permit photosynthesis to continue much as before in the older tissues of the plants, while limiting the accumulation of manganese and the rate of photosynthesis in the newly developed leaves. The greater the amount of dry matter already accumulated, the lower would be the rate of decline of the relative dry matter yield upon withdrawal of the manganese.

The manganese concentration in the youngest open leaf blades reflects the rate of photosynthesis in these blades at the time of measurement, and the relationship is probably fairly constant with time as new leaves are produced. The extent to which the new leaves characterize the photosynthesis by the plant as a whole, however, will vary with time. As the plant grows, the new leaves will represent a progressively smaller proportion of the total tissue, and the relationship between the rate of photosynthesis in the new tissue and in the older tissue will change according to the mobility of the nutrient in the plant, the external supply, and other factors. Thus, although an empirical relationship observed between a measurement of nutrient concentration in photosynthetic tissue and the rate of photosynthesis of a crop as a whole or the final yield of the crop may be useful for the circumstances it represents, the relationship must be expected to be different under other circumstances. The limitations will become clearer as the following sections are covered.

Bouma and Dowling (1980) developed a method in which net photosynthesis by detached leaves was measured by determining the increase in dry weight during a week under artificial light with the petioles immersed in water. Before the test period, the petioles were immersed for 7 to 8 hours in either water or a solution containing the nutrient to be tested. Fig. 3-10 shows the results they

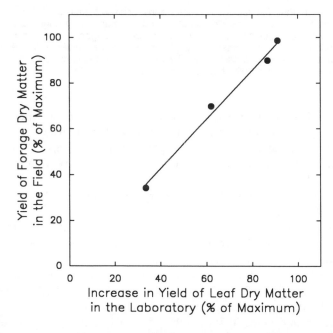

Fig. 3-10.   Yield of subterranean clover forage with different applications of fertilizer phosphorus as a percentage of the maximum yield in a field experiment in Australian Capital Territory versus the increase in weight of detached leaves under artificial light in the laboratory as a percentage of the increase in weight of comparable detached leaves that had been infiltrated with a solution containing orthophosphate. The phosphorus supplied to the treated leaves in the laboratory was intended to elicit the maximum response from deficient leaves during the test period. (Bouma and Dowling, 1980)

obtained with leaves from plants on field plots in an experiment in which different quantities of superphosphate had been applied and in which the detached leaves were immersed temporarily in water or a solution containing orthophosphate. The relative yield of forage dry matter in the field was directly proportional to the relative increase in yield of dry matter of detached leaves in the laboratory.

The Bouma and Dowling method provides more information when used with chemical analyses on the plants, but analyses are not essential. If used on plant samples taken during the growing season to estimate the relative yield at final harvest, the method would have the same disadvantage as do analyses for nutrient concentration to be discussed in subsequent pages, namely, that it cannot take into account the changes in nutrient supply or other conditions between the time of testing and the final harvest that may affect the results.

3-4. Concentration of Nutrients and Related Substances

3-4.1. Partial Analysis

Atterberg's principles, considered in Section 3-1.2, are based upon the concentrations of nutrients in the entire above-ground parts of oat plants. More recently, a great deal of work has been done on the composition of different plant parts and on different chemical fractions of the various nutrients in plants. Much of this research has been done with the idea of obtaining either a more sensitive measure of nutrient status, an easier method of assessing nutrient status, or both. Easier methods are especially important when one is dealing with large crop plants or trees. Obtaining an adequate sample of the total above-ground tissue of such plants for chemical analysis is difficult and expensive. See a paper by Jones and Steyn (1973) for a review of plant sampling, handling, and analysis procedures.

Two terms, *plant analysis* and *tissue testing*, are in common use. Both refer to chemical analyses, usually with the objective of evaluating the degree of deficiency or excess of one or more nutrients in the plants investigated. Sometimes the terms are used interchangeably. More commonly, *plant analysis* is used in connection with quantitative analyses made in the laboratory for the total concentration of a nutrient or specific forms of a nutrient, and *tissue testing* is used for roughly quantitative colorimetric tests made in the field for an inorganic fraction of a nutrient that reacts with the reagent employed.

3-4.1.1. Composition of Different Plant Parts

In general, the concentrations of a given nutrient in different plant parts are related to each other and to the concentration in the entire plant. An example of this behavior is found in Fig. 3-11, which shows the nitrogen concentrations in the leaves, stems, and roots of sweet-orange seedlings grown in sand cultures supplied with different quantities of nitrogen. Nonetheless, great differences in concentration may exist within individual classes of plant parts, and these differences are not consistent for different nutrients. Fig. 3-12 illustrates this fact with data on the composition of leaves from different positions on sugarcane

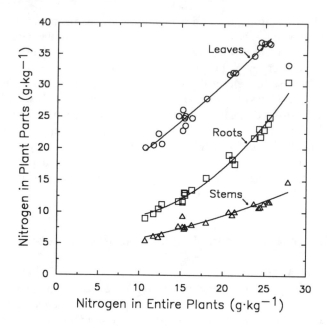

Fig. 3-11. Nitrogen concentrations in the dry matter of leaves, stems, and roots of sweet-orange plants versus the nitrogen concentrations in the total plants. The plants were grown in sand cultures through which nutrient solutions with different concentrations of nitrate were circulated. (Chapman and Liebig, 1940)

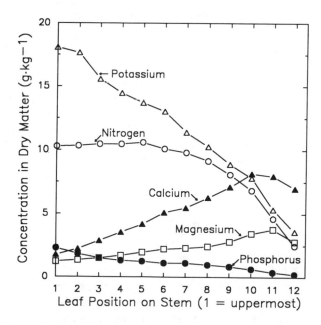

Fig. 3-12. Concentrations of certain nutrient elements in leaves of field-grown sugarcane plants in Hawaii. The leaves were numbered successively from the tip of the stem downward. (Data by Tanimoto quoted by Baver, 1960)

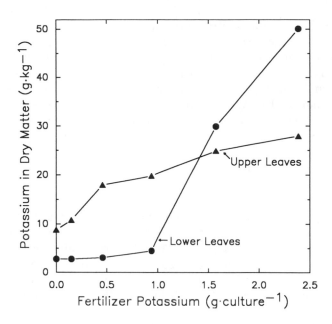

Fig. 3-13. Potassium concentrations in upper and lower leaves of tobacco with additions of different quantities of fertilizer potassium to cultures of potassium-deficient soil. (Drake and Scarseth, 1939)

plants. The concentrations of nitrogen, phosphorus, and potassium decreased in moving down the stem from the youngest leaves to the older leaves, but the calcium and magnesium concentrations increased.

In some instances, the relative sensitivity of a particular plant part to variations in nutrient supply changes with the supply of the nutrient. To judge from the slopes of the different lines in Fig. 3-11, leaves or roots would provide a more sensitive indication of the nitrogen concentration in the plant than would the stems. Taylor (1971) also noted that the nitrogen concentration in peach tree roots during the dormant season was a more sensitive indicator of previous nitrogen fertilization than was the nitrogen concentration in the 2 to 3 year old shoots, the current shoots (minus buds), the leaf and flower buds, or the leaves sampled during the following season when the nitrogen fertilization treatments were repeated.

Fig. 3-13 shows a marked difference in response of the potassium in lower and upper leaves of tobacco to differences in supply of fertilizer potassium. The potassium concentration increased more rapidly in the upper leaves than in the lower leaves with the smaller additions of potassium, and the reverse was true with the larger additions. The behavior of potassium in the leaves at different positions on the plant is related to the translocation of potassium from older leaves to younger leaves under conditions of deficiency, and this is indicated by the initial appearance of potassium deficiency symptoms on the older leaves. The older leaves thus are a more sensitive indicator of potassium deficiency than the younger leaves. Data similar to those in Fig. 3-13 were obtained by Bowling and Brown (1947) in a Maryland field experiment on tobacco. When an excess of potassium or of salts is absorbed by plants, it tends to accumulate in the lower leaves, and these leaves may be shed, ridding the plants of some of the excess.

Potassium is a highly mobile element in plants. Relatively immobile elements show substantially the reverse behavior. Thus, Loneragan and Snowball (1969) found that when they grew subterranean clover plants on a nutrient solution high in calcium and then transferred the plants to a solution low in calcium, the old leaves retained their calcium, and the new leaves developed calcium deficiency symptoms. Similarly, when subterranean clover plants were transferred from a nutrient solution containing manganese (a nutrient that is relatively immobile in plants) to one without manganese, the manganese that had accumulated in the cotyledonary and older leaf blades during the period with manganese tended to remain during the subsequent period without manganese. The manganese concentration in the youngest leaves, however, dropped rapidly (Nable and Loneragan, 1984). As will be noted later, Nable et al. (1984) followed up on this and related observations by proposing that the concentration of manganese in the youngest open leaves be used to provide an index of the current supply of manganese to subterranean clover plants.

Usually the composition of the vegetative portions of plants varies more strongly with the supply of a nutrient than does the composition of the fruit. An example is found in Fig. 3-14 for oat straw and grain. The contrast between grain and straw is most pronounced for potassium. The potassium concentration in the straw varied about fourfold from the lowest to the highest addition of potassium, but the potassium concentration in the grain was essentially inde-

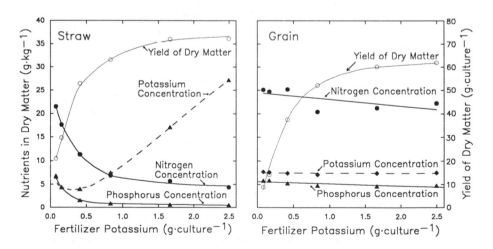

Fig. 3-14. Relative sensitivities of the nitrogen, phosphorus, and potassium concentrations in the straw (left) and grain (right) of oat plants to change, as illustrated by the results of an experiment in which oat plants were grown in sand cultures containing 1.5% peat and were treated with different quantities of potassium as potassium sulfate. (Alten et al., 1940)

pendent of the potassium supply. On the other hand, for calcium and boron, which are relatively immobile, the fruit may provide a better index of sufficiency than the leaves (Bould, 1966). For example, Askew and Chittenden (1936) found in New Zealand that with the Rome Beauty apple, trees receiving no borax or 0.4 kilogram of borax per tree both had 26 milligrams of boron per kilogram of leaf dry matter. But the concentration of boron per kilogram of the fruit dry matter was 7 milligrams in the control and 31 milligrams in trees receiving the borax treatment.

In research on the critical concentration of boron for toxicity to barley, Nable et al. (1990) reported a range from 15 to 125 milligrams per kilogram of dry matter from data in the literature. In their own experiments to investigate possible causes of the great range, they found that the concentration of boron increased greatly from the basal to the terminal portions of the leaves, increased with the evaporative demand, and decreased when the plants were subjected to artificial rainfall. They referred to their unpublished work indicating that for monitoring barley for possible boron toxicity, the boron concentration in the grain is more useful than the concentration in the foliage.

Information on the relative sensitivity of different plant parts for nutrient analyses is mostly scattered in the literature. A publication by Goodall and Gregory (1947) contains what is probably the most extensive summary available.

3-4.1.2. Nutrient Fractions

Plant nutrients are absorbed mostly as inorganic ions, and they are found in plants in the same forms. Some (nitrogen for example) are transformed mostly to organic forms in plants. Others (potassium for example) remain mostly in inorganic forms. The inorganic forms function as sources for the organic forms and as balancing ions to maintain electrical neutrality and a suitable pH and osmotic concentration in the plant sap. But as would be supposed, the greater the concentration of a nutrient in soluble inorganic form in a plant, the less likely it is that the nutrient will be deficient for its organic functions.

The inorganic fraction of nutrients in plants generally furnishes a more sensitive indication of nutrient status than does the organic fraction. To judge from the observed behavior, plants have more control over their organic components than they do over their inorganic components. The soil has a greater effect on the inorganic components than on the organic components.

Some authors have used the ratio of the concentration of an inorganic fraction of a nutrient to the total concentration of the nutrient in plants as an index of nutrient sufficiency. Scaife and Burns (1986) criticized the ratio of sulfate sulfur to total sulfur as an inferior index. They argued that (1) determining the ratio requires two analyses and twice as much analytical work as determining either sulfate sulfur or total sulfur alone, and (2) the ratio is likely to be less sensitive than either measurement alone because the numerator is the major variable in the denominator.

With nutrient fractions, as with plant parts, the concentrations of individual fractions normally increase with the total. The relative sensitivity of different

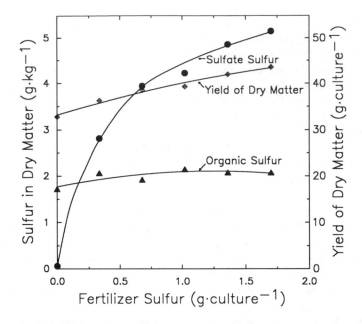

Fig. 3-15.   Yield of rape, and concentrations of organic and sulfate forms of sulfur in the plant tissue versus the quantity of sulfur added as ammonium sulfate. Complementary quantities of ammonium chloride were added to equalize the ammonium additions among cultures. (Saalbach et al., 1961)

fractions to differences in supply, however, may depend upon the circumstances. Fig. 3-15 shows that with an increase in the external supply of sulfur, the sulfate fraction varied more strongly with supply than did the organic sulfur. Similar data for alfalfa were published by Rendig (1956). Although the relative increases in inorganic sulfur, nitrogen, and phosphorus generally exceed the relative increases in organic sulfur, nitrogen, and phosphorus at high levels of supply, the reverse may occur at low levels. Data illustrating this behavior were published, for example, by Breimer (1982) for nitrogen in spinach and by Westermann and Kleinkopf (1985) for phosphorus in potato foliage.

Nitrate has received much attention, one reason being the fact that high concentrations may contribute to methemoglobinemia in human infants and ruminant livestock. A summary table prepared by Viets and Hageman (1971) from published literature showed a range of nitrate contents from 310 to 3,810 milligrams per kilogram of market spinach and a range from 930 to 8,060 milligrams per kilogram of market beet. These values are for nitrate (NO_3^-) on a fresh weight basis, which is the way analyses are usually expressed for human health purposes. On a dry-weight, nitrate-nitrogen basis, the ranges found by the author they quoted were 680 to 8,900 milligrams per kilogram for market spinach and 610 to 7,670 milligrams per kilogram for market beet. Lorenz (1978) reported the nitrate contents of various parts of different vegetable species (some with up to three cultivars) that had been grown together in the field in California with deficient, adequate, and excessive supplies of fertilizer nitrogen.

From the standpoint of plant metabolism, nitrate may be regarded as a storage form of nitrogen. For use by the plant, it is first reduced to nitrite and then to ammonium, after which it is promptly metabolized to glutamine. From glutamine, it is transferred to other organic compounds (Joy, 1988). Despite the seeming nonessential nature of the nitrate form, some plants require a relatively high concentration to avoid nitrogen deficiency symptoms and decreased growth. For example, Swiader et al. (1988) found that pumpkin developed deficiency symptoms if the nitrate-nitrogen concentration fell below 1,500 milligrams per kilogram in the petiole dry matter.

Critical values for human infants have not been established. Some limits have been proposed in The Netherlands and the United States, however, and a legal limit on the nitrate content of spinach is in effect in Switzerland (see Section 2-9).

Forages grown on high-nitrate soils, especially under conditions of drought, may contain enough nitrate to be toxic to ruminant livestock. According to Prins (1983), maximum grass yields generally occur when the content of nitrate-nitrogen is 680 to 1,360 milligrams per kilogram of dry matter. For cattle, the concentration of clinical significance depends upon the method of feeding, and may range up to 3,400 milligrams of nitrate-nitrogen or somewhat higher per kilogram of dry matter for grazed pasture.

In some plants, for example apple trees, nitrate normally is reduced in the roots, and its occurrence in the above-ground portions is a rarity. Ammonium normally occurs in plants only in very small quantities. The reason is that ammonium combines rapidly with organic metabolites to form organic nitrogen compounds.

For plants that do accumulate nitrate in the above-ground parts, chemical tests for nitrate often are made as an index of the nitrogen supply. Jakob et al. (1986) found a good correlation between the nitrate content of samples of the base of wheat stems as found by a colorimetric test (diphenylamine in concentrated sulfuric acid) and the values found by a laboratory test with a nitrate-sensitive electrode. The quick colorimetric test, which could be done in the field, provided a basis for predicting whether a second application of fertilizer nitrogen should be made. Schulz and Marschner (1986) made similar findings. They also compared the nitrate test with a quick test for soluble amino nitrogen (Schulz and Marschner, 1987). Scaife and Stevens (1983) reported work on cabbage in which the quick field test was done with a commercially prepared test paper impregnated with the appropriate chemicals. Binford et al. (1990) used the nitrate concentration in the base of cornstalks as an indication of excessive use of fertilizer nitrogen (see Section 3-4.2.3).

Fig. 3-16 shows the relation between the relative yield of spinach and the nitrate concentration in the plant tissue at harvest in a group of seven nitrogen fertilization experiments. These results indicate that a substantial concentration

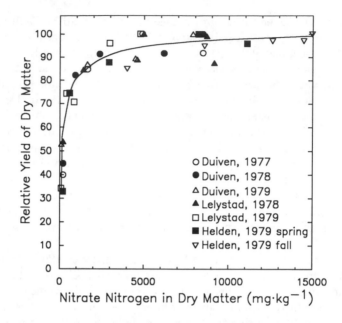

Fig. 3-16. Relative yields of spinach in seven field experiments in The Netherlands versus the concentrations of nitrate nitrogen in the plants. The experiments involved additions of different quantities of fertilizer nitrogen. (Breimer, 1982)

of nitrate in the plants was needed to produce the maximum yield. Nitrate must be reduced by the nitrate and nitrite reductase enzymes before it can be used to form organic nitrogen compounds, from which it would appear that spinach could use nitrogen more efficiently if the rate of nitrate reduction to ammonium were enhanced.

The results in Fig. 3-16 show a good relationship between the nitrate content of the spinach plants and the relative yield in the various experiments. The relationship is not always so close. For example, as a result of experiments in Alabama, Touchton et al. (1981) concluded that "Cotton growers of Alabama are not likely to benefit from a cotton-petiole-nitrate monitoring service because of the erratic behavior of nitrate levels during the period that is essential for identifying a deficiency of N."

The nitrate content of plants is not stable, but may vary with such factors as age of the plants, water supply, light intensity, and time of day. See reviews by Viets and Hageman (1971) and Maynard et al. (1976). Fig. 3-17 illustrates an unusually pronounced diurnal variation in nitrate content of red table beet plants. Such rapid and marked variation would seem to require diurnal movement of nitrate in and out of the plants, as indicated by Morgan's (1970) work. Variation of the nitrate content of plants with sampling time during the day has been noted for various crops, but the magnitudes of the changes with time have been far smaller than those in Fig. 3-17. Scarseth (1943) noted that with corn, the base of the stalk might contain nitrate in the early morning, but none later in the day.

The negative correlation of light intensity and nitrate concentration in plants mentioned by a number of investigators appears to be a consequence of an osmotic role of nitrate (Veen and Kleinendorst, 1986). When light intensity is high, the osmotic role is fulfilled by accumulated sugars, and nitrate then is not needed. At low light intensity or in darkness, the sugars are depleted, and the nitrate concentration then increases to maintain the osmotic concentration. Chloride, another readily absorbed anion, may substitute for nitrate.

Goodall and Gregory (1947) reviewed earlier literature on diurnal variations of concentrations of nutrients in general. Some of the most definitive work was done by Phillis and Mason (1942) on cotton leaves. Indications are that the effects are great enough to make it worth standardizing the time of sampling where feasible.

3-4.1.3. Organic Substances

Each essential mineral nutrient plays one or more indispensable roles in plant metabolism. Overall metabolism involves many distinct processes, all of which, however, are more or less interrelated. As a consequence, one may find that when a mineral element is deficient, certain of the organic substances undergoing metabolic transformation are present in unusually high or unusually low concentration because the transformations directly related to the deficient element are proceeding relatively slowly in comparison with the others.

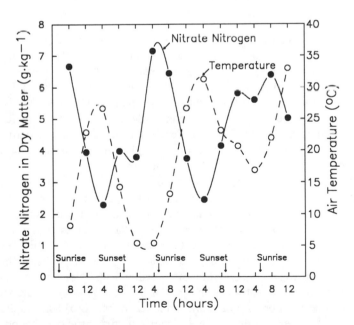

Fig. 3-17. Nitrate-nitrogen concentration in red table beet plants and air temperature at 4-hour intervals over a 52-hour period in a field experiment in New York. The plant samples analyzed included the leaves, petioles, and fleshy roots. (Minotti and Stankey, 1973)

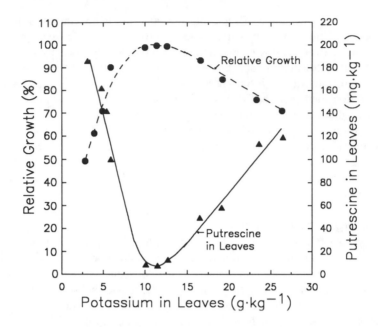

Fig. 3-18. Relative growth of the leaves of apple seedlings and putrescine concentration versus the potassium concentration in the leaves. Values for putrescine are on the fresh weight basis, and those for potassium are on the dry weight basis. (Hoffman and Samish, 1971)

Extensive investigations of soluble organic substances in relation to the mineral nutrition of plants have been made in connection with potassium. These followed the discovery by Richards and Coleman (1952) that potassium deficiency in barley caused a remarkable accumulation of putrescine, $NH_2(CH_2)_4NH_2$, and that infiltrating putrescine into cut leaves produced visual symptoms resembling those of potassium deficiency. Smith (1984) estimated that in barley plants suffering from extreme potassium deficiency, the concentration of putrescine in the dry matter of the dead tissue could be as high as 12 grams per kilogram and could account for at least 20% of the total nitrogen.

The biochemical connection between potassium deficiency and the accumulation of putrescine is not yet clear. Nevertheless, the accumulation of relatively high concentrations of putrescine under conditions of potassium deficiency in plants has been verified repeatedly, and use of the concentration of putrescine as an index of potassium deficiency has been suggested. There is limited evidence that the concentration of putrescine may be a more sensitive indicator of potassium deficiency than the concentration of potassium itself. Note in this connection Fig. 3-18, in which the plateau of the plot of relative growth of apple seedlings against the concentration of potassium in the leaves was broader than the valley in the plot of putrescine against potassium. In this instance, the putrescine concentration in the tissue increased when potassium was present in excess as well as when it was deficient. In most research, the additions of potassium have not reached the level at which they caused a depression in growth, and so it is not known what would have happened to the putrescine concentration in the presence of excess potassium.

Various factors other than the potassium supply affect the putrescine concentration in plants. In tobacco, Takahashi and Yoshida (1960) found that putrescine, which normally was not present in appreciable amount, occurred in plants deficient in potassium, phosphorus, calcium, magnesium, iron, manganese, sulfur, or boron; it accumulated in relatively large concentrations in plants deficient in potassium or phosphorus. Putrescine accumulates to different extents in different plants, but potassium deficiency seems invariably to result in its buildup.

As an indicator of water deficiency, proline accumulation in plants rivals putrescine for its specificity. In perennial ryegrass subjected to deficiency of water at different levels of nitrogen fertilization in Denmark, Jensen (1982) found that of 17 amino acids, proline was the only one for which the amino acid nitrogen as a percentage of the total nitrogen increased significantly as a result of water deficiency. The evidence for proline indicates that the accumulation of this amino acid is mostly a consequence of stimulated synthesis under water stress, inhibited oxidation, and impaired protein synthesis. Some proline is also released from pre-existing protein by proteolysis (Stewart and Hanson, 1980). Rutherford (1989) looked upon proline as an "osmoregulant."

Research has been done also on the response of free amino acids in plants to variations in nutrient supply. For example, Coïc et al. (1962) found that in the nonprotein-nitrogen fraction in barley leaves, deficiency of sulfur produced a relatively great increase in glycine and asparagine, and a considerable increase in serine and glutamine. In alfalfa, Rendig and McComb (1961) found similarly that asparagine increased where sulfur was deficient. Stewart (1962) found that arginine accumulated in citrus leaves where zinc was deficient, and tryptophan accumulated where magnesium was deficient.

Because different nutrients play different roles in metabolism, a set of analyses of a number of the soluble organic substances in plants in significant quantities probably would yield data that could be interpreted, with adequate calibration, in terms of the degree of sufficiency of the individual mineral nutrients concerned. Although the interpretation process probably could be computerized, this approach to diagnosis has not been developed for practical use, perhaps because of the amount of work involved in making the chemical analyses needed, and the possibility that the indirect values might be no better than the direct analyses for the nutrients.

The proportions of amino acids in the protein fraction respond to some degree to the supply of nitrogen. For example, the protein known as zein tends to accumulate in the grain of corn at high levels of nitrogen supply. This protein has lower nutritional value for nonruminants than the nonzein fraction because of deficiencies of essential amino acids. Nevertheless, the proportions of amino acids in the protein fraction are more nearly constant than the proportions of nitrogenous compounds in the nonprotein organic fraction. Variations in the amino acid composition of plant proteins thus do not appear to hold much potential for diagnosis of the supply of nitrogen or other mineral nutrients.

Additionally, work has been done on the activity of specific enzymes as indicators of mineral nutrient deficiencies. The basic concept is that specific enzymes catalyze the various metabolic processes in plants, that mineral nutrients are involved in some of the enzyme reactions, and that when a specific nutrient is deficient, the activity of one or more enzymes for which this nutrient is essential may be depressed, whereas the activity of some others may increase.

Essential metals known to participate in enzymatic reactions include iron, copper, zinc, molybdenum, manganese, nickel, magnesium, calcium, and potassium. Brown and Hendricks (1952) and Nason et al. (1952) introduced the concept of enzyme activity in relation to mineral deficiency, and presented some experimental work. More recently, a number of papers have been published, mostly on the possible use of the activity of certain enzymes as an index of adequacy of supply of certain nutrients.

Much of the enzyme assay work must be done in the laboratory, but efforts have been made to adapt certain tests for field advisory use. Bar-Akiva et al. (1978) developed a test for peroxidase activity that was more sensitive to iron deficiency than the determination of iron in plant tissue and could be carried out in the field. Delhaize et al. (1982) developed a test for copper deficiency based upon the activity of ascorbic acid oxidase in the youngest folded leaves of subterranean clover. In a field experiment in Australia in which different quantities of copper sulfate were added to a copper deficient soil, they found that the ascorbic acid oxidase activity as estimated by a field titration method was well related to the copper concentrations in the youngest open leaves. Loneragan et al. (1982) found that the ascorbic acid oxidase activity test was more sensitive to copper deficiency than was the copper concentration of the same leaves. The enzymic tests for copper and iron would have the advantage of responding to the active fraction of these nutrients, both of which tend to be inactivated to some degree in plants.

In another investigation, Kessler (1961) measured the activity of ribonuclease enzyme in the leaves of two varieties of healthy and zinc deficient apple trees in Israel. Zinc ions inhibit ribonuclease activity. His results indicated that the zinc deficient trees were separated from nondeficient trees a little better by the ribonuclease measurements than by the zinc concentrations. The samples from the zinc deficient trees included leaves showing zinc deficiency symptoms and leaves without symptoms. Samples from healthy trees included leaves from orchards without symptoms and leaves from trees without symptoms in orchards in which some trees did show symptoms.

As an application of the enzymic approach, Bar-Akiva et al. (1967) used a test for peroxidase activity on leaves from grapefruit trees in two orchards in Israel as an aid to diagnosing the deficiencies. In one orchard, the deficiency symptom was a chlorosis characteristic of iron deficiency. This deficiency was confirmed by the relatively low peroxidase activity in the chlorotic leaves, the higher peroxidase activity in the green leaves, and the disappearance of the symptom upon addition of an iron chelate compound to the soil. In the other orchard, the deficiency symptom suggested a combination of iron deficiency and zinc deficiency. In this orchard, the chlorotic leaves yielded a higher peroxidase value than the green leaves, which indicated that iron deficiency was not a cause of the symptom. Spraying the affected trees with a solution of zinc sulfate, manganese sulfate, and calcium hydroxide resulted in recovery, verifying that iron deficiency was not involved and suggesting that zinc, manganese, or both were deficient.

Considerable research has been done on nitrate reductase. A significant observation made in the course of this work was the finding that yields of grain and yields of nitrogen in the grain of a group of corn hybrids were correlated positively with nitrate reductase activities measured in the leaves (Deckard et al. 1973). This correlation suggests that differences among hybrids in efficiency of reducing nitrate were a factor in the productivity of the hybrids.

3-4.2. Critical Concentrations

The previous sections have emphasized qualitative concepts about nutrient supplies and their relation to plant behavior. This section on critical concentrations moves in the direction of quantitative evaluations. How does one determine from plant analysis when plants are adequately supplied with a nutrient?

3-4.2.1. Concepts and Nomenclature

When concentrations of nutrients in plants are to be used for diagnostic purposes, the usual procedure for obtaining the calibration data is to conduct experiments in which increasing quantities of the nutrient in question are supplied. The absolute or relative yields then are plotted against the concentrations of the nutrient in the variously treated plants to obtain curves such as those in Fig. 3-19.

Fig. 3-19A (a plot of some of the data from Fig. 3-1 in a different form) shows at the left a range in which the crop yield increased rapidly, but the phosphorus concentration remained very low (the *minimum concentration* range). The plot for old leaf blades in Fig. 3-19B shows a similar behavior. Next is a zone in which the yield increased less rapidly and the nutrient concentration increased more rapidly (the zone of *poverty adjustment*). This zone terminates with the *critical concentration* of the nutrient. Beyond the critical concentration is a zone in which the concentration of the nutrient continues to increase, but the yield remains about the same (the zone of *luxury consumption*). The terms italicized here were proposed by Macy (1936).

Beyond the plateau region, the concentration of the nutrient increases still further, but the crop yield decreases as a consequence of one or more toxic effects. This zone is evident only in Fig. 3-19A. The curve for young leaf blades in Fig. 3-19B illustrates one additional zone that sometimes occurs — a decrease in concentration of the nutrient in the plant with an increase in supply when the

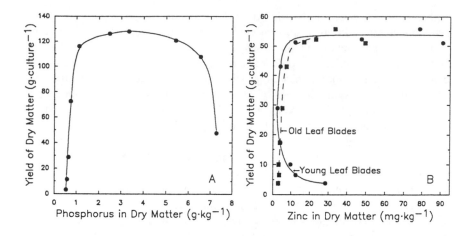

Fig. 3-19. A. Yields of above-ground parts of oat plants grown in sand cultures receiving different quantities of fertilizer phosphorus as monobasic sodium phosphate versus the phosphorus concentrations in the plants. (Mitscherlich, 1935) B. Yields of the tops of sugarbeet plants grown in nutrient solutions containing different concentrations of zinc versus the zinc concentrations in young and old leaf blades. (Rosell and Ulrich, 1964)

plant is initially very deficient. This zone was described in Section 3-1.2.1 as the Piper-Steenbjerg effect.

Added more recently has been the term *sufficiency range*, which Jones et al. (1990) defined as the range of concentrations of a given nutrient between the critical value and an excess or toxic concentration. As a second definition, they used the range in nutrient concentration in which no yield reduction occurs and no nutrient stress symptoms appear.

3-4.2.2. Definition

The critical concentration is the most commonly used concept in relating plant composition to nutrient sufficiency in plants, but Macy (1936) did not define it precisely in terms of the yield versus nutrient concentration function. The closest he came to a definition seems to be in the statement that "When sufficiency is reached at the critical percentage, and luxury consumption sets in, the Liebig law again holds, there being no appreciable further response."

In Macy's plots of experimental data from the literature, the critical concentration corresponded to a sharp decrease in slope of the yield versus nutrient concentration function. Sometimes the break appeared to be at the maximum yield, and sometimes it was a little below the maximum. The sharpness of the break was emphasized by his method of plotting. In subsequent use, the meaning of the critical concentration term generally has been kept within the bounds of Macy's original concept, but specific definitions have differed.

Macy (1936) and most others have looked at critical concentrations from the standpoint of supplying nutrients to eliminate deficiencies. Ohki (1984) and El-Gharably and Bussler (1985) took a broader view and defined a critical deficiency concentration that corresponded to 10% below the maximum yield on the low side and a critical toxicity concentration that corresponded to 10% below the maximum yield on the high side.

Selecting the concentration of a nutrient corresponding to the maximum crop yield or response as the critical concentration is logical in an instance such as that in Fig. 3-20, which shows a plot of the rate of photosynthesis of the youngest open leaf blades of subterranean clover plants against the manganese concentration in the leaf tissue. The data are well represented by a linear response followed by a level plateau, in which the intersection of the ascending and plateau segments of the model occurs at 22 milligrams of manganese per kilogram of dry matter in the youngest open leaf blades.

In other research, McGrath and Robson (1984) used a linear response and plateau function to represent a plot of the percentage of the maximum dry weight of Monterey pine seedlings versus the zinc concentration in the apical primary leaves. In their work, the plateau beyond the intersection of the two linear segments of the response function had a slight upward slope, and so it was logical to represent the point of intersection as the critical concentration even though a small increase in yield occurred with greater applications of zinc.

Burns (1986) defined the critical potassium concentration in lettuce plants as the concentration of potassium in the plant sap at the time the growth rate of

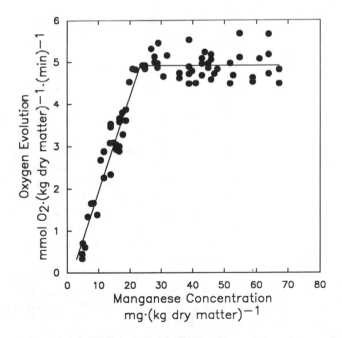

Fig. 3-20. Rate of photosynthetic oxygen evolution from the youngest open leaf blades of subterranean clover plants versus the manganese concentration in the dry matter of these leaves. (Nable et al., 1984)

the plant just starts to decrease due to potassium deficiency. He made the determination from the ratio of the fresh weight of the plants whose potassium supply was ample initially, but which had been cut off, to the fresh weight of comparable plants whose potassium supply was continuously ample. The ratio was plotted on a logarithmic scale against the reciprocal of the potassium concentration in the sap. When measurements were made at different times after cutoff of the potassium supply, the resulting straight line could be extrapolated to a ratio of 1.0, from which a value of the critical potassium concentration could be obtained.

Escano et al. (1981a) tried a different approach to establishing critical concentrations. In a series of factorial experiments on corn fertilization, they analyzed samples of the corn leaf below the uppermost ear at the time of 50% silking for each treatment. They also analyzed the grain yield data to determine for the plants receiving each submaximum quantity of the nutrient whether an increase in yield was or was not likely upon application of an additional quantity of the nutrient in question. Then they selected arbitrary values of the nutrient concentrations and used them as "trial critical concentrations" to determine whether they correctly predicted no increase or a likely increase in yield with each treatment evaluated. In a plot of numbers of correct predictions against trial values, the trial value associated with the maximum number of correct predictions was taken as the best estimate of the true critical concentration for the data analyzed. The authors found that values obtained in this way were similar to those obtained in the usual way by plotting the yield against the concentration of the nutrient in the plant tissue when increasing quantities of the nutrient were supplied by fertilization.

Illustrating the use of a property other than the concentration of a nutrient, Hannam et al. (1987) employed the fluorescence behavior of chlorophyll *a* in the youngest emerged leaf blades of barley as an index of the sufficiency of manganese. When manganese is deficient, the effectiveness of chlorophyll *a* in photosynthesis is impaired. Part of the light energy that normally would be used to split water is then re-emitted as fluorescence, which can be measured. The fluorescence measurements were closely related to the manganese concentration in the dry matter of the youngest expanded leaf blades of barley plants. Using the method of Smith and Dolby (1977) for estimating critical concentrations of manganese in milligrams per kilogram of dry matter of the selected leaf tissue, Hannam et al. derived a value of about 12 from data on yields of dry matter versus manganese concentrations and a value of about 14 from data on fluorescence versus manganese concentrations.

A different concept of critical concentrations was pioneered by Kenworthy (1961, 1967, 1973) and developed further by Walworth, et al. (1986). The basic idea is that if a crop produces a relatively high yield, the mineral composition must be such as to produce this yield.

Walworth, Letzsch et al. (1986) showed with extensive data on corn produced in various countries and in different years that scatter diagrams for grain yield against the concentrations of nitrogen, phosphorus, and potassium in the dry

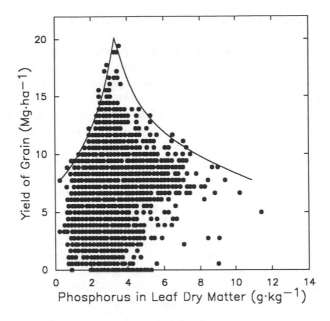

Fig. 3-21. Scatter diagram of corn grain yields versus the phosphorus concentrations in the leaf below the ear. The individual points were derived from measurements made in Africa, Canada, France, South America, and the United States. (Walworth, Letzsch, et al., 1986)

matter of the leaf below the ear resulted in peak yields at specific concentrations of the respective nutrients. Their plot for phosphorus is shown in Fig. 3-21.

They defined the nutrient concentrations corresponding to the peak yields as critical concentrations. The authors noted that if one has access to such a set of data, it becomes possible to develop a set of critical values that "should be diagnostically precise and more universally applicable" than the values obtained by the usual procedure described in the first paragraph of this section.

In contrast to the preceding examples, in which the criterion was physiological, Malavolta and Da Cruz (1971) defined the critical concentration "as the range of a given element in the leaf, below which the yield is limited and above which the use of fertilizer is no longer economical." Here the criterion is economics, and the critical concentration would vary with the price of the crop and the cost of fertilization.

The critical concentration is also defined in terms of the nature of the measurement and the response function from which this concentration is inferred. For example, Ware et al. (1982) fitted yield and nutrient concentration data by a modified Mitscherlich equation. They defined the concentration corresponding to 90% of the maximum yield as the critical concentration. Smith and Dolby (1977) represented the critical concentration as the concentration corresponding to the lower limit of the 95% probability level for the maximum yield.

In work on nitrogen fertilization of corn in Iowa, Binford et al. (1990) used a given set of data on relative corn grain yields versus (a) nitrate nitrogen concentration and (b) total nitrogen concentration, both in the lower portion of the stalks at the approximate time of physiological maturity of the grain, to obtain six different critical concentrations — three for nitrate nitrogen and three for total nitrogen. One pair of critical concentrations was obtained when the two sets of data were fitted by the linear response and plateau function. A second pair was obtained with use of a statistical procedure based on the "Cate-Nelson split" (Cate and Nelson, 1971) to decide which yields should be included in the linear response segment of the function and which should be included in the plateau segment (the Cate-Nelson split is discussed in Chapters 1, 2, and 4). A third pair was obtained with a quadratic response and plateau function, in which the concentrations of nitrate nitrogen and total nitrogen corresponding to the economic optimum application of fertilizer nitrogen were denoted as the critical concentrations. They reported average critical nitrate-nitrogen concentrations of 0.39 and 1.8 grams per kilogram of dry matter for the linear response and plateau function and the economic optimum, respectively. For total nitrogen, the corresponding critical concentrations were 3.0 and 4.3 grams per kilogram.

In Australia, Maier (1986) summarized the results of field experiments on potato to find the critical concentration of potassium. His results showed that the statistical approach used had a considerable effect on the outcome. The estimated critical values for potassium in the dry matter of leaf petioles in grams per kilogram were 90 for the Cate-Nelson split, 105 and 125 with the Mitscherlich equation for 90 and 95% of the maximum yield, and 113 for the Smith-Dolby bent hyperbola, which is a complex function that in this instance produced results equivalent to a linear response followed by a plateau.

In Colorado, Goos et al. (1982) used the chi-square method of Keisling and Mullinix (1979) to estimate the critical protein concentration in winter wheat grain as an index of the nitrogen status of the crop. This method is strictly statistical, and is closely related to the Cate-Nelson split.

The term "critical concentration" thus has been applied to a considerable range of values. Although it might be argued that the term should not be used so indiscriminately, the various approaches may be regarded as operational definitions as long as they lead to results within or close to the range described by Macy. The more important questions are why are the measurements made, and are the measurements useful? In this connection, for example, the description of the procedure used by Escano et al. (1981a) makes it clear that they were looking for a critical concentration that would separate plants that would respond to addition of more of a nutrient from those that would not respond. This sort of critical concentration, and in fact all critical concentrations of the Macy type, would be practically useful for situations in which the cost per unit of fertilizer is low relative to the price of the crop. For situations in which the fertilizer/crop price ratio is much higher, Macy's critical concentration concept does not provide a practical guide for decision making. Under such circumstances, it would be appropriate to apply fertilizer only when the nutrient concentration in the crop is well below Macy's critical concentration range. Then the Malavolta and Da

Cruz (1971) economic criterion would be more appropriate, and the numerical values obtained might better be called critical economic concentrations.

3-4.2.3. Behavior

Most of the available information on critical concentrations (and almost all of the information in the remainder of this section) is based upon the usual procedure, in which critical concentrations are found by analyzing samples of crops supplied with increasing quantities of a given nutrient. The Kenworthy concept is discussed in Section 3-4.3, however, and Section 3-5 on nutrient ratios is based primarily on scatter diagrams analogous to Fig. 3-21.

The usual change in concentration of nutrients in plants with time causes problems for the critical concentration approach. Moreover, it has required recognition of two different concepts of critical concentration. These are related to the way critical concentrations are used. One is a critical concentration that applies to the sufficiency of the nutrient at the time of measurement. The other is a critical concentration in which the concentration of the nutrient in the plant at the time of measurement applies to the sufficiency of the nutrient in question for producing the final yield.

The numerical values of the two critical concentrations are not the same. Especially for nutrients that must be supplied in fertilizers and hence usually decrease in availability as the season progresses, the concentrations in the crop at the time of preharvest sampling for nutrient analyses may need to be considerably in excess of the needs of the crop at the moment of sampling if the critical concentration is to apply to the final yield. The data in Table 3-2 represent an extreme case. They are critical concentrations of sulfate sulfur in subterranean clover derived from yields and sulfate analyses on plants grown on cultures of a sulfur-deficient soil at six different levels of sulfate fertilization. The concentration of sulfate sulfur per kilogram of leaf tissue required for 90% of the maximum yield at 133 days was 150 times greater when measured at 33 days than when measured at 133 days.

In addition to the change in critical concentration with time, another problem that may be encountered is that plant analyses made early in the season may not

Table 3-2. Critical Concentrations of Sulfate Sulfur for 90% of the Maximum Yield of Subterranean Clover Forage as Estimated From the Sulfate Sulfur Concentrations Measured in the Young Leaf Laminae at Different Times (Spencer et al., 1977)

Age of Plants at Time of Leaf Analysis (days)	Estimated Critical Concentrations of Sulfate Sulfur in the Leaf Dry Matter (mg·kg^{-1}) at Time of Leaf Analysis for Plants Harvested at Indicated Age		
	33 Days	61 Days	133 Days
33	200	700	1500
61	—	30	30
133	—	—	10

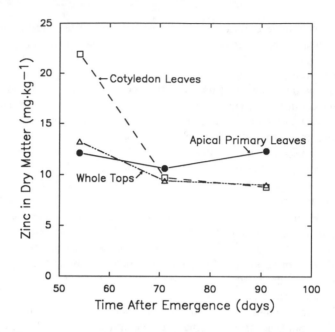

Fig. 3-22. Zinc concentrations in different parts of Monterey pine plants of different ages on a zinc-deficient soil in Australia. The soil had received zinc sulfate in various quantities, and at each sampling date the plants selected for analysis were from cultures that had received just enough zinc sulfate to produce the maximum yield. The cultures sampled at 54, 71, and 91 days had received 0.9, 1.2, and 2.1 milligrams of zinc, respectively. Note that the zinc concentrations were more nearly constant in the apical primary leaves than in the cotyledon leaves or the whole tops. (McGrath and Robson, 1984)

be sufficiently responsive to higher levels of a nutrient to be useful as indexes of final yields. Thus, in 14 experiments on nitrogen fertilization over a period of 4 years in Iowa, Binford et al. (1992) found that the range in total nitrogen concentrations in young corn plants sampled when their height was 15 to 30 centimeters was considerably less than the range in concentrations of nitrate in soil samples taken at the same time. In these experiments, the correlation between the soil nitrate and the nitrogen concentration in the plants was $r = 0.52$, and the correlations between the relative yields of corn and the nitrogen measurements were 0.57 for total nitrogen in the plants and 0.87 for soil nitrate concentration.

Researchers have dealt in various ways with the change of critical concentrations with time. For some plant species and some nutrient elements, it has been possible to circumvent much of the problem by selecting only young leaf tissue for analysis. The basis for this practice is that young leaf tissue present at different times should be physiologically similar and thus might be expected to have a relatively constant critical nutrient concentration. See Fig. 3-22 for illustrative data by McGrath and Robson (1984). These authors found that the critical concentrations of zinc in the apical primary leaves were more nearly constant with time than were the zinc concentrations in the cotyledon leaves or the total plant tops. Additionally, the zinc concentrations in the apical primary leaves provided a better estimate of the percentage of the maximum yield of plant dry matter than did the zinc concentration of the total above-ground parts of the plants. The proportions of the variance of the percentages of the maximum dry weight of the plants accounted for by regressions of these values on the zinc concentrations in the apical primary leaves and the whole plant tops were 85% and 63%, respectively.

Like many other matters relating to plant composition, the constancy of critical concentrations in young leaf tissue needs to be verified experimentally for the species and nutrient in question. Constancy is not assured. For example, Smith (1975) found that the critical concentration of phosphorus in the youngest expanded leaf of the pasture grass setaria ranged from about 5 grams per kilogram of dry matter at age 17 days to about 1.5 grams at age 45 days. Similar values were obtained for the whole plant tops. Bell et al. (1987) found that the concentration of potassium in the youngest fully expanded leaf blade of soybean plants corresponding to 95% of the maximum yield dropped from about 9 grams per kilogram of dry matter when the plants were 37 days old to about 4 grams when they were 55 days old.

In recognition of the general tendency for the concentrations of the most commonly deficient nutrients to decrease with time in the foliage of potato, Prummel and Von Barnau-Sijthoff (1984) used a statistical method of adjusting the data. By use of regression equations, they adjusted the phosphorus and potassium concentrations in the dry matter of potato tops to a standard value of 50 grams of nitrogen per kilogram in samples taken from fertilizer experiments in The Netherlands sampled from mid June to mid August. The relative yields of tubers then were plotted against the adjusted phosphorus and potassium concentrations to obtain estimates of the critical concentrations. Leece (1976) published a set of correction factors for use with analyses made on leaves of various fruit trees sampled at different times in Australia. Also in Australia, Williams and Maier (1990) calculated regressions of the nitrate concentration in the petioles of potato plants on the length of the longest tuber as a manifestation of growth stage.

Carter et al. (1971) had a different idea. They found that after irrigated sugarbeet in Idaho was fertilized with ammonium nitrate, the concentration of nitrate in the leaf petioles increased to a peak and then decreased approximately exponentially with time. They represented the decrease in concentration with time by the equation

$$N = N_0 e^{-ct},$$

where N_0 is the nitrate nitrogen concentration in the dry matter of the leaf petioles at the first sampling date following the peak when the time t in days is taken as zero, N is the nitrate nitrogen concentration in the dry matter of the leaf petioles at the second sampling date when the time in days is t, c is a constant, and e is the base of natural logarithms.

The constant c, which provides a measure of the rate of decrease of the concentration with time, can be evaluated by determining the concentration of nitrate nitrogen in the petioles at two times after the peak and substituting the values in the equation

$$c = (2.3/t)(\log_{10}N_0 - \log_{10}N).$$

This relationship fitted different sets of data relatively well, as indicated by coefficients of determination (r^2) ranging from 0.77 to 0.99 in 31 trials.

The approach suggested by Carter et al. (1971) represents a fundamental improvement over the usual practice of using the results of an analysis made at a particular time during the growing season for predicting the situation at the end. The reason is that it provides not only an estimate for a particular time, but also an estimate of the rate of change with time. Both factors influence the sufficiency of the nutrient at the end of the season. The rate of change with time varies with the circumstances. In a summary of their own work with different quantities of fertilizer nitrogen in different years, plus findings of other researchers in four states, they found a range of c values from 0.018 to 0.098.

The authors' objective was to predict whether the nitrate-nitrogen concentration in the petioles would remain above the critical level of 1,000 milligrams per kilogram of petiole dry matter until 4 to 6 weeks before harvest, as needed to optimize the yield of sugar with respect to nitrogen supply. Because the analyses were made 3 to 4 months before harvest, sufficient time remained to make an additional application of fertilizer nitrogen if needed. The fact that the crop was irrigated would make it possible for applied nitrogen fertilizer to become effective almost immediately.

Fig. 3-23 shows how the data from one experiment fitted the semilogarithmic plot. If a straight line drawn through the first two points for a given quantity of fertilizer nitrogen is extended downward to the point of intersection of the line with a concentration of 1,000 milligrams of nitrate nitrogen per kilogram, the estimated number of days the plants will have enough nitrogen to avoid growth limitation can be read off of the linear time axis.

Giles et al. (1977) conducted similar experimental work in Colorado and found that the rate of decrease in nitrate nitrogen concentration in the sugarbeet petioles could be expressed well by an exponential equation. Their findings pointed up the importance of precision in the two analyses from which the extrapolation was to be made. As may be inferred from the scattering of points around the lines in Fig. 3-23, substantial differences in the extrapolated values may result from small deviations one way or the other in nitrate analyses, and these effects are of increasing importance as the length of time between the two analyses decreases.

Field tests by colorimetric methods may be adequate for practical use of one-analysis methods (see Section 3-4.1.2), but more precise analyses are needed for the two-analysis extrapolation method. Carter et al. (1971) reported that the results obtained in separate analyses for nitrate made by a quick-test method were not as good as those with the regular laboratory quantitative method.

In Denmark, Østergaard (1989) reported that in 24 field experiments over a 3-year period the nitrate concentrations in the basal portion of the stems of spring barley plants decreased in regular fashion with growth stage approximating an exponential plot for each of several levels of nitrogen supply. For unknown reasons, however, the trends of the nitrate concentration curves in winter wheat did not follow a uniform course over the years.

Analogides (1988) found that the concept of an exponential rate of decrease in nitrate concentration in sugarbeet petioles with time applied also in Greece. His preference, however, was to categorize the nitrate concentrations in the petioles into limiting, normal, high, and excessive classes. In work on pumpkin, Swiader et al. (1988) found that the rate of decrease of nitrate concentration in the petioles with time sometimes followed an exponential pattern, forming curves that were concave upward, but sometimes formed curves that were concave downward for the greatest applications of fertilizer nitrogen. A similar tendency may be noted in Fig. 3-23.

Westermann and Kleinkopf (1985) applied the technique of Carter et al. (1971) to phosphorus in irrigated potato. In five experiments on a given soil type in Idaho, they estimated from analyses at the first two samplings the numbers of days required for the concentrations of soluble phosphorus in the petioles to drop

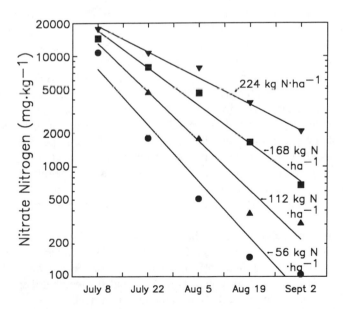

Fig. 3-23. Decrease in concentration of nitrate nitrogen with time in the dry matter of sugarbeet petioles with different quantities of fertilizer nitrogen in a field experiment in Idaho. Note the logarithmic scale of concentrations. (Carter et al., 1971)

to the critical value of 1,000 milligrams per kilogram of dry matter. The values obtained yielded a correlation of $r = 0.93$ with the actual times found from plotting the phosphorus concentration data for the entire growing season. The difference between the observed and estimated values averaged 9%. The authors found that the procedure worked well for other data they studied from published literature.

The precision of prediction offered by an evaluation of the exponential rate of decrease of nutrient concentration with time lends itself to a procedure for increasing fertilizer use efficiency. Soil testing or experience could be used as a basis for selecting a tentative but deficient quantity of the nutrient to apply at planting or early in the season, and subsequent tests could be used to indicate the time the nutrient concentration in the petioles would be expected to fall below the chosen level. The supply then could be adjusted by a second application. The quantity to apply would increase with the estimated number of days of deficiency before harvest, but the most appropriate quantity would have to be estimated from special experiments to develop the needed empirical relationships.

Two different views regarding the magnitude and significance of variations in critical concentrations may be found in the scientific literature. One view is that the variations in critical concentrations are not so great as to be of concern in their practical use. For example, Ulrich (1961) noted that sugarbeet had been found nitrogen deficient in field experiments and upon field inspection under different climatic and soil conditions in California, Utah, Washington, and Colorado when the nitrate-nitrogen concentration in the dry matter of the petioles of recently matured leaves fell below 1,000 milligrams per kilogram. The opposing view is that critical concentrations cannot be defined well enough and are affected by too many factors to be of much value.

Both views may be supported by reference to published field experimental data. Supporting the positive view, Binford et al. (1990) found in 15 experiments on nitrogen fertilization of corn in Iowa that the range in concentration of nitrate nitrogen in the dry matter of the basal portion of the stalks at physiological maturity of the grain was from zero to more than 7 grams per kilogram. The critical concentration of nitrate nitrogen in the 15 experiments as found by the linear response and plateau model covered less than 8% of the total range. Thus, the mean critical value (0.4 gram per kilogram) was a valuable statistic for determining whether the corn had received an excessive application of fertilizer nitrogen.

Supporting the negative view, the critical concentrations of nitrate nitrogen in the individual experiments ranged from 0.22 to 0.78 gram per kilogram. Their nearly fourfold range and their 46% coefficient of variation show that the overall critical concentration was not precisely defined and suggest that there may have been real differences in critical values among experiments.

In the work by Binford et al. (1990), the correlation between the maximum yield of grain in the various experiments and the critical concentration of nitrate nitrogen in the stalks was not significant. This observation indicates that if real

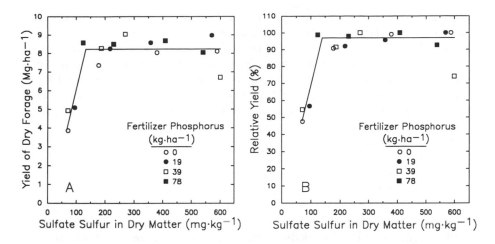

Fig. 3-24. Absolute yields (A) and relative yields (B) of subterranean clover forage versus sulfate sulfur in the plants in an experiment with different additions of sulfur as calcium sulfate and phosphorus as concentrated superphosphate in California. (Jones, 1962)

differences in critical values did exist among experiments, they were a consequence of factors other than those that were responsible for the differences in maximum yields.

Fig. 3-16 shows the results of seven field experiments on nitrogen fertilization of spinach. On the positive side, the relative yields obtained in all experiments could be fitted well by a single curve without appreciable segregation of the data for individual experiments. This observation implies that the nitrate-nitrogen concentration corresponding to a given percentage of the maximum yield could be selected as a critical concentration for all experiments. On the negative side, if the concentration of nitrate nitrogen in the crop dry matter from the highest yielding treatment in each experimental series is taken as an estimate of the critical concentration, the range in critical concentrations is from 4.8 to 15.0 grams per kilogram. The threefold range of the critical concentrations covers 69% of the total range of nitrate-nitrogen concentrations observed in the seven experiments, and the coefficient of variation of the critical concentrations is 46%.

The data in Fig. 3-24 were derived from a single field experiment involving applications of four levels each of fertilizer phosphorus and sulfur to a soil deficient in both nutrients. On the positive side, the relative yields could be fitted with a single linear response and plateau function without segregation of the values on the basis of the quantity of fertilizer phosphorus applied. On the negative side, if the highest observed yield obtained with increasing additions of fertilizer sulfur at each level of fertilizer phosphorus is taken as the maximum yield and the corresponding sulfate-sulfur concentration as the critical concentration, the range of critical concentrations covers 60% of the range of total concentrations. The coefficient of variation of the critical concentrations is 33%.

The existence of differences in numerical values of critical concentrations due to factors other than plant parts and growth stage seems to be generally accepted. But one perceives two points of view about their significance.

According to one school of thought, there is only one true critical concentration corresponding to a particular plant part at a particular growth stage, and this may be evaluated by adding increasing quantities of the nutrient of concern when all other growth factors (or at least all other nutrients) are present in adequate supply. Values observed under suboptimum conditions are not to be considered critical concentrations. This philosophy is consistent with the common practice of making a basal application of nutrients other than the one under test when the effect of applying a given nutrient is to be investigated. (But see Section 2-7.9 and Bouma et al. (1969) for some possible consequences.)

According to a second school of thought, the critical concentrations depend upon the supplies of other nutrients and perhaps on other factors as well. Prevot and Ollagnier (1961) advocated the use of factorial experiments to investigate the interactions of nutrients in determining the critical concentrations. Such experiments include a number of levels of each of the nutrients concerned, alone and in the presence of all combinations of levels of all the other nutrients being tested. An important limitation of factorial experiments is that they quickly become too large to handle if the numbers of nutrients, quantities of each nutrient, and replicates needed to establish independent response curves are included. As a consequence, adequate experiments are rare.

Böszörményi (1957) carried out a factorial experiment in which Scots pine seedlings were grown in sand cultures with five levels each of nitrogen and phosphorus. He used three replicates, which meant that there were 75 cultures in total. His results in Fig. 3-25 indicate a significant variation in critical concentration of nitrogen in the leaves, depending upon the concentration of phosphorus, and vice versa. In this experiment, the author selected the concentration of the nutrient in the highest-yielding member of each experimental series as the critical concentration, and the same was done in preparing Fig. 3-25. Perhaps four times more replicates as well as twice as many rates of application of the nutrients would have been needed to obtain the kind of smooth curves desired to obtain fairly precise critical concentrations. Nevertheless, the trends in Fig. 3-25 provide a clear enough indication that the critical concentrations of each of two nutrients were affected by the supply of at least one other nutrient.

Hylton et al. (1967) supplied Italian ryegrass with culture solutions containing eight different concentrations of sodium, and estimated that the critical concentration of potassium for 90% of the maximum yield of tops was 8 grams per kilogram in the dry matter of leaf blade 1 with high sodium in the nutrient solution (8 millimoles per liter) and 35 grams per kilogram with low sodium in the nutrient solution (1 millimole per liter). A difference in critical levels in the direction of that observed would be expected on the basis of the partial replacement of potassium by sodium. Smith (1974) and Burns (1986) similarly found that the critical concentration of potassium was reduced by addition of sodium.

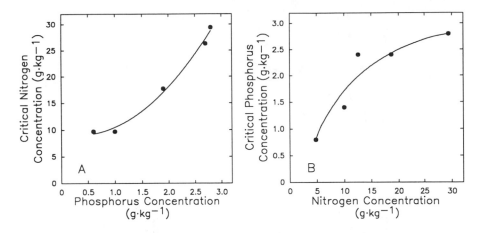

Fig. 3-25. Critical concentrations of nitrogen (A) and phosphorus (B) in Scots pine seedlings in sand cultures supplied with different quantities of fertilizer nitrogen and phosphorus. (Böszörményi, 1957)

The water supply was implicated by Ollagnier et al. (1987) as a cause of differences in the critical concentrations of potassium in the foliage of the oil palm, where the critical concentration was defined in economic terms. Their data in Fig. 3-26 are remarkable for correlating the results of experiments from four different countries. In 18 field experiments with siratro in Australia, White and Haydock (1970) found an upward trend in the critical phosphorus concentration with the rainfall received in the 28 days before sampling. In the experiments on siratro, the critical concentration was defined as the phosphorus concentration in the tip growth associated with 90% of the maximum yield.

Whether the differences in concentrations of a given nutrient at a given percentage of the maximum yield that are associated with environmental factors, including supplies of other nutrients, are or are not termed differences in critical concentrations is a matter of definition. The important point is that differences may exist and that erroneous interpretations may be made if an inappropriate value is used as a critical concentration.

Despite extensive use of critical concentrations, the theoretical background is still unsettled. The question of an operational definition has been raised by new methods of estimating critical concentrations and the fact that the values obtained by different methods may not be the same.

Referring again to the scatter diagram in Fig. 3-21, one may observe that peak corn yields were obtained at a particular phosphorus concentration in the leaf dry matter. Walworth et al. (1986) obtained plots with a similar appearance for nitrogen and potassium. The peak yields were associated with average concentrations of 32.9 grams of nitrogen, 3.3 grams of phosphorus, and 21.9 grams of potassium per kilogram of leaf dry matter. Sumner (1990) suggested that these values might be regarded as "the ultimate critical values, that is, critical values good for all conditions."

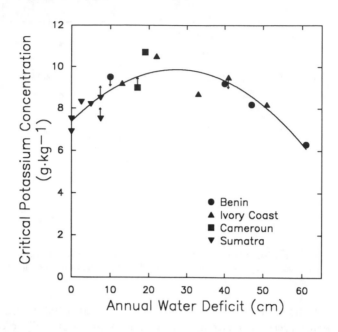

Fig. 3-26. Critical concentrations of potassium versus annual water deficit in oil palm in 17 field experiments in four countries. The arrows associated with certain points indicate that the critical concentration was equal to or greater than (downward pointing arrows), or equal to or less than (upward pointing arrows), the potassium concentrations recorded. The critical concentration was defined as the potassium concentration in leaf 17 associated with a rate of increase in value of bunches of nuts equal to five times the cost of 1 kilogram of potassium chloride. (Ollagnier et al., 1987)

Designating the nutrient concentrations associated with peak yields in the world data as the ultimate critical values or concentrations seems appropriate in that the observed peak yields are probably close to the ultimate maximum yields of the crop as it existed when the data were obtained. The nutrient concentrations associated with these peak yields are so precisely defined that they must have some fundamental significance. For example, Walworth, Letzsch, et al. (1986) found that the optimum nitrogen concentrations in corn leaves in the world data were 34.1 and 31.7 grams per kilogram as found by two independent methods.

Whether the nutrient concentrations in the peak yields may be regarded as the critical concentrations for normal conditions in which yields are much lower is as yet uncertain. The average value of 32.9 grams of nitrogen per kilogram found by Walworth, Letzsch, et al. (1986) is above the range of 21.9 to 32.0 grams that Walworth and Sumner (1988) found reported as critical values in the literature. Escano et al. (1981a) also published a table summarizing critical values from the literature. Most of the low values might be explained on the basis that the authors defined their critical values in terms of yields below the maximum. But some of the critical values appeared to be for maximum yields.

3-4.2.4. Conventional Systems

Numerous applications of the critical concentration concept have been made. Perhaps the most common is classifying nutrient concentrations into ranges. The critical concentration is located centrally in the optimum range. In some instances, crop yield may be the only consideration. In others, quality is important. When the amount of information derived from experimentation is limited, there may be only two ranges: adequate and deficient. Or a third "excess" range may be added if the crop is grown under circumstances in which excesses of nutrients such as boron or manganese may be common.

When more extensive calibrating information is available, additional ranges may be added. For example, in California, where much experimentation had been done on citrus fertilization, Embleton et al. (1978) classified data for citrus leaves into five ranges (deficient, low, optimum, high, and excess). They adjusted the widths of the ranges for the various nutrients, reflecting experience with responses to applications of individual nutrients. The narrowest optimum range was for nitrogen — from 24 to 26 grams per kilogram of the dry matter (a relative range from 1 to 1.08). The widest optimum range was for molybdenum — from 0.1 to 3.0 milligrams per kilogram of leaf dry matter (a relative range from 1 to 30). See Westerman (1990) for more information.

3-4.2.5. Lemaire System

Lemaire et al. (1989) employed a different concept, with results illustrated in Fig. 3-27. They were able to coordinate the data from four independent experiments on pasture grasses to a good approximation by expressing the relative yield of grass as a function of the relative concentration of nitrogen in the grass:

$$w/w_p = -0.82 + 3.07(N/N_c) - 2(N/N_c)^2,$$

where N and w were the nitrogen concentration in the dry matter and the yield of dry matter in a particular cutting in a particular experiment with a particular addition of fertilizer nitrogen, and where N_c and w_p were the corresponding concentration of nitrogen and yield of dry matter when nitrogen was nonlimiting. They called N_c the critical concentration of nitrogen. The entire set of data yielded an R^2 value of 0.95.

Although the data were fitted best by a curve, the relationship was not far from linear ($r^2 = 0.86$ for the linear relationship in Fig. 3-27), and the values of w/w_p were approximately equal to the values of N/N_c. These findings suggest the possible existence of a general approximate equality of w/w_p and N/N_c that could be useful in evaluating the adequacy of the nitrogen supply for crops. The authors noted that the relationship might be improved by modifying the nitrogen function to $(N - N_1)/N_c - N_1)$, where N_1 is the concentration of nitrogen in the crop at which no growth occurs (the minimum concentration); however, their data did not extend to such an extreme deficiency.

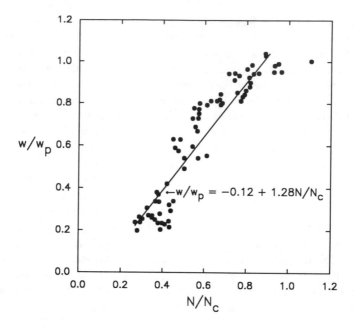

Fig. 3-27. Relative dry matter yield versus relative nitrogen concentration in pasture grasses in France. The data are from four experiments, two with each of two species of grass. One experiment was conducted in each of 4 years. Each experiment involved several cuttings of grass receiving 0, 50, 100, and 150 kilograms of fertilizer nitrogen per hectare. See the text for definitions of terms. (Lemaire et al., 1989)

Greenwood et al. (1991) incorporated the modified nitrogen function in a model they developed as an extension of the concept of Lemaire et al. (1989). In testing the model, they used data from French bean grown in England, potato in Scotland and The Netherlands, tall fescue in France, and winter wheat in Belgium and Sweden. They found, among other things, that the relationship between measured and calculated nitrogen concentrations in the assembly of data followed a 1:1 relationship to a good approximation, with 86% of the variance accounted for; however, they did not show any test of the correspondence of relative yields and relative nitrogen concentrations.

The experiments used for locating the critical concentrations provide information that is helpful as a guide to practice, but other information is needed. To be of maximum value in fertilizer practice, each range for each nutrient must be defined in terms of the kind and amount of fertilizer and the time and method of application needed to produce the desired response. To obtain this information requires a great deal of experimentation.

3-4.2.6. Møller Nielsen System

Møller Nielsen and colleagues in Denmark developed a system in which nutrient concentration data obtained on small plants at a standard yield of dry matter are used to judge the sufficiency of the nutrients for producing the final yield. The initial work used the concentrations measured in soil culture experiments, but later work used field data.

The small plants were used for two reasons. First, diagnosis of nutrient deficiencies early in the season would provide an opportunity to make a supplemental application of fertilizer to the crop. And second, the final yield of grain was found to be essentially directly proportional to the concentrations of nitrogen and potassium in the dry matter at this stage when different quantities of fertilizer nitrogen and potassium had been applied at the time of planting (Møller Nielsen and Friis-Nielsen, 1976).

The system is complex, and details of the procedures will not be given here. A summary progress report on the general approach was published by Møller Nielsen (1985) in a symposium on assessing the fertilizer nitrogen requirement of crops.

3-4.2.7. Application to the Mature Crop

The conventional justification for making analyses for nutrients in crops during the growing season is to permit the application of deficient nutrients, so that the final yield may be increased. For the most part, however, the logistics of making any but field tests and getting the information to the grower in time to make an effective application of fertilizer are discouraging.

The most widely reported plant analysis data in the literature are for the corn leaf just below the ear at the time of silking. By the time the plants have reached this stage, they are normally so large that they would be damaged excessively by the passage of common mechanical equipment for applying fertilizers, but fertilizers still could be added in irrigation water. Delayed applications are most effective with irrigation, sandy soils, and soluble nitrogen fertilizers. Elwali and Gascho (1988), for example, reported experimental work on the use of foliar analysis of corn at the earlier 10- to 12-leaf stage as a guide for supplemental applications of fertilizers to an irrigated sandy soil in Georgia.

If a basis for applying fertilizers to the current crop is not the objective of plant analysis for nutrients, one way to circumvent the problem of variations in the relationship of critical values determined during the growing season to the final yield of the crop is to make the nutrient analyses at final harvest. In many situations, the nutrient concentrations at that time may be expected to reflect more clearly the effect of nutrient supplies on final yield than do analyses made at any earlier time.

In an investigation involving collation of experimental data from Illinois, Indiana, Iowa, Missouri, Ontario, and Oregon, Pierre et al. (1977a) obtained the results in Fig. 3-28A relating the percentage of the maximum yield of corn in nitrogen fertilization experiments to the nitrogen concentration in the grain. The points in the figure represent values for individual nitrogen levels in the Iowa experiments and interpolated values for selected percentages of the maximum yields in the experiments from other states and Ontario. Some segregation of the data by locations is evident in that the Illinois and Ontario data tend to

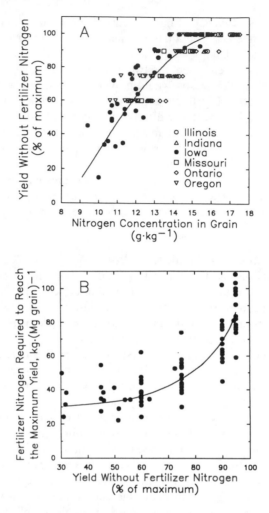

Fig. 3-28. A. Relative yields of corn grain versus nitrogen concentrations in the grain as found in nitrogen fertilization experiments at different locations, with different cultivars, and in different years. (Pierre et al., 1977a) B. Kilograms of fertilizer nitrogen required per megagram of corn grain to reach the maximum yield attainable with nitrogen fertilization when the yield without nitrogen fertilization represents different percentages of the maximum. (Pierre et al., 1977b) In both *A* and *B*, the data obtained with different quantities of fertilizer nitrogen are plotted directly for the Iowa experiments. Values corresponding to 60, 75, 90, and 100% of the maximum yields attainable with nitrogen fertilization are shown for experiments from Illinois, Indiana, Missouri, Ontario, and Oregon. These values were obtained from response curves for the individual experiments.

be displaced to the right. Nonetheless, an average line drawn through the points would permit a fairly good estimate of the percentage of the maximum yield from a measurement of the nitrogen concentration in the corn grain.

In a second paper, Pierre et al. (1977b) described a procedure for estimating the quantity of fertilizer nitrogen required to produce the maximum yield of corn on the basis of the current yield and the nitrogen concentration in the corn grain. The first step is to determine the relationship between the yield of corn as a percentage of the maximum and the nitrogen concentration in the corn grain, as in Fig. 3-28A.

The second step is to determine the maximum potential yield and potential yield increase. The maximum potential yield is obtained by dividing the current yield by the percentage of the maximum yield, which is derived from use of the nitrogen concentration in the corn grain and the function described in the first step. The potential yield increase is given by the difference between the maximum potential yield and the current yield.

The third step is to determine the "nitrogen requirement index," which is the kilograms of fertilizer nitrogen required per megagram increase in yield between the current yield and the maximum yield. This index is a function of the current yield as a percentage of the maximum. Fig. 3-28B shows this relationship for the data in Fig. 3-28A. (The authors noted some segregation of points representing well drained and poorly drained soils that is not indicated differentially in Fig. 3-28A.)

The fourth step is to determine the total quantity of nitrogen needed to produce the maximum yield. This quantity is the product of the potential yield increase in megagrams per hectare and the nitrogen requirement index for the initial percentage of the maximum yield.

In contrast to the findings by Pierre et al. (1977a,b), Cerrato and Blackmer (1990) observed little relationship between the relative yield of corn grain and the nitrogen concentration in the grain in 12 more recent experiments in Iowa. The poorer relationship observed by Cerrato and Blackmer may be in part a consequence of the smaller range of relative yields and the fact that the heavy applications of fertilizer nitrogen resulted in a considerable portion of the data being on the yield plateau in all but one experiment. An additional procedural matter that could affect the interpretations is the fact that Cerrato and Blackmer based their calculations on individual plot data, whereas Pierre et al. used the treatment means over all replicates, and for the experiments from locations other than Iowa they used data interpolated from response curves. The procedures of Pierre et al. would smooth the data.

In the work by Pierre et al. (1977a,b), the critical concentration of nitrogen in the corn grain (corresponding to the maximum yield) averaged 15.4 grams per kilogram for the 13 Iowa experiments and 16.1 grams per kilogram for the 24 experiments from other areas. In 42 similar experiments in New Zealand in which a statistically significant response to nitrogen fertilization was obtained in 22, Steele et al. (1982a,b) found that the nitrogen concentration in the grain corresponding to the maximum yield averaged 15.2 grams per kilogram. The weighted average of the critical concentration estimates from the two sets of investigations is thus 15.6 grams per kilogram.

From a plot of observed responses to individual increments of fertilizer nitrogen in their experiments, Steele et al. (1982a,b) developed the following empirical equation relating the overall increases in yield of grain y in kilograms per hectare per kilogram of fertilizer nitrogen to the relative yield of grain x without the nitrogen increments:

$$y = -12 + 7\sqrt{101 - x}.$$

To approximate a response curve for a particular site for which estimates of the yield of the control and the maximum yield are available, the relative yield of the control (in percent) is substituted in this equation. The value of y corresponding to the relative yield x of the control then is multiplied by a quantity of fertilizer nitrogen not to exceed 50 kilograms per hectare. The product is the total increase in grain yield expected from applying the number of kilograms of fertilizer nitrogen in the first increment. The value of the increase is added to the absolute yield of the control and is plotted against the number of kilograms of fertilizer nitrogen in the first increment. A new value of x is calculated by dividing the expected yield with the first increment of fertilizer nitrogen by the estimated maximum yield and multiplying by 100. The new value of x is substituted in the foregoing equation, and a new value of y is calculated. The new value of y is multiplied by a quantity of fertilizer nitrogen not to exceed 50 kilograms per hectare. The calculated increase in yield then is added to the yield obtained with the first increment of fertilizer nitrogen and is plotted against the sum of the first and second increments of nitrogen. The procedure is repeated as desired to obtain the estimated yield corresponding to additional increments of fertilizer nitrogen.

Steele et al. (1982a,b) did not compare the results obtained by their estimation procedure with those obtained by the Pierre et al. (1977a,b) procedure. The findings, however, seem as concordant as could be expected in view of the fact that the data were obtained at different times, with different cultivars, and on opposite sides of the world. For example, the estimated numbers of kilograms increase in corn yield per kilogram of fertilizer nitrogen in the respective investigations were 27 and 22 at a relative yield of 70, and 11 and 14 at a relative yield of 90.

Similar research on nitrogen fertilization of winter wheat grown in a wheat-fallow sequence under dryland conditions at various locations in eastern Colorado was reported by Goos et al. (1982). There were 30 experiments over a period of 3 years. They plotted the relative yield of wheat grain against the protein concentration in the grain as found by a dye adsorption method. In the 29 site-treatments in which the nitrogen supply was inadequate to produce the maximum yield, 28 were in the range between 60 and 95%, and the scatter was too great to provide any useful indication of a relationship between degree of nitrogen deficiency and protein percentage in the grain. The authors attributed much of the scatter to the wide diversity in water stress and temperature conditions in the experiments.

In another group of 15 experiments on corn, Binford et al. (1990) observed a close relationship between the relative grain yields and the total nitrogen

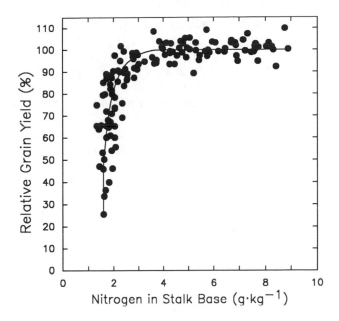

Fig. 3-29. Relative yield of corn grain versus concentration of total nitrogen in a basal segment of the stalk. The results are from 15 experiments on corn in Iowa, each experiment involving ten different quantities of fertilizer nitrogen. (Binford et al., 1990)

concentrations in a basal segment of the stalks at the time of physiological maturity of the grain. See Fig. 3-29. The basal portion of the stalks contains the highest concentration of nitrate. The nitrate form of nitrogen was only a minor component of the total nitrogen in the stalk segments low in total nitrogen, but was the major component in stalks high in total nitrogen.

The primary objective in the work by Pierre and coworkers (1977a,b) in analyzing corn grain for its nitrogen concentration was to assess the degree of nitrogen deficiency and develop a way to relate the nitrogen concentration to the quantity of fertilizer nitrogen needed to increase the yield to the desired level. The primary objective in the work by Binford et al., on the other hand, was to find a way to tell when corn had received more fertilizer nitrogen than was needed to produce the maximum yield. In principle, both the corn grain and the basal portion of the corn stalks are appropriate for both purposes. Examination of Fig. 3-29 indicates, however, that the nitrogen concentration in the basal portion of the stalks would be unsatisfactory for use on the deficiency side. Only in the narrow transition zone between the steep ascending part of the curve and the flat plateau could any substantial discrimination be made among relative yields on the basis of the nitrogen concentrations. The data published by Pierre et al. provide no basis for judging the suitability of grain analysis for determining whether excessive quantities of fertilizer nitrogen had been applied.

Other sources of information, however, indicate that the basal portion of the stalks would be a more satisfactory tissue for this purpose.

In principle, the procedure used by Pierre et al. (1977a,b) should apply to nutrients in general, and especially to those that are protein constituents. Randall et al. (1981) found in Australia that application of 50 kilograms of sulfur as calcium sulfate dihydrate (gypsum) per hectare increased the yield of wheat grain from 2.7 to 3.8 megagrams per hectare and increased the sulfur concentration in the grain from 1.0 to 1.8 grams per kilogram. These results suggest that sulfur concentration in the grain may be useful as an index of sulfur sufficiency. Rehm et al. (1983) reported the results of a field experiment on phosphorus fertilization of corn in Nebraska, however, in which the procedure would have been unsatisfactory. The experiment was repeated for 5 years in the same location with irrigation. Although the phosphorus concentration in the total aerial portion of the plants at silage harvest increased in all years with the quantity of fertilizer phosphorus applied, the phosphorus concentration in the grain increased in only 1 year and decreased in 4 years.

3-4.2.8. Crop Logging

Annual crops that occupy the land for a single growing season or part of a growing season usually receive fertilizer in a single application at the time of planting. Perennial crops, as well as annual crops that occupy the land for many months, absorb nutrients over a long time; this may provide an opportunity to apply fertilizers at intervals according to current needs instead of applying the total amount at one time. Repeated analyses on the crop can provide a check on the effect of fertilizer additions made previously as well as a basis for adjusting the next addition.

H. F. Clements popularized the concept of repeated analyses and fertilizer additions with his research on sugarcane. He called the system he developed *crop logging*, crediting the term to a colleague, P. T. Agee. The most recent reviews of the system apparently have been made by Clements (1961, 1980) and Bowen (1990). Baver (1960) reviewed the system from a somewhat different perspective. The log to which reference is made is a large graph used to plot the measurements on the crop, target values, treatments applied, and other pertinent observations as the crop develops. The procedures evolved as a result of much research designed to produce good crops efficiently.

The system for sugarcane is complex, and the details are not given here. As a guide to fertilizer applications, however, the primary purpose of the crop-logging procedure is to determine whether fertilization with a given nutrient will or will not produce an increase in yield. The appropriate quantities of fertilizer to apply are based upon experience.

From the standpoint of plant nutrients, principal emphasis is placed upon nitrogen in managing the crop. The reasons are the importance of nitrogen in controlling the balance between growth of the crop and sugar production and the fact that fertilizer nitrogen can be applied and can become available to the crop throughout its life.

Fig. 3-30 illustrates the application of the crop-logging principle to sugarbeet in California, where the growing season is long and where irrigation is used to control the water supply. In this instance, one sees that with no fertilizer nitrogen, the concentration of nitrate nitrogen in the petioles was below the critical level throughout the season (the authors defined the critical concentration as the concentration of nitrate nitrogen in the dry matter of the petioles of recently matured leaves below which nitrogen fertilization will produce a growth response). The application of 90 kilograms of fertilizer nitrogen per hectare at thinning raised the concentration of nitrate nitrogen in the petioles well above the critical level for a time, but by late July this application was exhausted, and the concentration in the petioles was below the critical level.

The figure shows the effect of late July applications of three different additional quantities of fertilizer nitrogen on the concentration of nitrate nitrogen in the petioles subsequently to the time of harvest in late October. The sugar yields are shown in the inset table. Although the highest sugar yield was obtained with the greatest total amount of fertilizer nitrogen (269 kilograms per hectare), the most profitable treatment was the one with the next greatest amount (180 kilograms per hectare).

With the heaviest application of fertilizer nitrogen, the concentration of nitrate nitrogen in the petioles apparently remained above the critical level to the time of harvest. Sugarbeet, like sugarcane, should run out of nitrogen toward the end

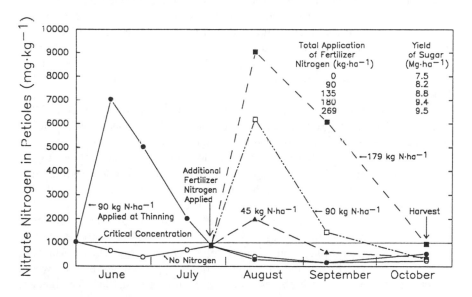

Fig. 3-30. Concentrations of nitrate nitrogen in the dry matter of petioles of sugarbeet with and without an application of fertilizer nitrogen at thinning and with three additional quantities of fertilizer nitrogen in late July in a field experiment in California. Yields of sugar corresponding to the total amounts of fertilizer nitrogen applied during the season are shown in the inset table. (Ulrich et al., 1959)

of the season, so that the carbohydrate produced by photosynthesis tends to be stored as sugar rather than being used to produce more top growth. In connection with the experimental data in Fig. 3-30, the authors noted that had the beets not been harvested until November or December (which would have been feasible at the Davis, California, location), the yield of sugar with the heaviest application of fertilizer nitrogen would have been increased, and that application then probably would have been the most profitable.

3-4.3. Kenworthy Nutrient Indexes

Kenworthy (1961, 1967, 1973) developed a system of leaf analysis for fruit trees in Michigan in which analyses eventually were made for 11 nutrient elements for diagnosing the status of various nutrients. The Kenworthy nutrient indexes are considered here in a section between critical concentrations and nutrient ratios because the theory is intermediate.

As an alternative to critical concentrations, Kenworthy developed a set of "standard values," which were the means of concentrations of nutrient elements found in leaves from high-performing trees showing good growth, production, and fruit quality. It is this method of obtaining the standard values that is responsible for the location of the Kenworthy approach between critical concentrations and nutrient ratios. The theory of the standard values will become clearer when the topic is considered in more detail in Section 3-5.4.1.

Initially, the analytical values for the concentrations of each nutrient in the leaf dry matter of samples being analyzed from growers' orchards were expressed as percentages of the corresponding standard values. Values from 83 to 117% of the standard were represented arbitrarily as normal, values from 50 to 83% as below normal, and values from 17 to 50% as shortage. Values from 117 to 150% of the standard were represented as above normal, and values of 150 to 183% as excess.

The initial procedure was found to be unsatisfactory because it often resulted in diagnoses of shortage or excess for trees which, in the case of some of the shortage diagnoses, were known not to respond to additions of the nutrient diagnosed as being in shortage. These faulty diagnoses were related to the scattering of the individual analyses from which the standard concentrations were developed, together with the fact that the breadth of the scatter relative to the standard concentration (the mean) differed considerably among nutrients.

In terms of the coefficient of variation (the standard deviation as a percentage of the mean), the extreme values reported for the scatter of analyses for tree fruits were 9% for nitrogen and 77% for copper (Kenworthy, 1961). Applying the ranges set initially for the different nutrient supply categories thus would indicate that many of the high-performing trees used to establish the standard value for copper were deficient in this nutrient.

To alleviate the problem created by differences in coefficients of variation in the analyses averaged to produce the standards, Kenworthy adjusted the values by the following procedure. If the concentration of the nutrient in the sample (*x*) is lower than the concentration in the standard (s),

$$\frac{100x}{s} = P,$$

$$(100 - P)(V/100) = I,$$

and

$$P + I = B,$$

where *P* = concentration of the nutrient in the sample as a percentage of the concentration in the standard, *V* = coefficient of variation (in percent) of the values averaged to obtain the standard, *I* = correction for the influence of the variation of the values averaged to obtain the standard, and *B* = balance index. If the concentration of the nutrient in the sample is greater than that in the standard,

$$\frac{100x}{s} = P,$$

$$(P - 100)(V/100) = I,$$

and

$$P - I = B.$$

When the relationships are expressed in the manner used by Kenworthy, the need to add the adjustment factor *I* when *x* < *s* and to subtract it when *x* > *s* is fairly clear. The use of different systems for the two situations is unnecessary in practice, however, because the equations may be combined to

$$Balance\ Index = B = \frac{100x}{s} + (V)\left(1 - \frac{x}{s}\right),$$

which applies whether *x* < *s* or *x* > *s*. According to this equation, the analytical value for the sample as a percentage of the standard will not be adjusted in either direction if *x*/*s* = 1. If *x*/*s* < 1, the second term on the right-hand side of the equation will cause *x*/*s* to be adjusted upward by a fraction of the coefficient of variation. The hypothetical maximum upward adjustment would be equal to the coefficient of variation if *x*/*s* were equal to zero. The second term on the right-hand side of the equation results in a downward adjustment ranging from zero where $x/s = 1$ to V where $x/s = 2$ and to $2V$ under the hypothetical circumstance in which $x/s = 3$.

As an example related to Kenworthy's findings, if $x/s = 0.4$ for both nitrogen and copper, the adjusted balance index would be $40 + 9(1 - 0.4) = 45.4$ for nitrogen and $40 + 77(1 - 0.4) = 86.2$ for copper. The unadjusted value $x/s = 0.4$ is in Kenworthy's shortage class. Because of the difference in the coefficient of variation (9 for nitrogen and 77 for copper), however, the adjusted balance index for nitrogen would remain in the shortage class, but the adjusted balance index for copper would rise into the normal class. In Kenworthy's modified system, the nutrient index values for samples are changed by including the coefficient of variation, but the numerical values of the various nutrient categories remain the same as they were originally.

Kenworthy's approach has the advantage that it could be put into operation for advisory purposes on a qualitative basis after only a year for collecting and analyzing samples from the test trees. Collecting and analyzing these samples would involve much less effort than conducting experiments to find critical values and making the needed analyses on the leaf samples. The question of whether the critical values obtained are sufficiently constant to adopt for use would require experiments in additional locations and additional years.

The "standard" values derived from analyses of leaves from the best trees in a single year of course are not the final answer either. Additional samples in other years are needed to provide a basis for adjustments that almost certainly will be required. Moreover, if all the trees sampled are from a limited area, such as a specialized fruit-producing area in a single state, the possibility exists that some of the standard values will be biased, no matter how many samples are analyzed from high-yielding trees, because the area may not include the range of conditions in which certain nutrients are present in optimum concentrations. This point will be discussed further in the next section.

As is true for the critical concentration approach, Kenworthy's nutrient indexes represent a system for rating nutrient supplies on a scale that bears no fundamental relationship to the quantities of nutrients needed to raise the concentrations to equal those in the standard or to any other level. These quantities must be found by empirical studies, and the results of the experiments probably will need some adjustment from one area to another.

3-5. Nutrient Ratios

3-5.1. Background
Plants take up and lose nutrients continuously, and the net uptake rate changes with time and differs among nutrients. The nutrients are translocated to different degrees from one tissue to another, and are diluted to different degrees in different plant parts by organic materials produced as plants grow. Despite this complexity in the behavior of plant nutrients, it seems intuitively that there should be a set

of optimum ratios among the nutrient elements within a given plant for promoting the growth of the plant.

Lagatu and Maume in France generally are credited with popularizing the concept of nutrient ratios as well as the term "foliar diagnosis" for evaluating the nutrient status of plants. A number of their papers are cited in an article by Thomas (1937), who had studied in France and who described the concepts in English.

The nutrient status of plants was considered by Lagatu and Maume (1934a,b) to have (1) a quantity or intensity aspect, which they represented by the sum of the calculated percentage concentrations of nitrogen, phosphorus pentoxide, and potassium oxide in the leaf dry matter, and (2) a quality aspect, which they represented by the ratios of the percentages of nitrogen, phosphorus pentoxide, and potassium oxide to each other. These ratios were depicted by plotting the percentages in an equilateral triangle analogous to the familiar soil texture triangle. In the nutrient triangle used by Lagatu and Maume, the sum of the percentage concentrations of nitrogen, phosphorus pentoxide, and potassium oxide in the leaves was taken as 100%, and the apexes were 100% N, 100% P_2O_5, and 100% K_2O. See the paper by Thomas (1937) for examples.

Although the triangle diagrams introduced by Lagatu and Maume did not become popular, the basic concept of quality and quantity aspects in the mineral composition of plants has endured. Interest in nutrient ratios has continued, and progress has been made in developing their usefulness.

One aspect of the physiological logic behind the theory of an optimum ratio of nutrients in a given plant is illustrated by the results of research by Dijkshoorn et al. (1960). In experimental work with perennial ryegrass, these authors found a content of about 0.027 mol of protein sulfur per mol of protein nitrogen in the foliage, despite variations in the concentrations of total sulfur and total nitrogen in the plant dry matter. See Fig. 3-31. The molar ratio of total organic sulfur to total organic nitrogen similarly was about 0.027 because the organic sulfur and organic nitrogen were present mostly in protein forms. Dijkshoorn and coworkers inferred from their findings that for normal growth, perennial ryegrass should contain 0.027 mol of organic sulfur per mol of organic nitrogen.

Dijkshoorn and Van Wijk (1967) later reviewed published data on nitrogen-sulfur relationships in plants and found that the molar ratio of organic sulfur to organic nitrogen ranged from 0.025 in legumes to 0.032 in grasses, and that this was the same as the range in the proteins that contained about 80% of the organic sulfur and nitrogen. In *Brassica* species, the ratio of organic sulfur to organic nitrogen is higher because a substantial proportion of the organic sulfur in these species is in forms other than proteins.

The reason for the approximate constancy of the ratios of protein sulfur to protein nitrogen is that nitrogen occurs in all amino acids, sulfur occurs in some of them, and the ratios of the various amino acids in plant proteins remain about the same except under extreme conditions. The ratio of protein sulfur to protein nitrogen in a plant as a whole of course might change with time if the composition

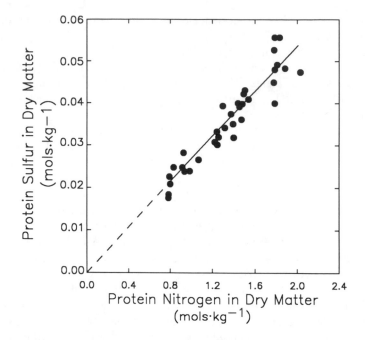

Fig. 3-31. Protein sulfur versus protein nitrogen in herbage of perennial ryegrass grown on nutrient solutions in seven experiments, in each of which sulfate was substituted for nitrate to different degrees. (Dijkshoorn et al., 1960)

of the proteins changes, as conceivably could be the case with the development of the fruit. In an investigation of ratios of sulfur to nitrogen in the protein of wheat grain, Randall et al. (1981) found that the values were variable, and the same was true for the ratios of total sulfur to total nitrogen.

The case of protein sulfur and protein nitrogen appears to be one in which in many instances there is a close, almost stoichiometric, relationship between two nutrients. When the soil supplies of sulfate and nitrate are ample, however, substantial proportions of the total sulfur and nitrogen in plants may be present as sulfate and nitrate, so that a range of ratios of total sulfur to total nitrogen may be found. Kelling and Matocha (1990) illustrated the implication of the ratio of nitrogen to sulfur for performance of crops in the field by a graph showing a decline in the relative yield of Coastal bermudagrass from 100 to about 50 as the ratio of nitrogen to sulfur in the forage increased from 12 to 36.

A second aspect of physiological logic behind the theory of an optimum ratio of nutrients is the tendency of additions of certain elements to induce a deficiency of another independently of the dilution that may accompany increased growth. Although the phenomenon is not understood, part of the effect seems to be a consequence of internal interference of one element with the functioning of another.

Table 3-3 illustrates the interaction of iron and zinc in young corn plants grown in nutrient solutions. Addition of zinc resulted in poor growth and iron

Table 3-3. Relationships of Iron and Zinc in Corn Plants Grown in Culture Solutions (Watanabe et al., 1965)

Concentration in Solution (μmol·l^{-1})		Yield of Dry Matter (g·culture^{-1})	Concentration in Plant Dry Matter (mg·kg^{-1})		
Iron	Zinc		Iron	Zinc	Iron/Zinc
40	0.75	20.0	44	15	2.9
40	2.25	11.9[a]	68	58	1.2
80	2.25	23.7	65	25	2.6

[a] Plants had iron deficiency symptoms.

deficiency that was related to a low ratio of iron to zinc in the plants, but not to a low concentration of iron.

Fig. 3-32 illustrates the reverse situation, observed in field studies in France, in which zinc deficiency was the problem. The zinc concentration in the plants is plotted against the ratio of iron to zinc, the plants with zinc deficiency symptoms being indicated by open circles and those without zinc deficiency symptoms by filled circles. The results in the figure indicate not only that zinc deficiency symptoms appeared when plants had a low concentration of zinc, but also that at a given concentration of zinc the proportion of the samples showing zinc deficiency increased with the ratio of iron to zinc.

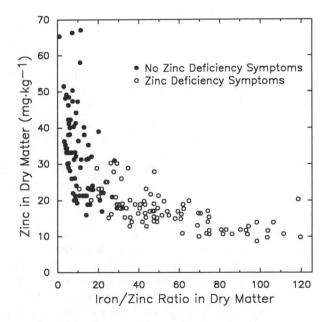

Fig. 3-32. Zinc concentration versus ratio of iron to zinc in the tissue of young corn plants on noncalcareous soils in France over a period of several years. The data are classified according to whether the plants did or did not exhibit zinc deficiency symptoms. (Lubet et al., 1983)

Sajwan and Lindsay (1988) found that as reduction became more pronounced and as iron solubility increased in flooded soil, the solubility of zinc decreased. Conversely, adding zinc depressed the solubility of iron. These findings indicate that at least part of the iron-zinc interaction observed in the field in rice production is a consequence of solubility effects in the soil.

Further evidence exists for a special interrelationship of magnesium and manganese in plants. In an experiment with wheat on an acid sandy soil that apparently was deficient in magnesium and toxic in manganese, Le Bot et al. (1990) added calcium carbonate and various quantities of magnesium sulfate. They observed that the relative yields of shoots were well related to the ratio of magnesium to manganese in the shoots, but not to the concentration of either magnesium or manganese alone. With tomato, they found that magnesium depressed the uptake of manganese and also increased the concentration of manganese in the plants required to induce toxicity symptoms.

3-5.2. Advantages

Aside from the theoretical aspects discussed in preceding paragraphs, the use of nutrient ratios has perhaps three principal advantages. First, when properly handled, nutrient ratios are not subject to the rigid requirement of the critical concentration system to sample a specific plant part at a specific stage of growth. Appropriate use of nutrient ratios makes possible a diagnosis of the nutrient status of a crop over a range of situations. As a result of the first advantage, nutrient ratios have a second advantage in making diagnoses possible at sampling times or growth stages for which information on critical concentrations may be inadequate. Third, they make possible an automatic ordering of nutrients in terms of relative sufficiency. The critical concentration approach also permits an ordering of nutrients in terms of relative sufficiency, but this adaptation is seldom used and is less satisfactory than the procedure developed for nutrient ratios.

The second and third advantages of nutrient ratios will be discussed in Section 3-5.4 in connection with Beaufils nutrient indexes. Especially the third advantage must be credited to the Beaufils system. The first advantage will be considered here.

The first advantage of nutrient ratios, that of relaxing the sampling requirements or broadening the range of conditions over which diagnoses can be made, is a consequence of two physiological principles, together with some mathematical juggling. The first principle has to do with the functions of nutrients in plants in relation to the manner of expressing the analytical values.

When quantitative chemical analyses on plants are made as a basis for inferences about the sufficiency of the various nutrients, the analyses almost invariably are expressed as concentrations in the dry matter of the plants or plant parts. The primary reasons for using the dry weight are that the concentration of water varies with the soil and atmospheric conditions and the stage of growth, and that it is inconvenient to maintain plants at their field water content in the usual situation in which the plants sampled are located in the field some distance from the laboratory.

Table 3-4. Proportion of the Variance of Nutrient Concentrations Associated With Plant Age, Hour of Sampling, Wind Force, Cloud Density, Temperature, and Plant Population in 350 Samples of Leaves From Healthy, Good-Yielding Corn Plants When the Concentrations Were Expressed on Different Bases (Beaufils, 1971)

Analytical Basis for Nutrient Concentration	Nutrient or Nutrient Ratio	Proportion of Variance Accounted for (R^2)
Dry weight	Nitrogen	0.44
	Phosphorus	0.28
	Potassium	0.57
Water weight	Nitrogen	0.07
	Phosphorus	0.06
	Potassium	0.03
Fresh weight	Nitrogen	0.09
	Phosphorus	0.09
	Potassium	0.11
(Nutrient ratio)	Nitrogen/Potassium	0.10
	Nitrogen/Phosphorus	0.04
	Potassium/Phosphorus	0.07

The activities of and requirements for most nutrients other than carbon, hydrogen, and oxygen are associated primarily with the protoplasm and not with the carbohydrates and lignin. Carbohydrates and lignin are by-products of protoplasmic activity, but they make up most of the dry weight, and the ratio of protoplasm to carbohydrates and lignin is variable.

Plant physiologists sometimes avoid the confusion associated with the dry-matter convention by expressing nutrient concentrations on the basis of the water content or the fresh weight of the plant tissue sampled. The theoretical justification for using these bases for expressing the analyses is that the protoplasm of plant cells is mostly water during the vegetative period when most analyses are made for evaluating nutrient status. The major component of the fresh weight of herbaceous plants is water during this period.

Table 3-4 shows that the nitrogen, phosphorus, and potassium concentrations in the dry matter of corn leaves were associated to a greater extent with plant age and other factors that would affect the water content of the plants than were the nutrient ratios or the nutrient concentrations calculated on the basis of the fresh weight or the weight of water in the tissues analyzed. These results indicate the potential value of the latter three modes for expressing the plant nutrient data as a means of alleviating the need to standardize the critical concentrations of the nutrients to correspond to a particular growth stage because of the marked variation of these concentrations with time of sampling. Although in Table 3-4 the nutrient ratios were not the ideal in terms of their independence of the other factors mentioned, they have the practical advantage that they can be derived from analyses of dried tissue, which is most convenient for plants grown in the field.

The data in Table 3-4 indicate that nutrient ratios provide a way to reduce the effect of accumulation of dry matter with plant age, which often has such a pronounced effect on critical concentrations of nutrients. Additional evidence supports this inference, but the situation is not as simple as Table 3-4 suggests. The use of nutrient ratios to deal with the effects of age and other factors will be discussed further in Section 3-5.4.7 under Beaufils nutrient indexes because the research has been done in that connection.

With the background just provided, it may be profitable to consider in more detail the research by Leigh and Johnston (1983a) referred to in Section 3-1.2.2. They found that nitrogen fertilization increased the potassium concentration in the dry matter of barley plants throughout most of the growing season (Fig. 3-33A) in a field experiment in which nitrogen fertilization increased the grain yield from 1.9 to 8.1 megagrams per hectare. This behavior is an exception to Atterberg's second principle.

The explanation for the effect of nitrogen fertilization on the behavior of potassium may be inferred from Fig. 3-33B and C. Fig. 3-33B shows that nitrogen fertilization increased the ratio of fresh weight to dry weight of the plants. That is, it increased the proportions of water and protoplasm in the plants. Nitrogen fertilization thus increased the requirement for other nutrients that occur in plant protoplasm per unit of dry matter.

Fig. 3-33C shows that the concentration of potassium in the tissue water was essentially unaffected by nitrogen fertilization. In this instance, potassium was not deficient. The soil had an ample supply of potassium, as judged by the usual chemical criteria, and equal additions of fertilizer potassium were made to all plots. In a companion paper (Leigh and Johnston, 1983b), the potassium concentration in the tissue water of barley plants was found to be lower in plants deficient in potassium than in those well supplied with potassium.

Johnston and Goulding (1990) suggested that the special value of expressing potassium concentrations on the basis of the content of tissue water results from the use of most of the potassium to generate osmotic pressure. In the experimental work they described, part of which was from the same paper that served as the source of Fig. 3-33, the concentration of potassium in the tissue water of barley remained approximately constant until senescence, when the crop was losing water. Then the concentration in the tissue water increased.

Because of the approximately constant concentration of potassium in the tissue water of barley and the other crops they studied, Johnston and Goulding (1990) concluded that potassium deficiency could be diagnosed at any time during the spring and summer if the concentrations of potassium are expressed on a tissue water basis. When the data are expressed on this basis, the critical concentration would be little affected by time of sampling. The behavior of potassium may be unique in this regard. In an analogous investigation of nitrogen behavior, Leigh and Johnston (1985) concluded that nitrogen concentrations expressed on the basis of tissue water are unlikely to be useful for determining the nitrogen requirement of barley.

3-5.3. Disadvantage

If the optimum ratio of, say, nitrogen to potassium in the leaf tissue of a particular crop is 1.5, the information that a given sample of the leaf tissue from this crop

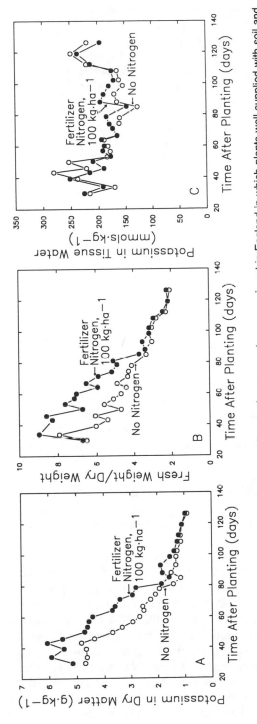

Fig. 3-33. Potassium behavior in barley plants during the growing season in an experiment in England in which plants well supplied with soil and fertilizer potassium received either no supplemental nitrogen or 100 kilograms of fertilizer nitrogen per hectare. A. Concentration of potassium in the dry matter. B. Ratio of fresh weight to dry weight of the above-ground portions of the plants. C. Concentration of potassium in the water present in the barley tissue. (Leigh and Johnston, 1983a)

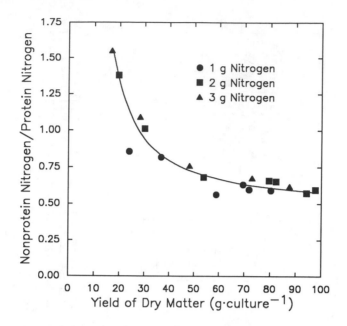

Fig. 3-34. Ratio of nonprotein nitrogen to protein nitrogen in oat plants in sand cultures 90 days after planting versus yield of dry matter. The cultures received 1, 2, or 3 grams of fertilizer nitrogen and various quantities of fertilizer potassium. (Alten et al., 1940)

has a nitrogen-to-potassium ratio of 2 merely says that the ratio exceeds the optimum to the measured degree. Conceivably, the supply of nitrogen could be adequate and potassium deficient. Potassium could also be adequate and nitrogen in excess. Or both nutrients could be deficient or present in excess. Similarly, ratios do not tell whether the total package of nutrients is deficient, adequate, or excessive.

One possible way to address this issue would be to use the ratio of nonprotein nitrogen to protein nitrogen as an internal standard for judging the sufficiency of nitrogen and, indirectly, the other nutrients. Nitrogen would seem to be an appropriate reference element because of the fundamental importance of proteins in protoplasm.

When new growth is inhibited by environmental factors, including deficiencies of nutrients other than nitrogen, the production of additional protoplasm and protein usually is reduced more sharply than uptake of nitrogen from the soil. The nitrogen absorbed thus tends to accumulate in forms other than protein. This principle is illustrated in part by Fig. 3-34 for oat plants grown in sand cultures with different quantities of fertilizer nitrogen and potassium.

The protein fraction of nitrogen often is estimated by the difference between total nitrogen and nitrogen soluble in alcohol or nitrogen soluble in trichloroacetic acid. Frequently it is called insoluble nitrogen. The nitrogen soluble in alcohol or trichloroacetic acid is partly organic and partly inorganic. The organic nonprotein nitrogen includes amides, amino acids, and peptides, and the inorganic forms include nitrate and ammonium. In the experiment in Fig. 3-34, the ammonium and nitrate nitrogen concentrations in the plant tissue were low in almost all treatments. The appearance of the plot of the ratio of nonprotein organic nitrogen to protein nitrogen against the yield thus had about the same appearance as the plot in Fig. 3-34.

Brooks and Reisenauer (1985) found that the relative growth rate of wheat plants increased with the ratio of soluble nitrogen to protein nitrogen in the leaf tissue up to about 0.11 and then decreased at higher ratios, which were associated with deficiencies of phosphorus or sulfur. In work on conifer seedlings, Kim et al. (1987) found that the ratio of glutamine nitrogen to protein nitrogen was more sensitive to the nitrogen supply than was the ratio of soluble nitrogen to protein nitrogen. Thus, at least for conifer seedlings, this ratio should be preferable to the ratio of soluble nitrogen to protein nitrogen.

Different methods were used, and different results were obtained in the work cited, which indicates that more research is needed to develop a suitable technique for judging the degree to which plant growth is limited by nutrient supplies. As a practical alternative that can be applied to existing data, the concentration of nutrients in the dry matter has been introduced as the equivalent of an additional nutrient in some work on nutrient ratios. This usage will be considered in Section 3-5.4.

3-5.4. Beaufils Nutrient Indexes

In studies on rubber trees in Vietnam and Cambodia in the 1950s, Beaufils developed a nutrient-ratio concept as part of an overall system for evaluating the nutritive and other factors limiting plant yields. He later applied the system to corn in South Africa (Beaufils, 1971, 1973). Originally called the *physiological diagnosis system*, it was later renamed the *diagnosis and recommendation integrated system* or *DRIS*.

A number of more recent scientific studies of the plant analysis aspects of the system have been made by other investigators, and the concept seems to be accepted increasingly as a useful way to evaluate the nutrient status of plants. Reviews have been published by Walworth and Sumner (1987, 1988).

The nutrient-ratio concept pioneered by Beaufils leads to special indexes, which here are called *Beaufils nutrient indexes* or merely nutrient indexes or indexes. Beaufils developed a way to use these indexes to rate the nutrients in order of their need by the plants analyzed.

The DRIS system involves the use of environmental factors and soil analyses in addition to the nutrient indexes. Although the term DRIS is commonly used by more recent authors, they usually deal only with the ratio technique that leads to nutrient indexes, and neglect the remainder of the system. The other parts of the system are more conventional.

In the Beaufils system, the index values for individual nutrients in a specific plant sample to be evaluated are derived in part from analytical values for the various nutrients being studied. The rationale takes into account the deviations of the nutrient ratios found in the sample to be evaluated from the corresponding optimum ratios of each nutrient to each other nutrient. The calculation process will be described in Section 3-5.4.3. But first, the evaluation of optimum ratios will be considered. Evaluating the required optimum ratios is a separate process.

3-5.4.1. Evaluating Optimum Nutrient Ratios

Beaufils plotted crop yields against the values of each ratio of nutrients found by analysis in many samples of a crop grown under various circumstances — some in fertilizer experiments, in which each plot or each treatment could provide a sample, and some on farmers' fields. When a distribution including many samples had been developed, he found that the highest yields were associated with certain ratios. These ratios are called here the optimum ratios. The term "norm," sometimes used, has other meanings.

Fig. 3-35A is an example from more recent work, showing a plot of yields of corn against the ratio of potassium to phosphorus in the leaf below the ear. This figure is a condensed version of the original data representing more than 8,000 observations.

Fig. 3-36 is an adaptation of an interpretive diagram by Sumner and Farina (1986) to explain the significance of a distribution of points such as that found in Fig. 3-35A. Points at the lower left represent situations in which the yield is limited by a deficiency of potassium, perhaps also by an excess of phosphorus, and by other factors. Points at the lower right represent situations in which the yield is limited by a deficiency of phosphorus, perhaps also by an excess of potassium, and by other factors. As one moves toward the central portion of the

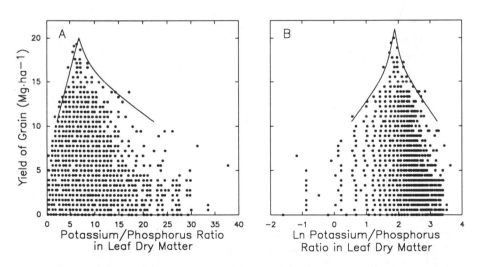

Fig. 3-35.   Distribution of yields of corn grain with different ratios of potassium to phosphorus in the leaf just below the ear. The individual points were derived from measurements made in Africa, Canada, France, South America, and the United States. A. Untransformed data as published by the authors. B. Data transformed by taking the natural logarithm of the ratios of potassium to phosphorus. (Walworth, Letzsch, et al., 1986)

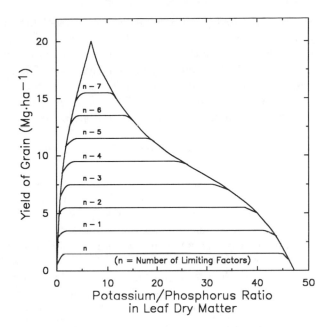

Fig. 3-36. Hypothetical response curves illustrating the concept of Sumner and Farina (1986) regarding the action of limiting factors in controlling the distribution of observations in the plot of observed corn yields against the potassium-to-phosphorus ratios in the leaf tissue in Fig. 3-35A. Successive elimination of limiting factors leads to higher and higher plateaus in the potassium-phosphorus response curves.

distribution of points, the deficiencies and possible excesses of phosphorus and potassium are eliminated. Yields are higher (and do not necessarily form flat plateaus as shown), but still are limited by other factors. As the limitations imposed by factors other than phosphorus and potassium are removed one by one, the yield rises, step by step. Supporting this interpretation are yield data, phosphorus-to-zinc ratios in the leaves, and visual deficiency symptoms of phosphorus and zinc reported by Takkar et al. (1976) in a factorial field experiment on phosphorus and zinc fertilization of corn in Punjab.

A given ratio of potassium to phosphorus in the central portion of the distribution may be found in a crop producing a low yield or a high yield, depending upon the importance of limiting factors other than potassium and phosphorus. The narrowing of the distribution of points at high yields, however, indicates that the higher the yields, the more critical is the ratio of potassium to phosphorus. The extremely narrow distribution at very high yields suggests that only a certain ratio or a narrow range of ratios corresponds to the optimum.

One cannot experimentally produce the highest possible yield and with it the optimum ratios of the nutrients, but the optimum ratio of each pair of two nutrients can be estimated in different ways. One of these is the "boundary line" method.

Although the boundary line concept was used without particular comment by Beaufils (1971, 1973), Webb (1972) wrote a paper in which he described explicitly the use of a line fitted to the best members of a population studied "as a standard against which to judge the remainder, on the assumption that there are reasons other than chance which account for the inferior performance of part of the population." Fig. 3-35A includes two boundary lines. One line ascends from the left and the other from the right as the yields obtained with a given ratio of potassium to phosphorus increase. The intersection of the two lines provides an estimate of the optimum ratio of the two nutrients.

One problem with the boundary line technique is the fact that the number of points thins out at the boundaries, so that only a few points may be available to estimate the location of the boundary lines. A second problem is that the number of points to be included in estimating the lines is somewhat arbitrary. Both of these problems are illustrated in Fig. 3-35A. Many points are available to estimate the boundary line at potassium to phosphorus ratios below the optimum, but few are available at ratios above the optimum. The location of the boundary line for ratios above the optimum is thus not estimated as precisely as that of the line for ratios below the optimum.

The commonly used method for estimating the optimum ratio between two nutrients is to use a direct average of the ratios found in the plants producing the highest yields, on the assumption that these yields are normally distributed. The maximum coincides with the mean in a normally distributed population.

The reason for averaging only the highest yields is that the total populations of nutrient ratios are skewed and not normally distributed. In a skewed population, the maximum does not coincide with the mean. The distribution of ratios for the highest-yielding members of the population is approximately normal, however, and so an average of these values gives a satisfactory estimate of the ratio corresponding to the maximum yield.

Walworth, Letzsch, et al. (1986) found that their estimates of optimum nutrient ratios were about the same when they used the boundary line method and the method of averaging the ratios associated with the highest-yielding members of the survey. And Letzsch and Sumner (1984) found that the cutoff value in selecting the highest yielding members to be averaged was not critical. As may be inferred from Fig. 3-35A, however, the yield level selected should be high enough to avoid an appreciable drift of the mean ratio as a consequence of skewness in the data. Letzsch (1985) published a computer program for making the calculations associated with finding the optimum ratios and the coefficients of variation.

The studies emphasize that to obtain good estimates of the optimum values, many observations representing a wide range of conditions and including exceptionally high yields are needed. If the number of observations is small, the boundary line approach will suffer from the availability of too few points to represent the boundary lines precisely or to obtain a mean that has good precision. If the high-yielding fields are not exceptionally high yielding, the estimated optimum ratio may be poorly defined. And if all the high-yielding fields are from a restricted area, the coefficient of variation of the optimum nutrient ratio may be underestimated.

To reduce the skewness in the survey data, Beverly (1987a,b) took the natural logarithm of the ratios, which stretches out the tail of the distribution on the left and compresses the tail on the right. This effect is illustrated by a comparison of Fig. 3-35B with Fig. 3-35A. The boundary lines for the upper part of the scatter of points are evidently more symmetrical in the transformed data than in the original data.

If a survey distribution is completely normalized by taking the natural logarithm of the nutrient ratios or by some other manipulation, the means of the nutrient ratios will be the same, within experimental error, whether derived from the highest-yielding 1, 10, 25, or 50% of the population or the total population. In a study of ratios based upon measured concentrations of nitrogen, phosphorus, potassium, calcium, and magnesium in the dry matter of Valencia orange leaves, Beverly (1987b) found that the means of transformed ratios for the higher-yielding 50% of the population ranged from 99 to 101% of the means for the total population. For the highest-yielding 25% of the population, the range was from 97 to 102% of the population mean. And for the highest-yielding 10%, the means ranged from 90 to 106% of the population mean. These results suggest that the logarithmic transformation effectively normalized the population and that the deviations with the highest-yielding members were a consequence of experimental error due to smaller numbers of observations.

In a companion paper, Beverly (1987a) used the natural logarithmic transformation on soybean tissue analysis data. In this instance, he found closer agreement between the mean nutrient ratios for the total population and the highest-yielding 1%, 5%, or 10% of the population than in the study on Valencia orange leaves. The logarithmic transformation thus satisfactorily normalized the distributions.

Visual examination of Fig. 3-35 suggests that although the symmetry of the peak of the distribution was improved by taking the natural logarithm of the ratios (Fig. 3-35B versus Fig. 3-35A), the logarithmic transformation did not normalize the total distribution. The natural logarithmic transformation of course is arbitrary, and other transformations might be more effective in normalizing some data sets.

The foregoing discussion of the way the optimum ratio between two nutrients is evaluated helps to explain the consequences of Kenworthy's use of high-yielding fruit trees as a basis for the standard values he used for the various nutrients (Section 3-4.3). His standard values would represent estimates of critical concentrations defined as the nutrient concentrations associated with the maximum yields (see Fig. 3-21). Use of leaf samples from the same high-yielding trees as sources of the standard values for all nutrients would be a step in the direction of assuring that the ratios of the standard values to each other were similar to the optimum ratios.

The usual concept of optimum nutrient ratios as estimated by the Beaufils procedure applies to crop yields. Hockman et al. (1989), however, applied the concept to crop quality. In work on Fraser fir grown as commercial Christmas trees in North Carolina, they substituted premium grade trees for the high yields employed in evaluating the optimum nutrient ratios. In a later paper, Rathfon and Burger (1991) returned to a criterion analogous to the yields of herbaceous crops, using the diameter of the trunk at ground level.

A radically different system of evaluating optimum ratios was employed by Hallmark et al. (1990a). Instead of using peak yields, they used trial and error. They started with a set of optimum ratios derived by the peak yield procedure from 3898 samples for which soybean grain yields and leaf composition were known. Then they made arbitrary adjustments in these values to obtain a new set of ratios that would yield the maximum numbers of correct diagnoses in an independent group of 706 experimental observations from fertilizer experiments in which phosphorus, potassium, calcium, manganese, and zinc were known to be either deficient or sufficient. By use of this approach, they were able to increase the proportion of correct diagnoses from 72% to 95%.

From the theoretical standpoint, both the traditional use of peak yields and the iterative approach pioneered by Hallmark et al. (1990a) are based upon the concept that optimum nutrient ratios exist. The traditional approach involves the additional assumption that the optimum ratios inferred in the usual way from plants producing peak yields apply also to plants producing much lower yields. The assumption seems reasonable, but its validity has not been established. The procedure proposed by Hallmark et al. (1990a) does not depend upon the second assumption, but merely finds by trial and error the ratios that are judged to be optimum because they give the best agreement between diagnosed and observed deficiencies. Although Hallmark et al. started with estimates of optimum ratios derived from an independent set of data, their procedure would have yielded the same result even if the initial ratios were mere guesses. That is to say, the experimentally estimated optimum ratios are not necessary.

The procedure proposed by Hallmark et al. (1990a) will produce better agreement between predictions and observations than the standard procedure because the optimum ratios obtained experimentally are estimated with experimental error. Although the trial-and-error procedure was highly successful in terms of improved diagnoses for the data on which it was tested, a disturbing observation is the fact that the optimum ratios found by the iterative procedure ranged from 8% to 1034% of the optimum ratios found by the standard procedure.

Further research will be needed to determine whether the optimum nutrient ratios Hallmark et al. (1990a) obtained by iteration are useful for independent sets of soybean data or whether individual sets of response data will need rather different sets of optimum ratios to obtain maximum agreement between diagnosed and observed deficiencies. If the optimum ratios they obtained by the iterative procedure turn out to be widely applicable, the indications will be that there were gross errors in the optimum ratios found initially. If substantial adjustments in the optimum ratios found by iteration are needed for different sets of data, the indications will be that the optimum ratios are still poorly defined as a consequence of the experimental error, that there are important differences in optimum ratios among the data sets, or both. The iterative procedure itself is perfectly valid. As a way of arriving at optimum ratios, it can be more economical than the standard procedure.

3-5.4.2. Variability of Optimum Nutrient Ratios

Evaluating a set of optimum nutrient ratios via the usual survey investigation is a major effort. A question of concern to a person choosing to use the survey approach is therefore whether it will be necessary to perform a new survey to obtain estimates of the optimum ratios or whether ratios developed elsewhere will be satisfactory.

For most crops, some information is available in published form, and additional unpublished information can be obtained to provide a start toward a survey from which optimum ratios may be inferred. For a smaller number of crops, estimates of optimum ratios based upon research by different investigators are available in published form. Walworth and Sumner (1987) reviewed this literature and summarized most of the data in tabular form.

The most extensive data are available for corn and are based upon analyses of the leaf just below the ear, usually at about the time the silks appear from the ear. Walworth and Sumner (1987) summarized more than 8,000 observations made throughout the world, and presented sets of optimum ratios for nitrogen, phosphorus, potassium, calcium, magnesium, and sulfur in a table in which the data were classified by locations. The estimated optimum nutrient ratios for the six individual areas listed in the table are well correlated with the corresponding overall optima, as shown in Fig. 3-37A. The correlation coefficients are 0.96 for Hawaii and 0.99 for each of South Africa, southeastern United States, northeastern United States and Canada, midwestern United States, and New Zealand. The overall correlation coefficient is 0.97.

The favorable results in Fig. 3-37A are in part a consequence of the fact that the optimum ratios in the combined data exhibit a 116-fold range. When the optimum values for individual areas are compared with the optimum values for the combined data on a percentage basis, the results shown in Fig. 3-37B are obtained. The use of percentages puts all nutrient ratios on the same basis and displays the results in a manner that is relevant to the way the values are used. The deviations of ratios for individual areas from the overall mean ratios in Fig. 3-37B range from +14 to −12% for N/P to +39 to −28% for K/Mg. The greatest deviations are for ratios involving calcium and magnesium, reflecting the relatively high coefficients of variation of the concentrations of these nutrients in the dry matter in the combined data. Walworth and Sumner (1987) reported coefficients of variation of 12, 30, 32, and 42% for the concentrations of nitrogen, phosphorus, potassium, and magnesium, respectively, in the dry matter.

The median deviations in Fig. 3-37B may be used as a basis for judging the possible effect of the departures of the optimum ratios for individual areas from the overall optimum ratios. The median positive and negative deviations are both 13%. Thus, if an optimum ratio of 100 with a 20% coefficient of variation were used from the overall data and a ratio of 150 were found in a specific sample,

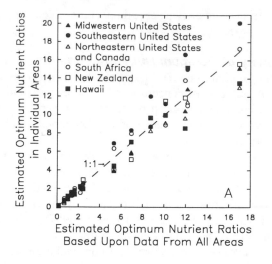

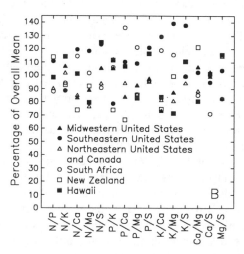

Fig. 3-37. Variability among geographic regions in the estimated optimum nutrient ratios found in the corn leaf below the ear. A. Plot of estimated optimum nutrient ratios for individual regions against estimated optima based upon data from all regions. B. Plot of estimated optimum nutrient ratios for individual regions as percentages of the estimated optima based upon data from all regions. (Walworth and Sumner, 1987)

the function value for that nutrient in the sample (see the next section for calculating nutrient functions) based on the overall optimum ratio would be 25. The corresponding values for the nutrient function would be 16 if the optimum ratio for the specific area is 113% of the overall mean or 36 if the optimum ratio for the specific area is 87% of the overall mean. Differences of this magnitude suggest that transferring optimum ratios from one area to another is appropriate only until enough information has been obtained in the area of concern to determine whether a different set of optima applies to the specific conditions. (The details of calculations for nutrient functions and nutrient indexes are found in the following section.)

An obvious possible cause of differences among areas is the fact that the ratios selected as optimum are those leading to the highest grain yields at the end of the growing season, whereas the plant composition data are obtained from analyses of leaves sampled at about midseason. As noted previously in connection with critical concentrations, differences in conditions between plant sampling and final harvest may affect the relationship between the composition data and the yields.

Experimental error of course is involved in all the measurements, and this no doubt is a cause of some of the differences Walworth and Sumner (1987) found among areas. The introduction of the coefficient of variation into the formula for calculating nutrient functions aids in reducing the effect of experimental error by reducing the weight given to ratios derived from relatively variable data when the nutrient functions are used to calculate the Beaufils nutrient indexes. This point will be explained in the following section.

The data from Hawaii that Walworth and Sumner (1987, 1988) found somewhat at variance with the combined data they collected from various parts of the world were published by Escano et al. (1981b). Analysis of their experimental data showed that local calibration increased the numbers of correct diagnoses based upon both critical concentrations and nutrient ratios, and that locally derived optimum nutrient ratios gave more correct diagnoses than did locally derived critical concentrations.

Escano et al. (1981b) tested the suitability of their locally derived critical concentrations and optimum nutrient ratios by applying them to the same experiments from which they were derived. In general, the most critical judgments about the performance of local calibrations relative to published values of critical concentrations and nutrient ratios may be made when the local calibrations are derived from experiments other than the ones to which they are applied.

In Canada, MacKay et al. (1987) derived estimates of optimum nutrient ratios from experimental data on potato obtained in Alberta and Nova Scotia, and found that the optima were similar. The ratios differed somewhat from the optimum ratios that had been reported previously from South Africa.

The requirement of large numbers of observations to obtain the best results need not be a hindrance to a useful beginning. Walworth et al. (1988) investigated experimentally the practicality of estimating optimum nutrient ratios from small numbers of observations. They published average values for the concentrations of nitrogen, phosphorus, potassium, calcium, magnesium, and sulfur in the ear leaf of corn at the early silk stage from ten areas yielding more than 18 megagrams of grain per hectare from a single fertilizer experiment in New Jersey. From these values, the 15 ratios corresponding to those from the world samples in Fig. 3-37 have been calculated. The percentage deviations from the ratios calculated in the same way from the world samples ranged from -22% to $+119\%$.

For the six ratios involving nitrogen, phosphorus, potassium, and sulfur, the values ranged from -22% to $+84\%$.

Walworth et al. (1988) found that their optimum ratios and those derived from the world samples were about equally useful in interpreting the results of a factorial experiment in the greenhouse in which different quantities of fertilizer nitrogen, potassium, and sulfur were applied to corn. This outcome indicates that at least for the experiment in question, differences in estimated optimum ratios of the magnitude observed were not of major importance for the qualitative evaluations they made of crop responses to applied nutrients.

Other investigators using small numbers of observations to estimate optimum nutrient ratios have included Langenegger and Smith (1978), Kim and Leech (1986), Ward et al. (1985), and Hockman et al. (1989). The respective crops were pineapple, hybrid poplar, eucalyptus, and Fraser fir.

In a further examination of the question of regional and sample-size effects, Hallmark et al. (1990b) used the accuracy of diagnoses of phosphorus and potassium deficiencies in a large experiment on soybean in Iowa as a basis for deciding among four sets of optimum nutrient ratios. Out of a total of 248 possible diagnoses of phosphorus and potassium sufficiency by the Beaufils system on the basis of leaf analyses, the percentages of correct diagnoses were 28, 75, 77, and 85 where the respective bases for evaluation were (a) ten soybean cultivars grown in one year at one site in Louisiana, (b) 137 samples from midwestern United States, (c) 137 samples from midwestern United States plus 313 samples from southeastern United States plus 3 samples from Brazil, and (d) 313 samples from southeastern United States.

The findings by Hallmark et al. (1990b) are not entirely consistent with the general principle that the effect of sampling error on the estimated optimum ratios should decrease with increasing numbers of samples used in developing the estimates, and they convey no clear indication of the significance of regional effects. Theoretically, the best estimates of optimum ratios should be derived from local data, followed in order by regional data, national data, and world data. A relatively simple way to circumvent the confounding influence of sample numbers associated with different sets of data for testing location effects would be to derive estimates of optimum ratios from equal numbers of samples selected at random within each of the sets of data being compared.

Walworth and Sumner (1987) did not give detailed consideration to the possible differences in optimum ratios among cultivars of a given crop. Cultivar effects sometimes are very important where deficiencies of certain nutrients are concerned. For example, Scott (1940) found in experimental work on a sandy soil in South Carolina that 11 grape cultivars were severely affected by boron deficiency, 19 were moderately affected, and 11 apparently were not affected. When otherwise susceptible cultivars were grafted on vigorous rootstocks, they showed no symptoms of boron deficiency. These findings typify the general observation that differences in sensitivity of cultivars and strains of a given species to the supply of a nutrient in the soil are most common and most pronounced with certain micronutrients.

Cultivars of a given species are not genetically identical, and Goodall and Gregory (1947) reviewed earlier evidence to the effect that the relationship between plant development and internal nutrient concentrations may differ among cultivars. They cited, for example, some work by Crowther in Egypt in which a particular cotton cultivar had a high nitrogen concentration but responded as well to nitrogen fertilizers as the other cultivars studied. More recent work by Huffaker and Rains (1978) suggests a possible explanation. They noted that when the wheat cultivars Anza and UC 44-111 were grown side by side in the field in California, Anza had up to twice the concentration of tissue nitrate found in UC 44-111. This difference was associated with a lower nitrate reductase activity in the leaves of Anza. The lower nitrate reductase activity in Anza indicates that this cultivar was slow in metabolizing nitrate to organic nitrogen, allowing an accumulation of nitrate. With an accumulation of nitrate, the process would proceed more rapidly. The nitrate concentration eventually decreased.

Varietal effects have not been found of significance in some work on nutrient ratios as indexes of the relative sufficiency of different nutrients. See papers by Sumner (1977b) and Payne et al. (1986) for specific tests on soybean cultivars grown together in the same experiments. Jones and Bowen (1981) stated that they had found some differences in optimum nutrient ratios between sugarcane cultivars in Hawaii; however, it was not certain whether the differences were due to the cultivars, the differences in environments under which they were grown, or both. Analyses of the ear leaf of inbred lines of corn grown under similar conditions (Kovacevic et al., 1987) have shown such large differences among lines in concentrations and ratios of mineral elements that the existence of real differences in optimum ratios seems likely.

3-5.4.3. Calculating a Nutrient Index

The first step in quantifying the nutrient balance in the Beaufils system is to calculate for the ratio of each pair of nutrients in the sample being evaluated the value of a function that represents a measure of the deviation of the observed ratio from the optimum. Two general mathematical formulations of the functions have been proposed. The "traditional" procedure will be described first, along with the associated explanations needed to obtain an understanding of the general concept and the way it has been used. The second mathematical formulation is discussed at the end of this section.

According to the traditional procedure, if the concentrations of two nutrients in a sample of a crop are A and B and their optimum ratio is a/b, the function $f(A/B)$ for the sample is given by

$$f\left(\frac{A}{B}\right) = 10\left(\frac{(A/B)}{(a/b)} - 1\right)\left(\frac{100}{V}\right),$$

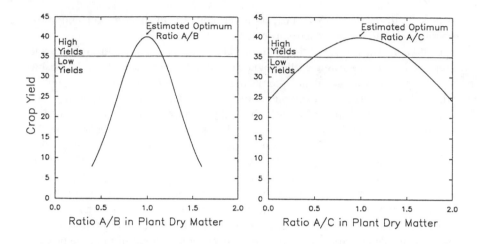

Fig. 3-38. Schematic representation of two normal distributions. On the left, the distribution for the ratio of nutrient *A* to nutrient *B* has a relatively low coefficient of variation. On the right, the distribution for the ratio of nutrient *A* to nutrient *C* has a relatively high coefficient of variation. See the text for the significance of the distributions in calculating Beaufils nutrient indexes.

where $A/B \geq a/b$, or by

$$f\left(\frac{A}{B}\right) = 10\left(1 - \frac{(a/b)}{(A/B)}\right)\left(\frac{100}{V}\right),$$

where $A/B < a/b$. All the nutrients are expressed on the same basis, conventionally their concentrations in the dry matter of the tissue analyzed.

In the foregoing expressions, V is the coefficient of variation in percent associated with the optimum ratio, that is, the coefficient of variation of ratios for the members of the high-yielding population averaged as an estimate of the optimum ratio. The high-yielding members of the population will be the same for the various nutrient ratios, but the ratios and their coefficients of variation will differ. The coefficient of variation was not included in the initial formulas for calculating nutrient indexes (Beaufils, 1971), but was added later (Beaufils, 1973).

The factor 100 is introduced because the standard deviations have been multiplied by (100/mean) to obtain the coefficients of variation. The factor 10 is arbitrary, and has been used to adjust the final nutrient index values to a convenient magnitude. Other factors could be used if desired, as long as they were the same for all the functions.

The reason for introducing the coefficient of variation (or, in effect, the standard deviation) of the estimated optimum ratio may be explained in connection with Fig. 3-38. This figure shows hypothetical normal distributions of survey observations for nutrient ratios A/B and A/C. The distribution for A/B has a relatively low coefficient of variation, and that for A/C has a relatively high coefficient of variation.

The purpose of the nutrient indexes, which will be discussed next, is to rate the supply of each nutrient relative to the others as a potential limiting factor. The more rapid decrease of crop yield with increasing deviation of A/B from its optimum than with deviation of A/C from its optimum in Fig. 3-38 signifies that the yield is more sensitive to variations in A/B than to variations in A/C. The value of A/B in the test sample thus provides more information about the effect of nutrient A on yield than does the value of A/C. When combining the two ratios to calculate the nutrient index, therefore, it is logical to give more weight to A/B than to A/C. Multiplying the expressions by the reciprocals of the coefficients of variation of the respective estimated optimum ratios provides such a weighting.

From the equations for the functions, it is evident that if the ratio of two nutrients in the sample exceeds the estimated optimum ratio, the value of the function will be positive. And if the ratio of two nutrients in the sample is less than the estimated optimum ratio, the value of the function will be negative. Moreover, if the ratio of two nutrients in the sample is equal to the estimated optimum ratio, the function will have a value of zero.

To obtain the Beaufils indexes for individual nutrients A through N, the weighted functions for the ratios are averaged in the following way:

$$A \ index \ = \ \frac{f\left(\frac{A}{B}\right) \ + \ f\left(\frac{A}{C}\right) \ + \ f\left(\frac{A}{D}\right) \ + \ \dots \ + \ f\left(\frac{A}{N}\right)}{z} \ ,$$

$$B \ index \ = \ \frac{-f\left(\frac{A}{B}\right) \ + \ f\left(\frac{B}{C}\right) \ + \ f\left(\frac{B}{D}\right) \ + \ \dots \ + \ f\left(\frac{B}{N}\right)}{z} \ , \dots \ ,$$

and

$$N \ index \ = \ \frac{-f\left(\frac{A}{N}\right) \ - \ f\left(\frac{B}{N}\right) \ - \ f\left(\frac{C}{N}\right) \ - \ \dots \ - \ f\left(\frac{M}{N}\right)}{z} \ ,$$

where A, B, C, ..., M, and N are different nutrients and z is the number of functions.

The Beaufils index for a given nutrient is the mean of expressions of the deviations of the ratios containing that nutrient from the corresponding optimum ratios. Each index includes all functions containing the nutrient in question. The sign preceding a function is positive if the nutrient occurs in the numerator, and negative if the nutrient occurs in the denominator. Thus, if the function for a particular pair of nutrients has a negative value, it will be entered with a positive value in the equation for calculating the index value for the nutrient in the denominator.

The assembly of indexes for the nutrients concerned provides a measure of the sufficiency of each measured nutrient relative to the others. The greater the index number in the positive direction, the greater is the relative supply.

To illustrate the calculations with an example from Walworth and Sumner (1987), the optimum values for nutrient ratios in corn leaves may be taken as $n/p = 10.04$, $n/k = 1.49$, and $k/p = 6.74$. The corresponding coefficients of variation of the high-yielding population from which the optimum ratios were estimated are $V_{N/P} = 14$, $V_{N/K} = 21$, and $V_{K/P} = 22$. The concentrations of the nutrients found by analysis in the dry matter of a corn leaf sample to be evaluated may be taken as N = 33.0, P = 2.0, and K = 12.0, all in grams per kilogram.

The nutrient ratios in the sample are then N/P = 33.0/2.0 = 16.50, N/K = 33.0/12.0 = 2.75, and K/P = 12.0/2.0 = 6.00, and the values for the nutrient functions are:

$$f\left(\frac{N}{P}\right) = (10)\left(\frac{16.50}{10.04} - 1\right)\left(\frac{100}{14}\right) = 45.96,$$

$$f\left(\frac{N}{K}\right) = (10)\left(\frac{2.75}{1.49} - 1\right)\left(\frac{100}{21}\right) = 40.27,$$

and

$$f\left(\frac{K}{P}\right) = (10)\left(1 - \frac{6.74}{6.00}\right)\left(\frac{100}{22}\right) = -5.61.$$

The values for the functions are averaged in the following way to obtain indexes for the individual nutrients:

$$\text{Nitrogen index} \quad = \frac{f\left(\frac{N}{P}\right) + f\left(\frac{N}{K}\right)}{2}$$

$$= \frac{(45.96 + 40.27)}{2} = 43.115,$$

$$\text{Potassium index} \quad = \frac{-f\left(\frac{N}{K}\right) + f\left(\frac{K}{P}\right)}{2}$$

$$= \frac{-40.27 - 5.61}{2} = -22.940,$$

and

$$Phosphorus\ index\ =\ \frac{-f\left(\frac{N}{P}\right)\ -\ f\left(\frac{K}{P}\right)}{2}$$

$$=\ \frac{-45.96\ +\ 5.61}{2}\ =\ -20.175.$$

An additional point, which may be verified by the numerical values just calculated, is that the sum of the positive indexes is equal to the sum of the negative indexes. The indexes usually are rounded off to whole numbers because their precision does not justify the number of significant figures shown here. A computer program for calculating the nutrient indexes is available from Letzsch and Sumner (1983). Interpreting the indexes will be discussed in Section 3-5.4.5.

Modifications of the foregoing method of calculating the Beaufils nutrient indexes have been suggested. Beverly (1987a) proposed simplified calculations for the circumstances in which transforming nutrient ratios normalizes the distributions. Elwali and Gascho (1984) modified the method of calculating nutrient indexes by considering any nutrient ratio within the range of the estimated optimum plus or minus the standard deviation as being in balance.

For the foregoing example, in which the optimum value of N/P is taken as 10.04 and the coefficient of variation as 14%, the standard deviation would be $(14)(10.04)/100\ =\ 1.41$. If the observed ratio of nitrogen to phosphorus in samples to be evaluated is in the range from $10.04\ -\ 1.41\ =\ 8.63$ to $10.04\ +\ 1.41\ =\ 11.45$, Elwali and Gascho (1984) would take the value for $f(N/P)$ as zero. If an observed value of N/P is outside the range of the estimated optimum plus or minus the standard deviation, however, they would calculate the expression for $f(N/P)$ in the regular way. Use of the Elwali and Gascho modification would reduce the absolute magnitude of the nutrient indexes, would increase the number of instances in which the nutrient supplies are indicated to be in balance, and would decrease the number of instances in which deficiencies would be indicated. The results obtained by Savoy and Robinson (1990) on white clover in Louisiana followed this pattern. In their work, a range of about 8/3 standard deviation appeared to be about the best compromise in terms of avoiding overestimates of the numbers of deficiencies and underestimates of the numbers of nondeficiencies. Rathfon and Burger (1991) found the Elwali and Gascho procedure useful for macronutrients in Fraser fir, but for micronutrients, which occurred over a relatively wide range of concentrations, they considered a wider range than the mean plus or minus one standard deviation to be in balance. Use of the estimated optimum ratio plus or minus some arbitrary deviation to represent a range of ratios over which the nutrients might be considered in balance is analogous to the "optimum range" used by Embleton et al. (1978) (Section 3-4.2.4) and the "normal range" proposed by Kenworthy (1961) (Section 3-4.3).

The second mathematical formulation referred to in the first paragraph of this section was proposed by Jones (1981). For simplicity, his approach is presented here in the same notation used in preceding paragraphs. Jones observed that the equations for the functions of the nutrient ratios may be written as

$$f\left(\frac{A}{B}\right) = (10)\left(\frac{(A/B) - (a/b)}{(a/b)}\right)\left(\frac{100}{V}\right),$$

where $(A/B) \geqslant (a/b)$, and as

$$f\left(\frac{A}{B}\right) = (10)\left(\frac{(A/B) - (a/b)}{(A/B)}\right)\left(\frac{100}{V}\right),$$

where $A/B < a/b$.

Noting that the denominator in the second equation is smaller than that in the first equation, Jones called attention to the fact that a given difference between A/B and a/b would have a greater effect on $f(A/B)$ where $A/B < a/b$ than where $A/B \geq a/b$. The nutrient index values thus would be affected. Taking advantage of the fact that the coefficient of variation V is the ratio of the standard deviation S to the mean a/b, Jones then simplified the first equation to

$$f\left(\frac{A}{B}\right) = (10)\left(\frac{(A/B) - (a/b)}{S}\right)$$

and proposed that this equation be used to calculate the nutrient functions irrespective of whether A/B exceeds or is less than a/b. In this equation, the denominator is fixed. The arbitrary multiplier 10 has been included for consistency with the preceding equations, although Jones did not use it.

Hallmark et al. (1987a) made calculations according to both the traditional method and the Jones (1981) method using hypothetical data and showed that for values of A/B greater than a/b, the numerical values of $f(A/B)$ increased linearly with both methods, and the numerical values obtained with the two methods were equal. For values of A/B less than a/b, the same linear rate of change of $f(A/B)$ persisted with the Jones method, but the rate of change of $f(A/B)$ with the traditional method became logarithmic, and the values of $f(A/B)$ decreased much more rapidly. For example, with $a/b = 1$ and $A/B = 0.1, 1.00$, and 1.90, the values of $f(A/B)$ with the traditional method were $-9,000$, 0, and 900, but the values of $f(A/B)$ with the Jones method were -900, 0, and 900, where the arbitrary values of the standard deviation and the coefficient of variation remained constant.

Hallmark et al. (1987a) then went on to test the two methods of calculation on data from an extensive fertilizer experiment on two soybean cultivars by calculating the nutrient indexes and comparing the resulting diagnoses with the yield response data. The results obtained by the Jones method were definitely superior to those obtained with the traditional method. In a later paper, Hallmark et al. (1991) noted that the apparent advantage of the Jones method disappeared when the nutrient index values were calculated in such a way that the nutrient under consideration always appeared in the numerator or the denominator when the nutrient functions were evaluated. With this method of calculation, simplicity thus appears to be the chief virtue of the Jones method.

See a paper by Hallmark and Beverly (1991) for a review of problems they perceived in calculating nutrient indexes. Whether the endpoint in modifications has been reached seems doubtful.

3-5.4.4. Selecting Nutrient Ratios

Beaufils (1973) suggested using only the ratios for which the yields tend to peak in the survey, as illustrated for the ratio of potassium to phosphorus in Fig. 3-35. To test for the existence of a peaking tendency, Beaufils proposed using the variance. According to this test, a peaking tendency exists if the variance of the nutrient ratios in members of the high-yielding population is lower than the variance of the members of the corresponding low-yielding population. In Fig. 3-35, for example, the variability of the ratios evidently decreases as the yields increase.

In studies of nine nutrients and 36 nutrient ratios in sugarcane, Jones (1981) found that the means of the ratios in the high-yielding populations differed significantly from those in the low-yielding populations for 33 of the 36 nutrient ratios, but that the variances differed significantly for only 19 of the 36 ratios. He suggested, therefore, that in deciding on the important factors that influence yields, one should include nutrients in which the high- and low-yielding groups differ significantly in ratio means as well as in ratio variances.

Two ratios are possible for each pair of nutrients, depending upon whether a given nutrient appears in the numerator or the denominator. The results are not the same with the two forms. The usual procedure for resolving this question in estimating the optimum ratios has been to formulate all possible ratios. The selection between the reciprocal ratios for each pair of nutrients then is made on the basis of the variance ratios between the low- and high-yielding populations. The form with the higher variance ratio is selected because it provides the better discrimination between the two populations.

In connection with work on soybean, Beverly (1987a) argued that the reason for the difference between the reciprocals of the ratios is the skewness of the distributions of the ratios. When he transformed the data by taking the natural logarithms of the ratios, his statistical test indicated that the skewness had been eliminated. The variances of the two ratios then were equal. In later work, Hallmark et al. (1990a) adopted the practice of using only the ratios in which the nutrient under consideration is placed in the numerator when calculating the nutrient functions, presumably after transforming the nutrient ratios by taking the natural logarithms.

Indications from studies by Walworth and Sumner (1987) are that, as a minimum, all nutrients that influence yield under the circumstances in question should be involved in calculating nutrient ratios, and an index value should be calculated for each nutrient. Failure to include all the important nutrients in developing the indexes may result in some erroneous diagnoses because one or more of the principal limiting factors may remain unknown. Walworth and Sumner (1987) illustrated this point by data from a factorial experiment on corn involving different quantities of fertilizer nitrogen, phosphorus, potassium, and sulfur.

On the other hand, including nutrients beyond those of importance in the immediate circumstances may add little or nothing in the way of diagnostic value, to judge from the findings reported by Escano et al. (1981b). In experiments on corn in Hawaii, they varied the value of the nitrogen index used as the maximum value to indicate a response, and compared the numbers of correct diagnoses obtained with nitrogen indexes derived from 36 ratios (from nitrogen, phosphorus, potassium, calcium, magnesium, sulfur, iron, copper, and zinc), 10 ratios (from nitrogen, phosphorus, potassium, calcium, and magnesium), and 3 ratios (from nitrogen, phosphorus, and potassium). Through most of the range of arbitrary cutoff values of the nitrogen index adopted for testing, the numbers of correct predictions were greatest with 10 ratios, intermediate with 36 ratios, and lowest with 3 ratios, but the differences were small. When the cutoff index was high enough, no difference was observed among the numbers of ratios used.

3-5.4.5. Interpreting Nutrient Indexes
Under the hypothetical conditions in which all the nutrient ratios are equal to their respective optima, the nutrient indexes will be equal to zero, and the nutrients will be "in balance." If the supply of the total package of nutrients is optimum, the crop should produce the maximum yield permitted by other limiting factors.

In practice, there are invariably positive and negative values for the nutrient indexes. If the index values are arranged in numerical order from the nutrient with the maximum positive value down to the nutrient with the maximum negative value, the implication is that each nutrient in the descending sequence is present in lower supply relative to the need than the nutrient above it. The greater the numerical difference between two indexes, the greater is the implied difference in degree of sufficiency. Thus, the greater the spread among the various nutrient indexes, the poorer will be the nutrient balance, and the lower is the yield that may be expected. Within a given fertilizer experiment, in which factors other than the nutrient supplies are fairly constant, a close negative correlation may be found between the crop yield and the sum of the absolute values of the Beaufils nutrient indexes for the crop receiving different treatments (Walworth and Sumner, 1987; Angeles et al., 1990).

The nutrient balance, however, is only one of the factors that may affect crop yields. Consequently, when the yields of a crop under a variety of conditions are plotted against the sums of the absolute values of the Beaufils nutrient indexes, the general concept may be preserved, but a wide scatter of points is observed. Illustrating this point is Fig. 3-39, a boundary-line plot of corn yields versus the

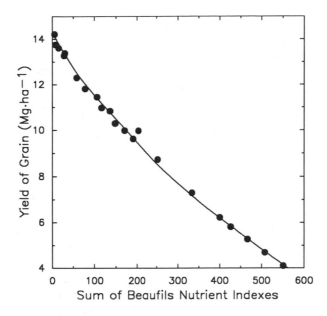

Fig. 3-39. Boundary-line plot of yield of corn grain versus the sum of the Beaufils nutrient indexes without regard for sign. The nutrient index values were based upon the composition of the leaf below the ear at tasseling or first silk as observed in France, Nigeria, Egypt, South Africa, and 14 states in the United States. Only the points used to represent the boundary line are shown in the figure. There were many points to the left of the boundary line, but no points to the right of those shown. (Sumner, 1977a)

sums of the absolute values of nutrient indexes for nitrogen, phosphorus, and potassium in the leaves. No points were found to the right of those plotted as boundary points in the figure, but the area to the left beneath the curve was well filled with points representing individual observations. The implication of Fig. 3-39 is that an optimum balance of nutrients is necessary for high yields, but that it is not a sufficient condition for high yields. The same inference may be made from Fig. 3-35.

Two important aspects of interpretation are consequences of the fact that the Beaufils nutrient indexes are expressions of the supplies of nutrients relative to each other. One has to do with the consequences of a divergent concentration of one nutrient. The concentration of each nutrient has an effect on the index value for each of the other nutrients. An abnormally high concentration of one or more nutrients, for example, will decrease the index values of other nutrients and perhaps will make some nutrients appear deficient when they are in ample supply.

Langenegger and Smith (1978) noted an instance in which the index values for nitrogen, phosphorus, and potassium were -9, -15, and -3, respectively, when their absolute concentrations in the dry matter of pineapple leaves were all in the high to excessive range. In this case, the index value for magnesium was $+22$ as a result of a magnesium concentration more than double the usual concentration reported in their work. If the magnesium concentration had been in the normal range, the index values for nitrogen, phosphorus, and potassium would have been much closer to zero than they were.

Jones and Bowen (1981) found that high concentrations of sulfur in sugarcane samples from fields belonging to one company in Hawaii resulted in two- to three-fold differences among plantations in the estimated optimum values of nutrient ratios involving sulfur. Rathfon and Burger (1991) had difficulty with manganese in applying the Beaufils system to Fraser fir. The relative range in manganese concentrations in the needles was so great (about 0.15 to 3.0 grams per kilogram) that it dominated the indexes for the macronutrients, which had relatively narrow ranges of possible concentrations, and masked the influence of these nutrients. Scrutiny of data for unusual values thus can be helpful in improving interpretations.

The second problem of interpretation is a consequence of the fact that nutrient ratios lack a device for relating them to a practical scale of plant needs and responses. In experiments on sugarcane in Hawaii, Jones (1981) obtained the best agreement with observed responses when he predicted responses from application of all nutrients with indexes less than zero. Thus, in his work, nutrients with positive indexes appeared to be in excess, and nutrients with negative indexes appeared to be deficient. In experiments on corn, also in Hawaii, Escano et al. (1981b) obtained the best agreement with observed responses when the cutoff point was a small positive value for the nutrient indexes.

No general answer can be provided on the cutoff point because of the effect of variation in concentration of the total package of nutrients, a factor that is likely to be of increasing importance as the range of conditions increases. For example, Langenegger and Smith (1978) noted that for two commercial pineapple plantings in South Africa, both yielding 86 tons of fruit per hectare, the leaf compositions in grams of nutrient per kilogram of dry matter were:

Planting *A*: 13.4 for N, 0.64 for P, 12.5 for K, 2.0 for Ca, and 1.4 for Mg

Planting *B*: 23.2 for N, 1.10 for P, 30.5 for K, 3.8 for Ca, and 3.8 for Mg

The nutrient balance was similar in the two plantings. The sums of the nutrient indexes without regard for sign were 46 in planting *A* and 41 in planting *B*. The much lower nutrient concentrations in the leaves of planting *A*, however, suggest that the cutoff index for applying fertilizers to this plantation would be higher than that for planting *B*.

A device that has been tested with available data for relating the nutrient index values to a practical scale of plant responses has been that of introducing dry matter into the calculations (Hallmark et al., 1987b; Walworth, Sumner, et al., 1986). Instead of using just the ratios N/P, N/K, and K/P for calculating the indexes for nitrogen, phosphorus, and potassium, for example, these investigators used the indicated ratios together with the concentrations of the nutrients in the dry matter to obtain a dry matter index, in which dry matter is entered in the sense of another nutrient. To use the Jones (1981) method for calculating the function for the ratio of nitrogen to dry matter, for example,

$$f(N/DM) = (10)\left(\frac{N/DM - n/dm}{S}\right),$$

where N/DM = ratio of nitrogen to dry matter in the sample, n/dm = optimum ratio of nitrogen to dry matter, S = standard deviation of the optimum ratio of nitrogen to dry matter, and the units are grams of nitrogen per kilogram of dry matter.

Where dry matter is included as the equivalent of a nutrient in an instance in which only indexes of nitrogen, phosphorus, potassium, and dry matter are calculated, the nitrogen index would be

$$N\ Index = \frac{f(N/P) + f(N/K) + f(N/DM)}{3},$$

and the dry matter index would be

$$Dry\ Matter\ Index = \frac{-f(N/DM) - f(P/DM) - f(K/DM)}{3}.$$

An example using a dry matter index is discussed in Section 3-6.

As a result of their work on alfalfa, Walworth, Sumner et al. (1986) concluded that a dry matter index was useful for judging whether the nutrients were deficient or in adequate supply. Nutrients with index values equal to or greater than the dry matter index were considered nondeficient, and those with index values below the dry matter index were considered deficient. They noted, however, that introducing a dry matter index could increase the sensitivity of the nutrient index values to the age of the plant tissue sampled for analysis. Some of the advantage conferred by the use of pure nutrient ratios, which eliminate the use of dry matter, thus could be lost.

In subsequent work with more extensive data on soybean leading to 95% accurate diagnoses by a modified procedure versus 72% by the regular procedure, Hallmark et al. (1990a) assumed that all nutrients with negative index values were deficient and all nutrients with positive index values were nondeficient. They calculated a dry matter index, but did not use it as a basis for discrimination, presumably because it was not found advantageous.

One final point worth mentioning is the fact that nutrient index values are affected by the estimated optimum ratios, the standard deviations of the observations from which the optimum ratios are estimated, and the value of the arbitrary constant that may have been used in calculating the nutrient ratio functions. Caution therefore is needed in making comparisons of nutrient index values for different circumstances.

3-5.4.6. Relation to Critical Concentrations

The concepts involved in the use of nutrient ratios, as outlined in the preceding sections, provide background for considering at this point the relationship between the critical concentration concept and Beaufils nutrient indexes. The critical concentration concept involves the action of nutrients singly. There is diagnostic value in determining the ratio of the concentration of at least nitrogen in grass to the critical concentration of nitrogen, as found by Lemaire et al. (1989) (see Fig. 3-27). The extent to which the behavior illustrated in Fig. 3-27 applies to other nutrients and other species needs investigation.

Attention is called once again to Fig. 3-21, which shows that in a plot of world data on yields of corn against the concentrations of phosphorus in the leaves, the yields peaked at a certain phosphorus concentration, which may be called a critical concentration. The authors (Walworth, Letzsch, et al., 1986) demonstrated a similar behavior in plots of yields against concentrations of nitrogen and potassium in the leaves.

The peaking of the yields at particular concentrations of the nutrients is so clear that it cannot be attributed to chance. It must have fundamental significance. Because the same yield data were used in all plots to the extent of availability of analytical data for the various nutrients, the critical concentrations of the various nutrients occurred in essentially the same samples. Therefore, if the critical concentration is defined as the concentration associated with the maximum yield, the ratios of these critical concentrations are also the optimum concentration ratios. The Beaufils nutrient ratio approach assumes in effect that the ratios of the critical concentrations found at peak yield levels are optimum ratios that apply as well to the suboptimum conditions that normally prevail.

The Beaufils nutrient ratio approach uses these experimentally determined optimum nutrient ratios as a basis for rating the sufficiencies of individual nutrients in a sample relative to each of the other nutrients included in the evaluation. The numerical values obtained are not necessarily equal to the ratios of the sample concentrations to the corresponding critical concentrations under the prevailing suboptimum conditions. But under the hypothetical circumstances in which the ratios of the critical concentrations are equal to the optimum ratios, the two sets of values are closely related. An illustration based upon hypothetical data is found in Table 3-5, and plots of the two sets of evaluations with nutrient indexes calculated by the traditional and Jones methods are shown in Fig. 3-40. The two methods used to calculate the Beaufils nutrient indexes yielded different numerical values. When the sample concentrations were equal to the critical concentrations, however, the Beaufils nutrient indexes were close to zero by both methods.

3-5.4.7. Effects of Plant Tissue and Age

For nutrient ratios to reduce or largely eliminate the influence of time of sampling, which is so troublesome with critical concentrations, the concentrations of the nutrients forming the numerator and denominator of the ratios must change in the same direction as the age of the tissue increases. Table 3-6 shows data on

Table 3-5. Comparison of the Critical Concentration Concept and Beaufils Nutrient Indexes for Evaluating the Relative Supplies of Five Nutrients Based Upon Hypothetical Data

Nutrient	Critical Concentration (y)	Sample Concentration (x)	x/y	Beaufils Nutrient Index[a] Traditional Method	Jones Method
A	0.2	0.12	0.60	−38.8	−21.3
B	2.0	1.60	0.80	−12.4	−6.1
C	0.1	0.09	0.90	−2.8	−5.0
D	1.6	1.70	1.06	10.8	7.1
E	0.4	0.60	1.50	43.2	25.3

[a] Calculated on the assumptions that (a) the optimum ratios of nutrients are equal to the ratios of the critical concentrations, (b) the coefficient of variation of all optimum ratios is 20%, and (c) the arbitrary multiplier used in calculating the functions of the various ratios is 10.

the nitrogen, phosphorus, and potassium concentrations in soybean leaves sampled at different dates and from different locations on the plants, together with the Beaufils nutrient indexes. The nutrient concentrations in the leaf dry matter varied considerably with both the time after emergence of the plants and the location of the leaves on the plants. The Beaufils nutrient indexes also varied with time, but they tended to change in the same direction. Moreover, in all samples the nutrient indexes indicated that potassium was in low supply relative

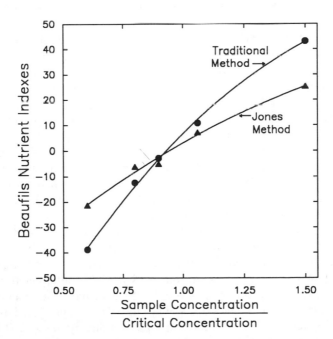

Fig. 3-40. Beaufils nutrient indexes calculated by two methods for five nutrients in a hypothetical plant sample versus the ratio of the sample concentrations to the critical concentrations. The data are derived from Table 3-5. See the table footnote for assumptions.

Table 3-6. **Nutrient Concentrations in Soybean Leaves From a Field Experiment in Iowa With Samples Taken at Different Times and From Different Locations on the Plants, Together With Beaufils Nutrient Indexes Calculated From the Concentrations[a] (Hanway and Weber, 1971; Sumner, 1977b)**

Days After Emergence (approx.)	Concentration in Leaf Dry Matter ($g \cdot kg^{-1}$)			Nutrient Indexes		
	N	P	K	N	P	K
Leaves from lowest 7 nodes						
52	48.0	2.4	10.0	35	−3	−32
73	40.0	2.4	10.0	21	3	−24
92	31.0	2.0	10.5	9	−1	−8
102	22.0	1.6	7.5	5	5	−10
Leaves from nodes 8 to 14						
52	59.0	3.9	15.0	18	8	−26
73	53.0	3.1	14.0	20	0	−20
92	40.0	2.3	13.0	13	−5	−8
102	29.5	2.2	11.5	1	3	−4
Leaves from nodes above 14						
73	64.5	4.2	17.5	15	6	−21
92	42.5	2.5	11.0	20	1	−21
102	29.0	2.0	9.5	8	3	−11

[a] Data published by Hanway and Weber, with nutrient indexes calculated by Sumner.

to nitrogen and phosphorus, and in 9 of 11 samples the phosphorus supply was indicated as low relative to the nitrogen supply. These findings illustrate the virtue of the nutrient-ratio approach in relaxing the sampling limitations required to obtain results that can be interpreted in a meaningful way. Sumner (1977c) reported similar data for leaves taken from different positions on corn plants.

In some instances, age does seem to be of significance. Barry et al. (1989) found that the standard procedure of analyzing the corn leaf below the ear at silking provided a less sensitive indication of the response of grain yield to fertilizer phosphorus in Ontario than did analyses made of the young shoots at the 4- to 6-leaf stage. This was true whether they used the phosphorus concentration as such or the Beaufils phosphorus index. Their results reflect the "starter effect" that seems especially important for corn grown in cooler areas. The pronounced responses of young plants to fertilizer phosphorus diminish as the plants grow, and responses often are no longer visually evident by silking time. The authors noted that paralleling the visual observations, the numerical values for the phosphorus index increased from − 18 at 32 days after planting, to − 11 at 39 to 47 days after planting, and to − 3 at 55 to 62 days after planting. Other findings indicating significant effects of age include those published by Amundson and Koehler (1987) in Washington for wheat and by Savoy and Robinson (1989) in Louisiana for dallisgrass. These examples are all for grass species that do not show the marked increase in calcium concentration with age that will be noted later in this section.

Preliminary experimental verification of course is always desirable to determine whether evaluations made from analyses on samples differing from those on which optimum ratios have been evaluated are valid. Langenegger and Smith (1978) reported findings on pineapple that illustrate the effect of plant parts with this crop. They estimated optimum ratios among certain nutrients from analyses on the "D leaf" of the pineapple plant and then applied these optima in calculating Beaufils nutrient indexes from analyses of the total leaf and various segments of the leaf. The estimated orders of nutrient requirements were:

> Entire leaf . . . . N>Mg>Ca>P>K
> Green section . . . . N>Mg>Ca>P>K
> Leaf tip . . . . N>P>Mg>Ca>K
> Transitional section . . . . N>Mg>Ca>K>P
> Basal white section . . . . Ca>Mg>N>K>P

Because all these evaluations were derived from the same plants, some were evidently inappropriate.

Similar ratios of certain nutrients in plant tissues often may be obtained over a range of times of sampling; however, constancy of nutrient ratios with time is not to be expected. Nutrient-availability ratios in soils are not necessarily constant with time. For example, the highly soluble sources of nutrients supplied in fertilizers at or before planting time may lose much of their potency as the season progresses because of utilization by the crop, reaction with the soil, or loss from the soil. Many experiments on fertilizers tagged with phosphorus-32 have shown that the percentage of the plant phosphorus derived from the fertilizers decreases with time during the season. Thus, for example, the ratio of phosphorus to magnesium in the tissue of plants grown on soil fertilized with a highly soluble phosphate fertilizer would be expected to decrease with time. Depending upon the conditions, the phosphorus component of the ratio might reflect mostly the supply of fertilizer phosphorus at the beginning of the season and the supply of soil phosphorus at the end.

Although the concentrations of nitrogen, phosphorus, and potassium usually decrease with time, complications arise because some nutrients do not behave in this way. Fig. 3-8 illustrates a situation in which the concentrations of calcium and magnesium in peach leaves increased with time, whereas the concentrations of nitrogen, phosphorus, and potassium decreased. The ratio of calcium to magnesium remained fairly constant with times of sampling; however, the ratio of calcium (which increased) to nitrogen (which decreased) varied more with time than did the concentration of either of these nutrients.

Beaufils (1973) developed a system for elements whose concentrations increase with time (he mentioned calcium, magnesium, and manganese) that was similar to the system for nutrients whose concentrations decrease with time. To connect the two classes of nutrients in a common system, he entered the nutrients whose concentrations increase with time as reciprocals. As examples, he gave

K/(1/Mn) and N/(1/Mn), where potassium and nitrogen are the nutrients whose concentrations decrease with time, and manganese is the nutrient whose concentration increases with time. The reciprocals decrease with time. Using the reciprocal of the manganese concentration in the foregoing examples is equivalent to entering the product of the concentrations of potassium and manganese and the product of the concentrations of nitrogen and manganese.

For the data in Fig. 3-8, where the concentrations of calcium and magnesium increased with time and concentrations of nitrogen, phosphorus, and potassium decreased, calcium and magnesium would be entered as $X = 1/Ca$ and $Y = 1/Mg$. The ratios of the concentrations of nitrogen to these "new" nutrients would be $N/X = N/(1/Ca) = N{\cdot}Ca$ and $N/Y = N/(1/Mg) = N{\cdot}Mg$ in calculating the functions of the nutrient ratios and the nutrient indexes. Sumner (1990) observed that after going through the calculations using N/X for $N/(1/Ca)$ and N/Y for $N/(1/Mg)$, one changes the sign of the X and Y indexes to obtain the calcium and magnesium indexes. This is because calcium and magnesium are really in the numerator instead of the denominator.

For the data in Fig. 3-8, the reciprocal device transformed the calcium and magnesium concentrations so they behaved about like the concentrations of the nutrients that did actually decrease with time. The ratios of nitrogen to the transformed calcium and magnesium concentrations were fairly stable with time. Thus, Fig. 3-41 shows that the coefficients of variation of the ratios of the

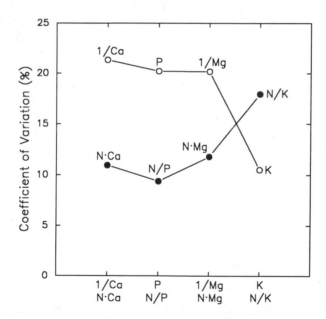

Fig. 3-41. Coefficients of variation for concentrations of single nutrients (upper line) and certain nutrient ratios and products (lower line) in peach leaves sampled 11 times from 32 to 183 days after full bloom in Washington. The calculations employed data shown in Fig. 3-8. The coefficient of variation for the nitrogen concentration (not shown) was 22.3%. This value exceeded all of those plotted in the figure. (Batjer and Westwood, 1958)

nitrogen concentration to the phosphorus concentration and the reciprocals of the calcium and magnesium concentrations with time of sampling were much lower than the corresponding coefficients of variation of the concentration of phosphorus and the reciprocals of the concentrations of calcium and magnesium. In this example, the coefficient of variation of the nitrogen/potassium ratio exceeded the coefficient of variation of the potassium concentration, an anomalous result that is related to the way the potassium concentration changed with time in the early samplings in Fig. 3-8. The more usual change of potassium concentration with time is a gradual decrease, as shown for apple leaves by Emmert's (1954) data.

With adequate knowledge of the way the concentrations of individual nutrients change with time in a given crop, it is possible to treat the data in a way that will yield fairly constant nutrient ratios. Thus, a nutrient balance diagnosis can be made over a range of growth stages, and an indicated nutrient deficiency can be treated by applying a fertilizer (if this is feasible), with checks made on subsequent samples to determine the effects of the treatment. In other words, the system lends itself to the crop-logging concept discussed previously in Section 3-4.2.8.

In preliminary work to investigate the suitability of seeds as a tissue for use with the Beaufils system, Hallmark et al. (1985) found that in two experiments on zinc fertilization of soybean in Alabama, the highly negative indexes for zinc increased with zinc fertilization and indicated that zinc was the most deficient of the nutrients tested. Seed analyses of course may provide an index of deficiencies to be expected in the future.

3-5.4.8. Use With Soil Analysis Data

The reason for the term "diagnosis and recommendation integrated system" developed by Beaufils is that the total system involves an integration of plant analyses with soil analyses and other measurable factors that may affect the growth and yield of a crop as a basis for recommending practices to improve the crop. As an example of the value of the joint use of plant and soil analyses, Sumner (1974) published the data in Table 3-7, derived from a factorial experiment in which fertilizer nitrogen, phosphorus, and potassium were applied to corn. Analyses of the leaf below the ear for nitrogen, phosphorus, potassium, sulfur, calcium, and magnesium were made at the time of tasseling, and these data are shown in the original paper. Table 3-7 includes only the analyses for nitrogen, phosphorus, and potassium, and the nutrient indexes calculated therefrom. In this instance, Sumner modified the usual system of representing the nutrient indexes by taking the one with the highest numerical value as zero and expressing the others relative to this value. This gives all nutrients negative signs except the one with the highest index.

In treatment *A*, which received a basal application of fertilizer nitrogen (N1), phosphorus (P1), and potassium (K1), the nutrient indexes indicated that phosphorus was the most limiting of the three nutrients. When extra phosphorus was applied in superphosphate (treatment *B*), the grain yield increased, and the nu-

Table 3-7. Grain Yields Obtained in an Experiment on Corn in Natal, Union of South Africa, as Related to Diagnoses of Nutrient Deficiencies Based Upon Nutrient Indexes Derived from Leaf Analyses (Sumner, 1974)

Treatment[a]		Concentration in Leaf Dry Matter (g·kg⁻¹)			Nutrient Indexes			Grain Yield
Letter	NPK	N	P	K	N	P	K	(Mg·ha⁻¹)
A	111	34.1	3.8	21.3	0	−41	−36	5.86
B	121	21.9	4.3	22.2	−12	0	−6	6.31
C	221	32.7	4.2	20.0	0	−17	−37	6.97
D	222	33.2	3.8	23.2	0	−36	−15	6.65

[a] Treatments as follows: N1 = 60 kg N·ha⁻¹ and N2 = 180 kg N·ha⁻¹, both as limestone ammonium nitrate containing 300 g N·kg⁻¹; P1 = 60 kg P·ha⁻¹ and P2 = 180 kg P·ha⁻¹, both as ordinary superphosphate containing 83 g P·kg⁻¹; K1 = 60 kg K·ha⁻¹ and K2 = 180 kg K·ha⁻¹, both as potassium chloride containing 500 g K·kg⁻¹.

trient indexes indicated that nitrogen was the most limiting of the three nutrients. When fertilizer nitrogen was applied (treatment *C*), the grain yield increased again, and the nutrient indexes indicated that potassium was the most limiting of the three nutrients. When potassium chloride was applied as a source of potassium in treatment *D*, however, the yield decreased, implying that something else that had not been taken into account was limiting the crop yield.

To judge from the soil analysis data, the problem caused when potassium chloride was added to the fertilizer was aluminum toxicity. The soil was strongly acid, with a pH of 4.3 in molar potassium chloride solution, and the extractable aluminum was more than doubled by the quantity of fertilizer potassium added. The plant analysis data provided no information on these matters.

The relevance of the Beaufils nutrient indexes to the response of the crop to fertilization was illustrated further in the same paper (Sumner, 1974) by information showing that when the nutrient indexes indicated phosphorus to be the most important limiting nutrient, applying fertilizer nitrogen caused the yield to decrease. The yield dropped even more if fertilizer potassium was applied.

3-5.4.9. Diagnostic Use

The preceding sections have emphasized the complex mechanics of the nutrient indexes developed by the Beaufils system. For practical purposes, the estimated optimum nutrient ratios derived from many measurements of crop yield and nutrient concentration may be used as a set of standards against which to judge the sufficiencies of various nutrients relative to each other in specific plant samples from growers' fields. The usefulness of the information is increased with the availability of other data that provide a more comprehensive numerical evaluation of the circumstances.

For research purposes, Beaufils nutrient indexes can be useful as a source of information on nutrient-availability effects of treatments that may not involve direct additions of nutrients. Bethlenfalvay et al. (1990) used the indexes to investigate the effects of inoculating soybean with different isolates of mycorrhizal fungi. Crop rotations and tillage systems are additional possibilities.

3-6. Comparing Evaluation Systems

Several criteria have been used to rate different systems for evaluating the nutrient status of crops. One is to determine which system results in the smallest sum of Beaufils nutrient indexes calculated without reference to sign. A second is to compare the increases in yield that result when different systems are used. A third is to compare the numbers of correct and incorrect diagnoses when different systems are used on a given set of experimental data showing responses to fertilization.

Walworth and Sumner (1987) reviewed the results that had been obtained in comparisons of systems and found the Beaufils nutrient index system superior. In some more recent work, however, the Beaufils system has not performed as well as the sufficiency range or critical concentration system (Hallmark et al., 1989).

Walworth and Sumner (1987) made their ratings on the basis of the systems as used, recognizing that in many instances the systems were not strictly comparable. The following two examples of comparisons of the Beaufils system with others are intended to provide some appreciation for the difficulties involved.

As an index of the acceptability or validity of the evaluation criteria, Hallmark et al. (1987b) used the degree of agreement between experimentally observed responses in a complex fertilizer experiment on soybean and predictions by the critical concentration and Beaufils systems. An evaluation was classed as correct or incorrect according to whether or not it agreed with the finding of a response or no response to phosphorus and potassium, which were the primary deficiencies. Table 3-8 summarizes the numbers of correct evaluations. Over all, the three methods made correct predictions in about two-thirds of the cases.

Table 3-8. Comparison of Yield Responses of Soybean to Fertilizer Phosphorus and Potassium in a Field Experiment in Iowa With the Phosphorus and Potassium Status of the Crop as Evaluated by Different Systems (Hallmark et al., 1987b)

| | | Number of Evaluations[a] | | | |
| Increase in Bean Yield From Fertilizer Phosphorus or Potassium (kg·ha^{-1})[c] | Primary Deficiency[c] | Total | Number of Correct Evaluations by Indicated System[b] | | |
			Critical Concentration	Beaufils	Modified Beaufils
>100	K	12	12	12	12
	P	5	0	3	0
14–100	K	5	2	4	1
	P	3	0	0	0
0	None	6	5	0	5

[a] Based upon trifoliate leaf samples taken at full bloom. The experiment included 31 treatments with seven levels each of limestone and fertilizer phosphorus and potassium.

[b] Nutrients were diagnosed as deficient if the concentration in the dry matter was below the critical concentration in the critical concentration system, if the nutrient index was more negative than the dry matter index in the modified Beaufils system, and presumably if the nutrient index was negative in the Beaufils system.

[c] Based upon the magnitude of the increases in yield from fertilization as estimated from the empirical equation used to represent the response surface.

Additional interpretation is needed to appreciate more fully the significance of the results. The basic limitation is that in the table, one has no real standard for comparison. The field experiment was not without experimental error. The results of the experiment were interpreted through use of an empirical response surface equation for which the R^2 value was 0.74. That is to say, the yield effects used as the basis for evaluation were the differences in yields estimated for the various treatments by the statistically fitted response surface and not by the differences in observed yields. This procedure would have the advantage of reducing the effect of random variations in treatment yields and responses due to experimental error. But if a different empirical model had been chosen, the estimated yield effects would have differed from those shown.

Because of the problem of experimental error, the primary deficiencies have been separated in the table according to whether they were associated with increases in yield exceeding 100 kilograms per hectare or with increases in yield ranging from 14 to 100 kilograms per hectare, or whether neither nutrient resulted in an increase in yield. Relatively large increases in yield were likely to be real, but small increases might have resulted from experimental error in the yield data or from an inappropriate fit of the statistical model. Classifying the data in this way showed that where the increases in yield from the fertilizer treatments exceeded 100 kilograms per hectare, the Beaufils system performed best, with the other two systems not far behind.

Where the increases in yield from the treatments ranged from only 14 to 100 kilograms per hectare, the Beaufils system gave the best performance with four correct diagnoses out of a possible eight. The critical concentration system had only two correct predictions, and the modified Beaufils system had only one.

Where no deficiency of phosphorus or potassium was indicated, the critical concentration and modified Beaufils system performed best, each with five correct diagnoses out of a possible six. The Beaufils system had no correct predictions. When the criterion of response is a negative nutrient index, the Beaufils system would invariably be expected to produce at least some inaccurate diagnoses because, on average, half of the indexes will be negative.

If the no-response category is eliminated, the Beaufils system was the best of the three, with 19 correct diagnoses out of 25 possibilities. The critical concentration system was second, with 14 correct evaluations, and the modified Beaufils system was third, with 13 correct evaluations.

The modified Beaufils system was tested as a means of estimating when the supply of all the nutrients evaluated was adequate. The modified Beaufils system was as successful as the critical concentration in indicating no deficiency of phosphorus or potassium where none was found, but it was less successful than the Beaufils system where the increase in yield from fertilization was only 14 to 100 kilograms per hectare. Whether this difference is meaningful is not known.

One other significant point of interpretation is needed to appreciate the difficulty in making valid comparisons between the different methods. All depend upon standard values that were determined independently of the experiment, and the accuracy of these values is not guaranteed. For example, the critical concentration method failed to diagnose all five phosphorus deficiencies that resulted in yield increases exceeding 100 kilograms per hectare. This result suggests that the value taken as the critical concentration for phosphorus was too high. If a lower value had been used for the critical concentration, some or all of the diagnoses based upon the critical concentration approach could have been correct.

Experimental work by Elwali and Gascho (1984) is described as a second example. They used yields and the sums of Beaufils nutrient indexes to compare soil tests, critical nutrient concentrations, and Beaufils nutrient indexes (calculated in the manner described in connection with their work in Section 3-5.4.3) as bases for fertilizing sugarcane in eight field experiments in Florida.

Soil samples were taken in May, and all the fertilizer phosphorus and potassium indicated by the soil test values was applied in May. No soil tests were made for micronutrients, and no micronutrients were applied in the soil test system, as is the accepted practice for ratoon sugarcane in Florida. No soil test was made for nitrogen. Fertilizer nitrogen attributed to the soil test system was applied to three fields according to Cooperative Extension Service recommendations on the basis of soil type (applications were made to two fields located on a sandy soil and a limited application was made to a field on a mucky sand, but no fertilizer nitrogen was applied to five fields on mucks.)

The critical concentrations for nitrogen and phosphorus were estimated from local experiments by plotting the yields against the concentrations of the nutrients individually and finding statistically the concentrations corresponding to 95% of the maximum yields. The same procedure was used for other nutrients, except that some adjustments were made to take into account values obtained elsewhere. Optimum nutrient ratios for calculating the Beaufils nutrient indexes were estimated from scatter diagrams of yields versus nutrient ratios (see Section 3-5.4.1). The data were derived from previous analyses on about 1,600 leaf samples from local commercial sugarcane fields and experimental plots and from the corresponding yields.

For diagnosing nutrient deficiencies from plant analyses in the experiments, leaf samples were taken in May and July, and fertilizers were applied in May and July on the basis of the findings. Another set of samples was taken in September to evaluate the results obtained with the different systems.

In none of the three systems did the indicated tests provide a basis for deciding how much of the nutrients indicated to be deficient should be applied. Fertilizer applications based upon soil test results were made according to calibration experiments conducted many years previously. Quantities recommended in the critical concentration and Beaufils nutrient index systems were standard applications published previously for conditions under which individual nutrients were indicated to be deficient. Fertilizer applications were made when the nutrient concentration was below the critical concentration or when the Beaufils nutrient index was negative. Generally, more fertilizer was used when the value was much below the critical concentration or when the Beaufils nutrient index was strongly negative. This practice was not followed for potassium applications to

Table 3-9. Nutrients Applied, Yields of Sugar, and Sums of Absolute Values of Beaufils Nutrient Indexes in Eight Field Experiments in Florida on Sugarcane Fertilized on the Basis of Soil Tests, Critical Nutrient Concentrations, and Beaufils Nutrient Indexes (Elwali and Gascho, 1984)

Basis for Fertilizer Application	Total Nutrients Applied (kg·ha⁻¹)	Yield of Sugar (Mg·ha⁻¹)	Sum of Beaufils Nutrient Indexes in September[a] Nine Nutrients[b]	NPK
Critical concentrations	107	7.7	36	3
Soil testing	262	7.9	46	8
Beaufils nutrient indexes	351	9.4	24	3
L.S.D.[c]	—	0.7	10	6

[a] Sum of absolute values without regard for sign.
[b] Nitrogen, phosphorus, potassium, calcium, magnesium, iron, manganese, zinc, and copper.
[c] Least significant difference at the 5% probability level.

the sands, where some of the potassium might be lost by leaching. In these instances, a second application was relied upon to supply potassium that otherwise would have been applied at the earlier date.

The most significant points to note from the results in Table 3-9 are: (1) The nutrient balance, as indicated by a low sum of nutrient indexes, was best where the Beaufils system was used for diagnosis and poorest where the soil testing system was used. (2) The yields of sugar increased with the total amounts of nutrients applied in connection with the different systems.

The relatively high sum of Beaufils nutrient indexes in the soil testing system is probably in part a consequence of the fact that the critical concentration and Beaufils systems involved diagnoses for nitrogen, phosphorus, and potassium plus six other nutrients, and the comparisons in the fourth column were based upon indexes for all nine of these nutrients. The soil testing system, however, involved diagnoses for only nitrogen, phosphorus, and potassium.

As a basis for discriminating among diagnostic methods, the correlation between sugar yields and nutrients applied must be discounted to an unknown degree because although all methods provide indications of relative deficiencies, none provide information on the quantities of nutrients needed. The quantities applied in the various systems were based upon different sets of criteria.

A significant auxiliary point noted by Elwali and Gascho (1984) in favor of plant analysis in their experiments on sugarcane is related to the fact that they studied "ratoon" fields (fields on which a sugarcane crop is produced from the shoots that develop from the stubble of a harvested crop without replanting). They noted that sugarcane farmers in South Africa and Florida had been discouraged from sampling the soil of such fields in favor of a single preplant soil sampling for an entire cycle of the first year "plant" crop and two to four succeeding ratoon crops. An important reason is that residues from banded applications of fertilizer to the preceding crop or crops cause problems in soil sampling and interpreting soil analyses. On ratoon crops, therefore, plant analysis has a theoretical advantage over soil analysis because the plants integrate the effects of the nonuniform distribution of nutrients in the soil.

As indicated by the descriptions of the procedures in the soil testing, critical concentration, and Beaufils systems, the three systems differed in a variety of ways. The causes of the observed differences in results are therefore uncertain.

3-7. Concluding Remarks

Qualitative indications of nutrient status are useful and often were considered acceptable in the past. In the current quest for quantitative evaluations, no generally satisfactory system of testing plants for nutrient deficiencies has emerged.

The critical concentration concept, although useful, has the disadvantage that critical concentrations are not constant. This lack of constancy is to be expected from the fact that critical concentrations are ratios of quantities of nutrients to the total mass of the plant tissue in question, and the major variable in the denominator is the organic substances evolved by the plants in response to the uptake of nutrients and to growing conditions. The quantities of organic substances produced are not determined uniquely by the quantity of a specific nutrient of interest.

Using ratios among nutrients is an improvement on the critical concentration concept in that the results are less variable with growth stage of the plant, and they make possible a relative ranking of nutrients. At the same time, the ratio concept loses the advantage the critical concentration concept does have in indicating whether the plant has enough of a particular nutrient or whether the supply is deficient or excessive.

The major need of the nutrient ratio concept is a way to regain the loss sustained by divorcing the consideration of the nutrients from the behavior of the plants. A way to judge unambiguously the sufficiency of the complete assembly of nutrients or of only one nutrient would make possible a further advance. The introduction of dry matter as the equivalent of another plant nutrient has been of value in some circumstances, but not all.

Thus far, the research papers on the nutrient ratio concept have depended mostly upon data already published or developed in other connections, in which the nutrient concentrations were expressed on a dry matter basis. Development of an improved way to evaluate nutrient sufficiency would very likely require a fresh start with new experiments so that the needed measurements could be made. Although the nutrient sufficiency evaluations should lead to improved predictions, they still would have limitations because they would respond to conditions such as water deficiency that change with time.

The findings made by Hallmark et al. (1990a) (Section 3-5.4.1) in diagnosing the nutrient status of soybean with the modifications they proposed to the Beaufils ratio system appear promising. Hopefully they can be confirmed in application to independent data on soybean and in application to other crops.

3-8. Literature Cited

Alten, F., E. Rauterberg, and H. Loofmann. 1940. Der Einfluss des Kalis auf den Stickstoffhaushalt der Pflanzen. *Bodenkunde und Pflanzenernährung* 19:22-55.

Amundson, R. L., and F. E. Koehler. 1987. Utilization of DRIS for diagnosis of nutrient deficiencies in winter wheat. *Agronomy Journal* 79:472-476.

Analogides, D. A. 1988. Time dependent interrelationships of plant nitrogen status and crop performance with reference to sugarbeet. Pp. 145-157. In D. S. Jenkinson and K. A. Smith (Eds.), *Nitrogen Efficiency in Agricultural Soils*. Elsevier Applied Science Publishers, London.

Angeles, D. E., M. E. Sumner, and N. W. Barbour. 1990. Preliminary nitrogen, phosphorus, and potassium DRIS norms for pineapple. *HortScience* 25:652-655.

App, F., V. Ichisaka, and T. S. Gill. 1956. The value of green manure crops in farm practice. *Better Crops With Plant Food* 40, No. 3:16-22, 41-43; No. 4:9-16, 40-46.

Asher, C. J., and J. F. Loneragan. 1967. Response of plants to phosphate concentration in solution culture: I. Growth and phosphorus content. *Soil Science* 103:225-233.

Askew, H. O., and E. Chittenden. 1936. The use of borax in the control of "internal cork" of apples. *Journal of Pomology and Horticultural Science* 14:227-245.

Atterberg, A. 1901. Die Variationen des Nährstoffgehalte bei dem Hafer. *Journal für Landwirtschaft* 49:97-172.

Bar-Akiva, A., M. Kaplan, and R. Lavon. 1967. The use of a biochemical indicator for diagnosing micronutrient deficiencies of grapefruit trees under field conditions. *Agrochimica* 11:283-288.

Bar-Akiva, A., D. N. Maynard, and J. E. English. 1978. A rapid tissue test for diagnosing iron deficiencies in vegetable crops. *HortScience* 13:284-285.

Barry, D. A. J., M. H. Miller, and T. E. Bates. 1989. Ear leaf and seedling P concentration and DRIS indices as indicators of P nutrition for maize. *Communications in Soil Science and Plant Analysis* 20:1397-1412.

Bates, T. E. 1971. Factors affecting critical nutrient concentrations in plants and their evaluation: A review. *Soil Science* 112:116-130.

Batjer, L. P., and M. N. Westwood. 1958. Seasonal trend of several nutrient elements in leaves and fruits of Elberta peach. *Proceedings of the American Society for Horticultural Science* 71:116-126.

Baver, L. D. 1960. Plant and soil composition relationships as applied to cane fertilization. *Hawaiian Planters Record* 56:1-86.

Beaufils, E. R. 1971. Physiological diagnosis — a guide for improving maize production based on principles developed for rubber trees. *Fertilizer Society of South Africa Journal* 1:1-30.

Beaufils, E. R. 1973. *Diagnosis and Recommendation Integrated System (DRIS): A General Scheme for Experimentation and Calibration Based on Principles Developed From Research in Plant Nutrition*. University of Natal, Soil Science Bulletin 1.

Beeson, K. C. 1941. *The Mineral Composition of Crops With Particular Reference to the Soils in Which They Were Grown. A Review and Compilation*. U.S. Department of Agriculture, Miscellaneous Publication 369.

Bell, R. W., D. Brady, D. Plaskett, and J. F. Loneragan. 1987. Diagnosis of potassium deficiency in soybean. *Journal of Plant Nutrition* 10:1947-1953.

Bergmann, E., and W. Bergmann. 1985. *Comparing Diagrams of Plant/Leaf Analysis Presenting by Rapid Inspection the Mineral Nutrient Element Status of Agricultural Crop Plants*. Potash Review, Subject 5, 52nd Suite, No. 2/1985.

Bergmann, W. 1983. *Ernährungsstörungen bei Kulturpflanzen*. VEB Gustav-Fischer-Verlag, Jena.

Bethlenfalvay, G. J., R. L. Franson, and M. S. Brown. 1990. Nutrition of mycorrhizal soybean evaluated by the diagnosis and recommendation integrated system (DRIS). *Agronomy Journal* 82:302-304.

Beverly, R. B. 1987a. Comparison of DRIS and alternative nutrient diagnostic methods for soybean. *Journal of Plant Nutrition* 10:901-920.

Beverly, R. B. 1987b. Modified DRIS method for simplified nutrient diagnosis of 'Valencia' oranges. *Journal of Plant Nutrition* 10:1401-1408.

Binford, G. D., A. M. Blackmer, and M. E. Cerrato. 1992. Nitrogen concentration of young corn plants as an indicator of nitrogen availability. *Agronomy Journal* 84:219-223.

Binford, G. D., A. M. Blackmer, and N. M. El-Hout. 1990. Tissue test for excess nitrogen during corn production. *Agronomy Journal* 82:124-129. (With additional information supplied by private communication.)

Borden, R. J. 1946. Variety differences in nitrogen utilization. *Hawaiian Planters Record* 50:39-49. (With additional data supplied by private communication.)

Böszörményi, Z. 1957. Leaf analysis investigations with scotch pine seedlings; the problem of the constancy of critical nutrient concentrations. *Acta Botanica Academiae Scientiarum Hungaricae* 4:19-44.

Bot, J. Le, M. J. Goss, G. P. R. Carvalho, M. L. van Beusichem, and E. A. Kirkby. 1990. The significance of the magnesium to manganese ratio in plant tissues for growth and alleviation of manganese toxicity in tomato (*Lycopersicon esculentum*) and wheat (*Triticum aestivum*) plants. Pp. 223-228. In M. L. van Beusichem (Ed.), *Plant Nutrition — Physiology and Applications*. Kluwer Academic Publishers, Dordrecht, The Netherlands.

Bould, C. 1966. Leaf analysis of deciduous fruits. Pp. 651-684. In N. F. Childers (Ed.), *Nutrition of Fruit Crops*. Horticultural Publications, Rutgers University, New Brunswick, NJ.

Bouma, D., and E. J. Dowling. 1980. Field evaluation of a test for phosphorus deficiency in pastures based on dry matter responses induced in detached subterranean clover leaves. *Communications in Soil Science and Plant Analysis* 11:861-872.

Bouma, D., K. Spencer, and E. J. Dowling. 1969. Assessment of the phosphorus and sulphur status of subterranean clover pastures. 3. Plant tests. *Australian Journal of Experimental Agriculture and Animal Husbandry* 9:329-340.

Bowen, J. E. 1990. Plant tissue analysis of sugarcane. Pp. 449-467. In R. L. Westerman (Ed.), *Soil Testing and Plant Analysis*. Third Edition. Soil Science Society of America, Madison, WI.

Bowling, J. D., and D. E. Brown. 1947. *Role of Potash in Growth and Nutrition of Maryland Tobacco*. U.S. Department of Agriculture, Technical Bulletin 933.

Braud, M. 1972. Le contrôle de la nutrition minérale du cotonnier par analyses foliaires. Pp. 469-487. In *3ᵉ Colloque Européen et Méditerranéen sur le Contrôle de la Nutrition Minerale et de la Fertilisation*. Publishing House of the Hungarian Academy of Sciences, Budapest.

Breimer, T. 1982. Environmental factors and cultural measures affecting the nitrate content in spinach. *Fertilizer Research* 3:191-292.

Brooks, S. L., and H. M. Reisenauer. 1985. Plant nitrogen fractions as indicators of growth rate. *Journal of Plant Nutrition* 8:63-71.

Brown, J. C., and S. B. Hendricks. 1952. Enzymatic activities as indications of copper and iron deficiencies in plants. *Plant Physiology* 27:651-660.

Burns, I. G. 1986. Determination of critical K concentrations in sap from individual leaves or from whole plants using K-interruption experiments. *Plant and Soil* 94:301-312.

Carter, J. N., M. E. Jensen, and S. M. Bosma. 1971. Interpreting the rate of change in nitrate-nitrogen in sugarbeet petioles. *Agronomy Journal* 63:669-674.

Cate, R. B., Jr., and L. A. Nelson. 1971. A simple statistical procedure for partioning soil test correlation data into two classes. *Soil Science Society of America Proceedings* 35:658-660.

Cerrato, M. E., and A. M. Blackmer. 1990. Relationships between grain nitrogen concentrations and the nitrogen status of corn. *Agronomy Journal* 82:744-749.

Chapman, H. D. (Ed.). 1966. *Diagnostic Criteria for Plants and Soils*. Division of Agricultural Sciences, University of California.

Chapman, H. D., and G. F. Liebig, Jr. 1940. Nitrate concentration and ion balance in relation to citrus nutrition. *Hilgardia* 13:141-173.

Clements, H. F. 1961. Crop logging of sugar cane in Hawaii. Pp. 131-147. In W. Reuther (Ed.), *Plant Analysis and Fertilizer Problems*. American Institute of Biological Sciences, Publication No. 8. Washington.

Clements, H. F. 1980. *Sugarcane Crop Logging and Crop Control Principles and Practices*. The University Press of Hawaii, Honolulu.

Coïc, Y., G. Fauconneau, R. Pion, C. Lesaint, and S. Godefroy. 1962. Influence de la déficience en soufre sur l'absorption des substances minérales et le métabolisme de l'azote et des acides organiques chez l'orge. *Annales de Physiologie Végétale* 4:295-306.

Dean, L. A., and M. Fried. 1953. Soil-plant relationships in the phosphorus nutrition of plants. Pp. 43-88. In W. H. Pierre and A. G. Norman (Eds.), *Soil and Fertilizer Phosphorus in Crop Nutrition*. Agronomy, Volume IV. Academic Press, NY.

Deckard, E. L., R. J. Lambert, and R. H. Hageman. 1973. Nitrate reductase activity in corn leaves as related to yields of grain and grain protein. *Crop Science* 13:343-350.

Delhaize, E., J. F. Loneragan, and J. Webb. 1982. Enzymic diagnosis of copper deficiency in subterranean clover. II A simple field test. *Australian Journal of Agricultural Research* 33:981-987.

De Villiers, J. I., and C. J. Beyers. 1961. Leaf analysis as a guide to fertilization in commercial orange growing. Pp. 107-119. In W. Reuther (Ed.), *Plant Analysis and Fertilizer Problems*. American Institute of Biological Sciences, Publication No. 8. Washington.

Dijkshoorn, W., J. E. M. Lampe, and P. F. J. van Burg. 1960. A method of diagnosing the sulphur nutrition status of herbage. *Plant and Soil* 13:227-241.

Dijkshoorn, W., and A. L. van Wijk. 1967. The sulphur requirements of plants as evidenced by the sulphur-nitrogen ratio in the organic matter. A review of published data. *Plant and Soil* 26:129-157.

Drake, M., and G. D. Scarseth. 1939. Relative abilities of different plants to absorb potassium and the effects of different levels of potassium on the absorption of calcium and magnesium. *Soil Science Society of America Proceedings* 4:201-204.

El-Gharably, G. A., and W. Bussler. 1985. Critical levels of boron in cotton plants. *Zeitschrift für Pflanzenernährung und Bodenkunde* 148:681-688.

Elwali, A. M. O., and G. J. Gascho. 1984. Soil testing, foliar analysis, and DRIS as guides for sugarcane fertilization. *Agronomy Journal* 76:466-470. (With additional information supplied by private communication.)

Elwali, A. M. O., and G. J. Gascho. 1988. Supplemental fertilization of irrigated corn guided by foliar critical nutrient levels and diagnosis and recommendation integrated system norms. *Agronomy Journal* 80:243-249.

Embleton, T. W., W. W. Jones, and R. G. Platt. 1978. Leaf analysis as a guide to citrus fertilization. Pp. 4-9. In H. M. Reisenauer (Ed.), *Soil and Plant-Tissue Testing in California*. University of California, Division of Agricultural Sciences, Bulletin 1879 (Revised June 1978).

Emmert, F. H. 1954. The soluble and total phosphorus, potassium, calcium, and magnesium of apple leaves as affected by time and place of sampling. *Proceedings of the American Society for Horticultural Science* 64:1-8.

Escano, C. R., C. A. Jones, and G. Uehara. 1981a. Nutrient diagnosis in corn grown on hydric dystrandepts: I. Optimum tissue nutrient concentrations. *Soil Science Society of America Journal* 45:1135-1139.

Escano, C. R., C. A. Jones, and G. Uehara. 1981b. Nutrient diagnosis in corn grown on hydric dystrandepts: II. Comparisons of two systems of tissue diagnosis. *Soil Science Society of America Journal* 45:1140-1144.

Giles, J. F., A. E. Ludwick, and J. O. Reuss. 1977. Prediction of late season nitrate-nitrogen content of sugarbeet petioles. *Agronomy Journal* 69:85-88.

Goodall, D. W., and F. G. Gregory. 1947. *Chemical Composition of Plants as an Index of Their Nutritional Status*. Imperial Bureau of Horticulture and Plantation Crops, Technical Communication 17.

Goos, R. J., D. G. Westfall, A. E. Ludwick, and J. E. Goris. 1982. Grain protein content as an indicator of N sufficiency for winter wheat. *Agronomy Journal* 74:130-133.

Greenwood, D. J. 1982. Nitrogen supply and crop yield: The global scene. *Plant and Soil* 67:45-59.

Greenwood, D. J., F. Gastal, G. Lemaire, A. Draycott, P. Millard, and J. J. Neeteson. 1991. Growth rate and %N of field grown crops: Theory and experiments. *Annals of Botany* 67:181-190.

Grunes, D. L. 1959. Effect of nitrogen on the availability of soil and fertilizer phosphorus. *Advances in Agronomy* 11:369-396.

Hallmark, W. B., J. F. Adams, and H. F. Morris. 1985. Detection of zinc deficiency in soybeans by the diagnosis and recommendation integrated system using seeds. *Journal of Fertilizer Issues* 2:11-16.

Hallmark, W. B., and R. B. Beverly. 1991. Review — An update in the use of the Diagnosis and Recommendation Integrated System. *Journal of Fertilizer Issues* 8:74-88.

Hallmark, W. B., R. B. Beverly, C. J. deMooy, and J. Pesek. 1991. Relationship of diagnostic nutrient expressions to soybean phosphorus and potassium diagnoses. *Agronomy Journal* 83:858-863.

Hallmark, W. B., R. B. Beverly, H. F. Morris, L. M. Shuman, D. O. Wilson, F. C. Boswell, J. F. Adams, and D. A. Wall. 1990a. Continued modification of the M-DRIS for soybean. *Communications in Soil Science and Plant Analysis* 21:1313-1328.

Hallmark, W. B., R. B. Beverly, M. B. Parker, J. F. Adams, F. C. Boswell, K. Ohki, L. M. Shuman, and D. O. Wilson. 1989. Evaluation of soybean zinc and manganese requirements by the M-DRIS and sufficiency range methods. *Agronomy Journal* 81:770-776.

Hallmark, W. B., R. B. Beverly, M. E. Sumner, C. J. deMooy, H. F. Morris, J. Pesek, and J. D. Fontenot. 1990b. Soybean phosphorus and potassium requirement evaluation by three M-DRIS data bases. *Agronomy Journal* 82:323-328.

Hallmark, W. B., C. J. deMooy, and J. Pesek. 1987a. Comparison of two DRIS methods for diagnosing nutrient deficiencies. *Journal of Fertilizer Issues* 4:151-158.

Hallmark, W. B., J. L. Walworth, M. E. Sumner, C. J. deMooy, J. Pesek, and K. P. Shao. 1987b. Separating limiting from non-limiting nutrients. *Journal of Plant Nutrition* 10:1381-1390.

Hammond, L. C., C. A. Black, and A. G. Norman. 1951. *Nutrient Uptake by Soybeans on Two Iowa Soils*. Iowa Agricultural Experiment Station, Research Bulletin 384. Ames.

Hannam, R. J., J. L. Riggs, and R. D. Graham. 1987. The critical concentration of manganese in barley. *Journal of Plant Nutrition* 10:2039-2048.

Hanway, J. J. 1962. Corn growth and composition in relation to soil fertility: II. Uptake of N, P, and K and their distribution in different plant parts during the growing season. *Agronomy Journal* 54:217-222.

Hanway, J. J., and C. R. Weber. 1971. N, P, and K percentages in soybean (*Glycine max* (L.) Merrill) plant parts. *Agronomy Journal* 63:286-290.

Hart, A. L., and D. H. Greer. 1988. Photosynthesis and carbon export in white clover plants grown at various levels of phosphorus supply. *Physiologia Plantarum* 73:46-51.

Hiatt, A. J., and H. F. Massey. 1958. Zinc levels in relation to zinc content and growth of corn. *Agronomy Journal* 50:22-24.

Hockman, J. N., J. A. Burger, and D. W. Smith. 1989. A DRIS application to Fraser fir Christmas trees. *Communications in Soil Science and Plant Analysis* 20:305-318.

Hoffman, M., and R. M. Samish. 1971. Free amine content in fruit tree organs as an indicator of the nutritional status with respect to potassium. *Recent Advances in Plant Nutrition* 1:189-206.

Huffaker, R. C., and D. W. Rains. 1978. Factors affecting nitrate acquisition by plants; assimilation and fate of reduced nitrogen. Pp. 1-43. In D. R. Nielsen and J. G. MacDonald (Eds.), *Nitrogen in the Environment*. Volume 2. *Soil-Plant-Nitrogen Relationships*. Academic Press, NY.

Hylton, L. O., A. Ulrich, and D. R. Cornelius. 1967. Potassium and sodium interrelations in growth and mineral content of Italian ryegrass. *Agronomy Journal* 59:311-314.

Jacob, W. C., C. H. Van Middelem, W. L. Nelson, C. D. Welch, and N. S. Hall. 1949. Utilization of phosphorus by potatoes. *Soil Science* 68:113-120.

Jakob, F., H.-P. Vielemeyer, and W. Podlesak. 1986. Untersuchungen zur Anwendung eines Nitrat-Schnelltests für die Bemessung der N-Düngung zu Winterweizen. *Archiv für Acker- und Pflanzenbau und Bodenkunde* 30:613-621.

Jensen, H. E. 1982. Amino acid composition of ryegrass in relation to nitrogen fertilization and soil water status. *Journal of Plant Nutrition* 5:1109-1120.

Johnson, V. A., P. J. Mattern, and J. W. Schmidt. 1969. The search for high protein wheat. *War on Hunger* 3, No. 3:14-17.

Johnston, A. E., and K. W. T. Goulding. 1990. The use of plant and soil analyses to predict the potassium supplying capacity of soil. Pp. 177-204. In *Development of K-Fertilizer Recommendations*. Proceedings of the 22nd Colloquium of the International Potash Institute held at Soligorsk/USSR. International Potash Institute, Worblaufen-Bern, Switzerland.

Jones, C. A. 1981. Proposed modifications of the diagnosis and recommendation integrated system (DRIS) for interpreting plant analyses. *Communications in Soil Science and Plant Analysis* 12:785-794.

Jones, C. A., and J. E. Bowen. 1981. Comparative DRIS and crop log diagnosis of sugarcane tissue analyses. *Agronomy Journal* 73:941-944.

Jones, J. B., Jr., H. V. Eck, and R. Voss. 1990. Plant analysis as an aid in fertilizing corn and grain sorghum. Pp. 521-547. In R. L. Westerman (Ed.), *Soil Testing and Plant Analysis*. Third Edition. Soil Science Society of America, Madison, WI.

Jones, J. B., Jr., and W. J. A. Steyn. 1973. Sampling, handling, and analyzing plant tissue samples. Pp. 249-270. In L. M. Walsh and J. D. Beaton (Eds.), *Soil Testing and Plant Analysis*. Revised Edition. Soil Science Society of America, Madison, WI.

Jones, M. B. 1962. Total sulfur and sulfate sulfur content in subterranean clover as related to sulfur responses. *Soil Science Society of America Proceedings* 26:482-484.

Joy, K. W. 1988. Ammonia, glutamine, and asparagine: a carbon-nitrogen interface. *Canadian Journal of Botany* 66:2103-2109.

Kamprath, E. J. 1987. Enhanced phosphorus status of maize resulting from nitrogen fertilization of high phosphorus soils. *Soil Science Society of America Journal* 51:1522-1526.

Keisling, T. C., and B. Mullinix. 1979. Statistical considerations for evaluating micronutrient tests. *Agronomy Journal* 43:1181-1184.

Kelling, K. A., and J. E. Matocha. 1990. Plant analysis as an aid in fertilizing forage crops. Pp. 603-643. In R. L. Westerman (Ed.), *Soil Testing and Plant Analysis*. Third Edition. Soil Science Society of America, Madison, WI.

Kenworthy, A. L. 1961. Interpreting the balance of nutrient-elements in leaves of fruit trees. Pp. 28-43. In W. Reuther (Ed.), *Plant Analysis and Fertilizer Problems*. American Institute of Biological Sciences, Publication No. 8. Washington.

Kenworthy, A. L. 1967. Plant analysis and interpretation of analysis for horticulture crops. Pp. 59-75. In G. W. Hardy, A. R. Halvorson, J. B. Jones, R. D. Munson, R. D. Rouse, T. W. Scott, and B. Wolf (Eds.), *Soil Testing and Plant Analysis. Part II, Plant Analysis*. SSSA Special Publication No. 2. Soil Science Society of America, Madison, WI.

Kenworthy, A. L. 1973. Leaf analysis as an aid in fertilizing orchards. Pp. 381-392. In L. M. Walsh and J. D. Beaton (Eds.), *Soil Testing and Plant Analysis*. Revised Edition. Soil Science Society of America, Madison, WI.

Kessler, B. 1961. Ribonuclease as a guide for the determination of zinc deficiency in orchard trees. Pp. 314-322. In W. Reuther (Ed.), *Plant Analysis and Fertilizer Problems*. American Institute of Biological Sciences, Publication No. 8. Washington.

Kiesselbach, T. A. 1948. Endosperm type as a physiologic factor in corn yields. *Journal of the American Society of Agronomy* 40:216-236.

Kim, Y. T., C. Glerum, J. Stoddart, and S. J. Colombo. 1987. Effect of fertilization on free amino acid concentrations in black spruce and jack pine containerized seedlings. *Canadian Journal of Forest Research* 17:27-30.

Kim, Y. T., and R. H. Leech. 1986. The potential use of DRIS in fertilizing hybrid poplar. *Communications in Soil Science and Plant Analysis* 17:429-438.

Kovacevic, V., L. J. Radic, and N. Vekic. 1987. Genetic differences in the ear-leaf nutrient content of inbred lines of corn (*Zea mays* L.). Pp. 399-402. In W. H. Gabelman and B. C. Loughman (Eds.), *Genetic Aspects of Plant Mineral Nutrition*. Martinus Nijhoff Publishers, Dordrecht, The Netherlands.

Krauss, H. H. 1965. Untersuchungen über die Melioration degradierter Sandböden im nordostdeutschen Tiefland. IV. Kalkungs- und Hilfspflanzenanbauversuche — Ernährung und Wachstum meliorerter Kiefernkulturen. *Archiv für Forstwesen* 14:499-532.

Lagatu, H., and L. Maume. 1934a. Action d'un engrais simple annuel sur l'alimentation NPK d'une même espèce végétale au cours de quatre années successives de culture dans le même sol. *Comptes Rendus Hebdomadaires des Séances de L'Académie D'Agriculture de France* 20:549-563.

Lagatu, H., and L. Maume. 1934b. Recherche, par le diagnostic foliaire, de l'équilibre optimum d'alimentation NPK chez une plante cultivée. *Comptes Rendus Hebdomadaires des Séances de l'Academie d'Agriculture de France* 20:631-644.

Langenegger, W., and B. L. Smith. 1978. An evaluation of the DRIS system as applied to pineapple leaf analysis. Pp. 263-273. In A. R. Ferguson, R. L. Bieleski, and I. B. Ferguson (Eds.), *Plant Nutrition 1978*. Proceedings 8th International Colloquium on Plant Analysis and Fertilizer Problems, Aukland, New Zealand. New Zealand Department of Scientific & Industrial Research, Information Series No. 134. Wellington.

Lawlor, D. W., M. Kontturi, and A. T. Young. 1989. Photosynthesis by flag leaves of wheat in relation to protein, ribulose *bis*phosphate carboxylase activity and nitrogen supply. *Journal of Experimental Botany* 40:43-52.

Le Bot, J., M. J. Goss, G. P. R. Carvalho, M. L. van Beusichem, and E. A. Kirkby. 1990. The significance of the magnesium to manganese ratio in plant tissues for growth and alleviation of manganese toxicity in tomato (*Lycopersicon esculentum*) and wheat (*Triticum aestivum*) plants. Pp. 223-228. In M. L. van Beusichem (Ed.), *Plant Nutrition — Physiology and Applications*. Kluwer Academic Publishers, Dordrecht, The Netherlands.

Leece, D. R. 1976. Diagnosis of nutritional disorders of fruit trees by leaf and soil analyses and biochemical indices. *Journal of the Australian Institute of Agricultural Science* 42:3-19.

Leigh, R. A., and A. E. Johnston. 1983a. Concentrations of potassium in the dry matter and tissue water of field-grown spring barley and their relationships to grain yield. *Journal of Agricultural Science (Cambridge)* 101:675-685.

Leigh, R. A., and A. E. Johnston. 1983b. The effects of fertilizers and drought on the concentrations of potassium in the dry matter and tissue water of field-grown spring barley. *Journal of Agricultural Science (Cambridge)* 101:741-748.

Leigh, R. A., and A. E. Johnston. 1985. Nitrogen concentrations in field-grown spring barley: an examination of the usefulness of expressing concentrations on the basis of tissue water. *Journal of Agricultural Science (Cambridge)* 105:397-406.

Lemaire, G., F. Gastal, and J. Salette. 1989. Analysis of the effect of nutrition on dry matter yield of a sward by reference to potential yield and optimum N content. Pp. 179-180. In *Proceedings of the XVI International Grassland Congress, Nice, France*.

Letzsch, W. S. 1985. Computer program for selection of norms for use in the Diagnosis and Recommendation Integrated System (DRIS). *Communications in Soil Science and Plant Analysis* 16:339-347.

Letzsch, W. S., and M. E. Sumner. 1983. Computer program for calculating DRIS indices. *Communications in Soil Science and Plant Analysis* 14:811-815.

Letzsch, W. S., and M. E. Sumner. 1984. Effect of population size and yield level in selection of diagnosis and recommendation integrated system (DRIS) norms. *Communications in Soil Science and Plant Analysis* 15:997-1006.

Loneragan, J. F. 1978. Anomalies in the relationship of nutrient concentrations to plant yield. Pp. 283-298. In A. R. Ferguson, R. L. Bielski, and I. B. Ferguson (Eds.), *Plant Nutrition 1978*. Proceedings 8th International Colloquium on Plant Analysis and Fertilizer Problems. Volume 1. New Zealand Department of Scientific and Industrial Research, Information Series No. 134.

Loneragan, J. F., E. Delhaize, and J. Webb. 1982. Enzymic diagnosis of copper deficiency in subterranean clover. I Relationship of acorbate oxidase activity in leaves to plant copper status. *Australian Journal of Agricultural Research* 33:967-979.

Loneragan, J. F., and K. Snowball. 1969. Calcium requirements of plants. *Australian Journal of Agricultural Research* 20:465-478.

Lorenz, O. A. 1978. Potential nitrate levels in edible plant parts. Pp. 201-219. In D. R. Nielsen and J. G. MacDonald (Eds.), *Nitrogen in the Environment*. Volume 2. *Soil-Plant-Nitrogen Relationships*. Academic Press, NY.

Lubet, E., J.-P. Soyer, and C. Juste. 1983. Appréciation de l'alimentation en zinc du maïs par la détermination du rapport Fe/Zn dans les parties aériennes du végétal. *Agronomie* 3:45-49.

Lundegårdh, H. 1941. Die Tripelanalyse. Theoretische und praktische Grundlagen einer pflanzenphysiologischen Methode zur Bestimmung des Düngerbedürfnisses des Ackerbodens. *Lantbrukshögskolans Annaler* 9:127-221.

MacKay, D. C., J. M. Carefoot, and T. Entz. 1987. Evaluation of the DRIS procedure for assessing the nutritional status of potato (*Solanum tuberosum* L.). *Communications in Soil Science and Plant Analysis* 18:1331-1353.

Macy, P. 1936. The quantitative mineral nutrient requirements of plants. *Plant Physiology* 11:749-764.

Maier, N. A. 1986. Potassium nutrition of irrigated potatoes in South Australia. 2. Effect on chemical composition and the prediction of tuber yield response by plant analysis. *Australian Journal of Experimental Agriculture* 26:727-736.

Malavolta, E., and V. F. Da Cruz. 1971. A meaning for foliar diagnosis. Pp. 1-3. In R. M. Samish (Ed.), *Recent Advances in Plant Nutrition*. Volume 1. Gordon and Breach Science Publishers, NY.

Martin-Prével, P., J. Gagnard, and P. Gautier (Eds.). 1987. *Plant Analysis as a Guide to the Nutrient Requirements of Temperate and Tropical Crops*. Lavoisier Publishing, NY.

Maynard, D. N., A. V. Barker, P. L. Minotti, and N. H. Peck. 1976. Nitrate accumulation in vegetables. *Advances in Agronomy* 28:71-118.

McGrath, J. F., and A. D. Robson. 1984. The distribution of zinc and the diagnosis of zinc deficiency in seedlings of *Pinus radiata*, D. Don. *Australian Forest Research* 14:175-186.

Minotti, P. L., and D. L. Stankey. 1973. Diurnal variation in the nitrate concentration of beets. *HortScience* 8:33-34.

Mitscherlich, E. A. 1935. Phosphorsäuredüngungsversuche in Sandkulturen. *Die Phosphorsäure* 5:517-526.

Møller Nielsen, J. 1985. Nitrogen fertilization system based on plant analysis, a progress report. Pp. 83-93. In J. J. Neeteson and K. Dilz (Eds.), *Assessment of Nitrogen Fertilizer Requirement*. Institute for Soil Fertility and Netherlands Fertilizer Institute, Haren, The Netherlands.

Møller Nielsen, J., and B. Friis-Nielsen. 1976. Evaluation and control of the nutritional status of cereals. I. Dry matter weight level. *Plant and Soil* 45:317-338.

Morgan, M. A. 1970. *Direct and Indirect Effects of Calcium and Magnesium on a Proposed Nitrate Absorption Mechanism.* Ph.D. Thesis, North Carolina State University, Raleigh. (Cited by Viets and Hageman, 1971)

Nable, R. O., A. Bar-Akiva, and J. F. Loneragan. 1984. Functional manganese requirement and its use as a critical value for diagnosis of manganese deficiency in subterranean clover (*Trifolium subterraneum* L. cv. Seaton Park). *Annals of Botany* 54:39-49.

Nable, R. O., and J. F. Loneragan. 1984. Translocation of manganese in subterranean clover (*Trifolium subterraneum* L. cv. Seaton Park). I. Redistribution during vegetative growth. *Australian Journal of Plant Physiology* 11:101-111.

Nable, R. O., J. G. Paull, and B. Cartwright. 1990. Problems associated with the use of foliar analysis for diagnosing boron toxicity in barley. *Plant and Soil* 128:225-232.

Nason, A., H. A. Oldewurtel, and L. M. Propst. 1952. Role of micronutrient elements in the metabolism of higher plants. I. Changes in oxidative enzyme constitution of tomato leaves deficient in micronutrient elements. *Archives of Biochemistry and Biophysics* 38:1-13.

Ohki, K. 1984. Zinc nutrition related to critical deficiency and toxicity levels for sorghum. *Agronomy Journal* 76:253-256.

Ollagnier, M., C. Daniel, P. Fallavier, and R. Ochs. 1987. Influence du climat et du sol sur le niveau critique du potassium dans le diagnostic foliare du palmier à huile. *Oléagineux* 42:435-445.

Østergaard, H. S. 1989. Analytical methods for optimization of nitrogen fertilization in agriculture. Pp. 224-235. In J. C. Germon (Ed.), *Management Systems to Reduce Impact of Nitrates.* Elsevier Applied Science, London.

Payne, G. G., M. E. Sumner, and C. O. Plank. 1986. Yield and composition of soybeans as influenced by soil pH, phosphorus, zinc, and copper. *Communications in Soil Science and Plant Analysis* 17:257-273.

Phillis, E., and T. G. Mason. 1942. On diurnal variations in the mineral content of the leaf of the cotton plant. *Annals of Botany (New Series)* 6:437-442.

Pierre, W. H., L. Dumenil, and J. Henao. 1977b. Relationship between corn yield, expressed as a percentage of maximum, and the N percentage in the grain. II. Diagnostic use. *Agronomy Journal* 69:221-226.

Pierre, W. H., L. Dumenil, V. D. Jolley, J. R. Webb, and W. D. Shrader. 1977a. Relationship between corn yield, expressed as a percentage of maximum, and the N percentage in the grain. I. Various N-rate experiments. *Agronomy Journal* 69:215-220.

Piper, C. S. 1942. Investigations on copper deficiency in plants. *Journal of Agricultural Science (Cambridge)* 32:143-178.

Prevot, P., and M. Ollagnier. 1961. Law of the minimum and balanced mineral nutrition. Pp. 257-277. In W. Reuther (Ed.), *Plant Analysis and Fertilizer Problems.* American Institute of Biological Sciences, Publication No. 8. Washington.

Prins, W. H. 1983. Effect of a wide range of nitrogen applications on herbage nitrate content in long-term fertilizer trials on all-grass swards. *Fertilizer Research* 4:101-113.

Prummel, J., and P. A. von Barnau-Sijthoff. 1988. Optimum phosphate and potassium levels in potato tops. *Fertilizer Research* 5:203-211.

Randall, P. J., K. Spencer, and J. R. Freney. 1981. Sulfur and nitrogen fertilizer effects on wheat. I. Concentrations of sulfur and nitrogen and the nitrogen to sulfur ratio in grain, in relation to the yield response. *Australian Journal of Agricultural Research* 32:203-212.

Rao, J. K., K. L. Sahrawat, and J. R. Burford. 1987. Diagnosis of iron deficiency in groundnut, *Arachis hypogaea* L. *Plant and Soil* 97:353-359.

Rathfon, R. A., and J. A. Burger. 1991. Diagnosis and recommendation integrated system modifications for Fraser fir Christmas trees. *Soil Science Society of America Journal* 55:1026-1031.

Rehm, G. W., R. C. Sorensen, and R. A. Wiese. 1983. Application of phosphorus, potassium, and zinc to corn grown for grain or silage: Nutrient concentration and uptake. *Soil Science Society of America Journal* 47:697-700.

Rendig, V. V. 1956. Sulfur and nitrogen composition of fertilized and unfertilized alfalfa grown on a sulfur-deficient soil. *Soil Science Society of America Proceedings* 20:237-240.

Rendig, V. V., and E. A. McComb. 1961. Effect of nutritional stress on plant composition. II. Changes in sugar and amide nitrogen content of normal and sulfur-deficient alfalfa during growth. *Plant and Soil* 14:176-186.

Reuter, D. J., and J. B. Robinson (Eds.). 1986. *Plant Analysis. An Interpretation Manual.* Inkata Press, Melbourne.

Richards, F. J., and R. G. Coleman. 1952. Occurrence of putrescine in potassium-deficient barley. *Nature* 170:460.

Roach, W. A. 1938. *Plant Injection for Diagnostic and Curative Purposes.* Imperial Bureau of Horticulture and Plantation Crops, Technical Communication No. 10.

Rosell, R. A., and A. Ulrich. 1964. Critical zinc concentrations and leaf minerals of sugar beet plants. *Soil Science* 97:152-167.

Rutherford, R. S. 1989. The assessment of proline accumulation as a mechanism of drought resistance in sugarcane. *Proceedings of the Annual Congress — South African Sugar Technologists' Association* No. 63:136-141. (*Soils and Fertilizers* 53, Abstract 13790. 1990.)

Saalbach, E., G. Kessen, and G. K. Judel. 1961. Über den Einfluss von Schwefel auf den Ertrag und die Eiweissqualität von Futterpflanzen. *Zeitschrift für Pflanzenernährung, Düngung und Bodenkunde* 93:18-26.

Sajwan, K. S., and W. L. Lindsay. 1988. Effect of redox, zinc fertilization and incubation time on DTPA-extractable zinc, iron and manganese. *Communications in Soil Science and Plant Analysis* 19:1-11.

Savoy, H. J., Jr., and D. L. Robinson. 1989. Development and evaluation of preliminary DRIS norms for dallisgrass. *Communications in Soil Science and Plant Analysis* 20:655-683.

Savoy, H. J., Jr., and D. L. Robinson. 1990. Norm range size effects in calculating Diagnosis and Recommendation Integrated System indices. *Agronomy Journal* 82:592-596.

Sayre, J. D. 1948. Mineral accumulation in corn. *Plant Physiology* 23:267-281.

Scaife, A., and I. G. Burns. 1986. The sulphate-S/total S ratio in plants as an index of their sulphur status. *Plant and Soil* 91:61-71.

Scaife, A., and K. L. Stevens. 1983. Monitoring sap nitrate in vegetable crops: Comparisons of test strips with electrode methods, and effects of time of day and leaf position. *Communications in Soil Science and Plant Analysis* 14:761-771.

Scarseth, G. D. 1943. Methods of diagnosing plant nutrient needs. *American Fertilizer* 98, No. 12:5-8, 22, 24, 26.

Schjørring, J. K., N. E. Nielsen, H. E. Jensen, and A. Gottschau. 1989. Nitrogen losses from field-grown spring barley plants as affected by rate of nitrogen application. *Plant and Soil* 116:167-175.

Schulz, R., and H. Marschner. 1986. Optimierung der Stickstoff-Spätdüngung zu Winterweizen mit dem Nitratschnelltest. *Schriftenreihe, Verband Deutscher Landwirtschaftlicher Untersuchungs- und Forschungsanstalten, Reihe Kongressberichte* (1986), No. 20:343-359. (*Soils and Fertilizers* 51, Abstract 1906. 1988.)

Schulz, R., and H. Marschner. 1987. Vergleich von Nitrat- und Amino-N Schnelltest zur Charakterisierung des Stickstoff-Versorgungsgrades von Winterweizen. *Zeitschrift für Pflanzenernährung und Bodenkunde* 150:348-353.

Scott, L. E. 1940. An instance of boron deficiency in the grape under field conditions. *Proceedings of the American Society for Horticultural Science* 38:375-378.

Shear, C. B., and M. Faust. 1971. Value of various tissue analyses in determining the calcium status of the apple tree and fruit. Pp. 75-98. In R. M. Samish (Ed.), *Recent Advances in Plant Nutrition.* Volume 1. Gordon and Breach Science Publishers, NY.

Smith, F. W. 1974. The effect of sodium on potassium nutrition and ionic relations in Rhodes grass. *Australian Journal of Agricultural Research* 25:407-414.

Smith, F. W. 1975. Tissue testing for assessing the phosphorus status of green panic, buffel grass and setaria. *Australian Journal of Experimental Agriculture and Animal Husbandry* 15:383-390.

Smith, F. W., and G. R. Dolby. 1977. Derivation of diagnostic indices for assessing the sulphur status of *Panicum maximum* var. *Trichoglume. Communications in Soil Science and Plant Analysis* 8:221-240.

Smith, P. F. 1962. Mineral analysis of plant tissues. *Annual Review of Plant Physiology* 13:81-108.

Smith, T. A. 1984. Putrescine and inorganic ions. Pp. 7-54. In B. N. Timmerman, C. Steelink, and F. A. Loewus (Eds.), *Phytochemical Adaptations to Stress.* Recent Advances in Phytochemistry 18.

Spear, S. N., C. J. Asher, and D. G. Edwards. 1978. Response of cassava, sunflower, and maize to potassium concentration in solution. *Field Crops Research* 1:347-361.

Spencer, K., M. B. Jones, and J. R. Freney. 1977. Diagnostic indices for sulphur status of subterranean clover. *Australian Journal of Agricultural Research* 28:401-412.

Sprague, H. B. (Ed.). 1964. *Hunger Signs in Crops.* New Third Edition. David McKay Co., NY.

Steele, K. W., D. M. Cooper, and C. B. Dyson. 1982a. Estimating nitrogen fertiliser requirements in maize grain production. 1. Determination of available soil nitrogen and prediction of grain yield increase to applied nitrogen. *New Zealand Journal of Agricultural Research* 25:199-206.

Steele, K. W., D. M. Cooper, and C. B. Dyson. 1982b. Estimating nitrogen fertiliser requirements in maize grain production. 2. Estimates based on ear leaf and grain nitrogen concentrations. *New Zealand Journal of Agricultural Research* 25:207-210.

Steenbjerg, F. 1945. Om kemiske Planteanalyser og deres Anvendelse. *Tidsskrift for Planteavl* 49:158-177. (*Soils and Fertilizers* 9:14. 1946.)

Stewart, C. R., and A. D. Hanson. 1980. Proline accumulation as a metabolic response to water stress. Pp. 173-189. In N. C. Turner and P. J. Kramer (Eds.), *Adaptation of Plants to Water and High Temperature Stress.* John Wiley & Sons, NY.

Stewart, I. 1962. The effect of minor element deficiencies on free amino acids in citrus leaves. *Proceedings of the American Society for Horticultural Science* 81:244-249.

Sumner, M. E. 1974. An evaluation of Beaufils' physiological diagnosis technique for determining the nutrient requirement of crops. Pp. 437-446. In J. Wehrmann (Ed.), *Plant Analysis and Fertilizer Problems.* Volume 2. German Society of Plant Nutrition, Hannover, Federal Republic of Germany.

Sumner, M. E. 1977a. Use of the DRIS system in foliar diagnosis of crops at high yield levels. *Communications in Soil Science and Plant Analysis* 8:251-268.

Sumner, M. E. 1977b. Preliminary N, P, and K foliar diagnostic norms for soybeans. *Agronomy Journal* 69:226-230.

Sumner, M. E. 1977c. Effect of corn leaf sampled on N, P, K, Ca and Mg content and calculated DRIS indices. *Communications in Soil Science and Plant Analysis* 8:269-280.

Sumner, M. E. 1990. Advances in the use and application of plant analysis. *Communications in Soil Science and Plant Analysis* 21:1409-1430.

Sumner, M. E., and M. P. W. Farina. 1986. Phosphorus interactions with other nutrients and lime in field cropping systems. *Advances in Soil Science* 5:201-236.

Swiader, J. M., J. G. Sullivan, J. A. Grunau, and F. Freiji. 1988. Nitrate monitoring for pumpkin production on dryland and irrigated soils. *Journal of the American Society for Horticultural Science* 113:684-689.

Takahashi, T., and D. Yoshida. 1960. Occurrence of putrescine in the tobacco plant with special reference to the nutrient deficiency. *Soil and Plant Food* 6:93.

Takkar, P. N., M. S. Mann, R. L. Bansal, N. S. Randhawa, and H. Singh. 1976. Yield and uptake response of corn to zinc, as influenced by phosphorus fertilization. *Agronomy Journal* 68:942-946.

Taylor, B. K. 1971. Soluble nitrogenous fractions of tissue extracts as indices of the nitrogen status of peach trees. Pp. 241-249. In R. M. Samish (Ed.), *Recent Advances in Plant Nutrition.* Volume 1. Gordon and Breach Scientific Publishers, NY.

Terry, N. 1980. Limiting factors in photosynthesis. I. Use of iron stress to control photochemical capacity *in vivo. Plant Physiology* 65:114-120.

Thomas, W. 1937. Foliar diagnosis: Principles and practice. *Plant Physiology* 12:571-599.

Timmer, V. R., and L. D. Morrow. 1984. Predicting fertilizer growth response and nutrient status of jack pine by foliar diagnosis. Pp. 335-351. In E. L. Stone (Ed.), *Forest Soils and Treatment Impacts.* Proceedings of the 6th North American Forest Soils Conference. University of Tennessee, Knoxville.

Timmer, V. R., and E. L. Stone. 1978. Comparative foliar analysis of young balsam fir fertilized with nitrogen, phosphorus, potassium, and lime. *Soil Science Society of America Journal* 42:125-130.

Touchton, J. T., F. Adams, and C. H. Burmester. 1981. *Nitrogen Fertilizer Rates and Cotton Petiole Analysis in Alabama Field Experiments.* Alabama Agricultural Experiment Station, Bulletin 528. Auburn.

Ulrich, A. 1961. Plant analysis in sugar beet nutrition. Pp. 190-211. In W. Reuther (Ed.), *Plant Analysis and Fertilizer Problems.* American Institute of Biological Sciences, Publication No. 8. Washington.

Ulrich, A., D. Ririe, F. J. Hills, A. G. George, and M. D. Morse. 1959. *Plant Analysis ... a Guide for Sugar Beet Fertilization.* California Agricultural Experiment Station, Bulletin 766, Part 1. Berkeley.

Veen, B. W., and A. Kleinendorst. 1986. The role of nitrate in osmoregulation of Italian ryegrass. *Plant and Soil* 91:433-436.

Viets, F. G., Jr., and R. H. Hageman. 1971. *Factors Affecting the Accumulation of Nitrate in Soil, Water, and Plants.* U.S. Department of Agriculture, Agriculture Handbook No. 413.

Villiers, J. I. de, and C. J. Beyers. 1961. Leaf analysis as a guide to fertilization in commercial orange growing. Pp. 107-119. In W. Reuther (Ed.), *Plant Analysis and Fertilizer Problems.* American Institute of Biological Sciences, Publication No. 8. Washington.

Wallace, T. 1961. *The Diagnosis of Mineral Deficiencies in Plants by Visual Symptoms.* Second Edition. Chemical Publishing Co., NY.

Walsh, L. M., and J. D. Beaton (Eds.). 1973. *Soil Testing and Plant Analysis.* Revised Edition. Soil Science Society of America, Madison, WI.

Walworth, J. L., W. S. Letzsch, and M. E. Sumner. 1986. Use of boundary lines in establishing diagnostic norms. *Soil Science Society of America Journal* 50:123-128.

Walworth, J. L., and M. E. Sumner. 1987. The diagnosis and recommendation integrated system (DRIS). *Advances in Soil Science* 6:149-188.

Walworth, J. L., and M. E. Sumner. 1988. Foliar diagnosis: A review. *Advances in Plant Nutrition* 3:193-241.

Walworth, J. L., M. E. Sumner, R. A. Isaac, and C. O. Plank. 1986. Preliminary DRIS norms for alfalfa in the southeastern United States and a comparison with midwestern norms. *Agronomy Journal* 78:1046-1052.

Walworth, J. L., H. J. Woodard, and M. E. Sumner. 1988. Generation of corn tissue norms from a small, high-yield data base. *Communications in Soil Science and Plant Analysis* 19:563-577.

Ward, S. C., G. E. Pickersgill, D. V. Michaelson, and D. T. Bell. 1985. Responses to factorial combinations of nitrogen, phosphorus and potassium fertilizers by saplings of *Eucalyptus saligna* Sm., and the prediction of the responses by DRIS indices. *Australian Forest Research* 15:27-32.

Ware, G. O., K. Ohki, and L. C. Moon. 1982. The Mitscherlich plant growth model for determining critical nutrient deficiency levels. *Agronomy Journal* 74:88-91.

Watanabe, F. S., W. L. Lindsay, and S. R. Olsen. 1965. Nutrient balance involving phosphorus, iron, and zinc. *Soil Science Society of America Proceedings* 29:562-565.

Waugh, J. G., and F. P. Cullinan. 1940. The nitrogen, phosphorus, and potassium content of peach leaves as influenced by soil treatment. *Proceedings of the American Society for Horticultural Science* 38:13-16.

Webb, R. A. 1972. Use of the Boundary Line in the analysis of biological data. *Journal of Horticultural Science* 47:309-319.

Weetman, G. F., and C. G. Wells. 1990. Plant analysis as an aid in fertilizing forests. Pp. 659-690. In R. L. Westerman (Ed.), *Soil Testing and Plant Analysis.* Third Edition. Soil Science Society of America, Madison, WI.

Westerman, R. L. (Ed.). 1990. *Soil Testing and Plant Analysis.* Third Edition. Soil Science Society of America, Madison, WI.

Westermann, D. T., and G. E. Kleinkopf. 1985. Phosphorus relationships in potato plants. *Agronomy Journal* 77:490-494.

White, R. E., and K. P. Haydock. 1970. Phosphate concentration in siratro as a guide to its phosphate status in the field. *Australian Journal of Experimental Agriculture and Animal Husbandry* 10:426-430.

White, W. C., and C. A. Black. 1954. Willcox's agrobiology: III. The inverse yield-nitrogen law. *Agronomy Journal* 46:310-315.

Willcox, O. W. 1948. Why some crop plants yield more than others. *Science* 108:38-39.

Willcox, O. W. 1949. The factual base of the inverse yield-nitrogen law. *Agronomy Journal* 41:527-530.

Williams, C. M. J., and N. A. Maier. 1990. Determination of the nitrogen status of irrigated potato crops I. Critical nutrient ranges for nitrate-nitrogen in petioles. *Journal of Plant Nutrition* 13:971-984.

CHAPTER 4

Soil Testing and Fertilizer Requirement

S OIL TESTING IS A GENERIC TERM that includes measurements made on soils for various purposes. In this chapter, which emphasizes the testing of soils for soil fertility evaluation and control, the usage of the term will be restricted to measurements made for that purpose.

Supplies of plant nutrients differ greatly among soils and among locations within soils. The supplies of nutrients in soils have a direct bearing upon the most suitable addition of fertilizers, and they influence other soil management practices; consequently, they receive much attention from scientists specializing in soil fertility. This chapter deals with the nutrient availability concept, biological indexes of availability, the adaptation of chemical methods to evaluate different aspects of availability, and the combined use of chemical and biological methods to estimate fertilizer requirements of crops.

Of special value for further reading on this subject are three books on *Soil Testing and Plant Analysis* published by the Soil Science Society of America (Soil Science Society of America, 1967; Walsh and Beaton, 1973; Westerman, 1990). Although the second was represented as a revised edition and the last as a third edition, the books are to a considerable extent individual treatises because of changes in authors and topics and the tendency of authors to emphasize different aspects of the subject matter. Related publications include those by Brown (1987), Jones (1985a), and Peck et al. (1977). Brief treatments include reviews by Cope and Evans (1985) and Jones (1988).

4-1. The Availability Concept

The words used most commonly to describe the supply of plant nutrients in soil are *available* and *availability*. These words are useful, but the precise meaning to be attached to them is not agreed upon, the basic problem being difficulties of experimental measurement. For purposes of discussion here, the word *available* will mean *susceptible to absorption by plants*. *Availability* will mean *effective quantity*.

Absolute availabilities cannot be evaluated. The objective of the biological methods is to estimate ratios of availabilities or to obtain values that hopefully are proportional to availabilities. The latter may be called *indexes of nutrient availability*.

The concept of nutrient availabilities in soils has some similarity to the concept of activities of ions in chemistry. In chemistry, the activity concept has the advantage of a standard state to which the activities in a given situation (designated as γx, where x is the concentration and γ is the activity coefficient) may be referred. No theoretical standard state has been developed for the availability concept in soil fertility. Although it may be of no practical value, a standard state for availability could be defined as a condition in which a nutrient is completely available. Complete availability is evidenced by complete absorption and an availability coefficient of unity. A real situation corresponding to the standard state exists to a good approximation where plants are grown in a nutrient solution containing initially a low concentration of a nutrient. See the data in Fig. 4-11, which is introduced later in a different context. In this experiment, the plants absorbed virtually all of the phosphorus present initially in the nutrient solution.

In Chapter 5 on fertilizer evaluation, the effective quantity is represented as the product of the quantity x that is susceptible to absorption by plants and an availability coefficient γ. The quantity x is taken arbitrarily as the total quantity of the nutrient or sometimes as the legally available quantity. The concept of availability as the product γx is an oversimplification in fertilizer evaluation, and the same is true for soil nutrients because of the complexity of the factors that contribute to what the plant sees as "effective."

4-2. Biological Indexes of Availability

The primary measurements to obtain availability ratios or availability indexes must be made using plants because the concept of availability is defined in terms of plants. Several different biological methods are used to provide the primary indexes. The chemical methods used in practice in soil testing work because of their economy and speed are calibrated against the biological measurements.

This section considers three classes of biological methods: internal standard, direct, and slope ratio. In internal standard methods, the nutrient availability index is derived from the comparative responses produced by the supply of the nutrient in the soil and by known additions of the nutrient in the form of fertilizer. In direct methods, no fertilizer is added, and the availability index is inferred directly from the magnitude of some plant response measurement. Slope-ratio methods are an adaptation of direct methods.

4-2.1. Internal-Standard Methods

4-2.1.1. Mitscherlich b Value

Mitscherlich early recognized that the equation he had proposed to represent the response of plants to addition of fertilizer could be used also to provide an index of nutrient availability in soils. He normally wrote the equation in the form

$$\text{Log}(A - y) = \text{Log}A - c(x + b),$$

where b is the availability index of the nutrient in the soil and seed when none is added in the fertilizer, y is the yield with the quantity x of the nutrient in the fertilizer, A is the maximum yield obtainable with increasing x, and c is the effect factor or efficiency factor that denotes the rate at which the yield approaches the maximum value with addition of x, as indicated in Section 1-2.1.

The significance of b and the method by which it is obtained may be visualized by reference to Fig. 4-1. In this figure, the solid curved line represents the Mitscherlich equation calculated from the experimental observations. Because the smallest quantity of fertilizer that can be added experimentally is zero, the shape of the response curve to the left of the y axis cannot be determined by direct observation. But if the response curve to the left of the axis may be represented by the Mitscherlich equation, the numerical values of y for various negative values of x will be those indicated by the broken line. The distance along the x axis from the point of intersection of that axis with the broken line to the y axis is b, the availability index of the nutrient in the soil. The units of b are the same as those of x.

Because b is a calculated value and not an experimentally determined value, and because the Mitscherlich equation has no fundamental significance, it is not clear without some kind of experimental evidence that b has the physical significance Mitscherlich attached to it. One way to investigate this question is to determine whether the values of b obtained with a range of supplies of a given nutrient are directly proportional to the values of some other availability index that is measured independently.

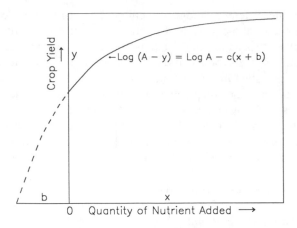

Fig. 4-1.   Use of the Mitscherlich equation to provide an estimate of the effective quantity *b* of a nutrient in soil on the basis of the yields of a test crop obtained with different quantities of the nutrient supplied in a fertilizer. The *b* value is the negative distance on the *x* axis from zero fertilizer to the intersection of the extrapolated yield curve with the *x* axis.

Finding a suitable independent index is problematical. Only if the two sets of index values are highly correlated is the validity of the *b* concept verified. A low correlation could result from unsuitability of either or both of the indexes and also from experimental error.

Perhaps the best independent index for the purpose is derived from chemical measurements on soil nitrogen. The boundary line between the nitrogen that plants can absorb from soils and the nitrogen they cannot absorb is generally sharp, and under certain conditions a good index of nitrogen availability can be obtained by chemical analysis.

Mitscherlich and Sauerlandt (1935) published some appropriate experimental data on nitrogen. In one experiment, they added different green manures to two soils and allowed the mixtures to incubate for several months. Then they analyzed the mixtures for ammonium plus nitrate nitrogen and grew a test crop to evaluate the *b* values in the variously treated samples of the soils. Fig. 4-2 shows that to a good approximation the values of *b* were directly proportional to the content of ammonium plus nitrate nitrogen in the soil. The *b* values exceeded the quantities of ammonium plus nitrate nitrogen, as would be expected from the fact that nitrogen mineralization would supply additional mineral nitrogen during the growth of the test crop. Data of this kind verify that both the chemical and biological methods were satisfactory and that they were performing in the intended manner.

A second way to test the validity of Mitscherlich *b* values is to determine whether the increases in *b* values due to addition of known quantities of nitrogen in a form available to plants agree numerically with the quantities of nitrogen added. This technique is analogous to that of determining the percentage recovery of known quantities of the substance for which an analysis is being made as a test of validity in quantitative chemical analysis.

Mitscherlich and Sauerlandt (1935) carried out an experiment of this kind by growing oat plants in sand cultures to which increasing quantities of fertilizer nitrogen were added and by determining the *b* values associated with different quantities of different chemical forms of nitrogen added as basal treatments. The results showed approximate 1 to 1 correspondence between the *b* values and the quantities of nitrogen added.

The two experimental tests just described verify that *b* has the significance Mitscherlich attached to it. They give one confidence that *b* may be used as an index of nutrient availability in soils in instances in which a clear experimental verification is not possible.

Mitscherlich applied his equation almost exclusively to data on crop yields versus quantities of nutrients applied in the form of fertilizers. But in one paper, in which he was commenting on findings by other investigators, Mitscherlich (1919) used his equation to fit phosphorus concentrations in plants produced with different quantities of fertilizer phosphorus.

Mitchell and Chandler (1939) made a different application of the Mitscherlich equation for nutrient percentages versus quantities of nutrient applied as fertilizer.

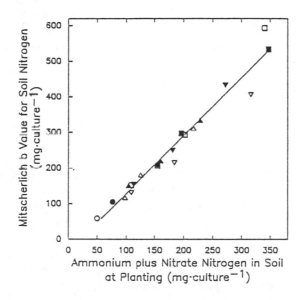

Fig. 4-2. Mitscherlich *b* values for soil nitrogen, as estimated by the Mitscherlich equation from the yields of oat plants, versus ammonium plus nitrate nitrogen in a sandy loam soil (open symbols) and a loam soil (filled symbols). The soil cultures had been treated in the autumn with different quantities of each of three leguminous green manures (serradella, triangle symbols; lupine, square symbols; and vetch, inverse triangle symbols), kept moist but subjected to outdoor temperatures during the winter, and planted to an oat test crop in the following May. The correlation coefficient is r = 0.98. (Mitscherlich and Sauerlandt, 1935)

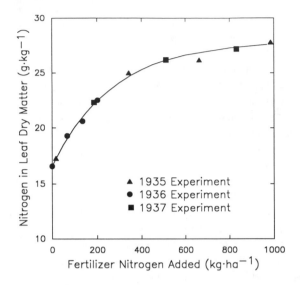

Fig. 4-3.   Nitrogen concentrations in the leaves of red oak trees versus quantity of fertilizer nitrogen added in an experiment in 1935 in New York State. The Mitscherlich equation was fitted to the data, resulting in a *b* value of 264 kilograms of nitrogen per hectare. Analogous data from experiments in 1936 and 1937 on nearby sites were plotted on the curve by spotting the nitrogen concentrations in leaves from the control plots on the calculated curve for the 1935 experiment, yielding *b* values of 247 kilograms for the 1936 experiment and 434 kilograms for the 1937 experiment. The quantities of nitrogen were adjusted mathematically by extending the curve downward to provide for the lower nitrogen concentration in the control in 1936 than in 1935, so that the control value for 1936 is the reference value for zero addition of fertilizer nitrogen. The adjustment involved adding 17 kilograms of nitrogen per hectare to all the other values. (Mitchell and Chandler, 1939)

They conducted several experiments in which trees of different species in mixed stands were fertilized with different quantities of nitrogen. The nitrogen concentrations in leaf samples then were determined. As illustrated for red oak in Fig. 4-3, all the data for a particular species could be correlated into a single Mitscherlich response curve if a suitable *b* value was assigned to the supply of nitrogen in each soil. These results signify that where the supply of nitrogen in the unfertilized soil was taken into account, the nitrogen concentration versus nitrogen applied curve was largely independent of the site and year.

Using the nitrogen concentration in the leaves to obtain an index of nitrogen availability in soils supporting forest trees is relatively convenient. Moreover, the nitrogen concentration probably provides a more precise availability index than could be obtained over a period of one or a few years from yield measurements in stands of large trees.

Although the Mitscherlich equation and others may be fitted to data on nutrient concentrations in plant tissue, extrapolating the curves to zero nutrient concentration, which in effect is done in deriving *b* values for the supply of a nutrient in the soil, is a less credible procedure for nutrient concentrations than for yields.

As noted in Chapter 3, nutrient concentrations do not drop continuously as the supply of nutrients decreases. Rather, they reach a minimum and then remain fairly constant over a wide range of yields. Thus, general success in applying the *b* value concept to nutrient concentration data is not to be expected.

4-2.1.2. Dean a Value

Dean (1954) made use of a linear relationship between the yield of phosphorus in plants and the quantity of fertilizer phosphorus added to obtain an index of the availability of phosphorus in soils. To obtain this index, one calculates the equation

$$y = mx + s$$

that represents the relationship for the experimental data at hand (where *y* is the yield of nutrient in the plants with the addition *x* of the fertilizer, *m* is the slope constant, and *s* [for soil] is the *y* intercept) and evaluates *x* at the point where *y* is equal to zero. Fig. 4-4 illustrates the procedure. The concept is analogous to that used in the preceding section to obtain Mitscherlich *b* values, but the response function is different.

Data obtained by Munson (1954) may be cited as verification of the validity of the approach. Fig. 4-5A shows that in a greenhouse experiment in which millet was grown on various soils with differing past management, the *a* values increased linearly with the sum of the initial nitrate nitrogen present in the soils and the nitrate nitrogen produced in the soils during a 6-week incubation in the laboratory. Fig. 4-5B shows that the *a* values found by extrapolating the yield-

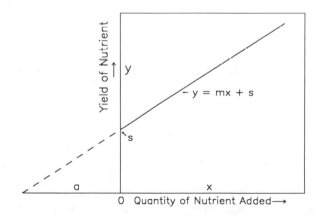

Fig. 4-4. Schematic representation of the *a* value method for deriving an index of nutrient availability by extrapolating the linear relationship often obtained between the yield of a nutrient in the crop and the quantity of the nutrient applied in the fertilizer. The *a* value is the negative distance on the *x* axis from zero fertilizer to the intersection of the extrapolated yield of nutrient with the *x* axis. (Dean, 1954)

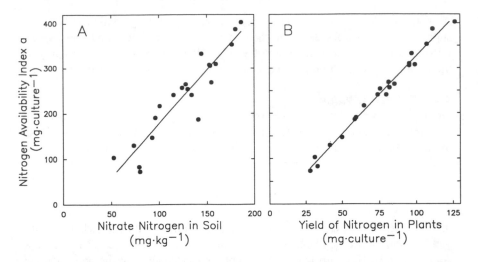

Fig. 4-5. Use of the Dean *a* value concept in a greenhouse experiment with 21 soils. A. Dean *a* values versus the sum of the initial nitrate-nitrogen content of the soils and the nitrate nitrogen produced during incubation of the soils for 6 weeks at 35° C. B. Dean *a* values versus the yield of nitrogen in the test crop of millet. The millet was grown for 50 days. (Munson, 1954)

of-nitrogen curves increased linearly with the yield of nitrogen in the test plants grown on the soil cultures that received no fertilizer nitrogen. This behavior indicates that in addition to validity of the *a* values, the differences among soils were not great enough to prevent equal uptakes of nitrogen by plants from equal supplies in the soil.

The quantity *a*, which is an index of the availability of the nutrient in the soil and seed in units of the quantity of the nutrient added in the fertilizer, is analogous to *b*, which is found by the Mitscherlich equation involving plant yields. The *a* value, however, has two advantages. First, the measurement is made directly on the nutrient itself and not on another characteristic of the plant. Yield response is less directly related to supply than is nutrient absorption. Second, the yield of a nutrient is more sensitive to changes in availability at high availabilities than is the yield of dry matter. Fig. 3-1 illustrates the second advantage.

A problem encountered in determining *a* values is that although the yield of nutrient may increase linearly with the smaller quantities of nutrient applied, a downward curvature is observed if the additions are great enough. In such instances, some arbitrariness is involved in determining how many points to use in calculating the linear regression, and the decision made will have some effect on the numerical value of *a*. Chien et al. (1990) suggested that an objective decision on the number of points to include may be made by fitting both first- and second-order polynomial regressions to the data. The number of points to include is one less than the number at which the quadratic coefficient first becomes significant at the 95% probability level.

A close correspondence of yield-of-nutrient curves to a linear relationship is obtained most often with nitrogen and potassium. The curves obtained with phosphorus deviate from linearity more frequently. Approximate linearity is observed more often in greenhouse experiments than in field experiments. In field experiments, the yields of nutrient in the control plots are often so high that the extrapolated *a* values have low precision. An analogous situation exists where Mitscherlich *b* values are calculated from field experimental data.

Both the *a* value and *b* value methods involve an assumption about the applicability or validity of the particular response function chosen. Sometimes neither relationship fits the facts.

4-2.1.3. Isotope Dilution

The isotope-dilution method is discussed here as a biological index of availability because biological indexes are considered first, and the method has been used to provide biological indexes. The method could be discussed equally well in Section 4-3 as a chemical index of availability because it has been used in that sense also, but in the interest of clarity, both the biological and chemical aspects are discussed here.

Basic Concept.   The isotope-dilution method provides a way to determine the quantity of a nontagged substance from analyses made with the aid of the same substance in tagged form. For example, if a known quantity q of a nutrient tagged with a nuclide or isotope and dissolved in water is thoroughly mixed with an unknown volume of nutrient solution containing the unknown quantity Q of the same nutrient in untagged form, the isotope-dilution equation can be used to estimate the quantity Q from an analysis of a small sample of the solution. If the sample is analyzed for the quantity q_s in tagged form and for the total quantity Q_s, where the subscript s represents the sample, one may write

$$\frac{q_s}{Q_s} = \frac{q}{Q}$$

and

$$Q = \frac{q Q_s}{q_s} \quad .$$

If, instead of removal of a sample of the solution for analysis, a plant is grown in the solution, it will absorb a mixture of the tagged and untagged forms of the nutrient. If the tagged and untagged forms are in the same chemical form, the plant will make no appreciable distinction between the two and will absorb them in the proportion in which they are present in the solution. Hence, if the plant is analyzed for the quantity q_s in tagged form and the total quantity Q_s, the quantity Q of the untagged nutrient originally present in the solution can be calculated as before by use of the equation

$$Q = \frac{qQ_s}{q_s} \; .$$

This formulation ignores the nontagged nutrient in the seed. Being untagged, like the nutrient in the soil or in the present instance the solution, the presence of the nutrient derived from the seed implies that the plants have absorbed more untagged nutrient than they really have. Brookes (1982) reviewed the attempts that had been made to correct for the nutrient supplied by the seed, and he proposed a method of his own. The usual procedure is to grow the plants long enough so the error will be small and to make no correction.

E Values, L Values, and A Values.   Applying the isotope-dilution concept to nutrients in soils is a more complex matter, not only because the nutrients may occur in various forms in the soil solution and in the soil solids, but also because the terms used by different investigators are inconsistent. There is also a difference in the sense in which certain observations are interpreted.

When a tagged nutrient is added to a soil, it mixes with the nontagged form of the nutrient in solution, and for most nutrients it also exchanges with the same form of the nutrient held by the soil solids. And especially tagged nitrogen becomes involved in the biological turnover that goes on in soils. Although for a short time the mineral nitrogen in a soil may remain essentially constant, tagged nitrogen added in solution in mineral form is taken up by microorganisms, which initially release only nontagged nitrogen.

The symbol E (for exchangeable) has been used to represent the quantity of a nutrient with which an added tagged nutrient exchanges in the soil solids when the measurements are made on a volume of solution separated from soil. For phosphorus, with which most of the research has been done, the quantity of nontagged phosphorus in solution is very small relative to the quantity of solid-phase phosphorus that exchanges with the tagged phosphorus added. The total soil phosphorus with which the added tagged phosphorus is diluted is consequently designated as E.

The symbol L (for Larsen) has been used to represent the sum of the quantities of a nutrient with which an added tagged nutrient mixes in solution and with which the tagged nutrient exchanges in the soil solids when the measurements are made on the sample of the nutrient removed from soil by plants. Sigurd Larsen (1952) apparently was the first to show that when a soluble tagged phosphate source was thoroughly mixed with soil, calculations based upon the measured quantities of soil and fertilizer phosphorus absorbed by plants grown on the soil indicated that the quantity of soil phosphorus with which the added phosphorus had equilibrated was independent of the quantity of phosphorus added, as would be expected from the isotope-dilution equation. In some instances, the L value has been called *labile* phosphorus. The term labile is applied also to chemical measurements of exchangeable phosphorus (E values).

Fried and Dean (1952) employed the isotope-dilution concept in a different sense. They proposed that plants absorb the soil and fertilizer sources of a nutrient in the same proportions as their availabilities or effective quantities. The equation they derived is identical with the isotope-dilution equation, but they assigned the meaning *availability* instead of *quantity* to the isotope-dilution measurements made on plants, and they used the symbol A instead of E or L. Fried (1964) later published a paper in which he discussed at length the terms E, L, and A.

Probably the most basic point involved in the difference between E and L values on the one hand and A values on the other has to do with the mixing or equilibration of the tagged nutrient added and the nutrient in the soil. E values and L values, which are applications of the classical isotope-dilution concept, involve the assumption that the added tagged source has become thoroughly mixed with the soluble and exchangeable soil sources, so that after equilibration a small sample of the nutrient taken at any place in the soil solution by either chemical or biological methods would have the same ratio of tagged and non-tagged nutrient as a small sample taken at any other place.

If the conditions for E values and L values are met, the numerical values obtained for E, L, and A theoretically will be identical for very small uptakes of the nutrient by the plants grown to obtain the L and A values. A values, however, do not involve the assumption of thorough incorporation and uniform distribution. A values thus may be calculated from data on uptake of a nutrient from nonuniform distributions, such as band applications of tagged fertilizer and applications of tagged fertilizer made in the field, where plants have access to the nutrient in locations that have not equilibrated with the tagged source. Here is where the availability concept arises. A values thus may be numerically rather different from E values or L values.

As indexes of availability, E values, L values, and A values have the advantage of sensitivity over a wide range of nutrient supply. They can produce measurements of supply of a soil nutrient with good precision even at high levels of availability at which crops would cease to respond in terms of yield.

As examples of the types of availability effects reflected in measurements of A values, Caldwell et al. (1952) found that mixing ammonium nitrate with phosphorus-32-tagged superphosphate in a localized application caused an increase in uptake of fertilizer phosphorus, a decrease in uptake of soil phosphorus, and a decrease in the A value for soil phosphorus. Such behavior would be expected if the extra nitrogen increased root growth and phosphorus uptake in the vicinity of the local application of tagged fertilizer. Pesek and Webb (1954) found that the A value calculated for phosphorus in a soil in field experiments with oat on adjacent locations in two successive years was about twice as great in a dry season as in a wet season. The higher ratio of soil phosphorus to fertilizer phosphorus absorbed by the plants in the dry season was presumably a consequence of dryness of the surface soil, which increased the ratio of root activity

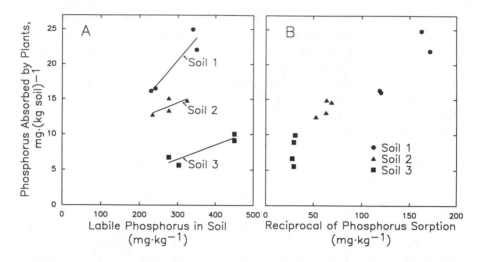

Fig. 4-6. Phosphorus uptake by rye plants from three soils that had been treated with soluble orthophosphate and incubated 118 days before planting. Two samples of each soil were incubated air-dry, and two were incubated at field capacity. Soil 1 was calcareous alluvium, pH 7.6. Soil 2 was a loam, pH 6.0. Soil 3 was a strongly phosphate-fixing loam derived from basalt, pH 5.8. A. Phosphorus uptake versus labile phosphorus in the soils by isotopic dilution. B. Phosphorus uptake versus the reciprocal of phosphorus sorption by the soils. (Russell et al., 1957)

in the lower layers of soil to that in the surface layer containing the tagged fertilizer.

A values for soil phosphorus have remained fairly constant with different quantities of fertilizer phosphorus in some experiments, but they usually change with fertilizer placement. Bullen et al. (1983) conducted an experiment in which both quantity and placement of fertilizer were changed. They added phosphorus-32-tagged monobasic calcium phosphate in solution in three quantities and five placements to a soil low in phosphorus and grew soybean as a test crop in a growth chamber. The placements included mixing the fertilizer phosphorus with all of the soil and with 50%, 25%, 12.5%, and 1% of the soil. In each of the placements, the top of the fertilized portion of the soil was located 2.5 centimeters below the seed.

From the uptake of soil and fertilizer phosphorus by the plants, the *A* values calculated for soil phosphorus in milligrams per culture were found to decrease from 186 to 149 as the proportion of the soil receiving the fertilizer decreased from 100% to 1% with addition of 100 milligrams of fertilizer phosphorus per culture, from 266 to 174 with 200 milligrams of fertilizer phosphorus per culture, and from 384 to 266 with 400 milligrams of fertilizer phosphorus per culture. The uptake of soil phosphorus by the plants increased with the plant yield, which suggests that an effect of fertilization on root growth was an important factor in the changes in *A* values with the various fertilizer treatments.

Problems With Phosphorus.   Correlations among measurements of *E*, *L*, and *A* are generally high (especially those between *E* and *L*). All these measurements generally correlate fairly well with biological indexes of nutrient availability and with the better chemical indexes involving nutrient extraction. Nevertheless, there are some problems.

For example, Gachon (1969) found that although the correlations between *E* values and uptake of phosphorus by plants in greenhouse cultures were 0.90 with 25 soils derived from crystalline rocks and 0.91 with 15 soils derived from volcanic ash, much less phosphorus was absorbed at a given *E* value from the latter group of soils than from the former. Soils derived from volcanic ash characteristically contain much amorphous material, react strongly with phosphorus, and have high requirements for fertilizer phosphorus. Similar observations were made by Russell et al. (1957). Fig. 4-6 illustrates the differences that may occur among soils, as well as the observation that the values for phosphorus uptake by plants were more closely related to the reciprocal of phosphorus sorption by the soils than to the *E* values. Thus, indications are that *E* values may provide a good index of soil phosphorus availability if the soils are similar, but that the availability of the phosphorus associated with a given *E* value decreases as the number of phosphorus sorption sites increases.

Problems With Exchangeable Cations.   For exchangeable cations, the complementary-cation effect is a cause of difficulty in relating quantities of the ions to availability. Table 4-1 shows the results of an experiment in which calcium-45-tagged calcium sulfate was mixed with samples of soil from a field experiment involving different applications of limestone. A test crop of oat was grown, and the *A* values (which in this instance are also *L* values because of the experimental method) were calculated from the data obtained on the tagged and untagged calcium present in the plants.

Table 4-1. *A* Values for Calcium Obtained in a Greenhouse Experiment With Oat Plants Grown on Soils to Which Different Quantities of Calcium Sulfate Tagged With Calcium-45 Had Been Added (Fried and Dean, 1952)

	A Value Found for Calcium, mg·kg^{-1}		
Calcium Sulfate Added, mg·kg^{-1}	Low-Calcium Soil	Medium-Calcium Soil	High-Calcium Soil
38	432	605	1230
75	486	576	1210
150	451	587	1250
300	491	597	1330
Average *A* Value, mg·kg^{-1}	465	591	1260
Exchangeable Calcium, mg·kg^{-1}	392	562	1070
A Value/Exch. Calcium	1.2	1.1	1.2

In this experiment, the calcium from the calcium sulfate should have equilibrated with the soluble and exchangeable calcium in the soil before the plants absorbed any appreciable amount of calcium because (1) the quantities of calcium sulfate were relatively small, (2) calcium sulfate has moderate solubility, and (3) other experiments have shown that tagged calcium added in solution equilibrates with exchangeable calcium in clay within an hour. The results indicate that the quantities of soil calcium with which the tagged calcium equilibrated were essentially independent of the quantity of calcium sulfate added and that they were a little greater than the values obtained in chemical measurements of exchangeable calcium in the soils.

Now, in a situation such as that in the experiment supplying the data in Table 4-1, adding increasing quantities of potassium to the soil would be expected to have only a small effect on the quantities of soil calcium with which the calcium-45-tagged calcium from the added calcium sulfate would equilibrate, and hence on the A values or L values. But the potassium could markedly decrease the uptake of soil calcium by the plants. In other experimental work, for example, Van Itallie (1938) found that in a soil containing 13.7 centimoles charge of exchangeable calcium per kilogram, adding 7 centimoles charge of potassium decreased the concentration of calcium in an Italian ryegrass test crop from 7.1 to 2.2 grams per kilogram of dry matter. The yield of the test crop was essentially unaffected. Thus, the addition of potassium could markedly reduce the calcium availability while having little or no effect on the A value or L value.

Problems With Nitrogen.   Stanford et al. (1973) obtained a correlation of $r = 0.94$ between the A values for nitrogen calculated for 39 soils cropped to sudangrass in the greenhouse and the sum of the initial mineral nitrogen plus the amounts of nitrogen mineralized during growth of the test crop, as estimated from measurements in the laboratory. In their work, the nitrogen-15-labeled fertilizer was sodium nitrate, which was applied in solution to the surface of the soil.

With nitrogen, the isotope-dilution method poses two problems that are different from those described for phosphorus. The first is that there are two principal forms of mineral nitrogen in soil — ammonium and nitrate, and they occur in varying ratios. Uniformly tagging the mineral nitrogen present in soil at a given moment would require analyzing the soil for these two forms and adding tagged ammonium and nitrate sources with nitrogen-15 concentrations adjusted to produce equal concentrations of nitrogen-15 in the soil ammonium and nitrate.

A second problem in using the isotope-dilution method with nitrogen is that although the usual interest is in the net effects of the fertilizers, the nitrogen-15 data do not necessarily reflect these effects. Löhnis (1926) is credited with being the first to observe that in some instances the increase in yield of nitrogen in a series of test crops upon adding a quantity of a green manure to soil may exceed the nitrogen added in the green manure. He inferred that the cause of his observation was "an intensified mineralization of the humus nitrogen." Subsequently, evidence of enhancement of soil organic nitrogen mineralization was obtained with additions of fertilizer nitrogen (see, for example, Fig. 1-11).

The later availability of nitrogen-15 as a tracer made possible an experimental distinction between soil nitrogen and added nitrogen. It also introduced new problems of interpretation because of the complexity of the processes taking place in the soil and the fact that a quantitatively significant part of the fertilizer nitrogen may become involved in these processes.

The most definitive recent analysis has been made by Jenkinson et al. (1985) and Hart et al. (1986). According to these investigations, when calculations based upon the results of experiments with tagged nitrogen indicate an increase in uptake of soil nitrogen by plants, the effect may be real, only apparent, or a combination of both. A real increase may arise if nitrogen fertilization enhances the rate of mineralization of soil nitrogen. A real increase may result also from absorption of extra soil nitrogen if addition of fertilizer nitrogen increases the volume of soil occupied by roots.

The apparent increase is an artifact resulting from the use of the tagged nitrogen. Under conditions suitable for microbial activity, organic nitrogenous substances are being formed continuously in new microbial tissue in soil, and organic nitrogen is being mineralized continuously. When tagged mineral nitrogen is added, it mixes with the untagged mineral nitrogen in the soil and reduces the amount of untagged nitrogen immobilized in the organic nitrogenous substances being synthesized. As a result, the amount of untagged mineral nitrogen in the soil available for plant uptake is greater than the amount that would have been available in the absence of the tagged nitrogen. A similar result would be obtained if some of the mineral nitrogen in the soil disappeared as a result of denitrification, except that in this case the nitrogen that disappeared would be lost from the soil.

The apparent increase in nitrogen mineralization upon addition of tagged nitrogen has implications for the use of isotope-dilution methods as indexes of nitrogen availability. Whether the observed effect represents a real increase, only an artifact, or a combination of the two, the consequence will be an increase in the value of E, L, or A, as the case may be.

Because some of the experimental data to be shown in Table 4-2 can be presented only as A values (Fried and Dean, 1952), the equation for calculating A, the effective quantity of the nutrient, is given here:

$$A = \frac{f(1 - y)}{y} \quad ,$$

where f is the quantity of the nutrient added in the fertilizer, and y is the proportion of the nutrient in the crop derived from the fertilizer. The units of A are the same as those of f.

A values calculated from data obtained in experiments with nitrogen-15-tagged fertilizer are shown in Table 4-2. Also shown is a second set of A values calculated

Table 4-2. *A* Values for Soil Nitrogen as Calculated From Data on Nitrogen-15 and From Increases in Yield of Nitrogen From Fertilization in Three Experiments With Wheat as the Test Crop (Hart et al., 1986)

Experiment	*A* Values[a]	
	Calculated From Nitrogen-15 Data	Calculated From Increase in Yield of Nitrogen From Fertilization
Greenhouse, fallowed soil	53	33
Greenhouse, grassland soil	248	66
Field	81	65

[a] Units of *A* values are kilograms of nitrogen per hectare for the field experiment and milligrams of nitrogen per kilogram of soil for the greenhouse experiment. Values are averages derived from applications of 24, 30.5, 121, and 242 milligrams of fertilizer nitrogen per kilogram of soil in the greenhouse experiment and 58.5, 116.9, 174.4, and 233.9 kilograms of fertilizer nitrogen per hectare in the field experiment. Only the above-ground plant material was used in the calculations.

from conventional data on increases in yield of nitrogen from fertilization. The latter values have been calculated in accordance with Fried's (1964) statement that "The use of labelled standard is not a necessary requirement for determining 'A' value, as the methodology is not stated in the definition. Thus if the amount of nutrient taken up by the plant is linearly related to the amount present in the medium, the amount of nutrient taken by the plant from a given standard can be obtained by deducting the amount obtained in a check treatment in the absence of added standard [from] the amount obtained by the plant in the presence of the added standard. This gives enough information to calculate an 'A' value without the use of radioactive isotopes." The required linear relation was obtained between the yield of nitrogen in the plants and the mineral nitrogen added in the experimental data from which the *A* values in the last column in the table were calculated.

The data in Table 4-2 show that the *A* values derived from the nitrogen-15 data were consistently higher than those derived from the total nitrogen data. The difference was more than threefold for the grassland soil. The closest agreement was found in the field experiment, where the difference was only 25%.

In the field experiment, the uptake of soil nitrogen was essentially unaffected by the addition of fertilizer nitrogen. The increases in yield of nitrogen from fertilization thus closely paralleled the yields of true fertilizer nitrogen derived from the nitrogen-15 data. True fertilizer nitrogen is defined here as the actual atoms of nitrogen added in the fertilizer. By using nitrate as the tagged source of nitrogen in the field experiment, the authors attempted to minimize the extent to which the fertilizer nitrogen would substitute for soil mineral nitrogen in the microbial immobilization process. They also applied the fertilizer as a topdressing to the existing crop of winter wheat in April, which would limit the mixing of fertilizer nitrogen with the mineral nitrogen in the soil.

In the greenhouse experiment, the uptake of soil nitrogen by the wheat plants increased when fertilizer nitrogen was added to both soils. In this experiment, the authors attempted to maximize the extent to which the fertilizer nitrogen would substitute for soil nitrogen in the microbial immobilization process by adding the tagged nitrogen as ammonium, which soil microorganisms utilize preferentially over nitrate as a source of nitrogen, and by thoroughly mixing the tagged fertilizer with the soil before the test crop of spring wheat was planted.

The authors devised a computer program to model the substitution of tagged nitrogen for soil nitrogen in the immobilization process, and found that their theoretical considerations accounted for the increase in uptake of soil nitrogen by the crop grown on the fallowed soil. For the crop on the grassland soil, where the increase in uptake of soil nitrogen with nitrogen fertilization was more marked, they found that the substitution of tagged nitrogen added in the fertilizer for mineral nitrogen in the soil in the immobilization process accounted for 56% of the increase in soil nitrogen uptake by the plants. The remaining 44% was accounted for by an increase in mineralization of soil nitrogen resulting from addition of the tagged fertilizer. (These figures were derived from a comparison of the control cultures with the cultures receiving the maximum addition of fertilizer nitrogen, which was 242 milligrams per kilogram of soil.)

The results of this experimental work thus show that the differences in the A values in Table 4-2 calculated from the nitrogen-15 data and the total nitrogen data were associated with the substitution of tagged fertilizer nitrogen for soil mineral nitrogen in the microbial immobilization of nitrogen that takes place continuously in warm, moist soils. In the field experiment, there was no significant substitution or enhanced soil nitrogen mineralization, and the A values were similar. In the greenhouse experiment on the fallowed soil, the substitution accounted for the increase in soil nitrogen uptake by plants when the soil was fertilized, and also for the increase in A value. In the greenhouse experiment on the grassland soil, the substitution accounted for part of the increase in soil nitrogen uptake and part of the increase in A value; the increase in soil nitrogen mineralization accounted for the remainder.

The apparent increase in availability of soil nitrogen that results from the substitution of tagged fertilizer nitrogen for soil mineral nitrogen in the microbial immobilization process is an artifact of the nitrogen-15 method. It is reflected in A values, but it is not an availability effect. The magnitude of the substitution evidently will vary with the circumstances in individual experiments, and so will the effect on the A values.

Fortunately, the theoretical problems with using the isotope-dilution method to obtain an availability index for nitrogen are not of much practical concern. Measurements of ammonium and nitrate in soils can be made more easily and inexpensively by chemical methods than by isotope-dilution methods.

Analogous biological interchange takes place with phosphorus, but here the process has been neglected because the behavior of phosphorus from the standpoint of plant nutrition generally is considered to be dominated by the inorganic fraction. Work by Helal and Dressler (1989) and Helal and Sauerbeck (1984), however, indicates that this may not be true for soils high in organic phosphorus. See also Section 4-3.6.2.

4-2.1.4. The Internal Standard Concept

All the internal standard methods depend upon some response of plants to the soil supply of a nutrient relative to their response to one or more known quantities of the nutrient applied in a fertilizer. Internal standard methods have the advantage that if the plant response differs among soils because of factors other than the availability of the nutrient being investigated, the extraneous factors will affect the uptake of the nutrient from both native and fertilizer sources.

The Achilles heel of the internal standard concept is the assumption that the standard is a standard in terms of availability. When fertilizers are added to soils, they interact with the soils to some degree, with the result that the availability of the standard source can vary from one situation to another. As the availability of the standard drops, the derived indexes of availability for the nutrient in the soils are artificially inflated. Additionally, the fertilizer may affect the availability of the nutrient supplied by the soil, as indicated in the preceding section. To date, no one has devised an internal standard method that can be said to lead to calculated availabilities of soil nutrients that are independent of the soil. As a substitute, adjustments for fertilizer efficiency, discussed in Section 4-4.3, have been developed for use in estimating the fertilizer requirement.

4-2.2. Direct Methods

In direct methods for obtaining nutrient availability indexes, no fertilizer is added. The magnitude of some plant response is used directly as the availability index of the nutrient. Direct methods are used principally, if not exclusively, in greenhouse work, where an important application is in evaluating chemical methods for obtaining indexes of nutrient availability in soils.

Direct methods find little application in field research. The probable reason is the view that field responses generally are affected too much by factors other than the nutrient of concern to be useful. This inference may not be valid for all situations, as will be brought out later.

4-2.2.1. Neubauer Method

The Neubauer method, proposed by Neubauer and Schneider (1923), and its numerous modifications probably have been used more than any other method. For a description in English, see Stewart (1932).

In the original method, 100 grams of soil are mixed with 50 grams of sand, the mixture is covered with 250 grams of sand, and 100 rye seeds are planted in the surface layer of sand. The rye seedlings are harvested on the 17th day after planting. The tops and roots are analyzed together to determine the quantities of phosphorus and potassium they contain. The quantities contained in the plants grown in parallel cultures without soil are subtracted to obtain the quantities absorbed from the soil. The resulting figures are taken as indexes of the availabilities of phosphorus and potassium in the soil.

In this method, nutrients other than phosphorus and potassium are not added. Because of the short period of growth, the seeds supply most of the nutrients needed. The effect of factors other than the supply of phosphorus and potassium in the soil is limited by dilution of the soil with sand and by the short period of growth. The plants on different soils produce about the same yield of dry matter. A technical difficulty encountered with some soils is that the roots cannot readily be separated from the soil for analysis.

4-2.2.2. Modifications of the Neubauer Method

The Neubauer method apparently has been used very little for many years, but a number of modifications have been published. In the modifications, the authors attempted to incorporate one or more advantages. The original idea of using sand as a diluent in one way or another has been used in most instances.

A method devised by Stanford and DeMent (1957) and DeMent et al. (1959a,b) for phosphorus, potassium, and nitrogen is described here because of its usefulness for various purposes. Other modifications will be neglected. In the Stanford and DeMent method, seeds of the test crop are planted in sand contained in a waxed cardboard carton with the bottom removed. This carton is nested in a similar container with bottom intact. A nutrient solution lacking the nutrient of concern is added, and the plants are grown 2 or 3 weeks, with periodic additions of nutrient solution and water, to produce seedlings that are deficient in the element for which the test is to be made. By the end of this time, the plants have developed a thick mat of roots on the bottom of the sand, and the inner bottomless containers can be lifted out of the outer containers without loss of sand. The inner cartons then are placed in other cartons containing 200 grams of soil, and growth is allowed to continue, with further additions of water. (In some instances, the soil was premoistened; in others, air-dry soil was used, and water was added to the overlying sand.) Extensive penetration of roots into the underlying soil normally occurs within 24 hours. After the desired period of contact between the sand culture and the soil, the tops of the plants, or the tops and the portion of the roots in the sand, are analyzed for the element under test.

The addition of nutrients other than the one under test is intended to reduce the effect of differences in these factors in the soils being tested. The relatively short time required for testing permits one to follow effects that may take place during only a few days, such as changes in availability after addition of fertilizers to soils. With the method as described, the uptake of the nutrient increases approximately linearly with the supply in the soil over a considerable range.

4-2.2.3. Direct Methods With an External Standard

Mitchell (1934) devised what might be called an external standard method for obtaining an index of nutrient availability in soils. His work was with pine seedlings. The external standard is supplied by the nitrogen concentrations in seedlings grown in sand cultures receiving different quantities of nitrogen fertilizer, but containing no soil. Alongside the standard cultures, pine seedlings are grown in the various soils under test with no nitrogen additions. Nutrients other than nitrogen are supplied to both the sand and soil cultures. At the end of the growth period, the nitrogen concentrations are determined in the seedlings in all the cultures.

To interpret the data obtained, the Mitscherlich equation that fits the nitrogen concentration versus nitrogen-added function for the sand cultures is calculated. The nitrogen concentration in the seedlings grown in the cultures of a given soil then is substituted in the equation, and the equation is solved for the nitrogen added. The solution of the equation for each soil represents the nitrogen availability index in the soil expressed in the same units as the nitrogen added to the sand cultures.

In comparison with internal standard methods requiring the construction of a response function for each soil, the principal advantage of the Mitchell method seems to be a reduction in the amount of work required to obtain an availability index. Only one response curve is needed.

The performance of the plants in the sand cultures that constitute the external standard is not affected by the differing conditions that may exist in the cultures of the various soils. The external standard does not modify these conditions, however, and their existence may affect the nutrient concentrations in the plants and the derived nutrient availability indexes.

The Wiessmann (1928) method [see Stewart (1932) for a description in English] is similar to the Mitchell method in that plant responses in sand cultures are used to interpret response data on soils in terms of nutrient availability indexes — again what might be called an external standard method. In the Wiessmann method as described for potassium, culture vessels are filled with 7200 grams of sand or with 5700 grams of sand and 1500 grams of soil. All nutrients except potassium are added in ample supply. The sand cultures receive potassium as potassium sulfate in quantities ranging from 0 to 1.24 grams per culture; the soil cultures receive 0 and 1.24 grams. The 1.24-gram addition is considered sufficient to produce the maximum yield of the oat test crop.

The yields of the test crop are determined on all cultures. To provide an adjustment for the effect of conditions other than the differences in potassium supply on the yields, the yields on the soil cultures without potassium are multiplied by the ratio of the yield on the sand cultures with 1.24 grams of potassium to the yield on the soil cultures with the same addition of potassium. The resulting corrected values for the soil culture controls then are spotted on the response curve for the sand cultures to find the addition of potassium that produced an equivalent yield.

The Wiessmann method uses yields of dry matter, whereas the Mitchell method uses nutrient concentrations. The Wiessmann method provides an overt adjustment for differences in conditions other than potassium supply in the various soils. This adjustment implies that the relative effect of factors other than the nutrient under test is independent of the supply of this nutrient. The adjustment is equivalent to Mitscherlich's assumption that the c value in his equation is constant.

Table 4-3.   Results of a Test of the Slope-Ratio Biological Assay Method for Availability Ratios in Which the Theoretical Ratio of the Slopes is 2 (McCants and Black, 1957)

	Value of $(b_2-b_0)/(b_1-b_0)$ for Indicated Response Criterion	
Subsoil	Yield of Dry Matter	Yield of Nitrogen
Tama	2.0	2.2
Shelby	1.9	2.2

4-2.3. Slope-Ratio Method

In the slope-ratio method, each of the preparations to be tested is added in increasing quantities to a medium suitable for the growth of the organism to be used to measure the biological responses. Conditions are arranged so that linear responses are produced. The results of the assay are represented graphically as a series of straight lines with a common control value corresponding to zero addition of the substance being assayed, and the relative quantities of the substance supplied by the various preparations are inferred from the ratios of the slopes of the lines. The slope-ratio method is now standard for biological assay of preparations of substances not readily determined by chemical analysis.

In an application of the slope-ratio method investigated by McCants and Black (1957), the "preparations" were soils. Each soil was added to sand cultures in several quantities. All cultures were supplied with nutrients other than the one under test, and plants were grown to determine the yields of dry matter and the nutrient being tested. (A different application of the slope-ratio method to fertilizer evaluation is described in Section 5-2.3.4.)

As a test of validity of the adaptation of the slope-ratio method for estimating nutrient availability ratios in soils, McCants and Black added 0, 20, and 40 milligrams of nitrate nitrogen per kilogram to each of two subsoils, and added each subsoil at each nitrate level at 0, 300, 600, and 900 grams per 4 kilograms of soil-sand mixture. Equal quantities of nitrate nitrogen were added to sand cultures with no soil. In this test, there were thus three equations for each soil, one equation for each quantity of nitrate nitrogen added:

$$y = a + b_0 x$$
$$y = a + b_1 x$$
$$y = a + b_2 x$$

where y is the response measurement, a is the response where the addition x of the preparation is zero (the Y intercept), and the b values are the slopes of the lines obtained with the different quantities of nitrate nitrogen added (b_0 for no addition, b_1 for 20 milligrams per kilogram, and b_2 for 40 milligrams per kilogram). The response of the test crop was linear for both yield of dry matter and yield of nitrogen, and the slopes led to the slope ratios shown in Table 4-3. The observed values are all close to the theoretical value 2. These results indicate that the method can be used to estimate the relative availabilities of a nutrient in different soils.

The slope ratios can be calculated and used in the same way as any of the availability indexes described in preceding pages. McCants and Black, however, used a special test to investigate the possibility that the observed responses under the artificial conditions employed to eliminate differences due to factors other than the availabilities of the nutrient under consideration had actually eliminated these differences, so that the slope ratios reflected only the relative availabilities. The test, described by De Wit (1953), consists of plotting the yield of dry matter against the yield of the nutrient and determining whether a single functional relationship exists. If the functional relationship is the same for two different soils, there is no evidence of an effect of factors other than the relative availabilities.

McCants and Black found that in their experiment with nitrogen, the results of which are given in Table 4-3, there was no evidence of a difference in the yield-of-dry-matter versus yield-of-nitrogen curves between the two subsoils. The yield-of-dry-matter versus yield-of-nitrogen curves for the soils, however, did differ from the corresponding curve obtained for sand. Similar results were obtained by McCants (1955) in another experiment with two surface soils to which no nitrogen was added. Although the yield-of-dry-matter versus yield-of-nitrogen curves provided no evidence of invalidity of the estimate of relative availability of nitrogen in the two soils, the difference in curves between the soils and the sand suggests that evidence of invalidity might be expected with some soils.

In related experiments with phosphorus, the yield-of-dry-matter versus yield-of-phosphorus curves for most soils did not coincide. These findings indicate that the responses obtained from adding increasing quantities of different soils were not measuring exactly the same thing in the different soils. Whether the differences were a consequence of factors other than phosphorus, differences in the course of phosphorus uptake with time that could be said to be an aspect of the availability phenomenon, or both cannot be inferred from the test. In any event, using only small quantities of soil and supplying nutrients other than the one under investigation was not enough to satisfy the test of validity.

4-2.4. Comparison of Internal-Standard and Direct Methods
As yet, not enough is known to permit categorical statements regarding the relative value of internal-standard methods and direct methods. In work done in the greenhouse, Dos Santos et al. (1960) found that direct methods were better than internal-standard methods. They compared yields of dry matter and phosphorus in the control cultures (as the direct methods) with Mitscherlich *b* values, Dean *a* values, and another extrapolation method (as the internal-standard methods). They found that the experimental error in making the measurements was

a major factor in determining the relative suitability of the different methods. The experimental error was relatively low in measuring the yields of dry matter and phosphorus. Another factor thought to be important was the arrangement of experimental conditions to make phosphorus strongly limiting, so that the sensitivity was good and the results were on the nearly linear portion of the response curves.

In field work with alfalfa involving experiments in a number of states and Ontario, Hanway et al. (1961) found that the Dean *a* value (an internal-standard method) gave the best results. The yield of potassium and the potassium percentage in the alfalfa (two direct methods) gave poorer results.

4-2.5. Application to Field Conditions

4-2.5.1. Applicability of Different Biological Methods

The first matter to be considered in applying indexes of nutrient availability under field conditions is the nature of the methods themselves. All the internal-standard methods and some of the direct methods can be applied under either field or greenhouse conditions, but some direct methods require artificial conditions. As direct indexes of nutrient availability, yields of dry matter and nutrients in control cultures are valuable under controlled conditions, but are of little use in the field because these quantities may be influenced too much by factors other than the availability of the nutrient that may differ from one field location to the next.

4-2.5.2. Timeliness and Cost

Indexes of nutrient availability for crops in the field provide information on nutrient supplies that can be used as a basis for planning soil management practices, particularly fertilization. The information thus must be in hand in time to permit the necessary management steps to be carried out. If the indexes of nutrient availability are to be determined directly in the field and are to be of use in making decisions about fertilizer practices, the measurements usually must be made at least one season before the information is to be used.

Except for measurements of plant composition, biological indexes of nutrient availability seldom are used in practical soil testing work on producers' fields. Field experiments and even biological tests in the greenhouse are relatively expensive. The biological methods discussed are used almost entirely to evaluate and calibrate the chemical tests that are used on producers' samples.

4-2.5.3. Mitscherlich Method

At one time, Mitscherlich operated a large-scale soil-testing operation using his biological method to obtain an index of availability of nutrients in soils. He avoided to some extent the cost of determining an availability index for nutrients under field conditions by bringing samples of soil to the greenhouse and determining the availability indexes (*b* values) there. Plants were grown on the soils in specially designed enameled metal containers in the greenhouse. The procedure was streamlined, but it was still an expensive, time-consuming operation.

Mitscherlich did not assume that the *b* values in the field would be the same as those he obtained in the greenhouse, where he had a layer of soil only 20 centimeters in thickness and other artificial conditions. Rather, he determined *b* in field experiments and in greenhouse experiments on samples of the same soils to determine how the two were related. According to Atanasiu (1954), Mitscherlich came to the conclusion that the *b* values in the field were on average about double those obtained in the greenhouse, and so he multiplied the greenhouse-determined *b* values by 2 for the conditions in East Prussia where his work was done.

With the estimated *b* value for the field in hand, the next step was to find from a table the percentage increase in yield to be expected from applying any desired quantity of fertilizer. The Mitscherlich equation could be used directly, but less conveniently, to obtain the same information. To put the results on an absolute basis, for purposes of estimating profitability, one could use an estimate of the yield obtainable without fertilization.

When Mitscherlich moved to West Germany during World War II, he headed another institute, and one of the subjects investigated was the relationship between *b* values obtained in the field and greenhouse. Results obtained in this work were published by Atanasiu (1954), and those for phosphorus are shown in Fig. 4-7. These results do not support the factor of 2 used previously by Mitscherlich, and indicate instead that a factor less than 1 was more suitable under the conditions of these experiments. The correlation coefficient $r = 0.76$ obtained in this work is about as good as could be expected in view of the experimental errors involved in estimating *b* values, especially where the *b* values are high.

In more recent research, Munk and Rex (1987) calculated Mitscherlich *b* values in a group of 82 long-term experiments on phosphate fertilization in Germany. They found that the *b* values were highly correlated with the economic optimum applications of phosphate fertilizer when the yield level was taken into account and the data were classified according to the Mitscherlich *c* values or efficiency factors. They used the data for economic optimum quantities of fertilizer to evaluate several chemical soil tests for phosphorus.

Although the Mitscherlich equation is still used extensively for calibrating chemical soil tests, emphasis has shifted from the original Mitscherlich concept of deriving a *b* value. Instead, the equation generally is used in a modified form in which *b* does not appear.

The Mitscherlich equation may be said to divide the total response curve from zero yield to the maximum yield into two parts: (1) the hypothetical portion from $y = 0$ to $y = y_0$, where y_0 is the control yield with none of the nutrient

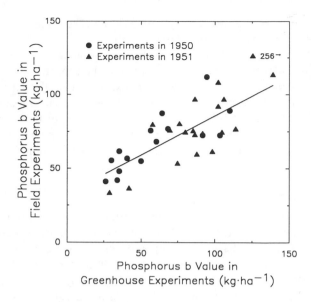

Fig. 4-7. Relation of *b* values obtained in field experiments in two years in Germany to *b* values obtained on samples of the same soils in the greenhouse. Crops represented in the summary include potato, fodder beet, sugarbeet, winter rye, winter barley, hemp, serradella, and summer barley. (Atanasiu, 1954)

added, and (2) the portion from y_0 to the maximum yield A. The second part can be observed experimentally, but the first part, from which b is derived, can only be calculated on the basis of certain assumptions about the response function. Current practice is to modify the Mitscherlich equation to eliminate b and to deal only with the observable portion of the response curve, which still involves the concept of a supply of the nutrient in the soil and seed, although not explicitly.

The Mitscherlich equation in exponential form with the base e is

$$y = A(1 - e^{-c(x+b)}) = A(1 - e^{-cx} \cdot e^{-cb}).$$

In a given experimental series, e^{-cb} is a constant. If this constant is represented by β, the equation becomes

$$y = A(1 - \beta e^{-cx}).$$

When $x = 0$, $y_0 = A(1 - \beta)$. Thus $\beta = 1 - y_0/A$. Then letting $e^{-c} = \rho$, the equation becomes

$$y = A(1 - \beta \rho^x),$$

which is a relatively convenient form for curve fitting. The value of c is calculated from the fitted curve by the relation $c = -ln \rho$.

The trend toward use of the Mitscherlich equation without calculating an explicit b value index of availability is an example of a general shift in emphasis. At present, yields of dry matter, relative yields, or yields of a nutrient usually are regressed on chemical soil test values according to a selected function without explicitly calculating availability indexes. This bypassing of the theoretical aspects of what biological measurements represent is of no direct practical consequence, although it may encourage some loss of appreciation for the theory and limitations.

4-2.5.4. Other Biological Methods

Some use has been made of the Neubauer method in relating biological values in the greenhouse or laboratory to biological values in the field. Although various microbiological methods have been proposed, even less use has been made of most of them. Measurements of nitrogen mineralized during incubation are an exception. Although the mineralization is a microbiological process, the actual measurements are made by chemical methods. As a consequence, measurements of nitrogen mineralization are not looked upon in the same sense as growing colonies of specific microorganisms under conditions in which the nutrient of concern in the soil is made limiting.

Except for purposes of evaluation and calibration, the biological methods now have been virtually displaced by chemical methods that can be applied to samples of soil on a mass-production basis at relatively low cost. Chemical methods will be discussed in the next section.

4-3. Chemical Indexes of Availability

4-3.1. Developing and Ranking Chemical Indexes

The approach used most commonly in chemical tests for nutrient availability in soils has been to obtain an extract of the soil by adding an excess of water or a solution containing certain chemical reagents, to measure the concentration of the nutrient in question in an aliquot of the extract, and to express the result in terms of the quantity of the nutrient extracted per unit mass or per unit volume of soil. The usual objective has been to find a combination of extractant and extracting procedure that removes quantities of the nutrient that are highly correlated with the biological indexes of nutrient availability obtained on a group of soils under test. There is almost always some theoretical basis for the choice of extractant.

Emphasis here is on general principles rather than procedural details. Recent reviews of the literature on soil test procedures proposed for the various nutrient elements are found in a book edited by Westerman (1990). The book also includes reviews on the use of plant analysis in general and for specific crops.

The usual procedure for ranking different methods of soil testing that have been developed as indexes of nutrient availability is to analyze a group of soils by each of the methods to be compared and to select as the most efficient the method that gives the highest correlation with biological indexes of nutrient availability found experimentally for the same group of soils. This research aspect of the soil testing rationale is often called *correlation* (Dahnke and Olson, 1990). Fixen and Grove (1990) summarized in tabular form the results of many such comparisons of soil phosphorus extractants.

The samples used in the evaluation may be derived from a group of field experiments or from a greenhouse experiment in which different soils have been brought together to obtain biological indexes of nutrient availability under uniform conditions. Although the ultimate aim of the evaluation of methods is to select the one that gives the best performance in terms of correlation with indexes of nutrient availability in the field on the soils of the area for which the selected method is to be used, greenhouse experiments often are employed for comparing soil testing methods because they can yield more critical comparisons of methods with less work and expense than can field experiments.

In the historical development of chemical indexes of nutrient availability in soils, the focus often was on a single nutrient, and the procedure was adjusted to provide the best correlation with the biological indexes employed. Multinutrient extractants now are becoming more important, and two of these will be discussed as illustrations. Although the illustrations logically should be covered here, the discussion is delayed to Section 4-3.8 because the examples involve some concepts that have not yet been explained.

4-3.2. Theory of Nutrient Supply

The theory behind the chemical methods would be of no concern from the practical standpoint if methods could be devised that would give consistently high correlations with appropriate biological indexes of nutrient availability. No method has been found that does this. The theory now available explains why such methods are unlikely to be found, and it helps to point the way to improvements in chemical methods and the way they are used. The basic problem is that the Mitscherlich *b* value and other biological indexes of nutrient availability discussed in this chapter are composite values that are the net result of plant integration of the factors contributing to the effective quantity of the nutrient of concern.

From the standpoint of plant nutrition, an ideal soil would supply continuously in the soil solution at root surfaces a concentration of each nutrient sufficient to prevent deficiency. This ideal is achieved to various degrees with different nutrients in different soils, and the mechanisms differ among nutrients, but the steps involved are the same:

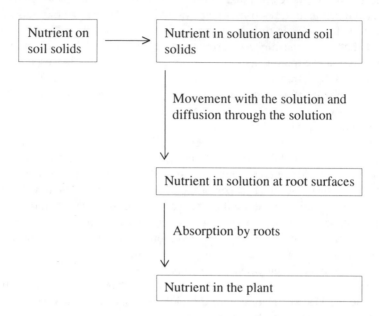

This concept has been described in terms of phosphorus in various papers, including those by Gunary and Sutton (1967), Dalal and Hallsworth (1976), and Moody et al. (1988), and discussed in relation to experimental data on different soils.

4-3.3. Rate of Release From Soil

The first step, equilibration of the nutrient on or in the soil solids with the nutrient in solution around the solids, is a two-way process. The nutrient is being released from the solids to the solution at the same time as it is returning from the solution to the solids.

Where phosphorus is concerned, the solid-phase forms generally are treated as adsorption complexes and sometimes as relatively insoluble compounds. Although some immobilization of inorganic phosphorus in organic forms and mineralization of organic phosphorus to inorganic forms take place, these processes generally are overlooked and, to some degree, are evaluated indirectly in the measurements made on the inorganic fraction. Measurements on bulk soil, however, underestimate the significance of the turnover between inorganic and organic forms that occurs in the rhizosphere (Helal and Dressler, 1989). See also Section 4-3.6.2. Other nutrients that would be viewed as adsorption complexes equilibrating with the solution phase would include molybdenum and copper. Ions considered to be adsorbed exchange with the corresponding ions in solution, but the quantities present in adsorbed form vary with the concentrations in solution.

Calcium, magnesium, potassium, ammonium, and sodium, which occur in exchangeable forms, are in a different class. These ions equilibrate with the solution phase by exchange, but the total number of charge positions holding these ions remains relatively constant at a given pH value in most soils.

Nitrogen also is transferred back and forth between the solid and solution phases, but except for soils containing much exchangeable and nonexchangeable ammonium as a result of recent fertilization, the transfers are mostly biologically mediated and are classified as mineralization and immobilization. Except for these circumstances, the equilibration is neglected as being too limited to be of practical concern.

The behavior of sulfur is intermediate between that of phosphorus and nitrogen. Sulfate occurs in adsorbed forms, especially in acid soils high in hydrous oxides, and in such soils the inorganic equilibration process may be of major importance as a source of sulfur for plants. Like nitrogen, sulfur is involved in organic cycling in the soil, and in many unfertilized soils this process probably plays an important role in the transfer of sulfur between the solid and solution phases.

The rate of transfer of nitrogen between the solid and solution phases has long been recognized as of major importance in determining the supply of soil nitrogen for plants. Estimates of the net rate of transfer may be made by incubating the soil under selected conditions and measuring the change in mineral nitrogen present with time.

The rate of transfer of phosphorus from the solution phase to the soil solids has been found important where inorganic orthophosphate is added to soils in soluble form. The soluble phosphorus decreases rapidly, but not so rapidly that plants do not benefit from it.

On the other hand, where release of phosphorus from the soil solids to the solution phase is concerned, the evidence suggests that it is fast enough to replenish the solution to the equilibrium value relatively rapidly and that other factors are more important in determining phosphorus uptake by plants. For example, Fried et al. (1957) found that in 2 to 3 hours, the soils they tested would release as much phosphorus to the solution as most crops absorb during a growing season. Thus, in studies of the factors affecting the uptake of phosphorus from soils by plants, most authors ignore the rate of release of phosphorus from the soil solids as being nonlimiting. The same may be said for potassium.

4-3.4. Rate of Absorption by Roots

The concentrations of nutrients at root surfaces in soils can be measured only indirectly under special experimental conditions. Experiments with plants grown in continuously stirred nutrient solutions, however, have shown that the last step in the foregoing scheme, absorption of nutrients by roots, takes place at rates that increase with the concentrations of the individual nutrients in solution up to values exceeding those needed to provide adequate nutrition. Practical interest from the standpoint of soil fertility evaluation thus is focused upon the prior steps that will affect the concentrations of nutrients at root surfaces.

4-3.5. Nutrient Movement Mechanisms

Three general mechanisms of movement of nutrients in solution from their initial location in soil to roots are recognized. The first is passive transport in the solution, mainly as a consequence of transpiration of water from plant tops and accompanying absorption of water by plant roots. This mechanism often is designated as mass movement and sometimes as convection.

The second mechanism is diffusion. This mechanism is a response to the decreases in concentration of nutrients at root surfaces that result from nutrient uptake by the roots. Diffusion is a consequence of random movements of solute molecules or ions. In a uniform solution, the diffusive movement is the same in all directions, but if concentrations are unequal, net movement takes place from higher concentrations to lower concentrations.

The third mechanism is biological transport via mycorrhizas, which are fungi that are attached to plant roots but extend outward into the surrounding soil, sometimes to distances of several centimeters. As discussed in Section 1-3.7.5, nutrients absorbed from the soil by mycorrhizal fungi are transported through the mycelia to roots, and the fungi are sustained by organic nutrients derived from the roots. Mycorrhizal fungi can greatly increase the effective extent of a plant's root system.

Although not a mechanism of movement of nutrients through soils, an additional essential aspect of the over-all process is the extension of roots, so that they are in position for the short-range mechanisms to be effective. Root extension and nutrient availability may be closely related, as indicated by the results of an experiment by Khasawneh and Copeland (1973), in which a soil very deficient in phosphorus was cropped to cotton after addition of 0, 50, 100, and 200 milligrams of phosphorus as concentrated superphosphate per kilogram of soil. The plants were harvested and analyzed for phosphorus 4, 6, and 8 weeks after planting. When the yields of phosphorus in the plant tops were plotted against the root lengths per culture, all the data could be represented to a good approximation by a single curve with increasing slope, indicating that phosphorus fertilization increased root length and that increased root length resulted in increased phosphorus uptake.

Visual evidence of both the diffusion mechanism and the mass movement mechanism was provided by Barber (1962), Barber et al. (1963), and Lavy and Barber (1964) in the form of autoradiographs of soil containing plant roots. Illustrating the importance of diffusion, the autoradiographs showed concentrations of radioactive rubidium (used as a tracer for potassium) in roots and relatively wide depleted zones in the soil around the roots. For phosphorus, the accumulation in the roots was not evident, but there was a narrow depleted zone around the roots, indicating that only a relatively small volume of soil was of importance in supplying phosphorus.

Illustrating the transport of nutrients to plant roots in the water absorbed by the roots, other autoradiographs showed the accumulation of radioactive sulfate sulfur and radioactive strontium around roots (the behavior of strontium is similar to that of calcium). After removal of the roots from the soil, the accumulated strontium gradually diffused back into the soil. An autoradiograph of a sandy soil treated with radioactive molybdenum showed accumulation of radioactivity around the roots in the upper portion of the culture to which water was added through glass tubes, indicating excessive transport to the roots in the water. In the lower portion of the same culture, where water movement and water absorption would have been less, zones depleted of radioactivity appeared around the roots, indicating diffusion as the dominant mechanism of molybdenum transport.

Indirect means must be used to estimate the relative significance of movement of nutrients with the water and diffusion through the water. The simplest is to calculate on the basis of the amount of water absorbed by plants and the concentrations of nutrients in the soil solution whether the quantities of dissolved nutrients delivered to the root surfaces by the water absorbed are sufficient to supply the quantities of the nutrients found by analysis in the plants. If the quantities brought to the root surfaces in the water exceed the quantities found in the plants, it is inferred that movement of the nutrients with the water was more than enough to account for the nutrients found in the plants. If the quantities brought to the root surfaces in the water are less than the quantities in the plants, diffusion of the nutrients through the soil water and transport by mycorrhizas are inferred to have been necessary to account for the additional uptake. Data summarized by Barber (1962) and Barber et al. (1963) for corn indicate that movement in the water generally would deliver more than adequate amounts of sulfur and calcium, often would deliver more than adequate amounts of magnesium and nitrogen, and rarely would deliver adequate amounts of potassium and phosphorus.

Diffusion of nutrients to root surfaces from their origin in the soil depends upon several factors that are important in the supply of nutrients to plants. Some of these factors are evident from Fick's diffusion equation, which was shown by Olsen and Kemper (1968) as

$$\frac{\Delta Q}{\Delta t} = -DA\frac{\Delta C}{\Delta x} \quad ,$$

where ΔQ is an amount of the nutrient (milligrams or millimoles) diffusing in time Δt, t is the time (seconds), D is the diffusivity or diffusion coefficient (square centimeters per second), A is the cross sectional area (square centimeters) through which diffusion is occurring, ΔC is the change in concentration (milligrams or millimoles per cubic centimeter), and Δx is the distance (centimeters) in the direction of net movement. Various units may be used. From the Fick equation, the rate at which diffusion delivers a nutrient to a root surface may be inferred to increase with the time, the diffusion coefficient, the area of root surface, and the difference between the maximum and minimum concentration, and to decrease with increasing distance through which diffusion must occur.

Numerous modifications of the basic equation are made to apply the equation to different conditions. Most are adaptations of equations for heat flow and have been discussed at length by Olsen and Kemper (1968), but will not be considered here.

The basic Fick equation applies to a steady-state condition in which the difference in concentration and the distance remain constant. In nutrient diffusion to roots in soils, primary interest is in transient conditions in which a root grows into a volume of soil in which the concentration is uniform and then proceeds to absorb the nutrient. In this situation, the concentration at the root surface decreases, and the distance x through which diffusion occurs increases with time. Moreover, the diffusion takes place radially to a central cylinder, the root. This situation requires a modification of the equation.

Not evident from the Fick equation is the fact that in soils the diffusivity or diffusion coefficient D does not have a constant value that is characteristic of the nutrient. Rather, it varies with the soil and with conditions within a given soil, principally the water content. The volume of water, the thickness of the water films, the tortuosity of the diffusion path, and the electrical field around the soil particles all affect the diffusion coefficient. The decreased effective diffusion coefficient of potassium in relatively dry soil has been related, for example, to the increased response of plants to fertilizer potassium under these conditions (Grimme, 1990; Barraclough, 1990).

One further very important condition in diffusion of certain nutrients in soils is the fact that diffusion is not occurring through an inert medium, as is supposed by the Fick equation. Rather, some of the nutrient is found in solution, and some (generally a considerably larger amount than that in solution) is attached to the soil solids in forms that equilibrate with the solution as the nutrient in solution slowly diffuses past. As a result of the presence of these solid-phase forms, the rate at which the concentration of the nutrient at the root surface would otherwise decrease is moderated, and the solid-phase forms are depleted to a degree that increases with proximity to the root surface. This subject will be discussed further in following sections.

4-3.6. Integrated Nutrient Availability Factors

The gradual evolution of concepts of nutrient availability factors has resulted in recognition of three aspects of availability that can be expressed in terms of chemical measurements on soils. The first is intensity, the second is quantity, and the third is buffer capacity. These concepts will be explained here and will be applied in Section 4-4 on estimating fertilizer requirement.

4-3.6.1. Intensity

The intensity factor of nutrient availability usually is associated with the nutrients in the soil solution in contact with root surfaces. Characterizing the soil solution, however, has represented a challenge to researchers from the very beginning.

The composition of the soil solution cannot be measured in place, and hence the composition at the surface of roots cannot be determined. Moreover, there is evidence that the composition varies on both a macro basis and a micro basis. The discussion here will start with the micro variation and will proceed to other aspects of understanding and practice.

Activities and Concentrations. As knowledge of the physical chemistry of soils developed, it became accepted that in addition to differences among soils and among macro locations in a given soil, the composition of the soil solution varies with distance from soil particles as a consequence of the atmospheres of exchangeable cations attached to the soil particles. Theoretically, the activity of a dissolved salt should be the same throughout a solution at equilibrium with the soil solids. Thus, if a represents activity, b the solution in bulk, and s the solution within the atmospheres of exchangeable cations on the soil solids, one may write for monobasic calcium phosphate

$$[a_{Ca_b^+}][a_{H_2PO_{4b}}]^2 = [a_{Ca_s^+}][a_{H_2PO_{4s}}]^2$$

or

$$[a_{Ca_b^+}]^{1/2}[a_{H_2PO_{4b}}] = [a_{Ca_s^+}]^{1/2}[a_{H_2PO_{4s}}].$$

For monobasic potassium phosphate, one may write

$$[a_{K_b^+}][a_{H_2PO_{4b}}] = [a_{K_s^+}][a_{H_2PO_{4s}}].$$

Similarly, for two salts in the same solution at equilibrium with the soil solids, the ratio of the activities in solution is equal to the ratio of the activities within the atmospheres of exchangeable cations. If the two salts have the same anion, the anion activities divide out, and one has an expression for the ratios of the cation activities. Thus, dividing the third of the foregoing expressions by the second yields

$$\frac{a_{K_b^+}}{[a_{Ca_b^+}]^{1/2}} = \frac{a_{K_s^+}}{[a_{Ca_s^+}]^{1/2}} \quad .$$

Analogous expressions may be written for other salts.

The valences of the ions have been included in the foregoing expressions to show how they are related to the exponents. Usually the valences are omitted. The symbol a for activity is often replaced by parentheses around the ion in question. Especially where the expressions are used in negative logarithmic form, the ion symbols are used without valences, parentheses, or the a notation.

One can evaluate the activities of ions in a bulk solution, but not in the ionic atmospheres on the surfaces of soil particles. The theory of equal activities in the bulk solution and ion atmospheres was verified in a qualitative way, however, by Bower and Goertzen (1955), who made chloride analyses on successive increments of solution removed from moist soils by use of a pressure-membrane apparatus. Because the exchangeable cations contribute to the cation activities within the ionic atmospheres around the soil particles, the chloride activity should be lower within the ionic atmospheres than in the bulk solution. In agreement with theory, Bower and Goertzen found that when a soil in which the cation-exchange positions had been saturated artificially with sodium was equilibrated with a solution containing 10 millimoles of sodium chloride per liter, the first increment of solution removed contained 16 millimoles of chloride per liter, but the last contained only 6. This experiment also verified the theory of variability of the composition of the soil solution on a micro basis.

Because the composition of the soil solution cannot be measured in place, the solution must be removed from the soil in some way for experimental study. The evident nonuniformity of composition of the soil solution on a micro basis raises two important questions. First, how is the composition of the solution to be expressed, and second, to what do plants respond when they are deriving their mineral nutrients from the soil solution?

Answering these two questions has required considerable effort, but fortunately the outcome seems simple enough. The soil solution generally has been extracted by displacing it, most often with water containing potassium thiocyanate as a tracer. The moist soil is packed in a vertical column, and the displacing solution is added at the top. As the soil solution is displaced, it forms a gradually thickening layer of saturated soil below the downward moving displacing solution. When the saturated zone reaches the bottom of the column, the displaced solution drips from the column under gravity. Successive increments of the displaced solution are tested with ferric chloride solution, which forms a red color with the thiocyanate ion, to find when collection should cease because of contamination of the displaced solution with the water added at the surface. See a publication by Adams (1974) for more details.

A method that has been proposed more recently (Mubarak and Olsen, 1976) is adding to moist soil a dense organic liquid that does not mix with water and centrifuging the mixture at high speed. The organic liquid enters the soil, and soil solution appears in the centrifuge tube above the organic liquid. This method leaves much of the soil solution in the soil, perhaps because the organic liquids do not dispace the water molecules that adhere strongly to the solid surfaces. Adams et al. (1980) found, however, that the ionic compositions of soil solutions were the same, whether obtained by displacement, simple centrifuging, or centrifuging with an immiscible dense organic liquid.

In light of the experiment by Bower and Goertzen (1955), the displacement procedure described by Adams (and the centrifuging method as well) appears to displace a solution whose composition represents that of the portion of the soil solution most distant from the soil particles. Calling the solution that drips from the bottom of the soil column the ''soil solution'' thus takes no account of the

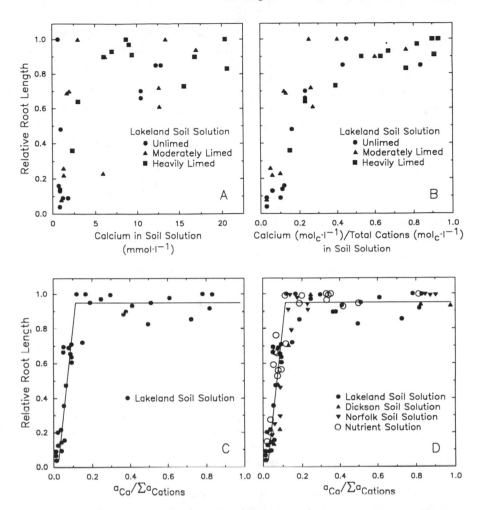

Fig. 4-8. Relative length of the tap root of cotton seedlings produced in 2 days in subsoils or nutrient solutions versus: A. Concentration of calcium in the soil solution from Lakeland subsoil with different liming treatments. B. Ratio of the concentration of calcium to the concentration of total cations in the soil solution from Lakeland subsoil with different liming treatments. C. Ratio of the activity of calcium to the sum of the activities of cations in the soil solution from Lakeland subsoil with no differentiation of liming treatments. D. Ratio of the activity of calcium to the sum of the activities of cations in nutrient solutions and in the soil solutions from Lakeland, Dickson, and Norfolk subsoils. (Howard and Adams, 1965; Adams, 1966, 1974)

fact that the concentrations of the salt constituents are greater in this solution than they are in the average solution as it occurs in the soil.

Indications are that plants respond to the portion of the soil solution that corresponds to the displaced solution and that the effects of the ion atmospheres can be neglected. Fig. 4-8 supplies important evidence. In this experimental work, the lengths of the tap roots of cotton seedlings were measured 2 days after the roots entered various subsoils that had received different treatments. The seedling roots must absorb almost all of their calcium from the soil because downward translocation of calcium from the seed is negligible. Thus, calcium deficiency affects the growth of the roots before it affects the tops.

Figs. 4-8A, B, and C show that to obtain a close relationship between the relative lengths of roots in the various experimental series and the measurements on the soil solution, it was necessary to express the data in an appropriate way. The relationship was better with the ratio of the activities of the calcium ion to the sum of the activities of all the major cations in solution (Fig. 4-8C) than with the ratio of the concentrations of the calcium ion to the sum of the concentrations of all the major cations in solution (Fig. 4-8B) or with the calcium concentration (Fig. 4-8A).

The most critical evidence is in Fig. 4-8D, which shows that when the solution composition was expressed as ratios of the activities of calcium to the sum of activities of all the major cations, the data for the soils did not segregate into separate plots, and the data for the nutrient solutions followed the same trend as the soil solutions. The logical inference from Fig. 4-8D is that within the limits of sensitivity of the experimental work, there was no evidence that the plants responded differently to the activity ratios in the soil solutions than they did to the activity ratios in the nutrient solutions. If the reduced concentrations or activities of salts within the exchangeable cation atmospheres surrounding the soil particles had been of sufficient importance, the plots of the soil solution data should have diverged from the plot of the nutrient solution data.

The findings in Fig. 4-8 provide important evidence on the soil solution as the nutrient medium of plants. They are of further significance because they indicate the superiority of ion activities over concentrations and the importance of considering the ratio of calcium ion activity to the activities of all the major cations.

Perhaps the most unequivocal support for the use of ion activities has been obtained in experiments with algae in nutrient solutions under conditions in which the depletion of the solutions was slight and the interpretation of the data was not complicated by possible interactions with solids. Sunda and Guillard (1976) determined the copper concentration in algal cells in seawater as affected by the concentrations of copper and tris, which is a copper-chelating agent. The concentration of copper per algal cell increased with the concentration of copper in solution, but decreased with the concentration of tris, resulting in a different plot of copper per cell versus the total concentration of copper for each concentration of tris. In an analogous plot of copper per cell against the calculated activity of the cupric ion, all the points fell very close to a single line, indicating that the uptake of copper was controlled by the copper ion activity and not the total copper concentration. Analogous data for zinc were published by Allen et al. (1980).

Similarly, experimental work by Parker et al. (1988) with wheat seedlings maintained for 2 days on culture solutions with different concentrations and activities of aluminum species showed that the length of the roots was related much more closely to the activities of the aluminum ion (Al^{+3}) than to total concentrations of aluminum in solution. The reason for measuring root length was that the primary obvious effect of aluminum toxicity on plants is on root elongation. The results of experiments on cadmium uptake by crop plants from culture solutions (Checkai et al., 1987) as well as soils (Sposito and Bingham, 1981) indicate that uptake was related to the cadmium activity, not the concentration, although the interpretation of the results is more complex than that of the experiments with algae.

Ion activities may be regarded as effective concentrations, where chemical reactivity is concerned. Mathematical procedures for converting the concentrations found analytically to activities are found in papers by Adams (1971, 1974) and in physical chemistry books, for example, Barrow (1979). Computer programs are available to calculate ion activities from the ionic strengths of solutions and the total concentrations of the ions in solution. The computations are complicated by the fact that in complex solutions, such as those in soils, salt components have a tendency to associate with other species, so that, for example, calcium may be present in solution not only as the Ca^{++} ion but also as the $CaHCO_3^+$ ion and as the neutral ion pair $CaSO_4^0$. The computation methods require that the total concentrations of all the ions be known for accurate results.

Ion-selective electrodes make possible direct measurements of the activities of individual ions for which such electrodes are available. Ideally, a specific-ion electrode for a particular ion will respond only to the activity of that ion in solution, and not to that of the associated charged or uncharged species containing that ion or to foreign ions. In practice, the selectivity is less than ideal, and there are usually interferences that cause problems.

Whether ion activities or concentrations should be used in soil fertility and plant nutrition studies is an issue that has not been settled, perhaps because the evidence does not indicate that the advantage is all in one direction or the other. There seems to be no question but that the activity of an ion in the soil solution should control the degree to which the ion saturates the carrier sites on plant roots that are specific for its uptake.

The situation is confused, however, by the results of experiments on higher plants in nutrient solutions with micronutrient metals supplied at constant activities but different concentrations by use of chelates. Higher concentrations have been associated with greater uptakes. In part this could be a consequence of a buffering effect of the chelates on the activities of the metal ions at root surfaces. There is also evidence of uptake of the chelates as such. See a paper by Bell et al. (1991) for experiments, discussion, and literature.

In an investigation of 21 samples of Queensland soils and subsoils, Bruce et al. (1988) found that the best index of the relative length of roots of soybean seedlings in the untreated soils to the length in the soils treated with calcium carbonate or other calcium salts was the percentage saturation of the effective cation exchange capacity of the soils with calcium. The ratio of the calcium ion activity to the sum of the activities of calcium, magnesium, potassium, and sodium in the soil solutions was a somewhat less effective index. The calcium ion activity was considerably less effective, and the calcium concentration was still less effective.

The nutrition of plants in soils is a more complex matter than the uptake of traces of copper or zinc by algae from a well stirred nutrient solution. For plants in soils, the question of activities versus concentrations may be reduced to one basic issue, namely, measurement. Although the soil solution may be regarded as the nutrient solution for plants in soils, the solution is not stirred, and the absorption of nutrients and water by plants changes the concentrations and activities of the solutes at root surfaces; most decrease, but some increase. It may be possible to calculate the concentrations and activities of ions at root surfaces as a result of some kind of experimental measurements, but measurements commonly are made on the bulk solution, which is remote from root surfaces, and the data obtained do not tell what the situation is at the root surfaces.

Moreover, measurements of concentrations and activities represent the situation at the point and time of measurement, but the nutrition of plants in soils is a dynamic process in which the plants interact with the soils and change the situation in the soils. Appropriate time-related measurements of concentrations and activities can provide information on the dynamic processes, but single measurements cannot. Activities are more difficult to deal with than concentrations, even where computers are used to make the calculations. The upshot has been that concentrations generally are used, even for research purposes.

Similarly, it is difficult to deal with soil solutions. Little research is done on them, and they are not used for practical soil testing work. As a substitute, so-called saturation extracts are in common use for saline soils. Enough water is added to saturate the soil, and after a period for equilibration a quantity of solution is removed by vacuum filtration for analysis. As indicated in the following section, the solution used for analysis in nonsaline soils is often obtained by equilibrating a sample of soil with a relatively large volume of a dilute calcium chloride solution (usually 0.01 molar).

Potentials. Related to ion activities is the concept of nutrient potentials, which has developed from Schofield's (1947) *ratio law*: "When cations in a solution are in equilibrium with a larger number of exchangeable ions, a change in the concentration of the solution will not disturb the equilibrium if the concentrations of all the monovalent ions are changed in one ratio, those of all the divalent ions in the square of that ratio and those of all the trivalent ions in the cube of that ratio." Although Schofield expressed his proposal in terms of concentrations, more recent practice has been to use ion activities because what Schofield had described was a physicochemical relation that is expressed more precisely in terms of activities.

The left side of the equality in the second equation in the preceding section on activities and concentrations has been termed the monocalcium phosphate potential, the calcium phosphate potential, the phosphate potential, or the Schofield phosphate potential. White and Beckett (1964) used the last of these terms to credit Schofield and also because there are two other physicochemical phosphate potentials defined in different ways. Usually the phosphate potential is written as $0.5pCa + pH_2PO_4$, where pCa and pH_2PO_4 are the negative logarithms of the activities of the respective ions.

By analogy with the calcium phosphate potential, the left side of the equality in the fourth equation in the preceding section on activities and concentrations might be called the potassium-calcium potential, but this term does not seem to be used. More commonly, it is called the potassium potential, which, like the phosphate potential, is somewhat of a misnomer. Beckett (1964a) modified the potassium-calcium activity ratio by adding the activity of magnesium to that of calcium in the denominator. For this expression, he used the term "activity ratio," which he abbreviated to AR^K.

Note that both the phosphate potential and the potassium potential involve calcium. The reason for this is that, harking back to Schofield's (1947) ratio law, calcium usually is the predominant exchangeable base. Related to this, the potentials are almost invariably inferred from measurements on dilute calcium chloride solutions (usually, but not always, 0.01 molar) that have been equilibrated with soils (Beckett, 1964a; White and Beckett, 1964).

Schofield (1955) was the first to propose "the chemical potential of monocalcium phosphate as the function most likely to give a numerical index of the condition in the soil which mainly controls the availability of phosphate." Beckett (1964b) credited Scheffer and Ulrich (1962) with the proposal that the equilibrium potassium potential is a satisfactory measure of the availability of potassium to plants.

No serious question seems to have been raised regarding the validity of the principle that the activity of a salt is equal throughout the soil solution at equilibrium. Extension of this concept to imply that an activity product or ratio represents an appropriate way to express the intensity factor of the supply of a nutrient or to provide an index of the availability of a nutrient to plants, however, is another matter.

From the mathematical formulation of the potentials, it is evident that the numerical value of the phosphate potential depends upon the activity of calcium as well as that of phosphate and that the numerical value of the potassium potential depends upon the sum of the activities of calcium and magnesium as well as the activity of potassium. Nevertheless, calling the formulations the phosphate and potassium potentials, neglecting to mention the other ions involved, and discussing the potentials in the context of availability has created some confusion.

Using measurements of the phosphate potential or the potassium potential as indexes of the intensity of supply of these nutrients would be satisfactory from the standpoint of correlating the results with biological indexes of phosphorus or potassium availability if the contribution of calcium or calcium plus magnesium to the potential were the same in all the soils being compared. Under such circumstances, inclusion of the calcium component of the potential would have no effect on the relative values of the indexes for the different soils because in effect it would amount to multiplying the phosphate or potassium activities in all soils by a constant factor.

Wild (1964) first tested the availability inference experimentally by growing plants in sand cultures supplied with solutions having three different concentrations of phosphorus, each at two different calcium concentrations, leading to different phosphate potentials. He found that wide differences in yield of the test plants resulted from the differences in phosphorus concentration, but not from differences in phosphate potential.

In a subsequent paper, Wild et al. (1969) pointed out that if the uptake of potassium is independent of the calcium and magnesium concentrations, the activity ratio might be misleading when comparing soils with different contents of calcium and magnesium. If two soils have the same activity ratio of potassium to the square root of the sum of calcium and magnesium, but the activity of calcium plus magnesium in solution in one soil is four times as great as in the other, the activity of potassium would be twice as great in the first soil as in the second, and the uptake of potassium by plants might also be rather different.

To test the validity of the activity ratio concept as a characterization of potassium availability, they grew ryegrass and flax in sand cultures supplied with solutions having three ratios of potassium activity to the square root of the calcium activity at three different concentrations of potassium. All solutions were prepared at two concentrations of magnesium. The results showed that the yield of dry matter and the potassium uptake depended upon the concentration of potassium and not upon the activity ratio of potassium to the square root of calcium. The higher magnesium concentration did not depress the uptake of potassium. Fig. 4-9 shows the average data on potassium uptake by ryegrass. Mengel (1963) used a wider range of compositions of culture solutions that produced nutrient deficiency symptoms in the extreme ranges, but his findings also led to the conclusion that the ion activities or concentrations were more important for the test plants than were the activity ratios.

Nutrient solutions were used in all the foregoing tests of the activity ratio concept, and the techniques employed were not adequate to prevent significant changes in composition of the solutions as the plants absorbed nutrients. The experiments thus do not answer the question of whether nutrient activities or concentrations are more appropriate where plants are concerned, but they still provide qualitative evidence on the ratio effect. The evidence indicates that where the intensity of supply of phosphorus and potassium is concerned, the concentration or activity of the ions is the important property, not their ratio to calcium or calcium plus magnesium.

As a contrasting example, Ozanne and Shaw (1967) published work on soil samples from 42 locations in Australia in which the phosphorus concentrations

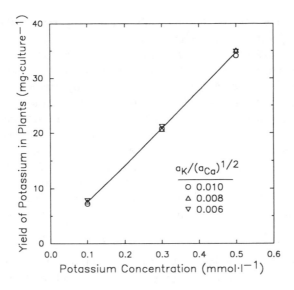

Fig. 4-9. Yield of potassium in ryegrass tops versus potassium concentration in three nutrient solutions added to sand cultures. The ratio of the activity of potassium to the square root of the activity of calcium (mol · l⁻¹)¹ᐟ² ranged from 0.006 to 0.010, corresponding to a threefold range in calcium concentrations. (Wild et al., 1969)

in 0.01 molar calcium chloride extracts of the soils were highly correlated ($r =$ 0.9999) with values of the phosphate potentials derived from the same extracts. In their investigation, the equilibrium concentrations of calcium seldom varied far from 0.01 mole per liter, and the pH values were mostly such that 98% or more of the phosphorus was in the $H_2PO_4^-$ form. They concluded that for the range of soils they examined there was no advantage in performing the laborious calculations needed to calculate the phosphate potentials.

For the range of soils investigated by Ozanne and Shaw (1967), the correlation between the phosphate potentials and the concentrations of phosphorus in solution would have been about the same if they had merely assumed that the activities of calcium in the solutions after equilibration were equal to the activity of calcium in the added solution. On the other hand, for the previously mentioned reason given by Wild et al. (1969), differences in the calcium component of the potential could degrade the correlation of the potential with phosphorus concentrations and biological indexes of availability in circumstances in which the contribution of the activities of calcium to the potential differed from one soil to another.

In research on 14 soils of Greece, Keramidas and Polyzopoulos (1983) characterized the phosphorus intensities by various conventional and physicochemical methods, and determined the correlations between the phosphorus uptake from the soils by ryegrass and the soil measurements to rate the soil measurements. The phosphorus concentrations and $H_2PO_4^-$ activities in 0.01 molar calcium chloride extracts of the soils were as effective as the Schofield phosphate potentials for the first cutting, but gradually became superior for the cumulative uptake of phosphorus in successive cuttings.

The evidence described on the significance of nutrient ratios thus is conflicting. For calcium (Fig. 4-8), the other ions needed to be taken into account, but for phosphorus and potassium, the ratios were mostly irrelevant. There seems to be no reason to question either side of the evidence. Rather, it seems appropriate to suggest that the situation may differ from one nutrient to another. Cations in the soil seem to have a special relation to cations in plants, a theme elaborated by Nielsen and Sørensen (1984) and Nielsen and Hansen (1984). They considered that plants have a finite cation uptake capacity and behave as saturated cation exchangers in equilibrium with the solution bathing the roots. Plants have long been known to exhibit marked selectivity in absorbing phosphorus and potassium from very dilute solutions.

4-3.6.2. Quantity

The quantities of most nutrients present in soil solutions are so small relative to the quantities taken up from soils by plants that the soil solids obviously must hold the major portions of the nutrients that are available to plants during a season. The methods commonly used to provide chemical indexes of nutrient availability extract far greater amounts of most nutrients than are present in solution in most soils. Thus, they may be said to provide indexes of the available quantities of the nutrients or some combination of quantity with other properties.

An example of a method that can reflect both intensity and quantity aspects of availability is the anion exchange resin method for extracting soil phosphorus (Amer et al., 1955). The shorter the equilibration time, the more nearly the amount of phosphorus extracted approaches the intensity aspect, and the longer the equilibration time, the more nearly the amount extracted approaches the quantity aspect.

A similar method as regards the effect of time on the balance between the intensity and quantity factors of phosphorus supply is the isotope dilution method of measuring the quantity of soil phosphorus that comes to equilibrium with radioactive orthophosphate added in solution (Russell et al., 1954). Measurements made on soils are often called E values (for exchangeable phosphorus — more commonly called labile phosphorus), and those made on plants are often called L values after Larsen (1952), who described this usage. Some authors use the symbol L in the equivalent sense of labile soil phosphorus.

Two theoretical advantages are associated with the isotope dilution method. First, the specific activity of the phosphorus taken up from soil by plants is equal to the specific activity of the phosphorus in the soil solution (Fardeau and Jappe, 1982). Second, when plants are grown for a longer time on soil uniformly tagged with phosphorus-32, the specific activity of the phosphorus in the plants is generally about the same as the specific activity of the soil phosphorus that has equilibrated with added phosphorus-32. That is, the estimates of labile soil phosphorus based upon measurements in the laboratory (E values) are approximately equal to those based upon measurements derived from uptake by plants (*L* values).

If a strictly 1:1 relationship were obtained between *L* values and *E* values, the implication would be that the plants were deriving their phosphorus from the fraction of soil phosphorus found by isotope dilution measurements in the laboratory. The data from three independent investigations summarized in Fig. 4-10 indicate that on average this was approximately the situation with the soils examined. The positive deviations somewhat exceeded the negative deviations, however, which suggests that a little of the phosphorus in the plants came from sources other than the labile phosphorus measured in the soil.

Both *E* values and *L* values increase with time. After an initial rapid increase, *E* values for phosphorus increase slowly for an indefinite length of time. To estimate the *E* values at a time corresponding to the end of the plant growth period, Fardeau and Jappé (1976) extrapolated a linear log-log plot of *E* values against time. Gachon (1966a) and Probert (1972) found that *L* values increased in successive cuttings of test plants grown on soils. And in experimental work with many soils of France, in which ryegrass was used as a test crop, Gachon (1966b, 1988) found that the *L* values definitely exceeded the *E* values.

In an investigation of possible reasons for the difference between *E* values and *L* values, Grinsted et al. (1982) and Hedley et al. (1982) grew rape plants in thin layers of soil so that after 14 days all the soil in effect became rhizosphere

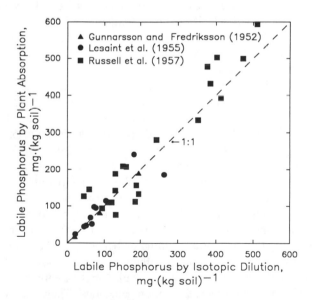

Fig. 4-10. Labile soil phosphorus as estimated from absorption of phosphorus-32 and phosphorus-31 from phosphorus-32-tagged soils by plants and as estimated from isotope dilution measurements on the soils in the laboratory. The data are from three independent investigations. The data from Russell et al. are from four different experiments. (Gunnarsson and Fredriksson, 1952; Lesaint et al., 1955; Russell et al., 1957)

soil. They found that during the first 14 days, when the E values and L values were equal, only the inorganic phosphorus fractions extractable from the cropped soil by an anion-exchange resin, 0.5 molar sodium bicarbonate at pH 8.5, and 0.1 molar sodium hydroxide were depleted by the plants. With extension of the period of plant growth, the soil pH values dropped as much as 2.4 units (from 6.5 to 4.1). (Separate tests showed that lowering the pH of the uncropped soil to this extent could result in at least a ten-fold increase in phosphorus concentration in solution. See a paper by Marschner et al. (1986) for color photographs illustrating by the use of indicator dyes the changes in pH that may occur in the root zone of plants.) Concurrently, the L values gradually exceeded the E values measured on the uncropped soil at the same times. After 35 days, the L values were approximately twice as great as the E values. Fractionation of the soil phosphorus showed that the acid-soluble and residual phosphorus fractions had decreased. The indications were, therefore, that these forms of phosphorus were not exchangeable initially, but that they were brought into play when the plants decreased the soil pH. No decrease in soil organic phosphorus was found.

Armstrong and Helyar (1992) used a culture technique like that employed by Grinsted et al. (1982) and Hedley et al. (1982), and they similarly found no significant decrease in soil organic phosphorus. The plant species tested depleted phosphorus from inorganic fractions, including an acid-soluble inorganic fraction, even though the rhizosphere pH increased in most instances. According to other investigators, however, the presence of plant roots stimulates a pronounced turnover (Helal and Sauerbeck, 1984) and net mineralization (Tarafdar and Junck, 1987) of organic phosphorus in the soil immediately adjacent to root surfaces. This involvement of plants in soil organic phosphorus transformations would help to account for L values exceeding E values.

Some investigations have shown that the values of E or labile phosphorus measured in different soils are well correlated with biological indexes of phosphorus availability to plants. In other work, the correlations have been poor.

Gachon (1969) found that the correlations between uptake of phosphorus by plants in greenhouse cultures and E values were $r = 0.90$ with 25 soils derived from crystalline rocks and $r = 0.91$ with 15 soils derived from volcanic ash. The regressions, however, were different. The phosphorus uptake was considerably lower from the latter group of soils than from the former. Soils derived from volcanic ash characteristically contain much amorphous material, react strongly with phosphorus, and have high requirements for fertilizer phosphorus.

Tran et al. (1988) found a good correlation between uptake of phosphorus by plants and E values measured after 1 minute of equilibration on 38 soils with low and medium phosphorus sorption capacities, but essentially no correlation on 20 soils with high phosphorus sorption capacities. Their use of a 1-minute equilibration time yielded E values that were greater by a factor of four than the phosphorus in a 1:10 soil:water extract, but were still lower than the values of E that would have been obtained in the usual equilibration for 24 hours or longer. Thind et al. (1989) found a small negative correlation between labile phosphorus and the yields of phosphorus in the control cultures in a greenhouse experiment with 12 alkaline soils from the Delhi region in India, whereas the five extraction methods tested all gave statistically significant positive correlations.

The findings made with tagged phosphorus have turned up the information that numerically equal quantities of phosphorus in equilibrium with the phosphorus in the soil solution from which plants absorb their phosphorus may have very different levels of availability. This information has increased the basic understanding of phosphorus availability. It provides evidence that the uptake of phosphorus from soils by plants is not necessarily governed primarily by the quantity factor.

The problem, then, is to characterize the way the quantity factor contributes to the availability of phosphorus and other nutrients to plants. The buffer capacity concept discussed in the next section is the current understanding.

4-3.6.3. Buffer Capacity

The buffer capacity concept is a combination of two more elementary principles. First, the availability of a nutrient decreases as plants reduce its concentration in the soil solution. Second, the degree to which the concentration decreases depends upon how much of the nutrient the plants absorb and how much is released by the soil solids in response to the drawdown of the concentration in the solution. The buffer capacity of a soil for a nutrient is the rate of change of quantity with respect to concentration, which usually is represented as *intensity* for theoretical purposes.

As an introduction, it may be helpful to consider the nutrient relations when plants are grown in a solution of constant volume with all the nutrients in solution. Each milligram of each nutrient absorbed by the plants then will result in a decrease in both the intensity (I) and quantity (Q) aspects of supply by 1 milligram per liter. That is, $\Delta Q = \Delta I$, and $\Delta Q/\Delta I = 1$. Fig. 4-11 illustrates this point with data from an experiment by Olsen (1950). If buffering had occurred, due for example to continuing passage of the nutrient solution through a column of soil, with release of phosphorus to the solution, the Q/I plot would lie above the 1:1 line in the figure and would be a curve that is concave downward. The slope of this curve would exceed unity, and the ratio of ΔQ to ΔI would represent the buffer capacity of the system at the intensity selected.

The buffer capacity of soils for plant nutrients is related to "fixation" of nutrients, such as phosphorus and potassium, but it is a newer term with different significance. Fixation refers to the reaction of a soluble nutrient with soils, with a consequent reduction in solubility and availability to plants. In buffer capacity terms, an addition of a soluble nutrient causes an increase in the concentration of the nutrient in solution and an increase in the quantity of the sorbed nutrient held by the soil solids. The increase in quantity of solid-phase nutrient in equilibrium with the solution per unit increase in concentration of the nutrient in solution is termed the buffer capacity at the particular concentration. Fixation and the increase in quantity of solid-phase nutrient in equilibrium with the solution

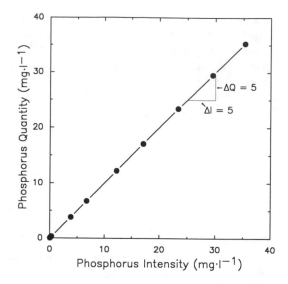

Fig. 4-11. Quantity versus intensity of phosphorus in a nutrient solution of constant volume as phosphorus was removed from solution by rye plants. (Olsen, 1950)

are not necessarily identical, however. Generally some of the nutrient that reacts with the soil does not remain in equilibrium with the solution, where it would contribute to the buffer capacity, but disappears into less soluble forms.

When some of the nutrient is removed from the solution by plants or otherwise, the removal causes a change in the opposite direction. The concentration in solution decreases, and the ensuing release of some of the nutrient from the solids to reestablish equilibrium replaces some of the nutrient that was removed from the solution. Then the buffer capacity is the decrease in quantity of the solid-phase nutrient in equilibrium with the solution per unit decrease in concentration in solution. Fixation does not apply directly to this important aspect of soil nutrient behavior.

In a given soil, the buffer capacity, quantity, and intensity of a nutrient are interrelated and vary together. As intensity increases, so also does the quantity. But the buffer capacity decreases. Buffer capacity curves are empirical and depend upon the methods employed. Nevertheless, they may be useful in providing improved estimates of fertilizer requirement, as will be explained in Section 4-4.

Phosphorus. The interrelationships and significance of the intensity, quantity, and buffer capacity factors are illustrated in Fig. 4-12 for three soils differing in phosphorus buffer capacity that had been adjusted to a range of supplies of phosphorus. Fig. 4-12A shows that the phosphorus uptake by corn seedlings increased with the phosphorus concentration supported in solution by the solids in each soil (the intensity of the phosphorus supply), but at a given initial

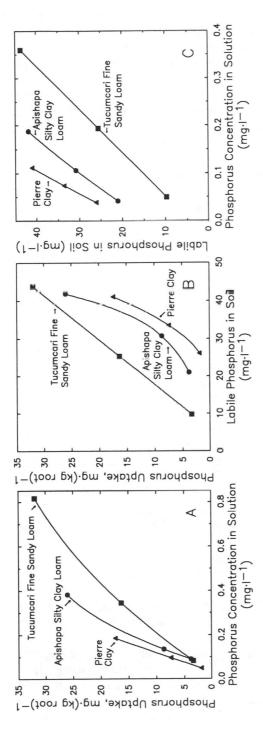

Fig. 4-12. Intensity, quantity, and buffer capacity factors in phosphorus uptake by corn seedlings in 24 hours from three calcareous soils. Each soil had been equilibrated with different quantities of concentrated superphosphate to vary the phosphorus supply. A. Phosphorus uptake versus initial phosphorus concentration in solution. B. Phosphorus uptake versus labile soil phosphorus. C. Labile soil phosphorus versus initial phosphorus concentration in solution. (Olsen and Watanabe, 1963)

concentration in solution the uptake increased with the clay content of the soil. Fig. 4-12B shows that the phosphorus uptake by the plants increased with the quantity of labile phosphorus in each soil, but that the uptake with a given quantity of labile phosphorus decreased with increasing content of clay in the soil.

Fig. 4-12C shows the relationships between the labile phosphorus and the phosphorus concentration in solution that help to explain the findings in Fig. 4-12A. The quantity of labile phosphorus held by the soil solids at a given concentration of phosphorus in solution increased with the clay content of the soil. Moreover, the slopes of the lines, representing the buffer capacities, increased with the clay content of the soil. Thus, the divergence of phosphorus uptakes from the different soils at a given initial phosphorus concentration in solution in Fig. 4-12A is related to differences in the phosphorus buffering capacities of the soils. As the buffer capacity decreased, uptake of phosphorus by the roots caused a greater drop in phosphorus concentration in solution and thus a drop in uptake. To put it another way, the uptake at a given initial concentration increased with the clay content of the soil and with the phosphorus buffer capacity because of lesser depletion of phosphorus from the solution.

For phosphorus, the buffer capacity has a maximum value as the concentration of phosphorus in solution approaches zero, and it decreases gradually as the concentration increases. See Fig. 4-13, which represents a *Q/I* or buffer capacity plot for one soil. The incidental construction showing the slopes at different points on the curve has been added to illustrate the significance of the plot in terms of phosphorus supply to plant roots.

The slopes of the lines shown at two different phosphorus concentrations in Fig. 4-13 indicate that for the sample from the field plots that had received fertilizer phosphorus at the rate of 142 kilograms per hectare some years previously, 760 milligrams of phosphorus would be released per kilogram of soil per milligram decrease in concentration of phosphorus per liter of solution in equilibrium with the soil. For the corresponding sample from the plots that had received 11 kilograms of fertilizer phosphorus per hectare at the same time, 1100 milligrams of phosphorus would be released per kilogram of soil per milligram decrease in phosphorus concentration per liter of solution. Extrapolating the curve to zero adsorbed phosphorus and zero concentration would give the maximum value of 2100 milligrams of phosphorus released per kilogram of soil per milligram decrease in phosphorus concentration per liter of solution.

Holford and Mattingly (1976a) designated the slope at the origin as the maximum buffer capacity. The other values were designated as the equilibrium buffer capacities for the samples of the soil that had been treated with different quantities of fertilizer phosphorus.

The units of buffer capacity described here may be abbreviated to liters per kilogram. These units are used in the figure. Unfortunately, it is difficult to relate the abbreviated units to the process they describe. Buffer capacity could also be defined more meaningfully for plants as milligrams of phosphorus released per liter of soil per milligram decrease in phosphorus concentration per

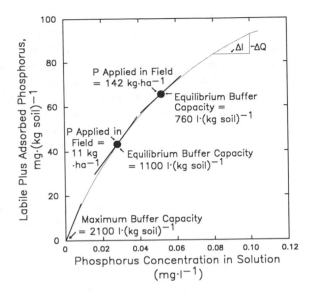

Fig. 4-13. The buffer capacity concept for phosphorus, illustrated by a plot of indigenous labile plus adsorbed phosphorus in a soil versus phosphorus concentration in solution. The two points represent samples taken from field plots that had received 11 or 142 kilograms of fertilizer phosphorus per hectare some years previously. The heavy solid lines through the points and at the origin are tangents to the curve. The slopes of the tangents represent the buffer capacities at the various locations on the curve, that is, the ratio of the change in labile plus adsorbed phosphorus to the change in phosphorus concentration in solution. (Holford and Mattingly, 1976a)

liter of solution in equilibrium with the soil (see Fig. 4-12C, where analogous units were used by the authors).

Where the buffer capacity varies continuously with the concentration of phosphorus in solution, one must face the question of how to deal with the variability in possible soil testing applications. This matter has not been resolved. Various "buffer capacities" and "buffer capacity indexes" are used.

Attention has been called to the fact that where one wishes to obtain an index of soil phosphorus availability to plants, the buffering process concerned is desorption of phosphorus held by the soil. For reasons of operational convenience, buffer capacity usually has been estimated by adding soluble phosphorus to soil and measuring the concentration in solution after a period of equilibration. Adsorption and desorption buffer capacities are not the same. Bowman and Olsen (1985) found that at a given concentration of phosphorus in solution, the desorption buffer capacities were smaller than the adsorption buffer capacities. That is, at a given concentration of phosphorus in solution, the quantity of phosphorus released per unit decrease in phosphorus concentration in solution was smaller than the quantity of phosphorus adsorbed per unit increase in phosphorus concentration in solution. The reason is that only a portion of the added phosphorus remains in labile form in rapid equilibrium with the solution.

Bowman and Olsen (1985) measured the desorption buffer capacities by equilibrating samples of soil with water and an anion exchange resin in the form of small spheres. The particles of resin were retained in a small wire basket that could be removed from the soil suspension to measure the phosphorus removed from the soil. The soil suspensions then were equilibrated for another 20 hours to allow an equilibrium phosphorus concentration to be established before the concentration of phosphorus was measured in the solution. Using a method similar to that employed by Bowman and Olsen, Holford et al. (1988) found that for correlating soil analysis data with uptake of phosphorus from soils in a greenhouse experiment, desorption buffer capacity measurements were superior to adsorption buffer capacity measurements.

Potassium.   Fig. 4-14 shows a *Q/I* or buffer capacity plot for potassium that is analogous to Fig. 4-13 for phosphorus. The horizontal line in Fig. 4-14 indicates an equilibrium condition in which the activity ratio of cations in the solution added to the soil resulted in no net uptake or release of potassium by the solid phase. Beckett (1964a) designated the equilibrium activity ratio of potassium to the square root of the sum of the activities of calcium and magnesium in the unchanged soil as AR_e^K. If similar construction had been used in Fig. 4-13 for phosphorus, there would be two horizontal lines, one through each of the

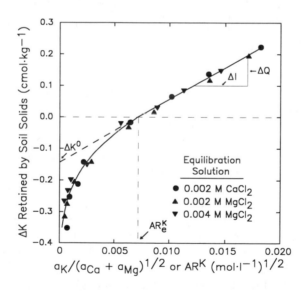

Fig. 4-14.   A *Q/I* plot for soil potassium, as found by equilibrating different quantities of soil with 50 milliliters of 0.002 molar calcium chloride, 0.002 molar magnesium chloride, or 0.004 molar magnesium chloride, each from 0 to 0.001 molar to potassium chloride. The equilibrium activity ratio AR_e^K corresponds to zero change in the potassium retained by the solid phase. The value of ΔK^0 found by extrapolating the linear portion of the plot to a zero activity ratio is designated as the labile potassium in the soil. The potential buffering capacity of the soil for potassium is indicated by the construction for $\Delta Q/\Delta I$. (Le Roux and Sumner, 1968)

two points, because the points represent equilibrium conditions for the two samples that had been differentiated by prior application of fertilizer phosphorus in the field. With phosphorus, the equilibrium condition would be no change in the concentration of phosphorus in the dilute calcium chloride solution added to the soil.

In both figures, the slope of the plot is a ratio of quantity to intensity, that is, the Q/I ratio. The Q/I terminology was introduced for potassium, and for the most part its use has been restricted to potassium. It is used only occasionally for phosphorus.

The construction showing ΔQ and ΔI in the plot for phosphorus yields an average ratio for the interval covered because the plot is not linear. For phosphorus, the slope of the plot gradually increases as zero concentration of phosphorus in solution is approached. This behavior indicates that the phosphorus buffer capacity is increasing as the concentration is decreasing and that the bonding energy with which the phosphorus is held to the soil is gradually increasing also.

In the figure for potassium, the portion of the plot above the equilibrium activity ratio is linear, but most of the part below this ratio is curved and behaves like the entire plot for phosphorus. As with phosphorus, the slope of the plot for potassium is connected with the bonding energy with which the potassium is held by the soil. The bonding energy is uniform in the upper linear portion and is thought to be a consequence of attachment of potassium in that range (resulting from uptake of potassium by the soil from the solution) in nonspecific cation exchange positions. The lower curved portion, which includes part of the exchangeable potassium, indicates that the potassium is bonded more and more strongly as the activity ratio approaches zero. Selective attachment of potassium ions in edge positions in micaceous minerals probably is responsible for some of the stronger bonding in this range. Bar et al. (1987) found that in a group of soils, mostly from Israel, the preferential adsorption of potassium by the method they employed was correlated ($r = 0.82$) with the percentage content of illite in the soils.

Although Fig. 4-14 represents the behavior of potassium in only one soil, quantity versus intensity plots for potassium in most soils are characterized by a linear upper portion and a curved lower portion. In soils that are very low in potassium, however, the equilibrium ratio may fall on the curved portion of the plot.

Incidentally, the close clustering of the different classes of points around the central line in Fig. 4-14 provides a verification of Schofield's ratio law for the soil on which the data were obtained. In a test of the ratio law with another soil and a different procedure, Beckett (1964a) found that the soil was at equilibrium with seven solutions ranging in concentrations of calcium plus magnesium from 0.00063 molar to 0.199 molar when the ratio of the activity of potassium to the square root of the activity of the sum of calcium plus magnesium averaged 0.0135 $(mol \cdot liter^{-1})^{1/2}$, with a standard deviation of 0.0006. Additional tests have been made on numerous other soils, and from these it may be concluded that the ratio law usually represents a good approximation where the solutions are dilute, the range of ion activities is limited, and the exchange sites are mostly saturated with calcium and magnesium.

The value of $-\Delta K$ obtained by extrapolating to an ion activity ratio of zero the linear portion of the plot of change in solid-phase potassium against the activity ratio has been termed "labile potassium" (Beckett, 1964b), denoted by the symbol ΔK^0 (see Fig. 4-14). $-\Delta K^0$ may be regarded as an estimate of the quantity of exchangeable potassium present in nonspecific exchange positions at the equilibrium activity ratio if this ratio lies on the linear portion of the Q/I plot.

The range between ΔK^0 and the intersection of the curve with the ΔK axis is a measure of the potassium released from potassium-selective exchange positions. Sometimes denoted by K_x, this portion of ΔK is seldom if ever determined directly because of difficulty in measuring the low activity ratios and the fact that the line approaches the ΔK axis asymptotically.

K_x is not to be viewed as the difference between exchangeable potassium and ΔK^0. The location of the boundary between exchangeable and nonexchangeable potassium on the ΔK axis is not fixed for a given soil, but depends upon the methods employed, as is true also for the intersection of the activity ratio line with the ΔK axis. Bradfield (1969) found that when he used molar ammonium acetate to displace the exchangeable potassium from five soils, the boundary line was below the extrapolated point of intersection of the curved line with the ΔK axis for each soil (average exchangeable potassium = 0.28 centimole per kilogram and average extrapolated intersection = 0.21 centimole per kilogram). If Bradfield had used as the displacing agent for exchangeable potassium the salt of a cation such as sodium that is not held selectively in the potassium-selective exchange positions, the value for exchangeable potassium would have increased. And if he had used a longer equilibration time for determining the activity ratios, the extrapolated point of intersection would have dropped to a more negative value of ΔK.

Other Nutrients.   The buffer capacity concept has been researched almost exclusively with phosphorus and potassium. Its potential applicability to other nutrients varies.

Except for ammonium, the buffer capacity concept would appear inapplicable to nitrogen. Nitrate, the principal form of nitrogen generally absorbed by plants in aerated soils, is not retained by the soil solids, except in soils (generally from the tropics) carrying positive charges, nor is it released from organic forms in response to uptake by plants. Like nitrate, chloride is not adsorbed appreciably by soil solids, and hence would not appear amenable to the buffer capacity concept.

The buffer capacity concept would appear inapplicable to iron. Slight release of iron from the great excesses present in soil solids in response to absorption of iron from solution by plants may occur by the passive desorption and solubility equilibria mechanisms described for phosphorus and potassium. The equilibrium solubility of ferric iron in neutral and alkaline solutions is vanishingly small, however, and plants would be unable to obtain enough iron from them. Other mechanisms are involved. In aerated soils, a deficiency of iron activates plant mechanisms that increase iron availability. In dicotyledonous plants, these mechanisms include extrusion of hydrogen ions and reduction of ferric iron to the ferrous form at root surfaces. In monocotyledonous plants, the mechanism appears to be synthesis and extrusion of chelating agents called phytosiderophores, which subsequently are absorbed in part after they have chelated ferric iron from external sources (Marschner et al., 1989). In flooded soils, microbial reduction of iron to the ferrous form greatly increases the availability of iron to plants, so that iron deficiency is normally not a problem.

The buffer capacity concept would be expected to apply to sulfur in soils in which adsorbed sulfate is the major source of sulfur used by plants. In soils that adsorb very little sulfate and in which the major sources of sulfur used by plants are atmospheric depositions and organic sulfur mineralization, the concept probably would be of no practical value.

The buffer capacity concept would be expected to apply to the other mineral nutrients, which are more numerous than the few mentioned in the preceding paragraphs. As noted by Schofield (1955), the concept applies also to water availability to plants.

4-3.6.4. Combination Indexes

Considered in this section is the joint use of two or more independent measurements on soils to provide an index of availability of a nutrient other than nitrogen to plants. Nitrogen is discussed separately in Section 4-4.6. Intensity, quantity, and buffer capacity measurements will be considered first to complement the coverage of the individual factors in the preceding three sections.

Intensity, Quantity, and Buffer Capacity. In some instances, a marked improvement in precision of estimating the uptake of certain nutrients by plants from different soils has been observed where the conventional soil test measurements have been classified according to the buffer capacity of the soils for the nutrients. Two examples are shown in Fig. 4-15. In Fig. 4-15A, the uptake of phosphorus by the test crop of corn in a greenhouse experiment with 26 soils from Australia is plotted against the phosphorus extracted by 0.5 molar sodium bicarbonate according to Colwell's (1963) modification of the method of Olsen et al. (1954). The overall correlation coefficient is $r = 0.54$, but the correlation coefficients obtained for the individual groups of soils classified according to buffer capacity are $r = 0.98, 0.84$, and 0.98. In Fig. 4-15B, the concentration of zinc in the youngest fully expanded leaf of the navy bean test crop is plotted against the zinc concentration in the soil solution in a greenhouse experiment with seven soils from Western Australia and six from Queensland. The overall correlation coefficient is $r = 0.29$, and the correlation coefficients for three

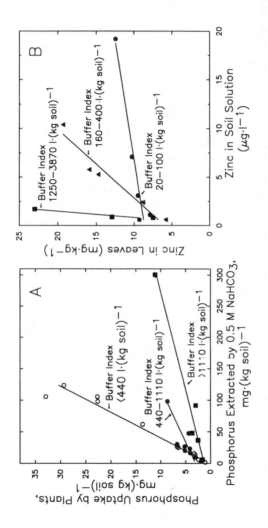

Fig. 4-15. The effect of classifying soils according to their nutrient buffer capacities on the precision of estimations of nutrient uptake from different soils by plants on the basis of soil test values for the nutrients. Each point represents a different soil. A. Data on phosphorus from a greenhouse experiment in Australia with corn as test crop. (Moody et al., 1988) B. Data on zinc from a greenhouse experiment in Australia with navy bean as test crop. (Armour et al., 1990)

groups of soils classified according to buffer capacity are $r = 0.94, 0.97$, and 0.97.

The two examples in Fig. 4-15 appear to be contradictory in that the slopes of the high and low buffer capacity plots are in reverse order. Fig. 4-15B shows the expected relationship of greater availability of the nutrient to plants with increasing buffer capacity at a given intensity of supply. The opposite behavior in Fig. 4-15A is probably a consequence of failure of the phosphorus extracted by the sodium bicarbonate method to reflect the intensity of phosphorus supply in the soil. Most of the phosphorus extracted would represent the quantity aspect of supply. Thus, the segregation of data into different classes according to buffer capacity would be a consequence of the way the extractable phosphorus values were related to the measurements of buffer capacity.

In the research on phosphorus in Fig. 4-15A there was no appreciable seg-regation of data by buffer capacity when soil solution phosphorus was plotted on the x axis. In the research on zinc in Fig. 4-15B, there was a slight indication of segregation of plots according to buffer capacity like that shown in Fig. 4-15A when the zinc extracted by the diethylaminetriaminepentaacetate method of Lindsay and Norvell (1978) was plotted on the X axis. With the zinc extracted by the ethylenediaminetetraacetate method of Fujii and Corey (1986), there was no appreciable indication of an effect of buffer capacity. Evidently much remains to be learned about the significance of the measurements of extractable nutrients and the way they relate to measurements of buffer capacity.

Work by a number of investigators has shown that the relative importance of the three factors — intensity, quantity, and buffer capacity — varies with the circumstances. Among the publications judged especially significant in this area are those by Barrow (1967a,b), Dalal and Hallsworth (1976), Holford and Mattingly (1976b), Holford (1980), Keramidas and Polyzopoulos (1983), Kumaragamage and Keerthisinghe (1988), Moody et al. (1988), Mendoza (1989), Warren (1990), and Soon (1990) for phosphorus. Corresponding publications for potassium include those by Zandstra and MacKenzie (1968), During and Duganzich (1979), and Wietholter (1984). Individual publications are not re-viewed here because of the diversity of methods and findings.

Perhaps three significant observations may be distilled from a review of the literature. One is that as the amount of a nutrient absorbed from soils by plants increases, the contribution of the initial intensity to the correlation between soil measurements and biological indexes of availability tends to decrease, and the contribution of the quantity and buffer capacity factors tends to increase.

A second observation is that according to theoretical considerations by Olsen and Watanabe (1963), the amount of phosphorus arriving at root surfaces by diffusion through soil increases in direct proportion to the intensity or concen-tration factor and in direct proportion to the square roots of the buffer capacity and the diffusion coefficient. Pointing to this theoretical relationship, Moody et al. (1988) noted that for the same proportional range in values, the intensity factor will have a greater effect on phosphorus uptake than the buffer capacity, an interpretation they found in agreement with their experimental data.

A third observation is that, like biological indexes of phosphorus availability, the quantities of phosphorus extracted by the methods that provide the better chemical indexes of phosphorus availability are composite values. That is, they integrate to some degree the intensity, quantity, and buffer capacity factors that contribute to phosphorus availability. Correlations between chemical and biological indexes thus depend in part upon how well the integration provided by the chemical indexes mimicks the integration provided by the biological indexes. Under such circumstances, a reasonable way to proceed in research designed to improve the precision of relationships between the chemical and biological measurements is to make independent measurements of the intensity, quantity, and buffer capacity factors.

Work on soil analysis methods is an essential part of the needed research because different researchers have used different methods, and the methods used yield different results. The most commonly used measure of the phosphorus quantity factor for research purposes has been the labile phosphorus by isotope dilution (equilibration of soil with radioactive orthophosphate), probably because this measurement seems logical in view of evidence that plants derive their phosphorus predominantly from this source. But there has been no corresponding de facto consensus on measurements of the intensity and buffer capacity factors. And the isotope dilution method unfortunately is unsuitable for routine soil testing purposes. Some winnowing is needed, and some new modifications may be found superior to the old.

Khasawneh and Copeland (1973) made a start toward a theoretical basis for integrating the intensity, quantity, and buffer capacity factors with the change in buffer capacity as a nutrient is removed from soil by plants. Brar et al. (1986) applied the Khasawneh and Copeland integrated supply formulation to data from an experiment in which wheat had been grown on soils after addition of a range of quantities of fertilizer phosphorus. In their work, the correlations between phosphorus uptake and the integrated supply parameter exceeded the correlations between phosphorus uptake and the individual supply factors (concentration of phosphorus in solution or intensity, E value or quantity, and buffer capacity).

Van Noordwijk et al. (1990) developed an integrated model of phosphorus availability using a mechanistic approach based upon the long-standing Dutch method for water-soluble phosphorus as an index of soil phosphorus availability. Although the specific method of measurement might be improved, avoidance of chemical extractants is desirable in models that attempt to integrate intensity, buffer capacity, and diffusion factors of soil phosphorus supply.

Diffusion Model.   In contrast to the foregoing instances, in which buffer capacity and extractable nutrients in effect were considered as separate variables, Baldwin et al. (1973) developed a theoretical equation in which the buffer capacity took the place of the usual extractable nutrient variable. From the standpoint of soil testing, their equation has the disadvantage that some of the plant factors cannot be measured until the plants are grown.

With the objective of eliminating the need to measure the plant factors, so that the availability indexes could be based upon analyses of soil samples taken before the crop is grown, Corey (1987) combined the plant factors in the following modification of the original equation by Baldwin et al. (1973):

$$U = C_{li}b\left\{1 - \exp\left[\frac{-2\pi A_1 P_n}{b\left(1 + \dfrac{A_1 P_d}{D_1 \theta f}\right)}\right]\right\} \quad ,$$

where U = uptake per unit volume of soil in unit time, C_{li} = initial concentration of nutrient in solution, b = buffer capacity of the soil for the nutrient, A_1 = fractional area of water contacting the roots, θ = volumetric water content of the soil, D_1 = diffusion coefficient of the nutrient ion in solution, f = conductivity of the soil for the nutrient (incorporating the effective path length, electrical charge, and water viscosity factors), P_n = product of plant factors in the numerator of the original equation (root absorbing power, root radius, and length of roots per unit volume of soil), and P_d = product of plant factors in the denominator (root absorbing power, root radius, and natural logarithm of [half-distance between roots/1.65 root radius]).

Corey (1987) noted that if $A_1 P_d/D_1 \theta f$ is much less than 1, the equation may be simplified to

$$U = C_{li}b\left[1 - \exp\left(\frac{-2\pi\theta P_n}{b}\right)\right] \quad ,$$

and if $A_1 P_d/D_1 \theta f$ is much greater than 1, the equation may be simplified to

$$U = C_{li}b\left[1 - \exp\left(\frac{-2\pi D_1 \theta f P_n}{b P_d}\right)\right] \quad .$$

In these equations, values for D_1 are available in handbooks. C_{li} and b can be measured, although the value of b for a given soil will vary with the amount of the nutrient removed. θ of course varies with time in a given soil, and this causes f to vary. Soil texture is an important determinant of θ and f. More information is needed, but estimates of f based upon soil texture and locally experienced values of θ may be made from data published by Porter et al. (1960) and Barraclough and Tinker (1981).

If $A_1 P_d/D_1 \theta f$ is approximately equal to 1, the foregoing simplifications cannot be made. To obtain the best-fitting values of P_n and P_d through a correlation approach, Corey (1987) suggested varying P_d systematically, so that $A_1 P_d/D_1 \theta f$ ranges from 0.1 to 10. At each selected value of P_d, P_n is varied, so that the values of the exponential term range from -0.01 to -5. The values of P_n and

P_d selected for use are those that result in the highest correlation between the calculated and measured uptakes of the nutrient by the plants.

Corey (1987) noted that the first of the two simplified equations applies if a single value of P_n yields the highest correlations irrespective of the value assigned to P_d. If the highest correlations are obtained at a specific P_n/P_d ratio irrespective of the absolute values, the second simplified equation applies. Otherwise, the first of the three equations must be used.

Corey's approach requires lengthy calculations, but the long calculations are needed only in developing the calibration. Once the appropriate calibration equation has been selected, application to specific soil samples is relatively simple.

Wietholter (1984), one of Corey's students at the University of Wisconsin, also used the diffusion equation by Baldwin et al. (1973), which brought into play both the buffer capacity and intensity factors of supply. His use of the equation was somewhat different from that described by Corey (1987). Wietholter (1984) tested his model by applying it to corn growing on soils with textures ranging from sand to silty clay loam at 51 sites. He found that the correlation between the potassium concentrations in the ear leaves of the corn plants and the potassium uptake predicted by the diffusion model was $r = 0.76$. The corresponding correlation with the potassium concentrations in a 0.004 molar strontium nitrate solution equilibrated with the soils was $r = 0.70$, and the correlation with exchangeable potassium was $r = 0.38$.

The diffusion approach has the advantage of dealing in a more realistic way with the dynamic situation in soils than do the conventional soil test methods and the physicochemical nutrient potential methods. The diffusion approach is built around the concept that plants alter the nutrient status in a portion of the soil adjacent to their roots, leaving more distant locations unaltered. The other approaches treat soil as a homogeneous medium analogous to a stirred nutrient solution, in which the nutrient quantity or potential is reduced uniformly as absorption proceeds.

For practical soil testing, Corey's approach has the advantage of the capability of being truly predictive. Wietholter's model required estimates of root length based upon the yields of above-ground dry matter, which could not be determined before the crop was grown, and which statistically would contain some of the information sought on potassium availability.

Diffusion Method. A newly developing technique should make possible a direct evaluation of the combined effects of the concentration, buffer capacity, diffusion coefficient, water content, and conductivity (*f*) factors that influence nutrient uptake by plants. The technique, attributed in part to a suggestion by H. A. Sissingh, involves imbedding specially treated filter paper strips in moist soil. The paper contains a nutrient adsorbent that adsorbs the nutrient in question from the soil solution. After an appropriate time, the strips are removed, and the adsorbed nutrient is extracted for determination.

Preliminary research has been published on the imbedding technique (Menon et al., 1990), which brings diffusion into play in a manner similar to that involved in uptake of nutrients from soils by plants. In most of the work, however, the adsorbent has been added to a suspension of soil in water or a dilute electrolyte, which has been shaken to expedite the transfer of the nutrient under test from the soil to the adsorbent. Van der Zee et al. (1987) and Menon et al. (1989a,b,c) used impregnated filter paper, but Saggar et al. (1990) and Schoenau and Huang (1991) used strips of an anion exchange resin membrane.

Information from research on the Sissingh technique and other aspects of the integration of the intensity, quantity, and buffer capacity factors should lead to improved understanding about the way in which plants integrate the factors and to the development of more efficient soil testing methods. Almost certainly, the methods that emerge will be more complex than those used at present. In the end, extensive and precise groups of field experiments will be required to adapt the findings to practical conditions in different areas.

The Sissingh technique is applicable to nutrients other than phosphorus. Searle (1988) used a strip of an anion exchange resin saturated with phosphate for extracting sulfate from soils. As a possible future development, Van Diest (1987) visualized a system of soil testing in which strips of cation and anion exchangers would be placed in the soil for a given time by the producer, who then would remove the strips and submit them to a soil-testing laboratory for analysis. The laboratory would make the analyses, recharge the exchangers, and return them to the producer for reuse.

Multiple Regression. The multiple regression technique of statistics has been used extensively in investigations of nutrient availability. The technique is flexible and has been used in various ways. Multiple linear regression has been used most often, partly because the experimental data usually do not justify more complex models.

From the soil fertility standpoint, a particularly valuable use of multiple regression has been to evaluate and relate separate measurements on a given nutrient. Some of these applications were discussed incidentally in the previous subsection on "intensity, quantity, and buffer capacity." Other applications have been to arbitrary measurements of a given nutrient by two or more methods.

Applications of the latter type may be highly empirical, providing nothing more than a statistical device for increasing the proportion of the variance of the response measurements accounted for. The applications may be more meaningful if the soil measurements represent distinct fractions of the nutrient. Fractionation of the nutrient by locations in the soil is discussed in Sections 4-4.2.5 and 4-4.2.6. Relevant here are applications to distinct chemical fractions of a nutrient at a given location. For example, if samples of a group of soils are brought together and cropped under uniform conditions in a greenhouse where the uptake of the nutrient of interest by the crop increases with the supply in the various soils, the simplest multiple regression model for two chemical fractions would be

$$y_{total} = y_1 + y_2 = k_1 x_1 + k_2 x_2,$$

where y_{total} is the total yield of the nutrient in the test crop, y_1 is the yield of the nutrient derived from chemical fraction x_1 in the soils, y_2 is the yield of the nutrient derived from chemical fraction x_2 of the nutrient in the soils, and k_1 and k_2 are proportionality factors. Usually provision is made for inclusion of a constant term on the right side of the equation. This term would include the contributions of the nutrient from the seed and from fractions of the nutrient not included in x_1 and x_2, and it would also help to compensate for lack of fit of the model and some experimental error.

A special value of applications of this type in which the fractions of the nutrient are distinct and do not overlap is that the ratios of the regression coefficients are estimates of the availability coefficient ratios of the nutrients in the different fractions measured. In an early application of this approach, Pratt (1951) measured the total yield of potassium in nine cuttings of alfalfa obtained from 13 soils in the greenhouse and used multiple linear regression to relate the yield of potassium (y) to the exchangeable potassium (x_1), the nonexchangeable potassium released to a cation exchange resin during moist incubation (x_2), and the nonexchangeable potassium not released to the exchange resin (x_3), obtaining the equation

$$y = 0.36x_1 + 0.099x_2 + 0.0011x_3 + 66,$$

where the units of x were milligrams per kilogram of soil. The calculations thus indicate that the ratio of the availability coefficient of fraction x_1 to x_2 was $0.36/0.099 = 3.6$ and that the ratio of the availability coefficient of fraction x_1 to x_3 was $0.36/0.0011 = 327$. The statistical analysis showed that the regression coefficient for fraction x_3 was not significantly different from zero. Many experiments of this type have been done on soils from various parts of the world.

Soil properties other than the content of a particular fraction of a nutrient often are measured and used in multiple regression equations with measurements on nutrients in an attempt to develop indexes of nutrient availability that yield improved correlations with biological indexes (Mortvedt, 1977). When improvements in correlations are obtained, the reasons, if any, that are put forward to account for the results are generally only speculative because of inadequate information.

The clay content of soils sometimes is used in an attempt to improve the predictability of measurements on soil nutrients. For example, the correlation of phosphorus sorption and release with clay content is well known. Lins and Cox (1989) used the clay content of soils as a substitute for the buffer capacity for phosphorus. The clay content of soils has numerous implications for nutrient supplies.

Soil pH probably is found useful as often as any of the other accessory soil properties. Data on manganese will be used here for illustration. In some instances, for example, findings by Rich (1956) and Mascagni and Cox (1985), the biological indexes of manganese availability may correlate as well or better with soil pH than with extractable soil manganese. The latter investigators ob-

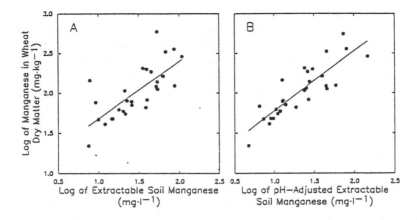

Fig. 4-16. Concentration of manganese in wheat plants grown on soils from 29 countries versus soil manganese extracted by the diethylenetriaminepentaacetic acid extractant of Lindsay and Norvell (1978). Each point represents the mean data for a given country. A. Unadjusted values of extractable soil manganese (r = 0.73). B. pH-adjusted values of extractable soil manganese (r = 0.86). (Sillanpää 1982)

tained coefficients of determination of 0.17, 0.28, and 0.59 when based upon extractable manganese alone, pH alone, and extractable manganese and pH as independent variables.

Fig. 4-16 summarizes the results obtained by Sillanpää (1982) in a study of samples of 3538 soils from 29 countries. Each point in each graph represents the mean data for a given country. As an index of plant manganese, the superiority of pH-adjusted extractable soil manganese (Fig. 4-16B) over extractable soil manganese (Fig. 4-16A) is evident from the differences in range of values and scatter of points. The ratios of the concentration of manganese in the test plants (wheat) to the concentration of extractable manganese in the soil ranged from 3.03 at pH 4.5 to a minimum of 0.48 at pH 6.7 and then increased to 1.62 at pH 8.5. This ratio reflects the difference in effectiveness of the plants relative to the extracting solution in removing manganese from the soils as a function of soil pH. In this instance, the extractant was the diethylenetriaminepentaacetic acid (DTPA) solution proposed by Lindsay and Norvell (1978).

Sillanpää (1982) found that adjusting measurements of extractable soil manganese, molybdenum, and zinc for soil pH made an important improvement in correlations between the soil measurements and the concentrations of these nutrients in the test plants. For copper, an adjustment for the organic carbon content of the soils was important. For boron, an adjustment for the cation exchange capacity of the soils was important. Brennan and Gartrell (1990) found that the critical level of DTPA-extractable zinc in soils of Western Australia for maximum growth of subterranean clover was affected by the soil pH and the clay content.

4-3.7. Potassium

Phosphorus has been emphasized in previous sections because of the relatively extensive research on that nutrient. Brief separate consideration of potassium therefore seems justified as an aid to understanding the problems with this nutrient. Several papers dealing with the intensity, quantity, and buffer capacity aspects of potassium supply have already been cited. Others that illustrate the range of findings made in research on the soil-plant relationships of the intensity, quantity, and buffer capacity factors include papers by Barrow (1966), Von Braunschweig and Mengel (1971), El Sarnagay et al. (1988), Koch et al. (1970), Mazumdar and Saxena (1988), Munns and McLean (1975), and Ross and Cline (1984).

Beckett (1972) reviewed the subject of cation activity ratios in relation to the nutrition of plants in soils, with emphasis on the ratio of the activity of potassium to the square root of the sum of the activities of calcium and magnesium. Because he considered that some authors had sought relationships under circumstances in which little or no relationship should be expected, he set forth the conditions necessary for an activity ratio to describe the availability of potassium or other cation in the numerator of the activity ratio. The necessary conditions are not met for plants growing in soils, and so the best that reasonably can be expected is a more or less close correlation between activity ratios and biological indexes of potassium availability. The arguments Beckett developed for activity ratios apply also to concentration ratios.

The fundamental problems with potassium are similar to those with phosphorus, but the mechanisms differ. Some understanding of the reasons for inconsistency of findings relative to ion activity ratios may be derived from the results of research on the behavior of potassium in soils in relation to potassium uptake by plants. Fig. 4-17A illustrates the first point, namely, that the concentration of potassium in the soil solution at root surfaces may differ markedly from that in the bulk soil. Barber (1984) pointed out that after a root has been absorbing potassium and calcium for only a few hours, the activity ratio in the solution at the root surface may differ by a factor of 20 from the activity ratio in the bulk solution that is analyzed to evaluate the activities of these ions. The change in the ratio is a consequence of the fact that potassium is absorbed preferentially from the solution and hence moves mostly by diffusion, whereas calcium is preferentially excluded and hence tends to accumulate at root surfaces as it is transported by mass movement in the water the roots absorb from the soil. In view of these considerations relative to Fig. 4-17A, a correlation of the equilibrium activity or activity ratio of potassium in the soil with a biological index of potassium availability is to be expected only because of some indirect relationship, such as a correlation between the activity of potassium in solution in the bulk soil and the rate of diffusion of potassium through the soil to the root surface.

The second point, illustrated by Fig. 4-17B, is that absorption of potassium by plant roots produces a drawdown in exchangeable potassium in the soil

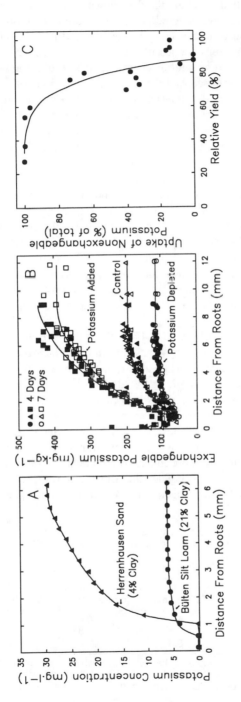

Fig. 4-17. Some soil-plant relationships of potassium. A. Concentration of potassium in the soil solution in two soils of Germany versus distance from corn roots. (Claassen and Jungk, 1982) B. Concentration of exchangeable potassium in a silt loam soil of Germany versus distance from rape roots. The control soil received no potassium. The potassium-depleted soil had been cropped previously to corn to reduce the exchangeable potassium content. The potassium-fertilized soil had received 391 milligrams of potassium as potassium chloride per kilogram. (Kuchenbuch and Jungk, 1984) C. Uptake of nonexchangeable potassium as a percentage of the total potassium uptake by oat plants on different soils of Germany versus relative yield. (Grimme, 1974)

adjacent to root surfaces that parallels the drawdown of potassium in solution shown in Fig. 4-17A. (The few scattered points well above the minimum level of exchangeable potassium at distances less than 0.5 millimeter from the root surfaces may have been a consequence of inclusion of root hairs in the soil samples.) When the potassium in solution decreased as a result of absorption by roots, the shifting activity ratio caused the exchange of cations other than potassium (principally calcium and magnesium) in solution for potassium, which then entered the diffusion stream leading to the roots. The fact that a difference between 4 and 7 days was evident only where the exchangeable potassium had been increased by adding potassium chloride suggests that depletion of potassium from greater distances would be very slow, and extension of roots into untapped soil would be the principal source of additional potassium.

In connection with Fig. 4-17B, Kuchenbuch and Jungk (1984) made some additional significant observations that are summarized in Table 4-4. Because of the increase in the effective diffusion coefficient with the concentration of exchangeable potassium, the distance to which potassium was depleted increased, the volume of soil supplying potassium increased, and the amount of potassium absorbed (designated by the authors as available potassium) increased. Note that although the ratio of exchangeable potassium in the potassium-fertilized soil (row 3) to the potassium-depleted soil (row 1) was 4.6, the corresponding ratio of available potassium was 22.

The third point is illustrated by Fig. 4-17C. Soils that are only slightly or moderately weathered generally contain potassium in nonexchangeable forms in layered minerals related to micas. This potassium is held more strongly than the conventional exchangeable potassium, and it does not become exchangeable until the activity or concentration of potassium in solution drops low enough. Then it is released rapidly. Fergus and Martin (1974) reported that they had observed little or no increase in "labile" (presumably exchangeable) potassium when soils from Australia that they had partly or completely exhausted of exchangeable potassium by intensive cropping were incubated in moist condition for 58 to 78 months. These results indicate that the equilibrium release of nonexchangeable potassium had already occurred before incubation began. Bray and

Table 4-4. Potassium Availability Factors in Soil in the Root Zone of Rape Plants as Calculated From Information Derived From Fig. 4-17B (Kuchenbuch and Jungk, 1984)

Exchangeable Potassium (mg·kg^{-1})	Depletion Zone[a]	Extent (mm)	Volume (cm^3 soil·cm^{-1} root)	Available Potassium[b] (µg·cm^{-1} root)
115	1.2	4.6	0.66	16
202	5.3	5.3	0.88	52
534	27.0	6.3	1.24	356

[a] Zone extending from the root surface to the distance at which the concentration of exchangeable potassium was 90% of that in the bulk soil uninfluenced by the plants.
[b] Potassium absorbed by the plants as inferred from the decrease in exchangeable potassium.

DeTurk (1939) had reported earlier that in the absence of additions of potassium to or removals of potassium from ten soils of Illinois they tested, the exchangeable potassium content was relatively stable, averaging 75 milligrams per kilogram before and after incubation in a moist condition for 5 years.

Fig. 4-17C shows that the importance of initially nonexchangeable potassium as a source of supply for plants grown in a greenhouse on a number of different soils increased as the relative yield of the test plants decreased. The use of some nonexchangeable potassium by the plants even at high relative yields where the activity or concentration of potassium in the bulk soil would be great enough to inhibit the release of potassium from nonexchangeable forms can be accounted for by the drawdown of soluble and exchangeable potassium by plants in the soil adjacent to root surfaces shown in Figs. 4-17A and B.

From Fig. 4-17 one may infer that although plants may derive potassium only from solution, the solution is resupplied by potassium from exchangeable and nonexchangeable forms. The proportions of the potassium derived from these three forms vary with time and potassium uptake from a given soil and also with the soil. Net changes in the field are much slower and less pronounced than those under greenhouse conditions because the volume of soil per plant is much greater, and with many crops most of the potassium removed is returned to the soil in crop residues. The root density in the field is less than that in the greenhouse, so that there is more bulk soil from which potassium is not removed, but the principles illustrated in Fig. 4-17 should still apply.

In view of the situation as elaborated in connection with Fig. 4-17, it is to be expected that the intensity, quantity, and buffer capacity factors of potassium supply have not been found in different investigations to behave consistently as indexes of potassium availability. The same is true for exchangeable potassium (a quantity factor), which is the more conventional measurement for soil testing purposes. In occasional investigations (for example, one by Spencer and Govaars, 1982), measurements of potassium on the soils may be of relatively little value if the soils differ sufficiently.

The supply of potassium to plants in soils is more appropriately related to the dynamics of potassium behavior illustrated in Figs. 4-17A and B, which must be simulated in a different way. Von Braunschweig and Mengel (1971) and Shaw et al. (1983) found that measurements of the diffusive flux were useful, but such measurements are difficult and inappropriate for soil testing purposes. The Sissingh technique described in Section 4-3.6.4 would come closer to practicality. The data obtained by Shaw et al. (1983) indicated that the root surface area was an additional factor that needed to be taken into account in comparing the quantities of potassium absorbed by plants from different soils (see also Barber, 1985).

A succinct summary of methods of evaluating the soil potassium supply was published in a paper by Johnston and Goulding (1990). Munson (1985) edited an extensive monograph on potassium.

4-3.8. Multinutrient Extractants

Multinutrient extractants logically should have been discussed following Section 4-3.1. They are considered here, however, because the examples involve some concepts that had not been explained in Section 4-3.1 and the preceding text.

Two general approaches have been used in developing chemical extractant procedures to obtain indexes of nutrient availability in soils. One is to adjust the procedures to obtain the highest possible correlation between the chemical and biological indexes of availability. The result of this approach has been a set of individual procedures for individual nutrients.

The second approach is to make analyses for more than one nutrient on a single extract. Multinutrient extractants have been in use for many years. Early bulletins describing soil testing systems using multinutrient extractants were published by Spurway (1933) and Morgan (1935). More recent publications include those by Soltanpour and Schwab (1977), Lindsay and Norvell (1978), Wolf (1982), and Mehlich (1984). Selected for illustration here are two examples of directed efforts to develop multinutrient extractants based upon theoretical concepts. One of these systems is due to Baker and Amacher (1981) and the other to Németh (1979). The illustrations will serve also to provide a picture of how similar basic concepts may be adapted to soil testing systems built around quite different methods. Jones (1990) reviewed the subject of multinutrient extractants and gave some procedures. Two powerful forces now promoting the use of multinutrient extractants are the continuing drive for increased economy in performing the analyses and the availability of instrumental methods of analysis that are capable of making analyses very quickly for a number of elements in a single solution.

4-3.8.1. Baker and Amacher Extractant

Instead of the usual extractant, the Baker and Amacher (1981) system uses a combination equilibrant and extractant. A sample of soil is equilibrated with a solution containing 2.5×10^{-4} molar potassium chloride, 10×10^{-4} molar magnesium chloride, 50×10^{-4} molar calcium chloride, and 4×10^{-4} molar diethylaminetriaminepentaacetic acid. The solution is buffered at pH 7.3 with triethanolamine.

According to Baker and Amacher, the concentrations of potassium, magnesium, and calcium "chosen for the test solution represent 'optimum' availability levels for the growth of most agronomic crops in the 'ideal' soil. The 'ideal' soil represents soils which supply the minimum availability of K, Ca, and Mg for optimum plant nutrition when the exchange complex is saturated with 90 percent bases, 2 percent K, and 10 percent Mg." The changes in the solution composition during equilibration with the soil provide information on the deviation of the soil from the ideal, and at the same time the changes are not extensive enough to affect the quantities of manganese, iron, copper, zinc, aluminum, lead, nickel, and cadmium brought into solution by the diethylamine-triaminepentaacetic acid, which is a metal-complexing reagent.

By means of extensive calculations made with the aid of a computer program available in machine-readable form from Pennsylvania State University, the activities of all the named ions in solution are calculated. Molar activities of phosphorus are calculated also. These values are taken to represent the intensity aspect of the supplies of the elements. As indexes of the quantity factors for sodium, sulfur, manganese, iron, copper, zinc, aluminum, lead, nickel, and cadmium, the quantities of these elements brought into solution by the equilibration-extraction solution are used. The quantity factors for potassium, magnesium, and calcium are taken as the quantities removed by 1 molar ammonium acetate at pH 7.0. The phosphorus extracted by the Bray and Kurtz (1945) 0.025 molar hydrochloric acid-0.03 molar ammonium fluoride extractant is used as an index of the quantity factor for that element.

If the concentration of potassium, magnesium, or calcium in the solution after equilibration is equal to or greater than the initial concentration, the availability of the ion in the soil is considered adequate. The interpretation is the reverse if the concentration after equilibration is less than the initial concentration.

The quantities of potassium and magnesium recommended for addition are those required to increase the potassium saturation to 2% of the cation exchange capacity and magnesium to 10% of the cation exchange capacity. If the saturation with potassium exceeds 5%, the corrective treatment for magnesium is that calculated to increase the magnesium saturation to twice the potassium saturation. The interpretation thus involves the cation saturation ratio concept described in Section 4-4.5.

In the Baker and Amacher program, buffer capacity (see Section 4-3.6.3) is not considered explicitly. The authors note, however, that the factors they use for estimating the potassium and magnesium requirements from the changes in concentrations of these nutrients in the equilibration-extraction solution include buffer capacity components that are average values for different soils. Other aspects of interpretation also are involved. They are described in brief in the bulletin by Baker and Amacher (1981). The quantities of nutrients recommended for addition are those needed to bring the soil up to the "ideal" level plus the amounts expected to be removed by the crops. The system thus involves the buildup and maintenance concept discussed in Section 4-4.5.1.

4-3.8.2. Németh Electro-Ultrafiltration

In electro-ultrafiltration, water is the extractant. Electro-ultrafiltration is a modification of electrodialysis. As the system was described by Németh (1979), a stirred 1:10 suspension of soil in water is contained in the center chamber of an assembly of three chambers connected by membranes that allow the passage of solutes but not solids. A direct current passed between electrodes in the outer chambers causes anions to migrate from the soil through the filter to the electrode maintained at a positive potential, where they form acids. Cations concurrently migrate from the soil through the other membrane to the electrode maintained at a negative potential, where they form hydroxides.

The electro-ultrafiltration modification involves continuous addition of water to the central chamber containing the soil to maintain the 1:10 ratio of soil to water. Suction is applied to the outer cells to remove the contents to collecting chambers, from which the accumulated volumes of solutions are collected at intervals for analysis.

Corresponding fractions containing cations and anions normally are combined for analysis. According to Németh, analyses may be made for calcium, magnesium, potassium, sodium, copper, iron, manganese, zinc, cadmium, nickel, ammonium, nitrate, sulfate, phosphate, borate, and other ions.

As the electro-ultrafiltration process continues, the ions transported into the outer chambers represent forms in the soil that support succeedingly lower concentrations in solution. For potassium, for example, these would be potassium salts in solution, potassium in nonselective exchange positions, potassium in potassium-selective exchange positions, and nonexchangeable potassium in interlayer positions in micaceous minerals present mostly in the clay fraction. Analyses of volumes of solution derived from the outer chambers at successive time intervals provide a quantitative measure of the changes in rates of release of ions from the soil between the time intervals selected.

The rate of removal of a given ion from a sample of soil in the central chamber depends upon the voltage applied between the electrodes in the outer chambers and the temperature. The ion extraction is started with a potential of 50 volts between the electrodes. Because of the design of the apparatus, this potential is equivalent to 11.1 volts per centimeter. The 50-volt potential is maintained for the first 5 minutes. According to Németh, potassium ions added in solution to the central chamber are removed almost quantitatively in 5 minutes.

In the interest of speed in completing the analysis, the potential then is raised to 200 volts. Németh noted that for clays such as kaolinite that have no potassium-selective exchange sites, the exchangeable potassium is essentially removed by the end of a 25-minute period at 200 volts. He included a table showing that the potassium removed by electro-ultrafiltration from a ''humic sand'' somewhat exceeded the exchangeable potassium. None of the other soils for which Németh gave data behaved in this way. Much of the conventional exchangeable potassium was not removed by the end of the 25-minute period at 200 volts. The electro-ultrafiltration method thus distinguishes differences in ease of removal of potassium within what is measured as one fraction in the extraction of exchangeable potassium by a salt solution. Németh's view was that the slow release of potassium in these soils was a consequence of the potassium-selective sites in which the potassium was held.

Although slow release of the selectively held potassium would continue at a potential of 200 volts, the presence of this fraction of potassium is made more evident at the end of the 25-minute period at 200 volts by raising the electrical potential to 400 volts and increasing the temperature to 80° C. The sum of the potassium released at the 200- and 400-volt potentials was sometimes less than and sometimes greater than the conventional exchangeable potassium in the data shown by Németh.

In an investigation involving eight soils from Germany, Grimme (1980) found that the calculated maximum cumulative amounts of potassium released with increasing time were independent of the electrical field strength in the range he tested (11.1 to 88.9 volts per centimeter or total potentials from 50 to 400 volts) and were almost equal to the initial quantities of exchangeable potassium in the soils. Data from two soils shown graphically indicated that the extraction of exchangeable potassium had not been completed in 50 minutes at a field strength of 88.9 volts per centimeter.

The value of the electro-ultrafiltration method for simulating the supplying power of soils for potassium is indicated in Fig. 4-18. The three soils tested all had about the same initial content of exchangeable potassium, but the uptake of potassium by ryegrass in the first cutting and the rate of decline with succeeding cuttings differed considerably among soils. Within a desorption time of 35 minutes, the electro-ultrafiltration method gave a good indication of the changes in uptake of potassium from the soils by the ryegrass test crop in four cuttings.

Nair and Grimme (1979) found that the potassium extracted by the electro-ultrafiltration method with an extraction time of 5 minutes at 50 volts and 5 minutes at 200 volts was more highly correlated with the yields of dry matter and grain of barley and the content of potassium in the plants grown on soils of Germany in a greenhouse experiment than were the equilibrium activity ratios for potassium in the same soils. (See Section 4-3.6.1 and Fig. 4-14, where the equilibrium activity ratio is designated as AR_e^K).

Increasing the rigor of the electro-ultrafiltration extractions improves the correlations with long-term uptake of potassium from soils by plants. Schubert et al. (1990) found in a group of eight widely different soils from Germany that when they modified the regular electro-ultrafiltration method by repeating the 5-minute extraction at 80° C and 400 volts five times, the proportion of nonexchangeable potassium in the total was about the same as that found by extracting potassium from the soils for 21 days with a fresh batch of cation exchange resin each day. The significance of the results was verified by growing ryegrass on the soils and measuring the potassium removed in 11 cuttings.

In research on nine soils of Ontario, Richards and Bates (1989) found that the electro-ultrafiltration method extracted amounts of potassium that were more highly correlated with the uptake of potassium from the soils in eight crops of alfalfa in a greenhouse experiment than were the potassium activity or the amounts of potassium extracted by any of the other methods tested. The coefficient of determination was $r^2 = 0.97$ with the electro-ultrafiltration method and $r^2 = 0.48$ with the more popular method involving extraction with boiling molar nitric acid. Their modification of the electro-ultrafiltration method used six 5-minute desorptions at 200 volts and 20° C followed by six 5-minute desorptions at 400 volts and 80° C.

To find the fertilizer potassium requirement, Németh first determined the quantities of potassium extracted by the electro-ultrafiltration method from soils that produced very high yields of sugarbeet, wheat, corn, and other crops. Then he determined the additions of fertilizer potassium required to increase the ex-

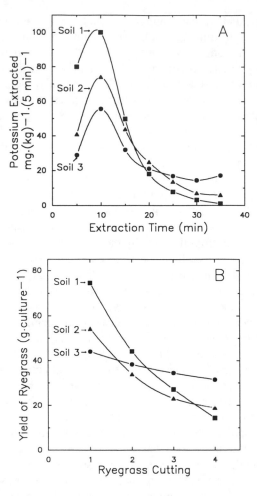

Fig. 4-18. A. Potassium extracted from three soils by the electro-ultrafiltration method in successive 5-minute intervals. In the first 5 minutes, the potential was maintained at 50 volts. From 5 to 30 minutes, the potential was maintained at 200 volts. From 30 to 35 minutes, the potential was maintained at 400 volts, and the temperature was increased to 80° C. The exchangeable potassium contents of the three soils were similar — 238 milligrams per kilogram in soil 1, 241 milligrams per kilogram in soil 2, and 262 milligrams per kilogram in soil 3. B. Yields of dry matter in four successive cuttings of ryegrass on the same three soils. (Németh, 1979)

tractable potassium to the determined levels for the various crops. For a given initial level of extractable potassium, the quantity of fertilizer potassium required to reach the target increased with the clay content of the soil. Accordingly, he made an adjustment for clay content by increments of 10% clay. The adjustment for clay content would include effects of both potassium fixation and the percentage of the exchange positions saturated with potassium.

The clay content was determined as a part of the electro-ultrafiltration process by weighing the membrane leading to the compartment that accumulated anions. Clay adheres tightly to this membrane, and Németh found that the weight of adhering clay was highly correlated with the clay content of the soils as found by conventional methods. The nature of the clay was also an important factor in the potassium requirement, and would lead to different sets of adjustment factors for clay content.

The requirement for fertilizer phosphorus was obtained by a procedure similar to that for potassium. As with potassium, the phosphorus requirement increased with the clay content, and an adjustment was made for the clay content.

Thus, it may be perceived that the fraction of a nutrient removed quickly at a low voltage by the electro-ultrafiltration method may be used as an index of the intensity factor. The sum of this fraction and the fraction removed over a longer time at a higher voltage represents a portion of the quantity factor. These two fractions plus the third fraction removed at a still higher voltage and temperature would represent a closer approximation to the quantity factor. Measurements of the quantities of a nutrient removed in successive equal time intervals at a given voltage and temperature would provide an indication of differences in buffer capacity. The adjustments for clay content in estimating the fertilizer requirement would represent corrections for buffer capacity effects.

Rex and Munk (1990) checked the relative value of the phosphorus extracted by the electro-ultrafiltration method, water, a calcium chloride solution, and two lactate methods by comparing the quantities of phosphorus extracted with the economic optimum quantities of fertilizer phosphorus found in a number of long-term fertilizer trials in Germany. In their work, the lactate methods proved superior, and the other methods, which extracted quantities of phosphorus more closely related to phosphorus concentrations in the soil solution, gave unsatisfactory correlations.

As explained by Németh (1979), there are some problems in use of the electro-ultrafiltration method for ammonium and nitrate nitrogen. Some 20 to 25% of the ammonium may be lost as ammonia. Moreover, some amino acids and other organic nitrogen compounds are extracted. They accumulate mainly in the anode chamber, and release some nitrogen as ammonium in the electro-ultrafiltration process. Although loss of ammonia is a disadvantage, the "problems" with organic nitrogen appear to be an advantage from the soil testing standpoint because these substances are regarded as easily mineralized forms of nitrogen.

Extraction of some organic, readily mineralizable nitrogen, however, is not unique to electro-ultrafiltration. Appel and Steffens (1988) found in experimental work with 25 soils from Germany that 0.01 molar calcium chloride also extracted some organic nitrogen along with mineral nitrogen. The correlation between organic nitrogen extracted and uptake of nitrogen from the soils by plants was as good with a calcium chloride extract ($r = 0.93$) as with electro-ultrafiltration ($r = 0.92$). Moreover, the coefficient of variation of total nitrogen extracted was 2.6% for calcium chloride and 6.2% for electro-ultrafiltration.

Magnesium hydroxide is relatively insoluble, and the precipitate it forms on the cathode must be dissolved to provide a figure for the total magnesium removed from the soil. A similar problem would be expected with some of the other polyvalent cations.

A general disadvantage that has been noted for the electro-ultrafiltration method is the high cost of the ultrafiltration units and the limited number of samples that can be processed per unit per day. See a paper by Novozamsky and Houba (1987) for information on the comparative performance of electro-ultrafiltration and conventional extraction methods.

4-4. Fertilizer Requirement

The fertilizer requirement is the quantity of fertilizer needed to supply the difference between the nutrient requirement of the crop and the nutrient supplied by the soil. This general concept encompasses various methods of measurement and various philosophies regarding the action that should be taken in a given situation. In practice, considerable individuality and subjectivity are involved in the concept of fertilizer requirement.

4-4.1. Critical Soil Test Values

According to one view, the objective is to find for each nutrient the *critical value* of the nutrient in the soil. The critical value is the soil test value that, depending upon the definition employed, produces the best separation of soils that gave a yield response, a statistically significant yield response, or an economically profitable yield response from those that did not in the group of soils or experiments used for calibration. The critical soil test value has also been defined as the minimum soil test value associated with the maximum crop yield.

If the critical-value concept were to perform ideally, there would be a unique critical value. In practice, the critical value may have a considerable uncertainty range. In some instances, the range of values that otherwise might be averaged as an estimate of a critical value can be shown to be dependent upon another property, from which it may be inferred that one is dealing with a range of critical values. For example, in a group of 38 field experiments on soybean in North Carolina, Mascagni and Cox (1985) found that the critical value for extractable soil manganese ranged from 1.9 milligrams per liter of soil at pH 5.5 to 8.4 milligrams at pH 7.1. In Western Australia, Brennan and Gartrell (1990) found in a greenhouse experiment that for the 54 soils they investigated the critical soil test value for zinc according to the diethylenetriaminepentaacetate method ranged from 0.14 to 0.54 milligram per kilogram. The critical value was influenced by the soil pH and the clay content.

Whether the experimental observations are taken to represent a single critical value or a range of critical values, the calibration data ideally separate the soils tested into a group in which application of the nutrient in a fertilizer would be economically unprofitable and another group in which it would be profitable. This qualitative interpretation of the significance of soil test data provides the needed practical information that the required quantity of the nutrient is zero for soils above the critical level, but for soils below the critical level it fails to supply guidance on the quantity of the nutrient that should be applied.

Critical values thus are of special practical significance in instances in which the cost of the treatment is small and in which unfavorable results of applying the treatment where it is not needed are rarely encountered. The critical value approach is appropriate also when a soil testing program is being started or when another nutrient is added to an existing program. A calibration based upon a relatively superficial sampling of the conditions involved may be more useful in a new program than an equal amount of effort expended on a few detailed experiments. The detailed work can come later.

4-4.1.1. Visual Inspection Method

A straight-forward but subjective way to locate a critical value for a soil test in data from a group of soils used in calibration is to plot the soil test values in order of increasing magnitude, indicating in some way the nature of the associated responses. Bar graphs are the usual alternative. The bars may be solid for responding sites and open for nonresponding sites or vice versa. Borderline responses may be indicated by crosshatching or some other means of indicating an intermediate condition. An example is shown in Fig. 4-21, introduced later in another connection.

Fig. 4-19 illustrates a different way of plotting the data in an investigation in which there were many sites. To avoid an unduly wide graph, the numbers of fields at each soil test value are plotted against the soil test values, with the different classes of responses indicated by different symbols.

Ideally, the soil samples from all the responsive fields would yield less than the critical value of the nutrient in question, and all the nonresponsive fields would yield more. But if the number of observations is great enough, some overlapping inevitably occurs as a consequence of failure of the test to provide the desired discrimination, together with experimental error. And this is where the difficulty and subjectivity arise in the visual inspection method. Should the critical value be located to correspond with the nonresponding site having the lowest test value, with the responding site having the highest test value, with equal numbers of misclassified observations on each side, or in some other place?

Once the decision is made on the criterion for selecting the critical value, finding the value is simple. In a data set such as that in Fig. 4-19, the upper and lower critical values are obvious, and the intermediate critical value could be ascertained very quickly. But the answer obtained may be troubling, especially when there is some scattering of responding and nonresponding sites or soils throughout most of the range of soil test values, as was the case with some of the zinc extraction methods tested by Brown et al. (1971).

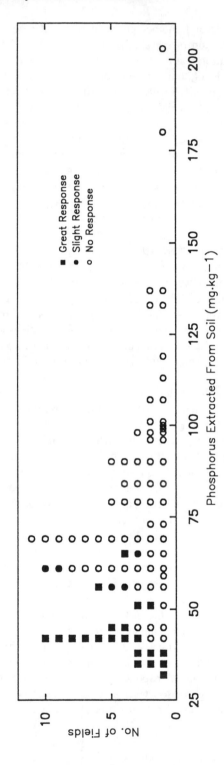

Fig. 4-19. Numbers of fields showing great, slight, or no response to phosphorus fertilization in Colorado versus phosphorus extracted from the surface 15 centimeters of soil by a 1% potassium carbonate solution. The data include observations on 12 different crops. (Hockensmith et al., 1933)

4-4.1.2. Maximum Accuracy Method

The method of maximum accuracy provides a simple way to obtain a critical value that does not involve the subjectivity of the visual inspection method. In the method of maximum accuracy, one finds by trial and error the soil test value (the critical value) for separating responsive and nonresponsive fields or soils for which the sum of the numbers of observations classified correctly as responding and nonresponding is a maximum. Hockensmith et al. (1933), who reported the data in Fig. 4-19, used this approach as the basis for selecting the potassium carbonate test over the other chemical soil tests they investigated. The potassium carbonate method gave the greatest number of correct diagnoses. On the basis of the method of maximum accuracy, they selected 55 milligrams of inorganic phosphorus per kilogram of soil as the critical value. This value is evidently well within the transitional range, with six nonresponding sites below and seven responding sites above.

4-4.1.3. Probability Adaptation

Fitts (1955) proposed that soils be divided into fertility categories on the basis of the probability of obtaining a profitable yield response from fertilization. Imaizumi and Yoshida (1958) observed an overlap of responding and nonresponding soils with similar test values in the results of 46 field experiments in Yamagata Prefecture, Japan, to determine the relation between the response of rice to applications of silicate slag and the extractable silica in the soils. They adopted the alternative of expressing their findings as probabilities. Classifying the results of their experiments in groups, they found that the probabilities of obtaining a profitable increase in yield were 0.91 with less than 75 milligrams of silica extracted per kilogram of soil by a 1 molar acetate buffer at pH 4, 0.89 with 75 to 125 milligrams of extractable silica, 0.25 with 125 to 175 milligrams, and 0 with more than 175 milligrams. According to this adaptation, the extractable soil silica associated with a probability of 0.5 would represent a critical value analogous to that used in the method of maximum accuracy. The data in Fig. 4-19 suggest that many experiments in the transition range of soil test values would be needed to provide a good estimate of the 0.5 probability level. Villemin (1987) used the probability concept for expressing responses to potassium fertilization on soils of France differing in content of exchangeable potassium. The probability concept was discussed again more recently by Dahnke and Olson (1990).

4-4.1.4. Cate-Nelson Methods

A procedure sometimes called the "Cate-Nelson split" (Cate and Nelson, 1965, 1971) for locating a critical test value has become popular in recent years. In a graphic version, the yields of the controls or the control yields as percentages of the maximum yields (relative yields) are plotted on the *Y* axis against the soil test values on the *X* axis. A transparent overlay with a vertical line and an intersecting horizontal line is positioned so as to maximize the numbers of points in the first and third quadrants and minimize the numbers in the second and fourth quadrants. The soil test value corresponding to the location of the vertical line is taken as the critical value that best separates the high-responding group of experiments on the left from the low-responding group on the right. As noted by the authors, this procedure amounts to maximizing the chi-square test of the null hypothesis that the numbers of observations in the four quadrants are equal, which would be expected if the data represented a single random population. The graphic technique is illustrated in Fig. 4-20A.

In a statistical technique described in the second publication by Cate and Nelson (1971), one calculates the corrected sum of squares of deviations of observed yields or other biological values on the Y axis from the means of the populations on the left and right of an arbitrarily placed vertical line representing a trial or postulated critical value. This information is used to calculate the proportion of the total variance of biological values (R^2) that is accounted for by dividing the observations into two groups at the postulated critical value. The numerical value of R^2 changes with the postulated critical value and reaches a maximum where the postulated critical value is equal to the statistical critical value by this method. See Fig. 4-20B for an example that uses the data shown in Fig. 4-20A.

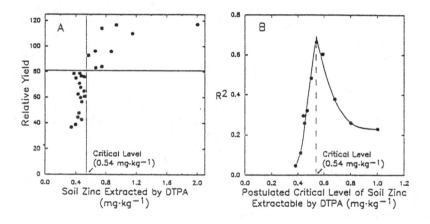

Fig. 4-20. The Cate-Nelson graphical and statistical techniques for locating a "critical" soil test value. A. Graphical technique. The crossed vertical and horizontal lines give the best separation of the soils into two groups, that is, with minimum numbers of observations in the second and fourth quadrants. The data are for relative yields of cluster bean on different soils in a greenhouse experiment versus zinc extracted from the soils by the diethylenetri-aminepentaacetate (DTPA) reagent. The yields obtained with addition of 5 milligrams of zinc per kilogram of soil were given a relative value of 100. B. Statistical technique. Plot of values of R^2 versus postulated critical values of extractable zinc obtained in the Cate-Nelson statistical technique for locating a critical soil test value. According to this test, the R^2 values peak at the critical value. The data used were derived from the plot in A. (Cate and Nelson, 1965, 1971; Singh, 1986)

Rodriguez et al. (1989) found that in a group of 14 field experiments on nitrogen and phosphorus fertilization of proso millet in Colorado, the graphical and statistical techniques yielded identical critical values for nitrogen and for three of four soil phosphorus tests. The values differed slightly for one phosphorus test.

If the data are sufficiently extensive and well enough defined, it may be preferable to recognize three soil test classes, such as low, medium, and high, on the basis of the response data obtained. The same statistical approach can be used for three groups as for two, but the required computations are greater because two boundaries instead of one must be shifted until the maximum among-groups sum of squares is obtained. The R^2 value then will represent the proportion of the variance of the total population of responses that is accounted for by fitting three means instead of two.

The purpose of the Cate-Nelson methods is to separate the data into groups with maximum statistical homogeneity within groups. The biological value associated with the critical soil test may vary. The same may be said for the zone of overlapping of responding and nonresponding sites or soils with similar soil test values. In the example shown in Fig. 4-20, the relative yield corresponding to the critical soil test value for zinc was 81. In an example given in the paper by Cate and Nelson (1971), the critical value was associated with a relative yield of about 78.

Where adequate response curve data are available, a common practice is to select the soil test value corresponding to 90 or 95% of the maximum yield as the breaking point for establishing a critical soil test value, not the approximately 80% that emerged in the two examples in the preceding paragraph. Thus, the two relatively homogeneous groups one obtains according to statistical criteria do not necessarily coincide with the separation desired from the standpoint of the magnitude of the responses or the economic effects.

The Cate-Nelson methods are used to find critical levels of both soil tests and plant tissue tests. In 87 experiments on nitrogen fertilization of corn in Pennsylvania, Fox et al. (1989) found that the segregation of relative grain yields on the basis of soil nitrate at the 5 to 7 leaf stage was much more clearcut than the segregation on the basis of the nitrate concentration in the base of the stalks at the same stage. Under other circumstances, the reverse might be expected.

4-4.1.5. Other Methods

A number of additional methods have been used to select critical levels of soil tests. Comerford and Fisher (1982) used discriminant analysis to classify forest soil sites as responsive or nonresponsive to fertilizer phosphorus on the basis of extractable soil phosphorus. Discriminant analysis is a more complex statistical operation than the Cate-Nelson split, but the results and limitations should be similar.

Havlin (1986) compared several methods for locating critical values for extractable soil iron using three sets of data. The methods included those discussed in the preceding sections and three others. The results obtained with one data

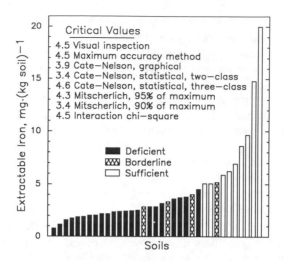

Fig. 4-21. Iron extractable by a diethylenetriaminepentaacetate reagent at pH 7.3 from 35 soils of Colorado classified according to their response to chelated iron added as fertilizer in a greenhouse test. The added iron produced a statistically significant increase in yield of sorghum grown as a test crop on the soils classed as deficient, and no significant increase on the soils classed as sufficient. On soils classed as borderline, the sorghum showed mild iron deficiency symptoms, but the yield response was not significant at the 5% level with the two replicates employed. The experimental work was done by Lindsay and Norvell (1978), and the critical values other than the 4.5 milligrams of extractable iron per kilogram of soil by visual inspection were calculated by Havlin (1986).

set are illustrated in Fig. 4-21. Here one sees that the critical values varied from 3.4 to 4.6 milligrams per kilogram of soil and that the method of visual inspection used by the authors of the research quoted was near the high end of the range. The two Mitscherlich figures apparently were based upon relative yields estimated by the Mitscherlich-Bray equation to be described in Section 4-4.2.1.

The chi-square method was described by Keisling and Mullinix (1979). In this method, interaction chi-square values are calculated using the formula given by the authors, with different arbitrarily selected soil test values as trial critical values for constructing a two-way contingency table of correctly and incorrectly classified responses versus deficient and sufficient (that is, responsive and non-responsive) indications from the plant assay tests on the sites or soils.

In a plot of the interaction chi-square values against the trial critical soil test values, the minimum chi-square corresponds to the critical soil test value. If the chi-square values are transformed to probabilities, as described by Steel and Torrie (1960, pp. 379-380), the probabilities peak at the minimum chi-square, and the width of the plot at the base of the peak is a measure of the width of the zone in which the soil test was not adequately discriminatory and classified some sites or soils incorrectly. The narrower the width, the more effective is the soil test in separating responding sites from nonresponding sites.

4-4.1.6. Evaluation of Methods

Mallarino and Blackmer (1992) devised an economic basis for evaluating the critical values obtained by different procedures. This approach to evaluation has the advantage of using an agriculturally practical criterion.

On the basis of 25 field experiments in Iowa on corn receiving different quantities of fertilizer phosphorus, and measurements of phosphorus extracted from the soils by three reagents, Mallarino and Blackmer (1992) fitted various response functions to the yield data. Then they calculated critical values for the soil tests on the basis of the individual response functions and certain assumptions about definitions of the critical level (for example, the percentage of the maximum yield). They assumed that advice on whether or not to apply fertilizer phosphorus was to be provided for 25 1-hectare fields, each representing one of the experiments. The fields for which the soil test values for phosphorus were below the critical level estimated by the particular response function and soil test then were assumed to be treated with 25 and 50 kilograms of fertilizer phosphorus per hectare (the two smallest quantities used experimentally). The remaining fields were assumed to receive no fertilizer phosphorus. The dollar values of the increases in yields from the fertilized fields were summed, and the magnitude of the sums was used as a basis for ranking.

In this investigation, the Cate-Nelson (1971) statistical procedure yielded the greatest return of the techniques tested (the Cate-Nelson graphical procedure was not tested). The linear response and plateau, quadratic response and plateau, Mitscherlich, and multivariable polynomial functions were less efficient.

4-4.2. Relative Yields From Soil Tests

4-4.2.1. Mitscherlich-Bray Equation

A step beyond the critical value concept is provided by an adaptation of the Mitscherlich equation to use of chemical soil test data that was pointed out by Bray (1944). He started with the Mitscherlich equation

$$\text{Log}(A - y) = \text{Log}A - c(x + b),$$

in which y is the yield with the quantity x of the nutrient applied in the fertilizer plus the effective quantity b of the nutrient in the soil in the same units as x, A is the maximum yield obtainable as the quantity x of the nutrient applied in the fertilizer increases indefinitely, and c is a measure of the effectiveness of the fertilizer in increasing the yield. See Chapter 1 for more information on this equation.

Bray noted that the foregoing equation may be written as follows when x is zero:

$$\text{Log}(A - y) = \text{Log}A - cb.$$

In this equation, y is the yield of the control, that is, the yield obtained with the effective amount b of the nutrient supplied by the soil, and it thus may be designated more explicitly as y_0.

In the field experimental data available to Bray, the average yields of the crops without potassium fertilization and the average yields with 46.5 kilograms of fertilizer potassium per hectare annually were known from long-term experiments at a number of locations in Illinois. On the assumption that no further yield increase would result from addition of more potassium, the numerical values for A and y_0 could be substituted into the simplified equation, leaving only c and b to be evaluated. (The validity of the assumption about substituting the yields of the fertilized plots for A was verified by doubling the application of potassium to the soils most deficient in potassium and obtaining no significant increase in yield.)

The exchangeable potassium in the surface layer of soil was substituted into the equation in place of b. This step involves the assumption that the quantity of exchangeable potassium was directly proportional to b, the potassium availability index that would be obtained by calculation following the usual procedure. If k is the constant of proportionality, the exchangeable potassium is equal to kb. This change requires that the value of c be divided by k to maintain equality in the equation

$$\text{Log}A - \text{Log}(A - y_0) = cb = (c/k)(kb).$$

If c/k is designated as c_1 and kb as b_1, the equation may be written

$$\text{Log}A - \text{Log}(A - y_0) = c_1 b_1.$$

This is the form in which the equation was used by Bray. Bray set A equal to 100 and expressed y_0 as a percentage of A. With all the values known except c_1, the value of this parameter could be calculated for each experiment. The average value of c_1 for corn over all experiments was 0.011 when b_1 was expressed in units of milligrams of potassium per kilogram of soil. That this method was of value in correlating the chemical analyses for exchangeable soil potassium with the responses of corn to potassium fertilization on the various experimental fields is evident from Fig. 4-22A.

The Mitscherlich-Bray equation has been widely used because it is relatively simple, and the experimental data requirements are so limited. The results obtained indicate that the approach is capable of providing fairly good to excellent correlations if the chemical soil test data are good.

Obtaining a good-fitting relationship between chemical soil test values and the crop yield data by use of the Mitscherlich-Bray equation or any other, however, is only the first step in making the desired prediction of the effects of different fertilizer additions. Once a body of data such as that obtained by Bray for potassium is available, the usual inference is that the average c_1 value may be applied for the same crop on other similar fields on which no experiments

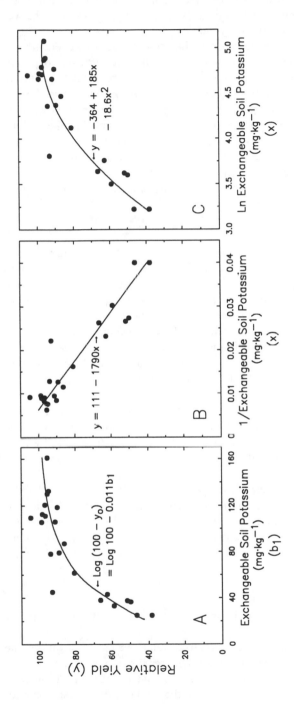

Fig. 4-22. Relative yields of corn in individual experimental fields in Illinois versus exchangeable potassium in the surface soil as represented in different ways: A. Substituting the exchangeable potassium for the potassium supplied by the soil (the *b* value) in the Mitscherlich equation (Bray, 1944). B. Using the reciprocal of the exchangeable potassium content (Semb and Uhlen, 1954). C. Using the natural logarithm of the exchangeable potassium content (Saunders et al., 1987a,b). The relative yields are the yields without potassium fertilization as percentages of the yields with applications of fertilizer potassium sufficient to produce the maximum yields obtainable with potassium.

have been conducted. Moreover, one may estimate the increase in exchangeable potassium or other nutrient needed to bring about any response between the current value and the maximum yield.

The increase in exchangeable potassium thus found, however, is not necessarily equal to the quantity of potassium needed as fertilizer to produce the calculated increase in relative yield. There are two reasons: First, the relative effectiveness of exchangeable and added potassium is not known. Some of the effectiveness of the added potassium may be lost by interaction with the soil. Second, the exchangeable potassium found in the portion of the soil sampled (normally only the plowed layer) is only part of the potassium available to plants and hence represents, in effect, an index of the availability of potassium in the entire soil profile, most of which is not measured.

To estimate the potassium fertilizer requirement, Bray used the equation

$$\text{Log}(A - y) = \text{Log}A - (c_1 b_1 + cx),$$

where c is the efficiency factor applying to the potassium fertilizer and x is the quantity of potassium added in the fertilizer. This equation makes provision for the possibility that the value of c for fertilizer potassium may be different from the value of c_1 for exchangeable potassium. If c_1, b_1, and c are known, the equation makes possible an estimation of the quantity of fertilizer potassium required to attain any desired percentage of the maximum yield that lies between the current yield and the maximum. Separate experiments were necessary to obtain the needed information, which is the c value in the Mitscherlich equation. Bray used an indirect method to estimate c for fertilizer potassium from the results of two field experiments.

The Mitscherlich-Bray equation may be expanded by adding more cb terms on the right-hand side for additional sources of a nutrient, such as those in an additional form in the soil or an additional layer of soil. Papers by Balba and Bray (1956, 1957) illustrated these adaptations.

4-4.2.2. Semb and Uhlen Equation

Semb and Uhlen (1954) used a different equation for relating chemical soil-test values to yield responses in the field. Like Bray (1944), they expressed the yields of control plots in field experiments as percentages of the maximum yields obtainable with the nutrient under test. These are the y values in their equation

$$y = c + b/x,$$

where c and b are constants and x is the chemical soil-test value for the nutrient in question.

Semb and Uhlen used their equation to correlate the results of various chemical soil tests for phosphorus and potassium with results from a large number of field experiments, but they gave only summaries and not the data from individual experiments. As an example to illustrate the method, Bray's data on potassium from Fig. 4-22A are replotted in Fig. 4-22B according to the Semb and Uhlen equation. The correlation r = 0.93 between relative yields and the reciprocal of the exchangeable potassium for this data set is higher than the correlation *R* = 0.79 between observed and estimated relative yields according to the Mitscherlich-Bray equation in Fig. 4-22A.

Semb and Uhlen pointed out that their equation was more convenient than the Mitscherlich-Bray equation because the results fall on a straight line, and the regression and correlation coefficients may be calculated readily by the method of least squares. From the standpoint of curve-fitting, the equation has the additional advantages that the curves need not pass through the point of origin, and the *y* values can exceed a relative yield of 100 at high levels of nutrient availability in the control plots. The Mitscherlich-Bray equation includes a *Log (A − y)* term, which causes trouble if *y* exceeds the value taken as *A* because one cannot take the logarithm of a negative number.

The advantages pointed out by Semb and Uhlen are bona fide improvements. Borlan et al. (1968) used a modified version of the equation, but the advantages seem generally to have been overlooked.

As with the Mitscherlich-Bray equation, the Semb and Uhlen equation may be modified to accommodate quantities of a nutrient supplied by a fertilizer, two or more forms in the soil, or additional layers of soil. These adaptations are made by adding *b/x* terms on the right-hand side of the equation.

4-4.2.3. Saunders, Sherrell, and Gravett Equation

Saunders et al. (1987a,b) used the equation

$$Relative\ Yield = a + b\ Ln\ Olsen\ P,$$

where *a* and *b* are constants to be fitted statistically, to relate the relative yields obtained in 135 field experiments on pasture vegetation in New Zealand to the phosphorus extracted from the soil by a slight modification of the 0.5 molar sodium bicarbonate extraction method proposed by Olsen et al. (1954). The data obtained by Bray (1944) used in the preceding two sections have been replotted in Fig. 4-22C to illustrate the performance of the equation.

As with the Semb and Uhlen (1954) equation in Section 4-4.2.2, the equation used by Saunders et al. may be expanded by adding extra terms on the right-hand side to represent the nutrient supplied by a fertilizer, different forms in the soil, or different layers of soil. And like the Semb and Uhlen equation, it is an improvement over the Mitscherlich-Bray equation in that relative yields exceeding 100 introduce no mathematical difficulties.

For the data set plotted in Fig. 4-22C, the linear correlation between relative yields and the values for the natural logarithm of exchangeable potassium is *r* = 0.91, which is intermediate between the correlations *R* = 0.79 between observed and estimated relative yields by the Mitscherlich-Bray method (Fig. 4-22A) and *r* = 0.93 between relative yields and the reciprocals of the exchangeable potassium by the Semb and Uhlen method (Fig. 4-22B). For this data set, a second-order polynomial provides a better fit than the linear equation used by Saunders et al. (1987a,b). Use of a second-order polynomial, as shown in Fig. 4-22C, increased the correlation to $R = 0.94$, which was a little better than the correlation $r = 0.93$ obtained by the Semb and Uhlen plot. The Semb and Uhlen plot is essentially linear, and the R value remained at 0.93 when a second-order polynomial equation was fitted to the data.

A special feature of the investigation by Saunders et al. (1987a,b) was the fact that the pastures were used for grazing all year. The growing conditions varied with the season. The relative yield associated with a given soil test value was higher during the spring flush, when the production rate was high and grass was dominant, than in the remainder of the year, when the production rate was lower and the pastures often were dominated by white clover. The overall correlation was poor. The relative yields diverged greatly when the extractable phosphorus was low, but they converged as the extractable phosphorus increased.

The authors noted that in the pastoral agriculture of New Zealand, the dry matter production rate during the period of slow growth is probably the major factor determining the maximum stocking rate for the year, so that if maximum stocking rate is the objective, phosphorus fertilization practice logically should emphasize the need for phosphorus during the period of slow growth. Accordingly, they adopted the convention of relating the soil test values for extractable phosphorus to the relative yields below which the observations would fall only 20% of the time. When they used this approach, the correlations between relative yields and extractable phosphorus were much improved, and the authors were satisfied that they had addressed the critical issue more effectively than they would have by using an overall correlation.

Superimposed on the variation among seasons within years, which could be taken into account by averaging the results over a period of years, was variation within seasons among years, which could not be taken into account. As an example, they included data from a group of experiments on one soil indicating that rainfall was a dominant controlling factor in herbage production during the autumn season. Within a 4-year period, when the autumn rainfall varied from 10 to 35 centimeters, the dry matter production increased from 5 to 74 kilograms per hectare per day, and the relative yield increased from 67 to 95.

One reason for discussing the results of these experiments in some detail is that the effect of season on the soil test calibration provides some insight into the results of soil test calibrations made under conditions in which the crop is grown only in a given season or is harvested for yield determination only at the end of its growth cycle, which in some instances may span more than one season in a given year. The experiments in New Zealand demonstrate the significance of seasonal conditions that may be expected to affect the results obtained under these other circumstances, in which calibrations usually are derived from experiments conducted over a period of years in various locations. Then the seasonal effects usually appear as experimental error, and they cannot be taken into account in the calibration because they cannot be predicted.

The calibration curves relating relative yields to soil test values also varied from one soil type to another in the experiments by Saunders et al. (1987a,b). This effect could be taken into account by using a calibration appropriate for the soil or perhaps by making other measurements on the soils. Differences in calibration among soil types, however, were smaller than the differences in calibrations among seasons.

4-4.2.4. Fitts Modification

The relative yield concept described in preceding sections has its basis in the assumption that the c value in the Mitscherlich equation is constant. In the International Soil Testing Program conducted by North Carolina State University under contract with the U.S. Agency for International Development during the 1960s and 1970s, J. W. Fitts (personal communication, 1991) found that in undeveloped areas where he was working, the relative yield concept did not provide the desired information on nutrient deficiencies. The basic problem he encountered was that some nutrient deficiencies were so extreme that unless that nutrient or those nutrients were added, there was little response to applications of other nutrients, which also were deficient.

He accordingly modified the approach in two ways. The first was to make an overt change in the experimental procedure from the Mitscherlich assumption that the relative yield of a crop with different additions of a given nutrient is independent of the supplies of other nutrients (the c value is constant),

$$Relative\ Yield\ =\ \frac{y_0}{y_n}$$

where y_0 is the yield of the crop without fertilizer and y_n is the yield with an adequate amount of the particular nutrient being tested. Fitts changed the meaning of y_0 to the yield of the crop with an adequate amount of all nutrients but the one under test, and the meaning of y_n to the yield with an adequate amount of all nutrients. The problem of determining what was an adequate amount led to the second modification of conducting laboratory tests on the soil samples to find the quantities of phosphorus, potassium, boron, sulfur, manganese, zinc, and copper that had to be added before the extractant used for soil testing would detect a significant increase. The information derived from the laboratory and greenhouse tests greatly improved the efficiency of the ensuing field experimentation designed to identify nutrient deficiencies and to determine the quantities of fertilizers needed to eliminate them. Alvarez and Fonseca (1990) found this approach useful in gauging the phosphorus needs of soils of Brazil that differed considerably in their tendency to react with soluble orthophosphate added in the laboratory.

4-4.2.5. Subsoil Nutrients and the Proportionality Assumption

The methods of Bray (Section 4-4.2.1), Semb and Uhlen (Section 4-4.2.2), and Saunders et al. (Section 4-4.2.3) all involve the assumption that the soil test value is proportional to the supply of the nutrient in the soil. This assumption is of significance because the soil test values almost invariably employed are measurements made on samples of only a surface layer of soil. Plants may derive a portion of their nutrients from soil below the sampled layer, which implies that the proportionality assumption is inappropriate to the extent that the subsoil contributions to the nutrient supply are not proportional to the values measured in soil from the sampled layer.

In experimental work on the principle of nutrient uptake from subsoils, Kuhl-mann et al. (1985) used rubidium as a tracer for potassium. Plants absorb potassium and rubidium in the ratio in which they are present in a nutrient solution, but because rubidium is held more strongly in exchangeable form than potassium, the authors grew wheat plants in 25 kilograms of rubidium-treated topsoil in 20-liter containers to calibrate the ratios of uptakes by the wheat test crop to the quantities in the soils. Similar containers without bottoms were placed on subsoil in the field. As a result of uptake of potassium from the subsoil, the ratio of rubidium to potassium in the plants was reduced, and the uptake of subsoil potassium at the six sites tested was calculated to be 22 to 49% of the total potassium absorbed. In this experimental work, the rubidium-treated surface soils were 30 centimeters in depth, and they replaced the surface 30 centimeters of soil. The soil below 30 centimeters was considered subsoil.

Mombiela et al. (1981) investigated the proportionality question by determining for two groups of field experiments on phosphorus fertilization of potato the functional relationship of soil test values for phosphorus to the *b* values calculated from the Mitscherlich equation. They found that the relationship could be represented better by a linear equation (including a constant with a positive value) than by a direct proportionality. Findings were similar when the soil supply of phosphorus was estimated using the quadratic response equation and the square root response equation, although the numerical values differed. The positive constant terms suggest a contribution of phosphorus not measured by the soil test to the nutrition of the crop.

The opposite inference may be made from Cope and Rouse's (1973) summary of data on response of corn, soybean, and cotton to potassium fertilization in Alabama. They used the Mitscherlich-Bray equation (Section 4-4.2.1) to represent the data, and found that with all crops the relative yields approached zero at a positive value of exchangeable potassium. They therefore modified the equation to the form $Log (A - y) = Log A - c(b - q)$, where q was a constant value that depended upon the crop and the group of soils in question. (They separated the soils into three groups.)

Under the conditions represented by the work of Cope and Rouse, the failure of the relative yields to approach zero at zero exchangeable potassium was made evident by the extreme potassium deficiencies they encountered (relative yields of about 10 for corn and 20 for cotton). If the minimum relative yields had been of the order of 50 or more, the discrepancy they noted probably could not have been verified.

Logically it would be expected that crops harvested for seed or lint would produce zero yields of these products at very low potassium availabilities even though they would make some vegetative growth. Cope and Rouse's observation thus does not mean that the crops derived no potassium from sources other than the exchangeable potassium in the surface layer sampled. Rather, if the research objective is finding whether such sources make a contribution to the supply of a nutrient used by a crop, the response curves should be for total dry matter instead of the harvested portion of the crop, where this represents a fruiting product that develops after considerable vegetative growth has taken place.

Hanway et al. (1961, 1962) made a more direct test of the assumption of direct proportionality of surface soil tests to soil nutrient supply in cooperative regional investigations including 51 field experiments on alfalfa in the North Central Region of the United States plus Alaska and Ontario and 31 field experiments on corn in the North Central Region. For both crops, the precision of estimates of the biological indexes derived from the field experiments was improved by measurements of the exchangeable potassium content of subsoil layers in addition to the usual measurements on samples of the surface soils. For alfalfa, the coefficients of determination (r^2 or R^2) for the yield of potassium in the plants on the control plots as estimated by the values for soil exchangeable potassium were 0.35, 0.38, 0.49, and 0.50 based upon exchangeable potassium in the first; first and second; first, second, and third; and first through sixth layer. Each layer of soil sampled was 15 centimeters thick, and the exchangeable potassium in each layer was entered as an independent variable.

Probably the most pronounced effects of subsoil nutrients are observed with nitrogen, and it is for nitrogen that the greatest need now exists for evaluating subsoil sources. In an investigation of the value of tests for subsoil nitrogen, Matar et al. (1990) conducted 40 field experiments with wheat in Syria and analyzed successive 20-centimeter layers of soil for nitrate. They found that the correlations of soil nitrate nitrogen with the yields of grain, total above-ground dry matter, and total above-ground nitrogen in the wheat plants increased as they added the amounts of nitrate nitrogen in soil layers to a depth of 80 centimeters.

Summing the amounts of a given nutrient found in various soil layers involves the assumption that a given amount of the nutrient has a given value to the crop, irrespective of the layer in which the nutrient occurs. This assumption may be appropriate for some circumstances but not for others. In The Netherlands, Neeteson and Zwetsloot (1989) reported data from 150 field experiments with sugarbeet and 98 with potato in which measurements of ammonium and nitrate nitrogen present in the soil in early spring were made on different layers of soil. Their results showed that the measured yield responses to fertilizer nitrogen could be estimated with significantly better precision as their yield response models included the mineral nitrogen in successively deeper soil layers. According to their analyses, 1 kilogram of mineral nitrogen in the 0-30, 30-60, and 60-100 centimeter layers of soil produced sugarbeet yields equivalent to 0.8,

0.85, and 0.15 kilogram of fertilizer nitrogen. For potato, 1 kilogram of mineral nitrogen in the 0-30 and 30-60 centimeter layers of soil produced yields equivalent to 0.67 and 0.33 kilogram of fertilizer nitrogen.

Determining the relative values of initial nitrate or mineral nitrogen in the surface soil and subsoil and the value of these quantities relative to current applications of fertilizer nitrogen under different conditions is a part of the calibration process needed to take into account the fact that all circumstances cannot be represented satisfactorily by an average relationship. Unfortunately, the calibrations for mineral nitrogen cannot be handled in the relatively simple way that is acceptable for most other nutrients. The problem with mineral nitrogen is that the ratios of concentrations in particular subsoil layers and in the subsoil as a whole to those in the surface soil are not stable. They vary with the time of sampling within a given year, weather conditions from one year to the next, and management factors related to nitrogen, especially previous fertilization. With nitrogen, what is needed is a way to take into account the effects of some of the salient factors that influence the relationships. Improvements in estimation can be accomplished by use of multiple regression equations based upon experimental measurements made over a period of years and by use of mechanistic models now being developed (see Section 4-4.6.4).

4-4.2.6. Arbitrary Soil Depths Versus Soil Horizons

Under natural conditions, and within a limited area, soil properties are more uniform horizontally than vertically. Thus, the use of superposed horizontal units of specified thicknesses that were uniform in different soils was a reasonable way for the investigators whose works were mentioned in the preceding section to proceed. In this simplified approach, the horizontal variation in availability coefficients brought about by differences in location with respect to the plants causes the modification of the availability coefficients for the nutrient in a given horizontal layer to an average value for that layer.

Using fixed thicknesses for the volume units involves the assumption that however the availability of the nutrient may vary within a given volume unit as defined, it does so in the same way in all the experimental areas upon which the multiple regression is based, and that for the usual linear model it is directly proportional to the quantity of the nutrient measured in that volume unit by the soil test. If, alternatively, the volume units were defined on the basis of soil horizons, the corresponding assumption would be that the availability of the nutrient in the individual volume units depends upon the horizon designations and is independent of the thickness and depth.

If the soils involved in the experimental work were very similar and differed mainly in the supply of the nutrient of concern, the system using horizon designations would seem preferable on the basis that the properties of the soil affecting the nutrient availability in individual horizons would be expected to correlate to some extent with the visually observable properties of the soil profiles. On the other hand, if the group of soils included in the experimental work were diverse, the system using horizon designations would lose much or all of the advantage it would seem to have if the soils were very uniform.

One of the problems is that a considerable element of arbitrariness may exist in designating the horizon names. A second difficulty is that soil horizon designations were not developed to characterize soils as substrates for plant growth, but rather to aid in describing and understanding the relationships among the various portions of individual profiles and to aid in soil classification. The horizons in a particular profile are named to a considerable extent in accordance with their properties in relation to each other. Hence, for example, if one were to remove the A3 horizon from a soil and examine it by itself, much of its property as an A3 horizon would disappear because of the loss of association with the other horizons. It might even pass for an A1 horizon in a different soil.

Except for soils that have very similar profiles, neither the fixed thickness system nor the horizon designation system provides the desired kind of data for analysis by multiple regression. To date, an experimental comparison between the fixed thickness system and the horizon designation system does not seem to have been made. The former system must have seemed preferable to the individuals who have investigated soil profile effects, however, because all have used that system. From the practical standpoint of collecting soil samples for testing, the fixed thickness system is definitely preferable because the individuals taking the samples are rarely if ever soil classification experts and cannot be expected to understand soil horizon designations or to take soil profile samples according to such a system.

4-4.3. Adjustments for Fertilizer Efficiency

Despite the considerable attention paid to evaluating the *effective supplies* of soil nutrients in the scientific literature, in this book, and in practical soil fertility evaluation, the fact is that the objective is to find the *additions* of nutrients needed to achieve the desired results. The quantity of a fertilizer source of a nutrient needed to achieve a selected return decreases as the effective supply of the nutrient in the soil increases, but there is no fixed relationship between the two. For a given effective supply of a nutrient in the soil, the additional quantity needed is influenced by the efficiency of the addition.

Two general approaches have been employed as a basis for adjusting the quantity of fertilizer for differences in efficiency among soils. One uses the relationship between the response curvatures and the buffer capacities of the individual soils for the nutrient in question. In the other, the soil is titrated to a critical nutrient intensity with a source of the nutrient.

4-4.3.1. Estimation of Response Curvature by Soil Buffer Capacity

As noted in Chapter 1, the c value in the Mitscherlich equation is a measure of the rate at which the yield approaches its maximum value when increasing quantities of a nutrient are added. For this reason, the c value often is called the efficiency factor. For given control and maximum yields, a high rate of increase

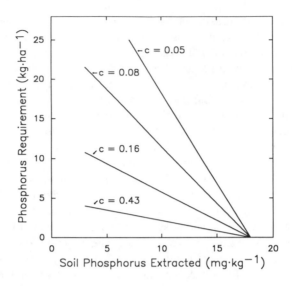

Fig. 4-23. Estimation of the phosphorus required to produce 90% of the maximum yield of wheat in New South Wales from the soil phosphorus extracted by the lactate method, showing the effect of different Mitscherlich c values, which represent the rate at which the yield approaches the maximum as fertilizer phosphorus is added. (Holford et al., 1985)

of the yield per unit of nutrient in the fertilizer (a relatively high c value in the Mitscherlich equation) results in a relatively low fertilizer requirement. And when the rate of increase of the yield per unit of nutrient in the fertilizer is low (a relatively low c value in the Mitscherlich equation), the fertilizer requirement is relatively high.

Mitscherlich adopted the convention of using a common c value for each nutrient. He found this convention useful, but fertilizer needs can be estimated more precisely from indexes of nutrient availability in soils if differences in c values among soils can be taken into account. The most definitive work on this subject on a field basis appears to have been done by Holford et al. (1985), Holford and Cullis (1985a), and Munk and Rex (1987).

Holford et al. (1985) illustrated the significance of the response curvature by calculating the required applications of fertilizer phosphorus associated with different values for extractable phosphorus measured by the lactate method of Egnér et al. (1960) in a group of 57 field experiments on wheat in New South Wales. As shown in Fig. 4-23, the estimates of fertilizer phosphorus requirement depended strongly upon the Mitscherlich c value when the phosphorus availability index was low, and all estimates converged at the critical value for the test, which was estimated at 18 milligrams of phosphorus per kilogram of soil. The various lines in the figure were derived from the calibration equation using the actual c values: *Fertilizer requirement in kg $P \cdot ha^{-1}$* $= (2.073 - 0.117x)/c$, where x is the soil test value.

Table 4-5. Comparative Value of Three Different Mitscherlich *c* Values for Estimating the Requirement of Fertilizer Phosphorus in 39 Field Experiments on Wheat in New South Wales When Used in Conjunction With Soil Phosphorus Availability Indexes Obtained by Two Methods of Phosphorus Extraction (Holford and Cullis, 1985a)

	Proportion of Variance of Fertilizer Requirement Accounted for (%)		
Soil Phosphorus Extraction Method[a]	Using the Average Mitscherlich *c* Value Over All Experiments	Using the *c* Values Estimated From the Phosphorus Buffer Capacity	Using the Actual *c* Values in the Individual Experiments
Best	32	75	93
Poorest	0	33	49

[a] The best of the six methods in terms of estimating the fertilizer requirement was the lactate method of Egnér et al. (1960), and the poorest was the sodium bicarbonate method of Colwell (1963).

For the same group of experiments, the value of including measurements of fertilizer efficiency was illustrated by using the Mitscherlich *c* value and the soil test values for phosphorus in the individual experiments to estimate the quantities of fertilizer phosphorus required to attain 90% of the maximum yield. When the average *c* value was used, only 23% of the variance of quantities of fertilizer phosphorus was accounted for. When the *c* values for individual experiments were used, 68% of the variance was accounted for.

Holford and Cullis (1985a) carried the analysis one step further by analyzing data from another set of 39 field experiments on wheat in New South Wales. They calculated the Mitscherlich *c* value for each experiment and also made measurements of the phosphorus buffer capacity and the soil phosphorus extracted by six methods. The central concepts in their findings are illustrated by data in Table 4-5 from the poorest and best of the six soil test methods they used. First, using the *c* values for individual experiments as opposed to using an average *c* value increased the proportion of the variance in fertilizer requirement accounted for from 32 to 93% with the lactate test and from 0 to 49% with the bicarbonate test. These increases illustrate the great significance of the response curvature. The second significant point is that for making predictions from the results of soil test values obtained on samples from producers' fields where the *c* values are not known, estimates of *c* based upon phosphorus buffer capacity measurements were useful. The buffer capacity estimates of *c* increased the percentage of the variance of fertilizer requirement accounted for from 32% to 75% with the lactate test and from 0 to 33% with the bicarbonate test. In terms of variance accounted for, the estimates of the *c* values derived from the buffer capacity data were of greater value than the indexes of phosphorus availability in the soils.

As with most practices in soil fertility control, adjustments are needed as the circumstances change. The two separate groups of field experiments on phosphorus fertilization of wheat grown on moderately acid to alkaline soils in New

South Wales (Holford et al., 1985; Holford and Cullis, 1985a) mentioned previously showed that the lactate method was the best of a number of soil analysis methods they used. The lactate method also proved superior in a study of another group of field experiments on wheat grown on mostly slightly acid to alkaline soils in the semiarid central plains of New South Wales (Holford et al., 1988). Phosphorus sorption (buffer capacity) index measurements made in one group of experiments markedly improved the precision of estimates of the quantities of fertilizer phosphorus required.

In work on phosphorus fertilization of wheat on strongly acid soils in New South Wales (Holford and Cullis, 1985b), a fluoride reagent (0.5 molar ammonium fluoride at pH 7) and the Truog (1930) reagent (0.001 molar sulfuric acid and 3 grams of ammonium sulfate per liter) emerged as the most effective, and the lactate method was less effective. In a study of 41 experiments on white clover pastures on acid soils in New South Wales, Holford and Crocker (1988) found that the Bray No. 1 extractant (0.03 molar ammonium fluoride and 0.025 molar hydrochloric acid) (Bray and Kurtz, 1945) was the most effective in estimating responsiveness and fertilizer requirement, and the lactate test was the least effective. In this work, the phosphorus sorption (buffer capacity) index was not of value in improving the estimates of fertilizer requirement by the most effective methods.

In a greenhouse experiment by Moody et al. (1988) with 26 soils of Queensland and corn as the test crop, a phosphorus intensity measurement (the equilibrium phosphorus concentrations in the soil samples from the control plots) was superior to the quantity index represented by the Colwell (1963) sodium bicarbonate extraction. The equilibrium phosphorus concentration and a phosphorus sorption index together accounted for 77% of the quantity of phosphorus required to attain 90% of the maximum yield. The equilibrium phosphorus concentration and the Mitscherlich c value together also accounted for 77% of the variance. Buffer capacity measurements were less efficient estimators of the variance of phosphorus requirement than were the sorption indexes. An additional observation of significance in this work was that at a given buffer capacity, soils with pH values above 7 generally had higher c values than did soils with pH values below 7. That is, the efficiency of the phosphorus applied as monobasic calcium phosphate was greater in alkaline soils than in acid soils.

The foregoing account of differences in results with changing experimental conditions, mostly from the work of one research group, illustrates the limited generalization that often is permissible. Moreover, it emphasizes the importance of making analyses of the intensity, quantity, and buffer capacity aspects of at least phosphorus availability, and of doing so by more than one method. One is then more likely to find the most effective combination than if the analyses are restricted to use of only one method on each of the three aspects of availability or to only one or two aspects of availability. In time, some standardization of methods may be possible, as the less effective methods are identified and dropped.

Another significant point is the importance of measurements of buffer capacity, which generally have not been used in the past. The evidence indicates that adding buffer capacity measurements to the usual chemical indexes of phosphorus availability sometimes does and sometimes does not improve the estimates of fertilizer requirement. Until experience has demonstrated that such measurements are unlikely to be helpful under specific circumstances, the potential benefit is worth the research time required to determine whether it is to be had.

This is probably the best place to introduce a concept that represents an alternative to the response curvature measured by the c value in the Mitscherlich equation. The c value is an appropriate measure of fertilizer effectiveness if one is comparing two or more fertilizers that supply the same nutrient when the fertilizers are compared in the same experiment, and when the yield of the control and the maximum yield are the same for all sources of the nutrient (see Chapter 5). The circumstances under consideration in the present section, however, are different. Experiments on different soils are being brought together. The control yields and the maximum yields will differ from one experiment to the next, and the magnitude of the response will vary.

As an alternative to the use of the Mitscherlich c value as an index of the effectiveness of the fertilizer, Bolland and Baker (1987) and Holford and Crocker (1991) have represented fertilizer effectiveness as the product $c(A - y_0)$, where c is the Mitscherlich efficiency factor, A is the maximum yield, and y_0 is the yield of the control. This expression is the derivative of the Mitscherlich equation at $x = 0$, which is the maximum rate of increase in a given experimental series.

Modifying the Mitscherlich efficiency factor by multiplying it by $(A - y_0)$ would make no difference for correlation purposes if the values of A and y_0 were the same for all soils because the $(A - y_0)$ term would be constant. On the other hand, when one is attempting to obtain an index of efficiencies of additions of a given fertilizer to different soils, $(A - y_0)$ is a variable, which means that the Mitscherlich efficiency factor c would be multiplied by a variable. Using $c(A - y_0)$ as a measure of effectiveness of a given fertilizer in different soils would be undesirable if c were independent of $(A - y_0)$. There is evidence, however, that c is not independent of $(A - y_0)$. Where a trend exists, the value of c seems invariably to decrease as $(A - y_0)$ increases, a behavior that is consistent with the requirement of an increasing quantity of the nutrient to produce additional yield. The inverse behavior of c and $(A - y_0)$ thus could lead to some advantage in using the product $c(A - y_0)$ as an index of effectiveness where different soils are being compared because multiplying the two terms would help to neutralize the effect of a downward trend in c values with an increase in $(A - y_0)$. This matter needs further investigation.

4-4.3.2. Titration to a Critical Intensity

Beckwith (1963, 1965) suggested a logical but radically different approach. To estimate the phosphorus fertilizer requirements of soils, he proposed bypassing

the usual procedure of obtaining an index of phosphorus availability and estimating the requirement directly by titrating the phosphorus sorbing capacity of
soil to a standard concentration of phosphorus in solution in equilibrium with
the soil.

Ozanne and Shaw (1967) tested Beckwith's idea. They estimated the quantity
of fertilizer phosphorus required in the field by equilibrating one part of soil in
the laboratory with ten parts of 0.01 molar calcium chloride solution containing
different concentrations of orthophosphate. By interpolation, they found the
phosphorus addition needed to increase the phosphorus concentration of the
equilibrium solution after 17 hours to 0.3 milligram per liter. Their preliminary
studies showed that the pastures with which they worked did not respond to
phosphorus fertilization if the surface 10 centimeters of soil did not adsorb
phosphorus from a 0.01 molar calcium chloride solution containing 0.3 milligram
of phosphorus per liter. In a group of 42 field experiments on phosphate fertilization of pastures in Western Australia, they added different quantities of fertilizer phosphorus as a broadcast topdressing, and from the response curves
estimated the quantities of fertilizer phosphorus required to produce 95% of the
maximum yields. These values were closely correlated ($r = 0.96$) with the
quantities of phosphorus sorbed by the soils in the laboratory tests.

In a subsequent paper, Ozanne and Shaw (1968) reported that the correlation
of phosphorus required with phosphorus sorbed was increased from $r = 0.87$
to $R = 0.91$ by including measurements of phosphorus buffer capacity in 27
field experiments on soils that had received phosphorus fertilizers previously.
Thus, there was some evidence of an effect of phosphorus buffer capacity in
their work, but the influence was not great. They did not report the phosphorus
buffer capacities of the soils. The direct titration method largely substituted for
the buffer capacity information.

Data obtained by Dear et al. (1992) in subsequent similar work showed that
the correlations between the quantities of phosphorus in superphosphate required
to produce 90% of the maximum yield of subterranean clover in Western Australia
and the additions of phosphorus in the laboratory required to obtain a concentration of 0.1 milligram of phosphorus per liter in a 1:10 soil:0.01 molar calcium
chloride extract were $r = 0.95$ in ten field experiments and $r = 0.87$ in 11
soils in a greenhouse experiment. The corresponding correlations of the phosphorus requirement for subterranean clover with the phosphorus buffer capacity
at 0.1 milligram of phosphorus per liter were $r = 0.71$ and $r = 0.79$. Thus,
the direct titration method appeared advantageous.

McLean, Adams, et al. (1982) adapted the direct titration concept in a different
way for estimating the quantities of fertilizer potassium required to obtain the
maximum yield of corn on soils of Ohio. They first used the soil test for potassium
(extraction with ammonium acetate) to determine the percentage recovery of a
fixed addition of potassium to a group of soils after a 2-hour equilibration period,
which was considered a short enough time to be used for routine soil testing.
The recovery ranged from 46 to 94%. Next, they used the same method for

potassium on a second group of samples that had been equilibrated with added potassium for 7 weeks. The recoveries after 7 weeks ranged from 20 to 75%. The relationship between the recoveries obtained in the 7-week equilibration and the recoveries in corresponding soils in the 2-hour equilibration then was fitted by a second degree polynomial equation. This relationship permitted the use of the recovery found in a 2-hour equilibration of a soil sample to estimate the recovery after 7 weeks and the quantity of fertilizer potassium needed to reach the maximum yield based upon the 7-week equilibration. In a later review, McLean and Watson (1985) changed the time of equilibration from 7 weeks to 60 days, and Moorhead and McLean (1985) changed the fertilizer potassium requirement to 95% of the maximum yield.

Analogous work for phosphorus was reported by McLean, Oloya, et al. (1982) and Mostaghimi and McLean (1983). Moorhead and McLean (1985) employed the general approach described in the preceding paragraphs, with some changes in methods, to measurements in the field.

As with the work on phosphorus by Ozanne and Shaw (1967, 1968), the direct titration approach employed by McLean and coworkers is a variation on the buffer capacity concept. It largely substitutes for measurements of buffer capacity and in a way is an improvement on the buffer capacity approach. The goal of McLean and coworkers was to improve the accuracy of estimates of fertilizer requirement made from soil tests by incorporating operationally practical measurements of the differential tendencies of individual soils to reduce the availability of nutrients added in fertilizers.

Further work, all on phosphorus, has been reported by Fox et al. (1974), Singh and Jones (1977), Peaslee and Fox (1978), Vander Zaag and Kagenzi (1986), Pew et al. (1988), Klages et al. (1988), and others. The direct titration concept is especially appropriate for nutrients that are retained by the soil solids and are released to the solution when the concentration in solution is decreased by plant uptake. Breimer et al. (1988) described an application of the direct titration concept for greenhouse crops. In their production system, the soil was used as a source of iron, manganese, zinc, copper, and molybdenum, but other nutrients were added periodically as a nutrient solution compounded from concentrated stock solutions on the basis of analyses made on a water extract of the soil.

4-4.4. Calibration

Fig. 4-24 is a scatter diagram showing yields of corn versus extractable soil phosphorus in the Coastal Plain region of southeastern United States. Although the yields of corn tended to a maximum of some 20 megagrams per hectare at about 40 milligrams of extractable phosphorus per kilogram of soil, there were also yields less than 1 megagram per hectare with similar values of extractable phosphorus. The data are a sobering reminder of the limited predictive value of a soil test with no associated information. Analogous findings for soybean yields versus extractable phosphorus, potassium, calcium, and magnesium in soils were reported by Evanylo and Sumner (1987). An important reason for the scattering

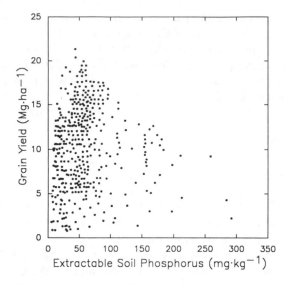

Fig. 4-24. Yield of corn grain versus phosphorus extracted by the Mehlich 0.05 molar hydrochloric acid, 0.0125 molar sulfuric acid method from the soils on which the crop was grown in farmers' fields and in experimental plots as found over a period of years in the Coastal Plain region of southeastern United States. (Sumner, 1987)

of points in Fig. 4-24 is of course the great influence of factors other than the phosphorus supply upon the yields of the crop. If all the factors causing yield differences were known and measured, and if they could be integrated properly with the measurements of extractable phosphorus, the scatter would disappear. The day when this can be done may never come.

In the meantime, the pragmatic answer to the scatter is *calibration*, which is the process of finding experimentally the relationships between soil test measurements and the crop responses to the nutrient in question under various sets of practical conditions. The two easiest of these conditions to record and allow for are the kind of crop and the kind of soil. From the relationships observed within the groups established are inferred the supplies of nutrients in soils and the quantities needed in fertilizers.

A weak link in applying the calibrations often results from the fact that producers do not know their soils by the terms used by soil scientists, nor do they know the legal description of the locations of their fields. Personnel in soil testing laboratories thus may be handicapped in attempting to relate the interpretation of the tests to the specific areas sampled. In Austria, a system has been developed to eliminate the information gap. Producers are supplied with an aerial photograph on which they are asked to mark the locations of the sites to be tested. In the laboratory, the aerial photograph is overlaid with a soil map of the same scale to identify the soil. The report that goes to the producer then contains information on the site characteristics as well as the laboratory results and the interpretation (Danneberg et al., 1990).

4-4.4.1. Soil Classification

From the standpoint of variation in soil properties, the ideal procedure would be to develop calibration curves for all nutrients of concern for all soil types and phases. Limited research resources, however, dictate that in practice a less ambitious approach must be used in which soils with similar properties are grouped for purposes of calibration and recommendation.

Andres (1990) published an example of a classical statistical situation illustrating the value of classifying soils from field experiments in Germany in relation to the quantities of potassium extracted from the soils by a calcium-acetate-lactate reagent. Over all experiments and all levels of extractable potassium, the long-term average yields of grain equivalent were negatively correlated with extractable potassium ($r = -0.58$), whereas the correlations within individual experiments were all positive.

As an example of the value of classifying soils according to type for interpreting the results of a soil phosphorus test, Wendt and Corey (1981) grew corn as a test crop in a greenhouse on 90 samples of Wisconsin soils. They found that the correlations between the uptake of phosphorus by the crop and the phosphorus extracted by the Bray and Kurtz (1945) reagent that contains 0.025 mole of hydrochloric acid and 0.03 mole of ammonium fluoride per liter ranged from 0.49 to 0.97 within five soil types (one group contained two soil types), all statistically significant, but the overall correlation for the 90 soils was only 0.13 and was not significant.

The soil type effects found by Andres (1990) and Wendt and Corey (1981) were a consequence of unknown correlations between soil properties, the plant responses, and the behavior of soil nutrients and the extracting agents. None of the properties directly responsible for the experimental results could be seen or taken into account in classifying the soils into types.

Some of the natural properties of the surface layer of cropped soils, often designated as the plowed layer or topsoil, have been obliterated by cultivation. For this and the additional reason that the plowed layer is only a minor part of most soil profiles, the usual classifications of soils according to their profile properties tend to emphasize the properties below the plowed layer. The upper portion of the soil, however, is of major importance to crops because it generally contains most of the roots and supplies the greater portion of the nutrients.

Referring to an extensive study in which many properties of profiles of soils in the Coastal Plain of North Carolina were correlated with corn yields, Sanchez et al. (1982) credited Sopher and McCracken (1973) with finding that 70% of the yield variability could be accounted for by soil properties that occurred in the plowed layer. This finding, which appears to have been derived from statistical analyses not included in the paper cited, emphasizes the need for a way of grouping soils according to properties that are important in crop production. More direct evidence of the importance of the surface layer of soil was provided by Engelstad and Shrader (1961). In a biological assay experiment on a permeable silt loam soil in Iowa, they found that the quadratic equivalent of the Mitscherlich *b* value for the effective supply of nitrogen in the soil was 134 kilograms per hectare for the normal soil and 45 kilograms where the surface layer of soil had been removed.

Although producing plants is the most important use of soils, the usual soil taxonomic systems are not designed to classify soils for this or any other specific use. As a result, a number of so-called technical classifications have been developed in which soil types are grouped in ways that are meaningful for specific uses (Bartelli, 1978). Clarke (1951) used measurements of soil texture, depth, and drainage as the basis for a numerical index of soil productivity for wheat. He found that the ratings of different soils correlated well with wheat yields under the limited range of conditions he investigated. A later and more comprehensive technical classification, called the *Fertility Capability Soil Classification System*, was proposed by Buol (1972), developed further by Buol and others, and reviewed by Sanchez et al. (1982).

The Fertility Capability Soil Classification System groups soils according to criteria that appear to have a direct influence on soil fertility management practices, especially fertilization. The system includes three category levels: texture of the surface soil, texture of the subsoil, and modifiers. According to Sanchez et al. (1982), 15 modifiers (gleying, dryness, low cation exchange capacity, aluminum toxicity, acidity, high phosphorus fixation by iron, x-ray amorphous clay, vertisol, low potassium reserves, basic reaction, salinity, sodium excess, cat clay, gravel, and slope) were recognized at the time of their publication. The properties thus include factors specific to phosphorus and potassium as nutrients, but they mostly represent factors that will modify the response of crops to a given supply of a nutrient. The system is intended to be used in connection with soil testing for nutrient supplies because the nutrient supplies will vary within each classification unit, as they do within individual soil types, as a consequence of differences in management practices.

Sanchez et al. (1982) reviewed the experimental results obtained with the Fertility Capability Soil Classification System, noting that use of the system improved relationships. For example, the returns from fertilization in 73 experiments on potato in Peru were $770 per hectare when based upon a single overall recommendation and $860 when based upon site-specific soil tests for phosphorus, potassium, and pH. When the experimental sites were grouped according to the system and the quantities of fertilizer were estimated from a generalized response curve for each category, the average return from fertilization was $920 per hectare. And when the sites were grouped according to the system, with the quantities of fertilizer estimated from a generalized response curve for each category and site-specific soil tests, the average return from fertilization was $965 per hectare. In this instance, grouping the experiments into categories based upon the properties recognized in the system was of greater value in terms of monetary returns from fertilization than were the soil tests for nutrient availability.

As a different kind of indication of the value of the classification, Sharpley and Buol (1987) summarized data from 19 literature sources worldwide on the relationship between the minimum exchangeable potassium content and the clay content of soils. The minimum exchangeable potassium content is the limiting content of exchangeable potassium after exhaustive cropping without addition of fertilizer potassium. The overall correlation for the total of 224 soils in all investigations was $r = 0.83$. But when the soils were divided into three groups on the basis of the Fertility Capability Soil Classification System, the correlations within groups were $r = 0.82, 0.91$, and 0.93. The classification system provided an indication of the potassium-release characteristics of the clay, which was advantageous in correlating the findings.

In a group of 13 experiments on potassium fertilization of tobacco conducted over a 3-year period in North Carolina, Denton et al. (1987) found that the classification system separated the soils into classes with markedly different responses and with economic optimum applications of fertilizer potassium ranging from 0 to 112 kilograms per hectare. Denton et al. (1986) also found the classification of value in segregating soils on the basis of the response of corn and soybean to deep tillage.

Smaling and Janssen (1987) used the qualitative Fertility Capability Soil Classification System to classify soils of the Kilifi Area in Kenya, to which they also applied a quantitative system described by Janssen et al. (1990) for evaluating the native nitrogen, phosphorus, and potassium fertility of tropical soils for corn. The quantitative system provided some information not supplied by the qualitative system. For example, in the area studied, it showed that nitrogen was the most limiting of the three major nutrients in almost all of the mapping units. Janssen et al. (1990) concluded that the two systems were supplementary.

4-4.4.2. Adjustments

The index of availability obtained by the soil test measurement in the laboratory is often adjusted for factors known to affect the nutrient requirement under practical conditions. For example, if the laboratory measurement is confined to the mineral nitrogen content of the soil, an adjustment may be made for the differences in nitrogen mineralization expected to occur as a consequence of differences in organic matter content of the soil and contributions of nitrogen from previous leguminous crops and applications of manure.

Where the soil samples represent only the surface layer, an adjustment may be made for differences in residual mineral nitrogen expected to remain in the portion of the rooting zone below the surface layer, depending upon the quantity of fertilizer nitrogen applied in the preceding year. Similarly, where soils are known to differ in availability of phosphorus, potassium, or sulfur in the subsoil, the figures obtained for these nutrients in the surface layer may be adjusted to reflect differences in the subsoil supplies known to be present in different soils. For nutrients other than nitrogen in the mineral form, the need for these latter adjustments will be reduced or eliminated if the calibration curves are based upon groups of similar soils.

Exchangeable potassium is the most widely used index of potassium availability, but the values obtained may be adjusted for other properties found to be of importance. One of these is the percentage saturation of the cation exchange capacity with potassium. The definitive work on the effect of degree of potassium saturation of the cation exchange capacity on potassium availability was done by Jenny and Ayers (1939). They found that the uptake of potassium by barley roots from equal quantities of exchangeable potassium in suspensions of clay saturated with potassium and calcium decreased with decreasing degree of potassium saturation. This effect was not found where the potassium and calcium were present in solution as the chlorides.

On a field basis, Cope and Rouse (1973) found that to obtain 95% of the maximum yields of cotton in Alabama required an average of 51, 73, and 94 milligrams of extractable potassium per kilogram of soils having cation exchange capacities of less than 5, 5 to 10, and more than 10 centimols of charge per kilogram, respectively. As an extractant, they employed the Mehlich reagent consisting of 0.05 molar hydrochloric acid and 0.0125 molar sulfuric acid. The solution was added to soil at a mass ratio of four to one, which would have been insufficient to replace all the exchangeable cations.

Adjustments for the degree of saturation often are expressed in terms of soil texture because texture can be estimated more readily than cation exchange capacity, with which it is correlated. See for example a paper by Gruber (1987). Soil texture or clay content may also carry some information on potassium release from nonexchangeable forms.

In France, the uptake of potassium by plants has been found to be correlated much more closely with the potassium extracted from the soils by sodium tetraphenylboron than with the smaller quantities of exchangeable potassium. Within soil types, however, the two soil potassium measurements are linearly related, with r^2 values of 0.9 or greater. Because the sodium tetraphenylboron reagent is expensive and the analytical procedure is more complex and time consuming than that for exchangeable potassium, the statistical relationships within soil types are used in practice to adjust the exchangeable potassium values to the sodium tetraphenylboron values (Collin, 1990).

4-4.4.3. Environmental Factors

Environmental factors are considered here to include all factors other than the nutrient in question that may affect the growth of the crop. Some of these factors are present in the soil, but others are not.

For a given nutrient availability index, the fertilizer requirement increases as the environmental conditions become more favorable for crop production, that is, as the *A* value in the Mitscherlich equation increases (see Figs. 1-3 and 1-10). Fig. 1-10 was based upon experimental work in greenhouses at different locations in Europe, but similar trends may be observed in field experiments. For example, Holford et al. (1988) found that in a semiarid region in New South Wales the response curvature or effectiveness of fertilizer phosphorus, as reflected by the *c* value in the Mitscherlich equation, was lower in the year that produced the highest yields than in other years. Rainfall from 3 weeks before flowering to 2 weeks after flowering was 11 centimeters in the year that produced the highest yields and from 3 to 8 centimeters in the other 4 years.

Water supply, one of the most important and most variable of the environmental factors, can be well controlled in dry regions by appropriate irrigation. The smallest deviations from average calibration curves for estimating fertilizer requirements can be expected under such conditions, where the water supply can be reproduced from one year to the next. For rainfed areas in which the variations in water supply from year to year result in large coefficients of variation in the mean fertilizer requirement (Fig. 2-6 shows an example), a calibration based upon average conditions may be considered the best that can be done in the absence of ability to predict the precipitation.

In another sense, however, a calibration based upon average conditions may be considered inappropriate because, in effect, it makes decisions for producers that will not be to their maximum financial advantage in most years. Because neither the researchers who develop a calibration nor the producers who use it can predict the deviations of specific years from the average, the alternative of preparing separate calibrations for, say, favorable, intermediate, and poor conditions, or for favorable and poor conditions may be considered. Then the producer can take into account his or her financial condition and outlook on future weather in deciding which calibration to use without being subjected to the bias associated with a calibration based upon the overall mean of a range of years.

Variations in requirement for an individual nutrient with the supplies of other nutrients were the subject of some of the earliest theories of plant response to limiting factors, as indicated in Chapter 1. Subsequent research has not disproved the basic concept, but rather has revealed more complexity than was appreciated initially.

The desirability of taking into account the supplies of all plant nutrients in making decisions about additions of individual nutrients is generally recognized. Comprehensive soil testing systems that provide availability indexes of a number of nutrients do take the supplies of various nutrients into account. Needs for individual nutrients usually are considered independently, however. Progress in dealing with joint effects has been inhibited by the complexity and experimental difficulty of dealing with many nutrients simultaneously, as well as the general lack of resources needed to address the situation adequately. Additionally, dealing with even one nutrient at a time has not proved simple, as evidenced by the subject matter of this chapter.

The problem is evaluating the way the supplies of individual nutrients interact to produce a net effect on the crop. From the physiological standpoint, an interaction among nutrients is almost always shown by the uptake of increasing quantities of nutrients not under test as increasing quantities of the nutrient under test are added. Significant statistical interactions of crop yield responses, however, are much less frequent when nutrient deficiencies are only moderate or slight than when they are marked, and this helps to support the use of relative yields in intensive agriculture, where extreme deficiencies are rare.

In calibration experiments, the standard practice is to measure the response to a given nutrient in the presence of a "blanket" application of other nutrients that might otherwise limit the magnitude of the response. Use of calibration curves obtained in this way for making inferences about expected responses from soil test values on producers' fields implies that the producers are applying other needed nutrients in quantities sufficient to avoid any significant deficiencies. This situation generally prevails where the ratio of fertilizer cost to crop value is low, but the calibration curves will overestimate the responses if producers apply only the given nutrient. See Section 2-7.9 for further discussion.

4-4.4.4. Field Experiments

Field experiments are necessary for most calibrations. Experiments conducted under greenhouse conditions are rarely a practical substitute where field crops are concerned, although they are appropriate for greenhouse crops. The field experimental approach to calibration varies.

Long-Term Experiments. Cope and Rouse (1973) and Cope and Evans (1985) rated field experiments in which the same treatments are repeated on the same plots for several years as the best for calibration purposes. They attributed three advantages to experiments of this type. First, such experiments provide information on the buildup or depletion of nutrients in the soil with time. Second, the years provide replications, increasing the precision of the measurements. Third, the experiments provide information on seasonal variations.

Gruber (1987) appeared to be of a similar persuasion. He noted that the use of results from short-term experiments has led to mistaken advice on fertilizer use, and argued that because farming is a long-term occupation, spanning generations, the experimental approach also should be long term.

On the other hand, the responses that occur after the first year are responses to a combination of effects of current fertilization and gradually changing residual effects of previous applications. With time, the deficiency of the nutrient under test increases in the control plots and decreases in the fertilized plots. Gruber (1987) gave an example in which potassium deficiency in grass was much greater in the second year of an experiment than in the first. In another experiment on grapes, a large response to potassium occurred only in the sixth and succeeding years. In long-term experiments, unknown deficiencies of other nutrients may increase in the fertilized plots to a greater extent than in the control plots. The response curves change.

Thus, whether the reflection of buildup or depletion of nutrients in years of repeated fertilizer applications on the same plots is to be considered an advantage in calibration depends upon the philosophy of interpretation and recommendations based upon the tests. If the information transmitted to a producer on the basis of a soil test is regarded as appropriate for the average of a period of years similar to that observed in the calibration experiment, the fact that the calibration data reflect cumulative effects is an advantage. If the calibration is intended to represent what the producer is to expect in the first year in which the information is implemented, a calibration based upon first-year results would be preferable. In this situation, the producer ideally would update the information by an annual soil test for a few years.

One-Year Experiments. Placing Gruber's (1987) views under long-term experiments may be inappropriate because he did not refer specifically to long-term experiments. His comment about inappropriate advice derived from short-term experiments is quite conventional if it is viewed in the context of the well known variation in results from one year to the next.

The significant distinction is between calibrations based upon 1-year experiments conducted over a period of years and calibrations based upon 1-year experiments conducted in a single year. A calibration derived from 1-year experiments conducted over a period of years is appropriate for circumstances in which the results are interpreted in terms of application to the first crop following the soil test. One-year experiments used for calibration are of various types. Unfortunately, all too many are experiments conducted for other purposes, with soil testing personnel using the results for what they are worth in the absence of enough other information from experiments designed specifically for calibration purposes.

Standard practice in calibration experiments is to record the kind of crop and the kind of soil because these factors are readily recognized, and they are known to have important effects on the results. This information provides two criteria for classifying the results and improving the precision of the calibrations derived from relating the plant responses in the groups of experiments to the soil test data obtained from applying a uniform test method to samples of soil from all experiments. Additional site-specific factors and seasonal factors, however, result in considerable variation that limits the precision of calibrations.

Statistical System. It is generally recognized that the plant response to supplies of soil and added nutrients at a given site in a given year depends upon a number of factors other than the kind of soil and crop. In theory, if these other factors could be measured and taken into account, it should be possible to improve the calibrations derived from identical sets of comprehensive measurements made at each experimental site. The problem has been how to implement the general concept in a practical way.

Statistical methods are used in treating data from both long-term and 1-year calibration experiments. The statistical computations required for handling the volume and types of information needed for characterizing sites in a comprehensive way, however, are detailed and complex, and so is the field and laboratory work. To keep the process manageable, some compromises must be made, and one must substitute the outcome in the form of a complex statistically fitted equation for the perceptions that may be derived from examining the raw data in the simpler systems.

Hanway (1973) discussed the theoretical background of the statistical system from the agronomic standpoint. Voss et al. (1970) described and illustrated a specific system using statistical methods to integrate soil, management, and environmental factors in a total of 23 1-year experiments in two successive years on two closely related soil types in Iowa. The experimental measurements included soil test values for nitrogen, phosphorus, and potassium in the surface soil and subsoil; response of the corn crop to application of six levels of each of these nutrients supplied by fertilizers; management factors, including past cropping, planting date, plant population, and grassy weed infestation; and a daily index of relative photosynthesis (water deficiency stress) based upon daily precipitation, evapotranspiration, and the available water content of the soil. The photosynthesis index was intended to integrate the weather factors that affect the supply of water to the crop. Measurements of surface soil test values, plant population and barrenness, and weed infestations were made on an individual plot basis.

The unconventional experimental design involved 25 plots per location, with only five treatments replicated. Each location was treated as a replicate in a single large experiment in which the number of replicates could be varied as circumstances permitted.

Analysis of the data from the overall experiment including the replicates at 23 locations showed that the nutrients nitrogen, phosphorus, and potassium accounted for only 20% of the grain yield variance, but that the overall equation accounted for 80% (the yield range was from about 1.4 to 8.3 megagrams per hectare). From the overall equation (see Voss, 1969) could be derived calibrations for the three nutrients plus estimates of the effects of management factors that usually are handled by adjustments of the calibration in the process of interpretating the results of a soil test for individual conditions.

In the analysis made by the authors, soil types and subdivisions due to slope and erosion did not appear in the final equation. Substituting for effects of these factors were estimates of the corn yield potential on the various sites that had been entered from independent soil survey sources.

Certain improvements in the system used by Voss et al. (1970) would seem desirable in further work. Adjustments in quantities of nutrients applied would be helpful. Moreover, under current conditions, in which fertilizer nitrogen is relatively economical and the supplies of phosphorus and potassium in many soils have been increased substantially as a consequence of repeated fertilization and residual effects, practical emphasis is focused upon the upper part of the response curve. The quadratic equation used by Voss et al. (1970) does not perform as well in this range as do some other equations, and so a better choice of equation probably could be made. The inverse polynomial equation proposed by Greenwood et al. (1980) is a possibility. Soil test values for nitrogen, phosphorus, and potassium can be entered in this equation, as in the others mentioned in Section 4-4.2. The Greenwood equation has been found to represent interaction effects to a good approximation.

4-4.5. Interpretation

The subjects discussed in the preceding section on calibration all represent objective aspects of relating soil test values to nutrient availability and crop response to fertilization, although subjectivity is involved in decisions on whether or not the practices are followed, and on the magnitude of the adjustments. Superimposed on the calibration process, however, are differences in interpretation that stem from philosophies about goals. These are subjective and can have a considerable effect upon the recommendations developed from given numerical values for soil tests. Olson et al. (1982, 1987) and Dahnke and Olson (1990) described the buildup and maintenance, sufficiency level, and optimum cation saturation ratio concepts as the three major philosophies used by U.S. soil testing laboratories.

4-4.5.1. Buildup and Maintenance Philosophy

According to the buildup and maintenance philosophy, supplies of nutrients in soils should be increased in 1 or 2 years to high soil test levels and then maintained by adding each year the quantities of nutrients the crop is expected to remove regardless of the soil test level. This philosophy sometimes is referred to as "fertilizing the soil." Although the philosophy applies to all nutrients, it seems to be applied in practice to only nitrogen, phosphorus, and potassium.

From the economic standpoint, fertilizing the soil with quantities of nutrients sufficient to produce the maximum yield would be the most beneficial policy for the producer if the fertilizers were free and could be applied without cost. Although the cost of fertilizers is generally low relative to crop values in developed countries, the optimum economic quantities are still below those that will produce the maximum yield. In less developed countries, fertilizers are more expensive relative to crop values, and the quantities that can be applied economically are considerably smaller. See Chapter 2 for a discussion of the economics of fertilization.

The idea of maintaining the supply of a nutrient in a soil by adding enough of a nutrient in the form of fertilizer to equal the amount of the nutrient removed from the soil in the harvested portion of a crop is an appealing concept. Consistency with the buildup philosophy of increasing the supply of nutrients in the soil to a high soil test level would suggest that the objective of the maintenance philosophy is to maintain the soil tests at the higher levels attained through the buildup. In direct field experimental tests by Fixen and Ludwick (1983) in Colorado and McCallister et al. (1987) in Nebraska, however, the apparent quantities of phosphorus and potassium required to maintain the soil test levels of these nutrients did not agree with the quantities removed in the crops.

Adding the quantities of nutrients removed in crops would be expected to fail to maintain the soil test levels under circumstances in which nutrients are lost from soils in ways other than crop removal. Moreover, the soil test levels for phosphorus and potassium may drop after an abrupt buildup even in the absence of removals from the soil. Under such circumstances, maintaining the soil test levels would be expected to require additions exceeding the quantities of nutrients removed in crops.

On the other hand, if the soil organic nitrogen is being mineralized fast enough to supply crop needs, adding more fertilizer nitrogen will contribute nothing to the crop-producing capability of the soil but will merely add to the possible losses from the soil. And in general if the level of a nutrient in the soil is high enough so that additions do not produce crop responses, an economic loss to the producer results from adding the nutrient even though all of it may remain in the soil.

Realistically, there is no fixed maintenance requirement. Nevertheless, the concept of maintenance applications is important and deserves careful attention because the supplies of phosphorus have been greatly increased in many soils of developed countries. In some situations, producers would be ahead financially to cease applications until the potential responses would justify the costs. Adding the kinds and quantities of fertilizers needed to maintain crop yields at an optimum economic level for the individual producer is a rational and practical objective for maintenance applications as well as for applications when greater deficiencies exist.

Theoretically, repeated fixed annual additions of a nutrient at a level below that corresponding to the maximum yield will cause crop yields to move gradually toward an equilibrium value relative to yields obtained with additions of other quantities of the same nutrient. When equilibrium has been attained, one can plot the yields (or relative yields) against the annual additions to obtain what Helyar and Godden (1977) termed a *maintenance requirement curve*. This curve is different from the initial response curve because the equilibrium yield associated with each annual addition (the maintenance requirement) on the X axis represents a summation of a first-year effect plus the accumulated residual effects of an equal addition in each previous year. Fig. 4-25 illustrates the concept using experimental data from Rossiter (1964) as reanalyzed by Helyar and Godden (1977). Note that the annual addition of fertilizer phosphorus required to produce a given relative yield at equilibrium (the maintenance requirement) is much smaller than the first-year requirement.

If desired, one could develop a second-year maintenance requirement, a third-year maintenance requirement, and so on. The curves for successive years would differ because of differences in cumulative residual effects. As the equilibrium yields were approached, the curves would draw more closely together.

The important principle to be derived from Fig. 4-25 and the equilibrium concept described by Helyar and Godden is that the maintenance requirement varies with the equilibrium yield or relative yield that is selected for maintenance. Furthermore, up to the time at which the equilibrium yield has been established, the maintenance requirement varies with the number of equal-quantity applications of fertilizer that have been made.

Cornforth and Sinclair (1982) applied the maintenance concept to the phosphorus requirement of permanent pastures in New Zealand that had received phosphorus-bearing fertilizers for a number of years and were in an approximately steady state such that the fertilizer phosphorus added for maintenance would be equal to the sum of the removal of phosphorus in the animals and the loss from the soil due to fixation, immobilization in organic forms, and erosion. A provisional soil loss factor of 0.1, 0.25, or 0.4 was assigned according to the soil group concerned. Losses through the animals were estimated by more detailed calculations. The loss that had to be replaced for maintenance increased with the herbage yield (carrying capacity) and stocking rate. Sinclair and Rodriguez Juliá (1993) developed an updated version of the model for presentation at the 1993 International Grasslands Congress.

The maintenance requirement concept illustrated in Fig. 4-25 for phosphorus applies to nutrients in general. The difference between the first-year response curve and the equilibrium maintenance requirement curve, however, will depend upon the factors that affect the retention of the nutrient by and its loss from the soil. For nitrogen, the curves usually will be much closer together than the curves for phosphorus. Curves for potassium will be intermediate.

Although a maintenance requirement curve cannot be found experimentally for each practical situation in which maintenance requirement information is needed, soil test values increase and decrease with the supplies of nutrients in

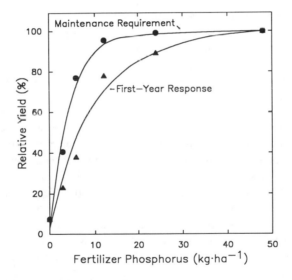

Fig. 4-25. Relative yields of pasture vegetation with different quantities of phosphorus applied annually as superphosphate in Australia. The yields obtained with 48 kilograms of fertilizer phosphorus per hectare are given a relative value of 100. The maintenance requirement curve represents the relative yields in years 9 to 13 of the experiment, by which time the relative yields with each quantity of annually applied fertilizer phosphorus had become approximately constant from one year to the next. The experimental data were published by Rossiter (1964), and the maintenance requirement curve and related theory were developed by Helyar and Godden (1977).

individual soils. For example, Karlovsky (1961) noted that after repeated annual topdressings of New Zealand pastures with constant quantities of superphosphate, the soil test values for phosphorus, like the relative yields, tended to remain constant with time, but to increase with the quantities of superphosphate supplied each year. Similarly, in analyzing the results of long-term field experiments on soils derived from loess in Germany, Munk and Rex (1990) found that soil test values for phosphorus increased with additions of fertilizer phosphorus and decreased with plant removal of phosphorus. Properly calibrated soil tests may be used with equal validity, whether the result is interpreted as a maintenance requirement or merely as a fertilizer requirement.

As with maintenance requirement curves based upon crop yields, the interpretation of soil test values in terms of maintenance requirements must be done with perception. Prior to attainment of equilibrium, the significance of a given soil test value in terms of maintenance applications will depend upon the number of years of continued constant annual applications. Proper calibrations are important, and periodic soil tests are needed to monitor the prognosis. Cornforth and Sinclair (1982) published a graph with their paper on the phosphorus maintenance requirement of New Zealand pastures showing how repeated soil test values for phosphorus could be used in conjunction with the relative yield level selected for maintenance as a guide to fertilization.

4-4.5.2. Sufficiency Level Philosophy

According to the sufficiency level philosophy, the objective of fertilization is to add enough nutrients to produce the economic or yield goal of the producer. No fertilizer is recommended if the soil test is at the level at which no economic response or no yield response is expected. This interpretation of soil tests sometimes is referred to as "fertilizing the crop."

Fig. 4-26 shows the numerical results obtained by following the recommendations made by five different soil testing laboratories on the basis of tests made on subsamples of composite samples from sites on three experimental fields in Nebraska. Laboratories 1 through 4 were commercial, and they handled about 80% of the soil testing business in the state at the time the work was done. Laboratory 5 was the Soil Testing Laboratory at the University of Nebraska. Although the research was done by scientists at the University, the samples were submitted as farmers' samples to all the testing laboratories, including the one at the University. The laboratory personnel thus were unaware that the consequences of their recommendations were being investigated and compared with others.

The irrigated corn used in the experiments responded strongly to fertilization at all sites, but the yields of corn fertilized according to the recommendations of the individual laboratories did not differ significantly except for one site, at which the yield obtained with the recommendations from Laboratory 3 was slightly below the others (Fig. 4-26A). Large differences were found, however, in the quantities (Fig. 4-26B) and costs (Fig. 4-26C) of the nutrients recommended. The costs of the nutrients recommended by Commercial Laboratory 3

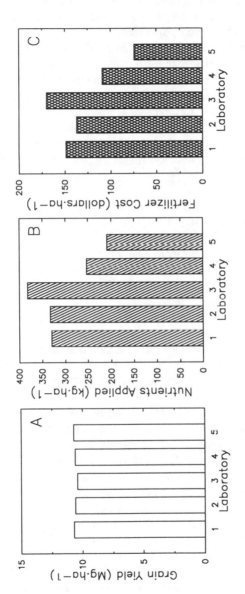

Fig. 4-26. Comparative results obtained by following the recommendations of five soil testing laboratories, as found experimentally over a period of years at three locations in Nebraska. Laboratory 5 was the Soil Testing Laboratory at the University of Nebraska, and the others were commercial laboratories. The crop was corn grown continuously. A. Average annual crop yields. The average control yield was 5.8 megagrams per hectare. B. Average annual applications of plant nutrients. C. Average annual fertilizer costs. (Olson et al., 1982)

were more than twice as great as the nutrients recommended by Laboratory 5, which was the University Laboratory.

According to Olson et al. (1987), all of the commercial laboratories appeared to be using the buildup and maintenance philosophy. The nutrient applications recommended by three of the laboratories suggested that they were also using the optimum cation saturation ratio philosophy.

The results of similar research were reported by Ewanek (1983) in Manitoba. In the Manitoba work, samples taken from ten locations were tested in the Provincial Laboratory and two commercial laboratories. Over all locations, the recommendations of the Provincial Laboratory resulted in the highest mean yield of wheat and barley, the lowest fertilizer cost, and the greatest return over the cost of the fertilizer. The range in yields, however, was only from 2.86 to 2.92 megagrams per hectare. The greatest relative range was in fertilizer cost, which was approximately half as great for the Provincial Laboratory as for one of the commercial laboratories.

4-4.5.3. Optimum Cation Saturation Ratio Philosophy

The optimum cation saturation ratio philosophy is credited to Bear et al. (1945), who suggested on the basis of nonspecific evidence from an experiment in which eight cuttings of alfalfa were taken from each of 20 soils in the greenhouse that in an ideal soil, "65 per cent of the exchange complex should be occupied by Ca, 10 per cent by Mg, 5 per cent by K, and 20 per cent by H." They noted that a New Jersey loam soil with these exchangeable cations would have a pH value of about 6.5. Graham (1959) later proposed the use of saturation ranges of 65 to 85% for calcium, 6 to 12% for magnesium, and 2 to 5% for potassium. Data supporting these ranges were not cited or included in the publication.

Although only two of 43 U.S. university soil testing laboratories surveyed by Eckert (1987) were using the optimum cation saturation ratio philosophy for potassium and five for magnesium, McLean (1977) cited a survey by Darst showing that essentially all of the commercial and industrial soil testing laboratories were using this philosophy. The great attractions of the concept of adjusting the exchangeable potassium, magnesium, and calcium in soils to an optimum ratio appear to be that as far as is known, the figures given by Bear et al. (1945) are indeed adequate for soils and crops in general and that no calibration against field experiments is needed. The drawback is the unnecessary cost to the producer who applies the recommendations.

If the originators of the concept of an optimum exchangeable cation ratio with specific numerical values had known the use that would be made of their statement, they might have rephrased it to eliminate the specificity. Their own previously published data (Hunter et al., 1943) showed that the highest yield of the first cutting of alfalfa was obtained with a ratio of exchangeable calcium to potassium of 32 to 1 on a charge basis and that over the range of ratios from 2 to 1 up to 16 to 1 the yields were only slightly lower and approximately equal. And Hunter (1949) later found that alfalfa yields did not differ significantly even though the ratio of exchangeable calcium to magnesium ranged from 0.25 to 1 up to 32 to 1. Moreover, a paper published in the same year as the bulletin containing the postulated optimum exchangeable cation ratios indicates that the authors held no illusions about the sanctity of the numbers they had suggested for the ratio of exchangeable calcium, magnesium, and potassium in an ideal soil. In that paper, which reported different parts of the data from which the optimum cation ratio concept apparently was derived, Bear and Prince (1945) argued that

> The evidence supports the belief that each of these cations has at least two functions in the plant, one specific and the other or others of the type that can be performed interchangeably by all three cations. Once the supply of each cation is adequate to meet the specific need for it, there can be a wide range in ratios in the remaining quantities that are absorbed by the plant to meet its total cation needs.
>
> . . . It would appear that the soil on which it [alfalfa] is to be grown should be fortified with an abundance of Ca and Mg in preparation for seeding, but that the K applications should be governed by the specific annual needs of the plant, a suitable application being made at seeding time and additional quantities being supplied each year the crop is allowed to continue on the same land.
>
> Because of alfalfa's tendency to accumulate K in excess of its critical need for it, difficulty is experienced in maintaining an adequate supply of this element in the soil. The annual application of K must be sufficient to maintain the K content of the plant at not less than 1%, but it should not be so large as to effect a substitution of K for Ca and Mg in the functions that are common to all three cations in the plant.

The last paragraph of the quotation contains elements of economics because it generally is less costly to supply calcium and magnesium in the form of dolomitic limestone than to supply potassium. To put the views of Bear and Prince (1945) in another way, it is of first importance to have enough of each of the three cations for the specific functions they perform. Once the plant has enough, the ratios are not of much importance. This last statement must be qualified where ruminant nutrition is concerned because a low magnesium content of forage may result in magnesium deficiency or "grass tetany" in the animals even though the content of magnesium in the plant tissues may be adequate for the plants.

Relatively little experimental work has been done to test the concept of an optimum ratio of exchangeable calcium, magnesium, and potassium that should be used for interpreting soil tests. The evidence available, including original data and literature citations in papers by McLean (1977), McLean et al. (1983), and Liebhardt (1981), supports the quotation just made from the paper by Bear and Prince (1945).

4-4.5.4. Other Philosophies

A philosophy not included in the major group by Olson et al. (1982, 1987) and Dahnke and Olson (1990) is that of increasing the recommended application of fertilizer beyond that indicated by the experimental data to compensate for the fact that losses to the grower from using too little fertilizer are greater than those from adding more fertilizer than is needed (Cope and Rouse, 1973). This philosophy derives from the fact that response curves are steeper below the economic optimum application than above. For example, Neeteson (1985) calculated from the results of an experiment on potato in the Netherlands that the financial loss from applying less than the quantity of fertilizer nitrogen required to yield the maximum net profit (230 kilograms per hectare) by 50 kilograms was five times as great as the loss from applying the same quantity in excess of the optimum. See also a graph by Bock and Hergert (1991, p. 162) illustrating the principle on the basis of data from experiments on nitrogen fertilization of corn in Nebraska. Cope and Rouse (1973) and Cope and Evans (1985) considered that recommending a little extra fertilizer would be advantageous as a safety factor to compensate for inaccuracies in sampling, analysis, and interpretation. It would also aid in maintaining high fertility because a part of the unneeded fertilizer would have residual value later.

Still another philosophy was described by Dahnke and Johnson (1990). For rainfed crops in semiarid to subhumid areas where loss of mineral nitrogen by leaching is not a problem, they suggested adding fertilizer nitrogen in a quantity that would be appropriate for the highest crop yield that has been observed in the past 5 to 10 years. This amount of nitrogen would help to assure that if the season is a good one, economic returns will not be sacrificed for lack of nitrogen. For the majority of years, which will not be good ones because of deficient precipitation, the nitrogen supplied will be in excess of crop needs. But most of the excess will remain in the soil and can be allowed for on the basis of a soil test for mineral nitrogen for the following crop.

Whatever philosophy is used, applications of phosphorus fertilizers probably will be great enough to increase the residual phosphorus supply in soils in general because of the low recovery of the added phosphorus in crops and the retention of phosphorus by soils. From a simulation study of the buildup of residual phosphorus in pastures in Australia, Barrow (1982) concluded that phosphorus reserves would accumulate if the annual addition of fertilizer phosphorus exceeded one-third of the phosphorus required to produce the maximum yield on a virgin soil.

An increase in residual potassium is less certain, and especially in humid regions an increase in residual nitrogen is still less certain. Nitrogen fertilization will increase the nitrogen returned to soils in crop residues, but this may not balance the loss of nitrogen from soil organic sources.

In their long-term field experiments with cotton and corn, Cope and Rouse (1973) found that when they applied the economic optimum quantities of fertilizer phosphorus and potassium, the residual effects increased the soil test values for phosphorus to the "high" category (no response to fertilizer phosphorus) in a few years on all soils. The same was true for potassium on soils with cation exchange capacities exceeding 5 centimols of charge per kilogram. In their area characterized by high winter rainfall, most soils with lower cation exchange capacities would not retain enough potassium from one year to the next to keep them in the high category. For such soils, most crops needed to receive fertilizer potassium each year.

4-4.6. Nitrogen

Although numerous brief references to nitrogen are made elsewhere in this chapter, this section is devoted specifically to nitrogen because of its importance and because the behavior of this nutrient is so different from the others, on which emphasis has been placed. The unique behavior of nitrogen leads to a unique set of problems where soil testing is concerned.

Most soil nitrogen occurs in organic forms in the soil solids. A small amount of organic nitrogen is found in the soil solution, but plants are not known to take up any significant part of it, and must depend upon the inorganic forms that are released when microorganisms decompose the compounds containing the organic forms.

The inorganic combined nitrogen in soils usually occurs mostly as nitrate, which behaves quite differently from phosphate. Phosphate is bound to soil particles, occurs only in very low concentrations in the soil solution, does not move downward to a significant degree except in the sandiest of soils, and generally moves no more than a millimeter or two to root surfaces. In most soils, nitrogen present as nitrate is found in the soil solution, leaches readily, frequently occurs in concentrations 100 or more times as great as phosphorus, and moves through the soil to roots from distances measured in centimeters. In some soils, typically strongly weathered acid soils of the tropics, positive charges hold some nitrate in exchangeable form, and this retards its movement and loss by leaching (Wong et al., 1990).

The importance of nitrogen and nitrogen fertilizers in agriculture should guarantee this element first priority among nutrients in terms of soil testing efforts, but the fact is that in some laboratories no tests are made for nitrogen. The reason is not a lack of capability to make chemical analyses for the mineral forms of nitrogen used by plants. These forms are well known, and their concentrations can be determined readily by chemical analysis. Rather, the reason is that the fleeting existence of mineral forms of nitrogen in soils of humid regions, the vagaries of net mineralization of nitrogen, and the time required to make the mineralization measurements decrease the utility of chemical soil tests. As a substitute, the content of soil organic matter as found by a rapid approximate method is used in some laboratories as an index of mineralization of nitrogen from the stable organic nitrogen in the soil. Keeney (1982) tabulated the procedures used in U.S. university soil testing laboratories, and Meisinger (1984) reviewed the literature on indexes of nitrogen availability.

Where no chemical tests are made, the recommendations for nitrogen fertilization are based upon a combination of experience developed from field experiments, knowledge of the nitrogen requirements of the crops, mineralization of organic nitrogen as influenced by the preceding crops and applications of animal manures, and other factors. A review by Becker and Aufhammer (1982) illustrates the wide range of factors that may be used to improve the prognosis.

4-4.6.1. Mineral Nitrogen

As a basis for an index of nitrogen availability, one ideally should have information on the amount and vertical distribution of mineral forms of nitrogen in the soil throughout the growing season of the crop. Measurements of this type can be made while the crop is growing, but much of the information would be available too late to be of greatest practical value. If information on the nitrogen supply is desired during crop growth, the usual alternative is to make the measurements on the crop, a subject discussed in Chapter 3. Winter cereals are an exception. For these crops, measurements of soil mineral nitrogen may be made in the spring before much growth occurs.

Before the use of fertilizer nitrogen in relatively large quantities became widespread, the principal emphasis in developing nitrogen availability indexes was on nitrogen mineralization. The amounts of mineral nitrogen in soils were generally low. Mineralization and fertilization were regarded as the major sources for crops. At present, fertilization and residual mineral nitrogen are often the major sources, and in soil testing the primary emphasis is on residual mineral nitrogen. This shift in emphasis is consistent with findings by Matar et al. (1990) in 40 field experiments on wheat in Syria. Where soil mineral nitrogen was low, organic nitrogen mineralization was the preferred index of nitrogen availability. Where soil mineral nitrogen was higher, mineral nitrogen was the preferred index.

Except for some question about the significance of exchangeable ammonium, which usually is only a minor portion of the mineral nitrogen in soils, measurements of mineral nitrogen do not present the problem of knowing what is being measured, as is true for most other nutrients. The principal concern has to do with positional availability and losses that may occur after the measurements have been made. This problem has been addressed by Burns, whose work is discussed in Section 7-6.

For practical soil testing purposes, measurements of mineral nitrogen in the soil before the crop is grown (or in the early spring for winter cereals) in semiarid regions often are made at present (Keeney, 1982; Hergert, 1987; Dahnke and Johnson, 1990) because, except for very sandy soils, loss of this nitrogen from the soil by leaching while the crop is growing is unlikely. This practice is also becoming standard in humid regions in a number of countries, except that the tests for mineral nitrogen often are made after the crop is in place — late enough so that the preseason loss of mineral nitrogen by leaching has already taken place, and late enough to reflect early-season nitrogen mineralization, but not so late that a topdressing or sidedressing of fertilizer nitrogen is impractical. Where feasible, such measurements can be more accurate than predictions from models involving measurements of conditions during the autumn, winter, and early spring that affect mineralization, leaching, and denitrification. Meijer and

Vreeke (1988) found that the economic optimum quantity of fertilizer nitrogen for ryegrass in the Netherlands was closely related to the content of mineral nitrogen in the soils to a depth of 90 centimeters in the early spring.

The importance of sampling time in humid regions is indicated by the results of experiments in Germany by Maidl and Fischbeck (1986), who found that the coefficients of determination (R^2) for the relationship between the nitrate contents of the surface 60 centimeters of soils and the yields of sugar in sugarbeet grown on the soils were 0.03, 0.12, 0.64, 0.59, 0.00, and 0.00 for soil samples taken in February, March, April, May, June, and August, respectively. Walther (1983) observed similarly that the value of soil mineral nitrogen measurements for predicting the nitrogen need of sugarbeet in Switzerland was greater if the measurements were made in May than if they were made in March.

Various alternatives are being used to accomplish the analyses and deliver the resulting information in the short time available for decision-making and fertilizer application. According to Magdoff et al. (1990), special delivery services to rush the samples to the laboratory are used in some states in northeastern United States. In one state, hand-held meters are used to measure the nitrate in solutions extracted from soils in Extension Service offices in counties. Commercial kits that can be used by producers to make their own tests are also being marketed.

Part of the predictive value of measurements of mineral nitrogen in the soil in the spring derives from the fact that some and sometimes almost all of the mineral nitrogen present at that time is a consequence of organic nitrogen mineralization. The mineral nitrogen found by analysis thus will be correlated to some extent with the nitrogen mineralization that will occur later during the growing season.

4-4.6.2. Organic Nitrogen Mineralization

Laboratory-derived indexes of mineralization of soil organic nitrogen are of two general types. One, discussed in this section, is tests in which the increase in content of mineral nitrogen in the soil is determined when a sample of soil is incubated in the laboratory under conditions of water supply and temperature favorable for nitrogen mineralization. A second type, discussed in the next section, is chemical measurements that do not depend upon the action of soil microorganisms on the soil organic nitrogen.

Conventionally, the mineral nitrogen that appears in exchangeable and soluble forms in soil during incubation or cropping is attributed to the release of ammonium from organic forms. No measurements are made to determine whether or to what extent ammonium has been released from or has disappeared into nonexchangeable forms. From the standpoint of plant nutrition, the source of a given ammonium ion is immaterial. From the theoretical standpoint, however, the inorganic process must be kept in mind along with the organic process. The fixation and release of ammonium may be significant in some circumstances and of no importance in others (Mengel and Scherer, 1981; Saha et al., 1982;

Mengel et al., 1990). Recently added ammonium that has become nonexchangeable is released more readily than nonexchangeable ammonium that has been in place for a long time (Wehrmann and Eschenhoff, 1986).

Where soil nitrogen mineralization is to be used as an index of nitrogen availability to plants, the best results would be expected from measurements of the sum of the mineral nitrogen present initially and the mineral nitrogen produced by soil microorganisms during incubation of a sample of moist soil in the laboratory. Incubation methods, however, have two principal practical drawbacks. One is that they are slow. Incubations generally last 1 or 2 weeks and sometimes longer. Tying up equipment and space for that length of time would be inconvenient for a soil testing laboratory processing many samples. A second disadvantage is that the values obtained are sensitive to the sample pretreatment and to the conditions of incubation. See a review by Bremner (1965) for an enumeration of the problems he perceived. These factors have some differential effects on different soils that decrease the correlations with the biological indexes of nitrogen availability measured in the field.

Laboratory measurements of soil nitrogen mineralization sometimes give poor results. In experiments on 53 soils in Germany, Wehrmann et al. (1988) found that they were unable to predict the amounts of nitrogen mineralized under field conditions. They tried four soil incubation methods, four extraction methods, and three modeling approaches, but none were satisfactory. The nitrogen mineralized under field conditions was estimated from plant analyses and from measurements of mineral nitrogen accumulation in fallow soils protected against leaching. Also in Germany, Müller and Feyerabend (1982) found that under bare fallow the increase in mineral nitrogen in the surface 90 centimeters of soil during the vegetative period averaged about 100 kilograms per hectare, but that about 80% more mineral nitrogen might accumulate in a favorable year. When the fallow soil was covered, an extra 30 to 40 kilograms of mineral nitrogen accumulated, presumably in the main because of reduced loss by leaching. In the Netherlands, Van der Meer and Van Uum-van Lohuyzen (1986) were unable to predict nitrogen uptake by grass from peat soils with a high and variable water table, and they attributed their failure to the effect of the water table level on nitrogen mineralization.

On the other hand, in a total of 18 field experiments on grassland conducted over a 3-year period, Van der Meer and Van Uum-van Lohuyzen (1986) obtained a correlation of $r = 0.84$ between the yield of nitrogen in the herbage and the mineral nitrogen in the root zone in early spring. When they added the nitrogen mineralized in the laboratory in 2 weeks at 30° C to the initial mineral nitrogen, the correlation was $r = 0.90$. Clement (1971) in the United Kingdom and Dalal and Mayer (1990) in Queensland also obtained good correlations between measurements of nitrogen mineralization in the laboratory and crop yield or nitrogen uptake in the field. Dahnke and Johnson (1990) cited a number of other researches in which the relationship between biological indexes of nitrogen availability in field or greenhouse and nitrogen mineralization in the laboratory was investigated.

Stanford and Smith (1972) developed a method for estimating the potentially mineralizable nitrogen in soils from the rates of nitrogen mineralization. The method, described more recently by Stanford (1982), involves measuring the nitrogen mineralized in successive incubations of soil samples with removal and determination of the accumulated mineral nitrogen at the end of each incubation period.

In their original paper on potentially mineralizable nitrogen, Stanford and Smith (1972) noted that with most of the soils they tested, the cumulative nitrogen mineralization was approximately proportional to the square root of time within the range from 4 to 30 weeks. If cumulative nitrogen mineralization were indeed proportional to the square root of time, the amount of nitrogen mineralized would increase indefinitely with the time, a behavior that is unrealistic. Stanford and Smith accordingly fitted to their data the equation for a first-order chemical reaction that represents the cumulative nitrogen mineralization under the hypo- thetical condition that a single source of nitrogen is undergoing mineralization at a rate proportional to the quantity of this source remaining in the soil. This is an exponential relationship analogous to the Mitscherlich equation:

$$N_t = N_p(1 - e^{-kt}),$$

where N_p is the nitrogen mineralization potential or potentially mineralizable nitrogen, N_t is the cumulative amount of nitrogen mineralized at time t, and k is a mineralization rate constant. Stanford and Smith (1972) showed the equation in the logarithmic form:

$$\text{Log}(N_p - N_t) = \text{Log}N_p - kt/2.303.$$

Working with previously published data on samples of 30 soils of Brazil, Parentoni et al. (1988) found that the functional relationship used by Stanford and Smith (1972) provided a good fit to the cumulative mineralization with time. The coefficients of correlation exceeded $R = 0.99$ for all soils. With 14 soils of Argentina, Sierra and Barberis (1983) obtained correlation coefficients ranging from 0.62 to 0.92 for the individual soils. Juma et al. (1984) and El Gharous et al. (1990) fitted both the exponential and a hyperbolic equation to minerali- zation data and concluded that both would fit their data well, although the estimates of potentially mineralizable nitrogen differed.

Others have used different equations. Molina et al. (1990) and Deans et al. (1988) used an exponential equation representing the sum of two first-order reactions. Richter et al. (1982) fitted an equation representing the sum of three reactions.

The problem Stanford and Smith (1972) encountered in fitting short-term mineralization to a long-term trend used to estimate the potentially mineralizable nitrogen is understood at present to result from the presence of fractions of organic matter that decompose at different rates (Nordmyer and Richter, 1985; Warren and Whitehead, 1988; Freytag, 1990), together with the standard practice

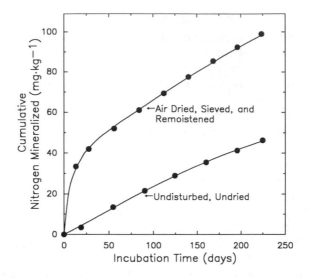

Fig. 4-27. Cumulative mineralization of nitrogen at 35° C in a soil from Kansas as measured on samples that were undisturbed and undried and on others that were air dried, sieved, and remoistened. (Cabrera and Kissel, 1988a)

of using soil samples that have been air dried, sieved, mixed, and remoistened for mineralization studies.

Nordmyer and Richter (1985) found that if they incubated undisturbed columns of soil that had been kept moist, the cumulative nitrogen mineralization increased almost linearly with time. The relatively rapid initial accumulation of mineral nitrogen that showed so markedly in air-dried and rewetted soil had almost disappeared. Cabrera and Kissel (1988b) similarly found that the mineralization rates in disturbed samples exceeded those in undisturbed samples manyfold during the first 4 weeks, but only slightly after that time (Fig. 4-27). The net mineralization in the undisturbed soil they described as a typical example increased almost linearly with time (that is, the rate was almost constant), and they referred to work of others with undisturbed, undried samples showing similar behavior.

When Cabrera and Kissel (1988a) omitted the mineral nitrogen produced during the first 4 weeks, they were able to estimate the nitrogen mineralized in the undisturbed samples to a good degree of precision from the nitrogen mineralized in the disturbed samples, together with the clay content of the soil (the overprediction of the rates of mineralization in the field increased with the ratio of clay to total nitrogen in the soils). In a preliminary test, they were able to estimate the mineralization rates in the field to a good approximation from values of mineralization found in the disturbed samples in the laboratory after subtracting the mineralization during the first 4 weeks, and after applying corrections for the clay content of the soil, the water content, and the temperature. This work extends the range of conditions within which rates of mineralization in undisturbed soil in the field may be estimated from rates of mineralization in disturbed samples in the laboratory, but it still leaves untouched the difference in short-term mineralization associated with drying and sieving, which appears to be the most important effect of all.

Raison et al. (1987) found that when they sampled field soil, sieved it at field moisture content, placed the sieved soil in polyethylene bags, and buried the bags in the field at the same depths from which the samples were collected, the rates of mineralization were far greater than those in adjacent undisturbed soil. In this instance, sieving moist soil without drying had a major effect on mineralization.

Hatch et al. (1990) described a method to estimate the total mineralization of nitrogen in soil under field conditions by incubating buried soil cores in the field at field moisture content in sealed jars containing some acetylene to inhibit ammonium oxidation and to inhibit denitrification beyond the nitrous oxide state. The method requires measurements of ammonium, nitrate, and nitrous oxide. Samples must be analyzed frequently and replaced with newly taken samples to limit the effects of changes in water content that take place due to precipitation, evaporation, and water use by plants. The authors used 2-week intervals. The method is too costly in time for practical purposes, but it could be useful for calibrating measurements made in more convenient ways.

Indications thus are that reliable estimates of mineralization rates in the field can be made from rates of mineralization in the laboratory on undisturbed, undried samples if appropriate adjustments are made for the effects of differences in temperature and water content. Undisturbed samples unfortunately are far more difficult to take and to deal with in the laboratory than are disturbed samples. Moreover, undisturbed samples must be incubated individually. They cannot be composited. Although the 30-week incubation period used in some of the experimental work to estimate the potentially mineralizable nitrogen is not necessary, the time required to obtain the information is still discouragingly long for practical soil testing work.

The most likely practical use of laboratory measurements of nitrogen mineralization on undisturbed soil samples appears now to be for obtaining improved estimates of nitrogen mineralization for use in computer simulations of the soil-plant relationships of nitrogen, such as the one by Addiscott and Whitmore (1987) that is being used for regionalized advisory purposes in the United Kingdom. The measurements could be justified for such a purpose, but they probably would be too time-consuming and expensive for use on individual field samples.

Problems thus remain in the use of mineralization rates in the laboratory to estimate mineralization rates in the field. The successes recorded for use of estimates based upon mineralization rates in air dried, sieved samples probably were obtained under circumstances in which differential effects of sample preparation upon mineralization were small. This situation would be expected to be approached most closely where the field measurements are confined to a single experiment on a cultivated sandy soil with low organic matter content, where the soil remains moist, and where the measurements are confined to the surface layer of soil, in which cultivation produces in the field some of the enhancement of mineralization rate that is associated with sieving of samples used for the mineralization measurements in the laboratory.

4-4.6.3. Chemical Indexes

In an attempt to circumvent the incubation time involved in conventional nitrogen mineralization studies, numerous methods and modifications of methods have been proposed for obtaining a chemical index of potentially mineralizable nitrogen. These have been reviewed by Bremner (1965), Keeney (1982), and Stanford (1982). Most methods have been highly empirical. In a more systematic approach, Keeney and Bremner (1964) looked for chemical forms of nitrogen in soils that might be readily attacked by microorganisms, but their results were disappointing. No fraction was found that disappeared much faster than the others.

Jenkinson (1966, 1968) obtained a lead from an experiment in which ryegrass labeled with radioactive carbon was allowed to decompose in soil for a year or longer. A small fraction of organic carbon that appeared to be present in the tissue of living microorganisms lost carbon faster than the remaining organic carbon. He used the amounts of polysaccharide (estimated as glucose) extracted from soils by 0.05 molar barium hydroxide, 0.01 molar sodium bicarbonate, and boiling water as indexes of the mass of microbial tissue in soils and as indexes of the rates at which the soils would produce mineral nitrogen in the absence of fresh organic material. He found that the amounts of polysaccharides extracted from soils by these reagents were correlated more highly with the nitrogen mineralized during incubation than were the amounts of nonnitrate nitrogen extracted by the same reagents. In 36 field experiments with barley, he found that the amounts of polysaccharides extracted by the barium hydroxide reagent were correlated more highly with the yields of the control plots and responses to added fertilizer nitrogen than were the values for nitrogen mineralized during incubation.

Jenkinson's methods eliminate the incubation for mineralizable nitrogen, which is undesirable as a routine operation in a soil testing laboratory, and in his work they appeared to do so without loss of information on mineralizable nitrogen. Some success with modifications of his methods has been reported by other investigators, but results have been inconsistent. Whitehead (1983) found that the polysaccharides measured by Jenkinson's method were more highly correlated with nitrogen mineralization in field experiments than with chemical indexes of nitrogen availability, but Smith et al. (1977) found that nitrogen mineralization in soils in the field was more highly correlated with laboratory measurements of potentially mineralizable nitrogen (N_p) than with the polysaccharides measured by Jenkinson's method. Stanford (1982), who reviewed the findings made by chemical methods in general, concluded that extractions with mild reagents, including those employed by Jenkinson (1968) showed the greatest promise.

Jenkinson (1968) pointed out that as indexes of nitrogen mineralization, estimates of microbial biomass or related entities, such as the microbial nitrogen, have an inherent weakness in that microbial tissue causes nitrogen immobilization as well as mineralization. In soils containing enough decomposable organic matter with a wide ratio of carbon to nitrogen, the net effect of a given amount of microbial tissue will be nitrogen immobilization rather than mineralization.

Fox and Piekielek (1978a,b) tested a number of chemical indexes of nitrogen availability by comparing them with the uptake of nitrogen by corn in the field in eight experiments conducted over a 2-year period in Pennsylvania. The highest correlation obtained was $r = 0.86$ with a method involving extraction of nitrogen by boiling 0.01 molar calcium chloride solution. Determination of the nitrogen in the calcium chloride extract by the Kjeldahl method is a slow process. The authors found, however, that when they measured the absorbance of a 0.01 molar sodium bicarbonate extract at 260 nanometers, which can be done rapidly, the correlation was equally as good as that obtained with use of the total nitrogen content of the 0.01 molar calcium chloride extract. In subsequent research involving 60 field experiments on nitrogen fertilization of corn in Pennsylvania, Fox and Piekielek (1983) obtained only low correlations between the yields of nitrogen in the plants on the control plots and the absorbance index of nitrogen availability, as well as several other chemical indexes of nitrogen availability. Whitehead (1983) similarly obtained poor results in a test of the absorbance method on samples of soils from 18 experiments on nitrogen fertilization of grassland in the United Kingdom. On the other hand, Sippola and Suonurmi-Rasi (1985) tested the absorbance of a 0.01 molar sodium bicarbonate extract as an index of nitrogen uptake by ryegrass from five peat soils and three mineral soils in a greenhouse experiment and found that the biological and chemical indexes of soil nitrogen availability were highly correlated. Appel and Mengel (1990) obtained higher correlations of plant uptake of nitrogen with the organic nitrogen extracted by 0.01 molar calcium chloride or electro-ultrafiltration than with nitrate nitrogen. This result is related to the fact that their soils initially contained only 1 to 9 milligrams of nitrate nitrogen but from 3 to 18 milligrams of soluble organic nitrogen per kilogram. Mengel (1991) reviewed the results of research on the electro-ultrafiltration method for extracting a potentially mineralizable fraction of organic nitrogen.

In further work, Gianello and Bremner (1986a,b) compared 12 chemical methods and five biological methods as indexes of mineralizable soil organic nitrogen on the basis of analytical results obtained on 30 soils. The highest correlation ($r = 0.95$) was obtained between the ammonium nitrogen released when soil was heated with 2 molar potassium chloride solution at 100° C for 4 hours and the nitrogen mineralized during incubation of soil under anaerobic conditions at 40° C for 1 week (Waring and Bremner, 1964). An equally high correlation was obtained between the chemical method and the potentially mineralizable nitrogen (N_p) discussed in the preceding section. Soil organic matter, which is used as an index of organic nitrogen mineralization in some soil testing laboratories, was less well correlated ($r = 0.75$ with nitrogen mineralized during anaerobic incubation and $r = 0.82$ with the potentially mineralizable nitrogen). Keeney and Bremner (1966) found that the better chemical methods they tested were essentially unaffected by drying and storage of soil samples in air-dry condition, which is a significant advantage over short-term incubation methods.

Hong et al. (1990) tested the 2 molar potassium chloride method of Gianello and Bremner with data from 49 field experiments on nitrogen fertilization of corn in Pennsylvania and found that the correlation $r = 0.48$ between quantities of nitrogen found by this method and the yields of nitrogen in mature plants on the control plots was significant at the 99% probability level but was still considerably lower than the correlation with the content of nitrate nitrogen in the soils at planting ($r = 0.75$). The correlation of yield of nitrogen in the plants with the total nitrogen content of the soils was $r = 0.47$, and the correlation with the organic matter content was $r = 0.22$. The correlation of yields of nitrogen with the absorbance of a 0.01 molar sodium bicarbonate extract at 200 nanometers was $r = 0.73$. At this wavelength the absorbance was well correlated with the nitrate content of the soils ($r = 0.85$), so that the measurement is to a considerable extent an index of nitrate. The authors favored this method because it is so easy and inexpensive.

Progress thus is being made, but it is slow. Hopefully further research will lead in time to an outcome more favorable to chemical indexes of nitrogen availability.

4-4.6.4. Models

For obtaining a perspective on the use of models for nitrogen fertilizer requirements, especially useful papers include those by Godwin et al. (1989), Greenwood (1982, 1983), Myers (1984), Sylvester-Bradley (1985), and Neeteson and Van Veen (1988). The papers by Greenwood emphasize the theoretical aspects of dynamic modeling of nitrogen requirements, the paper by Godwin ct al. describes the application of a dynamic model to nitrogen in cropping systems, the papers by Myers and Sylvester-Bradley describe the details of static modeling for soil testing purposes, and the paper by Neeteson and Van Veen bridges the dynamic and static approaches.

N_{min} **Method.**   The simplest model for the requirement for fertilizer nitrogen has been popularized in Europe as the N_{min} method. A particular crop is supposed to have a certain requirement for nitrogen, which can be met by soil nitrogen or fertilizer nitrogen. As the amount of mineral nitrogen present in the soil in the spring increases, the need for fertilizer nitrogen decreases. This model is formulated as

$$N_f = a - bN_m \quad ,$$

where N_f = fertilizer nitrogen required, N_m = mineral nitrogen present in the soil in the spring to the depth of rooting in the particular soil or to an approximate depth used for many soils in an area, and a and b are constants. N_m usually is considered to represent ammonium plus nitrate, but sometimes only nitrate is determined because in most soils almost all the mineral nitrogen is present as nitrate if no fertilizers supplying ammonium have been added recently. The constant a is the sum of the soil mineral nitrogen and the fertilizer nitrogen needed for the crop, and the constant b is the value of 1 kilogram of soil mineral nitrogen relative to that of 1 kilogram of fertilizer nitrogen. Thus, the model may be said to represent the difference between the mineral nitrogen required by the crop and the amount actually supplied by the soil.

In an application of the N_{min} method for winter wheat in Germany, Wehrmann and Scharpf (1986) considered the total nitrogen supply needed for the crop to be 200 kilograms per hectare. Fig. 4-28 shows the way they used the model to estimate the optimum early spring application of fertilizer nitrogen from the mineral nitrogen in the soil to a depth of 90 centimeters in February/March. For the application at that time, they considered the fertilizer nitrogen and the soil mineral nitrogen to be interchangeable (the line in the graph indicates a 1:1 relationship). The objective was to add enough fertilizer nitrogen to bring the total up to 120 kilograms per hectare. A top dressing of 70 to 100 kilograms of fertilizer nitrogen per hectare was given later, and this would provide the opportunity to adjust the final application as needed on the basis of plant or soil analyses. The model makes no explicit provision for the nitrogen supplied by mineralization, but both additions by mineralization and losses by leaching, volatilization, or denitrification would have an effect on the nitrogen test at the time of the late top dressing, and an allowance could be made for the net effect of these factors. Schulz and Marschner (1986) found that a test for nitrate on

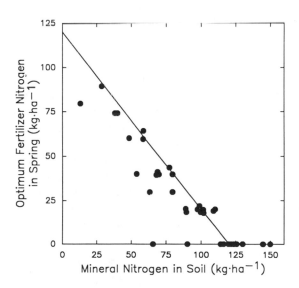

Fig. 4-28. Optimum quantity of fertilizer nitrogen for spring application to winter wheat in Germany versus mineral nitrogen in the soil to a depth of 90 centimeters. A topdressing of 70 to 100 kilograms of fertilizer nitrogen per hectare is applied later. (Wehrmann et al., 1988)

the base of winter wheat stems provided a convenient and effective way of gauging the adequacy of the soil nitrogen supply for carrying the crop to maturity and for adjusting the quantity of nitrogen to be applied in the final topdressing. Neeteson (1990) published a summary table giving for each of a number of crops the values of *a*, *b*, and sampling depths used in making nitrogen fertilizer recommendations in the Netherlands.

Boon (1983) added more details by incorporating a "humus factor" and a factor based upon prior applications of organic fertilizers. Both of these factors would aid in estimating the mineralization during the season. The amount of nitrogen that had been absorbed by the young cereal plants at the time of sampling was determined also.

In Denmark, Østergaard (1989) found that in 3 years the mineral nitrogen in the surface 100 centimeters of soil in the spring was equivalent to 32 to 39% of the quantity of nitrogen taken up by spring barley when the nitrogen supply was optimum. The correlation between mineral nitrogen in the surface 100 centimeters of soil and the yield of spring barley without nitrogen fertilization on many soils was $r = 0.74$, which implies that the measurement was valuable for prediction, but still that nearly half the variance of yields was accounted for by factors other than the supply of mineral nitrogen measured in the spring. Differences among soils in nitrogen mineralization during the season would be one of these factors. In some instances, the correlation between nitrogen mineralization and mineral nitrogen is good enough that no independent estimates of nitrogen mineralization are needed, as in an experiment by Lindén (1985) in Sweden. But in some instances, information on mineralization is needed. The more complex models generally make some provision for an independent contribution of mineralization that occurs after the measurement of residual mineral nitrogen.

Balance Method.   Stanford (1973) introduced what now is called the *balance method* or *balance sheet method* for estimating the requirement for fertilizer nitrogen. He proposed that the nitrogen content of the crop be represented as the sum of the nitrogen supplied by soil and fertilizer, and that the quantity of nitrogen derived from the fertilizer be equal to the product of the quantity of fertilizer nitrogen added and the fractional recovery of the nitrogen by the crop. The basic concept has been applied in a variety of ways, mostly by dividing the soil supply of nitrogen into discrete compartments and in some instances by adding a routine dealing with leaching of mineral nitrogen below the root zone.

Østergaard et al. (1985) used the Stanford approach to estimate the requirement of fertilizer nitrogen for barley following cereals in Denmark. Their formulation was

$$N_{f\ opt} = \frac{N_c - N_s}{N_{rf}} \quad ,$$

where $N_{f\,opt}$ = economically optimum quantity of fertilizer nitrogen, N_c = total amount of nitrogen absorbed by the crop (tops + roots) when $N_{f\,opt}$ is applied, N_s = nitrogen absorbed from the soil supply (estimated by the nitrogen in the tops and roots of the crop without nitrogen fertilization), and N_{rf} = fractional recovery of fertilizer nitrogen. $N_{f\,opt}$, N_c, and N_s were expressed in units of kilograms per hectare.

The value of N_c was estimated by the equation

$$N_c = 19x + 16,$$

where x = grain yield at 16% water content in megagrams per hectare. The value for N_s was found by the equation

$$N_s = 0.80N_m + 20,$$

where N_m = mineral nitrogen in the soil in March to the depth to which rooting is expected to occur. N_{rf} was estimated by the equation

$$N_{rf} = (0.75)(1 - e^{-0.015(N_c - N_s)}) + 0.08.$$

Østergaard et al. (1985) noted that they could estimate the nitrogen content of the unfertilized crop from the mineral nitrogen in the soil in March to ±15 to 20 kilograms per hectare and that they could not improve on this estimate by using an index of nitrogen mineralization. The numerical values in their equations of course apply to the specific conditions under which they were obtained and would be expected to be different under other conditions.

In France, Meynard et al. (1982) used for the nitrogen requirement of wheat the equation

$$N_f = by + N_{mh} - (N_m + N_{ms} + N_{mr}),$$

where N_f = fertilizer nitrogen required, b = total nitrogen per unit mass of grain, y = expected yield of grain with application of N_f, N_{mh} = mineral nitrogen in the soil at harvest (presumably to the same depth used for N_m), N_m = mineral nitrogen in the soil at the end of winter to the depth to which rooting is expected, N_{ms} = nitrogen mineralized from the soil organic matter during crop growth, and N_{mr} = nitrogen mineralized from the residues of the previous crop. In use of this formulation for predicting N_f, N_m would be measured directly, and b, N_{mh}, N_{ms}, and N_{mr} would have to be obtained from other experiments.

The authors tested the model in 91 field experiments conducted over a period of 5 years. In 40 experiments the observed yield was within ±5% of the expected yield (y), and in 75 experiments the observed yield was within ±15% of the expected value.

Also in France, Remy and Viaux (1982) used an equation for nitrogen fertilizer requirement (N_f) similar to that employed by Meynard et al. (1982):

$$N_f = \frac{by}{N_r} - (N_m + N_{ms} + N_{mr} + N_{mo}),$$

where b = total nitrogen per unit mass of grain, y = expected yield of grain with application of N_f, N_m = mineral nitrogen in the soil at the end of winter, N_{ms} = nitrogen mineralized from the soil organic matter, N_{mr} = nitrogen mineralized from residues of the previous crop, N_{mo} = nitrogen mineralized from organic manures, and N_r = efficiency of nitrogen utilization (fractional recovery in the grain, all sources being assumed to have equal efficiency). This formulation lacks N_{mh} but includes two terms (N_r and N_{mo}) not used in the equation by Meynard et al. The authors noted that in previous work by Viaux, the equation had been used without N_r. Although it had been found satisfactory for wheat in northeastern France, it was not satisfactory in other regions. The paper by Remy and Viaux provided information on additional factors that cause deviations from target yields, but none on the validation of the equation. A somewhat different equation was used for winter wheat in the Netherlands and Belgium by Neeteson et al. (1984).

In Canada and the United States, the expected yield that appears in the equations used by Meynard et al. (1982) and Remy and Viaux (1982) has been termed the target yield or yield goal. As in Europe, the concept has been applied for nitrogen. See a review paper by Peterson and Frye (1989).

Fehr (1970) described in some detail the procedure and supporting experimental data employed in the balance method as applied to the grain farming system used in Manitoba. The yield of grain was first determined as a function of the nitrogen content of the crop:

$$y = 0.0538e^{(2.83 + 0.0223N_c)},$$

where y = yield of grain in megagrams per hectare and N_c = kilograms of nitrogen per hectare in the above-ground portion of the crop. This equation is for barley, for which the proportion of the variance of grain yield accounted for by the nitrogen content of the crop was $R^2 = 0.86$.

The fractional recovery of fertilizer nitrogen in the above-ground portion of the barley crop was found to be 0.52 ± 0.14 in experiments with different quantities of ammonium nitrate applied broadcast. Residual nitrate was used as an index of the nitrogen supplied by the soil. In studies of sampling depth, the best relationship ($R^2 = 0.84$) between soil nitrate and nitrogen content of the above-ground portion of the crop was obtained with soil samples taken to a depth of 60 centimeters. Greater or lesser depths produced poorer agreement.

Combining the information on the relationships between grain yield and nitrogen content of the crop, the recovery of fertilizer nitrogen in the crop, and the relationship between the residual nitrate in the soil (N_m) to a depth of 60 centimeters, together with the quantity of fertilizer nitrogen (N_f), and solving the equation for the quantity of fertilizer nitrogen produced the following final model:

$$N_f = \frac{112[e^{(2.83 + 0.465y)} - 0.875N_m - 22.80]}{0.52} .$$

In this equation, the target grain yield or yield goal (y) is entered as megagrams per hectare, and the units of N_f and N_m are kilograms of nitrogen per hectare.

According to Power and Broadbent (1989), the University of Nebraska Soil Testing Laboratory applies the yield goal concept for corn production through the equation

$$N_f = \frac{14.4y}{1 - 0.0128y} + 55 - N_m - N_{mw} - N_{ml} - N_{mo} ,$$

where N_f = fertilizer nitrogen required, N_m = nitrate nitrogen in the soil to a depth of 60 centimeters before planting, N_{mw} = nitrate nitrogen added in the irrigation water, N_{ml} = nitrogen mineralized from previous legume crops, and N_{mo} = nitrogen mineralized from organic manures. All the N values are in kilograms per hectare, and y is in megagrams of grain per hectare. Nitrogen supplied in the irrigation water is included in the accounting because some of the irrigation water is pumped from shallow groundwater that contains enough nitrate to constitute a significant source. The values entered for N_{mw}, N_{ml}, and N_{mo} must be derived from other experiments and are not measured in individual instances. Vigil and Kissel (1991) provided a basis for estimating the mineralization of nitrogen from crop residues during the first season from laboratory data on the residue composition.

An advantage of the yield goal approach is that the farm operator should know better than the soil testing laboratory personnel the yields that can be produced. If the producer supplies the yield goal, the soil testing laboratory can provide an estimate of the quantity of fertilizer nitrogen needed to produce that yield. On the other hand, producers may not fully appreciate the distinction between the yields they can produce and the yields they would like to produce. Moreover, they may hold some unwarranted optimism about the yields they can produce. In a large project involving many producers of irrigated corn in Nebraska, Schepers et al. (1991) reported that in the 4 years from 1980 through 1983, producers provided yield goals that exceeded their measured production by an average of 2.0 megagrams of grain per hectare. This optimism would be reflected in excessive estimates of fertilizer nitrogen requirements. The continuing educational work associated with the project apparently resulted in increased producer awareness because the authors reported an overestimate of only 0.9 megagram of corn grain yield per hectare in 1988 and a reduction of fertilizer nitrogen application from about 280 kilograms per hectare before 1980 to 164 kilograms per hectare in 1988.

Fehr (1970) explained that in the information supplied to producers in Manitoba, a range of target yields was given, along with the quantity of fertilizer nitrogen to add for each yield. This procedure should have the advantage of stimulating more perceptive decisions by some producers.

Except for the work by Meynard et al. (1982) and Remy and Viaux (1982), the foregoing equations do not take into explicit account the nitrogen mineralized from soil organic matter independently of that supplied by the organic manures and previous crops of legumes. In effect, the nitrogen mineralized from the soil organic matter is considered to be constant (independent of location), and its contribution is hidden in the empirical equations.

In experimental work on sugarbeet in Idaho, Carter et al. (1976) added a laboratory-determined index of nitrogen mineralization. They noted that to obtain the maximum yield of sucrose, the nitrogen requirement under their conditions was 5.5 ± 0.5 kilograms per megagram of fresh beet roots. For the work on commercial fields to be described, they used the figure of 6 kilograms because the farm managers in that area generally applied more irrigation water than needed, which would cause some loss of nitrogen by leaching below the root zone. The total net nitrogen available to the crop N_t required to obtain the yield y of fresh beet roots associated with the maximum yield of sucrose (determined from farm records) is thus $6y$ kilograms.

N_t is given by

$$N_t = N_{rf}N_f + N_{rm}N_m + N_{rms}N_{ms} + N_{mr} \quad ,$$

where N_f = nitrogen added in the fertilizer, N_m = nitrate nitrogen in the soil to the depth sampled (1.5 meters in their experiments), N_{ms} = nitrogen mineralized in the soil to the depth sampled as determined by a 3-week aerobic incubation at 30° C, N_{rf} = fractional recovery of fertilizer nitrogen in the crop (increase in nitrogen in the crop from fertilization/nitrogen added in the fertilizer), N_{rm} = fractional recovery of soil nitrate nitrogen in the crop (crop-extractable nitrate nitrogen)/(nitrate nitrogen in the soil depth sampled), N_{rms} = fractional recovery in the crop of the nitrogen mineralized in the soil [(crop-extractable mineralizable nitrogen)/(field mineralizable nitrogen in the soil depth sampled)][(field mineralizable nitrogen)/(laboratory mineralizable nitrogen)], and N_{mr} = nitrogen mineralized from or immobilized by organic residues added to the soil. In this formulation, the quantities of nitrogen are in units of kilograms per hectare. N_{mr} has no recovery coefficient because the effect of the residue of the preceding crop on the nitrogen supply was determined in separate experimental work in which the net effect was measured.

Solving the foregoing equation for N_f and substituting $6y$ for N_t,

$$N_f = \frac{6y - (N_{rm}N_m + N_{rms}N_{ms} + N_{mr})}{N_{rf}} \quad .$$

Numerical values used in their work from previous investigations were $N_{rf} =$ 0.65, $N_{rm} = 1.2$, and $N_{rms} = 0.95$. The only figure they gave for N_{mr} was for wheat straw. The value obtained from prior biological assay experimental work involving determination of sugarbeet yields with different applications of straw and fertilizer nitrogen (Smith et al., 1973) was that a megagram of straw was equivalent to -7.5 kilograms of fertilizer nitrogen; that is, the straw immobilized some nitrogen, so that where straw was added as an organic residue, extra fertilizer nitrogen would be needed to compensate for the effect. The value of N_{mr} for 5 megagrams of straw in their work was $(5)(-7.5) = -37.5$ kilograms.

Carter et al. (1976) found that using the approach described, they obtained the best results with measurements of both nitrate nitrogen and mineralizable nitrogen in individual experiments. They were able to estimate the quantity of fertilizer nitrogen needed to produce the maximum yield of sucrose within 56 kilograms per hectare in 83% of the sites using the nitrate nitrogen and mineralizable nitrogen values measured in individual experiments, in 67% of the sites using the nitrate nitrogen values measured in individual experiments and average values for mineralizable nitrogen, and in 12.5% of the sites using the recommendations made by the fieldmen.

Stanford et al. (1977) extended this work in a somewhat different direction. Their approach was based upon earlier research by Stanford and Smith (1972) showing that the rate of nitrogen mineralization during a 30-week aerobic incubation gradually decreased with time. The rates of decrease in almost all soils tested followed the pattern that would be expected if a single source of mineralizable nitrogen was present and was decomposing at a rate proportional to the amount remaining. From the data obtained, they were able to calculate values for the hypothetical quantity of potentially mineralizable nitrogen, designated here as N_p.

Applying the N_p concept to samples of soils from 40 field experiments on nitrogen fertilization of sugarbeet in Idaho, Stanford et al. (1977) estimated an N_p value for each soil from mineralization data obtained in a 30-week incubation. Then, for each of the 6 months the crop was in the field (April through September), they calculated the average air temperature from weather records and assumed the soil would have the same temperature. Then they adjusted the rate constant for the temperature. The average rate constant k was 0.0058 per day at 35° C, 0.0029 at 25° C, 0.0015 at 15°, and 0.0007 at 5° C. Additionally, they adjusted the rate constant for the water content of the soil. The effect of the water content of the soils on the rate constant for nitrogen mineralization was determined in laboratory experiments, and the results were used to estimate the effect of water content in the field experiments in which the trends in water content were inferred from records of rainfall, irrigation, and estimated evapotranspiration.

The results showed that the correlation between the calculated uptakes of nitrogen and the observed uptakes on the plots receiving no nitrogen in the various experiments was $r = 0.80$. This is a relatively good correlation, considering the fact that each experiment was at a different location. Although the water content of the soil was variable, all the experiments were on irrigated land, and adequate irrigation water was supplied. In most situations, the crop yield will vary from one season to the next as a consequence of differences in water supply.

A model by Myers (1984) is discussed here in brief, with certain modifications made in consultation with the author. The Myers model was developed for crops grown under semiarid conditions, where the water stored in the soil at planting time is often used as a basis for deciding whether or not to plant. Under such conditions, the maximum yield obtainable with an adequate supply of nitrogen is determined primarily by the water supply. The nitrogen requirement of the crop thus will increase with the water supply. For grain sorghum, the equation obtained was

$$y_{g\ max} = -770 + 7.79(w_s + w_p),$$

where $y_{g\ max}$ = maximum grain yield in kilograms per hectare, w_s = millimeters of available stored water to a depth of 90 centimeters at the time of planting, and w_p = millimeters of precipitation between planting and physiological maturity of the crop. This equation, derived from experiments in Kansas, Nebraska, and northwestern Australia, had a coefficient of determination of $R^2 = 0.39$.

The quantity of nitrogen in the grain and stover ($N_{gs\ max}$) associated with the maximum yield of grain sorghum was given by

$$N_{gs\ max} = 0.0233y_{g\ max} - 17.5,$$

where $N_{gs\ max}$ has units of kilograms per hectare. The data for nitrogen content of the grain and stover at the maximum yield obtainable with nitrogen were derived from a published source. In this case, an R^2 value of 0.99 was obtained.

The next step was to represent the nitrogen supplied by the soil as the sum of the initial mineral nitrogen in the upper 90 centimeters of soil (N_m) and the nitrogen mineralized during crop growth (N_{ms}). Values of N_m were obtained by direct measurements. N_{ms} was assumed to be proportional to the product of the kilograms of total nitrogen per hectare in the upper 20 centimeters of soil (N_{ts}), the precipitation between planting and physiological maturity of the grain w_p, and the radiant energy flux as affected by the latitude (l). The equation obtained was

$$N_{ms} = 0.656 \cdot N_{ts} \cdot w_p \cdot l \cdot 10^{-4},$$

where the latitude factor was 0.5 for latitudes ≥ 49, 0.75 for latitudes < 49 and ≥ 23.5, and 1.0 for latitudes < 23.5.

Drawing on the empirical observation that when water is deficient the efficiency of recovery of soil mineral nitrogen in the grain and stover increases with the water supply, and assuming that $N_{rm} = N_{rms} = N_{rm,rms}$, where N_{rm} is the fractional recovery of the initial mineral nitrogen, N_{rms} is the fractional recovery of the nitrogen mineralized during crop growth, and $N_{rm,rms}$ is the fractional joint recovery of these two fractions, Myers used the linear equation

$$N_{rm,rms} = 0.000471(w_s + w_p) + 0.50$$

to represent the efficiency of recovery of the two fractions of soil mineral nitrogen. On the basis that about 20% of fertilizer nitrogen is lost in balance-sheet studies, Myers assumed that the amount of fertilizer nitrogen available for recovery is 80% of the amount added. On the additional assumption that the recovery of the remaining fertilizer nitrogen is the same as the recovery of the soil mineral nitrogen and the nitrogen mineralized from soil organic matter during growth of the crop, Myers represented the fractional recovery of fertilizer nitrogen (N_{rf}) in the grain and stover as

$$N_{rf} = 0.8 \cdot N_{rm,rms} \quad .$$

The addition of fertilizer nitrogen required to produce the maximum yield of grain ($N_{f\ max}$) was represented by the equation

$$N_{f\ max} = \frac{N_{gs\ max} - (N_m + N_{ms}) \cdot N_{ms,rms}}{N_{rf}}$$

$$= \frac{N_{gs\ max} - (N_m + N_{ms}) \cdot N_{rm,rms}}{0.8N_{rm,rms}} \quad .$$

To estimate the quantity of fertilizer nitrogen needed to produce the maximum net profit, Myers started with the following equation:

$$\frac{y_g}{y_{g\ max}} = \frac{2N_{gs}}{N_{gs\ max}} - \left(\frac{N_{gs}}{N_{gs\ max}}\right)^2,$$

where y_g = yield of grain obtained with the yield N_{gs} of nitrogen in the grain and stover, and $y_{g\ max}$ and $N_{gs\ max}$ have the same meaning as before. This empirical equation relating $N_{gs}/N_{gs\ max}$ to $y_g/y_{g\ max}$ was devised from some data Myers had published previously.

The rate at which the yield y_g in the foregoing equation increases with the quantity of nitrogen in the grain and stover (N_{gs}) is obtained by differentiating y_g with respect to N_{gs}:

$$\frac{dy_g}{dN_{gs}} = \frac{2y_{g\ max}}{N_{gs\ max}} - \frac{2y_{g\ max} \cdot N_{gs}}{(N_{gs\ max})^2} \quad .$$

The maximum net profit from application of fertilizer nitrogen is obtained when $dy_g/dN_{gs} = c_{Ngs}/p$, where c_{Ngs} = cost of one unit of nitrogen in the sorghum grain and stover and p = price of one unit of sorghum grain. From a previous

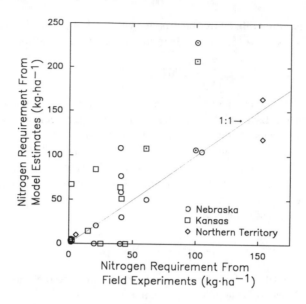

Fig. 4-29. Quantity of fertilizer nitrogen required to yield the maximum net profit in grain sorghum production in Nebraska and Kansas (U.S.A.) and Northern Territory (Australia), as estimated from a model and from field experiments. Each point represents an individual experiment. The points with a center dot represent experiments in which the response curve did not appear to have reached a maximum. (Myers, 1984)

relationship, $c_{Ngs} = c_{Nf}/N_{rf} = c_{Nf}/0.8 \cdot N_{rm,rms}$. Substituting $c_{Nf}/0.8 \cdot N_{rm,rms} \cdot p$ for dy_g/dN_{gs} and solving for N_{gs}, which then becomes the economic optimum quantity of nitrogen in the grain and stover $N_{gs\ opt}$,

$$N_{gs\ opt} = N_{gs\ max} - \frac{C_{Nf} \cdot (N_{gs\ max})^2}{1.6 \cdot p \cdot N_{rm,rms} \cdot y_{g\ max}} \ .$$

The quantity of fertilizer nitrogen to give the maximum net profit ($N_{f\ opt}$) is given by

$$N_{f\ opt} = \frac{N_{gs\ opt} - (N_m + N_{ms}) \cdot N_{rm,rms}}{0.8 \cdot N_{rm,rms}} \ .$$

When $N_{f\ opt}$ is negative, it is taken as zero.

Fig. 4-29 illustrates the results obtained in a test of the model using published data from Nebraska and Kansas (U.S.A.) and Northern Territory (Australia). Each point represents a single experiment. The model accounted for 60% of the variance of observed values. The total nitrogen concentrations in the soils were estimated from separate published data. The author reported that data from approximately one-third of the experiments were discarded because "no sensible response curve could be drawn through the data points." This observation suggests that the experiments that were used may not have been as precise as desired for correlation with the model.

In any event, the procedure outlined illustrates the way in which the model was developed from underlying relationships involving a few measured values. The data inputs required include (1) the available soil water at or just before planting, (2) the precipitation during the growing season (adjusted for runoff if necessary), (3) the nitrate in the soil profile to a depth of 90 centimeters at or just before planting, and (4) the total nitrogen concentration in the surface layer or plowed layer of soil. Because Myers was using the model on completed experiments, he had the precipitation during the growing season from the nearest weather station. If the model were to be used for prediction, the model could be used to provide estimates of the requirement of fertilizer nitrogen for specified amounts of precipitation. The use of a more complex nitrogen model for semiarid conditions in Kenya, without details, was described by Keating et al. (1991).

In the foregoing discussion of models, a consistent set of symbols has been used to reduce the complexity. As indicated, however, some of the symbols were defined in somewhat different ways by different authors to fit their purposes or the data available. Moreover, although the general objective of the models was the same, the models were not identical. None of the models took into account losses of nitrate by leaching. In most instances, no leaching loss would occur, but it would need to be taken into account in models for certain environments. Myers noted that he modified the data for sorghum in Northern Territory (Fig. 4-29) because leaching losses were known to occur there.

More sophisticated models have been developed for use in humid regions. Examples include models by Richter et al. (1985), Greenwood et al. (1987), Neeteson et al. (1987), Addiscott and Whitmore (1987) (see Whitmore and Addiscott (1987) for a more practically oriented account), and Burns (1980a,b). The model by Burns is discussed in Section 7-6. Ballard (1984) developed a simpler model for predicting the cumulative increase in volume growth of marketable timber to nitrogen fertilization that substituted site quality for soil nitrogen measurements and other growth factors. Inputs to this model were time, quantity of fertilizer nitrogen applied, stand composition, and site quality.

Models for use in humid regions become more complex than those for dry regions if they take into account the loss of nitrate by leaching. Incorporating the loss of nitrate by leaching requires simulation of water movement within the soil and loss of water by drainage, and this in turn requires information on the water-retaining properties of soils at different depths. Usually the soil is represented as a series of hypothetical thin uniform horizontal layers.

Another problem addressed by some models is that except when water movement in the soil occurs very slowly, the movement does not occur as a solid front. Rather, water moves faster through large pores than through small pores, and the movement of dissolved nitrate follows the same pattern. To take this into account requires measurements on the soil that provide information on pore size distribution. Addiscott and Whitmore (1987) used just two size-classes of pores in their model, and they found that the results were not highly sensitive to the proportions of the two sizes. Their work was done in England, where the rainfall characteristically is of low intensity. Pore size would have a more important effect in areas where high-intensity rains occur (Priebe and Blackmer, 1989). Under such conditions, the water movement would be more rapid, and a greater proportion of the moving water would be carried by the large pores.

Some of the models do not stop with the mineral nitrogen present in the soil in the spring, but continue through the growing season to simulate the accumulation and movement of mineral nitrogen in the soil as well as any loss that may occur by leaching. The growth of the crop also is brought into the picture. During development of the crop, uptake of mineral nitrogen by the crop dominates the changes in mineral nitrogen in the soil. The rate at which the supply of mineral nitrogen disappears due to crop uptake depends upon the location of the nitrogen in the soil in relation to crop roots and the demand by the developing crop, as affected by the weather conditions. Because of the complexity of the processes and their mutual effects, the models that have the greatest capability for precise simulation are *dynamic*, in that they continually incorporate new data on weather (such as rainfall, evaporation, and temperature) and recalculate the outputs on a daily basis.

When a model has been developed to the stage at which the calculated outputs are close enough to measured values to be acceptable, the model can be used with some assurance to simulate, for example, the expected nitrogen requirement of a crop if weather conditions for other years are entered or if a range of conditions is specified. It can also be used to provide continually updated advisory information on a regional basis, with breakdowns following according to soil types and management, as permitted by the availability of data on these factors.

4-4.7. Fertilizer Prescriptions

The term *fertilizer prescription* would be appropriate whenever quantities of fertilizer are adjusted to fit individual conditions. Thus far, however, the term does not seem to have been used in connection with models for requirement of fertilizer nitrogen (Section 4-4.6.4); rather, its use has been restricted to situations in which nitrogen, phosphorus, and potassium were involved.

Like the models for requirement of fertilizer nitrogen, the fertilizer prescription approach attempts to provide the nutrients needed for a specified quantity of crop on the upper portion of the response curve. The procedure incorporates aspects of soil tests and fertilizer efficiency introduced in Section 4-4.3, together with crop requirements for nutrients. It also is related to the procedure described in Section 3-4.2.7 for estimating the fertilizer nitrogen need of corn on the basis of the nitrogen content of the grain.

Truog (1961) has been credited as the originator of the fertilizer prescription concept. He noted that he had been asked to write a fertilizer prescription for a corn yield of 100 bushels per acre (6.3 megagrams per hectare) and that he eventually had done it on the basis of (1) the nutrients contained in the crop, (2) the nutrients supplied by the soil, and (3) the nutrients supplied by manure and fertilizers. Item (2) was derived from soil tests. Approximate efficiency factors were used with both items (2) and (3). Although it was not presented in the form of an equation, Truog's fertilizer prescription concept was a balance sheet method, preceding the one more formally proposed by Stanford (1973) (Section 4-4.6.4). The fertilizer prescription concept has developed in the direction of the original use of the balance sheet by Truog.

Truog reported that over a 5-year period, 2,766 farmers in Wisconsin had used the prescriptions in a demonstration project. Although the observed yields varied widely, the average was 106 bushels per acre (6.6 megagrams per hectare). In a test of the procedure by corn seed growers, prescriptions were made for yields up to 200 bushels per acre (12.5 megagrams per hectare). Actual yields approximated the prescription yields up to 125 bushels per acre (7.8 megagrams per hectare), but increased only a little beyond that level even though enough fertilizer was added for much higher yields.

A number of papers on the fertilizer prescription concept have been published in India, of which one by Ramamoorthy et al. (1967) appears to be the first. This paper described a method of calculation, represented by the following equation, for estimating the requirement of fertilizer phosphorus to obtain a target crop yield:

$$P_f = \frac{P_t \cdot y - P_0 \cdot P_s}{P_r} \quad ,$$

where P_f = fertilizer phosphorus in kilograms per hectare to attain the target yield, P_t = kilograms of phosphorus in the grain and straw of the fertilized crop per megagram of grain (determined in a calibration experiment to be described in following paragraphs), y = target grain yield in megagrams per hectare, P_0 = kilograms of phosphorus in grain and straw of the unfertilized crop per hectare per kilogram of soil test phosphorus per hectare, P_s = soil test phosphorus in the unfertilized control in kilograms per hectare, and P_r = fractional recovery of fertilizer phosphorus in grain and straw, averaged over treatments in the calibration experiment with different quantities of fertilizer phosphorus.

For use of the equation in calculating fertilizer prescriptions, P_t and P_0 are divided by P_r, producing a linear equation from which P_f can be calculated upon insertion of the value of soil test phosphorus P_s and the target yield y for the specific area in question:

$$P_f = \frac{P_t}{P_r} \cdot y - \frac{P_0}{P_r} \cdot P_s.$$

Analogous formulations may be written for nitrogen and potassium.

A novel feature of the fertilizer prescription concept as applied in India has been the calibration process. Ramamoorthy et al. (1967) performed the entire calibration for different fertility levels with a given crop and soil type in a single experiment at one location. According to Ramamoorthy and Velayutham (1971), the reason for adopting this procedure was to assure homogeneity in soil, management, and weather, which are causes of poor correlations in calibrations resulting from experiments conducted at different locations and in different seasons.

As the calibration procedure was described in a private communication from Dr. Rani Perumal, of Tamil Nadu Agricultural University, fertilizer nitrogen is added to four large plots in quantities of 0, 75, 150, and 300 kilograms per hectare. To these same plots are applied both fertilizer phosphorus and potassium in relative quantities corresponding to those of nitrogen. The quantities of phosphorus and potassium to be applied to correspond to 150 kilograms of nitrogen are found from the phosphorus and potassium fixation curves for the soil. The phosphorus applied is equal to the phosphorus fixation capacity as measured by a method proposed by Waugh and Fitts (1966), and the potassium applied is sufficient to produce 150 kilograms of exchangeable potassium per hectare. The quantities of phosphorus and potassium are calculated for the surface 15 centimeters of soil with a standard bulk density of 1.5.

A nonexperimental crop then is grown to provide time for the fertilizers to react with the soil, so that the different residual levels may be considered to have an effect similar to that of areas that have been brought to different nutrient levels by treatments made over a period of years. The nonexperimental crop is followed by the calibration experiment. Each of the four large plots is subdivided into subplots. In the design described by Ramamoorthy and Velayutham (1971), there were 43 subplots in each of the four large plots. Of these, 22 were control plots, and 21 received different quantities and combinations of fertilizer nitrogen, phosphorus, and potassium. In a calibration experiment for which treatments were tabulated in detail by Murugappan et al. (1989), there were 24 subplots in each large plot, and they received five quantities of fertilizer nitrogen, three of phosphorus, and four of potassium in a fractional factorial design. There were four control plots without fertilizer.

As may be seen from the foregoing equations, a prescription for fertilizer phosphorus P_f (or its counterparts N_f and K_f) requires five items of information. One of these is the soil test value P_s for the site to which the prescription is to be applied, and one — the target yield y — is supplied by the producer with the counsel of soil testing laboratory personnel. The other three are supplied by the calibration experiment.

The first item of information derived from the calibration experiment is the quantity of nutrient required by the crop per unit of grain (or other economic part) produced. A procedure for selecting the value of P_t and its counterparts N_t and K_t for use in the foregoing equation was described by Ramamoorthy et al. (1967), and at least some others have followed this approach. In their calibration experiment with wheat, Ramamoorthy et al. (1967) determined the yields of nitrogen, phosphorus, and potassium in the grain plus straw from the treatment that produced the next to highest yield of grain, and divided these values by the yield of grain. The treatment used for calculating these values for nutrient requirement per unit of produce was considered to have the best balanced supply of nutrients because it represented the highest yield obtained with the level of nitrogen used (99 kilograms per hectare) and only a slightly lower yield than was produced with more nitrogen and potassium.[1]

The second item of information required is the contribution of nutrients by the soil. To obtain this information, the quantities of nitrogen, phosphorus, and potassium in the grain and straw on the control plots of the calibration experiment are divided by the soil test values for these nutrients in the layer of soil upon which the soil tests were made. The quotients (P_0, N_0, and K_0) are efficiency coefficients for the soil test values of the nutrients.

The third item of information required is the fractional recovery in the crop of the nutrients applied by the fertilizer (P_r, N_r, and K_r). To obtain this information, the yields of phosphorus, nitrogen, and potassium in the grain plus straw on the control plots are subtracted from the yields of the corresponding nutrients on the fertilized plots. The differences are divided by the quantities of the respective nutrients added in the fertilizers, and the quotients are averaged by nutrients. The quotients represent efficiency coefficients for the fertilizer sources of the respective nutrients.

Although the equations for the various nutrients place no explicit upper limit on the target yield y, excessive values of y will lead to prescriptions of relatively large quantities of fertilizers, but the resulting yields will tend to fall below the target yields. Note in this connection the data in Fig. 4-30 derived from several independent trials of the fertilizer prescription concept. In each instance, there is a tendency for the observed yields to drop, relative to target yields, as the

---

[1]In a personal communication, Dr. Rani Perumal of Tamil Nadu Agricultural University described two other methods for determining values for P_t, N_t, and K_t. In the *maximum yield method*, the yields of grain and straw for the subplots in each of the four large calibration plots are arranged in groups. Within each group, the level of a given nutrient is constant. For example, in a grouping according to nitrogen level, a group might include treatment (a) N_1, P_0, and K_0, treatment (b) N_1, P_1, and K_0, and treatment (c) N_1, P_1, and K_1. The total yield of nitrogen in the grain and straw from the treatment producing the highest yield in this group is found by chemical analysis. The value obtained is divided by the corresponding yield of grain (both in the same units, for example, kilograms per hectare), and the quotients from all nitrogen groups in all four large plots are averaged to obtain N_t. The values of P_0, N_0, and K_0 are obtained by analyzing the grain and straw from four control subplots in each of the four large calibration plots. The four subplots selected represent the highest yield, the lowest yield, and two intermediate yields. The total yields of phosphorus, nitrogen, and potassium in grain and straw on each selected plot are divided by the corresponding soil test values, and the quotients are averaged to obtain P_0, N_0, or K_0, as the case may be.

In the *response method*, the subplots within each of the large calibration plots are arranged in groups such that within each group the level of one nutrient increases while the others remain constant. Within each group, the grain and straw from each treatment yielding more than the treatment with the smallest addition of the nutrient in question are analyzed if the response is more than marginal. The total yield of the nutrient in grain plus straw in each selected subplot then is divided by the corresponding yield of grain (both in the same units), and the quotients from all groups in all four of the large calibration plots are averaged to obtain P_t, N_t, or K_t, as the case may be. The values of P_0, N_0, and K_0 are obtained from six control subplots selected from each of the four large calibration plots. The selected subplots cover a wide range of yields and soil test values for nitrogen, phosphorus, and potassium. The total yields of phosphorus, nitrogen, and potassium in grain and straw on each selected subplot are divided by the corresponding soil test values, and the quotients are averaged over all selected subplots in all four of the large calibration plots to obtain P_0, N_0, or K_0, as the case may be.

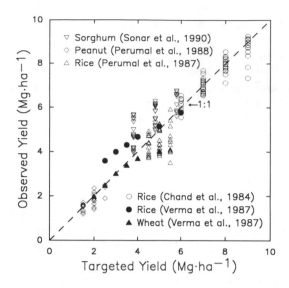

Fig. 4-30. Targeted and observed yields of crops in tests of the fertilizer prescription procedure. The data for peanut were obtained by Perumal et al. (1988) in 11 experiments in
Tamil Nadu on three related soil series with two or three target yields per experiment. The
data for rice by Perumal et al. (1987) represent the results of 16 individual experiments with
two or three target yields per experiment on closely related soils in Tamil Nadu. The data
for rice and wheat by Verma et al. (1987) in Himachel Pradesh are averages of the results
of 13 experiments with each crop over a 5-year period. The data for rice by Chand et al.
(1984) in Punjab are for a total of 25 individual experiments conducted over a period of 4
years, with apparently two or three target yields per experiment. The data for sorghum by
Sonar et al. (1990) in India are for a total of 19 experiments with two or three target yields
per experiment. The data have been displaced to target yields 0.2 megagram per hectare
below the actual target yields to avoid covering other points at the same target yields.

target yields are increased. A similar trend was noted in the original work by
Truog (1961). In that case, the highest targeted yields were well above the general
yield levels, and so the shortfall of observed yields could be accounted for by
maximum yield limitations. The highest targeted yields in the investigations in
Fig. 4-30 apparently were more realistic.

The fact that in some applications of the fertilizer prescription procedure the
uptake of a nutrient by the crop has exceeded the soil test value has proved
troubling to certain investigators. Such findings are not necessarily anomalies
or the effects of unsuitable tests, but may be regarded as consequences of the
empirical nature of the tests. Soil test phosphorus, for example, conceivably
could represent any value in the complete range from the quantity of phosphorus
in solution in the plowed layer of soil to the total phosphorus in the soil in the
entire root zone of the crop. The uptake of phosphorus from the soil by the crop
would far exceed the former quantity but would be far smaller than the latter.
The issue of importance is not the magnitude of the soil test relative to the uptake
of the nutrient by the unfertilized crop, but rather the product of P_0 and P_s in
the foregoing equations. The effect of a very small value for soil test phosphorus

P_s in the equation will be compensated by a very large value of phosphorus in the grain and straw of the unfertilized crop per unit of soil test phosphorus P_0.

Similar concerns have been expressed regarding apparent recoveries of an applied nutrient by the crop that exceed the quantities added in the fertilizer. Such results are possible, although they do not appear to be common. A more likely cause is the nature of the calibration procedure, in which the percentage recovery of a given nutrient appears to be calculated by the conventional procedure except that possible effects of other nutrients applied concurrently are ignored.

4-4.8. Some Problems

To judge from the fact that approximately 4 million soil samples are tested annually in the United States (Jones, 1985b, cited by Jones et al., 1990), many producers must consider the results a useful aid in production agriculture. From both the scientific and practical standpoint, however, there are some problems.

4-4.8.1. Soil Variability

One of the important problems in sampling soils for soil testing purposes is the heterogeneity of soils within both small areas and large areas. From the world literature on soil variability, Beckett and Webster (1971) developed the summary shown in Table 4-6 for topsoil variability within individual fields.

Variability on a micro scale was demonstrated by Qureshi and Jenkins (1987), who analyzed samples scraped from vertical planar faces of prismatic structural units in four soils from the United Kingdom. The excess of total phosphorus in the surface millimeter relative to that in the insides of the structural units ranged from 1 to 28% in the four soils. The enrichment of sulfur in the surface millimeter ranged from 11 to 14%.

Soil variability increases with management of soils for agricultural purposes, including grazing and the production of cultivated crops. Fertilizer residues are an important source of variability in plant nutrients. A bulletin by James and Dow (1972) illustrates several aspects of variability that are of general significance.

Table 4-6. Coefficients of Variation for Certain Properties of Topsoils Within Areas Differing in Size (Beckett and Webster, 1971)

Property[a]	Range of Coefficients of Variation[b] for Samples Taken Within Indicated Areas		
	1 Square Meter	100 Square Meters	One Field
Available potassium	11-40	11-112	21-142
Available phosphorus	7-45	13-121	11-131
Calcium	—	—	20-260
Magnesium	—	—	15-71

[a] Methods of analysis differed among the literature sources cited. To judge from the text, values for calcium and magnesium were for the exchangeable forms.
[b] The coefficients of variation were expressed as percentages.

Fig. 4-31A shows the results of measurements of phosphorus extracted by the 0.5 molar sodium bicarbonate method from closely spaced samples of soil from a field planted to potato. The samples were taken in a transect across two rows after harvest of the potato crop. The fertilizer had been applied in bands below and to both sides of the seed pieces at planting time. Although the digging equipment had caused some mixing of the soil when the potatoes were harvested, the location of the rows is made obvious by the relatively high values for residual phosphorus reflected in the soil test values. In research by Kitchen et al. (1990) in Colorado and Kansas, the distribution of extractable phosphorus in soil samples taken at right angles to fertilizer bands peaked sharply at band sites and dropped off rapidly and uniformly on both sides. Most of the increase in extractable phosphorus occurred within 10 centimeters on either side of the band. In their work, the soil had not been disturbed in the period of 12 to 23 months between fertilization and sampling. In an investigation involving three soils in Nebraska, Eghball et al. (1990) calculated from short-term measurements that if ammonium polyphosphate were applied in solution in bands and left undisturbed, the concentration of water-soluble phosphorus in these soils at the center of the band location would still be five times that in the original soil from 2.6 to 7.4 years after application, depending upon the soil and the concentration of fertilizer in the original bands.

A second type of heterogeneity due to fertilization is illustrated in Fig. 4-31B. In this experiment, granular concentrated superphosphate was broadcasted on the surface and plowed under for alfalfa planted in the spring. Soil samples were taken in the fall. The values for extractable phosphorus reflected the increasing quantities of fertilizer phosphorus applied, but the 95% confidence limits also increased. Some of the soil samples at each rate of application had values for extractable phosphorus comparable to those for the untreated control plots, indicating that these volumes of soil did not contain any of the phosphorus added in the fertilizer.

Fig. 4-31C illustrates a third type of heterogeneity associated with tillage methods. The fertilizer was broadcasted on the surface. Where moldboard plowing was practiced, the vertical distribution of extractable phosphorus was fairly uniform, but in the no-tillage system the concentration was greatest near the surface. The vertical distributions of extractable potassium in the same tillage systems (not shown) were similar to those for phosphorus. Nitrate did not show the marked accumulation at the surface.

Fig. 4-32 shows the values for extractable phosphorus and potassium in each of two commercial fields sampled in the direction of the long dimension by a grid system with grid lines 30.5 meters apart. In the field represented by Fig. 4-32A, the extractable potassium was relatively low and fairly uniform. Extractable phosphorus was more variable and showed a distribution pattern. In the field represented by Fig. 4-32B, both phosphorus and potassium were variable, but the distribution patterns did not coincide. These data illustrate the principle that the two nutrients may have rather different patterns of variability in a given field.

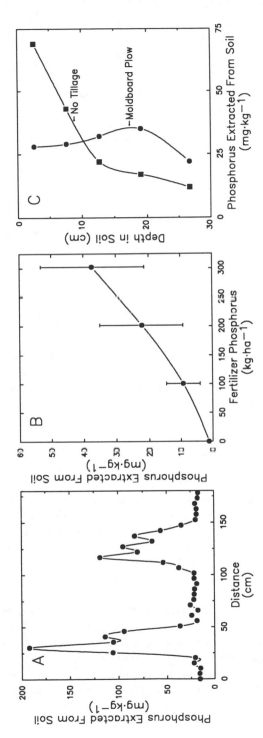

Fig. 4-31. Fertilizer use and soil variability. A. Phosphorus extracted by the 0.5 molar sodium bicarbonate method from samples of soil from a potato field in Washington. The samples were taken along a line perpendicular to two rows after the potatoes were dug in the fall. The fertilizer, which included phosphorus, had been banded below and to both sides of the seed pieces at the time of planting in the spring. (James and Dow, 1972) B. Phosphorus extracted by the 0.5 molar sodium bicarbonate method from samples of soil taken in the fall from an alfalfa field in Washington that had received different quantities of granular concentrated superphosphate in the spring. The vertical lines indicate the range within which the true mean value for soil test phosphorus would be expected to lie with 95% probability. (James and Dow, 1972) C. Vertical distribution of phosphorus extracted by the 0.025 molar hydrochloric acid-0.03 molar ammonium fluoride method from samples of a soil in Minnesota after 8 years of corn production with no tillage and moldboard plowing. The fertilizer, including 49 kilograms of phosphorus per hectare, was broadcast on the surface each year. (Randall, 1980)

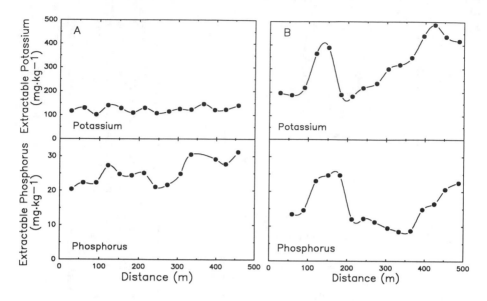

Fig. 4-32. Trends in extractable phosphorus and potassium in the long dimension of two commercially cropped fields (*A* and *B*) in Washington. (James and Dow, 1972)

For most characteristics, the variability is greater in surface soil samples than in subsoil samples. Thus, the number of samples from a given area required to attain a given degree of precision of the mean is usually smaller with subsoil samples than with surface soil samples.

4-4.8.2. Soil Sampling

Reviews of soil sampling written especially for soil testing include those by Peck and Melsted (1973), Sabbe and Marx (1987), and James and Wells (1990). These publications include citations to many other papers.

Abrupt discontinuities in nutrient supplies in soils generally affect crops over a distance exceeding the discontinuities because of the spreading of root systems beneath the surface and the fact that the root systems of adjacent plants generally overlap to some extent. Standard practice in field experiments with fertilizers is thus to confine measurements of yield and plant composition to the interiors of plots, where the results are not affected by the discontinuities and mixing that may occur at the plot borders.

Under conditions of no mechanical mixing due to tillage implements, Follett et al. (1991) found with use of nitrogen-15 as a tracer that border effects with wheat were recognizable to a distance of 46 centimeters from the boundary between tagged and nontagged fertilizer, but not beyond that distance. In further work with nitrogen-15, Sanchez et al. (1987) found that corn roots absorbed some nitrogen almost a meter from the base of the plants. Thus, if the distribution of roots was symmetrical, a given plant was absorbing nitrogen from the soil occupying an area of approximately 3 square meters. Individual soil cores, however, may correspond to a surface area no more than 5 square centimeters. In theory, 6,000 such cores could be taken within 3 square meters. Individual soil cores thus will reflect relatively small-scale discontinuities, the effects of which crop plants integrate with the nutrient supplies in the many other zones within the area covered by their root systems. The term *soil core* is used here to represent an individual sampling unit, which may properly be called a core if taken by a sampling tube, but also may be taken in other ways.

Sampling Procedures.   The standard statistical approach to sampling soils for chemical soil tests assumes that the objective is to estimate the true mean analysis for the area sampled. One takes a number of cores at random, analyzes the cores individually, and then determines the mean x_m and standard deviation $s = [\Sigma(x - x_m)^2/(n - 1)]^{1/2}$ for the sample, where x refers to the analyses on the individual cores, n is the number of cores, and Σ indicates a summation. The estimated number of cores (n_{est}) needed to assure that the population mean will be within the range of the sample mean plus or minus a specified confidence interval (d) with specified probability then is given by the formula (Petersen and Calvin, 1986)

$$n_{est} = t_p^2 s^2/d^2,$$

where t_p is Student's t value for a specified probability p with the number of degrees of freedom associated with the sample (number of individual cores minus one).

Petersen and Calvin (1986) illustrated the process of obtaining n_{est} with an example based upon measurements of exchangeable potassium in ten individual soil cores taken at random. The mean was 57.0 milligrams per kilogram, and the standard deviation was 9.627. For an allowable estimated population mean range of 57.0 ± 5 milligrams per kilogram, $d = 5$. The value of t for a probability of 95% is 2.262 for nine degrees of freedom, as found from a table of t values. Substituting these values in the formula yields

$$n_{est} = \frac{(2.262)^2(9.627)^2}{5^2} = 19.$$

Thus, the number of cores should have been 19 to keep the estimated population mean within the range 57.0 ± 5 with 95% probability. For the ten cores actually taken, the estimated population mean would be within the range 57.0 ± 5 with probability between 80 and 90%.

The smaller the permitted deviation and the greater the specified assurance that the population mean lies within the permitted range, the greater is the estimated number of cores required. For example, if the permitted range is 57.0 ± 2 and the probability is 99% (corresponding to $t = 3.250$ for nine degrees of freedom),

$$n_{est} = \frac{(3.250)^2(9.627)^2}{2^2} = 245.$$

Statisticians emphasize random sampling because samples consisting of cores taken at random permit an estimate of the standard deviation of the population. Random sampling theory is useful for estimating the number of cores needed to attain a given degree of precision of the mean even though the distribution of analyses of soil cores over an area is not random.

For soil sampling with the objective of estimating the true population mean, a given number of cores generally will yield better population estimates if the sampling is *stratified* or done by subdivisions of the area, so that the samples are well distributed over the area to be sampled and are not taken completely at random. The individual strata might be soil types. In fields where erosion has occurred or in fields leveled for surface irrigation, the strata might represent different thicknesses of topsoil. In other instances, the strata might be equal areas. Cores within strata might be taken at random, by a Latin square system in which a sample is taken in each hypothetical row and column, or in other ways. In taking samples, it is important to avoid an alignment of soil core locations, especially in the direction of machinery operations, where the alignment might result in picking up a series of samples that are unusually high or low in one or more nutrients because of some machine operation that has not treated all the soil uniformly.

In a soil sampling study in North Dakota, Swenson et al. (1984) found that 20 cores per field would estimate the field mean with 90% confidence or probability to ±17%, ±18%, and ±9% for soil test nitrate, phosphorus, and potassium, respectively. Samples for nitrate were taken to a depth of 60 centimeters, and those for phosphorus and potassium to a depth of 15 centimeters. The variability for nitrate was greater in a sampling study in Colorado by Reuss et al. (1977). They found that 20 cores per field would estimate the nitrate content of the surface 60 centimeters of soil to ±26% with 90% confidence. To estimate the nitrate content to ±15% with 90% confidence would require 82 samples per field.

Directions issued by soil testing laboratories often suggest that about 20 individual cores be taken from each area in a field that appears to be different or is known to have been differentiated by past management. In a North Central Regional Publication, Ellis and Olson (1986) recommended 20 to 30 cores from each sampling area up to 8 hectares and 30 or more cores from larger areas.

Although a soil testing laboratory may issue directions for sampling, the actual sampling almost always is done by someone else. The laboratory has no way to compensate for nonrepresentative samples, but must make the analyses and provide the interpretations on the samples as received.

The usual procedure in sampling soils for chemical analysis is to composite the cores taken to represent a given area, to mix the composited cores, and to take a sample of the mixture for analysis. Individual cores are not analyzed. Soil variability is an important reason for using samples derived from composites of numerous cores. Another reason is that collecting additional cores is a more economical way of increasing the precision of the soil test values than is making additional analyses. The coefficient of variation of repeated analyses on a given core is almost always much smaller than the coefficient of variation of analyses on different cores.

Compositing cores from an area such as an experimental plot or a field evidently obscures the effects of discontinuities that occur on a scale of distances exceeding the extent of the root systems of individual plants as well as the way individual plants respond to discontinuities within their root zones. Moreover, there may be some interaction among the segments of a composite sample, especially during equilibration of the extracting solution with the components of the composite. The net result may be an analytical value that does not reflect the way the effects of the variations would be integrated by a plant even if all the individual cores came from the root zone of a single plant. Only a beginning has been made in resolving these problems of sampling soils for chemical tests.

Depending upon the conditions, uneven application of fertilizers may have only a small effect or it may have a large effect upon the number of cores required to obtain a given degree of precision of the mean of chemical soil tests made subsequently for plant nutrients. The results of a sampling study on a field experiment in Oregon (Horneck et al., 1990) showed that to estimate the mean extractable phosphorus with a given level of precision required about 1.4 times as many samples per plot from plots receiving phosphorus-bearing fertilizer in bands as from plots on which the fertilizer was applied broadcast. In this instance, the percentage increase in extractable phosphorus from fertilization was small, and so only a minor effect of uneven application would be expected. Illustrating the opposite extreme, Hooker (1976) found in a soil sampling study in Nebraska that to estimate the mean extractable phosphorus with a given level of precision required 15 times more cores at two sites where phosphorus-bearing fertilizer had been placed in the row for corn for several years than at three sites on the same soil type where the fertilizer had been broadcast.

Where fertilizer banding is coordinated with crop rows, the location of the most recently applied bands of fertilizer is known when samples are taken in fields of established crops. The locations from which soil cores are obtained can be similarly coordinated to improve the makeup of composite samples. In an investigation of sampling soils for phosphorus tests in three such experiments in Colorado and Kansas, Kitchen et al. (1990) summarized their results by the formula that for each core taken in the fertilizer band, the number of cores that should be taken between bands to estimate the mean value of extractable phosphorus for the soil should be (0.27)(band spacing in centimeters). To accomplish the same objective, Blackmer et al. (1991) recommended taking samples in sets of eight cores oriented relative to the rows. The first core is taken one-eighth of the distance between any two rows. The second is taken from another spot two-eighths of the distance between any two rows in the same direction from the row as the first. The third through seventh cores are taken similarly at locations three-eighths through seven-eighths of the distance between rows, and the eighth is taken in a row. Except for the restriction regarding rows, the cores are taken at random locations within the area being sampled. The authors recommended taking two or three sets of eight cores from an area not exceeding 2.5 hectares and compositing the cores for submission as a sample for analysis. They recommended also that areas having different soil types or management histories should be sampled separately. Smith (1977) recommended a similar procedure. In an investigation of procedures for sampling wheat land after harvest where the wheat had received banded fertilizer beneath the rows, Mahler (1990) tried a procedure about like the one described by Blackmer et al. (1991). Mahler, however, decided that, all things considered, random sampling was the best of the procedures he tried.

Ashworth (1990) described a method for sampling soil containing banded fertilizer in which the soil picked up by a chain saw in a cut made perpendicular to the band is caught in a special container. This method would automatically mix the fertilized and unfertilized soil within the zone of saw blade action.

The objective of soil sampling and testing is of course to provide a basis for guidance regarding the degree of deficiency and the quantity of fertilizer needed. The soil test value is only a means to an end. Obtaining an estimate of the true mean soil test value is emphasized from the practical standpoint because in the calibration process crop responses are related to mean soil test values. Moreover, economy in testing is increased by compositing soil cores for analysis.

On the other hand, Kitchen et al. (1990) were concerned about the need for an evaluation that takes into account the potential effects of the irregular distribution of a nutrient associated with fertilizer banding as opposed to estimating a mean soil test value. They suggested that for situations in which fertilizer phosphorus is banded in the same location year after year, a soil sampling procedure designed to yield a mean value for the soil as a whole would be inappropriate. Rather, the sampling scheme should be designed to evaluate the residual fertilizer phosphorus in the bands.

The theoretical understanding of the consequences of residual bands as reflected in soil tests has not been developed as yet. Further theoretical and experimental work is needed to provide a guide for soil sampling, analytical practices, and interpretations where residual fertilizer has been banded.

Sampling Time.   Fig. 4-33 illustrates the effects of sampling time on extractable soil phosphorus and potassium, as found in two independent investigations in the United States. Except for the fact that the monthly trends in the data for corn would be subdued because the authors presented the data as 3-month moving averages, one can only speculate about the causes for the differences observed. The important point, however, is the fact that in both investigations the seasonal variations were great enough to influence the interpretation of the results.

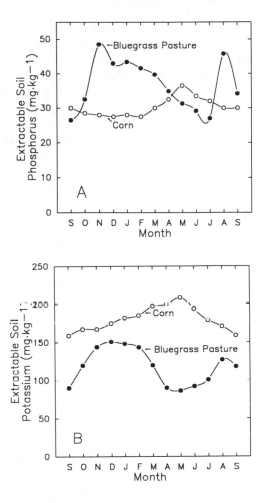

Fig. 4-33. Seasonal patterns of extractable phosphorus (A) and extractable potassium (B) in independent investigations on corn in Ohio (Lockman and Malloy, 1984) and bluegrass pasture in West Virginia (Childs and Jencks, 1967). The corn was fertilized in March or April before plowing, which incorporated the cornstalks from the crop in the previous year. The bluegrass pasture received no fertilizer. Samples were taken from the surface 20 centimeters of soil under corn and the surface 5 centimeters under grass.

Friesen et al. (1985) investigated the fluctuations in phosphorus extracted by two widely used soil test methods in monthly samples taken over a period of 2 years from pastures in New South Wales, where the areas were grazed regularly the year around. The results obtained with the two methods were correlated, and the quantities of phosphorus extracted increased with the amounts of phosphorus applied as superphosphate. Significant variations in extractable phosphorus with time were found, and the fluctuations were great enough to affect the interpretation of results, but the authors found no obvious relation of the results to the seasons.

The three investigations cited here are consistent in showing the importance of time of sampling. The inconsistency of the time effect and the limited understanding of the processes involved, however, indicate the importance of determining the effect of sampling time under the specific conditions for which tests are made. Sampling time may be a significant factor in calibration. Unless tests indicate otherwise, a consistent sampling time is preferable for both calibration experiments and samples from producers' fields.

If sampling time is found to be important, and if the seasonal trends are fairly consistent, the value of taking soil samples at an appropriate time can be stressed in information supplied to persons who take soil samples. But because a soil testing laboratory must make analyses on samples whenever they are taken, a part of the calibration process should be to determine the seasonal variation and to develop a way to adjust the analyses as needed for the time at which the samples were taken. If fluctuations in soil test values occur independently of season or other readily determined conditions, the fluctuations must be accepted as part of the experimental error.

In cropped soils, relatively high values of soil tests are to be expected after fertilization, and relatively low values are to be expected late in the growing season after crops have absorbed nutrients and after additions of fertilizer phosphorus have equilibrated with the soil. The decrease in phosphorus tests due to equilibration continues after crop removal, although by that time it should be slow.

Fertilizer potassium equilibrates with the soil more rapidly than fertilizer phosphorus. Because some potassium absorbed by grain crops may return to the soil through the roots before the grain is harvested, the minimum soil test value for potassium would be expected to coincide with the time of maximum content of potassium in above-ground parts.

Most of the potassium absorbed by crops harvested for grain is present in the vegetative portions of the plants and generally is returned to the soil. Being in soluble inorganic form, the potassium in the crop residues quickly replenishes the potassium supply in the soil. In the Ohio research on corn in Fig. 4-33, the extractable soil potassium reached a minimum in September and increased thereafter, but the cornstalks were not plowed under until the following spring. In the West Virginia data on bluegrass pasture, the extractable potassium reached a minimum in May, which would be close to the end of the period of rapid growth.

Particularly in humid regions, the time of sampling is more critical for mineral forms of nitrogen than for extractable phosphorus and potassium. Fig. 2-12 provides one illustration of the seasonal variability of mineral nitrogen. The peak concentration of mineral nitrogen was more than 15 times the minimum. More information related to sampling time for mineral nitrogen may be found in Section 4-4.6.

Statistical Mapping.   In recent years, many publications have appeared that deal with sampling and statistical procedures that are variously referred to as geostatistics, regionalized variable theory, and kriging (see Warrick et al, 1986; Nielsen and Alemi, 1989). The basic concept is that analyses on samples taken close together are correlated to some extent and that the correlation decreases with increasing distance. The theory makes it possible to estimate analytical values between sampling locations, so that a map can be drawn with contour lines to represent the distribution of estimated analytical values within a sampled area.

The technology is available to adapt this technique to improve the uniformity of nutrient supplies for crops in individual fields. By coupling a computer to a map of soil test values, adjustments in quantities of individual nutrients applied can be made automatically as a fertilizer distributor crosses a field. Whether the benefit in profitability of crop production would justify the extra costs of equipment and soil testing remains to be determined. Individual analyses on many soil samples would be required to prepare the map.

For a field experiment, the mapping technique could be used to characterize the preexisting levels of nutrients in individual plots to which fertilizers are to be applied. The map showing the various levels of a nutrient in a plot would have to be interpreted in some way to yield a single value for use in a covariance analysis of the experimental data to improve the precision by adjusting the yields or plant compositions on individual plots for the preexisting levels within those plots. Mean values of a soil characteristic of interest for individual plots no doubt could be obtained as an extension of computer programs now available to make the calculations and draw the maps. Whether values for individual plots obtained in this way would be superior to those obtained by the simpler procedure of analyzing samples composited by plots is not known.

4-4.8.3. Sample Treatment

In preparing quantities of soils for greenhouse experiments and associated laboratory analyses, it is convenient to dry the individual samples in air at room temperature, pass them through a relatively coarse sieve, and mix them thoroughly before placing equal weights or equal volumes of soil in the containers for growth of the test crop. Except for additional crushing to permit passing the soils through a finer sieve to promote uniformity among small samples, the soils used for laboratory analyses generally receive the same treatment as the soils on which the test crop is grown in the greenhouse, and the results of the analyses thus correspond to the initial conditions in the cropped cultures. (Procedures such as passing samples through a hammer mill can produce marked changes and generally are avoided.) Samples of soils submitted by growers also are normally dried before analysis. The treatment of the samples used for correlating chemical analyses made in the laboratory with the biological assays made in the greenhouse thus is consistent with the procedure used to test samples submitted by growers.

Air drying of samples used for testing is a satisfactory procedure for those nutrients and soils for which the effect of drying on nutrient availability and extractability is relatively small or where the effects are consistent among soils.

Research has shown, however, that these conditions are by no means universal. The air drying of samples where the limiting conditions do not prevail can be a source of error in estimating field performance from measurements of extractable nutrients in the laboratory.

Most of the research has been on nitrogen and potassium, but drying of soils affects the behavior of other nutrients as well. For nitrogen and phosphorus, the release from microorganisms killed by the desiccation is an important cause of an increase in availability. See a paper by Jenkinson (1988) for a review and discussion of methods to estimate the microbial biomass in soils.

The occurrence of a flush of nitrogen mineralization in soils that are re-moistened following drying has long been known from scientific evidence, and practices in native agriculture in some parts of the world have evolved to take advantage of it (see, for example, a paper by Birch, 1958). Marumoto et al. (1982) found a close relationship between the decrease in microbial biomass carbon and the increase in ammonium plus nitrate nitrogen in soils during incubation for 56 days after the soils had been dried at 70° C or fumigated with chloroform and then reinoculated with soil that had not received either of these treatments. They estimated that about 77% of the flush of mineral nitrogen caused by heat drying (70° C) and 55% of that caused by air drying were derived from nitrogen in the killed microbial tissue, and that the remainder came from other sources.

In a group of 18 soils from New Zealand, Sparling et al. (1985) found that biomass carbon declined from 11 to 68% as a result of drying. And Anderson and Domsch (1980) found that for a group of 29 soils from Germany, the average quantities of nutrients in the microbial biomass in kilograms per hectare were 108, 83, and 70 for nitrogen, phosphorus, and potassium, respectively.

Sparling et al. (1985) found that air drying the 18 soils from New Zealand increased the phosphorus extracted by the 0.5 molar sodium bicarbonate method by up to 184% (two sandy soils showed a decrease). The release of inorganic phosphorus from microbial tissue upon drying was estimated to account for 3 to 76% of the observed increase in extractable phosphorus. In subsequent work with 32 soils, Sparling et al. (1987) estimated that release of microbial phosphorus on drying accounted for up to 95% of the increase in sodium bicarbonate-extractable phosphorus and that the microbial phosphorus comprised 4 to 97% of the inorganic phosphorus extracted.

From the standpoint of calibration, which requires field experiments, these findings suggest that the conventional use of air-dry samples for laboratory analysis can be a significant source of error because the plants in the field respond to the conditions in the soils on which they grow. Soils in the field rarely if ever become air dry to the depth to which samples are taken for analysis. Sparling et al. (1985, 1987) found that the error due to phosphorus released from microbial cells upon drying was greatest in soils that were subjected to the least drying in the field and in soils having a relatively high content of organic matter and hence of microbial tissue. In some soils, dehydration of hydrous oxides of iron and aluminum upon drying decreases the phosphorus sorption capacity and increases the concentration of phosphorus in solution.

Air-drying of soil no doubt causes release of potassium from the killed microbial tissue, although this does not seem to have been investigated directly. Marumoto et al. (1982) found that fumigating soils with chloroform increased the exchangeable potassium for up to 3 days, but after that the exchangeable potassium fluctuated, and the amounts were not related to the microbial biomass.

With potassium, the principal effect of drying appears to result from inorganic processes involving the retention of potassium in interlayer positions in micaceous minerals in the clay and silt fractions of soils. Fig. 4-34 shows the effect of air-drying and oven-drying on the exchangeable potassium content of profile samples of four soils. Suggesting the importance of inorganic processes, the

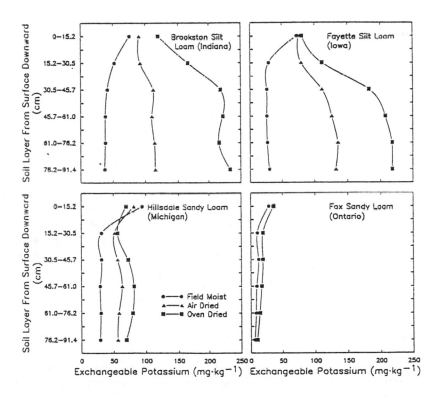

Fig. 4-34. Exchangeable potassium found by the ammonium acetate method in four soil profiles as affected by drying. (Hanway et al., 1961)

greatest effects of drying occurred in the subsoils, where the content of microbial biomass should be relatively low.

In most studies, including two extensive regional investigations (Hanway et al., 1961, 1962), the biological indexes of potassium availability have been correlated more highly with measurements of exchangeable potassium made on moist samples than on air-dried samples. The reverse was found in one investigation (Jones et al., 1961).

Although the effects of drying are well known and can be large, Haby et al. (1990) reported that field-moist samples are not being routinely analyzed at any public laboratory in the United States. The laboratory in Iowa, which used field-moist samples for potassium tests for more than 20 years, is now drying all samples.

Drying also affects the amounts of micronutrient cations extracted from soils by currently used extractants. Martens and Lindsay (1990) reviewed research on this subject, noting that the explanations advanced for the changes due to drying include dehydration, oxidation-reduction effects, and alteration of functional groups that bind the cations.

Field-moist samples submitted by producers are of course by no means moist to the same degree. For some nutrients this may be important. But for potassium at least, the water content does not appear to be of much importance as long as the samples are not actually air dried because the most significant changes occur as air dryness is approached (Luebs et al., 1956).

Attempting to return the exchangeable potassium content of air-dried soils to the original field-moist status by adding water and allowing the samples to stand is a slow process. Haby et al. (1988) were able to accomplish the desired conditioning by adding an excess of water, freezing the samples, and then allowing the samples to thaw and dry 48 hours. For the six soils they studied that retained less than 5% water after this time, the values for exchangeable potassium did not differ significantly from those measured in the original field-moist samples. Further work is needed to determine whether the procedure is generally applicable. The additional steps represent an inconvenience that must be balanced against the inconveniences of dealing with field-moist samples. Another possible substitute for moist samples is the use of adjustment factors on analyses of air-dried samples if warranted by experimentally determined relationships between the effects of soil drying and soil type or some other measured property of the samples.

The most satisfactory alternative is direct investigation of the effects of sample preparation procedures for the specific soils and conditions involved. The experience recorded by Daughtrey et al. (1973) is a case in point. They found that the phosphorus extracted by the Mehlich 0.05 molar hydrochloric acid-0.0125 molar sulfuric acid reagent gave better estimates of the biological indexes of phosphorus availability where measurements were made on air-dried samples than on field-moist samples. They attributed the superiority of the dried samples to the mineralization of different amounts of organic phosphorus when the various soils were dried, the extraction of part of the mineralized phosphorus by the acid reagent, and the mineralization of organic phosphorus that occurred while the test crop was growing on the soils, which had been maintained in a field-moist condition. The inference about the importance of organic phosphorus was verified by correlations of measured organic phosphorus mineralization with the amounts of inorganic phosphorus extracted by the acid reagent, as well as with the biological indexes of phosphorus availability.

For some soil tests, the time of extraction is important and is characteristically specified in the soil test procedures and observed in the soil testing laboratory by use of a timing device. Other procedural details, such as the nature and setting of the shaking machine, are seldom specified and, if specified, tend to be disregarded because of differences in available equipment. Although differences in the shaking procedure affect the results obtained by some tests, the consequences probably are not important if the calibration has been made with the equipment actually used.

The temperature of extraction is rarely specified in soil test procedures. Especially for phosphorus, the temperature effect is significant and may influence the interpretation of measurements of extractable phosphorus made at different times of the year if the effect is not known and allowed for. An approximate allowance for the temperature effect can be made as a part of the overall calibration, but the adjustment must be recognized as an approximation because the temperature coefficient differs among soils. As an average of five soils of Bulgaria, Ivanov et al. (1989) found that the amounts of phosphorus extracted by the sodium bicarbonate method of Olsen et al. (1954) ranged from 39 milligrams per kilogram of soil at 10° C to 46 milligrams at 20° C and 55 milligrams at 30° C.

4-4.8.4. Volume Versus Mass Basis for Analysis

Whether chemical tests on different soils to obtain indexes of nutrient availability and fertilizer needs should be made on the basis of equal weights or masses of soil or on the basis of equal volumes would seem to be a question that should have been settled years ago. The fact that both systems are currently in use indicates that the issue is still open.

Traditionally the data obtained in chemical analysis are expressed as a ratio of the mass of the constituent for which the analysis is made to the mass of the sample analyzed. The analytical chemistry tradition has carried over into soil analysis for research purposes and also into soil testing.

From the soil testing standpoint, the problem with the mass basis is that fertilizers are applied per unit of land area, and not per unit mass of soil. The mass of soil to any given depth per unit of land area varies with the bulk density of the soil, and the field measurements of bulk density needed to convert the analytical data to quantities per unit area to a fixed depth are not made. When a given bulk density is used for all soils tested, the quantity of fertilizer corresponding to a particular test value applies to unit land area only when the bulk density of the soil is equal to the assumed value.

In contrast to the mass basis for expressing soil tests, values expressed on the volume basis are directly proportional to area, and thus are theoretically preferable. Although the usual method of preparing soil samples for analysis changes the bulk densities, correlations of biological indexes of availability with soil tests indicate that analytical values for equal volumes are superior to analytical values for equal masses in both field experiments (Saunders et al., 1987a) and greenhouse experiments (Daughtrey et al., 1973). Shickluna's (1962) analyses of samples by both methods in the Soil Testing Laboratory at Michigan State University showed that duplicate analyses on scooped samples agreed almost as well as those made on weighed samples.

According to Jones (1985a), most soil testing laboratories measure samples by volume for analysis because of the saving of time over weighing. Some laboratories, however, employ the assumption that the samples measured by volume for analysis all have the same mass and that the soil in the field also has a certain bulk density. Laboratories using this procedure are actually using a volume basis even though they may not be thinking in those terms.

Mehlich (1972) and Van Lierop (1989) have published thoughtful papers on the mass versus volume issue. Van Lierop discussed the issue in terms of lime requirement. The philosophy of the lime requirement test differs from that of the usual soil tests for plant nutrients in that except when calcium deficiency is involved, the immediate purpose of liming is to treat the soil, not the crop. For cultivated crops, the intent is to mix the liming material with the plowed layer of soil so that it will react with that volume.

4-4.8.5. Research Support

The precision of available calibrations is never as good as desired, and sometimes is grossly inadequate. Despite the great sums of money invested in fertilizers each year, and the potential financial benefits to be derived from properly adjusting fertilizer applications to the many specific circumstances of crops, soils, environment, and management that exist, administrative support for calibration experiments has been and remains poor. An additional part of the problem of inadequate calibration is the limited scientific reward associated with the performance of the research.

Much more can be learned about optimizing fertilizer applications from multinutrient experiments, but these are more expensive than single-nutrient experiments. Support generally has been inadequate to conduct the needed multinutrient experiments over enough years to develop them into bases for practical soil test interpretations. The apparent success achieved with the more economical multinutrient fertilizer prescription procedure described in Section 4-4.7 justifies more extensive testing of this approach.

Coupled with inadequate calibration is the inadequate use of scientific advances that may improve the predictions made from soil tests. A well organized soil testing operation is complex, and considerable time and money are invested in the mechanics of the procedures, including computerization, which generally involves special programs developed on a contract basis by outside experts

(Donohue and Gettier, 1990). Needed to complement the systematics is continuing research on in-house aspects of the operation, as well as evaluation, testing, and adaptation of new approaches developed elsewhere. Most of the research function of soil testing must be performed by publicly supported laboratories, which have the responsibility of passing the information along to laboratories in the private sector. Because the performance of the soil testing service must take precedence, the usual limitation of funds operates to the disadvantage of research and updating. No ready solution to this problem is in sight.

Inadequate research support is related further to inadequate verification of results of soil testing programs. Crop responses to fertilizers are visually obvious when they are large, but as the responses decrease they are no longer obvious and must be measured. The smaller the differences, the less reliable are simple measurements, and the more important it becomes to replicate and randomize the fertilized and unfertilized locations to separate the effects sought from the effects caused by nonuniformities in the area and other sources of experimental error. Producers rarely make precise measurements of the effects of the fertilizers they apply, and the laboratories that make the soil tests do not measure the results either. In consequence, the only kinds of direct checks that are effective are those provided by large positive or negative responses.

Indirect checks sometimes are made by growers who submit duplicate soil samples to different laboratories. The differences in recommendations they are likely to receive naturally make the recipients skeptical of the validity of soil tests in general.

The obscurity of the results derived from soil testing tends to promote the status quo and to limit the drive for improvements. Soil testing laboratories, their patrons, and the supporters of soil testing research would be much less complacent than they are if the errors in estimating fertilizer requirements were as obvious as the errors made by meteorologists in predicting tomorrow's weather.

A related problem is created by the philosophy that in agriculture, information should be provided free of charge. Most soil tests are not done free of charge, but they are done for such low charges that the emphasis must be on mass production by streamlined procedures. This situation is not conducive to the use of advances that could improve the quality of the output.

Perhaps because of the limited support for research, a procedure sometimes followed when a new and seemingly advantageous method is developed is to analyze a group of soil samples by the old method and the new method and to determine the correlation between them (Fixen and Grove, 1990). If the methods are well correlated, the new method may be substituted for the old after due adjustment for the regression relation between methods without actually calibrating the new method. This shortcut procedure carries some risk because the combination of two correlation coefficients may be deceiving. Fixen and Grove reproduced a table from Schulte and Hodgson (1987) showing, for example, that if the correlation between the old method and the biological indexes of availability is 0.85, which is good for field data, a correlation of 0.95 between measurements by the two methods could mean that the correlation of the new method with the original biological indexes might range from 0.64 to 0.97.

4-5. Literature Cited

Adams, F. 1966. Calcium deficiency as a causal agent of ammonium phosphate injury to cotton seedlings. *Soil Science Society of America Proceedings* 30:485-488.

Adams, F. 1971. Ionic concentrations and activities in soil solutions. *Soil Science Society of America Proceedings* 35:420-426.

Adams, F. 1974. Soil solution. Pp. 441-481. In E. W. Carson (Ed.), *The Plant Root and Its Environment*. University Press of Virginia, Charlottesville.

Adams, F., C. Burmester, N. V. Hue, and F. L. Long. 1980. A comparison of column-displacement and centrifuge methods for obtaining soil solutions. *Soil Science Society of America Journal* 44:733-735.

Addiscott, T. M., and A. P. Whitmore. 1987. Computer simulation of changes in soil mineral nitrogen and crop nitrogen during autumn, winter and spring. *Journal of Agricultural Science (Cambridge)* 109:141-157.

Allen, H. E., R. H. Hall, and T. D. Brisbin. 1980. Metal speciation. Effects on aquatic toxicity. *Environmental Science and Technology* 14:441-443.

Alvarez V., V. H., and D. M. da Fonseca. 1990. Definição de doses de fósforo para determinação da capacidade máxima de adsorção de fosfatos e para ensaios em casa de vegetação. *Revista Brasileira de Ciência do Solo* 14:49-55.

Amer, F., D. R. Bouldin, C. A. Black, and F. R. Duke. 1955. Characterization of soil phosphorus by anion exchange resin adsorption and P^{32}-equilibration. *Plant and Soil* 6:391-408.

Anderson, J. P. E., and K. H. Domsch. 1980. Quantities of plant nutrients in the microbial biomass of selected soils. *Soil Science* 130:211-216.

Andres, E. 1990. Soil fertility data banks as a tool for site-specific K-recommendations. Pp. 291-305. In *Development of K-Fertilizer Recommendations*. Proceedings of the 22nd Colloquium of the International Potash Institute held at Soligorsk/USSR. International Potash Institute, Worblaufen-Bern, Switzerland.

Appel, T., and K. Mengel. 1990. Importance of organic nitrogen fractions in sandy soils, obtained by electro-ultrafiltration or $CaCl_2$ extraction, for nitrogen mineralization and nitrogen uptake of rape. *Biology and Fertility of Soils* 10:97-101.

Appel, T., and D. Steffens. 1988. Vergleich von Elektro-Ultrafiltration (EUF) und Extraktion mit 0.01 molarer $CaCl_2$-Lösung zur Bestimmung des pflanzenverfügbaren Stickstoffs im Boden. *Zeitschrift für Pflanzenernährung und Bodenkunde* 151:127-130.

Armour, J. D., A. D. Robson, and G. S. P. Ritchie. 1990. Prediction of zinc deficiency in navy beans (*Phaseolus vulgaris*) by soil and plant analyses. *Australian Journal of Experimental Agriculture* 30:557-563.

Armstrong, R. D., and K. R. Helyar. 1992. Changes in soil phosphate fractions in the rhizosphere of semi-arid pasture grasses. *Australian Journal of Soil Research* 30:131-143.

Ashworth, J. 1990. Method and an apparatus for taking soil samples. *United States Patent US 4,922,763.* (*Soils and Fertilizers* 54, Abstract 3885. 1991.)

Atanasiu, N. 1954. Die Gefäss- und die Felddüngungsversuche und ihre Vergleichbarkeit unter Zugrundelegung des Wirkungsgesetzes. *Zeitschrift für Pflanzenernährung Düngung Bodenkunde* 66:118-124.

Baker, D. E., and M. C. Amacher. 1981. *The Development and Interpretation of a Diagnostic Soil-Testing Program*. Pennsylvania State University, Agricultural Experiment Station Bulletin 826. University Park.

Balba, A. M., and R. H. Bray. 1956. New fields for the application of the Mitscherlich equation: I. A quantitative measure of the relative effectiveness of nutrients. *Soil Science* 82:497-502.

Balba, A. M., and R. H. Bray. 1957. New fields for the application of the Mitscherlich equation: 2. Contribution of soil nutrient forms to plant nutrient uptake. *Soil Science* 83:131-139.

Baldwin, J. P., P. H. Nye, and P. B. Tinker. 1973. Uptake of solutes by multiple root systems from soil. III. A model for calculating the solute uptake by a randomly dispersed root system developing in a finite volume of soil. *Plant and Soil* 38:621-635.

Ballard, T. M. 1984. A simple model for predicting stand volume growth response to fertilizer application. *Canadian Journal of Forest Research* 14:661-665.

Bar, A., A. Banin, and Y. Chen. 1987. Adsorption and exchange of K in multi-ionic soil systems as affected by mineralogy. Pp. 155-170. In *Methodology in Soil-K Research*. Proceedings of the 20th Colloquium of the International Potash Institute held in Baden bei Wien/Austria 1987. International Potash Institute, Worblaufen-Bern, Switzerland.

Barber, S. A. 1962. A diffusion and mass-flow concept of soil nutrient availability. *Soil Science* 93:39-49.

Barber, S. A. 1984. *Soil Nutrient Bioavailability. A Mechanistic Approach.* John Wiley & Sons, NY.

Barber, S. A. 1985. Potassium availability at the soil-root interface and factors influencing potassium uptake. Pp. 309-324. In R. D. Munson (Ed.), *Potassium in Agriculture*. American Society of Agronomy, Madison, WI.

Barber, S. A., J. M. Walker, and E. H. Vasey. 1963. Mechanisms for the movement of plant nutrients from the soil and fertilizer to the plant root. *Journal of Agricultural and Food Chemistry* 11:204-207.

Barraclough, P. B. 1990. Modelling K uptake by plants from soil. Pp. 217-230. In *Development of K-Fertilizer Recommendations*. Proceedings of the 22nd Colloquium of the International Potash Institute held at Soligorsk/USSR. International Potash Institute, Worblaufen-Bern, Switzerland.

Barraclough, P. B., and P. B. Tinker. 1981. The determination of ionic diffusion coefficients in field soils. I. Diffusion coefficients in sieved soils in relation to water content and bulk density. *Journal of Soil Science* 32:225-236.

Barrow, G. M. 1979. *Physical Chemistry*. Fourth Edition. McGraw-Hill Book Company, NY.

Barrow, N. J. 1966. Nutrient potential and capacity II. Relationship between potassium potential and buffering capacity and the supply of potassium to plants. *Australian Journal of Agricultural Research* 17:849-861.

Barrow, N. J. 1967a. Relationship between uptake of phosphorus by plants and the phosphorus potential and buffering capacity of the soil — an attempt to test Schofield's hypothesis. *Soil Science* 104:99-106.

Barrow, N. J. 1967b. Effects of the soil's buffering capacity for phosphate on the relation between uptake of phosphorus and the phosphorus extracted by sodium bicarbonate. *Journal of the Australian Institute of Agricultural Science* 33:119-121.

Barrow, N. J. 1982. Soil fertility changes. Pp. 197-220. In A. B. Costin and C. H. Williams (Eds.), *Phosphorus in Australia*. Centre for Resource & Environmental Studies, Canberra. (*Soils and Fertilizers* 48, Abstract 2261. 1985.)

Bartelli, L. J. 1978. Technical classification system for soil survey interpretation. *Advances in Agronomy* 30:247-289.

Bear, F. E., and A. L. Prince. 1945. Cation-equivalent constancy in alfalfa. *Journal of the American Society of Agronomy* 37:217-222.

Bear, F. E., A. L. Prince, and J. L. Malcolm. 1945. *Potassium Needs of New Jersey Soils*. New Jersey Agricultural Experiment Station, Bulletin 721. New Brunswick.

Becker, F. A., and W. Aufhammer. 1982. Nitrogen fertilization and methods of predicting the N requirements of winter wheat in the Federal Republic of Germany. Pp. 33-36. In *Symposium on Fertilisers and Intensive Wheat Production in the EEC*. Fertiliser Society (London) Proceedings No. 211.

Beckett, P. 1972. Critical cation activity ratios. *Advances in Agronomy* 24:379-412.

Beckett, P. H. T. 1964a. Studies on soil potassium I. Confirmation of the ratio law: measurement of potassium potential. *Journal of Soil Science* 15:1-8.

Beckett, P. H. T. 1964b. Studies on soil potassium II. The 'immediate' *Q/I* relations of labile potassium in the soil. *Journal of Soil Science* 15:9-23.

Beckett, P. T., and R. Webster. 1971. Soil variability: A review. *Soils and Fertilizers* 34:1-15.

Beckwith, R. S. 1963. Chemical extraction of nutrients in soils and uptake by plants. *Agrochimica* 7:296-313.

Beckwith, R. S. 1965. Sorbed phosphate at standard supernatant concentration as an estimate of the phosphate needs of soils. *Australian Journal of Experimental Agriculture and Animal Husbandry* 5:52-58.

Bell, P. F., R. L. Chaney, and J. S. Angle. 1991. Free metal activity and total metal concentrations as indices of micronutrient availability to barley [*Hordeum vulgare* (L.) 'Klages']. *Plant and Soil* 130:51-62.

Birch, H. F. 1958. The effect of soil drying on humus decomposition and nitrogen availability. *Plant and Soil* 10:9-31.

Blackmer, A. M., T. F. Morris, D. R. Keeney, R. D. Voss, and R. Killorn. 1991. *Estimating Nitrogen Needs for Corn by Soil Testing*. Iowa State University Extension PM-1381. Ames.

Bock, B. R., and G. W. Hergert. 1991. Fertilizer nitrogen management. Pp. 139-164. In R. F. Follett, D. R. Keeney, and R. M. Cruse (Eds.), *Managing Nitrogen for Groundwater Quality and Farm Profitability*. Soil Science Society of America, Madison, WI.

Bolland, M. D. A., and M. J. Baker. 1987. Increases in soil water content decrease the residual value of superphosphate. *Australian Journal of Experimental Agriculture* 27:571-578.

Boon, R. 1983. Avis de fumure en azote basé sur l'analyse du profil pour céréales d'hiver sur sols limoneux et sablo-limoneux profonds. *Agro* 2:46-61. (*Soils and Fertilizers* 47, Abstract 8968. 1984.)

Borlan, Z., C. Bardeiau, C. Hera, and T. Tbranu. 1968. The use of soil chemical testing data in agriculture. *Transactions 9th International Congress of Soil Science* 4:733-741.

Bower, C. A., and J. O. Goertzen. 1955. Negative adsorption of salts by soils. *Soil Science Society of America Proceedings* 19:147-151.

Bowman, R. A., and S. R. Olsen. 1985. Assessment of phosphate-buffering capacity: 1. Laboratory methods. *Soil Science* 140:287-291.

Bradfield, E. G. 1969. Quantity/intensity relations in soils and the potassium nutrition of the strawberry plant (*Fragaria* sp.). *Journal of the Science of Food and Agriculture* 20:32-38.

Brar, B. S., A. C. Vig, and G. S. Bahl. 1986. Relationship of growth and P uptake by wheat to quantity, intensity and buffering capacity factors of P in bench mark soils of Punjab. *Journal of Nuclear Agriculture and Biology* 15:60-66.

Braunschweig, L. C. v., and K. Mengel. 1971. Der Einfluss verschiedener den Kaliumzustand des Bodens charakterisierender Parameter auf den Kornertrag von Hafer. *Landwirtschaftliche Forschung, Sonderheft* 26/I:65-72.

Bray, R. H. 1944. Soil-plant relations: I. The quantitative relation of exchangeable potassium to crop yields and to crop response to potash additions. *Soil Science* 58:305-324.

Bray, R. H. 1945. Soil-plant relations: II. Balanced fertilizer use through soil tests for potassium and phosphorus. *Soil Science* 60:463-473.

Bray, R. H., and E. E. DeTurk. 1939. The release of potassium from nonreplaceable forms in Illinois soils. *Soil Science Society of America Proceedings (1938)* 3:101-106.

Bray, R. H., and L. T. Kurtz. 1945. Determination of total, organic, and available forms of phosphorus in soils. *Soil Science* 59:39-45.

Breimer, T., C. Sonneveld, and L. Spaans. 1988. A computerized programme for fertigation of glasshouse crops. *Acta Horticulturae* 222:43-50.

Bremner, J. M. 1965. Nitrogen availability indexes. Pp. 1324-1345. In C. A. Black (Ed.), *Methods of Soil Analysis. Part 2. Chemical and Microbiological Properties*. Agronomy 9. American Society of Agronomy, Madison, WI.

Brennan, R. F., and J. W. Gartrell. 1990. Reaction of zinc with soil affecting its availability to subterranean clover. I. The relationship between critical concentrations of extractable zinc and properties of Australian soils responsive to applied zinc. *Australian Journal of Soil Research* 28:293-302.

Brookes, P. C. 1982. Correction for seed-phosphorus effects in L-value determinations. *Journal of the Science of Food and Agriculture* 33:329-335.

Brown, A. L., J. Quick, and J. L. Eddings. 1971. A comparison of analytical methods for soil zinc. *Soil Science Society of America Proceedings* 35.105-107.

Brown, J. R. (Ed.). 1987. *Soil Testing: Sampling, Correlation, Calibration, and Interpretation*. SSSA Special Publication Number 21. Soil Science Society of America, Madison, WI.

Bruce, R. C., L. A. Warrell, D. G. Edwards, and L. C. Bell. 1988. Effects of aluminium and calcium in the soil solution of acid soils on root elongation of *Glycine max* cv. Forrest. *Australian Journal of Agricultural Research* 38:319-338.

Bullen, C. W., R. J. Soper, and L. D. Bailey. 1983. Phosphorus nutrition of soybeans as affected by placement of fertilizer phosphorus. *Canadian Journal of Soil Science* 63:199-210.

Buol, S. W. 1972. Fertility capability classification. Pp. 44-50. In *Agronomic-Economic Research on Tropical Soils. Annual Report for 1971*. Soil Science Department, North Carolina State University, Raleigh.

Burns, I. G. 1980a. A simple model for predicting the effects of leaching of fertilizer nitrate during the growing season on the nitrogen fertilizer needs of crops. *Journal of Soil Science* 31:175-185.

Burns, I. G. 1980b. A simple model for predicting the effects of winter leaching of residual nitrate on the nitrogen fertilizer need of spring crops. *Journal of Soil Science* 31:187-202.

Cabrera, M. L., and D. E. Kissel. 1988a. Potentially mineralizable nitrogen in disturbed and undisturbed soil samples. *Soil Science Society of America Journal* 52:1010-1015.

Cabrera, M. L., and D. E. Kissel. 1988b. Evaluation of a method to predict nitrogen mineralized from soil organic matter under field conditions. *Soil Science Society of America Journal* 52:1027-1031.

Caldwell, A. C., A. Hustrulid, and R. F. Holt. 1952. The effect of nitrogen and potassium on the uptake of phosphorus by corn, oats, and alfalfa in the greenhouse. Pp. 91-94. In *Compilation of Experimental Work Reported at the Fourth Annual Phosphorus Conference of the North Central Region, December 4-5, 1952*. Issued by the Mineral Deficiencies Committee of the North Central Region. Iowa State University, Ames.

Carter, J. N., D. T. Westermann, and M. E. Jensen. 1976. Sugarbeet yield and quality as affected by nitrogen level. *Agronomy Journal* 68:49-55.

Cate, R. B., Jr., and L. A. Nelson. 1965. *A Rapid Method for Correlation of Soil Test Analyses With Plant Response Data*. North Carolina Agricultural Experiment Station, International Soil Testing Series, Technical Bulletin No. 1. Raleigh.

Cate, R. B., Jr., and L. A. Nelson. 1971. A simple statistical procedure for partitioning soil test correlation data into two classes. *Soil Science Society of America Proceedings* 35:658-660.

Chand, M., A. C. Vig, G. Dev, and A. S. Sidhu. 1984. Evaluation of fertilizer recommendation based on soil tests for targetting rice yield on cultivators' fields. *Punjab Agricultural University, Journal of Research* 21:20-28.

Checkai, R. T., R. B. Corey, and P. A. Helmke. 1987. Effects of ionic and complexed metal concentrations on plant uptake of cadmium and micronutrient metals from solution. *Plant and Soil* 99:335-345.

Chien, S. H., P. W. G. Sale, and D. K. Friesen. 1990. A discussion of the methods for comparing the relative effectiveness of phosphate fertilizers varying in solubility. *Fertilizer Research* 24:149-157.

Childs, F. D., and E. M. Jencks. 1967. Effects of time and depth of sampling upon soil test results. *Agronomy Journal* 59:537-540.

Claassen, N., and A. Jungk. 1982. Kaliumdynamik im wurzelnahen Boden in Beziehung zur Kaliumaufnahme von Maispflanzen. *Zeitschrift für Pflanzenernährung und Bodenkunde* 145:513-525.

Clarke, G. R. 1951. The evaluation of soils and the definition of quality classes from studies of the physical properties of the soil profile in the field. *Journal of Soil Science* 2:50-60.

Clement, C. R. 1971. Effect of cut and grazed swards on the supply of nitrogen to subsequent arable crops. Pp. 166-171. In *Residual Value of Applied Nutrients*. Ministry of Agriculture, Fisheries and Food, Technical Bulletin No. 20. Her Majesty's Stationery Office, London.

Collin, D. 1990. Soil and plant test data in computerized fertilizer recommendations. Pp. 279-290. In *Development of K-Recommendations*. Proceedings of the 22nd International Colloquium of the International Potash Institute held at Soligorsk/USSR. International Potash Institute, Worblaufen-Bern, Switzerland.

Colwell, J. D. 1963. The estimation of the phosphorus fertilizer requirements of wheat in southern New South Wales by soil analysis. *Australian Journal of Experimental Agriculture and Animal Husbandry* 3:190-197.

Comerford, N. B., and R. F. Fisher. 1982. Use of discriminant analysis for classification of fertilizer-responsive sites. *Soil Science Society of America Journal* 46:1093-1096.

Cope, J. T., and C. E. Evans. 1985. Soil testing. *Advances in Soil Science* 1:201-228.

Cope, J. T., Jr., and R. D. Rouse. 1973. Interpretation of soil test results. Pp. 35-54. In L. M. Walsh and J. D. Beaton (Eds.), *Soil Testing and Plant Analysis*. Soil Science Society of America, Madison, WI.

Corey, R. B. 1987. Soil test procedures: Correlation. Pp. 15-22. In J. R. Brown (Ed.), *Soil Testing: Sampling, Correlation, Calibration, and Interpretation*. SSSA Special Publication Number 21. Soil Science Society of America, Madison, WI.

Cornforth, I. S., and A. G. Sinclair. 1982. Model for calculating maintenance phosphate requirements for grazed pastures. *New Zealand Journal of Experimental Agriculture* 10:53-61.

Dahnke, W. C., and G. V. Johnson. 1990. Testing soils for available nitrogen. Pp. 127-139. In R. L. Westerman (Ed.), *Soil Testing and Plant Analysis*. Third Edition. Soil Science Society of America, Madison, WI.

Dahnke, W. C., and R. A. Olson. 1990. Soil test correlation, calibration, and recommendation. Pp. 45-71. In R. L. Westerman (Ed.), *Soil Testing and Plant Analysis*. Third Edition. Soil Science Society of America, Madison, WI.

Dalal, R. C., and E. G. Hallsworth. 1976. Evaluation of the parameters of soil phosphorus availability factors in predicting yield response and phosphorus uptake. *Soil Science Society of America Journal* 40:541-546.

Dalal, R. C., and R. J. Mayer. 1990. Long-term trends in fertility of soils under continuous cultivation and cereal cropping in southern Queensland. VIII Available nitrogen indices and their relationships to crop yield and N uptake. *Australian Journal of Soil Research* 28:563-575.

Danneberg, O. H., P. Nelhiebel, W. Schneider, and H. Amman. 1990. Standortsbezogene Probenahme ein Weg zur optimalen Düngeempfehlung. *Verband Deutscher Landwirtschaftlicher Untersuchungs- und Forschungsanstalten Reihe Kongressberichte* No. 30:625-630. (*Soils and Fertilizers* 54, Abstract 1239. 1991.)

Daughtrey, Z. W., J. W. Gilliam, and E. J. Kamprath. 1973. Soil test parameters for assessing plant-available P of acid organic soils. *Soil Science* 115:438-446.

Dean, L. A. 1954. Yield-of-phosphorus curves. *Soil Science Society of America Proceedings* 18:462-466.

Deans, J. R., J. A. E. Molina, and C. E. Clapp. 1986. Models for predicting potentially mineralizable nitrogen and decomposition rate constants. *Soil Science Society of America Journal* 50:323-326.

Dear, B. S., K. R. Helyar, W. J. Muller, and B. Loveland. 1992. The P fertilizer requirements of subterranean clover, and the soil P status, sorption and buffering capacities from two P analyses. *Australian Journal of Soil Research* 30:27-48.

DeMent, J. D., G. Stanford, and B. N. Bradford. 1959a. A method for measuring short-term nutrient absorption by plants: II. Potassium. *Soil Science Society of America Proceedings* 23:47-50.

DeMent, J. D., G. Stanford, and C. M. Hunt. 1959b. A method for measuring short-term nutrient absorption by plants: III. Nitrogen. *Soil Science Society of America Proceedings* 23:371-374.

Denton, H. P., G. C. Naderman, S. W. Buol, and L. A. Nelson. 1986. Use of a technical soil classification system in evaluation of corn and soybean response to deep tillage. *Soil Science Society of America Journal* 50:1309-1314.

Denton, H. P., G. F. Peedin, S. N. Hawks, and S. W. Buol. 1987. Relating the Fertility Capability Classification System to tobacco response to potassium fertilization. *Soil Science Society of America Journal* 51:1224-1228.

De Wit, C. T. 1953. *A Physical Theory on Placement of Fertilizers.* Verslagen van Landbouwkundige Onderzoekingen No. 59.4. Staatsdrukkerij en Uitgeverijbedrijf, Wageningen.

Diest, A. van. 1987. Coordinator's report on the 2nd working session. Pp. 211-212. In *Methodology in Soil-K Research.* Proceedings of the 20th Colloquium of the International Potash Institute held in Baden bei Wien/Austria 1987. International Potash Institute, Worblaufen-Bern, Switzerland.

Donohue, S. J., and S. W. Gettier. 1990. Data processing in soil testing and plant analysis. Pp. 741-755. In R. L. Westerman (Ed.), *Soil Testing and Plant Analysis.* Third Edition. Soil Science Society of America, Madison, WI.

Dos Santos, A. F., M. T. Eid, A. van Diest, and C. A. Black. 1960. Evaluation of biological indexes of soil phosphorus availability. *Soil Science* 89:137-144.

During, C., and D. M. Duganzich. 1979. Simple empirical intensity and buffering capacity measurements to predict potassium uptake by white clover. *Plant and Soil* 51:167-176.

Eckert, D. J. 1987. Soil test interpretations: Basic cation saturation ratios and sufficiency levels. Pp. 53-64. In J. R. Brown (Ed.), *Soil Testing: Sampling, Correlation, Calibration, and Interpretation.* SSSA Special Publication Number 21. Soil Science Society of America, Madison, WI.

Eghball, B., D. H. Sander, and J. Skopp. 1990. Diffusion, adsorption, and predicted longevity of banded phosphorus fertilizer in three soils. *Soil Science Society of America Journal* 54:1161-1165.

Egnér, H., H. Riehm, and W. R. Domingo. 1960. Untersuchungen über die chemische Bodenanalyse als Grundlage für die Beurteilung des Nährstoffzustandes der Böden II. Chemische Extraktionsmethoden zur Phosphor- und Kaliumbestimmung. *Kungl. Landbrukshögskolans Annaler* 26:199-215.

El Gharous, M., R. L. Westerman, and P. N. Soltanpour. 1990. Nitrogen mineralization potential of arid and semiarid soils of Morocco. *Soil Science Society of America Journal* 54:438-443.

El-Sarnagawy, N. M., A. A. Lotfy, M. Mohey El-Din, and M. A. Maatouk. 1988. Estimation of available potassium for corn by soil analysis. *Egyptian Journal of Soil Science* 28:447-460.

Ellis, B. G., and R. A. Olson. 1986. *Economic, Agronomic and Environmental Implications of Fertilizer Recommendations.* North Central Regional Research Publication No. 310. Agricultural Experiment Station, Michigan State University, East Lansing.

Engelstad, O. P., and W. D. Shrader. 1961. The effect of surface soil thickness on corn yields: II. As determined by an experiment using normal surface soil and artificially-exposed subsoil. *Soil Science Society of America Proceedings* 25:497-499.

Evanylo, G. K., and M. E. Sumner. 1987. Utilization of the boundary line approach in the development of soil nutrient norms for soybean production. *Communications in Soil Science and Plant Analysis* 18:1379-1401.

Ewanek, J. 1983. Observations by Manitoba Agriculture relative to fertilizer recommendation from several labs. Pp. 108-119. In *Technical and Scientific Papers, 1983 Manitoba Agronomists' Proceedings*. University of Manitoba, Winnipeg.

Fardeau, J.-C., and J. Jappé. 1976. Nouvelle méthode de détermination du phosphore du soil assimilable par les plantes: extrapolation des cinétiques de dilution isotopique. *Comptes Rendus Hebdomadaires des Séances de L'Académie des Sciences, Série D,* 282:1137-1140.

Fardeau, J. C., and J. Jappe. 1982. Intérêt des cinétiques d'échanges isotopiques pour la caractérisation du phosphore assimilable des sols. *Science du Sol. Bulletin de l'AFES* 2:113-124. (Cited by Morel, C., and J. C. Fardeau. 1991. Phosphorus bioavailability of fertilizers: a predictive laboratory method for its evaluation. *Fertilizer Research* 28:1-9.)

Fehr, P. 1970. Interpretation of target yield fertilizer recommendations. Pp. 116-127. In *Manitoba Soil Science Society, Papers Presented at the Fourteenth Annual Manitoba Soil Science Meeting, December 9 and 10, 1970*. Economics and Publications Branch, Manitoba Department of Agriculture, Winnipeg.

Fergus, I. F., and A. E. Martin. 1974. Studies on potassium. IV Interspecific differences in the uptake of non-exchangeable potassium. *Australian Journal of Soil Research* 12:147-158.

Fitts, J. W. 1955. Using soil tests to predict a probable response from fertilizer application. *Better Crops With Plant Food* 39, No. 3:17-20.

Fixen, P. E., and J. H. Grove. 1990. Testing soils for phosphorus. Pp. 141-180. In R. L. Westerman (Ed.), *Soil Testing and Plant Analysis*. Third Edition. Soil Science Society of America, Madison, WI.

Fixen, P. E., and A. E. Ludwick. 1983. Phosphorus and potassium fertilization of irrigated alfalfa on calcareous soils: I. Soil test maintenance requirements. *Soil Science Society of America Journal* 47:107-112.

Follett, R. F., L. K. Porter, and A. D. Halvorson. 1991. Border effects on nitrogen-15 fertilized winter wheat microplots grown in the Great Plains. *Agronomy Journal* 83:608-612.

Fox, R. H., and W. P. Piekielek. 1978a. Field testing of several nitrogen availability indexes. *Soil Science Society of America Journal* 42:747-750.

Fox, R. H., and W. P. Piekielek. 1978b. A rapid method for estimating the nitrogen-supplying capability of a soil. *Soil Science Society of America Journal* 42:751-753.

Fox, R. H., and W. P. Piekielek. 1983. *Response of Corn to Nitrogen Fertilizer and the Prediction of Soil Nitrogen Availability With Chemical Tests in Pennsylvania*. Pennsylvania Agricultural Experiment Station, Bulletin 843. University Park.

Fox, R. H., G. W. Roth, K. V. Iversen, and W. P. Piekielek. 1989. Soil and tissue nitrate tests compared for predicting soil nitrogen availability to corn. *Agronomy Journal* 81:971-974.

Fox, R. L., R. K. Nishimoto, J. R. Thompson, and R. S. de la Pena. 1974. Comparative external phosphorus requirements of plants growing in tropical soils. *Transactions of the 10th International Congress of Soil Science* 4:232-239.

Freytag, H. E. 1990. Beziehungen zwischen N-Gehalten und N-Mineralisierungspotentialen in Bodenproben aus zwei Praxisschlägen (Versuche ohne Eingriff). *Archiv für Acker- und Pflanzenbau und Bodenkunde* 34:165-172.

Fried, M. 1964. 'E,' 'L' and 'A' values. *8th International Congress of Soil Science Transactions* 4:29-39.

Fried, M., and L. A. Dean. 1952. A concept concerning the measurement of available soil nutrients. *Soil Science* 73:263-271.

Fried, M., C. E. Hagen, J. F. Saiz del Rio, and J. E. Leggett. 1957. Kinetics of phosphate uptake in the soil-plant system. *Soil Science* 84:427-437.

Friessen, D. K., G. J. Blair, and M. Duncan. 1985. Temporal fluctutations in soil test values under permanent pasture in New England, N.S.W. *Australian Journal of Soil Research* 23:181-193.

Fujii, R., and R. B. Corey. 1986. Estimation of isotopically exchangeable cadmium and zinc in soils. *Soil Science Society of America Journal* 50:306-308.

Gachon, L. 1966a. Observations sur la mesure du phosphore isotopiquement diluable des sols. *Comptes Rendus Hebdomadaires des Séances de l'Académie d'Agriculture de France* 52:1103-1108.

Gachon, L. 1966b. Phosphore isotopiquement diluable et pouvoir fixateur des sols en relation avec la croissance des plantes. *Comptes Rendus Hebdomadaires des Séances de l'Académie d'Agriculture de France* 52:1108-1115.

Gachon, L. 1969. Les méthodes d'appréciation de la fertilité phosphorique des sols. *Bulletin de l'Association Française pour l'Etude du Sol* No. 4:17-31.

Gachon, L. 1988. Comparaison des valeurs E et L: application à l'examen de l'enrichissement de différents types de sol par divers engrais. Pp. 197-203. In L. Gachon (Ed.), *Phosphore et Potassium dans les Relations Sol-Plante. Conséquences sur la Fertilisation.* Institut National de la Recherche Agronomique, Paris.

Gianello, C., and J. M. Bremner. 1986a. A simple chemical method of assessing potentially available organic nitrogen in soil. *Communications in Soil Science and Plant Analysis* 17:195-214.

Gianello, C., and J. M. Bremner. 1986b. Comparison of chemical methods of assessing potentially available organic nitrogen in soil. *Communications in Soil Science and Plant Analysis* 17:215-236.

Godwin, D. C., U. Singh, G. Alagarswamy, and J. T. Ritchie. 1989. Simulation of N dynamics in cropping systems of the semiarid tropics. Pp. 119-128. In C. B. Christianson (Ed.), *Soil Fertility and Fertilizer Management in Semiarid Tropical India.* International Fertilizer Development Center, Special Publication SP-11. Muscle Shoals, AL.

Graham, E. R. 1959. *An Explanation of Theory and Methods of Soil Testing.* Missouri Agricultural Experiment Station, Bulletin 734. Columbia.

Greenwood, D. J. 1982. Nitrogen supply and crop yield: The global scene. *Plant and Soil* 67:45-59.

Greenwood, D. J. 1983. Quantitative theory and the control of soil fertility. *New Phytologist* 94:1-18.

Greenwood, D. J., T. J. Cleaver, M. K. Turner, J. Hunt, K. B. Niendorf, and S. M. H. Loquens. 1980. Comparison of the effects of potassium fertilizer on the yield, potassium content and quality of 22 different vegetable and agricultural crops. *Journal of Agricultural Science (Cambridge)* 95:441-456.

Greenwood, D. J., A. Draycott, and J. J. Neeteson. 1987. Modelling the response of diverse crops to nitrogen fertilizer. *Journal of Plant Nutrition* 10:1753-1759.

Grimme, H. 1974. Potassium release in relation to crop production. Pp. 131-136. In *Proceedings of the 10th Congress of the International Potash Institute held in Budapest.* International Potash Institute, Worblaufen-Bern, Switzerland.

Grimme, H. 1980. The effect of field strength on the quantity of K desorbed from soils by electro-ultrafiltration. *Zeitschrift für Pflanzenernährung und Bodenkunde* 143:98-106.

Grimme, H. 1990. Soil moisture and K mobility. Pp. 117-131. In *Development of K-Fertilizer Recommendations*. Proceedings of the 22nd Colloquium of the International Potash Institute held at Soligorsk/USSR. International Potash Institute, Worblaufen-Bern, Switzerland.

Grinsted, M. J., M. J. Hedley, R. E. White, and P. H. Nye. 1982. Plant-induced changes in the rhizosphere of rape (*Brassica napus* var. Emerald) seedlings I. pH change and the increase in P concentration in the soil solution. *New Phytologist* 91:19-29.

Gruber, P. 1987. Methodology of potash fertilizer recommendations. Pp. 409-421. In *Methodology in Soil-K Research*. Proceedings of the 20th Colloquium of the International Potash Institute held in Baden bei Wien/Austria 1987. International Potash Institute, Worblaufen-Bern, Switzerland.

Gunary, D., and C. D. Sutton. 1967. Soil factors affecting plant uptake of phosphate. *Journal of Soil Science* 18:167-173.

Gunnarson, O., and L. Fredriksson. 1952. Méthode pour déterminer, au moyen de P^{32}, la teneur du sol in phosphore "assimilable par la plante." *Bulletin de Documentation, Association Internationale des Fabricants de Superphosphate* 11:16-20.

Haby, V. A., M. P. Russelle, and E. O. Skogley. 1990. Testing soils for potassium, calcium, and magnesium. Pp. 181-227. In R. L. Westerman (Ed.), *Soil Testing and Plant Analysis*. Third Edition. Soil Science Society of America, Madison, WI.

Haby, V. A., J. R. Sims, E. O. Skogley, and R. E. Lund. 1988. Effect of sample pretreatment on extractable soil potassium. *Communications in Soil Science and Plant Analysis* 19:91-106.

Hanway, J. J. 1973. Experimental methods for correlating and calibrating soil tests. Pp. 55-66. In L. M. Walsh and J. D. Beaton (Eds.), *Soil Testing and Plant Analysis*. Revised Edition. Soil Science Society of America, Madison, WI.

Hanway, J. J., S. A. Barber, R. H. Bray, A. C. Caldwell, L. E. Engelbert, R. L. Fox, M. Fried, D. Hovland, J. W. Ketcheson, W. M. Laughlin, K. Lawton, R. C. Lipps, R. A. Olson, J. T. Pesek, K. Pretty, F. W. Smith, and E. M. Stickney. 1961. *North Central Regional Potassium Studies I. Field Studies With Alfalfa*. Iowa Agricultural and Home Economics Experiment Station, Research Bulletin 494. Ames.

Hanway, J. J., S. A. Barber, R. H. Bray, A. C. Caldwell, M. Fried, L. T. Kurtz, K. Lawton, J. T. Pesek, K. Pretty, M. Reed, and F. W. Smith. 1962. *North Central Regional Potassium Studies III. Field Studies With Corn*. Iowa Agricultural and Home Economics Experiment Station, Research Bulletin 503. Ames.

Hart, P. B. S., J. H. Rayner, and D. S. Jenkinson. 1986. Influence of pool substitution on the interpretation of fertilizer experiments with ^{15}N. *Journal of Soil Science* 37:389-403.

Hatch, D. J., S. C. Jarvis, and L. Philipps. 1990. Field measurement of nitrogen mineralization using soil core incubation and acetylene inhibition of nitrification. *Plant and Soil* 124:97-107.

Havlin, J. L. 1986. Comparison of statistical methods for separating iron deficient from sufficient soils. *Journal of Plant Nutrition* 9:241-249.

Hedley, M. J., R. E. White, and P. H. Nye. 1982. Plant-induced changes in the rhizosphere of rape (*Brassica napus* var. Emerald) seedlings III. Changes in *L* value, soil phosphate fractions and phosphatase activity. *New Phytologist* 91:45-56.

Helal, H. M., and A. Dressler. 1989. Mobilization and turnover of soil phosphorus in the rhizosphere. *Zeitschrift für Pflanzenernährung und Bodenkunde* 152:175-180.

Helal, H. M., and D. R. Sauerbeck. 1984. Influence of plant roots on C and P metabolism in soil. *Plant and Soil* 76:175-182.

Helyar, K. R., and D. P. Godden. 1977. The biology and modelling of fertilizer response. *Journal of the Australian Institute of Agricultural Science* 43:22-30.

Hergert, G. W. 1987. Status of residual nitrate-nitrogen soil tests in the United States of America. Pp. 73-88. In J. R. Brown (Ed.), *Soil Testing: Sampling, Correlation, Calibration, and Interpretation*. SSSA Special Publication Number 21. Soil Science Society of America, Madison, WI.

Hockensmith, R. D., R. Gardner, and J. Goodwin. 1933. *Comparison of Methods for Estimating Available Phosphorus in Alkaline Calcareous Soils*. Colorado Agricultural Experiment Station, Technical Bulletin No. 2. Fort Collins.

Holford, I. C. R. 1980. Greenhouse evaluation of four phosphorus soil tests in relation to phosphate buffering and labile phosphate in soils. *Soil Science Society of America Journal* 44:555-559.

Holford, I. C. R., E. J. Corbin, C. L. Mullen, and J. Bradley. 1988. Effects of rainfall variability on the efficacy of soil tests for wheat on semi-arid soils. *Australian Journal of Soil Research* 26:201-209.

Holford, I. C. R., and G. J. Crocker. 1988. Efficacy of various soil phosphate tests for predicting phosphate responsiveness and requirements of clover pastures on acidic tableland soils. *Australian Journal of Soil Research* 26:479-488.

Holford, I. C. R., and G. J. Crocker. 1991. Residual effects of phosphate fertilizers in relation to phosphate sorptivities of 27 soils. *Fertilizer Research* 28:305-314.

Holford, I. C. R., and B. R. Cullis. 1985a. Effects of phosphate buffer capacity on yield response curvature and fertilizer requirements of wheat in relation to soil phosphate tests. *Australian Journal of Soil Research* 23:417-427.

Holford, I. C. R., and B. R. Cullis. 1985b. An evaluation of eight soil phosphate extractants on acidic wheat-growing soils. *Australian Journal of Soil Research* 23:647-653.

Holford, I. C. R., and G. E. G. Mattingly. 1976a. A model for the behaviour of labile phosphate in soil. *Plant and Soil* 44:219-229.

Holford, I. C. R., and G. E. G. Mattingly. 1976b. Phosphate adsorption and plant availability of phosphate. *Plant and Soil* 44:377-389.

Holford, I. C. R., J. M. Morgan, J. Bradley, and B. R. Cullis. 1985. Yield responsiveness and response curvature as essential criteria for the evaluation and calibration of soil phosphate tests for wheat. *Australian Journal of Soil Research* 23:167-180.

Hong, S. D., R. H. Fox, and W. P. Piekielek. 1990. Field evaluation of several chemical indexes of soil nitrogen availability. *Plant and Soil* 123:83-88.

Hooker, M. L. 1976. *Sampling Intensities Required to Estimate Available N and P in Five Nebraska Soil Types*. M.S. Thesis, University of Nebraska, Lincoln.

Horneck, D. A., J. M. Hart, and D. C. Peek. 1990. The influence of sampling intensity, liming, P rate and method of P application on P soil test values. *Communications in Soil Science and Plant Analysis* 21:1079-1090.

Howard, D. D., and F. Adams. 1965. Calcium requirement for penetration of subsoils by primary cotton roots. *Soil Science Society of America Proceedings* 29:558-562.

Hunter, A. S. 1949. Yield and composition of alfalfa as affected by variations in the calcium-magnesium ratio in the soil. *Soil Science* 67:53-62.

Hunter, A. S., S. J. Toth, and F. E. Bear. 1943. Calcium-potassium ratios for alfalfa. *Soil Science* 55:61-72.

Imaizumi, K., and S. Yoshida. 1958. Edaphological studies on silicon supplying power of paddy fields. *Bulletin of the National Institute of Agricultural Sciences (Japan), Series B (Soils and Fertilizers)* No. 8:261-304.

Itallie, T. B. van. 1938. Cation equilibria in plants in relation to the soil. *Soil Science* 46:175-186.

Ivanov, P., E. Markov, and R. Jendova. 1989. [Effect of temperature on the amount of labile phosphorus and potassium contained in the soil, determined after different agrochemical methods.] (Translated title.) *Pochvoznanie i Agrokhimiya* 24, No. 4:49-55.

James, D. W., and A. I. Dow. 1972. *Source and Degree of Variation in the Field: the Problem of Sampling for Soil Tests and Estimating Soil Fertility Status.* Washington Agricultural Experiment Station, Bulletin 749. Pullman.

James, D. W., and K. L. Wells. 1990. Soil sample collection and handling: Technique based on source and degree of field variability. Pp. 25-44. In R. L. Westerman (Ed.), *Soil Testing and Plant Analysis.* Third Edition. Soil Science Society of America, Madison, WI.

Janssen, B. H., F. C. T. Guiking, D. van der Eijk, E. M. A. Smaling, J. Wolf, and H. van Reuler. 1990. A system for quantitative evaluation of the fertility of tropical soils (QUEFTS). *Geoderma* 46:299-318.

Jenkinson, D. S. 1966. Studies on the decomposition of plant material in soil. II. Partial sterilization of soil and the soil biomass. *Journal of Soil Science* 17:280-302.

Jenkinson, D. S. 1968. Chemical tests for potentially available nitrogen in soil. *Journal of the Science of Food and Agriculture* 19:160-168.

Jenkinson, D. S. 1988. Determination of microbial biomass carbon and nitrogen in soil. Pp. 368-386. In J. R. Wilson (Ed.). *Advances in Nitrogen Cycling in Agricultural Ecosystems.* C.A.B. International, Wallingford, Oxon, United Kingdom.

Jenkinson, D. S., R. H. Fox, and J. H. Rayner. 1985. Interactions between fertilizer nitrogen and soil nitrogen — the so-called 'priming' effect. *Journal of Soil Science* 36:425-444.

Jenny, H., and A. D. Ayers. 1939. The influence of the degree of saturation of soil colloids on the nutrient intake by roots. *Soil Science* 48:443-459.

Johnston, A. E., and K. W. T. Goulding. 1990. The use of plant and soil analyses to predict the potassium supplying capacity of soil. Pp. 177-204. In *Development of K-Fertilizer Recommendations.* Proceedings of the 22nd Colloquium of the International Potash Institute held at Soligorsk/USSR. International Potash Institute, Worblaufen-Bern, Switzerland.

Jones, J. B., Jr. 1985a. Soil testing and plant analysis: Guides to the fertilization of horticultural crops. *Horticultural Reviews* 7:1-68.

Jones, J. B., Jr. 1985b. Recent survey of number of soil and plant tissue samples tested for growers in the United States. In *Proceedings 10th Soil Plant Analyst's Workshop.* Council on Soil Testing and Plant Analysis, Athens, Georgia. (Cited by Jones et al., 1990.)

Jones, J. B., Jr. 1988. *Soil testing and plant analysis: Procedures and use.* ASPAC Food & Fertilizer Technology Center, Technical Bulletin No. 109. Taipei City, Taiwan.

Jones, J. B., Jr. 1990. Universal soil extractants: Their composition and use. *Communications in Soil Science and Plant Analysis* 21:1091-1101.

Jones, J. B., Jr., H. V. Eck, and R. Voss. 1990. Plant analysis as an aid in fertilizing corn and grain sorghum. Pp. 521-547. In R. L. Westerman (Ed.), *Soil Testing and Plant Analysis.* Third Edition. Soil Science Society of America, Madison, WI.

Jones, J. B., Jr., H. J. Mederski, D. J. Hoff, and J. H. Wilson. 1961. Effect of drying some Ohio soils upon the soil test for potassium. *Soil Science Society of America Proceedings* 25:123-125.

Juma, N. G., E. A. Paul, and B. Mary. 1984. Kinetic analysis of net nitrogen mineralization in soil. *Soil Science Society of America Journal* 48:753-757.

Karlovsky, J. 1961. Phosphate utilisation and phosphate maintenance requirements. Pp. 142-151. In *Proceedings of the Ruakura Farmers' Conference.* New Zealand Ministry of Agriculture and Fisheries. (Cited by Sinclair, A. G., and I. S. Cornforth. 1982. A model for calculating phosphate fertiliser requirements to maintain grazed pastures. Pp. 625-630. In A. Scaife (Ed.), *Plant Nutrition 1982.* Proceedings of the Ninth International Plant Nutrition Colloquium. Commonwealth Agricultural Bureaux, Slough, United Kingdom.)

Keating, B. A., D. C. Godwin, and J. M. Watiki. 1991. Optimising nitrogen inputs in response to climatic risk. Pp. 329-358. In R. C. Muchow and J. A. Bellamy (Eds.), *Climatic Risk in Crop Production: Models and Management for the Semiarid Tropics and Subtropics.* C.A.B. International, Wallingford, United Kingdom.

Keeney, D. R. 1982. Nitrogen availability indexes. Pp. 711-733. In A. L. Page (Ed.), *Methods of Soil Analysis. Part 2. Chemical and Microbiological Properties.* Second Edition. Agronomy Number 9 (Part 2). American Society of Agronomy and Soil Science Society of America, Madison, WI.

Keeney, D. R., and J. M. Bremner. 1964. Effect of cultivation on the nitrogen distribution in soils. *Soil Science Society of America Proceedings* 28:653-666.

Keeney, D. R., and J. M. Bremner. 1966. Comparison and evaluation of laboratory methods of obtaining an index of soil nitrogen availability. *Agronomy Journal* 58:498-503.

Keisling, T. C., and B. Mullinix. 1979. Statistical considerations for evaluating micronutrient tests. *Soil Science Society of America Journal* 43:1181-1184.

Keramidas, V. Z., and N. A. Polyzopoulos. 1983. Phosphorus intensity, quantity, and capacity factors of representative Alfisols of Greece. *Soil Science Society of America Journal* 47:232-236.

Khasawneh, F. E., and J. P. Copeland. 1973. Cotton root growth and uptake of nutrients: Relation of phosphorus uptake to quantity, intensity, and buffering capacity. *Soil Science Society of America Proceedings* 37:250-254.

Kitchen, N. R., J. L. Havlin, and D. G. Westfall. 1990. Soil sampling under no-till banded phosphorus. *Soil Science Society of America Journal* 54:1661-1665.

Klages, M. G., R. A. Olsen, and V. A. Haby. 1988. Relationship of phosphorus isotherms to $NaHCO_3$-extractable phosphorus as affected by soil properties. *Soil Science* 146:85-91.

Koch, J. T., E. R. Orchard, and M. E. Sumner. 1970. Leaf composition and yield response of corn in relation to quantity-intensity parameters for potassium. *Soil Science Society of America Proceedings* 34:94-98.

Kuchenbuch, R., and A. Jungk. 1984. Wirking der Kaliumdüngung auf die Kaliumverfügbarkeit in der Rhizosphäre von Raps. *Zeitschrift für Pflanzenernährung und Bodenkunde* 147:435-448.

Kuhlmann, H., N. Claassen, and J. Wehrmann. 1985. A method for determining the K-uptake from subsoil by plants. *Plant and Soil* 83:449-452.

Kumaragamage, D., and G. Keerthisinghe. 1988. Quantity and intensity relationships in predicting P availability of soils in Sri Lanka. *Zeitschrift für Pflanzenernährung und Bodenkunde* 151:395-398.

Larsen, S. 1952. The use of P³² in studies on the uptake of phosphorus by plants. *Plant and Soil* 4:1-10.

Lavy, T. L., and S. A. Barber. 1964. Movement of molybdenum in the soil and its effect on availability to the plant. *Soil Science Society of America Proceedings* 28:93-97.

Le Roux, J., and M. E. Sumner. 1968. Labile potassium in soils I: Factors affecting the quantity-intensity (Q/I) parameters. *Soil Science* 106:35-41.

Lesaint, M., E. Tyszkiewicz, and G. Barbier. 1955. Validité de la détermination physique, par dilution isotopique, de l'acide phosphorique assimilable du sol. *Comptes Rendus Hebdomadaires des Séances de l'Académie d'Agriculture de France* 41:350-353.

Liebhardt, W. C. 1981. The basic cation saturation ratio concept and lime and potassium recommendations on Delaware's Coastal Plain soils. *Soil Science Society of America Journal* 45:544-549.

Lierop, W. van. 1989. Effect of assumptions on accuracy of analytical results and liming recommendations when testing a volume or weight of soil. *Communications in Soil Science and Plant Analysis* 20:121-137.

Lindén, B. 1985. Mineral nitrogen present in the root zone in early spring and nitrogen mineralized during the growing season — their contribution to the nitrogen supply of crops. Pp. 37-49. In J. J. Neeteson and K. Dilz (Eds.), *Assessment of Nitrogen Fertilizer Requirement*. Institute for Soil Fertility and Netherlands Fertilizer Institute, Haren, The Netherlands.

Lindsay, W. L., and W. A. Norvell. 1978. Development of a DTPA soil test for zinc, iron, manganese, and copper. *Soil Science Society of America Journal* 42:421-428.

Lins, I. D. G., and F. R. Cox. 1989. Effect of extractant and selected soil properties on predicting the optimum phosphorus fertilizer rate for growing soybeans under field conditions. *Communications in Soil Science and Plant Analysis* 20:319-333.

Lockman, R. B., and M. G. Molloy. 1984. Seasonal variations in soil test results. *Communications in Soil Science and Plant Analysis* 15:741-757.

Löhnis, F. 1926. Nitrogen availability of green manures. *Soil Science* 22:253-290.

Luebs, R. E., G. Stanford, and A. D. Scott. 1956. Relation of available potassium to soil moisture. *Soil Science Society of America Proceedings* 20:45-50.

Magdoff, F. R., W. E. Jokela, R. H. Fox, and G. F. Griffin. 1990. A soil test for nitrogen availability in the northeastern United States. *Communications in Soil Science and Plant Analysis* 21:1103-1115.

Mahler, R. L. 1990. Soil sampling fields that have received banded fertilizer applications. *Communications in Soil Science and Plant Analysis* 21:1793-1802.

Maidl, F. X., and G. Fischbeck. 1986. Veränderung des bodenbürtigen Mineralstickstoffs im Frühjahr und seine Bedeutung für die Bemessung der N-Düngung von Zuckerrüben. *Journal of Agronomy and Crop Science* 156:1-12.

Mallarino, A. P., and A. M. Blackmer. 1992. Comparison of methods for determining critical concentrations of soil test phosphorus for corn. *Agronomy Journal* 84:850-856.

Marschner, H., V. Römheld, W. J. Horst, and P. Martin. 1986. Root-induced changes in the rhizosphere: Importance for the mineral nutrition of plants. *Zeitschrift für Pflanzenernährung und Bodenkunde* 149:441-456.

Marschner, H., M. Treeby, and V. Römheld. 1989. Role of root-induced changes in the rhizosphere for iron acquisition in higher plants. *Zeitschrift für Pflanzenernährung und Bodenkunde* 152:197-204.

Martens, D. C., and W. L. Lindsay. 1990. Testing soils for copper, iron, manganese, and zinc. Pp. 229-264. In R. L. Westerman (Ed.), *Soil Testing and Plant Analysis.* Third Edition. Soil Science Society of America, Madison, WI.

Marumoto, T., J. P. E. Anderson, and K. H. Domsch. 1982. Mineralization of nutrients from soil microbial biomass. *Soil Biology and Biochemistry* 14:469-475.

Mascagni, H. J., Jr., and F. R. Cox. 1985. Calibration of a manganese availability index for soybean soil test data. *Soil Science Society of America Journal* 49:382-386.

Matar, A. E., M. Pala, D. Beck, and S. Garabet. 1990. Nitrate-N test as a guide to N fertilization of wheat in the Mediterranean region. *Communications in Soil Science and Plant Analysis* 21:1117-1130.

Mazumdar, S. P., and S. N. Saxena. 1988. Calibration of critical limits of various soil potassium parameters for wheat crop in Torri-Psamment soils of western Ghat (India). *Anales de Edafologia y Agrobiologica* 47:955-971.

McCallister, D. L., C. A. Shapiro, W. R. Raun, F. N. Anderson, G. W. Rehm, O. P. Engelstad, M. P. Russelle, and R. A. Olson. 1987. Rate of phosphorus and potassium buildup/decline with fertilization for corn and wheat on Nebraska Mollisols. *Soil Science Society of America Journal* 51:1646-1652.

McCants, C. B. 1955. *A Biological Slope-Ratio Method for Evaluating Nutrient Availability in Soils.* Ph.D. Thesis, Iowa State University. Ames.

McCants, C. B., and C. A. Black. 1957. A biological slope-ratio method for evaluating nutrient availability in soils. *Soil Science Society of America Proceedings* 21:296-301.

McLean, E. O. 1977. Contrasting concepts in soil test interpretation: Sufficiency levels of available nutrients versus basic cation saturation ratios. Pp. 39-54. In M. Stelly (Ed.), *Soil Testing: Correlating and Interpreting the Analytical Results.* ASA Special Publication Number 29. American Society of Agronomy, Crop Science Society of America, Soil Science Society of America, Madison, WI.

McLean, E. O., J. L. Adams, and R. C. Hartwig. 1982. Improved corrective fertilizer recommendations based on a two step alternative usage of soil tests: II. Recovery of soil-equilibrated potassium. *Soil Science Society of America Journal* 46:1198-1201.

McLean, E. O., R. C. Hartwig, D. J. Eckert, and G. B. Triplett. 1983. Basic cation saturation ratios as a basis for fertilizing and liming agronomic crops. II. Field studies. *Agronomy Journal* 75:635-639.

McLean, E. O., T. O. Oloya, and S. Mostaghimi. 1982. Improved corrective fertilizer recommendations based on a two-step alternative usage of soil tests: I. Recovery of soil-equilibrated phosphorus. *Soil Science Society of America Journal* 46:1193-1197.

McLean, E. O., and M. E. Watson. 1985. Soil measurements of plant-available potassium. Pp. 277-308. In R. D. Munson (Ed.), *Potassium in Agriculture.* American Society of Agronomy, Crop Science Society of America, and Soil Science Society of America, Madison, WI.

Meer, H. G. van der, and M. G. van Uum-van Lohuyzen. 1986. The relationship between inputs and outputs of nitrogen in intensive grassland systems. Pp. 1-18. In H. G. van der Meer, J. C. Ryden, and G. C. Ennik (Eds.), *Nitrogen Fluxes in Intensive Grassland Systems.* Martinus Nijhoff Publishers, Dordrecht, The Netherlands.

Mehlich, A. 1972. Uniformity of expressing soil test results. A case for calculating results on a volume basis. *Communications in Soil Science and Plant Analysis* 3:417-424.

Mehlich, A. 1984. Mehlich 3 soil test extractant: A modification of Mehlich 2 extractant. *Communications in Soil Science and Plant Analysis* 15:1409-1416.

Meijer, W. J. M., and S. Vreeke. 1988. Nitrogen fertilization of grass seed crops as related to soil mineral nitrogen. *Netherlands Journal of Agricultural Science* 36:375-385.

Meisinger, J. J. 1984. Evaluating plant-available nitrogen in soil-crop systems. Pp. 391-416. In R. D. Hauck (Ed.), *Nitrogen in Crop Production*. American Society of Agronomy, Madison, WI.

Mendoza, R. E. 1989. Different performances of soil phosphate tests for reflecting the effects of buffering capacity on plant uptake of native phosphate with time. *Plant and Soil* 113:13-19.

Mengel, K. 1963. Untersuchungen über das "Kalium-Calciumpotential." *Zeitschrift für Pflanzenernährung Düngung Bodenkunde* 103:99-111.

Mengel, K. 1991. Available nitrogen in soils and its determination by the 'Nmin-method' and by electroultrafiltration (EUF). *Fertilizer Research* 28:251-262.

Mengel, K., D. Horn, and H. Tributh. 1990. Availability of interlayer ammonium as related to root vicinity and mineral type. *Soil Science* 149:131-137.

Mengel, K., and H. W. Scheier. 1981. Release of nonexchangeable (fixed) soil ammonium under field conditions during the growing season. *Soil Science* 131:226-232.

Menon, R. G., S. H. Chien, and L. L. Hammond. 1989a. Comparison of Bray I and P_i tests for evaluating plant-available phosphorus from soils treated with different partially acidulated phosphate rocks. *Plant and Soil* 114:211-216.

Menon, R. G., S. H. Chien, L. L. Hammond, and B. R. Arora. 1990. Sorption of phosphorus by the iron oxide-impregnated filter paper (P_i soil test) embedded in soils. *Plant and Soil* 126:287-294.

Menon, R. G., S. H. Chien, L. L. Hammond, and J. Henao. 1989b. Modified techniques for preparing paper strips for the new P_i soil test for phosphorus. *Fertilizer Research* 19:85-91.

Menon, R. G., L. L. Hammond, and H. A. Sissingh. 1989c. Determination of plant-available phosphorus by the iron hydroxide-impregnated filter paper (P_i) soil test. *Soil Science Society of America Journal* 53:110-115.

Meynard, J. M., J. Boiffin, and W. Sebillotte. 1982. Prevision of nitrogen fertilizer for winter wheat. Test of a model. Pp. 390-395. In A. Scaife (Ed.), *Plant Nutrition 1982*. Proceedings of the Ninth International Plant Nutrition Colloquium, Volume 2. Commonwealth Agricultural Bureaux, Farnham Royal, Slough, United Kingdom.

Mitchell, H. L. 1934. *Pot Culture Tests of Forest Soil Fertility With Observations on the Effect of Varied Solar Radiation and Nutrient Supply on the Growth and Nitrogen Content of Scots and White Pine Seedlings*. Black Rock Forest Bulletin 5. Cornwall-on-the-Hudson, NY.

Mitchell, H. L., and R. F. Chandler, Jr. 1939. *The Nitrogen Nutrition and Growth of Certain Deciduous Trees of Northeastern United States*. Black Rock Forest, Bulletin No. 11. Cornwall-on-the-Hudson, NY.

Mitscherlich, E. A. 1919. Zum Gehalt der Haferpflanze an Phosphorsäure und seinen Beziehungen zu der durch eine Nährstoffzufuhr bedingten Ertragserhöhung. *Journal für Landwirtschaft* 67:171-176.

Mitscherlich, E. A., and W. Sauerlandt. 1935. Salpeter- und Ammoniak-Stickstoff im Boden und die pflanzenphysiologisch wirksame Stickstoffmenge "b." *Landwirtschaftliche Jahrbücher* 81:623-654.

Molina, J. A. E., C. E. Clapp, and W. E. Larson. 1980. Potentially mineralizable nitrogen in soil: The simple exponential model does not apply for the first 12 weeks of incubation. *Soil Science Society of America Journal* 44:442-443.

Mombiela, F., J. J. Nicholaides, III, and L. A. Nelson. 1981. A method to determine the appropriate mathematical form for incorporating soil test levels in fertilizer response models for recommendation purposes. *Agronomy Journal* 73:937-941.

Moody, P. W., R. L. Aitken, B. L. Compton, and S. Hunt. 1988. Soil phosphorus parameters affecting phosphorus availability to, and fertilizer requirements of, maize (*Zea mays*). *Australian Journal of Soil Research* 26:611-622.

Moorhead, K. K., and E. O. McLean. 1985. Improved corrective fertilizer recommendations based on two-step alternative usage of soil tests: 4. Studies of field plot samples. *Soil Science* 139:131-138.

Morgan, M. F. 1935. *The Universal Soil Testing System*. Connecticut Agricultural Experiment Station Bulletin 372. New Haven.

Mortvedt, J. J. 1977. Micronutrient soil test correlations and interpretations. Pp. 99-117. In M. Stelly (Ed.), *Soil Testing: Correlating and Interpreting the Analytical Results*. ASA Special Publication Number 29. American Society of Agronomy, Crop Science Society of America, and Soil Science Society of America, Madison, WI.

Mostaghimi, S., and E. O. McLean. 1983. Improved corrective fertilizer recommendations based on a two-step alternative usage of soil tests: III. The Bray-1 test on soils with concretions. *Soil Science Society of America Journal* 47:966-971.

Mubarak, A., and R. A. Olsen. 1976. Immiscible displacement of the soil solution by centrifugation. *Soil Science Society of America Journal* 40:329-331.

Müller, A., and I. Feyerabend. 1982. N_{min} im Bodenprofil und N-Aufnahme durch die Zuckerrübe. *Mitteilungen der Deutschen Bodenkundlichen Gesellschaft* 34:39-44.

Munk, H., and M. Rex. 1987. Zur Eichung chemischer Bodenuntersuchungsmethoden auf Phosphat an mehrjährigen Feldversuchen. Pp. 275-297. In *Leistungsförderer in der Tierproduktion Möglichkeiten und Grenzen*. VDLUFA Schriftenreihe 20. Kongressband 1986. VDLUFA-Verlag, Darmstadt.

Munk, H., and M. Rex. 1990. Betrachtungen zur Frage der Ausnutzung von Phosphatvorräten im Boden. *Verband Deutscher Landwirtschaftlicher Untersuchungs- und Forschungsanstalten. Reihe Kongress Berichte* 30:229-235. (*Soils and Fertilizers* 54, Abstract 1996. 1991.)

Munns, D. A., and E. O. McLean. 1975. Soil potassium relationships as indicated by solution equilibrations and plant uptake. *Soil Science Society of America Proceedings* 39:1072-1076.

Munson, R. D. 1954. *Soil Nitrification Rate as Correlated With Crop Response to Nitrogen Fertilization*. M.S. Thesis, Iowa State College. Ames.

Munson, R. D. (Ed.). 1985. *Potassium in Agriculture*. American Society of Agronomy, Madison, WI.

Murugappan, V., G. V. Kothandaraman, S. P. Palaniappan, and T. S. Manickam. 1989. Fertilizer requirements for specified yield targets. II. Field verification of mathematical models for the estimation of soil and fertilizer nutrient efficiencies. *Fertilizer Research* 18:127-140.

Myers, R. J. K. 1984. A simple model for estimating the fertilizer requirement of a cereal crop. *Fertilizer Research* 5:95-108.

Nair, P. K. R., and H. Grimme. 1979. Q/I relations and electroultrafiltration of soils as measures of potassium availability to plants. *Zeitschrift für Pflanzenernährung und Bodenkunde* 142:87-94.

Neeteson, J. J. 1985. Effectiveness of the assessment of nitrogen fertilizer requirement for potatoes on the basis of soil mineral nitrogen. Pp. 15-24. In J. J. Neeteson and K. Dilz (Eds.), *Assessment of Nitrogen Fertilizer Requirement*. Institute for Soil Fertility and Netherlands Fertilizer Institute, Haren, The Netherlands.

Neeteson, J. J. 1990. Development of nitrogen fertilizer recommendations for arable crops in the Netherlands in relation to nitrate leaching. *Fertilizer Research* 26:291-298.

Neeteson, J. J., K. Dilz, and G. Wijnen. 1989. N-fertilizer recommendations for arable crops. Pp. 253-264. In J. C. Germon (Ed.), *Management Systems to Reduce Impact of Nitrates*. Elsevier Applied Science, London.

Neeteson, J. J., D. J. Greenwood, and A. Draycott. 1987. *A dynamic model to predict yield and optimum nitrogen fertiliser application rate for potatoes*. Fertiliser Society (London) Proceedings No. 262.

Neeteson, J. J., and J. A. van Veen. 1988. Mechanistic and practical modelling of nitrogen mineralization-immobilization in soils. Pp. 145-155. In J. R. Wilson (Ed.), *Advances in Nitrogen Cycling in Agricultural Ecosystems*. C.A.B. International, Wallingford, Oxon, United Kingdom.

Neeteson, J. J., and H. J. C. Zwetsloot. 1989. An analysis of the response of sugar beet and potatoes to fertilizer nitrogen and soil mineral nitrogen. *Netherlands Journal of Agricultural Science* 37:129-141.

Németh, K. 1979. The availability of nutrients in the soil as determined by electro-ultrafiltration (EUF). *Advances in Agronomy* 31:155-188.

Neubauer, H., and W. Schneider. 1923. Die Nährstoffaufnahme der Keimpflanzen und ihre Anwendung auf die Bestimmung des Nährstoffgehalts der Böden. *Zeitschrift für Pflanzenernährung und Düngung* 2A:329-362.

Nielsen, D. R., and M. H. Alemi. 1989. Statistical opportunities for analyzing spatial and temporal heterogeneity of field soils. *Plant and Soil* 115:285-296.

Nielsen, N. E., and E. M. Hansen. 1984. Macro nutrient cation uptake by plants II. Effects of plant species, nitrogen concentration in the plant, cation concentration, activity and activity ratio in soil solution. *Plant and Soil* 77:347-365.

Nielsen, N. E., and C. B. Sørensen. 1984. Macro nutrient cation uptake by plants I. Rate-determining steps in net inflow of cations into intact and decapitated sunflower plants and intensity factors of cations in soil solution. *Plant and Soil* 77:337-346.

Noordwijk, M. van, P. de Willigen, P. A. J. Ehlert, and W. J. Chardon. 1990. A simple model of P uptake by crops as a possible basis for P fertilizer recommendations. *Netherlands Journal of Agricultural Science* 38:317-332.

Nordmyer, H., and J. Richter. 1985. Incubation experiments on nitrogen mineralization in loess and sandy soils. *Plant and Soil* 83:433-445.

Novozamsky, I., and V. J. G. Houba. 1987. Critical evaluation of soil testing methods for K. Pp. 177-197. In *Methodology in Soil-K Research*. Proceedings of the 20th Colloquium of the International Potash Institute held in Baden bei Wien/Austria 1987. International Potash Institute, Worblaufen-Bern, Switzerland.

Olsen, C. 1950. The significance of concentration for the rate of ion absorption by higher plants in water culture. *Physiologia Plantarum* 3:152-164.

Olsen, S. R., C. V. Cole, F. S. Watanabe, and L. A. Dean. 1954. *Estimation of Available Phosphorus in Soils by Extraction With Sodium Bicarbonate.* U.S. Department of Agriculture, Circular No. 939. Washington.

Olsen, S. R., and W. D. Kemper. 1968. Movement of nutrients to plant roots. *Advances in Agronomy* 20:91-151.

Olsen, S. R., and F. S. Watanabe. 1963. Diffusion of phosphorus as related to soil texture and plant uptake. *Soil Science Society of America Proceedings* 27:648-653.

Olson, R. A., F. N. Anderson, K. D. Frank, P. H. Grabouski, G. W. Rehm, and C. A. Shapiro. 1987. Soil testing interpretations: Sufficiency vs. build-up and maintenance. Pp. 41-52. In J. R. Brown (Ed.), *Soil Testing: Sampling, Correlation, Calibration, and Interpretation.* SSSA Special Publication Number 21. Soil Science Society of America, Madison, WI.

Olson, R. A., K. D. Frank, P. H. Grabouski, and G. W. Rehm. 1982. Economic and agronomic impacts of varied philosophies of soil testing. *Agronomy Journal* 74:492-499.

Østergaard, H. S. 1989. Analytical methods for optimization of nitrogen fertilization in agriculture. Pp. 224-235. In J. C. Germon (Ed.), *Management Systems to Reduce Impact of Nitrates.* Elsevier Applied Science, London.

Østergaard, H. S., E. K. Hvelplund, and D. Rasmussen. 1985. Assessment of optimum nitrogen fertilizer requirement on the basis of soil analysis and weather conditions prior to the growing season. Pp. 25-36. In J. J. Neeteson and K. Dilz (Eds.), *Assessment of Nitrogen Fertilizer Requirement.* Institute for Soil Fertility and Netherlands Fertilizer Institute, Haren, The Netherlands.

Ozanne, P. G., and T. C. Shaw. 1967. Phosphate sorption by soils as a measure of the phosphate requirement for pasture growth. *Australian Journal of Agricultural Research* 18:601-612.

Ozanne, P. G., and T. C. Shaw. 1968. Advantages of the recently developed phosphate sorption test over the older extractant methods for soil phosphate. *9th International Congress of Soil Science, Transactions* 2:273-280.

Parentoni, S. N., G. E. França, and A. F. C. Bahia Filho. 1988. Avaliação dos conceitos de quantidade e intensidade de mineralização de nitrogênio para trinta solos do Rio Grande do Sul. *Revista Brasileira de Ciência do Solo* 12:225-229.

Parker, D. R., T B. Kinraide, and L. W. Zelazny. 1988. Aluminum speciation and phytotoxicity in dilute hydroxy-aluminum solutions. *Soil Science Society of America Journal* 52:438-444.

Peaslee, D. E., and R. L. Fox. 1978. Phosphorus fertilizer requirements as estimated by phosphate sorption. *Communications in Soil Science and Plant Analysis* 9:975-993.

Peck, T. R., J. T. Cope, Jr., and D. A. Whitney (Eds.). 1977. *Soil Testing: Correlating and Interpreting the Analytical Results.* ASA Special Publication No. 29. American Society of Agronomy, Crop Science Society of America, Soil Science Society of America, Madison, WI.

Peck, T. R., and S. W. Melsted. 1973. Field sampling for soil testing. Pp. 67-75. In L. M. Walsh and J. D.Beaton (Eds.), *Soil Testing and Plant Analysis.* Revised edition. Soil Science Society of America, Madison, WI.

Perumal, R., P. Duraisamy, S. Baskaran, and S. Chellamuthu. 1987. Rationalised fertiliser prescription for rice-alfisol alluvium based on soil test-crop response studies. *Madras Agricultural Journal* 74:312-319.

Perumal, R., P. Duraisamy, C. Jayaraman, and S. Mani. 1988. Rationalised fertiliser prescription for groundnut based on soil test crop response studies. *Madras Agricultural Journal* 75:164-172.

Pesek, J., and J. R. Webb. 1954. Unpublished data. *Iowa Agriculture and Home Economics Experiment Station*. Ames.

Petersen, R. G., and L. D. Calvin. 1986. Sampling. Pp. 33-51. In A. Klute (Ed.), *Methods of Soil Analysis. Part 1. Physical and Mineralogical Methods*. Second Edition. Agronomy Number 9 (Part 1). American Society of Agronomy and Soil Science Society of America, Madison, WI.

Peterson, G. A., and W. W. Frye. 1989. Fertilizer nitrogen management. Pp. 183-219. In R. F. Follett (Ed.), *Nitrogen Management and Ground Water Protection*. Elsevier Science Publishers B.V., Amsterdam, The Netherlands.

Pew, W. D., J. L. Abbott, B. R. Gardner, and T. C. Tucker. 1988. Determining phosphorus needs for cabbage grown on calcareous soils. *Journal of Plant Nutrition* 11:1701-1713.

Porter, L. K., W. D. Kemper, R. D. Jackson, and B. A. Stewart. 1960. Chloride diffusion in soils as influenced by moisture content. *Soil Science Society of America Proceedings* 24:460-463.

Power, J. F., and F. E. Broadbent. 1989. Proper accounting for N in cropping systems. Pp. 159-181. In R. F. Follett (Ed.), *Nitrogen Management and Ground Water Protection*. Elsevier Science Publishers B.V., Amsterdam, The Netherlands.

Pratt, P. F. 1951. Potassium removal from Iowa soils by greenhouse and laboratory procedures. *Soil Science* 72:107-117.

Priebe, D. L., and A. M. Blackmer. 1989. Preferential movement of oxygen-18-labeled water and nitrogen-15-labeled urea through macropores in a Nicollet soil. *Journal of Environmental Quality* 18:66-72.

Probert, M. E. 1972. The dependence of isotopically exchangeable phosphate (*L*-value) on phosphate uptake. *Plant and Soil* 36:141-148.

Qureshi, R. H., and D. A. Jenkins. 1987. Concentration of phosphorus and sulphur at soil ped surfaces. *Journal of Soil Science* 38:255-265.

Raison, R. J., M. J. Connell, and P. K. Khanna. 1987. Methodology for studying fluxes of soil mineral-N *in situ*. *Soil Biology and Biochemistry* 19:521-530.

Ramamoorthy, B., R. L. Narasimham, and R. S. Dinesh. 1967. Fertilizer application for specific yield targets of Sonora 64. *Indian Farming* 17, No. 5:43, 45.

Ramamoorthy, B., and M. Velayutham. 1971. Soil test crop response correlation work in India. *Food and Agriculture Organization of the United Nations, World Soil Resources Report* No. 41:96-102.

Randall, G. W. 1980. Fertilization practices for conservation tillage. *32nd Annual Fertilizer and Ag Chemical Dealers Conference*. Iowa State University Extension, EC-1498k. Ames.

Remy, J. C., and P. Viaux. 1982. The use of nitrogen fertilisers in intensive wheat growing in France. Pp. 67-92. In *Symposium on Fertilisers and Intensive Wheat Production in the EEC*. Fertiliser Society (London) Proceedings No. 211.

Reuss, J. O., P. N. Soltanpour, and A. E. Ludwick. 1977. Sampling distribution of nitrates in irrigated fields. *Agronomy Journal* 69:588-592.

Rex, M., and H. Munk. 1990. Eichung von Bodenuntersuchungsmethoden auf Phosphat ein Methodenvergleich (CAL, DL, P-H$_2$O, EUF, CaCl$_2$). *Verband Deutscher Landwirtschaftlicher Untersuchungs- und Forschungsanstalten Reihe Kongressberichte* 30:209-214. (*Soils and Fertilizers* 54, Abstract 1273. 1991.)

Rich, C. I. 1956. Manganese content of peanut leaves as related to soil factors. *Soil Science* 82:353-363.

Richards, J. E., and T. E. Bates. 1989. Studies on the potassium-supplying capacities of southern Ontario soils. III. Measurement of available K. *Canadian Journal of Soil Science* 69:597-610.

Richter, J., H. Nordmyer, and K. C. Kersebaum. 1985. Simulation of nitrogen regime in loess soils in the winter half-year: comparison between field measurements and simulations. *Plant and Soil* 83:419-432.

Richter, J., A. Nuske, W. Habenicht, and J. Bauer. 1982. Optimized N-mineralization parameters of loess soils from incubation experiments. *Plant and Soil* 68:379-388.

Rodriguez, J. B., G. A. Peterson, and D. G. Westfall. 1989. Calibration of nitrogen and phosphorus soil tests with yield of proso millet. *Soil Science Society of America Journal* 53:1737-1741.

Ross, G. J., and R. A. Cline. 1984. Potassium exchange characteristics in relation to mineralogical properties and potassium uptake by grapes of selected soils in the Niagara Peninsula of southern Ontario. *Canadian Journal of Soil Science* 64:87-98.

Rossiter, R. C. 1964. The effect of phosphate supply on the growth and botanical composition of annual type pasture. *Australian Journal of Agricultural Research* 14:61-76.

Roux, J. le, and M. E. Sumner. 1968. Labile potassium in soils I: Factors affecting the quantity-intensity (Q/I) parameters. *Soil Science* 106:35-41.

Russell, R. S., J. B. Rickson, and S. N. Adams. 1954. Isotopic equilibria between phosphates in soil and their significance in the assessment of fertility by tracer methods. *Journal of Soil Science* 5:85-105.

Russell, R. S., E. W. Russell, and P. G. Marais. 1957. Factors affecting the ability of plants to absorb phosphate from soils. I. The relationship between labile phosphate and absorption. *Journal of Soil Science* 8:248-267.

Sabbe, W. E., and D. B. Marx. 1987. Soil sampling: Spatial and temporal variability. Pp. 1-14. In J. R. Brown (Ed.), *Soil Testing: Sampling, Correlation, Calibration, and Interpretation*. SSSA Special Publication Number 21. Soil Science Society of America, Madison, WI.

Saggar, S., M. J. Hedley, and R. E. White. 1990. A simplified resin membrane technique for extracting phosphorus from soils. *Fertilizer Research* 24:173-180.

Saha, D., R. Mukherjee, and A. K. Mukhopadhyay. 1982. Availability of fixed NH_4^+ to crops I. Native fixed form. *Plant and Soil* 65:345-349.

Sanchez, C. A., A. M. Blackmer, R. Horton, and D. R. Timmons. 1987. Assessment of errors associated with plot size and lateral movement of nitrogen-15 when studying fertilizer recovery under field conditions. *Soil Science* 144:344-351.

Sanchez, P. A., W. Couto, and S. W. Buol. 1982. The Fertility Capability Soil Classification System: Interpretation, applicability and modification. *Geoderma* 27:282-309.

Santos, A. F. dos, M. T. Eid, A. van Diest, and C. A. Black. 1960. Evaluation of biological indexes of soil phosphorus availability. *Soil Science* 89:137-144.

Saunders, W. M. H., C. G. Sherrell, and I. M. Gravett. 1987a. A new approach to the interpretation of soil tests for phosphate response by grazed pasture. *New Zealand Journal of Agricultural Research* 30:67-77.

Saunders, W. M. H., C. G. Sherrell, and I. M. Gravett. 1987b. Calibration of Olsen bicarbonate phosphorus soil test for pasture on some New Zealand soils. *New Zealand Journal of Agricultural Research* 30:387-394.

Scheffer, F., and B. Ulrich. 1962. Considerations regarding the availability to plants of soil potassium. *Potash Review*, October 1962. (Cited by Beckett, 1964b.)

Schepers, J. S., M. G. Moravek, E. E. Alberts, and K. D. Frank. 1991. Maize production impacts on groundwater quality. *Journal of Environmental Quality* 20:12-16.

Schoenau, J. J., and W. Z. Huang. 1991. Anion-exchange membrane, water, and sodium bicarbonate extractions as soil tests for phosphorus. *Communications in Soil Science and Plant Analysis* 22:465-492.

Schofield, R. K. 1947. A ratio law governing the equilibrium of cations in the soil solution. *Proceedings of the XIth International Congress of Pure and Applied Chemistry* 3:257-261.

Schofield, R. K. 1955. Can a precise meaning be given to "available" soil phosphorus? *Soils and Fertilizers* 18:373-375.

Schubert, S., R. Paul, and K. Uhlenbecker. 1990. Charakterisierung des nachlieferbaren Kaliums aus der nichtaustauschbaren Fraktion von acht Böden mittels einer Austauschersmethode und EUF. *Verband Deutscher Landwirtschaftlicher Untersuchungs- und Forschungsanstalten Reihe Kongressberichte* 30:329-334. (*Soils and Fertilizers* 54, Abstract 1275. 1991.)

Schulte, E. E., and P. R. Hodgson. 1987. Suitability of the Mehlich-3 extractant for multielement analysis in soils of the North-Central states. In T. R. Peck (Ed.), *Proceedings of the 11th Soil-Plant Analyst's Workshop, St. Louis. 27-28 October.* University of Illinois, Urbana. (Cited by Fixen and Grove, 1990.)

Schulz, R., and H. Marschner. 1986. Optimierung der Stickstoff-Spätdüngung zu Winterweizen mit dem Nitratschnelltest. *Schriftenreihe, Verband Deutscher Landwirtschaftlichen Untersuchungs- und Forschungsanstalten, Reihe Kongressberichte* No. 20:343-359. (*Soils and Fertilizers* 51, Abstract 1906. 1988.)

Searle, P. L. 1988. The determination of phosphate-extractable sulphate in soil with an anion-exchange membrane. *Communications in Soil Science and Plant Analysis* 19:1477-1493.

Semb, G., and G. Uhlen. 1954. A comparison of different analytical methods for the determination of potassium and phosphorus in soil based on field experiments. *Acta Agriculturae Scandinavica* 5:44-68.

Sharpley, A. N., and S. W. Buol. 1987. Relationship between minimum exchangeable potassium and soil taxonomy. *Communications in Soil Science and Plant Analysis* 18:601-614.

Shaw, J. K., R. K. Stivers, and S. A. Barber. 1983. Evaluation of differences in potassium availability in soils of the same exchangeable potassium level. *Communications in Soil Science and Plant Analysis* 14:1035-1049.

Shickluna, 1962. A comparison of weighed and volume measurements of soil for determining exchangeable soil potassium and calcium. *Michigan State University, Agricultural Experiment Station, Quarterly Bulletin* 44:534-540. East Lansing.

Sierra, J. O., and L. A. Barberis. 1983. Analisis de un modelo de mineralizacion de nitrogeno en suelos del oeste de la Provincia de Buenos Aires. *Universidad de Buenos Aires, Revista de la Facultad de Agronomia* 4:309-315.

Sillanpää, M. 1982. *Micronutrients and the Nutrient Status of Soils: a Global Study.* FAO Soils Bulletin 48. Food and Agriculture Organization of the United Nations, Rome.

Sinclair, A. G., and M. Rodriguez Juliá. 1993. *Defining Economically Optimal Phosphorus Fertiliser Rates for Permanent Grass-Clover Pastures in Spain and New Zealand.* Paper prepared for presentation at the 1993 International Grasslands Congress.

Singh, B. B., and J. P. Jones. 1977. Phosphorus sorption isotherm for evaluating phosphorus requirements of lettuce at five temperature regimes. *Plant and Soil* 46:31-44.

Singh, K. 1986. The critical level of zinc in soil and plant for predicting response of cluster bean to zinc fertilization. *Plant and Soil* 94:285-288.

Sippola, J., and R. Suonurmi-Rasi. 1985. Simple extraction methods as indicators of available soil nitrogen supply. *Annales Agriculturae Fenniae* 24:125-129.

Smaling, E. M. A., and B. H. Janssen. 1987. Soil fertility. Pp. 107-117. In H. W. Boxem, T. de Meester, and E. M. A. Smaling (Eds.), *Soils of the Kilifi Area, Kenya.* Agricultural Research Reports 929. Centre for Agricultural Publishing and Documentation (Pudoc), Wageningen.

Smith, C. M. 1977. Interpreting inorganic nitrogen soil tests: Sample depth, soil water, climate, and crops. Pp. 85-98. In T. R. Peck, J. T. Cope, Jr., and D. A. Whitney (Eds.), *Soil Testing: Correlating and Interpreting the Analytical Results.* ASA Special Publication Number 29. American Society of Agronomy, Crop Science Society of America, Soil Science Society of America, Madison, WI.

Smith, J. H., C. L. Douglas, and M. J. LeBaron. 1973. Influence of straw application rates, plowing dates, and nitrogen applications on yield and chemical composition of sugarbeets. *Agronomy Journal* 65:797-800.

Smith, S. J., L. B. Young, and G. E. Miller. 1977. Evaluation of soil nitrogen mineralization potentials under modified field conditions. *Soil Science Society of America Journal* 41:74-76.

Soil Science Society of America. 1967. *Soil Testing and Plant Analysis.* SSSA Special Publication Series Number 2. Soil Science Society of America, Madison, WI.

Soltanpour, P. N., and A. P. Schwab. 1977. A new soil test for simultaneous extraction of macro- and micro-nutrients in alkaline soils. *Communications in Soil Science and Plant Analysis* 8:195-207.

Sonar, K. R., B. D. Tamboli, S. Y. Daftardar, and N. D. Patil. 1990. Fertilizer requirements of sorghum in vertisols based on yield goal approach. *Communications in Soil Science and Plant Analysis* 21:1245-1255.

Soon, Y. K. 1990. Comparison of parameters of soil phosphate availability for the northwestern Canadian Prairie. *Canadian Journal of Soil Science* 70:227-237.

Sopher, C. D., and R. J. McCracken. 1973. Relationships between soil properties, management practices, and corn yields on South Atlantic Coastal Plain soils. *Agronomy Journal* 63:595-599.

Sparling, G. P., J. D. G. Milne, and K. W. Vincent. 1987. Effect of soil moisture regime on the microbial contribution to Olsen phosphorus values. *New Zealand Journal of Agricultural Research* 30:79-84.

Sparling, G. P., K. N. Whale, and A. J. Ramsey. 1985. Quantifying the contribution from the soil microbial biomass to the extractable P levels of fresh and air-dried soils. *Australian Journal of Soil Research* 23:613-621.

Spencer, K., and A. G. Govaars. 1982. *The Potassium Status of Pastures in the Moss Vale District, New South Wales.* Commonwealth Scientific and Industrial Research Organization, Australia, Division of Plant Industry Technical Paper No. 38. Canberra.

Sposito, G., and F. T. Bingham. 1981. Computer modeling of trace metal speciation in soil solutions: Correlation with trace metal uptake by higher plants. *Journal of Plant Nutrition* 3:35-49.

Spurway, C. H. 1933. *Soil Testing. A Practical System of Soil Diagnosis.* Michigan Agricultural Experiment Station, Technical Bulletin No. 132. East Lansing.

Stanford, G. 1973. Rationale for optimum nitrogen fertilization in corn production. *Journal of Environmental Quality* 2:159-166.

Stanford, G. 1982. Assessment of soil nitrogen availability. Pp. 651-688. In F. J. Stevenson (Ed.), *Nitrogen in Agricultural Soils.* Agronomy 22. American Society of Agronomy, Crop Science Society of America, and Soil Science Society of America, Madison, WI.

Stanford, G., J. N. Carter, D. T. Westermann, and J. J. Meisinger. 1977. Residual nitrate and mineralizable soil nitrogen in relation to nitrogen uptake by irrigated sugarbeets. *Agronomy Journal* 69:303-308.

Stanford, G., and J. D. DeMent. 1957. A method for measuring short-term nutrient absorption by plants: I. Phosphorus. *Soil Science Society of America Proceedings* 21:612-617.

Stanford, G., J. O. Legg, and S. J. Smith. 1973. Soil nitrogen availability evaluations based on nitrogen mineralization potentials of soils and uptake of labeled and unlabeled nitrogen by plants. *Plant and Soil* 39:113-124.

Stanford, G., and S. J. Smith. 1972. Nitrogen mineralization potentials of soils. *Soil Science Society of America Proceedings* 36:465-472.

Steel, R. G. D., and J. H. Torrie. 1960. *Principles and Procedures of Statistics With Special Reference to the Biological Sciences.* McGraw-Hill Book Co., NY.

Stewart, R. 1932. *The Mitscherlich, Wiessmann and Neubauer Methods of Determining the Nutrient Content of Soils.* Imperial Bureau of Soil Science, Technical Communication 25.

Sumner, M. E. 1987. Field experimentation: Changing to meet current and future needs. Pp. 119-131. In J. R. Brown (Ed.), *Soil Testing: Sampling, Correlation, Calibration, and Interpretation.* SSSA Special Publication Number 21. Soil Science Society of America, Madison, WI.

Sunda, W., and R. R. L. Guillard. 1976. The relationship between cupric ion activity and the toxicity of copper to phytoplankton. *Journal of Marine Research* 34:511-529.

Swenson, L. J., W. C. Dahnke, and D. D. Patterson. 1984. *Sampling for Soil Testing.* North Dakota State University, Department of Soil Science, Research Report No. 8. Fargo.

Sylvester-Bradley, R. 1985. Possibilities for improving prediction of fertilizer nitrogen requirement on chalky boulder clay in East Anglia from measurements of soil mineral nitrogen. Pp. 51-63. In J. J. Neeteson and K. Dilz (Eds.), *Assessment of Nitrogen Fertilizer Requirement.* Institute for Soil Fertility and Netherlands Fertilizer Institute, Haren, The Netherlands.

Tarafdar, J. C., and A. Jungk. 1987. Phosphatase activity in the rhizosphere and its relation to the depletion of soil organic phosphorus. *Biology and Fertility of Soils* 3:199-204.

Thind, S. S., A. K. Rishi, and N. N. Goswami. 1989. Comparison of soil test methods for available phosphorus. *Journal of Nuclear Agriculture and Biology* 18:48-51.

Tran, T. S., J. C. Fardeau, and M. Giroux. 1988. Effects of soil properties on plant-available phosphorus determined by the isotopic dilution phosphorus-32 method. *Soil Science Society of America Journal* 52:1383-1390.

Truog, E. 1930. The determination of the readily available phosphorus of soils. *Journal of the American Society of Agronomy* 22:874-882.

Truog, E. 1961. Fifty years of soil testing. *Transactions of 7th International Congress of Soil Science* 3:46-53.

Van der Meer, H. G., and M. G. van Uum-van Lohuyzen. 1986. The relationship between inputs and outputs of nitrogen in intensive grassland systems. Pp. 1-18. In H. G. van der Meer, J. C. Ryden, and G. C. Ennik (Eds.), *Nitrogen Fluxes in Intensive Grassland Systems*. Martinus Nijhoff Publishers, Dordrecht, The Netherlands.

Van der Zee, S. E. A. T. M., L. G. J. Fokkink, and W. H. van Riemsdijk. 1987. A new technique for assessment of reversibly adsorbed phosphate. *Soil Science Society of America Journal* 51:599-604.

Van Diest, A. 1987. Coordinator's report on the 2nd working session. Pp. 211-212. In *Methodology in Soil-K Research*. Proceedings of the 20th Colloquium of the International Potash Institute held in Baden bei Wien/Austria 1987. International Potash Institute, Worblaufen-Bern, Switzerland.

Van Itallie, T. B. 1938. Cation equilibria in plants in relation to the soil. *Soil Science* 46:175-186.

Van Lierop, W. 1989. Effect of assumptions on accuracy of analytical results and liming recommendations when testing a volume or weight of soil. *Communications in Soil Science and Plant Analysis* 20:121-137.

Van Noordwijk, M., P. de Willigen, P. A. J. Ehlert, and W. J. Chardon. 1990. A simple model of P uptake by crops as a possible basis for P fertilizer recommendations. *Netherlands Journal of Agricultural Science* 38:317-332.

Verma, T. S., R. S. Minhas, R. C. Jaggi, and P. K. Sharma. 1987. Efficiencies of nitrogen, phosphorus and potassium for rice and wheat and their verifications for prescription based fertiliser recommendations. *Journal of the Indian Society of Soil Science* 35:421-425.

Vigil, M. F., and D. E. Kissel. 1991. Equations for estimating the amount of nitrogen mineralized from crop residues. *Soil Science Society of America Journal* 55:757-761.

Villemin, P. 1987. Translation of laboratory K-data into K fertilizer recommendations. Pp. 199-210. In *Methodology in Soil-K Research*. Proceedings of the 20th Colloquium of the International Potash Institute held in Baden bei Wien/Austria 1987. International Potash Institute, Worblaufen-Bern, Switzerland.

Von Braunschweig, L. C., and K. Mengel. 1971. Der Einfluss verschiedener den Kaliumzustand des Bodens charakterisierender Parameter auf den Kornertrag von Hafer. *Landwirtschaftliche Forschung, Sonderheft* 26/I:65-72.

Voss, R. E. 1969. *Response by Corn to NPK Fertilization of Marshall and Monona Soils as Influenced by Management and Meteorological Factors*. Ph.D. Thesis, Iowa State University, Ames.

Voss, R. E., J. J. Hanway, and W. A. Fuller. 1970. Influence of soils, management, and climatic factors on the yield response by corn (*Zea mays* L.) to N, P, and K fertilizer. *Agronomy Journal* 62:736-740.

Walsh, L. M., and J. D. Beaton (Eds.). 1973. *Soil Testing and Plant Analysis*. Revised Edition. Soil Science Society of America, Madison, WI.

Walther, U. 1983. Die N_{min}-Methode Entscheidungshilfe zu einer gezielten N-Düngung im Zuckerrübenbau? Pp. 521-532. In *Nitrogen and Sugar Beet*. International Institute for Sugar Beet Research, Brussels, Belgium.

Waring, S. A., and J. M. Bremner. 1964. Ammonium production in soil under water-logged conditions as an index of nitrogen availability. *Nature* 201:951-952.

Warren, G. P. 1990. Understanding P availability in tropical soils by the use of sorption parameters for P-32. *Transactions 14th International Congress of Soil Science* IV:104-109.

Warren, G. P., and D. C. Whitehead. 1988. Available soil nitrogen in relation to fractions of soil nitrogen and other soil properties. *Plant and Soil* 112:155-165.

Warrick, A. W., D. E. Myers, and D. R. Nielsen. 1986. Geostatistical methods applied to soil science. Pp. 53-82. In A. Klute (Ed.), *Methods of Soil Analysis. Part 1. Physical and Mineralogical Methods.* Second Edition. Agronomy Number 9 (Part 1). American Society of Agronomy and Soil Science Society of America, Madison, WI.

Waugh, D. L., and J. W. Fitts. 1966. *Soil Test Interpretation Studies: Laboratory and Potted Plant.* North Carolina State University Agricultural Experiment Station, International Soil Testing Series, Technical Bulletin No. 3. Raleigh.

Wehrmann, J., and C.-Z. Eschenhoff. 1986. Distribution of nitrate, exchangeable and non-exchangeable ammonium in the soil-root interface. *Plant and Soil* 91:421-424.

Wehrmann, J., and H.-C. Scharpf. 1986. The N_{min}-method an aid to integrating various objectives of nitrogen fertilization. *Zeitschrift für Pflanzenernährung und Bodenkunde* 149:428-440.

Wehrmann, J., H.-C. Scharpf, and H. Kuhlmann. 1988. The N_{min}-method — an aid to improve nitrogen efficiency in plant production. Pp. 38-45. In D. S. Jenkinson and K. A. Smith (Eds.), *Nitrogen Efficiency in Agricultural Soils.* Elsevier Applied Science, London.

Wendt, R. C., and R. B. Corey. 1981. Available P determination by equilibration with dilute $SrCl_2$. *Communications in Soil Science and Plant Analysis* 12:557-568.

Westerman, R. L. (Ed.). 1990. *Soil Testing and Plant Analysis.* Third Edition. Soil Science Society of America, Madison, WI.

White, R. E., and P. H. T. Beckett. 1964. Studies on the phosphate potential of soils. Part I - The measurement of phosphate potential. *Plant and Soil* 20:1-16.

Whitehead, D. C. 1983. Prediction of the supply of soil nitrogen to grass. Pp. 318-320. In J. A. Smith and V. W. Hays (Eds.), *Proceedings of the XIV International Grassland Congress.* Westview Press, Boulder, Colorado.

Whitmore, A. P., and T. M. Addiscott. 1987. Applications of computer modelling to predict mineral nitrogen in soil and nitrogen in crops. *Soil Use and Management* 3:38-43.

Wiessmann, H. 1928. Bestimmung des Nährstoffgehaltes der Böden durch den Gefäss-versuch. *Die landwirtschaftlichen Versuchs-Stationen* 107:275-293.

Wietholter, S. 1984. Predicting potassium uptake by corn in the field using the strontium nitrate soil testing method and a diffusion-controlled uptake model. *Dissertation Abstracts International* 44, 8B:2302.

Wild, A. 1964. Soluble phosphate in soil and uptake by plants. *Nature* 203:326-327.

Wild, A., D. L. Rowell, and M. A. Ogunfowora. 1969. The activity ratio as a measure of the intensity factor in potassium supply to plants. *Soil Science* 108:432-439.

Wit, C. T. de. 1953. *A Physical Theory on Placement of Fertilizers.* Verslagen van Landbouwkundige Onderzoekingen No. 59.4. Staatsdrukkerij en Uitgeverijbedrijf, Wageningen.

Wolf, B. 1982. An improved universal extracting solution and its use for diagnosing soil fertility. *Communications in Soil Science and Plant Analysis* 13:1005-1033.

Wong, M. T. F., R. Hughes, and D. L. Rowell. 1990. Retarded leaching of nitrate in acid soils from the tropics: measurement of the effective anion exchange capacity. *Journal of Soil Science* 42:655-663.

Zandstra, H. G., and A. F. MacKenzie. 1968. Potassium exchange equilibria and yield responses of oats, barley, and corn on selected Quebec soils. *Soil Science Society of America Proceedings* 32:76-79.

Zee, S. E. A. T. M. van der, L. G. J. Fokkink, and W. H. van Riemsdijk. 1987. A new technique for assessment of reversibly adsorbed phosphate. *Soil Science Society of America Journal* 51:599-604.

CHAPTER 5

Fertilizer Evaluation

T HE ASPECT OF FERTILIZER EVALUATION to be considered here is the experimental comparison of two or more fertilizers supplying a common nutrient to find the relative biological value of the nutrient in the different sources. Chemical analyses can be made to determine the concentrations of the nutrient. But because the nutrient may occur in different chemical compounds, in different mixtures of compounds, and in different physical conditions, such as particle sizes, equal biological values often are not associated with equal quantities of the nutrient from different sources. Fertilizer evaluation is an important aspect of soil fertility research.

Qualitative biological evaluations of fertilizers usually are made by applying equal quantities of the common nutrient in the sources to be compared, with due compensation for any differences that may exist in contents of other nutrients. The ranking of biological values then is given by the ranking of the yields of the test crop, the yields of the nutrient in the crop, or some other response criterion. When evaluations are made in this way, the biological values of the various sources may appear to differ considerably if the quantity of the nutrient applied causes the responses to fall on the steeply ascending portion of the response curve. But as the quantity of the nutrient increases, the similarity of performance of the fertilizers generally increases, and if the quantity is great enough so that all the sources supply enough of the nutrient to produce the maximum yield, the differences between fertilizers may disappear. If the quantity applied is sufficient to produce one or more responses on the descending arm of the response curve, the most effective source may appear to be the least effective because it produces the lowest yield. This behavior of qualitative evaluations is largely a consequence of the nonlinear relation between biological responses and quantity of nutrient.

5-1. The Biological Assay Concept

The problem of nonlinearity of biological responses to the dose of biologically active substances was addressed many years ago by scientists endeavoring to characterize the potency of preparations containing certain vitamins and other organic substances for which chemical analyses could not be made. They developed what now would be called the basic biological assay model, namely, if two preparations containing the same active ingredient in an inert matrix are tested on appropriate biological subjects, the ratio of the quantities of the preparations that produce equal biological responses is the inverse of the ratio of the concentrations of the active ingredient. The absolute concentration is not known in either case, only the ratio.

The German medical scientist Paul Ehrlich is regarded as the father of biological assay. Ehrlich is credited with the first bona fide assay standard, a preparation of diphtheria antitoxin that he kept in his laboratory. The potency of other preparations of diphtheria antitoxin in his and other laboratories was checked against his standard, which was used from the late 1800s until 1914.

5-2. Biological Assay of Fertilizers

5-2.1. Application of Biological Assay Theory

As a hypothetical example of classical biological assay in terms of fertilizers, consider a standard fertilizer consisting of 90% ammonium nitrate and 10% quartz sand, and a test fertilizer consisting of 45% ammonium nitrate and 55% quartz sand. These two fertilizers are applied in different quantities to separate plots in a field experiment or to separate containers of soil in a greenhouse experiment, and the responses of a test crop to the fertilizers are measured. The reason for using a range of quantities of each fertilizer is to obtain response curves, which are used to find the relative quantities of the two fertilizers that produce equal responses.

Within the limits of error, the experiment with the two ammonium nitrate fertilizers would show that if the quantities of the test fertilizer containing 45% ammonium nitrate and 55% quartz sand are plotted in a graph along with the results obtained with the standard fertilizer containing 90% ammonium nitrate and 10% quartz sand, the two response curves coincide if the data for the test fertilizer are plotted at half the quantity of fertilizer applied. One kilogram of the test fertilizer is biologically equivalent to 0.5 kilogram of the standard fertilizer. The value 0.5 applies to all pairs of quantities of the two fertilizers leading to equal responses, whether the responses are yields of dry matter, yields of nitrogen in the crop, concentrations of nitrogen in the crop, or any other measured response.

If the experimenter knew the nitrogen concentrations in the two fertilizers, the responses could be plotted against the quantity of nitrogen applied. Within experimental error, the two response curves then would coincide. In effect, the experiment would show that unit quantity of nitrogen as ammonium nitrate has a crop-producing effect equivalent to unit quantity of nitrogen as ammonium nitrate.

Another most significant point about the evaluation of the two fertilizers containing different concentrations of ammonium nitrate in quartz sand is that there would be no need to repeat the experiment in different years, at different locations, or with different crops. Within limits of error, the response curve for the test fertilizer could always be made to coincide with the response curve for the standard fertilizer by plotting the responses against the additions of nitrogen or against the quantity of the standard fertilizer and half the quantity of the test fertilizer.

A situation such as the two ammonium nitrate fertilizers is rarely if ever encountered in practice. Considerable information usually is available on the chemical forms of nutrients in fertilizers, and the concentrations of individual plant nutrients can be determined by chemical analysis. Chemical analyses are more precise and more economical than biological assays. Biological evaluation of fertilizers thus is rarely, if ever, biological assay in the classical sense.

In fertilizer evaluation, the desired information is generally the biological value of a particular nutrient in a test fertilizer relative to that of the same nutrient in another fertilizer, which serves as the standard. The chemical form, the physical condition, and the accompanying ingredients may differ between the standard and test fertilizers. These differences and the way they interact with soils and plants cause problems in making the assays and in interpreting the findings.

In general terms, when two fertilizers containing a given nutrient are used in an evaluation experiment, the response function for the fertilizer serving as the standard of comparison (the standard fertilizer) may be represented as

$$y = f_s(x_s),$$

where y is the yield or other response, x is the quantity of the nutrient applied, and the subscript s refers to the standard source. Similarly, the response function for the fertilizer being compared with the standard (the test fertilizer) may be represented as

$$y = f_t(x_t),$$

where x is the quantity of nutrient as before, and the subscript t applies to the test fertilizer.

In the classical biological assay model, the function relating response to dose is the same in the test substance as in the standard after multiplying the quantity of the test substance by a coefficient that represents the concentration of the active ingredient in the test substance relative to that in the standard. Thus,

$$y = f_s(\lambda x_t),$$

where λ is the concentration of the active ingredient in the test substance expressed as a decimal fraction of the concentration in the standard.

The situation in fertilizer evaluation is different. The concentrations of the nutrient in the test and standard fertilizers are known before the biological assay is made, and as a result, the additions of the nutrient sources almost invariably are expressed in units of nutrient rather than units of fertilizer. When this is the situation, the coefficient λ is a measure of the biological value of unit quantity of the nutrient in the test fertilizer expressed as a decimal fraction of the biological value of unit quantity of the same nutrient in the standard fertilizer.

Numerous terms have been used in the fertilizer evaluation literature to describe λ. These include availability coefficient ratio, availability ratio, availability, efficiency factor, relative efficiency, coefficient of assimilability, relative agronomic effectiveness, and substitution rate. A number of authors who were evaluating various phosphate fertilizers against superphosphate as a standard have called λ (in percent) the percentage superphosphate equivalent. No standard usage has emerged. Here the term availability coefficient ratio will be used because the parameter λ being estimated is a coefficient representing a ratio that reflects the quality commonly termed availability in soil science and plant nutrition.

As in classical biological assay, where absolute concentrations are not known and only relative concentrations can be determined, so also in fertilizer evaluation the absolute availabilities are not known and only relative availabilities can be determined. A properly conducted experiment will yield an evaluation that is a characteristic of the fertilizers and is independent of the quantities applied, the soil, the test plants, and the environmental conditions as long as the fertilizers satisfy the biological assay model of different concentrations of the same active ingredient in an inert matrix.

The availability or effective quantity of the nutrient in the test fertilizer may be represented as the product of the quantity and an availability coefficient. The same is true for the standard fertilizer. Accordingly, if the quantities of the nutrient in the test and standard fertilizers that produce equal responses are x_t and x_s, and if the corresponding availability coefficients are γ_t and γ_s, one may write

$$\gamma_t x_t = \gamma_s x_s,$$

from which

$$\frac{\gamma_t}{\gamma_s} = \frac{x_s}{x_t} \ .$$

Thus, the availability coefficient ratio, which is equal to λ, is the inverse of the quantities of the nutrient that produce equal responses.

In the following text, the term availability coefficient ratio often will be used. It is the ratio that is measured. The convention of expressing the results of the assay as a ratio of the test fertilizer to a standard (generally because the standard is widely used, relatively soluble, and highly effective) has the effect of representing the availability coefficient of the nutrient in the standard source as 1.00. This does not mean that it is 1.00 or that it remains the same under different conditions. Rather, the convention is simply a useful way of treating the data from a given experiment.

In some instances, the following text will refer to certain availability coefficient ratios as availability coefficients applying to the test fertilizers. In such instances, it must be remembered that the numerical values are nominal and are based upon an arbitrarily assigned availability coefficient of 1.00 for the standard fertilizer. The standard fertilizer generally will be indicated.

Where fertilizers that do not satisfy the classical biological assay model are compared, the ideal of a universally applicable ratio must be sacrificed. If the availability coefficient ratios obtained are properly validated, they are appropriate measures of the relative availabilities of unit quantities of the nutrient in the different sources under the conditions employed, but they do not have universal application. Their study yields further insight into the causes of the observed behaviors.

The availability coefficient ratio λ is evaluated most simply by graphic analysis from appropriate response curves. The curves for the test and standard fertilizers are plotted in the same graph. By trial and error, one finds the value of λ by which the quantity of the nutrient in the test fertilizer must be multiplied to cause the response curve for the test fertilizer to coincide with that for the standard. With care, graphic analyses will produce satisfactory evaluations. They are somewhat subjective, however, and the greater the experimental error, the greater is the subjectivity.

Barrow (1985) pointed out that for the circumstances in which the test fertilizer and the standard fertilizer behave like different dilutions of the same active ingredient, the response functions may be combined in a mathematical relation that is equivalent to

$$y = f(x_s + \lambda x_t),$$

where the symbols have the same meaning as before. This relationship is the basic concept employed in statistical procedures for evaluating λ.

Statistically, the evaluation is made by fitting simultaneously the responses to all levels of all fertilizers to the response model by the method of least squares. White et al. (1956) described statistical procedures for three types of response models: (1) concurrent straight lines, in which the response of the control is the same for all fertilizers (the value obtained by the fitting procedure is not necessarily equal to the observed control value) and in which the response is linearly related to the quantity of fertilizer, (2) parallel straight lines with responses and rates of application in logarithms (see Section 5-6.3), and (3) concurrent Mitscherlich curves, in which the response of the control is the same for all fertilizers (the value obtained by the fitting procedure is not necessarily equal to the observed control value) and the responses to all fertilizers approach the same maximum, but at different different rates.

Although White et al. (1956) described their procedures in words different from those used by Barrow (1985), the basic concepts were the same, amounting to fitting equations after substituting Barrow's $(x_s + \lambda x_t)$ for x in whatever the response functions might be. For one standard fertilizer and n test fertilizers, the substitutions would be $(x_s + \lambda_1 x_{t1} + \lambda_2 x_{t2} + \ldots + \lambda_n x_{tn})$.

If there were only one test fertilizer and one standard fertilizer, the equation for linear responses (the concurrent straight lines model) would be

$$y = a + b(x_s + \lambda x_t) = a + bx_s + b\lambda x_t.$$

The equation for quadratic responses would be

$$y = a + b(x_s + \lambda x_t) + c(x_s + \lambda x_t)^2$$
$$= a + bx_s + b\lambda x_t + c(x_s + \lambda x_t)^2.$$

And the equation for Mitscherlich responses would be

$$\text{Log}(A - y) = \text{Log}A - c[(x_s + \lambda x_t) + b]$$
$$= \text{Log}A - cx_s - c\lambda x_t - cb.$$

In these equations, y is the yield or other response; x_s and x_t are the quantities of the standard and test fertilizers or the quantities of the nutrient added in these fertilizers; and $a, b, c, \lambda,$ and A are constants found by fitting. When the equations have been calculated, λ is given by the ratio of the coefficient of x_t to the coefficient of x_s.

Papers emphasizing the theoretical aspects of fertilizer evaluation include those by Barrow (1985), Barrow and Bolland (1990), Black and Scott (1956), Black et al. (1956), Bouldin and Black (1960), Terman et al. (1962), and White et al. (1956). Books on biological assay include those by Bliss (1952), Emmens (1948), and Finney (1978). Although all the books were written by statisticians, the one by Emmens is more easily understood by nonstatisticians than are the others mentioned.

5-2.2. Tests of Validity

An apparent availability coefficient ratio λ can be calculated for any situation in which one can estimate the quantity of a test fertilizer that produces a yield equal to that obtained with a known quantity of a standard fertilizer. The question to be considered here is the conditions under which the calculated ratios may justifiably be called availability coefficient ratios.

5-2.2.1. Statistical Test

Significant deviations of measured responses from a theoretical biological assay model indicate that the model is inappropriate and hence that the calculated value of λ is biased. In the absence of a significant deviation from the response function, one infers that the coefficient λ is an estimate of the availability coefficient ratio of the nutrient in the two fertilizers.

Colwell and Goedert (1988) used the test-of-significance procedure in connection with a field experiment in which a partially acidulated phosphate rock and concentrated superphosphate were compared as sources of phosphorus for soybean. Fig. 5-1A shows their evaluation. Using the concurrent square root function to fit the data (see Section 1-5.1), they found that the data did not deviate significantly from the biological assay model. That is, the concurrent square root function and the model were acceptable. They found that $\lambda = 0.32$.

Fig. 5-1B shows the fit of the data to the $\lambda = 0.32$ outcome. For plotting, the quantities of phosphorus have been multiplied by $\lambda = 0.32$ for the partially acidulated phosphate rock and 1.00 for concentrated superphosphate. The authors calculated that the 95% confidence interval for $\lambda = 0.32$ was from 0.20 to 0.43.

Figs. 5-1C and D show the same data as Figs. 5-1A and B, but here the data are fitted to the concurrent straight lines function. In this instance, λ is estimated at 0.36. The fit of the concurrent straight lines function is not as good as that of the concurrent square root function, but the estimated availability coefficient ratio is about the same as with the better concurrent square root function, and it is well within the 95% confidence interval for that function.

The statistical calculations behind Fig. 5-1A and B are valuable for both the test of significance and the estimate of the availability coefficient ratio they make possible. Figs. 5-1C and D, however, illustrate the important point that the experimental error in the assay data is usually so great that a variety of models may prove acceptable. Improving the precision of the data is of first importance in improving the outcome.

Terman (1961) found that in 107 field experiments on fertilization of corn in the United States, the coefficient of variation of the grain yields ranged from about 4% to 24%. Failure to appreciate fully the implications of such uncontrolled variability of crop yields has led to the conduct of many fertilizer evaluation experiments that included too few rates of application and too few replicates to expect that differences in effectiveness among different sources of a common nutrient could be measured quantitatively or even detected. See a paper by Johnstone and Sinclair (1991) for statistical considerations of experimental design requirements.

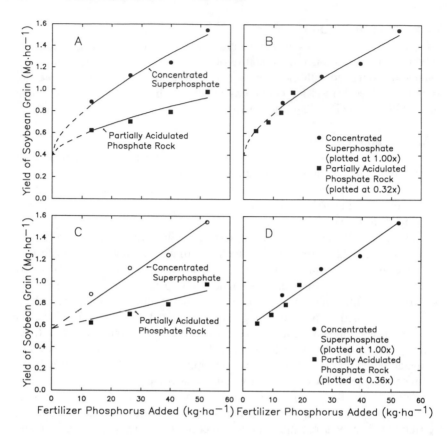

Fig. 5-1. Results of a field experiment in Brazil comparing partially acidulated phosphate rock and concentrated superphosphate as sources of phosphorus for soybean. In *A*, the lines represent the concurrent square root response function, with availability coefficients 0.32 for the phosphorus in partially acidulated phosphate rock and 1.00 for the phosphorus in concentrated superphosphate. In *B*, the yields for the same function are plotted against 0.32x for the phosphorus in partially acidulated phosphate rock and 1.00x for the phosphorus in concentrated superphosphate. In *C*, the lines represent the concurrent straight lines function, with availability coefficients 0.36 for the phosphorus in partially acidulated phosphate rock and 1.00 for the phosphorus in concentrated superphosphate. In *D*, the yields for the function in *C* are plotted against 0.36x for the phosphorus in partially acidulated phosphate rock and 1.00x for the phosphorus in concentrated superphosphate. (Colwell and Goedert, 1988)

Terman's (1960, 1961) summaries of the results of fertilizer evaluation experiments sponsored by the Tennessee Valley Authority bear testimony to the disastrous effects of uncontrolled variation. In a study of the results of 174 experiments in southeastern United States, he reported that if the availability coefficient of the water-soluble and citrate-soluble phosphorus in the test fertilizers investigated had been within the range of 0.5 to 1.5 relative to the value of 1.0 arbitrarily assigned to the phosphorus in the concentrated superphosphate standard, it would have been possible to find the differences from the standard significant at the 95% probability level in only a few experiments. In 124 experiments from northern and western states, he found that in fewer than 10% of the experiments showing response to phosphorus fertilization would it have been possible to find availability coefficients in the range of 0.5 to 1.5 significantly different at the 95% level from the value of 1.0 assigned to the phosphorus of superphosphate.

Limited response to the nutrient being evaluated is an important cause of experimental error in fertilizer evaluation. A narrow spread between control yields and maximum yields provides only a small range in which the effects of differences among fertilizers can be evaluated.

One way to increase the level of response and thus to increase the probability of obtaining useful results is to use preliminary soil tests as a basis for selecting potentially responsive sites. For experiments on nitrogen, cropping potential sites to nonlegumes for 2 or 3 years without additions of fertilizer nitrogen will usually result in a large response suitable for evaluating nitrogen sources.

Maintaining conditions for growth at a practical maximum is a second important way of increasing the level of response. The usual procedure is to add nutrients that are suspected to be limiting. Adding water may be of equal or greater value, but this is often impractical. Although some have suggested that failure to eliminate deficiencies of nutrients other than the one under test may result in incorrect estimates of availability coefficient ratios, a theoretical explanation for why this should be true is as yet lacking.

5-2.2.2. Physiological Test

A qualitative test to determine whether it is appropriate to calculate an availability coefficient ratio from the data in a particular experiment may be made by plotting the yield of dry matter against the yield of the nutrient or the concentration of the nutrient, and determining whether the functions for the test and standard fertilizers differ. This test, proposed by De Wit (1953) for another purpose, will be discussed further in Chapter 7 in connection with his work on fertilizer placement.

The basic theory behind the plots of yield of dry matter against yield of nutrient is that the primary effect by which the availability or effective quantity of a nutrient may be recognized is absorption of the nutrient by plants. The total amount of the nutrient absorbed, however, is determined only in part by the availability. Influencing both the absorption and the yield of the plants are all the other growth factors.

The total yield of the plants and the yield of the nutrient are mutually dependent variables. If the availability of the nutrient is varied by changing the quantity of a particular fertilizer, therefore, the change in yield of the plants will be related to the change in yield of the nutrient in the plants. The functional relationship between the two variables will be determined by the availability of the nutrient and all other attributes of the fertilizer, environment, and plants that affect either variable. As a result, if a plot of the yield of the plants against the yield of the nutrient or against the concentration of the nutrient in the plants for a test fertilizer does not diverge from the corresponding plot for the standard fertilizer (that is, if the internal efficiency of the nutrient in the plants does not differ between fertilizers), there is no evidence that the fertilizers are affecting the plants in ways other than the way they would if they represented different dilutions of the same nutrient in an inert matrix. If desired, a statistical test may be made to determine whether the plots diverge significantly.

A plot of yield of dry matter versus yield of the nutrient provides a means of finding whether the difference between the test fertilizer and the standard fertilizer is qualitative or merely quantitative. A qualitative difference is indicated by the emergence of curves that differ in vertical location in part or all of the range of the data. Divergence is most common in the upper range, where the yield is least limited by the nutrient being tested. If the plots diverge, the evidence indicates that something besides the availability of the nutrient is affecting the results, and the mechanism of action of the nutrient supplied by the test and standard fertilizers is not the same. A quantitative difference is indicated by the location of the data points for the test fertilizer to the left or right of the data points for corresponding quantities of the standard fertilizer, but on the same curve. The points for test fertilizers usually are located to the left of the points for the standard because the most effective fertilizer usually is chosen as the standard.

The physiological test sometimes is applied to nutrient concentrations instead of yields of the nutrient. For example, Fig. 5-2 gives three plots from data on copper fertilization of soil, with wheat as a test crop. Copper sulfate was used as the source of the various additions of copper in all instances, and the additional variables were time and temperature of incubation of the soil with or without the copper before the test crop was grown.

In each instance, the yield of dry matter is plotted on the vertical axis. In A, the X axis represents the yield of copper in the plants. In B, the X axis is the concentration of copper in the plants, and in C, the X axis is the concentration of copper in the youngest fully expanded leaf blade of the wheat test crop. In this instance, the scattering of points is greater in B than in A or C.

The reason for the relatively marked scatter of points in B is apparently that a plot of yield against copper concentration in the plants is in effect a plot of the square of the yield of dry matter against the yield of copper. The squaring process causes the points in A that are a little high to become much higher in B. The same principle applies to a comparison of C and A, but the fact that the concentration of copper in the young leaf tissue is a superior index of the current sufficiency of copper may be responsible for the favorable showing of C. Copper is a relatively immobile nutrient in plants. Little translocation from old tissue to developing tissue occurs.

The physiological test may be insensitive if applied to grain. Fig. 5-3A shows an instance in which two plots of yield of dry matter in oat grain versus the yield of potassium in the grain show good agreement. In this instance, the two "potassium fertilizers" employed were actually the same — potassium sulfate,

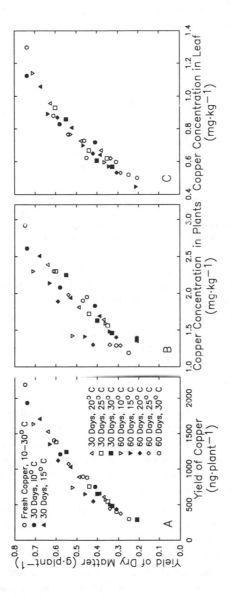

Fig. 5-2. Yields of copper in wheat plants versus: A. Yields of dry matter in the above-ground portions of the plants. B. Copper concentrations in the total above-ground dry matter. C. Copper concentrations in the dry matter of the youngest fully expanded leaf blade. Copper was applied in 5 milliliters of dilute copper sulfate solution in quantities of 0, 0.1, 0.4, and 0.6 milligram per culture. The soil then was moistened to field capacity and incubated for 30 or 60 days at different temperatures before the crop was planted. In the "fresh copper" series, the copper was added after the soil had been incubated. Incubation temperature had no significant effect on the results with this series, and the data were averaged. The legend in A applies to *B* and *C* as well. (Brennan et al., 1984)

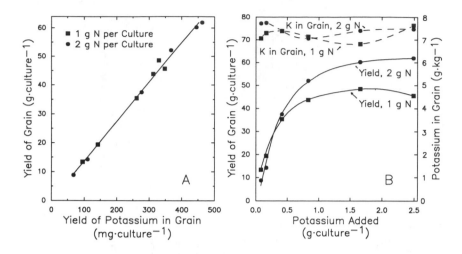

Fig. 5-3. A. Yields of dry matter versus yields of potassium in oat grain in an experiment in which oat plants were grown on sand-peat cultures with different quantities of potassium sulfate and two levels of nitrogen as calcium nitrate. B. Yield response curves and potassium concentrations in the grain in the experiment described in A. (Alten et al., 1940)

but the nitrogen supply differed. The explanation for the good agreement of the two plots is that both are mainly a function of the yield of dry matter because the potassium concentration in the oat grain was little affected by either the potassium supply or the nitrogen supply. See Fig. 5-3B. The principal effect of potassium supply on plant composition is in the vegetative parts.

Fig. 5-4 illustrates the divergence of plots of yield of dry matter versus yield of nutrient, a condition indicating that the plants are responding to differences in conditions other than quantitative variations in the supply of a common nutrient. Here the two curves are definitely different functions, and the cause is readily understandable. In this instance, the two ''potassium fertilizers'' again were actually the same — potassium sulfate, but one series of cultures was supplied with more nitrogen than the other. Hence, a given amount of potassium in the plants produced more dry matter in one series than the other because it had the help of extra nitrogen. Calculation of availability coefficient ratios would be inappropriate in such an instance, but if calculated, the values would vary with the quantities of the fertilizers added in the range where both response curves had numerical responses. No values would exist in the range above the maximum yield for the lower-yielding fertilizer.

The two experimental series illustrated in Fig. 5-4 had different maximum yields. Where a standard and test fertilizer produce different maximum yields, plotting the yields of the test fertilizer after multiplying the quantities of the nutrient by a constant factor will not cause the response curve for the test fertilizer to coincide with that for the standard. Sometimes, but not always, this means that the responses are not a consequence of differences in supply of the common nutrient. The explanation for the difference, however, is generally not so obvious as the explanation given for the observations in Fig. 5-4. Additional experiments usually are needed.

Bolland et al. (1988) found that different phosphate fertilizers did not give the same maximum yields even though the yield of dry matter versus yield of phosphorus curves did not segregate by fertilizers. Other experimental work, which will be reviewed in Section 5-2.3.4, provides a theoretical explanation.

Pesek and Webb (1957) reported the results of field experiments with corn in Iowa, in which the calculated maximum yields increased with the proportion of the citrate-soluble phosphorus in the fertilizers that was soluble in water. The fertilizers were applied to the side of and below the seed of corn, which was planted in hills. Thus, the fertilizers were highly concentrated. In subsequent experiments in which the fertilizers were broadcast (Webb and Pesek, 1959), the degree of water solubility had little or no effect on the responses, and there was no evidence that the maximum yields differed among sources. In this instance, the concentrated placement was associated with the failure of the response curves to behave according to the biological assay model of different dilutions of the same active ingredient. The explanation may be a combination of inhibition of the dissolution of water-insoluble phosphorus by the water-soluble phosphorus in the fertilizers plus inhibition of root penetration into the fertilizer zones by the soluble nitrogen and potassium components that were mixed with the phosphates.

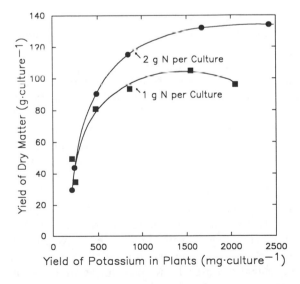

Fig. 5-4. Yields of oat plants in sand-peat cultures versus the yields of potassium in the plants with different quantities of potassium as potassium sulfate and with two levels of nitrogen as calcium nitrate. (Alten et al., 1940)

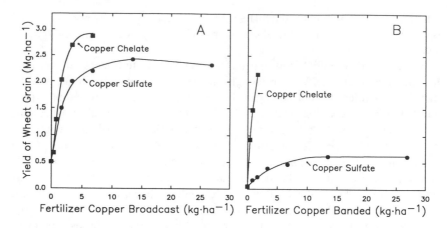

Fig. 5-5. Response of wheat on organic soils to *A* broadcast and *B* banded applications of copper sulfate and copper chelate (HEDTA). The data are average values from four field experiments in Minnesota. (Varvel, 1983)

Fig. 5-5 is an example with copper, in which placement again played a part. In these experiments, different quantities of copper were applied broadcast and in bands as copper sulfate and copper chelate, and the test crop was spring wheat. Here the maximum yields were higher with the chelate than with the sulfate, and the difference appeared to be greater with the banded application than with the broadcast application.

The experiments were conducted on organic soils, which are notable for their capacity to immobilize copper by chelation. The findings thus suggest that the difference in the functional relationships implied by the difference in maximum yields was related to interactions between the added copper and the organic soils. Additional experiments are needed, however, to investigate this possibility.

In interpreting the results of experimental work on different nitrogen sources for cabbage, Smith and Hadley (1988) suggested that osmotic stress from high concentrations of salts present in the soil with some sources may have been responsible for differences they noted in the response functions. The estimated maximum yields with the quick-acting sources (ammonium nitrate and dried blood) were lower than those with the slower-acting sources (feathermeal and a proteinaceous material derived from the activated sewage sludge process), and they were obtained with smaller quantities of nitrogen.

If the evidence for divergence is not significant in plots of yield of dry matter versus yield of nutrient, equal yields of dry matter or equal yields of the nutrient may be inferred to reflect equal availabilities or effective quantities of the nutrient. Where the yield of dry matter versus yield of nutrient functions for a test fertilizer and a standard fertilizer do not diverge, at least four theoretical possibilities appear to exist. In the following enumeration, the phrase "form or mixture of forms" refers to both physical and chemical forms, and only the first possibility listed corresponds to the classical biological assay model. In all instances, the values of λ are calculated on the basis of the quantity of the nutrient being evaluated.

1. If the value of λ is constant and equal to 1, indications are that the same form or mixture of forms of the nutrient occurs in the test and standard fertilizers.
2. If the value of λ is constant but not equal to 1, indications are that the form or mixture of forms of the nutrient is not the same in the test and standard fertilizers.
3. If the value of λ is constant and not equal to 1, but differs between evaluations based upon yield of dry matter and yield of the nutrient, a possible cause is a difference in form or mixture of forms of the nutrient in the test and standard fertilizers, with an associated difference in rates of change of availabilities of the nutrient during growth of the test crop.
4. If the value of λ varies with the quantity of fertilizer, indications are that the form or mixture of forms of the nutrient differs between fertilizers and that some characteristic of one of the fertilizers inhibits its action.

5-2.3. Examples
Included here are examples of the four theoretical possibilities outlined in the preceding section, in which the physiological test of validity provides no evidence that the yield of dry matter versus yield of nutrient functions differ for the test and standard fertilizers.

5-2.3.1. Availability Coefficient Ratio Unity
The behavior of test and standard fertilizers consisting of different dilutions of ammonium nitrate with quartz sand discussed in Section 5-2.1 is an example of the first possibility. As mentioned previously, this is essentially a hypothetical situation because such evaluations rarely if ever are made in practice.

5-2.3.2. Availability Coefficient Ratio Constant But Not Unity
Fig. 5-6A shows the results of an experiment in which three copper compounds were applied in different quantities as foliar sprays for wheat. The response curves approach similar maximum yields, as would be expected if the mechanism of the effect was the same for all copper sources. When the most effective compound (copper ethylenediaminetetraacetate) is taken as the standard and the quantities of copper are multiplied by the nominal availability coefficient of 1.00, multiplying the quantities of copper supplied by copper sulfate by 0.48 and the quantities supplied by copper oxychloride by 0.21 caused all the grain yield data to come together to form a single response curve. That is, a single availability coefficient for each fertilizer was adequate to express the relative potency of the copper supplied by the three compounds irrespective of the quantities applied.

Fig. 5-6 represents the results of one of three identical experiments on different soils in different years. The yields differed among experiments, but the availability coefficient ratios were similar. If the copper added as the ethylenediaminetetraacetate is assigned an availability coefficient of 1.00 in each experiment, the mean availability coefficient ratio of the copper in copper sulfate was 0.49 plus or minus a standard error of the mean equal to 0.05, and the mean availability coefficient ratio of the copper in copper oxychloride was 0.23 plus or minus a standard error of the mean equal to 0.03. These results indicate that the behavior of the three copper sources in the three experiments was close to that of the classical model of different dilutions of a common active ingredient, for which the relative effectiveness would correspond to the relative quantities of the active ingredient applied. In all experiments, the sprays were applied on cool, cloudy days and at a given growth stage. Less similarity in results would be expected if the conditions at the time of application had differed and if the copper had been applied to the soils.

A second example is found in Fig. 5-7. Fig. 5-7A shows a plot of yields of dry matter in rice plants grown on flooded soil versus the yield of phosphorus from seven different sources. The data from all sources can be represented to a good approximation by a single line, indicating that the differences in yield of dry matter were a consequence of differences in the uptake of phosphorus from the different sources.

In Fig. 5-7B, the yields of dry matter obtained with the seven sources are plotted against the quantities of phosphorus applied. Wide differences in availability of the phosphorus supplied by the various phosphate rocks are evident.

In Fig. 5-7C, the yields of dry matter obtained with the various fertilizers are plotted against the product of the quantities of phosphorus applied and an avail-

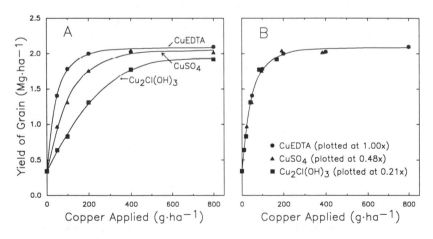

Fig. 5-6. A. Response of wheat to foliar sprays supplying different quantities of copper as copper ethylenediaminetetraacetate, copper sulfate, and copper oxychloride in a field experiment in Western Australia. B. Data from *A* in which the copper applied as copper ethylenediaminetetraacetate is taken as the standard source and assigned an availability coefficient of 1.00. The points representing yields obtained with copper sulfate and copper oxychloride are plotted after multiplying the quantities of copper added by estimated availability coefficient ratios of 0.48 and 0.21, respectively. (Brennan, 1990)

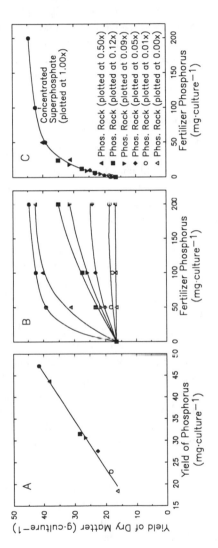

Fig. 5-7. Evaluation of six phosphate rocks against concentrated superphosphate as a standard fertilizer for rice in flooded soil cultures. A. Yields of dry matter versus yields of phosphorus in the plants supplied by the various sources. The data are averages of values obtained with 50, 100, and 200 milligrams of fertilizer phosphorus per culture. B. Yields of dry matter versus fertilizer phosphorus added by sources. C. Yields of dry matter versus fertilizer phosphorus added, with the yields plotted against 1.00x for concentrated superphosphate and values ranging from 0.00x to 0.50x for the various phosphate rocks. From top to bottom in the legend, the sources of the phosphate rocks are North Carolina, northern Florida, central Florida, Idaho, Tennessee, and Missouri. The legend in *C* applies to *A* and *B* as well. (Engelstad, 1978)

ability coefficient, the value of which ranges from 0.00 for Missouri phosphate rock to 1.00 for the concentrated superphosphate standard. The close cluster of the points for the phosphate rocks around the line for concentrated superphosphate shows that the data could be represented to a good approximation by the classical biological assay model in which the availability coefficient ratio λ remains the same for all pairs of values of the standard and test fertilizer producing equal yields. Although the λ values probably would have been found variable for some and perhaps all of the phosphate rocks, had great enough quantities been applied, this cannot be verified by the data from the experiment. The same may be said for a greenhouse experiment by Khasawneh (Khasawneh and Doll, 1978) in which the same group of fertilizers was compared using corn as the test crop.

5-2.3.3. Availability Coefficient Ratios Differing Among Response Criteria

Differences in availability coefficient ratios among response criteria may occur if the yield of dry matter versus yield of nutrient function differs among nutrient sources. If there is very little scattering of points in a plot of yields of dry matter against yields of the nutrient, however, the availability coefficient ratios should be almost the same for different response criteria, provided that appropriate response functions are used.

Fig. 5-8 is an example in which two phosphate sources were compared upon addition to a soil that had been treated with four different quantities of Fe_2O_3 as hydrous iron oxide. Fig. 5-8A shows that the points in a plot of yields of dry matter versus yields of phosphorus could be fitted well by a single line except for two points, and Fig. 5-8B shows that the availability coefficient ratios based upon yields of phosphorus were a little higher than those based upon yields of dry matter for two of the three iron oxide treatments. The addition (1.60 grams of Fe_2O_3 equivalent per kilogram of soil) for which the deviation between response criteria was greatest involved response curves using the point in Fig. 5-8A with the greatest deviation. Inspection of the original data suggests that the value for yield of phosphorus in the deviating point was too high, which would have inflated the estimated availability coefficient ratio based upon yields of phosphorus. Each response curve was defined by only three points, which means that one deviating point could have a considerable effect on the derived availability coefficient.

Fig. 5-9 illustrates a different situation, in which definite differences were observed between the availability coefficient ratios derived from different response criteria. Urea was the chemical form of nitrogen applied in all the four nitrogen sources tested, and paddy rice was the test crop.

In Fig. 5-9A, all the availability coefficient ratios based upon yields of nitrogen exceeded those based upon yields of grain except for the standard fertilizer, which was given a value of 1.0 with both yields of grain and yields of nitrogen. The choice of fertilizer-grade urea in a split application as the standard was appropriate from the practical standpoint because it represented the form of urea

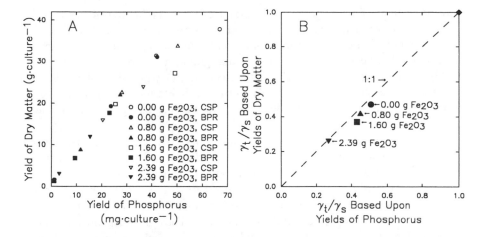

Fig. 5-8. Evaluation of Bayovar phosphate rock as a phosphorus source for corn against concentrated superphosphate as a standard fertilizer on an acid silt loam soil. Additions of 0, 0.8, 1.6, and 2.39 grams of Fe_2O_3 as hydrous ferric oxide per kilogram of soil were made to increase the phosphorus retention capacity in a greenhouse experiment. A. Yields of dry matter versus yields of phosphorus in the test crop. B. Ratios of the availability coefficients of phosphorus in the phosphate rock to that of phosphorus in concentrated superphosphate (γ_t/γ_s) derived from yields of dry matter versus the ratios derived from yields of phosphorus. The location of the diamond shaped symbol in the upper right corner of the graph represents the availability coefficient of 1.00 arbitrarily assigned to concentrated superphosphate for evaluations based upon both the yields of dry matter and the yields of phosphorus. (Hammond et al., 1986a)

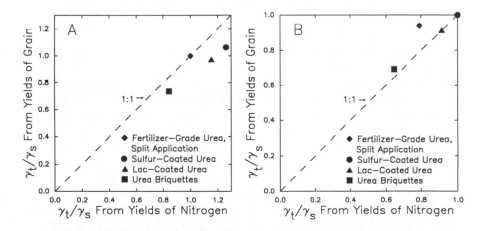

Fig. 5-9. Availability coefficient ratios (γ_t/γ_s) of various sources of urea based upon yields of grain versus ratios based upon yields of nitrogen in paddy rice in a field experiment in Punjab. All sources of urea were applied at the time of transplanting except for the fertilizer-grade urea, which was applied in a split application. A. Fertilizer-grade urea in a split application as the standard fertilizer. B. Sulfur-coated urea as the standard fertilizer. (Rana et al., 1984)

and the method of application used in practice. From the biological standpoint, however, the fertilizer-grade urea treatment was not the best standard because it was the only source applied in a split application, which could have affected the relationship between nitrogen uptake and grain production.

Fig. 5-9B shows the results when sulfur-coated urea was used as the standard fertilizer. Here the availability coefficient ratios based upon the two different response criteria were about the same except for the fertilizer-grade urea applied in a split application.

A possible cause of a difference in availability coefficient ratios calculated from yields of dry matter and yields of a nutrient is a lag of growth behind uptake, coupled with different rates of change of the availability of the nutrient when the standard and test sources are added to soil. A growth lag effect is shown to an extreme degree in the short-term uptake test of phosphorus availability proposed by Stanford and DeMent (1957). They grew seedlings of three species of test plants without an external supply of phosphorus and then placed the roots in contact with soil containing fertilizer phosphorus. Three days later, the average dry weights of the plants were 0.54 gram without phosphorus and 0.54 gram with phosphorus, but the phosphorus concentration in the plant tops had increased from an average of 2.4 grams per kilogram in the controls to 4.0 grams per kilogram with the maximum addition of fertilizer phosphorus.

5-2.3.4. Availability Coefficient Ratio Variable

In some instances, the crop may appear to be responding to a particular nutrient in the standard and test fertilizers, but yet the responses do not behave like those described in Section 5-2.1 for an experiment with the two mixtures of ammonium nitrate with sand. The examples found most frequently in the scientific literature involve phosphate rock. Investigations of the value of various phosphate rocks have been stimulated by the findings that phosphate rocks may differ greatly in value for direct application as sources of phosphorus and that some deposits of phosphate rock found in lesser developed countries are more economical sources of phosphorus than the highly soluble processed phosphates under local economic conditions. Many experiments have been done with various phosphate rocks in recent years.

Fig. 5-10 shows the results of a field experiment in which concentrated superphosphate and one of the phosphate rocks used in Fig. 5-7 were compared. Here one sees that with the lower additions of phosphorus in phosphate rock, multiplying the quantities of phosphorus added by 0.2 produced data that fell on the curve for concentrated superphosphate. The yields for larger additions fell well below the line for concentrated superphosphate. These findings indicate that the availability coefficient of the phosphorus of the phosphate rock relative to that of the concentrated superphosphate standard varied with the quantity of phosphorus added.

In Fig. 5-10, only two data points for the Tennessee phosphate rock plotted at $0.2x$ deviate from the line. Fig. 5-11 shows the results of a special experiment

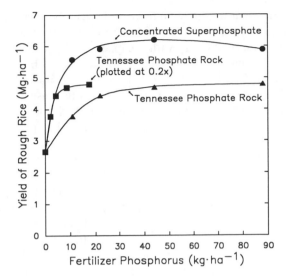

Fig. 5-10. Yield of rough unhulled flooded rice in a field experiment in Thailand versus the quantities of phosphorus applied in concentrated superphosphate and Tennessee phosphate rock. The yields obtained with the phosphate rock are also plotted against two-tenths of the quantity of phosphorus added. (Engelstad et al., 1974)

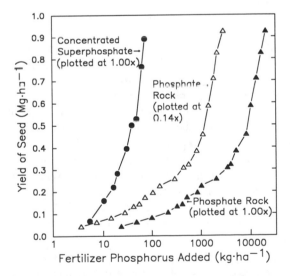

Fig. 5-11. Yields of slender serradella seed with different quantities of phosphorus applied as ordinary superphosphate and phosphate rock in a field experiment in Australia. Note the use of a logarithmic scale for the X axis. The response data plotted in the central line are the same as those for phosphate rock on the right, but the quantities of phosphorus have been multiplied by 0.14. (Bolland and Barrow, 1988)

in which 20 different quantities of both ordinary superphosphate and a phosphate rock were applied to make a more critical test of the hypothesis that the availability coefficient ratio may vary with the quantity of phosphorus applied. The heaviest addition of phosphate rock was equivalent to 20 megagrams of phosphorus per hectare. (Data for the heaviest applications of ordinary superphosphate have been omitted because they were excessive.) In this instance, the maximum attainable yield appeared to be about the same with both sources of phosphorus.

Because of the extreme range of quantities of fertilizer phosphorus applied, the yield response data have been plotted against the logarithms of the quantities of phosphorus added. The yield response curves thus have an unconventional appearance. The logarithmic scale, however, facilitates the demonstration that the availability coefficient ratio did change with the quantity of phosphate rock applied. Because of the logarithmic scale, multiplying all the additions of phosphorus in phosphate rock by a number less than 1 moves all the points on the response curve an equal distance to the left. When the quantities were multiplied by 0.14 (the central curve), the lowest yield moved to the left far enough to coincide approximately with the response curve for superphosphate, but the upper part of the phosphate rock curve remained far to the right. This behavior implies a marked decrease in availability coefficient ratio with an increase in rate of application. At the highest rate of application of phosphate rock, the availability coefficient ratio was well below 0.01. In effect, the phosphate dissolution per unit of phosphate rock decreased as the soil solution approached saturation with respect to the phosphate rock. Kanabo and Gilkes (1988) found experimentally by measuring the increases in exchangeable calcium and extractable phosphorus in soil treated with phosphate rock that the percentage dissolution decreased with increasing application rates. Up to 19% of the phosphate rock dissolved with an application of 200 milligrams of phosphorus per kilogram of soil, but only 5% dissolved when the application rate was increased tenfold.

Phosphate rocks with sufficiently low solubility are incapable of producing maximum yields similar to those obtainable with soluble phosphates. An extreme example is the igneous Missouri phosphate rock (see Fig. 5-7), the apatite mineral of which contained only 0.2% carbon dioxide compared with 1.5 to 5.0% carbon dioxide in the sedimentary phosphate rocks used in experiments by Engelstad et al. (1974). Evidence indicates that a carbonate ion and a fluoride ion may substitute for phosphate in the apatite structure and that the effectiveness of phosphate rocks as a source of phosphorus for plants increases with the degree of substitution.

If the quantities of phosphorus applied are great enough, variability in availability coefficient ratios is probably the rule rather than the exception in comparisons of phosphate rocks with soluble sources. In experiments with a wide range of quantities of fertilizer phosphorus, Bolland et al. (1988) and Weatherley et al. (1988) found that the Mitscherlich equation fitted the results, but as mentioned previously, the maximum yields usually differed. To evaluate the sources, the authors adopted the alternative of calculating Mitscherlich equations for the sources individually. Then they used the ratio of the slopes of the curves at zero application of fertilizer phosphorus as estimates of the availability coefficient ratios. The result was a fixed value for each availability coefficient ratio instead of a range of values that would depend upon the quantities of fertilizers applied.

Theoretically, using the ratios of initial slopes in individual Mitscherlich equations as estimates of the availability coefficient ratios is equivalent to using the ratios of Mitscherlich equations in differential form at zero fertilizer (see Section 1-2.1). In differential form, the Mitscherlich equation says that the rate of increase of crop yield (dy) with respect to quantity of fertilizer (dx) is proportional to the difference between the current yield (y) and the maximum yield (A) obtainable as the quantity of nutrient or fertilizer (x) increases indefinitely:

$$\frac{dy}{dx} = c(A - y),$$

where c is the proportionality factor. Numerical values for A and c are estimated in the curve-fitting process. If $x = 0$ is substituted in the fitted Mitscherlich equation, the corresponding value of y (that is, y_0) and $(A - y)$ can be calculated.

In the concurrent Mitscherlich response function, all fertilizers are fitted to the same control yield y_0 and the same maximum yield A. Thus, when the ratio of two initial slopes is calculated, the $(A - y)$ terms divide out, and the result is the ratio of the c values or Mitscherlich efficiency factors. When used to estimate availability coefficient ratios from the initial slopes of individually calculated Mitscherlich curves, the $(A - y)$ values generally will differ somewhat among fertilizers. The apparent availability coefficient ratios thus will represent some combination of the ratios of the efficiency factors c for the different fertilizers (which theoretically represent the availability coefficient ratios when the values of A and y_0 are the same for all fertilizers) with values of A and y_0, which have other meanings. As suggested in Section 4-4.3.1, however, the disadvantage just described is alleviated to some degree by the opposite trends in values of c and A and the compensating effect of multiplying them together. Despite the theoretical shortcomings, the ratio of the initial slopes may provide a good practical evaluation for the relatively small quantities of phosphate rocks that could be applied economically.

Barrow and Bolland (1990) developed a solution to the deficiencies of the evaluations made by the classical biological assay model and the ratio of the limiting slopes at zero applications of the fertilizers. Their proposal was, in effect, to recognize reality by modifying the classical biological assay model to accommodate a series of response curves in which the control yield is the same for all fertilizers, but the maximum yield may differ among fertilizers. For responses represented by the Mitscherlich equation, the ratio of the efficiency factors (the c values) then is no longer the ratio of the availability coefficients. Rather, the ratio of the availability coefficients changes with the quantity of the nutrient added, which it certainly does in comparisons of at least some phosphate rocks with superphosphate. The outcome thus is realistic, but the model is more complex than those considered previously.

The Barrow and Bolland (1990) model is

$$y = a - be^{-c(x_s + kx_t)},$$

where y is the yield of the crop or the nutrient, x_s and x_t are the quantities of the nutrient added in the standard and test fertilizers, and a, b, c, and k are parameters to be fitted. This basic equation is a transformation of the exponential form of the Mitscherlich equation to the base e, where a is equivalent to the maximum yield A produced by the standard fertilizer (which gives the highest yield — superphosphate in the usual comparisons between superphosphate and phosphate rocks), and $b = A - y_0$, where y_0 is the yield of the control without phosphorus fertilization.

The coefficient k is the function

$$k = (1 + mx_t)^{-n},$$

where m and n are parameters to be fitted. This equation gives the availability coefficient of the nutrient in the test fertilizer relative to that of the nutrient in the standard fertilizer as the quantity of the nutrient in the test fertilizer varies. According to this function, the availability coefficient of the test fertilizer cannot exceed 1.00, which is the value conventionally assigned to the standard fertilizer.

Additional test fertilizers may be accommodated by including additional kx_t terms. Thus, there are five adjustable parameters to fit for the standard fertilizer and one test fertilizer, and two more parameters for each additional test fertilizer.[1]

The authors tested the model on the data from three field experiments, each with a range of quantities of phosphorus as superphosphate and a calcined phosphate rock. In each instance, the maximum yield produced with the calcined phosphate rock was well below that produced with superphosphate. The model produced good fits to the yield data, and the curves for the calcined phosphate rock could be superposed on the corresponding curves for superphosphate when the changes in the values of k with x_t were taken into account.

Using the Barrow and Bolland (1990) model for fertilizer evaluation in the sense considered in this chapter still involves the restriction that the yield of dry

---

[1] A computer package to fit curves to responses to several sources of a given nutrient by four different response functions has been prepared by N. J. Barrow. It is possible to hold the relative effectiveness of each source constant or to let it decrease with increasing quantity of the nutrient. The package contains the source program in Turbobasic, which can be modified by users if desired; a compiled version of the program that should run on any MS DOS machine; several sets of test data, with outputs; and written instructions. To obtain a copy of the package, send one or two blank formatted disks to either (a) Dr. N. J. Barrow, CSIRO Private Bag, P.O. Wembley, WA 6014, Australia, or (b) Dr. Charles A. Black, Department of Agronomy, Iowa State University, Ames, Iowa 50011. The total information can be held by a single 3.5-inch disk, either double density or high density, and by a single 5.25-inch high-density disk. Two 5.25-inch double-density disks are required.

matter versus yield of nutrient curves for the various sources of the nutrient do not segregate. That is to say, the differences in crop yield obtained with the various fertilizers result from differences in supply of the nutrient being tested and not from differences in other effects. Where the yield of dry matter versus yield of nutrient curves segregate as a consequence of other differences between fertilizers, application of the model of equal control yields and differing maximum yields could be satisfactory for describing the results, but not as a basis for estimating availability coefficient ratios.

Figuring significantly in the recent literature on evaluation of phosphate rocks as sources of phosphorus for plants has been an adaptation of the relative-increase-in-yield method, which was one of the original methods of fertilizer evaluation. The basic relative-increase-in-yield evaluation is calculated as follows:

$$Relative\ Yield\ Increase = \frac{y_t - y_0}{y_s - y_0} ,$$

where y_s = yield obtained with a given quantity of nutrient in the standard fertilizer, y_t = yield obtained with an equal quantity of the nutrient in the test fertilizer, and y_0 = control yield without application of the nutrient. The values of y may be yields of plant substance or yields of the nutrient.

Chien et al. (1990) called the method of comparing increases in yield from fertilization a *vertical comparison* because in a plot of yield versus quantity of nutrient added, the comparison is between increases in yield plotted vertically where equal quantities of the nutrient have been applied in all the fertilizers being compared. They called the classical biological assay approach a *horizontal comparison* because it involves determining the relative quantities of the nutrient in different fertilizers required to obtain a given yield.

When linear responses are obtained under conditions that meet the requirements for classical biological assay, the ratio of the availability coefficients of the nutrient in the test and standard fertilizers, denoted by γ_t/γ_s in Section 5-2.1, is numerically equal to the relative effectiveness of the fertilizers as represented by the relative increases in yield in the foregoing equation. This is because $(y_t - y_0)/(y_s - y_0) = x_s/x_t$. In practice, the linear model is applicable only to the initial steeply ascending portion of the response curve. Beyond this range, the responses invariably become curvilinear.

For use with curvilinear responses to sources of a nutrient that lead to different maximum yields, and hence do not conform to classical biological assay theory, Chien et al. (1990) called attention to two equations that transform the x values representing the quantities of the nutrient so that the biological responses to the transformed quantities of the different sources of the nutrient are approximately linear. The slope-ratio concept in Section 4-2.3 (model of concurrent straight lines) then can be used to provide an estimate of the relative effectiveness of the *transformed* values of x. (Although the equations do not lead to maximum yields, this is not of much practical concern because the indicated upward trend at the plateau level is very slight.)

One equation involves a logarithmic transformation of x values:

$$y = y_0 + b\ln x,$$

where y is the yield with the addition x of the nutrient, y_0 is the yield with $x = 1$ (or a value close to the control at $x = 0$), and b is an effectiveness coefficient for the nutrient. Fig. 5-11 is a plot of logarithmic transformations of x values for one set of experimental data.

The increase in yield obtained with a given quantity of nutrient supplied by a test fertilizer relative to that obtained with the same quantity of nutrient supplied by the standard fertilizer then is given by

$$\frac{y_t - y_0}{y_s - y_0} = \frac{b_t \ln x_t}{b_s \ln x_s}.$$

Because $\ln x_t = \ln x_s$ in the vertical comparison method,

$$\frac{y_t - y_0}{y_s - y_0} = \frac{b_t}{b_s}.$$

Thus, because b_t and b_s are constants, the relative increase in yield with the two fertilizers is independent of the addition of nutrient at which the calculation is made.

If instead of the vertical comparison involving equal quantities of the nutrient in the test and standard fertilizers one uses the equation to make a horizontal comparison according to the classical biological assay theory and makes the comparison between fertilizers at equal values of y_t and y_s,

$$b_s \ln x_s = b_t \ln x_t.$$

Then

$$\ln x_t = \frac{b_s}{b_t} \ln x_s$$

and

$$x_t = x_s^{b_s/b_t}.$$

The quantities of the nutrient in the test fertilizer x_t and standard fertilizer x_s required to produce equal yields thus are not related by a constant factor, and the logarithmic equation would be expected to apply to the classical biological assay situation only where both the active ingredient and the dilution are the same for the fertilizers being compared.

The second equation discussed by Chien et al. (1990) for use in situations in which the maximum yield differs between nutrient sources involves an exponential transformation of x values:

$$y = y_0 + bx^{1/m},$$

where y_0 is the yield with zero addition of the nutrient, y is the yield with the quantity x of the nutrient, and b and m are constants. The parameter b is an effectiveness coefficient. The relative increases in yield produced by application of a nutrient in a test fertilizer and standard fertilizer then are given by

$$\frac{y_t - y_0}{y_s - y_0} = \frac{b_t x_t^{1/m}}{b_s x_s^{1/m}}.$$

For a vertical comparison at a given addition of the nutrient in the test and standard fertilizers, $x_t^{1/m} = x_s^{1/m}$, and

$$\frac{y_t - y_0}{y_s - y_0} = \frac{b_t}{b_s}.$$

The relative increases in yield thus are independent of the quantity of nutrient applied.

If instead of the vertical comparison involving increases in yield at equal values of x, the comparison is made horizontally at equal values of y in the classical biological assay mode, $y_t = y_s$ and

$$b_s x_s^{1/m} - b_t x_t^{1/m}.$$

Then

$$\frac{x_s^{1/m}}{x_t^{1/m}} = \frac{b_t}{b_s},$$

and

$$\frac{x_s}{x_t} = \left(\frac{b_t}{b_s}\right)^m.$$

The relative effectiveness of the fertilizers thus is b_t/b_s in the vertical mode and $(b_t/b_s)^m$ in the horizontal mode. Only when $m = 1$ (that is, when the untransformed responses are linear) will the relative effectiveness in the vertical and horizontal modes be equal. Otherwise, the horizontal evaluations will be equal to the vertical evaluations raised to the power m. For example, Chien et al. (1990) gave some results derived from data by Bationo et al. (1990) in which the transformation used was $y = y_0 + bx^{1/m}$ and m was taken as 2. The square root transformation provides a good linearization of many response curves. The horizontal evaluations were equal to the vertical evaluations squared. With concentrated superphosphate taken as the standard source having a relative effectiveness of 1.00, the two phosphate rocks compared with the superphosphate in a field experiment had relative effectiveness values of 0.90 and 0.49 in the vertical comparison mode, but the square of these values, 0.81 and 0.24, in the horizontal mode.

It is important to keep in mind the fact that the vertical and horizontal comparisons are not to be considered alternative methods for measuring a given property of the fertilizers being evaluated. As explained by Chien et al. (1990), the vertical comparison ranks ''a series of test fertilizers with respect to a standard fertilizer according to their agronomic potential to produce a yield response at the same rate of P applied.''

The relative effectiveness values obtained when the logarithmic or exponential transformations are applied to the x values make use of the classical biological assay theory. Because of the transformations, however, the relative effectiveness values obtained are not availability coefficient ratios.

Chien et al. (1990) emphasized the point that vertical comparisons involving phosphate rocks could be correlated with such characteristics as the citrate-soluble phosphorus in the rocks. The same is true of horizontal comparisons.

As noted by Chien et al., horizontal comparisons have the advantage that they can be used to provide information on the economics of fertilizer use. This is an important issue with phosphate rocks. A farmer with limited resources who is operating on the steep portion of the response curve may find phosphate rock a better buy than superphosphate, whereas a farmer with greater resources who is operating in the upper reaches of the response curve may find superphosphate the better buy because of the large quantity of phosphate rock required to produce the response or because the phosphate rock is incapable of producing the response.

These economic implications cannot be represented accurately by horizontal comparisons made by the classical biological assay model for comparisons of phosphate rock with superphosphate when the maximum yields differ because fitting the classical model to the data results in an apparently constant availability coefficient ratio even though the value may vary with the quantities of the fertilizers added. The Barrow and Bolland (1990) model makes allowance for the change in availability coefficient ratio with the quantities of fertilizers and could be used to make appropriate economic analyses.

5-2.4. Compensation Procedures

Differences among fertilizers in content of nutrients other than the one being evaluated can cause problems. For example, in an experiment to find whether ammoniation has affected the availability coefficient of the phosphorus in concentrated superphosphate, the evaluation would be invalidated if the plant response to the nitrogen in the ammoniated superphosphate were not compensated by an equal response to ammonium in the nonammoniated concentrated super-phosphate used as the standard fertilizer.

Problems of this kind generally are dealt with in one of two ways. In the more precise compensation procedure, the concentrated superphosphate would be mixed before application with an ammonium salt such as ammonium chloride that would provide an equivalent quantity of ammonium but would not interact with the concentrated superphosphate to produce an effect on the phosphorus similar to that of ammoniation. Equal additional quantities of fertilizer nitrogen might be applied with both sources if more nitrogen were needed to eliminate nitrogen as a limiting factor. Use of the more precise procedure is of special importance in instances in which an attempt is being made to compensate for a nutrient present in quantities so great that a uniform application to all treatments might produce an excess of this nutrient and a decrease in yield of the crop or uptake of the nutrient being studied.

In the less precise procedure, an equal "blanket" application of fertilizer nitrogen (preferably an ammonium source) would be added to all plots. The objective would be to add enough fertilizer nitrogen so that the plants would be on the plateau portion of the nitrogen response curve, and the yield would not be affected appreciably by the relatively small quantity of nitrogen supplied by the ammoniated superphosphate.

Sometimes the attempted compensation is successful in eliminating extraneous effects, and sometimes it is not. The degree of success usually can be judged by the degree of similarity of the maximum yields obtained with the various sources. As an example of a successful compensation, Mitscherlich and Sauer-landt (1935) grew oat plants as a test crop on two soils, to each of which had been added four different quantities of each of three different legumes. With complete fertilization, including nitrogen, the highest yields were obtained with the maximum addition of legume residues (100 grams per culture). The yield range with these additions was from 79.8 to 80.0 grams on one soil and from 72.3 to 75.9 grams on the other. Where nitrogen was omitted from the complete fertilization, the highest yields varied with the residue used, but were of the order of half those obtained with added nitrogen.

De Haan (1981) used a different procedure, which amounts to an adjustment for differences in plant response caused by factors other than the nutrient being evaluated. He described the procedure in connection with an experiment in which five different quantities of phosphorus supplied as sewage sludge and monobasic calcium phosphate were compared. Nitrogen, potassium, and magnesium were supplied in mineral form to the soil in all cultures. For the first cutting of the English ryegrass test crop, the amounts of these nutrients supplied by the sludge were subtracted from the additions made to the sludge-treated cultures. Only the mineral nitrogen in the sludge was taken into account on the basis that in most instances the effect of the organic part is negligible. (His experiment included different quantities of fertilizer nitrogen to permit an adjustment if needed for uptake of nitrogen originally present in organic form in the sludge.)

With equal yields of phosphorus in the ryegrass test crop, the sewage sludge yielded more dry matter than did the monobasic calcium phosphate. The indications thus were that despite the compensation for the principal nutrients, the sewage sludge was having some favorable effect on growth that was not a phosphorus effect. The extra dry matter yield should have increased the phosphorus yield also. To adjust for the nonphosphorus effect of the sludge, De Haan plotted the yield of phosphorus against the yield of dry matter, and converted the yields of phosphorus from the sludge to yields of phosphorus from the monobasic calcium phosphate at equal yields of dry matter. The set of adjusted yields of phosphorus from the sludge then presumably could be compared with the observed yields of phosphorus from the standard source to estimate the ratio of the availability coefficient of the phosphorus in the sludge to that of the phosphorus in the monobasic calcium phosphate standard. Further work is needed to investigate the proposed procedure.

5-3. Chemical Evaluation of Fertilizers

Biological evaluations provide the basic justification for the chemical methods of fertilizer analysis that are designed to discriminate among chemical forms of nutrients that are highly effective and those that are less effective or of little value. Chemical methods are used routinely for regulatory purposes. The numbers of such evaluations performed annually far exceed the numbers of fertilizer evaluations made by biological methods for all purposes.

The methods of analysis used for regulatory purposes in the United States are described by Rund (1984) in the *Official Methods of Analysis of the Association of Official Analytical Chemists*. The fourteenth edition was published in 1984, and that edition refers to earlier editions for some of the methods.

5-3.1. Nitrogen

For nitrogen, recognized fractions include water-soluble nitrogen, water-insoluble organic nitrogen soluble in neutral permanganate, water-insoluble organic nitrogen distilled from alkaline permanganate but not soluble in neutral permanganate, and residual nitrogen. The water-soluble fraction is designated as the "active" fraction.

For urea-formaldehyde compounds, the fractions include cold-water-soluble, cold-water-insoluble but soluble in a hot phosphate buffer solution, and residual. The difference between the cold-water-insoluble nitrogen and the hot-water-insoluble nitrogen expressed as a percentage of the cold-water-insoluble nitrogen is considered to provide an "activity index" of the nitrogen.

Additionally, methods are prescribed for nitrate, ammonium, urea, and biuret. These methods are not considered to be availability indexes.

5-3.2. Potassium

Potassium is extracted by boiling a quantity of the test fertilizer in an ammonium oxalate solution (additional methods are prescribed for organic materials and ash from organic materials). This procedure extracts water-soluble potassium plus the readily soluble and exchangeable portion of the potassium that might be present in the form of finely ground minerals, such as potassium feldspar. No provision is made, however, for determining the residual insoluble potassium, which means that the method does not recognize such residues as being of importance to plant nutrition.

5-3.3. Phosphorus

Three fractions of fertilizer phosphorus are recognized: soluble in water, insoluble in water but soluble in neutral ammonium citrate, and insoluble in neutral ammonium citrate. The first two fractions are considered to be available for absorption by plants. Other extractants used in some countries are an alkaline ammonium citrate solution and 2% citric acid.

The neutral ammonium citrate method extracts only a minor portion of the phosphorus in phosphate rocks. Following the discovery that phosphate rocks from different sources may differ greatly in the availability of their phosphorus to plants, much research has been done on the biological evaluation of phosphate rocks, and some has been done on chemical methods.

Chien (1978) found in experiments with seven phosphate rocks that the phosphorus extracted by ammonium citrate at pH 3 and by 2% formic acid gave relatively good correlations with the yields of plants obtained with application of these rocks to acid soils. The phosphorus extracted by the neutral ammonium citrate method gave a poor correlation, mainly because of a relatively low value of extractable phosphorus for one of the phosphate rocks. This particular rock contained calcium carbonate. The calcium carbonate dissolved readily in the ammonium citrate solution, and the calcium depressed the solubility of the calcium phosphate mineral in the citrate solution. The effect of the calcium carbonate on the availability of the phosphorus to plants in the acid soil was relatively small, presumably because the small quantities of calcium carbonate associated with individual phosphate rock particles dissolved in a short time with only a minor effect on the pH of the surrounding soil.

When samples of the various phosphate rocks were extracted a second time with neutral ammonium citrate, the phosphorus extracted from the phosphate rock originally containing calcium carbonate increased more than fourfold, whereas the phosphorus extracted from the others remained about the same. The correlations obtained between yields of test plants and the phosphorus in the second neutral ammonium citrate extract then ranked along with those obtained with the phosphorus extracted by ammonium citrate at pH 3 and by 2% formic acid.

For chemically solubilized phosphate fertilizers, such as superphosphate, solubility in neutral ammonium citrate solution is an effective way of discriminating between the phosphorus of phosphate rocks that has been changed to forms of greater solubility and availability to plants by processing, and the phosphorus in the unreacted residual phosphate mineral (see a review by Terman et al., 1964). Normally, the residual unreacted phosphate rock represents only a very small percentage of the total phosphorus, and it appears to be less soluble and available to plants than the portion of the phosphate rock dissolved by the acid treatment (Charleston et al., 1989; Junge and Werner, 1989; Resseler and Werner, 1989). Thus, the contribution of the phosphorus dissolved in the citrate solution from the unreacted phosphate rock is usually negligible.

For phosphate rocks, solubility in neutral ammonium citrate provides an index of the reactivity of the phosphate sources. The values obtained, however, do not have the same significance for plants as comparable values for the more soluble compounds that result from chemical processing of phosphate rock (Chien et al., 1987). For example, Engelstad et al. (1974) found that the proportions of the phosphorus of Idaho, Florida, and North Carolina phosphate rocks that were soluble in neutral ammonium citrate solution were 0.12, 0.20, and 0.26, respectively. But the plant response data obtained when rice plants were grown on an acid soil treated with different quantities of each of the phosphate rocks indicated that the availability coefficient ratios of the phosphorus of the phosphate rocks with superphosphate as the standard fertilizer were approximately 0.05, 0.1, and 0.5.

One of the problems with solubility in neutral ammonium citrate as a chemical index of the availability of the phosphorus of phosphate rocks is that the significance of a given solubility value depends upon the properties of the soil, principally the pH value. Soluble, chemically processed fertilizers are effective sources of phosphorus in alkaline soils as well as acid soils, but phosphate rocks are of little or no value in alkaline soils. Their solubility and the availability of their phosphorus to plants decreases with increasing soil pH (Rajan et al., 1991).

An evaluation approach geared more specifically to the soil may be derived from data by Khasawneh and Sample (1978). They grew cowpea as a test crop on cultures of an acid soil that had been treated with different quantities of various phosphate rocks and anhydrous dibasic calcium phosphate. At the end of 6 weeks, when the crop was harvested, they determined the concentrations of phosphorus in water extracts of samples of the cropped soil that had received the individual phosphate treatments.

Fig. 5-12 shows that the yields of dry matter obtained with the four different sources of phosphorus could be represented to a good approximation as a function of the concentration of phosphorus in the water extracts of the variously treated soils. This finding signifies that the plants were responding primarily to differences in phosphorus supplied by the various fertilizers, and that a given concentration of phosphorus in the extracts was associated with approximately equal values of fertilizer phosphorus, irrespective of the fertilizer source. Fig. 5-12 serves as a proxy for a plot of yields of dry matter against the yields of phosphorus in the plants, which could not be derived from the published data.

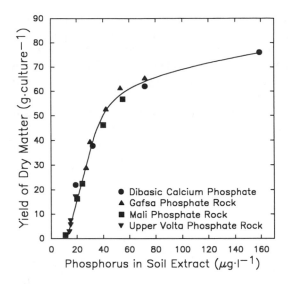

Fig. 5-12. Yields of dry matter of cowpea plants versus concentrations of phosphorus in 1:1 soil:water extracts obtained at the time of harvest from soil cultures treated with various quantities of phosphorus as dibasic calcium phosphate and three phosphate rocks. (Khasawneh and Sample, 1978)

Figs. 5-13A and B show plots of yields of plant dry matter versus the quantities of phosphorus added in the individual sources in the research by Khasawneh and Sample (1978). The data indicate that the ratios of the quantities of phosphorus applied in the Gafsa and Mali phosphate rocks to the quantities of dibasic calcium phosphate phosphorus producing equal yields were not constant. Because of Fig. 5-12, the interpretation would be that the numerical values could be inferred to be availability coefficient ratios, but that these ratios varied with the quantity of phosphorus applied.

Figs. 5-13C and D contain plots of the concentrations of phosphorus in the water extracts against the quantities of fertilizer phosphorus added. Although these response curves differ in shape from those in Figs. 5-13A and B, the results agree with the findings in 5-13A and B in that the ratios of the extractable phosphorus in soil treated with the Gafsa and Mali phosphate rocks relative to the extractable phosphorus in soil treated with dibasic calcium phosphate were not constant, but varied with the quantities of the phosphate rocks applied.

When the availability coefficient ratio of a nutrient supplied by two different fertilizers depends upon the soil or other conditions, one cannot expect that applying a chemical extraction method to the fertilizers will give results that apply well to conditions in general. Additional data are needed. The work by Khasawneh and Sample (1978) suggests one way in which evaluations more specific to individual soils may be obtained.

Hammond et al. (1986a) published similar work, in which measurements were made of the phosphorus extracted from treated soils by the widely used

Bray method to obtain indexes of soil phosphorus availability. In this work, no appreciable segregation of data for concentrated superphosphate from data for a phosphate rock was evident in a plot of yield of dry matter against either yield of phosphorus or phosphorus extracted from the fertilized soils by the soil testing method, but there was some evidence of a difference in response functions between the two phosphorus sources.

The success thus far achieved suggests that shifting research emphasis on chemical evaluation of fertilizers in the direction of chemical analysis of fertilizer-treated soils may be advantageous. Needed are methods of soil analysis producing data that correlate highly with the yields of a nutrient or the yields of dry matter

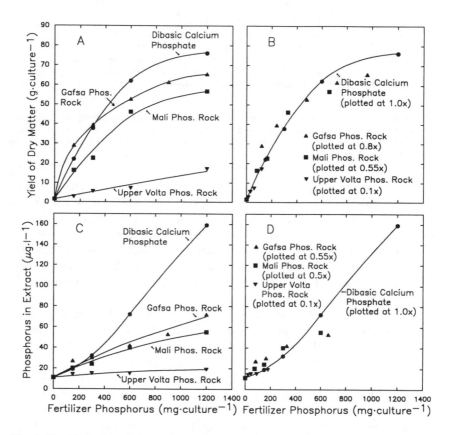

Fig. 5-13. Evaluation of three phosphate rocks as sources of phosphorus for cowpea in cultures of an acid soil, with anhydrous dibasic calcium phosphate as a standard source. A. Yields of dry matter versus additions of fertilizer phosphorus. B. Yields of dry matter versus additions of fertilizer phosphorus, with quantities of phosphorus in each source multiplied by an apparent availability coefficient. C. Concentrations of phosphorus in 1:1 soil:water extracts at harvest versus additions of fertilizer phosphorus. D. Concentrations of phosphorus in 1:1 soil:water extracts at harvest versus additions of fertilizer phosphorus, with quantities of phosphorus in each source multiplied by an apparent availability coefficient. (Khasawneh and Sample, 1978)

where increasing quantities of two or more sources of the nutrient have been added individually, without segregation of data by sources. Making analyses by such a method on individual soils to which increasing quantities of the test and standard fertilizers have been added would yield curves showing how individual soils respond to the fertilizer additions. If the extractability coefficient ratio of a test fertilizer relative to the standard is constant according to the soil analysis procedure, the same would be expected in a biological test to estimate the availability coefficient ratio, provided that any compensation needed for other fertilizer differences is successful. If the extractability coefficient ratio depends upon the quantities of the fertilizers added, the same would be expected in a biological test.

Research probably will be needed in most instances to find an extraction method that behaves like the one used by Khasawneh and Sample (1978) to obtain the data plotted in Figs. 5-12 and 5-13. The relative amounts of a nutrient removed from different sources by an extractant often differ from the relative amounts absorbed by plants. Hammond et al. (1986b) cited examples. Although extractants with the desired properties may be found in experiments in which the nutrient sources are dispersed throughout the soil, they cannot be expected to answer all the questions where placement effects are involved.

Direct estimates of the comparative reactivities of different phosphate rocks in specific soils may be made by incubating the phosphate rocks with the soils and then measuring the increase in exchangeable calcium, the increase in phosphorus extractable by sodium carbonate or sodium hydroxide, or the increase in labile phosphorus by isotopic dilution. These measurements are less artificial than measurements of the dissolution of phosphate rocks in solutions containing chemical extractants. In a review of the literature emphasizing the use of phosphate rocks as sources of phosphorus for plants grown on acid soils, Sinclair et al. (1991) noted further that the release of phosphorus from phosphate rocks to soil provides a more direct and meaningful index of reactivity than does the yield of plants or the uptake of phosphorus by plants from the soil. Chemical data on reactivity and biological data on availability of various phosphate rocks on different soils both contribute to understanding.

As an alternative approach, Barber and Ernani (1990) adapted the mechanistic concept of nutrient availability that had been developed by Barber and coworkers (see Section 7-4). They described an experiment in which four phosphate fertilizers ranging in solubility from phosphate rock to monobasic calcium phosphate were incubated for 30 days with each of two moist soils. From measurements on the incubated fertilized soils, they derived figures for the initial phosphorus concentrations in the soil solution, the buffering capacities of phosphorus on the soil solids for phosphorus in the soil solution, and the effective diffusion coefficients for phosphorus diffusion through the soils. Combining these values with data from a previous experiment (not described) for phosphorus uptake from soil by corn, they estimated the uptake of phosphorus by corn plants. Their results showed that the correlation between the initial concentration of phosphorus in the soil solution and the estimated phosphorus uptake by plants was $r = 0.98$.

5-4. Biological Assay of Fertilizer-Related Factors

The biological assay concept described for comparing different fertilizer sources of a common nutrient has applications to related factors of concern in soil fertility control. Fig. 5-14 is an example taken from research by Malzer and Graff (1984, 1985) on nitrogen fertilization of irrigated corn on a sandy soil in Minnesota. In this work, anhydrous ammonia was injected into the soil as a source of nitrogen before planting (with and without a chemical nitrification inhibitor) and also at the 10-leaf stage. Ammonia applied before planting was much less effective than ammonia applied at the 10-leaf stage. The increase of the availability coefficient ratio from 0.15 to 0.48 as a result of adding a nitrification inhibitor indicates that an important cause of the inefficiency of the preplant application was loss by leaching of some of the nitrate produced in the soil from the ammonia.

Evaluation of placement effects comes to mind as another possibility for applying the classical biological assay concept to fertilizer use. Experiments have shown, however, that although yield of dry matter versus yield of nutrient plots may provide no evidence of invalidity of comparisons of banded and broadcast fertilizer, the availability coefficient ratio of broadcast and band applications may vary with the quantity of fertilizer applied because the maximum yields are not the same. See Fig. 7-5 for a generalized representation of plant responses to broadcast and banded applications of fertilizer, and see Chapter 7 for an alternative assay concept.

5-5. Biological Assay of Nutrients in Organic Substances

The plant nutrients in the crop residues, animal manures, and sewage sludges used in agriculture do not occur in the same forms and do not behave in exactly the same way as the nutrients in the fertilizers used as standard sources for evaluation purposes. The nutrients do not occur singly. A number are present in at least small quantities. In experiments to evaluate nutrients in organic materials, the compensation procedures discussed in Section 5-2.4 for fertilizers normally are applied. In addition to the nutrient effects, however, organic substances may have physical effects and biological effects, such as modification of disease incidence, that are more difficult to deal with.

Theoretical and practical difficulties in evaluation notwithstanding, information on the availability coefficient ratios of the various nutrients supplied by crop residues, animal manures, and sewage sludges is needed. These materials are sources of significant amounts of plant nutrients. Whether they are added alone or in combination with fertilizers, estimating the effective amounts of

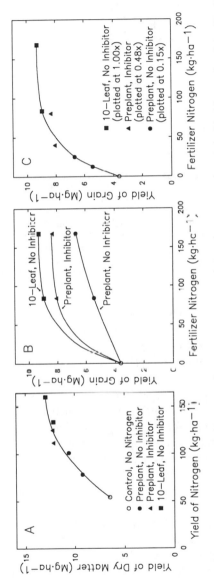

Fig. 5-14. A 2-year comparison of anhydrous ammonia applications for irrigated corn grown on a sandy soil in Minnesota where the ammonia was injected with and without a nitrification inhibitor before planting (preplant), and without a nitrification inhibitor at the 10-leaf stage. A. Yields of aboveground dry matter versus yield of nitrogen in the dry matter. B. Yields of corn grain obtained with different quantities of fertilizer nitrogen. C. Yields of corn grain obtained with different quantities of fertilizer nitrogen where the quantities of nitrogen applied at the 10-leaf stage are plotted at 1.00x (availability coefficient set at 1.00), those applied before planting with a nitrification inhibitor are plotted at 0.48x, and those applied before planting without a nitrification inhibitor are plotted at 0.15x. (Malzer and Graff, 1984, 1985)

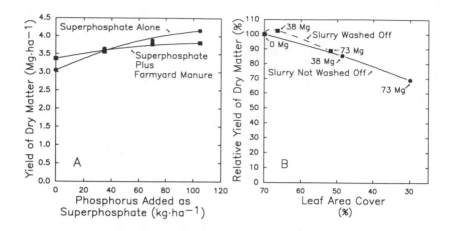

Fig. 5-15. A. Yields of perennial ryegrass topdressed with increasing quantities of phosphorus as superphosphate, with and without a topdressing of 40 megagrams of farmyard manure per hectare in The Netherlands. Basal dressings of nutrients other than phosphorus from mineral fertilizers were 160 kilograms of nitrogen and 199 kilograms of potassium per hectare distributed during the season. B. Relative yields of dry matter versus leaf area cover by perennial ryegrass topdressed with cattle slurry, with and without an immediately following vigorous spraying with 4 millimeters of water to wash off the slurry. Applications of slurry were 0, 38, and 73 megagrams per hectare. "Leaf area cover" is interpreted as the percentage of the land area covered by leaves that were not coated by the slurry dry matter. (Schechtner et al., 1980)

nutrients they convey is a part of the process of gauging the extent to which plant needs will be met by known applications of specific organic materials.

Illustrating a physical effect, Fig. 5-15 contains some results of research that grew out of an attempt to evaluate the phosphorus in a sample of manure. In the experiment that provided the data for Fig. 5-15A, different quantities of phosphorus were applied to grassland as a superphosphate standard, with and without a topdressing of 40 megagrams of farmyard manure per hectare.

If the compensation supplied for the nitrogen and potassium added in the manure had been adequate to cause the response to the manure to be a response to phosphorus, the response curves would have converged with increasing additions of phosphorus in the superphosphate. It would have been possible to make the curves coincide, within experimental error, by moving the farmyard manure curve to the right until the yield with manure but with no phosphorus as superphosphate fell on the curve for superphosphate.

The experimental results in Fig. 5-15A do not behave in this way. Rather, the shapes of the curves suggest that some beneficial factor was associated with the superphosphate that was not supplied by the manure, or conversely that some detrimental factor was associated with the manure that was not associated with the superphosphate.

Further observations and additional experiments provided evidence for what appeared to be a major cause of the difference in plant response to the phosphorus in the farmyard manure and superphosphate, namely, a reduction in photosynthesis as a consequence of interception of some of the light by a coating of manure on the grass. The authors noted that the negative effect seemed to be essentially confined to the first cutting of grass after application of the manure. Moreover, the negative effect did not appear when the manure had been dried and ground before application. It was most marked in a year of low rainfall between the time of manure application and the beginning of grass growth, and it did not appear in a year when there was heavy rainfall after the manure was applied.

The implication of these observations was reinforced by the results of a field experiment in which fertilizer nitrogen, phosphorus, and potassium were added to all plots, followed by application of cattle slurry (containing about 10% dry matter), with and without a hard spray of 4 millimeters of water to wash the slurry off of the grass. Fig. 5-15B shows that the yield of grass reflected the leaf area not coated by slurry and hence capable of unimpaired photosynthesis.

The dominant nutrient effect of manure in recent experiments usually has been that of nitrogen. The principal reason is probably a combination of the continuing need of plants for large quantities of available nitrogen, together with a fairly good supply of phosphorus and potassium already present in the soil as a result of the residual effects of previous additions of mineral fertilizers and manures.

To evaluate the nitrogen in surface-applied and injected cattle slurry, Van der Meer et al. (1987) conducted a number of experiments in The Netherlands. To compensate for possible effects of the additional nutrients in the slurry, they made frequent applications of fertilizer phosphorus, potassium, and magnesium to the experimental areas based upon the results of regular soil analyses. The data cited here are average values obtained from an experiment repeated 5 years at the same location.

Fig. 5-16 shows a plot of the yield of dry matter of the grass herbage against the yield of nitrogen in the herbage receiving nitrogen in the form of fertilizer and cattle slurry, topdressed and injected. The absence of an appreciable segregation of plots for the different nitrogen sources indicates that to a good approximation, the yield effects were nitrogen effects. That is, the differences in the response curves for yield of dry matter or yield of nitrogen versus nitrogen applied should be mostly a consequence of differences in the availability of the nitrogen in the different sources.

Fig. 5-17A shows an evaluation of the availability coefficient ratio of the nitrogen in topdressed cattle slurry, with fertilizer nitrogen as the standard source assigned an availability coefficient of 1.00. Fig. 5-17B shows a comparable evaluation based upon the yields of nitrogen. In this experiment, the maximum quantity of slurry applied was the same as the quantity of farmyard manure used as a topdressing in Fig. 5-15A, but the slurry contained only 10% dry matter, so that the quantity of solid material applied was much smaller than in the experiment in Fig. 5-15A.

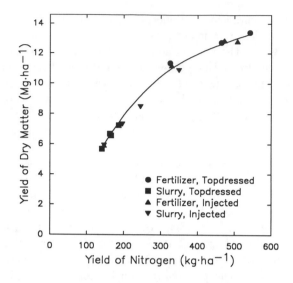

Fig. 5-16. Average annual yields of dry matter versus yields of nitrogen in grassland herbage in a 5-year experiment in The Netherlands on topdressing and injection of cattle slurry and fertilizer. To keep the supplies of phosphorus, potassium, and magnesium in the nonlimiting range, these nutrients were added frequently as needed in mineral fertilizers on the basis of soil tests. (Van der Meer et al., 1987)

In the same experiment with the nitrogen evaluations described in the preceding paragraph, the authors included an evaluation of the nitrogen in injected slurry. In this portion of the work, they used larger quantities of slurry because the injected slurry would not cover the grass. With a wider range in quantities of nitrogen, the experimental error of the evaluations should be reduced. Figs. 5-17C and D show the results of these evaluations. The availability coefficients of nitrogen in the injected slurry relative to those of fertilizer nitrogen are higher than the analogous values for topdressed slurry, presumably reflecting the conservation of the ammonium nitrogen in injected slurry by the soil.

With both the topdressed and injected slurry, the availability coefficient ratios based upon yields of nitrogen exceed those based upon yields of dry matter. Whether this observation is accounted for by a difference in rate of nitrogen uptake with time between the fertilizer and slurry sources (see the discussion in Section 5-2.3.3) is not known.

In a field experiment in Punjab, Y. Singh et al. (1987) added different quantities of nitrogen as urea and cattle manure to flooded rice, and found that the responses to manure nitrogen could be made to fall on the response curves for urea by plotting the manure nitrogen at 0.45N for yield of dry matter and 0.40N for yield of nitrogen. If the concurrent linear response function is fitted to data the same authors obtained in similar work with poultry manure (B. Singh et al., 1987), the availability coefficient ratios of the manure nitrogen relative to the urea nitrogen as the standard source may be estimated at 0.85 from the yield of grain data and at 0.81 from the yield of nitrogen data.

On a short-term basis, the effective nitrogen in cattle manure corresponds approximately to the ammonium nitrogen, and most of this is derived from the urine (Castle and Drysdale, 1966). Ernst (1987) reported the results of a 7-year experiment in Germany, in which the sum of the yield of nitrogen in grassland herbage and the change in mineral nitrogen in the soil was closely correlated with the sum of the nitrogen added in the fertilizer and the ammonium nitrogen added in a manure slurry. In a field experiment in Belgium, Barideau and Falisse (1982) added ammonium nitrate and sewage sludge in different quantities. On the basis of their finding that the percentage recovery of the fertilizer nitrogen was 50%, they assumed that the same was true for the ammonium nitrogen supplied in the sludge. Subtracting the equivalent of half the ammonium nitrogen in sludge from the nitrogen recovered in the perennial ryegrass test crop over a season (four cuttings), they estimated that the recovery in the crop of the nitrogen present in organic forms in the sludge was 6 to 8%.

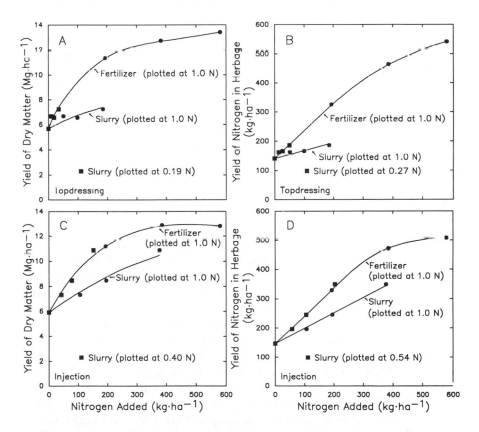

Fig. 5-17. Evaluation of nitrogen in cattle slurry relative to nitrogen in fertilizer for grassland in The Netherlands. A. Application by topdressing, evaluation based upon yields of dry matter. B. Application by topdressing, evaluation based upon yields of nitrogen. C. Application by injection, evaluation based upon yields of dry matter. D. Application by injection, evaluation based upon yields of nitrogen. (Van der Meer et al., 1987)

Research by Whitehead et al. (1989) on the availability of the nitrogen in various components of manure slurries adds understanding to the findings in overall evaluations. These investigators fractionated slurries from ten different sources into water-soluble and water-insoluble components, and further separated the water-insoluble components according to particle size. They found that the positive effect of the slurries on yields of dry matter and nitrogen in perennial ryegrass was attributable mostly to the water-soluble nitrogen, especially at the first cutting.

When the water-insoluble fraction of the slurries was added to soil, the yield of ryegrass herbage and the nitrogen uptake by the grass were increased to a small degree by pig slurries, but were decreased by cattle slurries, particularly at the first cutting, indicating immobilization of nitrogen by the water-insoluble fraction of the cattle slurries. A further indication of immobilization was provided by the finding that the percentage recovery of the water-soluble nitrogen in the herbage decreased with an increase in the carbon to nitrogen ratio of the water-insoluble material.

The foregoing experiments on evaluating the nitrogen supplied by organic fertilizers against that supplied by inorganic fertilizers have emphasized short-term effects. The results of short-term experiments normally show organic fertilizers to be inferior sources of nitrogen because the usual sources of nitrogen in commercial fertilizers are completely available to plants in the year of application, whereas the organic nitrogen in crop residues, manures, and sewage sludges is not.

The reduced availability of organic sources on a short-term basis leads to a greater buildup (or lesser decline) in total nitrogen content of the soil on a long-term basis than occurs with inorganic sources. Along with a buildup of total nitrogen is a gradual increase in nitrogen mineralization due to the accumulated residues. Thus, Suzuki et al. (1990) found that in a 60-year experiment in Japan in which approximately equal quantities of nitrogen were applied annually as ammonium sulfate and organic fertilizer (straw compost plus soybean cake), the yields of rice grain and straw with the organic fertilizer were below those with ammonium sulfate for the first decade. In the second and third decades, the yields were about the same, and in the fourth through sixth decades the organic fertilizer produced the higher yields. Analyses of the soil for total nitrogen and mineralizable nitrogen near the end of the experiment showed large differences that were related to the rice yields. Soil acidification due to the ammonium sulfate did not appear to be an important factor.

Bromfield (1961) conducted a plant culture test in which samples of manure from sheep fed on different natural and improved pastures in Australia were added to perlite in quantities to supply 3 milligrams of total phosphorus. To separate cultures, monobasic potassium phosphate was added in different quantities to obtain a response curve. Nutrients other than phosphorus were supplied from a nutrient solution added initially and at intervals.

To estimate the ratios of the availability coefficients of the manure phosphorus to that of phosphorus supplied by the soluble source, the quantities of phosphorus

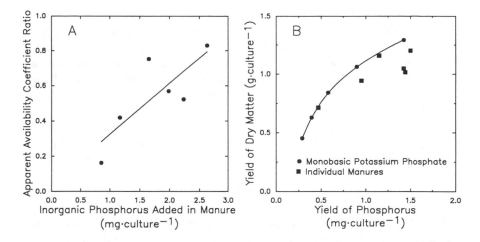

Fig. 5-18. A. Apparent availability coefficients of the phosphorus in six samples of sheep manure (with the phosphorus in monobasic potassium phosphate assigned an availability coefficient of 1.00) versus the inorganic phosphorus added in the manures. The data were derived from an experiment in which wheat plants were grown on perlite supplied with a minus-phosphorus solution and in which the manures were added in quantities to supply 3 milligrams of total phosphorus per culture. B. Yields of dry matter in the wheat plants plotted against the yields of phosphorus. The data points for the monobasic potassium phosphate standard are connected by a solid line. (Bromfield, 1961)

on the standard response curve that were associated with yields of dry plant material equal to those obtained with the individual samples of manure were divided by the quantity of manure phosphorus applied (3 milligrams per culture). The results in Fig. 5-18A show that the apparent availability coefficient ratios ranged from 0.16 to 0.83 and that they increased with the quantities of inorganic phosphorus supplied by the various manures.

Because the total phosphorus in the manures was the sum of the inorganic and organic fractions, the apparent availability coefficient ratios of the manure phosphorus decreased with an increase in the organic phosphorus (plot not shown). The findings regarding the organic phosphorus in the plant growth test were verified by incubation studies in which there was little or no increase in inorganic phosphorus in manures incubated for 4 or 8 weeks. In most of the samples investigated, the organic phosphorus after incubation exceeded that before incubation. Kaila (1950) also found little evidence of mineralization of the organic phosphorus in manure during incubation periods up to 3 months. If, on the basis of the comparative inactivity of the organic phosphorus, the inorganic fraction is assumed to be responsible for the total phosphorus contribution of the manures in the work by Bromfield (1961) in Fig. 5-18, the ratio of the apparent availability coefficients of the inorganic phosphorus in the manures to the availability coefficient of the phosphorus of the monobasic potassium phosphate standard averaged 0.92.

The plots of yields of dry matter against yields of phosphorus in the plants in Fig. 5-18B show that the values for most of the manure samples were below the curve representing the soluble phosphorus source, indicating that these samples had some unfavorable effect on the yield of dry matter and hence that the availability coefficient ratios for these samples were not entirely valid. These deviations are the reason for the use of "apparent" in reference to the ratios calculated in this experiment.

The inorganic phosphorus in manures is found by analysis as orthophosphate, and apparently occurs in that form. Bromfield (1961) found that the inorganic phosphorus of sheep manure appeared in successive water extracts far more slowly than would the phosphorus in superphosphate, for example, suggesting that at least part of it occurs in less soluble forms. Fordham and Schwertmann (1977) identified magnesium ammonium phosphate in manure by x-ray diffraction, and obtained indirect evidence for the existence of several other calcium and magnesium phosphates of intermediate solubility, including dibasic calcium phosphate. Phosphorus present in such compounds would not leach readily into the soil from manure on the soil surface, but it would be available to plants when the manure was incorporated.

Johnston and Beckett (1981) approached the evaluation of organic sources of nutrients as a statistical problem. They were dealing with a sewage sludge in which the issue was separating effects due to nitrogen and phosphorus from those due to heavy metals.

To provide a basis for the statistical tests of significance, they applied different combinations of nutrients from inorganic sources to describe a nutrient response surface, and they applied different quantities of sludge to describe the sludge response surface to the extent this could be done in view of the fixed ratios of the components. Polynomial regression equations then were fitted to the surface generated by the inorganic sources of nutrients (nitrogen and phosphorus in this case) to see if either variable could be eliminated. In their experiment, phosphorus could be eliminated as nonsignificant.

The question then was whether the decrease in yield of the barley test crop at high levels of sludge (up to 326 metric tons of dry solids per hectare) was due to the excess of nitrogen with associated lodging, to the heavy metal content (up to 2240 kilograms of zinc equivalent per hectare), or a combination of the two. The statistical tests made on the response surfaces for inorganic nutrients and sludge showed no significant difference, from which it was concluded that in the instance investigated, the effect was due to nitrogen.

5-6. Biological Assay Using Tracers

5-6.1. Sensitivity Advantage

The relative biological value of different sources of a common nutrient is arguably evaluated most appropriately from data on the yields of the harvested portions of crops. The best evaluations are obtained from experiments conducted on soils that are extremely deficient in the nutrient being evaluated because a wide range of responses permits the manifestation of differences in effectiveness if such exist, and it increases the precision of evaluations. Many experiments designed to use crop yields for fertilizer evaluation have failed, however, at the expense of much time and economic resources, because the increases in yield from fertilization were too small and the experimental error was too great to evaluate the relative effectiveness of the different nutrient sources with useful precision.

For commonly used soluble fertilizers, the yields of nutrients in plants continue to increase with applications in the plateau region of response curves based upon yields of the harvested portion of crops. Yields of the nutrient being evaluated thus represent a more sensitive basis for fertilizer evaluation at higher levels of nutrient availability than do yields of the harvested portion of crops. This is especially true for nitrogen and potassium. As an example of the operation of the sensitivity principle, Christensen and Meints (1982) obtained a multiple correlation coefficient of $R = 0.85$ between the quantities of fertilizer nitrogen applied and the quantities of total nitrogen in a winter wheat crop in Oregon. The corresponding multiple correlation coefficient between the quantity of fertilizer nitrogen applied and the yield of wheat grain was $R = 0.72$.

Absorption of a nutrient from the soil is the primary biological indication of availability of a nutrient, whereas the yield of the harvested portion of the crop is only a secondary indication. Thus, for evaluating differences in availability coefficient ratios of a nutrient supplied by different fertilizers at relatively high levels of availability, the yield of the nutrient has some scientific advantage as well as the practical advantage of sensitivity.

The sensitivity of evaluations may be increased still further by using tracers. For example, in the wheat experiment by Christensen and Meints (1982) just mentioned, the multiple correlation between tagged fertilizer nitrogen applied and tagged nitrogen in the crop was $R = 0.96$. Tracers reflect differences in availability of nutrient sources even if there is no increase in yield of the harvested portion of the crop or yield of the nutrient.

Fig. 5-19 illustrates both the sensitivity principle and a tracer technique useful for fertilizer evaluation. The figure shows three sets of response measurements made in an experiment by Gunary (1968), in which the soil phosphorus was tagged by mixing it thoroughly with a nutrient solution containing, per kilogram of soil, 15.7 milligrams of phosphorus as monobasic sodium phosphate tagged with ^{32}P — a radioactive nuclide. Also mixed with the soil in separate cultures were different quantities of superphosphate to provide a response curve and a single quantity of phosphorus (131 milligrams per kilogram of soil) as sheep manure. These two phosphorus sources had been powdered, so that they too would have an opportunity to equilibrate uniformly with the phosphorus-32 added and the labile phosphorus throughout the soil.

The shapes of the curves show how the sensitivity increased from left to right. In the yield of dry matter curve in *A*, the yield with the manure occurred on the

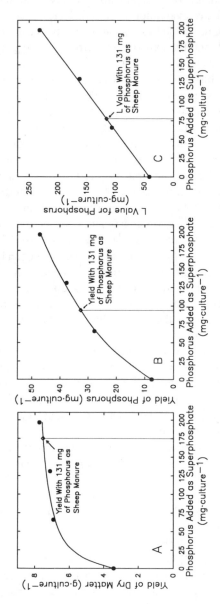

Fig. 5-19. Evaluation of the effectiveness of the phosphorus in a sample of sheep manure by three techniques differing in sensitivity, with super- phosphate as a standard source and ryegrass as a test crop. The superphosphate and manure were added to separate cultures containing 1 kilogram of soil. A. Evaluation on the basis of the yields of dry matter. B. Evaluation on the basis of the yields of phosphorus. C. Evaluation on the basis of the L values for phosphorus in the soil. (Gunary, 1968.)

relatively flat portion of the response curve. The scatter of points in the response curve indicates that the plateau portion was not precisely defined, and accordingly that the quantity of superphosphate phosphorus corresponding to the yield obtained with the application of manure was similarly not precisely defined. A small difference in the height of the response curve or in the yield obtained with the manure addition could cause a large difference in the apparent effectiveness of the manure phosphorus.

Gunary (1968) did not estimate the availability coefficient ratio of the phosphorus in the manure from the data on yield of dry matter (Fig. 5-19A) because he considered the response curve too poorly defined. The fact that the equivalent quantity of phosphorus as superphosphate exceeded the quantity of phosphorus applied in the manure, according to the response curve employed, is an indication of poor precision. The estimated availability coefficient ratio based upon the yield of phosphorus data in Fig. 5-19B is $94/131 = 0.72$, and the ratio based upon the L values for phosphorus in Fig. 5-19C is $77/131 = 0.59$. The evaluations in Figs. 5-19B and 5-19C are fairly similar.

As explained in Section 4-2.1.3, the L-value data in 5-19C are estimates of the theoretical quantities of phosphorus that have equilibrated with the tagged phosphorus added to the soil. Sometimes called Larsen values or labile soil phosphorus, they are derived from measurements of tagged and untagged phosphorus in plants grown on soil that has been mixed thoroughly with a known quantity of a soluble tagged source of phosphorus. They are calculated from the relation

$$L = \left(\frac{c_f}{c_p} - 1 \right) x,$$

where c_f is the specific activity or concentration of the tracer in the tagged nutrient applied in the fertilizer, c_p is the concentration of the tracer in the same nutrient present in the plant, and x is the quantity of the tagged nutrient added to the soil.

In Gunary's experiment, the tagged phosphorus would equilibrate with portions of the phosphorus supplied by the soil and superphosphate in the superphosphate-treated cultures and with portions of the phosphorus supplied by the soil and manure in the manure-treated cultures. Evaluations of this kind must be made under circumstances in which the total medium is tagged with the tracer and in which the fertilizers to be evaluated are uniformly dispersed throughout the total medium. Thus, the technique is suitable for experiments in soil cultures, but not for experiments in which plants are grown as they are normally grown in the field with the fertilizer added to the upper portion of the soil.

Fig. 5-20 illustrates again the sensitivity principle in connection with a different application of the tracer technique. In this experimental work, the fertilizers were specially prepared for the purpose and were tagged with phosphorus-32 during processing. The fertilizers were broadcast on the surface and disked into the soil.

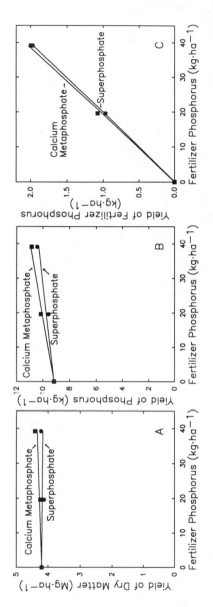

Fig. 5-20. Evaluation of the effectiveness of phosphorus in calcium metaphosphate by three techniques differing in sensitivity, with superphosphate as a standard source. The data are an average of results obtained in two field experiments in Iowa, with oat plants as the test crop. A. Evaluation on the basis of the yields of dry matter. B. Evaluation on the basis of yields of phosphorus. C. Evaluation on the basis of yields of fertilizer phosphorus. (Black, 1950)

The range from the control to the highest yield of dry matter amounted to only 5% of the control yield, which was insufficient to make a precise evaluation. The corresponding range for the yield of phosphorus data was 18%, which was better, but still not great enough for a good evaluation. Results such as those shown in the figure for the yield of dry matter and yield of phosphorus data suggest that if the experimental work had been done on a phosphorus-deficient soil that gave a large response, calcium metaphosphate would have been found far more effective than superphosphate. The data in Fig. 5-20C on yields of fertilizer phosphorus, calculated with the aid of the radioactive tag, however, indicate that the calcium metaphosphate and superphosphate were of essentially equal value. In interpreting the results, these data must be given the major weight because the control value is zero and the sensitivity is excellent.

5-6.2. Problems in Tagging

Work by McAuliffe and Peech (1949) and McAuliffe et al. (1949) illustrates a potential problem in direct tagging of fertilizers to be evaluated. They tested different techniques for labeling the phosphorus in manure, so that the plant phosphorus derived from the manure could be estimated directly from the total phosphorus in the test plants and the specific activities of the phosphorus in the manure and the test plants.

Their data indicated that the labeling of the organic phosphorus in the manure was 47% where the labeling was done by incubating radioactive monobasic potassium phosphate with the manure for 22 days (with addition of 1% soluble starch to enhance microbial activity and increase the labeling of the organic fraction) and 89% where the phosphorus in the manure was labeled by feeding a sheep radioactive monobasic potassium phosphate for 10 days and collecting the feces from the last 5 days of feeding for the experimental work. They found that the total recovery of labeled phosphorus in three cuttings of the ryegrass test crop was 10.7% with superphosphate and 9.6% with manure tagged by the feeding method, indicating that the manure phosphorus was somewhat less effective than that supplied by superphosphate.

If calculations of phosphorus uptake from the incompletely labeled manure were made on the assumption that the phosphorus was labeled uniformly, the uptake would be overestimated because the label would be concentrated in the tagged portion from which the plants were absorbing phosphorus. In the samples of sheep manure used by McAuliffe and Peech (1949) and McAuliffe et al. (1949), the organic phosphorus fraction amounted to about 15% of the total. Thus, if 47% of the organic phosphorus was labeled, the assumption of uniform labeling would result in uptake estimates 8% too high. And if 89% of the organic phosphorus was labeled, the uptake estimates would be 2% too high.

In producing ^{32}P-tagged superphosphate by adding a solution containing ^{32}P when phosphate rock is being processed by sulfuric acid, any unreacted phosphate rock would be essentially unlabeled. If the citrate-soluble phosphorus is the basis upon which the phosphorus content of the superphosphate is calculated, the unreacted and unlabeled phosphate rock probably would cause no significant error in the estimates of phosphorus uptake by the test crop. There would be small counterbalancing errors of dissolution of part of the unreacted phosphate rock by the citrate extractant and of absorption of some phosphorus from the unreacted phosphate rock by the plants.

5-6.3. Tagging Different Sources

The most straightforward way to use tracers in fertilizer evaluation is to tag each source and measure the uptake of the tagged nutrient from each source. This method is also the one most commonly used. The fertilizers are applied in different quantities, so that a response curve for yield of plant substance, yield of the nutrient, and yield of the tagged nutrient can be plotted for each fertilizer. Experimental work by Terman and Allen (1969) is a specific example.

Morel and Fardeau (1990b) proposed a simplified procedure. They suggested dispensing with the response curve concept and evaluating phosphate fertilizers by determining the uptakes of tagged phosphorus by a test crop from a given quantity of phosphorus added to the soil in the standard fertilizer and each of the test fertilizers. To estimate the availability-coefficient ratios, the uptake from each test fertilizer is divided by the uptake from the standard fertilizer. This approach is applicable when the uptakes of the nutrient from all fertilizers are proportional to the quantities applied.

Two other ways of using tracers for fertilizer evaluation are illustrated in Fig. 5-21. In this experimental work, the phosphate fertilizers to be evaluated were not tagged, but a small amount of tagged superphosphate was added to all plots. In such an arrangement, the greater the quantity of the untagged nutrient absorbed from the soil plus the fertilizers to be evaluated, the smaller is the quantity absorbed from the tagged fertilizer.

The experimental system used to obtain the data in Fig. 5-21 is similar to that used to obtain the data in Fig. 5-19, except that the fertilizers were mixed with only a portion of the soil from which the plants absorb the nutrient, and only regular field equipment was used to mix the fertilizers into the soil. (Mainly because the cost limited the quantity of tagged fertilizer available, the tagged fertilizer was applied to only a small portion of each of the field plots. The phosphorus data were obtained from the small areas treated with the radioactive fertilizer, and the crop yield data were obtained from the larger nonradioactive portions of the plots.)

As shown in Fig. 5-21A, the soil was strongly deficient in phosphorus. The approximately fourfold increase in yield of grain with superphosphate provided a good range for evaluating the various phosphorus sources. In Fig. 5-21C, the percentages of the plant phosphorus derived from the fertilizer (calculated from the radioactive phosphorus data) are plotted against the quantities of fertilizer phosphorus. Here there is a finite and substantial control value, and the relative range of data available for evaluation is actually less than the relative range in the yield of grain. In this method of using tracers, the greater the quantity of

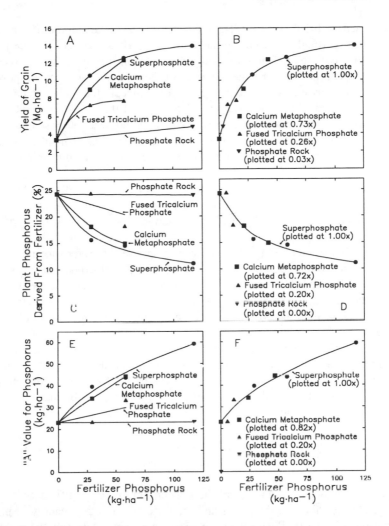

Fig. 5-21. Evaluation of the effectiveness of three phosphate fertilizers by different response criteria with superphosphate as the standard source in a field experiment on a calcareous soil in Iowa. *A, C,* and *E* show plots of the yield of oat grain, the percentage of the total phosphorus in the mature plants derived from superphosphate tagged with phosphorus-32 that was applied to all plots, and the "*A*" values for soil plus fertilizer phosphorus, all plotted against the quantities of phosphorus added in the untagged fertilizers being tested. *B, D,* and *F* show the same plots for the superphosphate standard, assigned an availability coefficient of 1.00, but with the quantities of phosphorus applied in each of the test fertilizers multiplied by an availability coefficient. (Webb, 1952)

tagged fertilizer added uniformly to all plots to make the evaluation, the lower the sensitivity would become.

To obtain values that increase with the effectiveness of the fertilizers being evaluated, one may use the data on percentage of the plant phosphorus derived from the fertilizer to calculate a quantity called the "*A*" value (see Chapter 4).

As explained by Fried (1964), "The 'A' value . . . is the amount of available nutrient in a particular source measured in terms of a fertilizer standard and based on the assumed definition that if a plant is confronted by two sources of a nutrient it will take up nutrient from each of these sources in direct proportion to the amounts available." According to this concept, the standard fertilizer source is assumed to have a certain availability, and the ratio of the uptake of the nutrient from the "particular source" (which would be all other sources from which the plant was absorbing the nutrient) to the uptake from the standard fertilizer source would be equal to the ratio of the amounts available in the two sources.

Fig. 5-21E shows plots of "*A*" values against the quantities of fertilizer phosphorus added. The relative range of data available for evaluation is similar to that for the percentage of the plant phosphorus derived from the fertilizer (Fig. 5-21C).

Sections *B*, *D*, and *F* in Fig. 5-21 show the response curves for superphosphate from the corresponding sections *A*, *C*, and *E*, with the superphosphate standard plotted at $1.00x$ and the other fertilizers plotted as the product of the quantity applied (x) and an appropriate availability coefficient relative to the standard. The estimated availability coefficients of the three test fertilizers are similar with all three bases for evaluation.

In the experiment described in Fig. 5-21, a single quantity of tagged fertilizer was used. A question of concern where different quantities of tagged phosphate fertilizers are applied is that the numerical values of "*A*" commonly show a trend with the quantities of the tagged fertilizer employed. To eliminate the effect of the trend, the empirical function

$$\frac{c_p}{c_f - c_p} = \alpha x^\beta$$

has been used, where c_p and c_f are the concentrations or specific activities of the tagged phosphorus in the plants and the fertilizer, x is the quantity of phosphorus added in the fertilizer, and α and β are constants. Adapting the function for comparing a standard source and a test source, as in Section 5-2.1, one may write

$$\frac{c_{ps}}{c_{fs} - c_{ps}} = \alpha(x_s)^\beta$$

for the standard fertilizer and

$$\frac{c_{pt}}{c_{ft} - c_{pt}} = \alpha(\lambda x_t)^\beta$$

for the test fertilizer, where c_{ps} and c_{pt} are the concentrations or specific activities of the tag from the standard and test fertilizers in the plants, c_{fs} and c_{ft} are the concentrations or specific activities of the tag in the standard and test fertilizers, α and β are constants, and λ is the ratio of the availability coefficient of the phosphorus in the test fertilizer to that of the nutrient in the standard fertilizer. The standard fertilizer is arbitrarily assigned an availability coefficient of 1.00.

The logarithm of the equation for the standard fertilizer is

$$\mathrm{Log}\left(\frac{c_{ps}}{c_{fs} - c_{ps}}\right) = \mathrm{Log}\alpha + \beta\mathrm{Log}x_s,$$

and the logarithm of the equation for the test fertilizer is

$$\mathrm{Log}\left(\frac{c_{pt}}{c_{ft} - c_{pt}}\right) = (\mathrm{Log}\alpha + \beta\mathrm{Log}\lambda) + \beta\mathrm{Log}x_t.$$

These are linear equations that have the same slope β. They plot as parallel straight lines. The intercept for the standard fertilizer is Log α, and the intercept for the test fertilizer is (Log α + βLog λ). The value of λ is determined by subtracting Log α from (Log α + βLog λ), dividing βLog λ by β, and taking the antilogarithm of Log λ. White et al. (1956) devised a statistical procedure for fitting the biological assay model with the function described.

5-6.4. Interpretation Problems

Perhaps another manifestation of the cause of changes in "*A*" values with quantity of tagged fertilizer added is a systematic difference in the numerical values of the availability coefficient ratios shown in Fig. 5-22 between evaluations based upon the yield of phosphorus and those based upon the yield of true fertilizer phosphorus (where true fertilizer phosphorus is defined as the atoms actually applied in the fertilizer and found by analysis of the plants for total phosphorus and the ^{32}P derived from the tagged fertilizer). In this experiment, the different fertilizers were ^{32}P-tagged granules of monobasic sodium phosphate of four particle sizes, each added in different quantities to six soils. The two soils with points close to the 1:1 line in Fig. 5-22A are the same as the ones in Fig. 5-22B with the smallest quantities of labile phosphorus by isotopic dilution. (This phosphorus includes the inorganic orthophosphate in solution and the inorganic orthophosphate in the soil solids that exchanges with ^{32}PO$_4$ added to make the analysis. Also included is a relatively small amount of organic phosphorus that is released by biological interchange.)

The soil phosphorus that equilibrates with added ^{32}PO$_4$ is thought to be involved in the difference in numerical values of evaluations derived from yield of phosphorus data and from ^{32}PO$_4$ data. The fertilizer may affect root growth and absorption of soil phosphorus from the portion of the soil that contains no fertilizer as well as the portion that contains fertilizer.

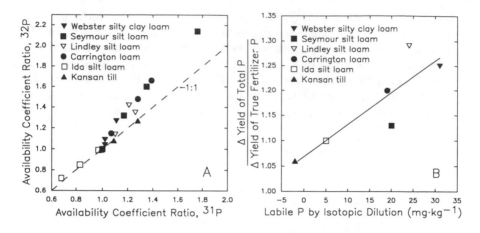

Fig. 5-22. Comparative evaluations of ^{32}P-tagged monobasic sodium phosphate of four particle sizes on the basis of the increases in yields of total phosphorus and yields of true fertilizer phosphorus in oat plants grown on cultures of six soils. Average particle diameters ranged from 2.03 millimeters in the coarsest separate to 0.59 millimeter in the finest separate, and the 0.59 millimeter particles were assigned an availability coefficient of 1.00. A. Availability coefficient ratios based upon yields of true fertilizer phosphorus versus availability coefficient ratios based upon the increases in yield of phosphorus. B. (Δ Yield of total phosphorus)/(Δ yield of true fertilizer phosphorus) versus the labile phosphorus in the soils by isotopic dilution. The Δ values are averages over all four granule sizes. (Bouldin and Black, 1960)

Within the fertilizer-affected soil, there is an additional complication. During movement of the phosphorus from the source particle or band, the spreading stream of phosphorus becomes progressively depleted of the tagged form by interaction with the soil and progressively enriched with the untagged form picked up from the soil. Thus, although the increase in concentration of the phosphorus present at a specific site removed from the source may be said to be derived from the fertilizer, only part of the nontagged atoms were derived from the atoms added in the fertilizer. The others were derived from the soil.

The loss of the tag from a solution moving through soil was demonstrated by Larsen (1952) in one of the early experiments on the behavior of tagged phosphorus in soil. He added a solution containing known quantities of $^{32}PO_4$ and $^{31}PO_4$ in the equivalent of 20 centimeters of water to the top of a column of soil saturated with water. When the added solution had entered the soil and drainage had ceased, he sectioned the column and analyzed the segments for total phosphorus and radioactive phosphorus. The results in Fig. 5-23 show that the percentage recovery of the $^{32}PO_4$ in the uppermost layer of soil exceeded that of the $^{31}PO_4$, but that the ratio of the recoveries decreased with depth. Below about 2.5 centimeters, the apparent recovery of the $^{31}PO_4$ added (actually the $^{31}PO_4$ added plus that picked up from the soil) exceeded the recovery of the "true fertilizer phosphorus" atoms, which were those added in the tagged solution at the surface. Bouldin (1953) found that the same was true where the movement of the tagged phosphorus occurred by diffusion only, and Burns (1962) found that when anion exchange resin particles were placed in soil at different distances from tagged monobasic sodium phosphate granules, the resin particles registered uptakes of soil phosphorus greater than those from the corresponding unfertilized soils.

The reader usually is left to infer that evaluations of tagged phosphate fertilizers by the tracer technique represent the comparative performance of the fertilizers in terms of differences in phosphorus uptake or yield response that would have occurred, had conditions of soil and environment led to readily measurable responses. Although the approximation may be acceptable for practical purposes, it is not necessarily correct because, as indicated by the preceding discussion, the measurements of the tracer in the plant do not necessarily reflect the total fertilizer effects. It is appropriate, therefore, to distinguish between two different kinds of estimates of uptake of fertilizer phosphorus that may be made in the presence of tracers.

The first, calculated without the benefit of the tracer, is

$$y_n - y_T - y_0,$$

where y_n is the net increase in yield of phosphorus from fertilization, and y_T and y_0 are the total yields of phosphorus in the plants with and without the fertilizer. This evaluation reflects the total fertilizer effects.

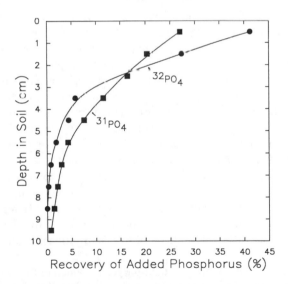

Fig. 5-23. Percentage recovery of $^{32}PO_4$ and $^{31}PO_4$ at different depths in soil after a solution containing known quantities of the two nuclides had been added to the surface and allowed to drain through. (Larsen, 1952)

The second, calculated using the tracer data, is

$$y_t = \left(\frac{c_p}{c_f}\right)y_T,$$

where y_t is the "yield of true fertilizer phosphorus," y_T is the total yield of phosphorus in the fertilized plants, and c_p and c_f are the concentrations or specific activities of the phosphorus in the plants and in the fertilizer. Calling y_t the yield of true fertilizer phosphorus involves the assumption that there is no discrimination between the fertilizer-derived ^{32}P atoms and ^{31}P atoms in any of the processes the fertilizer phosphorus undergoes, and hence that the ratio of the two types of atoms in the plants is the same as the ratio in the original fertilizer before addition to the soil. The plants of course contain additional ^{31}P atoms derived from the seed and soil.

To bring together the foregoing considerations in a specific example related to fertilizer evaluation, Morel and Fardeau (1989) found that application of tagged dibasic ammonium phosphate to a soil from France depressed the uptake of soil phosphorus, whereas tagged North Carolina phosphate rock had no consistent effect on soil phosphorus uptake. As a consequence of this behavior, the increase in total phosphorus uptake with increasing quantities of phosphorus as dibasic ammonium phosphate was smaller than it would have been if uptake of soil phosphorus had remained constant. The effectiveness of the phosphorus supplied by phosphate rock relative to that supplied by dibasic ammonium phosphate thus would appear greater if the evaluation were based upon data for total phosphorus than upon data for true fertilizer phosphorus. In another paper, Morel and Fardeau (1990a) reported that in a greenhouse experiment with 14 soils from France, addition of tagged phosphorus as dibasic ammonium phosphate increased the uptake of phosphorus from soils with low uptakes of soil phosphorus and decreased the uptake from soils with high uptakes of soil phosphorus. What is the significance of these findings in terms of methods of fertilizer evaluation?

The first question is whether an estimate of the ratio of the availability coefficient of phosphorus in the phosphate rock to that of phosphorus in the dibasic ammonium phosphate is valid when based upon total phosphorus uptake or uptake of tagged phosphorus. If both methods are interpreted as applications of the classical biological assay concept, the answer would be that the evidence indicates both estimates are invalid. The reason is that the differential effect of the two phosphorus sources on uptake of soil phosphorus shows that the sources were not behaving as different dilutions of the same active ingredient in an inert matrix.

The next question then is, for practical purposes, which of the two evaluations is the better? The answer here would be that if the sensitivity of both methods is good, the better evaluation would be the one that provides the kind of information one is seeking. If the interest is in finding the relative uptakes of the actual atoms of fertilizer phosphorus applied in the two sources, uptakes of tagged phosphorus would be an appropriate basis for evaluation. If the interest is in finding the net effects of the two sources, which would be applicable to the economics of the effects of applying the two fertilizers, uptakes of total phosphorus would be an appropriate basis for evaluation.

On the other hand, if the sensitivity of the evaluations based upon total phosphorus uptake is poor because the uptake of total phosphorus changes only a little with fertilization, the more desirable basis for evaluation very likely would be the uptake of tagged phosphorus regardless of the information desired. Evaluations based upon tagged phosphorus uptake would provide the better evaluation of the relative uptakes of tagged phosphorus, and they probably would provide a better estimate of relative net effects than the uptakes of total phosphorus.

The foregoing discussion and examples of the use of tracers in fertilizer evaluation have been restricted to phosphorus. Most of the work has been done in this area. Menzel and Smith (1984) reviewed uses in general.

With phosphorus, only the orthophosphate species need be considered, although Bouldin (1953) found that the phosphorus in the concentrated ^{32}P-tagged solution obtained from the U.S. Oak Ridge National Laboratory was not all in orthophosphate form. Special processing in the laboratory was required to convert the phosphorus to orthophosphate before it could be used as a valid tracer for orthophosphate.

The use of ^{15}N as a tracer presents a complication in that both the ammonium and nitrate species are important. Either or both can be tagged. These ions do not behave the same in either inorganic reactions or biological processes. Moreover, ammonium may be converted to nitrate, and nitrogen added to soil as nitrate may reappear later as ammonium.

Nitrogen-15, a stable nuclide, has been used extensively in soil nitrogen research, but not to a great extent for nitrogen fertilizer evaluation. Nitrogen behaves differently in the soil than does phosphorus, and this is a source of confusion (see Chapter 4). With phosphorus, the interchange of $^{32}PO_4$ with soil $^{31}PO_4$ is mostly an inorganic equilibration process, but with nitrogen the interchange is mostly biological and not an equilibration process.

Relatively large differences commonly are found between the uptake of soil nitrogen in the presence and absence of nitrogen fertilizer (Jansson and Persson, 1982; Bouldin, 1986). Fig. 5-24 shows the results of similar field experiments at two locations, in which a difference between the yields of true fertilizer nitrogen and the increases in yield of nitrogen from fertilization may be observed.

The true yields of nutrients derived from fertilizers thus are not necessarily equal to the net increases in yield of the nutrients from fertilization, and the differences depend upon the soil and other circumstances that as yet have not been quantified. Although further investigation will be needed to clarify the issue, the inconsistencies among soils provide further evidence that differences in the way a given nutrient supplied by a given pair of fertilizers interacts with

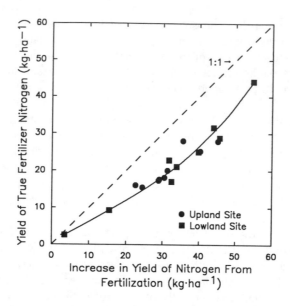

Fig. 5-24. Yields of true fertilizer nitrogen and increases in yield of nitrogen from fertilization in the first cutting of ryegrass in two similar field experiments in Ireland with three sources of nitrogen and three dates of application. (Stevens and Laughlin, 1988)

soils can affect the ratio of the availability coefficients of the nutrient from the two sources. Because the net increase in yield of a nutrient from fertilization is of practical concern, whereas the true yield of a nutrient supplied by a fertilizer is not, Jansson and Persson (1982) expressed the view that the conventional use of the net increase in nitrogen uptake from fertilization is still to be recommended for fertilizer evaluation.

The problem of differential effects of two sources of a nutrient on uptake of the nutrient from the soil might seem to be eliminated if the assay of the relative effectiveness of a common nutrient could be made in the presence of both sources. McLaughlin and Alston (1986) addressed the problem of distinguishing the contributions of two sources of phosphorus by a double-tagging procedure in which an organic source (medic) was tagged with ^{33}P and the standard source (monobasic calcium phosphate) was tagged with [32]P. Although this experiment was not designed to provide an assay in the sense being considered here, the two sources were found to have large mutual effects. Addition of monobasic calcium phosphate increased the recovery of the ^{33}P from the medic in the wheat crop from 8% to 16%. And conversely, addition of the medic decreased the recovery of ^{32}P from the monobasic calcium phosphate from 24% to 18%. The dry matter yield of the wheat plants was markedly decreased by addition of the medic residues and markedly increased by addition of the phosphate fertilizer. In the presence of the phosphate fertilizer, medic decreased the dry matter yield to only a small degree. The findings of the double-tagging experiment by

McLaughlin and Alston (1986) indicate that the evaluation of the phosphorus in the medic should be made in the absence of the standard source if the medic residues are to be used alone, or in the presence of the standard source if the medic residues are to be used with the standard source or another similar source.

The use of tracers thus increases the sensitivity of the evaluations in most instances, while introducing a new set of theoretical concepts. The advantage of improved sensitivity in fertilizer evaluation must be balanced against whatever importance is attached to loss of information on net effects of fertilization.

5-6.5. Practical Considerations

Along with the scientific aspects must be considered the practical aspect of cost in money and time. The cost of nitrogen-15 and phosphorus-32 tracers is not a problem for laboratory or greenhouse experiments, but it is for field experiments. In field experiments, the tracers commonly are applied only to "microplots" too small for good yield evaluations. Larger plots with untagged nitrogen or phosphorus are normally used to obtain the yield data.

Special equipment is required to make the analyses needed. The cost of purchasing and maintaining the equipment to make nitrogen-15 measurements is so great that an extensive research program requiring many analyses is needed to justify the purchase. This problem may be mitigated to some extent by shipping samples to a central laboratory for analysis. The equipment to make radioactivity assays is less costly, and it can be allowed to stand for long periods without use.

Much tracer work with tagged phosphate fertilizers was done when the concept was new. At that time, the U.S. Department of Agriculture devoted a special laboratory and skilled personnel to producing the fertilizers, which were distributed free to researchers (Hill et al., 1949). Eventually the laboratory was transferred to the Tennessee Valley Authority. The Chemical Development Department, National Fertilizer and Environmental Research Center, Tennessee Valley Authority, Muscle Shoals, Alabama, is continuing the service to scientists wishing to do research with fertilizers tagged with calcium, nitrogen, phosphorus, or sulfur. The nuclide desired by a user is purchased in the market, and the user is asked to pay for the nuclide and shipping charges plus a nominal fee for preparing the product.

At present, most of the stable nuclide products (tagged with ^{15}N) produced by the National Fertilizer and Environmental Research Center go to U.S. scientists, but most of the radioactive fertilizer shipments are to scientists in countries other than the United States. Requests from researchers in other countries generally are submitted to the International Atomic Energy Agency in Vienna, which contracts with the U.S. laboratory. The U.S. laboratory then ships the fertilizers direct to the users.

The currently limited use of radioactive fertilizers by U.S. scientists appears to be due in large part to the requirements and paper work detailed in some 38 pages of federal regulations (Anonymous, 1990) plus the restrictive policies adopted by the universities for public relations reasons. These barriers naturally encourage U.S. researchers to find what they need to know in ways that do not involve radioactive tracers.

Chemical sources of common fertilizer nitrogen compounds enriched with ^{15}N are commercially available direct to potential users. Although the physical form may not be identical with those in commercial nitrogen fertilizers, this probably is not of importance in most instances because the compounds are completely soluble in water. Where the physical form is of significance, the National Fertilizer and Environmental Research Center is equipped to produce the desired forms.

5-7. Literature Cited

Alten, F., E. Rauterberg, and H. Loofmann. 1940. Der Einfluss des Kalis auf den Stickstoffhaushalt der Pflanzen. *Bodenkunde und Pflanzenernährung* 19:22-55.

Anonymous. 1990. *Code of Federal Regulations*. Part 20 — Standards for Protection Against Radiation. Pp. 292-330. Office of the Federal Register. U.S. Government Printing Office, Washington.

Barber, S. A., and P. R. Ernani. 1990. Use of a mechanistic uptake model to evaluate phosphate fertilizer. *Transactions 14th International Congress of Soil Science* 2:136-140.

Barideau, L., and A. Falisse. 1982. La disponibilité de l'azote dans les boues résiduaires. *Bulletin des Recherches Agronomiques de Gembloux* 17:227-235.

Barrow, N. J. 1985. Comparing the effectiveness of fertilizers. *Fertilizer Research* 8:85-90.

Barrow, N. J., and M. D. A. Bolland. 1990. A comparison of methods for measuring the effect of level of application on the relative effectiveness of two fertilizers. *Fertilizer Research* 26:1-10.

Bationo, A., S. H. Chien, J. Henao, C. B. Christianson, and A. U. Mokwunye. 1990. Agronomic evaluation of two unacidulated and partially acidulated phosphate rocks indigenous to Niger. *Soil Science Society of America Journal* 54:1772-1777.

Black, C. A. 1950. Source of phosphate experiments on oats using P^{32}. Pp. 49-54. In *Compilation of Field Fertilizer Experiments Using Radiophosphorus*. Phosphorus Subcommittee of the North Central Region. Department of Agronomy, Iowa State University, Ames.

Black, C. A., and C. O. Scott. 1956. Fertilizer evaluation: I. Fundamental principles. *Soil Science Society of America Proceedings* 20:176-179.

Black, C. A., J. R. Webb, and O. Kempthorne. 1956. Fertilizer evaluation: III. Availability coefficient of water-soluble and citrate-soluble phosphorus in acidulated phosphate fertilizers. *Soil Science Society of America Proceedings* 20:186-189.

Bliss, C. I. 1952. *The Statistics of Bioassay With Special Reference to the Vitamins*. Academic Press, NY.

Bolland, M. D. A., and N. J. Barrow. 1988. Effect of level of application on the relative effectiveness of rock phosphate. *Fertilizer Research* 15:181-192.

Bolland, M. D. A., A. J. Weatherley, and R. J. Gilkes. 1988. Residual effectiveness of superphosphate is greater than that of rock phosphate fertilisers for lateritic soils in south-western Australia. *Australian Journal of Experimental Agriculture* 28:83-90.

Bouldin, D. R. 1953. *Phosphate Diffusion in Soils.* M.S. Thesis, Iowa State College, Ames.

Bouldin, D. R. 1986. The chemistry and biology of flooded soils in relation to the nitrogen economy in rice fields. *Fertilizer Research* 9:1-14.

Bouldin, D. R., and C. A. Black. 1960. Fertilizer evaluation: IV. Use of P^{32}-labeled fertilizers. *Soil Science Society of America Proceedings* 24:491-496.

Brennan, R. F. 1990. Effectiveness of some copper compounds applied as foliar sprays in alleviating copper deficiency of wheat grown on copper-deficient soils of Western Australia. *Australian Journal of Experimental Agriculture* 30:687-691.

Brennan, R. F., J. W. Gartrell, and A. D. Robson. 1984. Reactions of copper with soil affecting its availability to plants. III Effect of incubation temperature. *Australian Journal of Soil Research* 22:165-172.

Bromfield, S. M. 1961. Sheep faeces in relation to the phosphorus cycle under pastures. *Australian Journal of Agricultural Research* 12:111-123.

Burns, G. R. 1962. *Particle-Size Effects of Water-Soluble Phosphate Fertilizer.* Ph.D. Thesis, Iowa State University, Ames.

Castle, M. E., and A. D. Drysdale. 1966. Liquid manure as a grassland fertilizer. V. The response to mixtures of liquid manure (urine) and dung. *Journal of Agricultural Science (Cambridge)* 67:397-404.

Charleston, A. G., L. M. Condron, and I. W. M. Brown. 1989. The nature of the residual apatites remaining after partial acidulation of phosphate rocks with phosphoric and sulphuric acids. *Fertilizer Research* 18:257-273.

Chien, S. H. 1978. Dissolution of phosphate rocks in solutions and soils. Pp. 97-129. In *Seminar on Phosphate Rock for Direct Application.* International Fertilizer Development Center, Muscle Shoals, AL.

Chien, S. H., F. Adams, F. E. Khasawneh, and J. Henao. 1987. Effects of combinations of triple superphosphate and a reactive phosphate rock on yield and phosphorus uptake by corn. *Soil Science Society of America Journal* 51:1656-1658.

Chien, S. H., P. W. G. Sale, and D. K. Friesen. 1990. A discussion of the methods for comparing the relative effectiveness of phosphate fertilizers varying in solubility. *Fertilizer Research* 24:149-157.

Christensen, N. W., and V. W. Meints. 1982. Evaluating N fertilizer sources and timing for winter wheat. *Agronomy Journal* 74:840-844.

Colwell, J. D., and W. J. Goedert. 1988. Substitution rates as measures of the relative effectiveness of alternative phosphorus fertilizers. *Fertilizer Research* 15:163-172.

De Haan, S. 1981. Sewage sludge as a phosphate fertiliser. Pp. 149-162. In T. W. G. Hucker and G. Catroux (Eds.), *Phosphorus in Sewage Sludge and Animal Waste Slurries.* D. Reidel Publishing Company, Dordrecht, The Netherlands.

De Wit, C. T. 1953. *A Physical Theory on Placement of Fertilizers.* Verslagen van Landbouwkundige Onderzoekingen No. 59.4. Staatsdrukkerij en Uitgeverijbedrijf, Wageningen, The Netherlands.

Emmens, C. W. 1948. *Principles of Biological Assay.* Chapman & Hall, London.

Engelstad, O. P. 1978. Relative agronomic and economic effectiveness values for phosphate rocks. Pp. 291-303. In *Seminar on Phosphate Rock for Direct Application.* International Fertilizer Development Center, Muscle Shoals, AL.

Engelstad, O. P., A. Jugsujinda, and S. K. De Datta. 1974. Response by flooded rice to phosphate rocks varying in citrate solubility. *Soil Science Society of America Proceedings* 38:524-529.

Ernst, P. 1987. Utilization of nitrogen from cattle slurry applied to permanent grassland. Pp. 283-285. In H. G. van der Meer, R. J. Unwin, T. A. van Dijk, and G. C. Ennik (Eds.), *Animal Manure on Grassland and Fodder Crops. Fertilizer or Waste?* Martinus Nijhoff Publishers, Dordrecht, The Netherlands.

Finney, D. J. 1978. *Statistical Method in Biological Assay.* Third Edition. Macmillan Publishing Co., NY.

Fordham, A. W., and U. Schwertmann. 1977. Composition and reactions of liquid manure (Gülle), with particular reference to phosphate: II. Solid phase components. *Journal of Environmental Quality* 6:136-140.

Fried, M. 1964. 'E', 'L' and 'A' values. *Transactions 8th International Congress of Soil Science* 4:29-39.

Gunary, D. 1968. The availability of phosphate in sheep dung. *Journal of Agricultural Science (Cambridge)* 70:33-38.

Haan, S. de. 1981. Sewage sludge as a phosphate fertiliser. Pp. 149-162. In T. W. G. Hucker and G. Catroux (Eds.), *Phosphorus in Sewage Sludge and Animal Waste Slurries.* D. Reidel Publishing Company, Dordrecht, The Netherlands.

Hammond, L. L., S. H. Chien, and G. W. Easterwood. 1986a. Agronomic effectiveness of Bayovar phosphate rock in soil with induced phosphorus retention. *Soil Science Society of America Journal* 50:1601-1606.

Hammond, L. L., S. H. Chien, and A. U. Mokwunye. 1986b. Agronomic value of unacidulated and partially acidulated phosphate rocks indigenous to the tropics. *Advances in Agronomy* 40:89-140.

Hill, W. L., E. J. Fox, and J. F. Mullins. 1949. Preparation of radioactive phosphate fertilizers. *Industrial and Engineering Chemistry* 41:1328-1334.

Jansson, S. L., and J. Persson. 1982. Mineralization and immobilization of soil nitrogen. Pp. 229-252. In F. J. Stevenson (Ed.), *Nitrogen in Agricultural Soils.* Agronomy 22. American Society of Agronomy, Crop Science Society of America, and Soil Science Society of America, Madison, WI.

Johnston, N. B., and P. H. T. Beckett. 1981. Crop response to sewage sludge: Minimizing the risk of misinterpretation caused by heavy metals and nitrogen. Pp. 198-201. In *Heavy Metals in the Environment.* CEP Consultants, Edinburg, United Kingdom.

Johnstone, P. D., and A. G. Sinclair. 1991. Replication requirements in field experiments for comparing phosphatic fertilizers. *Fertilizer Research* 29:329-333.

Junge, A., and W. Werner. 1989. Investigations on interactions of phosphorus compounds in partially acidulated phosphate rock and fertilizer effectiveness. *Fertilizer Research* 20:129-134.

Kaila, A. 1950. Karjanlanta kasvien fosforin lähteenä. *Maataloustieteellinen Aikakauskirja* 22:107-121.

Kanabo, I. A. K., and R. J. Gilkes. 1988. The effect of the level of phosphate rock application on its dissolution in soil and on bicarbonate-soluble phosphorus. *Fertilizer Research* 16:67-85.

Khasawneh, F. E., and E. C. Doll. 1978. The use of phosphate rock for direct application to soils. *Advances in Agronomy* 30:159-206.

Khasawneh, F. E., and E. C. Sample. 1978. Phosphorus concentration in soil solution as a factor affecting phosphate rock effectiveness. Pp. 130-146. In *Seminar on Phosphate Rock for Direct Application.* International Fertilizer Development Center, Muscle Shoals, AL.

Larsen, S. 1952. The use of P^{32} in studies on the uptake of phosphorus by plants. *Plant and Soil* 4:1-10.

Léon, L. A., W. E. Fenster, and L. L. Hammond. 1986. Agronomic potential of eleven phosphate rocks from Brazil, Colombia, Peru, and Venezuela. *Soil Science Society of America Journal* 50:798-802.

Malzer, G. L., and T. Graff. 1984. Influence of nitrogen form, nitrogen rate, timing of nitrogen application and nitrification inhibitors for irrigated corn Becker, MN 1983. Pp. 8-13. In *A Report on Field Research in Soils*. University of Minnesota, Agricultural Experiment Station, Miscellaneous Publication 2 (revised) — 1984. St. Paul.

Malzer, G. L., and T. Graff. 1985. Influence of nitrogen form, nitrogen rate, timing of nitrogen application and nitrification inhibitors for irrigated corn — Becker, MN 1984. Pp. 16-21. In *A Report on Field Research in Soils*. University of Minnesota, Agricultural Experiment Station, Miscellaneous Publication 2 (revised) — 1985. St. Paul.

McAuliffe, C., and M. Peech. 1949. Utilization by plants of phosphorus in farm manure: I. Labeling of phosphorus in sheep manure with P^{32}. *Soil Science* 68:179-184.

McAuliffe, C., M. Peech, and R. Bradfield. 1949. Utilization by plants of phosphorus in farm manure: II. Availability to plants of organic and inorganic forms of phosphorus in sheep manure. *Soil Science* 68:185-195.

McLaughlin, M. J., and A. M. Alston. 1986. The relative contribution of plant residues and fertiliser to the phosphorus nutrition of wheat in a pasture/cereal system. *Australian Journal of Soil Research* 24:517-526.

Meer, H. G. van der, R. B. Thompson, P. J. M. Snijders, and J. H. Geurink. 1987. Utilization of nitrogen from injected and surface-spread cattle slurry applied to grassland. Pp. 47-71. In H. G. van der Meer, R. J. Unwin, T. A. van Dijk, and G. C. Ennik (Eds.), *Animal Manure on Grassland and Fodder Crops. Fertilizer or Waste?* Martinus Nijhoff Publishers, Dordrecht, The Netherlands.

Menzel, R. G., and S. J. Smith. 1984. Soil fertility and plant nutrition. Pp. 1-34. In M. F. L'Annunziata and J. O. Legg (Eds.), *Isotopes and Radiation in Agricultural Sciences. Volume 1. Soil-Plant-Water Relationships*. Academic Press, London.

Mitscherlich, E. A., and W. Sauerlandt. 1935. Salpeter- und Ammoniak-Stickstoff im Boden und die pflanzenphysiologisch wirksame Stickstoffmenge "b." *Landwirtschaftliche Jahrbücher* 81:623-654.

Morel, C., and J. C. Fardeau. 1989. Native soil and fresh fertilizer phosphorus uptake as affected by rate of application and P fertilizers. *Plant and Soil* 115:123-128.

Morel, C., and J. C. Fardeau. 1990a. Uptake of phosphate from soils and fertilizers as affected by soil P availability and solubility of phosphorus fertilizers. *Plant and Soil* 121:217-224.

Morel, C., and J.-C. Fardeau. 1990b. Agronomical evaluation of phosphate fertilizer as a nutrient source of phosphorus for crops: isotopic procedure. *Fertilizer Research* 24:115-122.

Pesek, J. T., and J. R. Webb. 1957. Economic interpretation of the importance of water solubility in phosphorus fertilizers when used as hill fertilizer for corn. Pp. 15-28. In E. L. Baum, E. O. Heady, J. T. Pesek, and C. G. Hildreth (Eds.), *Economic and Technical Analysis of Fertilizer Innovations and Resource Use*. Iowa State College Press, Ames.

Rajan, S. S. S., R. L. Fox, W. M. H. Saunders, and M. Upsdell. 1991. Influence of pH, time and rate of application on phosphate rock dissolution and availability to pastures I. Agronomic benefits. *Fertilizer Research* 28:85-93.

Rana, D. S., B. Singh, M. L. Kapur, and A. L. Bhandari. 1984. Relative efficiency of new urea based nitrogen fertilizers for rice grown in a light textured soil. *Journal of the Indian Society of Soil Science* 32:284-287.

Resseler, H., and W. Werner. 1989. Properties of unreacted rock residues in partially acidulated phosphate rocks affecting their reactivity. *Fertilizer Research* 20:135-142.

Rund, R. C. (Ed.). 1984. Fertilizers. Pp. 8-37. In S. Williams (Ed.), *Official Methods of Analysis of the Association of Official Analytical Chemists*. Fourteenth Edition. Association of Official Analytical Chemists, Arlington, VA.

Schechtner, G., H. Tunney, G. H. Arnold, and J. A. Keuning. 1980. Positive and negative effects of cattle manure on grassland with special reference to high rates of application. Pp. 77-93. In W. H. Prins and G. H. Arnold (Eds.), *The Role of Nitrogen in Intensive Grassland Production*. Centre for Agricultural Publishing and Documentation, Wageningen, The Netherlands.

Sinclair, A. G., S. S. S. Rajan, and J. H. Watkinson. 1991. The standardization of phosphate rocks. In *Workshop on Phosphate Sources for Acid Soils in the Humid Tropics of Asia, 6-7 November 1990*. Kuala Lumpur, Malaysia. (In Press)

Singh, B., Y. Singh, M. S. Maskina, and O. P. Meelu. 1987. Poultry manure as a N source for wetland rice. *International Rice Newsletter* 12, No. 6:37-38.

Singh, Y., B. Singh, M. S. Maskina, and O. P. Meelu. 1987. Availability to wetland rice of nitrogen from cattle manure. *International Rice Research Newsletter* 12, No. 6:35-36.

Smith, S. R., and P. Hadley. 1988. A comparison of the effects of organic and inorganic nitrogen fertilizers on the growth response of summer cabbage (*Brassica oleracea* var. *capitata* cv. Hispi F_1.) *Journal of Horticultural Science* 63:615-620.

Stanford, G., and J. D. DeMent. 1957. A method for measuring short-term nutrient absorption by plants: I. Phosphorus. *Soil Science Society of America Proceedings* 21:612-617.

Stevens, R. J., and R. J. Laughlin. 1988. The effects of times of application and chemical forms on the efficiencies of ^{15}N-labelled fertilisers for ryegrass at two contrasting field sites. *Journal of the Science of Food and Agriculture* 43:9-16.

Suzuki, M., K. Kamekawa, D. Sekiya, and H. Shiga. 1990. Effect of continuous application of organic or inorganic fertilizer for sixty years on soil fertility and rice yield in paddy field. *Transactions 14th International Congress of Soil Science* IV:14-19.

Terman, G. L. 1960. Yield response in experiments with phosphorus fertilizers in relation to: I. Meaningful differences among sources on acid soils of the southeastern states. *Soil Science Society of America Proceedings* 24:356-360.

Terman, G. L. 1961. Yield response in experiments with phosphorus fertilizers in relation to: II. Variability and differences among sources on soils of northern and western states. *Soil Science Society of America Proceedings* 25:49-52.

Terman, G. L., and S. E. Allen. 1969. Fertilizer and soil P uptake by maize, as affected by soil P level, granule size, and solubility of phosphate sources. *Journal of Agricultural Science (Cambridge)* 73:417-424.

Terman, G. L., D. R. Bouldin, and J. R. Webb. 1962. Evaluation of fertilizers by biological methods. *Advances in Agronomy* 14:265-319.

Terman, G. L., W. M. Hoffman, and B. C. Wright. 1964. Crop response to fertilizers in relation to content of "available" phosphorus. *Advances in Agronomy* 16:59-100.

Van der Meer, H. G., R. B. Thompson, P. J. M. Snijders, and J. H. Geurink. 1987. Utilization of nitrogen from injected and surface-spread cattle slurry applied to grassland. Pp. 47-71. In H. G. van der Meer, R. J. Unwin, T. A. van Dijk, and G. C. Ennik (Eds.), *Animal Manure on Grassland and Fodder Crops. Fertilizer or Waste?* Martinus Nijhoff Publishers, Dordrecht, The Netherlands.

Varvel, G. E. 1983. Effect of banded and broadcast placement of Cu fertilizers on correction of Cu deficiency. *Agronomy Journal* 75:99-101.

Weatherley, A. J., M. D. A. Bolland, and R. J. Gilkes. 1988. A comparison of values for initial and residual effectiveness of rock phosphates measured in pot and field experiments. *Australian Journal of Experimental Agriculture* 28:753-763.

Webb, J. R. 1952. Evaluation of the soil phosphorus levels of fertility and source-of-phosphate plots by means of "A" values. Pp. 113-118. In *Compilation of Experimental Work Reported at the Fourth Annual Phosphorus Conference of the North Central Region.* Mineral Deficiencies Committee of the North Central Region, Department of Agronomy, Iowa State University, Ames.

Webb, J. R., and J. T. Pesek, Jr. 1959. An evaluation of phosphorus fertilizers varying in water solubility: II. Broadcast applications for corn. *Soil Science Society of America Proceedings* 23:381-384.

White, R. F., O. Kempthorne, C. A. Black, and J. R. Webb. 1956. Fertilizer evaluation: II. Estimation of availability coefficients. *Soil Science Society of America Proceedings* 20:179-186.

Whitehead, D. C., A. W. Bristow, and B. F. Pain. 1989. The influence of some cattle and pig slurries on the uptake of nitrogen by ryegrass in relation to fractionation of the slurry N. *Plant and Soil* 117:111-120.

Wit, C. T. de. 1953. *A Physical Theory on Placement of Fertilizers.* Verslagen van Landbouwkundige Onderzoekingen No. 59.4. Staatsdrukkerij en Uitgeverijbedrijf, Wageningen, The Netherlands.

CHAPTER 6

Residual Effects

L AWES AND GILBERT (1884), the original leaders of the Rothamsted Experimental Station in England, wrote that "The recent legislative enactments, giving the cultivator of the soil a claim for the manure ingredients possessing a pecuniary value, which he has applied to the land, add greatly to the interest of all investigations which have a bearing upon this important subject." The system of compensation to which Lawes and Gilbert referred distinguished between the natural fertility of a soil and the accumulated fertility due to purchased livestock feed, manures, and fertilizers. The natural fertility, which was less readily exhausted, was considered the property of the landlord, whereas the accumulated fertility, which was more readily exhausted, was considered the property of the tenant.

The legislation did indeed stimulate scientific interest in residual effects. Summaries of information emphasizing conditions in the United Kingdom include those by Cooke (1967), Johnston (1970), and a symposium published by the Ministry of Agriculture, Fisheries and Food (1971). Tables listing standard compensation values were published by the Ministry of Agriculture, Fisheries and Food (1953) for fertilizers and by Berryman (1971) for animal manures. Much recent research has been done on phosphate fertilizers in Australia. A review of this work was published by Bolland et al. (1988) as a part of a paper on the relative effectiveness of phosphate rock and superphosphate.

This chapter starts with brief comments on residual effects in general, but it emphasizes principally the evaluation and interpretation of residual effects. Evaluation is of concern in connection with the compensation discussed by Lawes and Gilbert, but at present it is probably more important as a basis for adjusting current practices.

Residual effects of plant nutrients are effects on crops following the one to which the nutrients are added. In some situations, the quantities of residual nutrients can be determined to a good approximation by appropriate chemical analyses. The quantities, however, generally cannot be equated to quantities of freshly applied nutrients in terms of effects on plants. The effect on a given crop per unit quantity is generally lower for residual nutrients than for freshly applied nutrients; that is, the availability coefficient of a residual nutrient is usually lower than that of the same nutrient applied currently. On the other hand, the effect on successive crops per unit quantity generally decreases less rapidly for a residual nutrient than for the same nutrient freshly applied. Thus, if the residual quantity of a nutrient has a residual effect on a given crop equivalent to that of 25 kilograms of the freshly applied nutrient, the total effect on the current and future crops may be expected to be greater with the residual nutrient than with the freshly applied nutrient.

Integration of residual effects over time has received little consideration in research. The focus in research has been on determining the value of a residual nutrient in a given year as a proportion or percentage of the value of the same nutrient applied currently. Those interested in applying the research data to find the proportion of the total value of a nutrient remaining after various lengths of time then estimate these values as best they can on the basis of the scattered information available.

Various ways may be used to obtain an indication of residual effects, and reference will be made to several in the examples discussed in this chapter. Emphasis will be placed, however, upon quantitative evaluation by biological methods. Quantitative evaluation of the residual value of previously applied nutrients involves the biological assay principles applied to fertilizer evaluation in the preceding chapter, but there are several additional problems. First, the precision of measurement may be relatively poor because residual effects are usually considerably smaller than the effects on the first crop. Second, the residual effects occur at a different time than the effects on the first crop, so that some means is needed to separate residual effects from chance differences associated with the time. Third, the residual responses represent a combination of the effects of the remaining supplies of the nutrients added, together with various other effects. Fourth, the economic value of a given residual response decreases with an increase in the time required to obtain it. This chapter will be devoted primarily to these matters as they relate to evaluation, but the first section discusses differences among nutrients to provide that important perspective.

6-1. Nutrient Differences

Although the principles of evaluation are independent of the nutrient of concern, there are important differences among nutrients. Each nutrient has its own characteristic behavior. Some of the differences among nutrients are discussed first because the emphasis on phosphorus in the examples that follow might otherwise tend to bias the general perception of residual effects.

Phosphorus, which receives most attention in this chapter because of the availability of data, is a long-lasting nutrient because of its strong interactions with soil. Another nutrient with long-lasting effects is copper. Copper is strongly retained by soil, especially the organic fraction. Once accumulated, it remains indefinitely. The current toxicity of copper to citrus in certain areas in Florida has been related to excessive use of copper as a fertilizer and as a fungicide many years ago.

Hannam et al. (1982) used a combination of field and greenhouse experiments to make a qualitative test of the residual effect of applications of copper to three pastures that had been established 14 to 23 years previously on newly cleared land in Australia. At the time the pastures were established, 1 kilogram of copper per hectare had been applied to one pasture, and 2 kilograms had been applied to the other two. Although adjacent untreated areas were still deficient, field experiments in each of 3 consecutive years on each of the three previously treated pastures gave no increases in yield from applications of copper, indicating that the original applications were still supplying the plants with enough copper.

Nitrogen is at the other extreme. The end product of nitrogen metabolism in soils under aerobic conditions is nitrate, which remains in the soil solution and is readily used by plants and soil microorganisms, and which also moves with the soil solution as it is displaced downward by water added to the surface of the soil or upward in response to loss of water from the soil surface.

The residual effect of nitrogen may be very small if conditions are suitable for its removal from the soil. For example, Lozano and Abruña (1981) found in Puerto Rico that the yield of rough rice was increased from 3.83 to 8.36 metric tons per hectare by application of 224 kilograms of nitrogen as ammonium sulfate per hectare, but that the corresponding yields in the following year when no fertilizer nitrogen was applied were 3.92 and 3.95 metric tons. In the central valley of California, which is characterized by winter rainfall and mild winter temperatures, Krantz et al. (1968) found that sugarbeet gave no response to ammonium nitrate applied in the preceding fall in quantities supplying 34, 67, and 134 kilograms of nitrogen per hectare, but gave a 68% response to ammonium nitrate applied in the spring. The fall-applied nitrogen apparently was lost by leaching during the winter. In Scotland, Smith et al. (1988) found that only 0 to 12% of the nitrogen applied in the autumn to winter barley was present in the plants at harvest.

The residual effect of nitrogen may also be long. R. F. Finn (personal communication) observed that a single application of fertilizer nitrogen to a forest site in New York was still increasing the width of tree rings more than 10 years later, despite the occurrence of enough precipitation to cause loss of nitrate from the soil in drainage water. In this situation, the fertilizer nitrogen presumably was retained in part in the wood from year to year and recycled in part through leaf fall.

Where there is no leaching, residual effects of heavy applications of nitrogen fertilizers may last for a number of years. For example, a 10-year experiment was conducted at Swift Current, Saskatchewan (Read and Winkleman, 1982), where normally the clay loam soil remains dry below 90 centimeters and all available moisture is removed each year. Quantities of 0, 50, 100, 400, and 800 kilograms of nitrogen as ammonium nitrate per hectare were applied to a stand of crested wheatgrass at the beginning of the experiment, with none later. The applied nitrogen was soon depleted where only 50 or 100 kilograms had been applied. At the end of 10 years, the amounts of ammonium plus nitrate nitrogen remaining in the soil to the 120-centimeter depth in excess of the control were -5, 1, 21, and 134 kilograms per hectare with initial applications of 50, 100, 400, and 800 kilograms per hectare, respectively. During the 10-year period, the nitrogen removed in the forage on the fertilized plots in excess of the control plots ranged from 54 kilograms per hectare with the smallest application of nitrogen up to 284 kilograms with the greatest. The corresponding increases in yield of dry forage ranged from 3,160 to 11,700 kilograms per hectare. See papers by Ukrainetz and Campbell (1988), Ukrainetz et al. (1988), and Zentner et al. (1989) for similar work on bromegrass, also in Saskatchewan.

Pratt et al. (1973, 1976) used what they called a decay series to reflect the residual value of the nitrogen in animal manures. For conditions in southern California, they estimated that in the year of application, the fraction of the total nitrogen of dry dairy cow manure that appeared in mineral form was approximately 0.45. In the second year, the fraction of the remaining nitrogen that appeared in mineral form was 0.10. The corresponding fractions in the third and fourth years were 0.05 and 0.05. For liquid manure, the decay series was 0.75, 0.15, 0.10, and 0.05. Applying these series to the results of a 4-year field experiment in which 633 to 2,532 kilograms of manure nitrogen were added per hectare annually to simulate the practices of the dairy farmers in the area, they found that the increase in nitrogen content of the soil due to manuring, as estimated from the nitrogen content of the manure and the decay series, was essentially equal to the increase in nitrogen content of the soil found by analysis.

The close agreement found by Pratt et al. (1976) between observed and predicted values for the increase in nitrogen content of manure-treated soils indicates that the decay series employed were appropriate for the manures used and the conditions under which they were used. Other decay series might fit other circumstances. See a review of this subject by Bouldin et al. (1984).

Potassium is intermediate between phosphorus and copper on the one hand and nitrogen on the other. Potassium is held in the exchangeable form in soils, in which it resists loss by leaching but yet is readily available to plants. In some soils, almost all the fertilizer potassium added remains in exchangeable form, and the availability to plants remains high.

As a specific example for potassium, six field experiments in Mauritius showed that applying fertilizer potassium at the rate of 900 kilograms per hectare once in 6 years for sugarcane produced about the same yields of sugar as applying fertilizer potassium at 150 kilograms per hectare each year (Ng Kee Kwong and Deville, 1989). At one site, where the rainfall was relatively heavy (370 centimeters annually), the single 900-kilogram application produced less sugar than the smaller repeated annual applications. Analysis of the soil for potassium soluble in 1 N nitric acid (the standard potassium test in that area) showed that the content of extractable potassium in the 45- to 75-centimeter layer of soil, which had a cation exchange capacity of only 1 centimole of charge per kilogram, had been increased threefold by the heavy fertilization, suggesting that some of the potassium may have been lost by leaching. Another example of a residual effect of potassium observed experimentally is illustrated later in this chapter in Fig. 6-13.

In most soils, some of the added potassium disappears from the exchangeable form. An important cause of this loss is the entry of potassium into interlayer positions in micaceous minerals, where it exchanges less freely with ions in solution and where its availability to plants is reduced to some degree. Usually the loss of exchangeability of some of the potassium, as measured by the usual ammonium acetate extraction method, is little hindrance in recovery of the potassium by plants, but with some soils the loss is severe. According to Cassman et al. (1990), an area of 120,000 hectares of strongly potassium-fixing soils is found in the San Joaquin Valley in California. Applications as great as 18,800 kilograms of fertilizer potassium per hectare are needed to eliminate potassium deficiencies.

6-2. Increasing Measurement Precision

In most instances, the relatively poor precision of measurements of residual effects on crops is not adequately taken into account in designing experiments. Measuring residual effects with precision approaching measurements of direct effects requires more replications than usually are used for measuring direct effects.

Some have turned to the use of tracers to improve the precision of evaluations. Nitrogen-15, the stable tracer for nitrogen, has the advantage that if the enrichment or depletion of the fertilizer nitrogen with the tracer is great enough, an experiment with tagged nitrogen can follow the residual nitrogen for years. Although the values obtained for residual effects indicate the uptake of true fertilizer nitrogen, their significance in terms of the net residual effect of the fertilizer nitrogen is dubious, for reasons discussed in Section 5-6.

The half-lives of radioactive forms of phosphorus are so short that a tagged phosphate fertilizer is useful only in the season in which it is applied. The effects of nonradioactive phosphate fertilizers applied currently and in preceding years, however, may be estimated by making a uniform application of a tagged phosphate fertilizer and evaluating the dilution of the phosphorus derived from this fertilizer with nontagged soil and fertilizer phosphorus. The technique is the same as that described in Section 5-6 for evaluating different sources of a nutrient. The sensitivity of this use of a tracer is considerably less than that provided by the direct tagging feasible with stable tracers.

6-3. The Time Effect

Residual effects of a fertilizer occur at a different time than current effects. Measurements made at the different times thus reflect the environmental conditions at the times of measurement as well as the residual effects. This problem has been addressed in different ways.

In experimental work by Pommel (1981a,b), the residual effect of phosphorus in sewage sludge was measured by a test crop produced in a growth chamber maintained under constant conditions. Good control of the time effect can be achieved in this way, but the conditions are artificial, and it is unlikely that environmental control equipment will function continuously over the long period of time in which residual effects are of concern in the field.

In Pommel's (1981a) experiment, both sewage sludge and monobasic calcium phosphate were applied in three different quantities as sources of phosphorus for ryegrass grown on sand cultures supplied with a minus-phosphorus nutrient solution. The test crop permitted repeated cuttings to evaluate the changes with time until phosphorus recovery by the plants had essentially ceased. Fig. 6-1A shows the cumulative increases in yield of phosphorus observed with the maximum addition of 112.5 milligrams of phosphorus in each source per culture.

To obtain availability coefficient ratios of the phosphorus sources, Pommel fitted the concurrent straight lines function to the cumulative yields of phosphorus for each quantity of added phosphorus at each successive cutting with each of the two phosphorus sources. Fig. 6-1B shows how the availability coefficient of the sludge phosphorus relative to that of the monobasic calcium phosphate phosphorus increased with time.

To represent the change in the availability coefficient ratio of sludge phosphorus relative to monobasic calcium phosphate phosphorus with time, Pommel (1981a) used the equation

$$E = kt^c,$$

where E is the ratio of the availability coefficient of the phosphorus in the sludge to the availability of the phosphorus in the standard, taken as 1.00. The time in

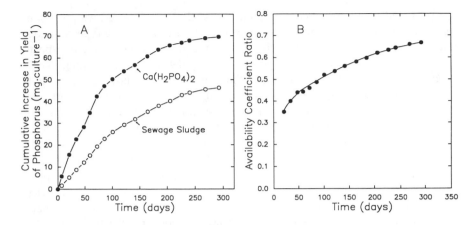

Fig. 6-1. A. Cumulative increases in yield of phosphorus in 17 successive cuttings of ryegrass grown on sand cultures with 112.5 milligrams of phosphorus as monobasic calcium phosphate and sewage sludge applied at the beginning of the experiment. B. Change with time in the ratio of the availability coefficient of phosphorus in the sludge to that of phosphorus in the monobasic calcium phosphate. An availability coefficient ratio was estimated for each cutting except the first from the cumulative yields of phosphorus obtained from the control and three quantities of phosphorus. (Pommel, 1981a)

days is t, and k and c are constants that must be found experimentally. In Fig. 6-1B, $k = 0.166$ and $c = 0.245$.

In a second paper, Pommel (1981b) used the same equation to express the results obtained with seven organic wastes. In all instances a plot of log E against log t fitted a straight line ($r \geq 0.98$), as would be predicted by the equation. Values of k ranged from $-784 \cdot 10^{-7}$ to 0.57, and values of c ranged from 0.1 to 1.9. The closer the k value for a waste to 1.00, the closer is the initial efficiency of the phosphorus relative to that of the standard. And the higher the value of c, the greater is the rate of increase in the efficiency of the phosphorus in the waste relative to that in the standard with succeeding cuttings.

In analyzing the results of a field experiment on residual effects of various phosphate fertilizers on wheat in Australia, Bolland (1985) made an adjustment for the time effect by expressing the results from different years as percentages of the maximum yield obtained with currently applied concentrated superphosphate. Concentrated superphosphate produced the highest yields and was used as the standard fertilizer.

In a field experiment in Australia in which grain sorghum was grown each year as the test crop, Arndt and McIntyre (1963) used the control yields instead of the maximum yields as the basis for adjusting for seasonal effects. For each year of the experiment, they multiplied the yields of grain on all plots by the ratio of the average yield on the control plots without phosphate fertilizer over all the 7 years of the experiment to the average yield on the control plots in the year in question.

Arndt and McIntyre (1963) fitted the adjusted yields obtained with different quantities of currently applied superphosphate to a Mitscherlich-type equation, and estimated the phosphorus equivalents of the superphosphate and phosphate rock treatments applied in prior years by interpolation in the resulting curve. Their data in Fig. 6-2 are analogous to those in Fig. 6-1 in showing, on the left, a plot of cumulative adjusted increases in yield of grain in successive years with two sources of phosphorus and, on the right, a plot of the availability coefficient ratio of the phosphate rock and superphosphate phosphorus against the time.

Conformance of Arndt and McIntyre's response data to the equation used by Pommel (1981a) was tested by plotting the logarithm of the availability coefficient ratio against the logarithm of the time. The points for the successive years could be fitted well by a straight line with the exception of the point for the first year. This point was somewhat off the line. The curve in Fig. 6-2B has been calculated by the Pommel equation, ignoring the point for the first year. The first cutting of ryegrass in Pommel's experiment (not plotted in Fig. 6-1) similarly deviated somewhat from the straight line that fitted the other cuttings well. Deviations of points at the beginning of the series are to be expected because the effectiveness of the test fertilizer relative to the standard usually does not approach zero at zero time (although an organic nitrogen source could show this behavior), but the equation has this property.

Figs. 6-1 and 6-2 show the relative cumulative values of two sources of a common nutrient. Other ways of representing the data are also revealing. When

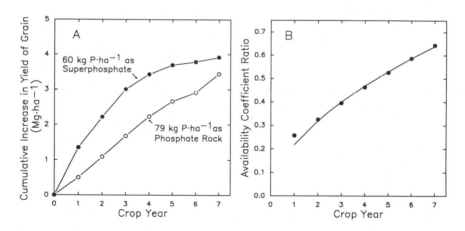

Fig. 6-2. A. Cumulative increases in yield of grain sorghum in seven successive crops in a field experiment in Australia with similar quantities of phosphorus applied as superphosphate and phosphate rock at the beginning of the experiment. B. Change with time in the ratio of the availability coefficient of phosphorus in phosphate rock to that of phosphorus in superphosphate, where the first-year availability coefficient of the superphosphate standard is given a value of 1.00. The availability coefficient ratios are averages of results with four different quantities of phosphate rock. (Arndt and McIntyre, 1963)

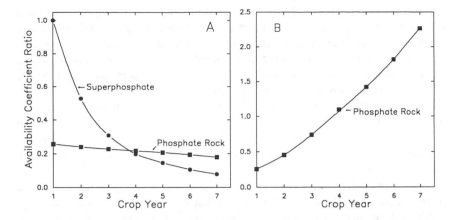

Fig. 6-3. Residual effects of superphosphate and phosphate rock in a field experiment in Australia. A. Change with time in the ratios of the availability coefficients of phosphorus in initial applications of superphosphate and phosphate rock to the phosphorus in superphosphate applied each year and assigned an availability coefficient of 1.00 in each year. B. Change with time in the ratio of the availability coefficient of phosphorus in an initial application of phosphate rock to that of phosphorus in superphosphate applied in the first year and given a value of 1.00 in that year and in each of the following years when the residual effect was being measured. (Arndt and McIntyre, 1963)

the data obtained by Arndt and McIntyre (1963) shown in Fig. 6-2 are recalculated and plotted against the crops (or years), with currently applied superphosphate as the standard each year, the availability coefficient ratios for both superphosphate and phosphate rock applied only in the first year decrease with time (Fig. 6-3A). And when the responses to superphosphate applied only in the first year are taken as the standard in the first year and succeeding years, the availability coefficient of the phosphorus in phosphate rock increases with time and eventually exceeds the availability coefficient of the phosphorus in superphosphate (Fig. 6-3B).

Colwell (1985a) was especially critical of the use of relative yields to adjust data for differences in the time effect. In two 3-year experiments he reported on phosphorus fertilization of wheat in Australia, calculating relative yields was of little or no value in correlating the results for the different years. His experiments were conducted under semiarid conditions, where the water supply may differ markedly among years and where dryness of the surface soil in some years may increase the dependency of the crop on subsoil phosphorus and decrease the dependency on both the currently applied and residual fertilizer phosphorus, which usually will be confined to the surface soil. Sometimes, as in Colwell's experiments, expressing the responses as relative yields is not helpful. In instances in which the water supply is better and less variable among seasons, relative yields can be more useful (see Fig. 6-11 for data obtained under such circumstances).

Barrow and Campbell (1972) emphasized the importance of the design used in experiments by Arndt and McIntyre (1963), Bolland (1985), and others as a means of avoiding the complicating effect of time of evaluation in the investigation of residual effects. The design includes extra control plots to which fertilizer can be applied in successive years to provide a current application for evaluating the responses to earlier applications.

To evaluate the residual effects of a single application of fertilizer according to this system, certain plots receive the requisite quantities of the fertilizer in the first year but none thereafter. An estimate of the 1-year residual effect is derived from second-year measurements of the response on these plots relative to the controls that have received no fertilizer and another set of plots that served as controls in the first year, but to which a fresh application of fertilizer is made in the second year. In the third year, another evaluation of a 1-year residual effect may be made from plots that were fertilized in the second year only, relative to the controls and plots with freshly applied fertilizer. Additionally, a 2-year residual effect may be derived from measurements of the response on the plots that received fertilizer in the first year but none since, relative to control plots and to plots that served as controls in the first and second years but were fertilized in the third year.

The design may be modified to provide data for more years and to provide data on residual effects of applications made in two or more consecutive years or in other patterns as desired. The essential components are plots with the residues of a prior application or applications to be evaluated, control plots that have not received fertilizer during the experiment, and plots that have received freshly applied fertilizer and no prior applications or prior applications that provide the desired comparison for the residual evaluation.

The rate of change of residual effects depends upon the circumstances. The data in Fig. 6-2 for superphosphate represent an intermediate rate for soluble phosphate fertilizers. In northern Scotland, where the soils are acid and strongly retentive of phosphorus, Williams and Reith (1971) found that in eight field experiments the availability coefficient of superphosphate broadcast in the fall after plowing for swede had dropped to one-third of that of superphosphate broadcast the following spring after cultivation but before planting. In ten field experiments in which swede was grown in rotation with other crops, the ratio of the availability coefficient of the residual superphosphate to that of currently applied superphosphate was evaluated when the rotation again returned to swede 6 to 9 years later. At that time, all values were less than 0.10, and the average was only 0.03.

Fertilizer phosphorus applied in soluble forms, such as superphosphate, is noted for the decrease in availability it undergoes due to reaction with soils. But in some soils, a rapid decrease is followed by a relatively long period of sustained availability. Smith (1956) conducted a 3-year experiment on two acid soils in Iowa in which the availability coefficient of phosphorus supplied as granulated concentrated superphosphate decreased with time at rates similar to that in the experiment shown in Fig. 6-3A. In an analogous experiment on a calcareous soil, however, the ratio of the availability coefficient of the residual phosphorus to that of currently applied phosphorus increased in each of the first 2 years after application. If a value of 1.00 is assigned to the availability coefficient in the year the fertilizer was applied, the values for the first and second residual years were approximately 1.1 and 1.9.

The increase in residual effect with time in the calcareous soil may be attributed primarily to the behavior of the reaction product formed in the soil. The acid solution emanating from the particles of monobasic calcium phosphate in superphosphate or concentrated superphosphate dissolves some of the calcium carbonate in calcareous soils. The extra calcium in solution and the higher pH value result in almost immediate precipitation of the phosphate. Dibasic calcium phosphate is the principal reaction product. Dibasic calcium phosphate is far less soluble than the monobasic calcium phosphate in the fertilizer, but it is still a good source of phosphorus for plants. Because of the very rapid reaction of the soluble phosphorus with the soil, only a small sphere of fertilizer-affected soil forms around the site of each granule. The total increase in the amount of dissolved orthophosphate in the soil is small because of the low solubility of dibasic calcium phosphate. For the same reason, the reversion of the phosphorus to less soluble forms is slow.

In Smith's (1956) experimental work, the cultivation done in preparation for the second crop stirred the soil and dispersed the very small dibasic calcium phosphate crystals more widely, so that more of the phosphorus dissolved. The same process was repeated before the third crop. In time, of course, the removal of phosphorus by the crops and the reversion of the phosphorus to forms less soluble than dibasic calcium phosphate would reduce the residual effect. Smith did not continue the experiment long enough to observe the turn-around when the effectiveness began to decrease.

Devine et al. (1968) obtained results consistent with those of Smith in an experiment in which dibasic calcium phosphate was applied in powdered and granular forms to three soils (two acid and one calcareous), and the relative effectiveness of the two sources was estimated on the basis of the uptake of phosphorus by a test crop of barley. Their data showed that the availability coefficient of the phosphorus in the granular product was much below that of the powdered product in the year the fertilizers were applied, a little below in the first residual year, and a little above in the second and third residual years. This experiment also involved mixing the soil in each culture each year. The granulated product consisted of the powdered product that was bound together by sucrose, which means that the individual crystals composing the granules would be subject to dispersion when the soil was mixed. The results are found in Fig. 6-4, introduced later in another connection.

The scientific literature on the behavior of phosphorus in soils emphasizes the reactions of soluble orthophosphate with soil constituents, commonly termed phosphorus fixation. The reactions are rapid at first and become progressively

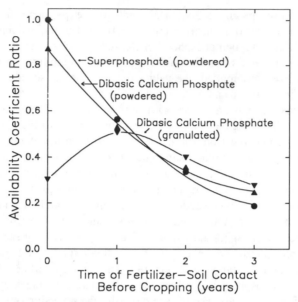

Fig. 6-4.   Change in availability coefficient of the phosphorus in three fertilizers over 3 years in the absence of crops, as assessed at the end of the experiment by a test crop of barley fertilized with different quantities of phosphorus as powdered superphosphate, with the phosphorus in this source assigned an availability coefficient of 1.00. The data are average values for three soils in the United Kingdom. (Devine et al., 1968)

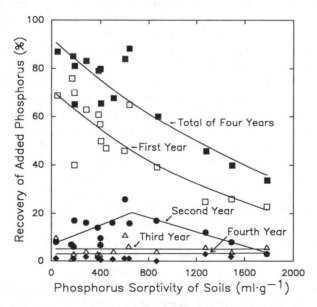

Fig. 6-5.   Recovery by ladino clover of phosphorus added to 15 soils in a greenhouse experiment versus the phosphorus sorptivity of the soils. The phosphorus was added as monobasic potassium phosphate at the beginning of the experiment, but the clover was grown in 4 successive years. (Holford, 1982)

slower as time goes on. These reactions decrease the availability of the phosphorus to plants (Suntheim et al., 1987), but it is questionable whether the phosphorus ever becomes completely unavailable.

6-4. Soil Effects

The effective amount remaining from previous applications of a nutrient represents the combined consequences of prior additions to and removals from the supply of nutrients in the soil, together with changes resulting from the processes the nutrients undergo in the soil. The latter may include reactions with the soil, changes in location, and volatilization. Considered in this section are biological evaluations of soil effects on the residual value of nutrients.

Devine et al. (1968) described a technique to evaluate the changes in availability of fertilizer phosphorus that take place over time in the absence of removal of phosphorus by plants. They added an equal quantity of various sources of phosphorus to quantities of three soils 0, 1, 2, and 3 years before a test crop of barley was planted in outdoor cultures. At planting, different quantities of powdered superphosphate were mixed with separate cultures to provide a response curve. Thus, the test crop was grown concurrently on all cultures at the end of several different lengths of time during which interactions of the fertilizers with the soils could proceed. This experiment was a truncated version of an ideal design in which addition of several quantities of each source would provide a response curve for each source.

Some of the data obtained in the experiment by Devine et al. (1968) are shown graphically in Fig. 6-4. The findings provide understanding about soil processes uncomplicated by the absorption of phosphorus by plants and the recycling that occurs with each crop when part of the absorbed phosphorus is returned to the soil in crop residues.

Widjaja-Adhi et al. (1985) used a similar technique in field experiments in which different quantities of phosphorus as concentrated superphosphate were allowed to react with soil for 8 months in Hawaii and 15 months in Indonesia. The residual value of the previously applied fertilizer phosphorus then was evaluated by comparing it with the response curve obtained with the same fertilizer freshly applied for a test crop.

Holford (1982) reported the results of an experiment in which the residual effect of the phosphorus supplied by monobasic potassium phosphate was followed for 4 years in 15 soils in a greenhouse. Bringing the soils together and carrying out the work under similar environmental conditions permitted improved precision in relating the results to the reduction in availability of the fertilizer phosphorus due to reaction with the soils, which was an important objective of the investigation.

Differences among soils in the extent of reduction in availability due to interactions with the soil are illustrated in Fig. 6-5 by plots of Holford's (1982) data on the percentage recovery of the added phosphorus against the phosphorus sorptivity of the soils. (See Section 1-3.2, footnote 3, for an explanation of the units for phosphorus sorptivity.) Note that the overall recovery and the recovery in the first year decreased with increasing sorptivity, but that in the third and fourth years the recovery was very low and not affected by sorptivity. In the second year, the recovery actually increased to some extent through part of the sorptivity range, apparently as a consequence of the relatively high recovery of added phosphorus from soils with the lowest sorptivity in the first year.

In the experiment in Fig. 6-5, the recovery of added phosphorus in the crop was unusually high. The high recovery may be attributed to a combination of conditions, including the low sorptivity of some of the soils, applying the mono-basic potassium phosphate in solution to the surface of the soil (which would result in placement near the surface), restriction of the roots to a small volume of soil, and growth of the crop in a greenhouse with good watering. Recoveries are lower under field conditions.

For phosphorus, retention of the added source by the soil, with a reduction in availability and a long-lasting residual effect, is the norm. With nitrogen, quite a different effect has been observed under some circumstances.

The accumulation of a surface mat of partially decayed organic matter is a characteristic of many soils developed under forest vegetation, particularly in colder areas. As the mats develop, they contain increasing amounts of nutrients and become a principal source of nutrients for the trees. In 1935, plots in the Black Rock Forest in New York received a one-time application of 0, 323, 645, and 968 kilograms of fertilizer nitrogen per hectare in an experiment on the effects of nitrogen fertilization (Mitchell and Chandler, 1939). Cooperating in this work was R. F. Finn (now deceased), who reported in a personal communication that the organic mat on the surface of the soil gradually disappeared under the influence of nitrogen fertilization.

In a more quantitative report, Turner (1977) found that in the first year after application of urea to a stand of Douglas fir in Washington, the biomass on the forest floor decreased 7% on the control, 18% with 220 kilograms of nitrogen per hectare, and 32% with 880 kilograms of nitrogen per hectare. The enhanced rate of decomposition would increase the seeming residual effect of the fertilizer nitrogen by augmenting it with soil nitrogen mineralized as a consequence of adding the fertilizer nitrogen. Along with the nitrogen would be released some phosphorus and other nutrients contained in the organic layer.

The residual effect of fertilization is distributed differently with Douglas fir and some other trees than with annual plants. A response in wood growth may be detected in the first year after nitrogen fertilization, according to Gessel et al. (1965), but greater responses may be expected in the second and third years. This effect may result in part from the stimulated release of nutrients from the organic mat on the surface of the soil, but the main factor is probably the fact that the potential for the annual growth flush is determined mostly by the number of cells present in the buds from which the new growth occurs. Improved growth conditions in a given year increase the number of cells laid down in the buds, and this translates into greater potential for growth from the buds in the following year.

6-5. Crop Effects

For nutrients in general, crops influence the residual value by absorbing some of the nutrient applied. The proportion of the added nutrient recovered in the crop generally decreases with an increase in the quantity applied. Differences among soils in reaction with the soil or loss from the soil influence the percentage recovery by the crop. The crop residues return to the soil a portion of the nutrient absorbed, and release of the nutrient from the residues constitutes a portion of the residual effect.

Fig. 6-6 illustrates the crop residue effect in an experiment on residual effects of phosphorus fertilization. As a result of continued reaction of the fertilizer phosphorus with the soil and incomplete release of the residue phosphorus, the residual value of the phosphorus declined rapidly even where the ladino clover herbage was returned to the soil. In this experiment, the only phosphorus removed

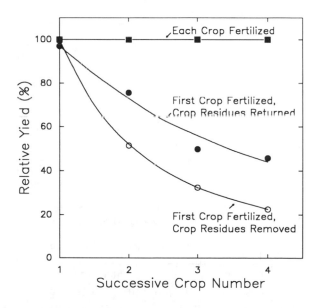

Fig. 6-6. Residual effect of phosphorus fertilization on four successive crops of ladino clover in a greenhouse experiment on a strongly weathered soil from Australia showing the effect of crop residue return and removal. The yields of the crops that received soluble phosphate fertilizer before the first crop only are expressed as percentages of the yields obtained when a quantity of fertilizer equivalent to that producing the maximum yield in the first year was added in each succeeding year to additional replicates. (Holford, 1982)

from the soil in the "crop residues returned" treatment was the small amounts in the herbage samples required to determine the phosphorus content.

The importance attached to modification of the residual effect brought about by crops is greatest with nitrogen. In addition to the universal crop effect of reducing the supply of soil nitrogen residual from fertilization by absorption, crops have other effects. Residues of some crops temporarily reduce the residual availability by inducing the immobilization of mineral nitrogen into organic forms by microorganisms during decomposition, whereas residues of others augment the supply of mineral nitrogen, acting as fertilizers. Crop residues may also contain nitrate, which serves directly as a fertilizer.

Illustrating the immobilization effect, Smith et al. (1973) found using biological assay procedures in field experiments in Idaho that when sugarbeet was grown following wheat, beet yields increased with nitrogen fertilization, but decreased with straw application. The deleterious effect of the straw on the beet crop could be compensated by adding an extra 7.5 kilograms of fertilizer nitrogen per megagram of straw.

Illustrating the mineralization effect, Fig. 6-7 shows that the yield of nitrogen in the grain of two successive wheat crops was lower following ten different nonleguminous crops than following eight different leguminous crops (points

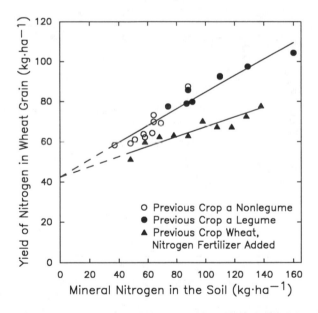

Fig. 6-7.   Total yields of nitrogen in wheat grain in two seasons following (a) wheat and 17 other crops and (b) wheat, but with different quantities of fertilizer nitrogen. The yields of nitrogen are plotted against the total amounts of nitrogen present in the soil as ammonium and nitrate to a depth of 1.2 meters before the first crop of wheat was planted. The experiment was conducted under relatively dry conditions in Australia. (Strong et al., 1986)

for two of the leguminous crops are superposed). The overall crop effect could be expressed approximately as a function of the mineral nitrogen found by analysis in the soil before the first crop of wheat was planted. The mineral nitrogen found at that time did not represent the whole of the residual value, as suggested by the difference in slopes of the lines representing the effect of the crops and the effect of fertilizer nitrogen, but the differential effects of the preceding crops appeared to be essentially dissipated during the growth of the two wheat crops. After the first crop, the soil mineral nitrogen ranged from 20 to 113 kilograms per hectare, with all crops falling in the range from 20 to 58 kilograms except for the legume showing the highest residual nitrogen before the first crop. After the second crop, the total range in mineral nitrogen was from only 17 to 28 kilograms. This experiment was conducted under semiarid conditions in Queensland.

A quite different analysis of crop effects, again focused on nitrogen, was made by Shrader et al. (1966). Their objective was to test the hypothesis that the differences in corn yields among the various crop sequences in a crop rotation experiment in Iowa could be attributed to differences in the supply of nitrogen in the soil. In this experiment, each corn crop received a range of quantities of fertilizer nitrogen, so that a response curve was available to estimate the supply of nitrogen in the soil in the absence of fertilizer nitrogen (the Mitscherlich b value — see Section 1-2.1).

The statistical model fitted to the data involved the assumption that the corn in each of the crop sequences was responding to the sum of the soil nitrogen and the fertilizer nitrogen, that the horizontal location of the individual response curves on the overall response curve depended upon the supply of nitrogen in the soil as affected by the preceding crops, and that the responses fitted the Mitscherlich equation. That is, the b value depended upon the cropping sequence, but the maximum yield A and the Mitscherlich nutrient efficiency factor c were the same for all crop sequences.

The overall response curve for the years 1958 to 1964 is shown in Fig. 6-8A. Statistically, no significant deviation from the model was found, which means that the hypothesis about control of yields by the nitrogen supply was not disproved and also that deviations from the Mitscherlich equation were not significant. As might be expected, continuous corn without supplemental fertilizer nitrogen produced the lowest yield and had the lowest supply of soil nitrogen ($b = 50$ kg·ha^{-1}), and corn following a year of an established legume meadow crop produced the highest yield and had the highest supply of soil nitrogen ($b = 327$ kg·ha^{-1}).

The data in Fig. 6-8A were obtained in the early years of the rotation experiment. Fig. 6-8B shows that after some 25 years the maximum yield level was higher, and the scatter of points at the high yield levels was greater than that observed in the early years. In Fig. 6-8B, the yields of the control plots without fertilizer nitrogen for each rotation were merely spotted on the response curve for continuous corn, as extended to fit the data for corn following legume

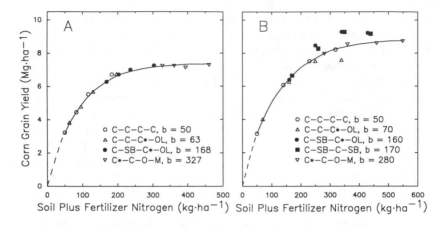

Fig. 6-8. Residual effect of crop rotations on yields of corn in an experiment in Iowa. A. Data obtained from 1958 to 1964 during the early years of the experiment. Each corn crop received 0, 34, 67, and 134 kilograms of fertilizer nitrogen per hectare. A Mitscherlich-type equation was fitted to the yield data on the assumption that a single response function would fit all the data when the statistically estimated b values representing the supply of nitrogen in the unfertilized plots were added to the quantities of fertilizer nitrogen. The b values in the legend are the effective quantities of nitrogen supplied by the soil as estimated from the Mitscherlich equation. The capital letters in the legend refer to corn (C), oat with mixed forage legume seeding (OL), soybean (SB), and meadow (M, consisting of mixed forage legumes). In instances in which data for some of the corn crops were not used, an asterisk indicates the corn crop in the rotation from which yield data were taken. (Shrader et al., 1966) B. Data obtained from 1984 to 1989. The additions of fertilizer nitrogen had been increased to 0, 90, 179, and 269 kilograms per hectare. The corn-soybean rotation was a continuation of a previous corn-soybean-corn-oat (legume) rotation. In *B*, the control yield for each rotation was spotted on the response curve for continuous corn, and the corresponding quantities of nitrogen were taken as the b values. (Duffy and Webb, 1990)

meadow. The results indicate that the efficiency of fertilizer nitrogen and the maximum yield were greater for the two rotations including soybean than for the others. Data for these rotations are distinguished in the figure by the use of solid points. The corn-soybean-corn*-oat (legume) rotation (filled circles) was a continuation of the rotation for which data are given in Fig. 6-8A. (The asterisk designates the corn crop to which the yield data correspond.) The corn-soybean-corn-soybean rotation (filled squares) in Fig. 6-8B, however, was originally a corn-soybean-corn-oat (legume) rotation that was modified by substituting soybean for the oat (legume) crop.

Compaction of the soil during the years and its reversal by the well known soil-loosening effect of the soybean crop has been suggested as an explanation for the favorable response of corn to fertilizer nitrogen in the corn-soybean rotation. The soil-loosening effect of the soybean crop could be classed as an indirect effect that modifies the behavior of a succeeding crop but is not a residual effect of the nutrient nitrogen. See a paper by Hanson et al. (1988) for experimental work on grain sorghum grown continuously and in rotation with soybean on claypan soils in Missouri.

Various kinds of crop-related indirect effects may exist that modify the behavior of a test crop and cause difficulty in interpretation. Greater use of water and other nutrients during previous crop growth on fertilized plots may result in a downward bias of the response to the residual nutrient in the year it is compared with the freshly applied nutrient. Such indirect effects are responsible for a portion of the responses used to evaluate the direct effects of the supplies of the nutrient remaining in the soil from the prior application or applications. They may be expected to have a greater effect on crop yields than on uptake of the nutrient being studied.

Interactions of crop effects and fertilizer effects are of special importance in some instances. Experiments by Holford and Crocker (1991) on phosphorus fertilization of clover-based pastures in Queensland are an example of one such interaction. They found that the percentage recovery of the fertilizer phosphorus in the herbage decreased more rapidly in succeeding years after fertilization than did the response in terms of yield of herbage. They attributed this difference to the effect of phosphorus in enhancing nitrogen fixation by the clover and the effect of the extra nitrogen in increasing the growth of grass, which responds strongly to nitrogen but is less sensitive to phosphorus deficiency than the clover species present in the sward.

A quite different type of interaction was described by Nelson et al. (1946). They found that where an oat-legume hay seeding was fertilized with nitrogen, the oat yield increased, but the yield of legume hay in the following year usually decreased. The decrease in hay yield was thought to be a consequence of an indirect effect of nitrogen fertilization, namely, a depression in growth of the hay seeding understory as a result of competition with the oat plants (and weeds) that responded better to the nitrogen fertilizer and made more rapid growth than the seeding. In nine field experiments, the depression in yield of legume hay with application of nitrogen to the oat crop tended to increase with the response of the oat crop to the nitrogen fertilizer. A similar depression of yields of alfalfa planted with wheat fertilized with nitrogen was reported by Bittman et al. (1991) in northeastern Saskatchewan.

A third type of interaction was described by Jourdan and Villemin (1990). In a factorial field experiment in France, they found that the exchangeable soil potassium residual from applications of different quantities of fertilizer potassium over a period of 21 years was greatly reduced by nitrogen fertilization. See Fig. 6-9. Chemical interactions may contribute to some effects of this type, but in the instance in question the predominant influence was probably the uptake of additional potassium by the crop when more nitrogen was supplied.

In general, the cause of indirect effects can only be surmised in the absence of special experiments. Theoretical difficulties notwithstanding, residual effects of nutrients are of practical importance in crop rotations, but the differences in growth usually are recognized as a combination of the consequences of residual effects, differences in responsiveness of individual crops to the residual nutrient, and whatever indirect effects may have resulted from the response of the preceding crop or crops to the nutrient additions.

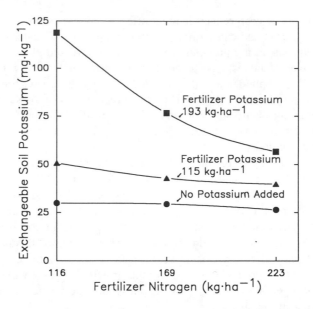

Fig. 6-9. Residual exchangeable potassium in a calcareous silt loam soil in France after 21 years of fertilization. The applications of fertilizer nitrogen and potassium varied with the crop, and the values given are the average annual applications. (Jourdan and Villemin, 1990, with additional information supplied by the senior author)

The experimental design described in Section 6-3 for estimating residual effects approaches the ideal design used in fertilizer evaluation, in which quantities of two or more sources of a common nutrient are applied concurrently to field plots that have not been differentiated by previous treatments. One deficiency remains. Although the control plots for the freshly applied fertilizer will be comparable to the control plots for the residual fertilizer, the plots that did receive the residual fertilizer may have been changed in respects other than the nutrient of concern as a result of the differences in growth of the test crop produced by the fertilizer in previous years.

Indirect effects associated with crops do not disappear in experiments in which the same crop is used in all years. Nonetheless, the fact that the indirect effects are not evaluated independently is justification for using the same crop in all years of an experiment when the objective is to evaluate the value of a residual nutrient in terms of the effect of the same nutrient applied currently. When the same crop is used, whatever bias from indirect effects does exist is at least related to the effects of a single crop.

6-6. Fertilizer Effects

Under practical conditions, the residual value of applied fertilizers is invariably influenced by crops, which remove some of the nutrients supplied. Some effects, however, are described more appropriately as fertilizer effects.

The residual value may differ from one source of a nutrient to another. The initial value of the phosphorus in superphosphate is considerably greater than that in most phosphate rocks, but the residual value declines much more rapidly. Figs. 6-2 and 6-3 illustrate the general behavior that has been observed. See a paper by Bolland et al. (1988) for a review.

Although interaction with crops and soils is involved, another situation that may described as a fertilizer effect is differences in residual value due to fertilizer placement. For example, Sander et al. (1990) found in an experiment in Nebraska that placement of fertilizer phosphorus for winter wheat affected the yield and phosphorus content of the subsequent crop of grain sorghum. Fertilizer supplying 33 kilograms of phosphorus as water-soluble polyphosphate per hectare that was "knifed" into the soil for winter wheat below the depth of tillage increased the yield of sorghum grain by about 50%. Equal quantities of fertilizer phosphorus applied to the wheat broadcast or with the seed increased the yield of sorghum grain by only 10%. For the wheat crop, to which the fertilizer was applied, placement with the seed was the most advantageous of the placements tested. Knife placement below tillage depth was a close second, and broadcasting was a poor third. Between the wheat crop and the sorghum crop, the soil was disked twice and cultivated once. These tillage treatments would have mixed the broadcast and seed applications with the soil.

6-7. Cumulative Effects

Principal emphasis in this chapter is on evaluating the residual effectiveness of a single application of fertilizer made at some time in the past. Such assays are valuable for advisory and compensatory purposes, but they differ from the usual practical situation in which repeated applications have been made.

If the quantities of nutrients added are great enough, the residual value of repeated applications shows a cumulative effect over time. In developed countries, the usual situation has been an accumulation of soil phosphorus as a residual effect of applications exceeding removals. For example, at an annual average application of about 11 kilograms of fertilizer phosphorus per hectare, Ireland is below the median for countries in the European Community, but Tunney (1990) estimated that the input of phosphorus to the soils in 1988 still exceeded the output of phosphorus by 47%. Concurrent with the accumulation of residual phosphorus has been an increase in soil phosphorus extractable by the ammonium acetate-acetic acid reagent used in Ireland to provide an index of phosphorus availability. The average extractable phosphorus in about 80,000 samples tested per year increased almost linearly from about 1.3 milligrams per kilogram of soil in 1955 to 8 milligrams in 1985.

Conversely, when the applications are discontinued, the accumulation ceases, and the residual value declines. Fig. 6-10 illustrates both the increasing and declining phases with data on phosphorus from field experiments in Alabama. Fig. 6-11 is one of the classic examples of residual effects resulting from the

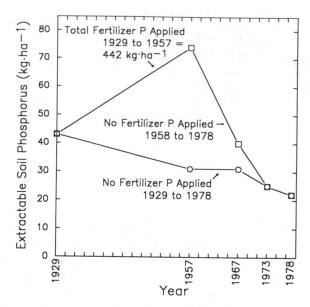

Fig. 6-10. Trends in extractable soil phosphorus in control plots and in plots receiving repeated applications of fertilizer phosphorus from 1929 through 1957 and none from 1958 through 1978. The phosphorus extractant was 0.05 molar hydrochloric acid and 0.0125 molar sulfuric acid. The data are average values from identical field experiments at six locations in Alabama. (Cope, 1981)

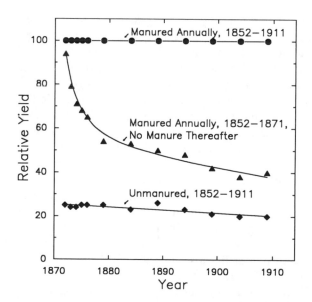

Fig. 6-11. Residual effect of manure on barley at the Rothamsted Experimental Station in the United Kingdom. The treatments were 31.4 megagrams of farmyard manure per hectare annually from 1852 through 1911, 31.4 megagrams annually from 1852 through 1871, and no manure from 1852 through 1911. (Hall, 1917)

accumulation of nutrients in the soil with repeated applications. In this experiment, 20 annual applications of 31.4 megagrams of farmyard manure per hectare were still doubling the yield of barley 40 years later.

The cumulative effect of repeated additions of a fertilizer has perhaps three significant consequences from the theoretical standpoint. First, the residual effect observed can be attributed only to the aggregate of the past treatments. Second, when different quantities of a fertilizer are applied annually or at other intervals to generate a response curve, the responses to applications after the first are responses to the quantities applied currently plus the cumulative and indirect effects of prior additions, and not responses to the quantities applied currently. And third, when the schedule of repeated applications is suspended on a given set of plots and continued on another for evaluation purposes, one is measuring a cumulative effect of repeated applications in the past against a cumulative effect plus a current effect. If the plots receiving continued additions have received more than one periodic addition of fertilizer after suspension of additions to the residual plots, the cumulative component will no longer be the same in the two sets of plots.

Cumulative effects do not invalidate the use of biological assay principles in evaluation, but the interpretations need to be consistent with the consequences mentioned in the preceding paragraph. Illustrating the second consequence of cumulative effects, Fig. 6-12 shows the yields of potato tubers with different quantities of fertilizer potassium applied annually in a 4-year rotation. Superimposed on the response curve are the yields with 62.5 kilograms of potassium per hectare applied at 3-year intervals 0, 1, and 2 years previously. This application is numerically equivalent to 20.8 kilograms annually. The results indicate, though, that the triennial application of 62.5 kilograms produced a yield equivalent to only 35 kilograms of fertilizer potassium applied annually even when the 62.5-kilogram application was made directly to the crop whose yield was measured. The difference in tuber yield between 35 and 62.5 presumably reflects an accumulation of residual potassium with annual additions. The plots receiving 62.5 kilograms annually had actually received three times that amount in the 3-year cycle. In this experiment, the mean effective quantity of annually applied potassium corresponding to the triennial additions of 62.5 kilograms 0, 1, and 2 years previously was 17.5 kilograms, as judged from the crop-producing effect, which is close to the average annual application of 20.8 kilograms. These results indicate that the crop-producing effect of the potassium applied once in 3 years was not much less than the effect when the potassium was applied annually.

6-8. Evaluating Accumulated Residues

The concepts of fertilizer evaluation may be applied directly to the evaluation of accumulated residues by soil analysis. If a method of analyzing soil can be found that reflects the relative availabilities to plants of current applications of a standard fertilizer and the residues of a nutrient applied previously in fertilizers

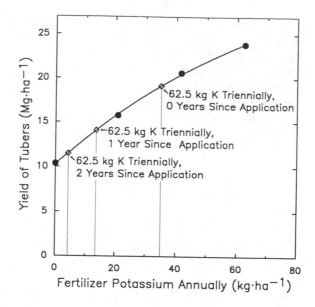

Fig. 6-12.   Residual effect of fertilizer potassium on yields of potato tubers in three experiments over two 4-year periods in the United Kingdom. The response curve shows yields with annual applications of the quantities indicated on the *X* axis. Spotted on the curve are the yields obtained with triennial applications of 62.5 kilograms of fertilizer potassium per hectare made 0, 1, and 2 years before the current crop. (Russell and Batey, 1971)

or manures, the values obtained from analyses made on soils by this method can be used to estimate the availability coefficient ratios of the nutrient supplied by the various sources. Such use of the method depends upon prior validation by biological assays.

The most widely used and most successful evaluations of accumulated residues are those for nitrate. The use of such evaluations has increased greatly as a consequence of the sometimes considerable residual effect of previously applied nitrogen fertilizers. In earlier years, when applications of fertilizer nitrogen were relatively small, emphasis in soil testing for nitrogen was on the release of mineral nitrogen from organic forms by mineralization because this was generally the major source. Now the much greater quantities of fertilizer nitrogen applied have greatly increased the importance of residual nitrate (Hergert, 1987). Soil nitrate can be determined readily, and it is absorbed readily by plants. Now the principal questions are where is the nitrate in the soil profile, what is the relative value of nitrate found at different depths, and how much of the nitrate is lost before it can be absorbed by crops? These matters are considered in Chapter 7 on fertilizer placement.

The isotope dilution method discussed in Section 4-2.1.3 may be adapted for evaluating the residual effect of previously added sources of a nutrient. This is done by equilibrating a known quantity of a tagged source of the nutrient with samples of soil that have received one or more additions of the untagged source of the nutrient in the past and with the control soil that has not received the untagged source. The quantity of the nutrient in the soil with which the tagged source equilibrates can be calculated from the quantity and specific activity of the tagged source and the specific activity of the nutrient in solution in equilibrium with the soil or in plants that have grown on the soil and absorbed some of the tagged and untagged forms.

Morel and Fardeau (1991) used the isotope dilution method to find the relative residual effects of concentrated superphosphate and phosphate rock that had been applied annually in equal quantities (presumably equal quantities of phosphorus) to a soil from France for 15 years. They did not give the quantity of soluble tagged phosphorus they had added, and so the increases in residual quantities of labile phosphorus in the soils from the two fertilizers could not be calculated. Instead, they gave the percentages of the phosphorus in plants grown on the soils and in solution in equilibrium with the soils that were derived from the residual fertilizer phosphorus. The increases due to the applied fertilizer were obtained by subtracting the corresponding values found for the control soil. In their experiment, they found that the percentages of the phosphorus derived from the residual labile fertilizer phosphorus in the soil were 67 from superphosphate and 4.7 from phosphate rock where the measurements were made on plants and 56 and 5.5 where the measurements were made on solutions in equilibrium with the soil. The residual value of the phosphorus of phosphate rock relative to that of superphosphate thus was $4.7/67 = 0.07$ as estimated from the measurements on plants and $5.5/56 = 0.10$ from the measurements on the soils.

The experiment by Morel and Fardeau (1991) described in the preceding paragraph involved samples of soil taken from the field. The measurements leading to the percentages of the phosphorus in the plants or in solution that were derived from the residual fertilizers were made after thorough mixing of the tagged phosphorus with the soil. Although the method described would be expected to produce ratios of residual effects that would correlate well with the corresponding ratios calculated from yield response curves or nutrient uptake curves by biological assay methods within individual soils, one must bear in mind that, as may be inferred from Section 4-2.1.3 and as illustrated in Fig. 4-6, the absolute values are related to quantities, and the availability of a given quantity may differ considerably from one soil to another.

Expressing dissatisfaction with the biological evaluation of residual effects of phosphorus fertilization because of seasonal variability in experiments in which the effects in residual years were not evaluated concurrently with those of freshly applied fertilizer, Colwell (1985a) suggested the use of soil analysis data as an alternative. He found that the phosphorus extracted by a modification of the 0.5 molar sodium bicarbonate method from samples of soil taken after the third crop in two 3-year experiments on phosphorus fertilization in Australia could be represented by a regression equation of the form

$$P_t = P_0 + b(P_c + P_{r1} + P_{r2}),$$

where P_t = total phosphorus extracted, P_0 = phosphorus extracted from the unfertilized control, P_c = fertilizer phosphorus added in the current year, P_{r1} = fertilizer phosphorus added in the first residual year, P_{r2} = fertilizer phosphorus added in the second residual year, and b is a regression coefficient. The model yielded an R^2 value of 0.74 for one experiment and 0.61 for the other.

The foregoing formulation implies that under the conditions of the experiments, the increase in extractable phosphorus depended upon the total amount of fertilizer phosphorus applied, but that when it was applied and how much of the applied phosphorus had been removed in the crops were not of importance. A more realistic model for estimating the total extractable phosphorus would provide for effects of quantity of fertilizer phosphorus applied, the year in which it was applied, and the quantity recovered in the grain. As a first approximation, these factors could be expressed in linear form. As a second approximation, curvilinearity could be investigated because it is known that with the weak extractants normally used to provide indexes of soil phosphorus availability, the regression coefficients for added phosphorus would increase with the quantity of phosphorus added (see Fig. 1-12). Colwell tried these modifications, but in his experiments the benefit of the refinements was apparently small enough and the experimental error great enough that little or no improvement in fit resulted. Possibly contributory to the limited value of the more refined models was the fact that the fertilizer phosphorus was drilled with the seed, whereas the analyses for extractable phosphorus were made on mixed samples from the surface 10 centimeters of soil.

From the standpoint of residual effects, the primary objective in using chemical analysis of soils as a substitute for bioassay is to find a method that will provide a precise index of the effective quantities of the nutrient supplied by the currently applied and residual sources. Eik et al. (1961) investigated the suitability of the soil phosphorus extractability ratios found by each of three soil testing methods as indexes of the relative availabilities of the phosphorus residues in three soils from fertilizers that had been applied in field experiments in prior years. Quantities of the variously treated soils were brought together for cropping under uniform conditions in a greenhouse. The results showed that for each of the soil testing methods the ratios of extractable phosphorus in soils with the various residual phosphorus treatments were well correlated with the availability ratios found with the test crop, but none of the soil testing methods satisfied the ideal of a 1:1 relationship between extractable phosphorus ratios and availability ratios. Similar results were obtained by Bolland (1992) for three soils with a modification of one of the methods (0.5 molar sodium bicarbonate) used by Eik et al. (1961).

In a field experiment in South Africa, Farina and Channon (1987) found that to a good approximation the yields of corn grain could be expressed as a function of the phosphorus extracted from the soil, irrespective of whether the phosphorus was residual from earlier applications or whether the applications were continued to the year in which the soil analyses were made and the yields reported. Field plots with residual phosphorus produced lower yields of corn than did those on which the applications were continued. The data in Fig. 6-13A and B show that the relative extractabilities of residual and currently applied phosphorus were similar to the relative availabilities estimated from the grain yields. Once a soil analysis method has been found that will lead to results such as these, it is evidently preferable to evaluate residual effects by soil analysis and forego the extra time and expense connected with the use of biological assay in field experiments.

On the other hand, problems may be encountered where the soils, fertilizers, or fertilizer placement are sufficiently different. For example, Ghosh and Gilkes (1987) found in a greenhouse experiment on the residual value of superphosphate and different phosphate rocks that the crop following the one to which the phosphorus sources were applied responded to the residual phosphorus, and that the yield of dry matter versus yield of phosphorus plot showed no segregation of points by phosphorus sources. A given amount of phosphorus extractable by sodium bicarbonate solution, however, corresponded to higher yields with phosphate rock residues than with superphosphate residues. In this instance, the relative availabilities of the different sources to plants were not accurately reflected by the relative extractabilities.

Fig. 6-13C shows data from a field experiment using subterranean clover as the test crop. In this instance, the data segregated into two functions, one for superphosphate, and the other for two calcined phosphate rocks. The percentage of the maximum yield associated with a given amount of extractable phosphorus depended upon the fertilizer that supplied the phosphorus. Response curves for the percentage of the maximum yield versus the quantity of phosphorus supplied as the two calcined phosphate rocks could be compared to estimate the relative effectiveness of the phosphorus in the two sources, but a comparison of the phosphate rocks with superphosphate would not be appropriate.

Similar divergence of relationships of yield versus phosphorus extracted by the same sodium bicarbonate reagent was observed in a field experiment in New Zealand in which Rajan et al. (1991b) were comparing phosphate rock with monobasic calcium phosphate. In both instances, the results indicated that compared with the availability of the soluble phosphorus source, which should have reacted completely with the soil, the availability of the phosphorus added as phosphate rock was underestimated by the sodium bicarbonate extractant.

The availability of the phosphorus of phosphate rock to plants is low in neutral and alkaline soils, but increases as the soil pH drops, as was well illustrated by the data by Rajan et al. (1991a). The increase in availability with decreasing pH parallels an increase in solubility. Although the sodium bicarbonate reagent is one of the best for providing an index of phosphorus availability to plants in acid soils as well as in alkaline soils, the findings in the experiments in which phosphate rock was applied to acid soils thus suggest that the plant roots were interacting directly with the surfaces of the phosphate rock particles to obtain phosphorus in a way that was not simulated by the sodium bicarbonate reagent at pH 8.5.

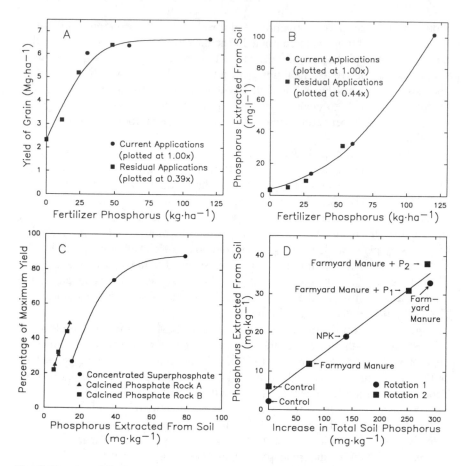

Fig. 6-13. A and B. Residual effects of concentrated superphosphate on corn in an experiment in South Africa, as estimated by yields of grain and by phosphorus removed from the soil by the Bray (0.025 molar HCl, 0.03 molar NH_4F) extractant. Residual plots received fertilizer phosphorus annually from 1972 through the 1980-1981 season and none thereafter. Annual additions were continued through 1983 to the plots with current applications. Yield and extractable phosphorus measurements were made for the 1983-1984 crop. (Farina and Channon, 1987) C. Percentage of maximum yield of subterranean clover versus soil phosphorus extracted by a modification of the Olsen et al. (1954) 0.5 molar sodium bicarbonate method in a field experiment in Australia with three phosphate fertilizers. Values shown were obtained in the year following application of the fertilizers. The yield produced in that year by a current application of 300 kilograms of phosphorus as superphosphate per hectare was taken as the maximum yield. Phosphate rock *A* was calcined at 500° C, and *B* was calcined at 900° C. (Bolland and Bowden, 1982). D. Phosphorus extracted from soil in a field experiment in the United Kingdom versus the increase in total soil phosphorus that had resulted from prior additions. In Rotation 1, the additions had been made annually by 13.4 megagrams of farmyard manure and 250 kilograms of superphosphate per hectare from 1899 to 1957. In Rotation 2, the additions had been made once in 4 years by 22.4 megagrams of farmyard manure, 941 kilograms of superphosphate (P_1), and 1882 kilograms of superphosphate (P_2) per hectare from 1899 to 1964, except for the P_2 treatment, which was discontinued in 1952. The soil phosphorus measurements were made on soil samples taken from Rotation 1 in 1957 and from Rotation 2 in 1964. (Mattingly et al., 1970)

A second possibly contributing factor may have been an interaction of placement of the fertilizers with the processes that went on during extraction of the fertilized soils with the sodium bicarbonate solution used. The fertilizers were broadcast on the surface. The layer of soil sampled for analysis would contain virtually all of the residual phosphorus, but the proportion of this layer occupied by fertilizer phosphorus at the time of soil sampling was undoubtedly greater for the soluble phosphates than for the phosphate rocks. The mixing of the soil within samples and the extraction by shaking the mixed soil with a volume of the extracting solution would result in some equilibration of the phosphorus from the fertilizers with the entire layer of soil sampled even though it had not been thus equilibrated in the field under the conditions of plant growth, and this would affect the amount of phosphorus found in the extract. Soil extractants give the most meaningful results when the soil being extracted is uniform. To provide a basis for estimating placement effects, the various phosphorus zones would need to be sampled and extracted individually.

Fig. 6-13D, derived from experimental work in the United Kingdom, shows a different application of soil analysis to investigation of residual effects. Here the extractable phosphorus is plotted against the increase in total soil phosphorus resulting from applying phosphorus in farmyard manure and superphosphate over periods of many years. Of special interest in this figure is the fact that the extractable phosphorus does not appear to depend upon whether the soil phosphorus had accumulated from applications of manure or inorganic fertilizer.

The observation in Fig. 6-13D for the phosphorus of farmyard manure does not necessarily apply to all organic materials or times of application. Johnston (1981) found that, of the phosphorus added to soil over a period of 18 years in a field experiment in the United Kingdom, 13.5% of the farmyard manure phosphorus was soluble in 0.5 molar sodium bicarbonate solution, but only 3.2% of the sewage sludge phosphorus was soluble. He attributed the difference mainly to the removal of water-soluble phosphorus at the sewage treatment plant. Sewage sludges may also contain substantial quantities of metals that react with phosphate to produce relatively insoluble compounds.

A second point that does not show in Fig. 6-13D is that the increase in total soil phosphorus was more highly correlated with the phosphorus extracted by the sodium bicarbonate solution than with the labile phosphorus by isotopic dilution or the phosphate potential ($0.5pCa + pH_2PO_4$). A third point is the marked increase in total soil phosphorus that had resulted from the long-continued fertilization. The phosphorus contents of the soil from the control plots without phosphate fertilization were 460 milligrams per kilogram in Rotation 1 and 434 milligrams per kilogram in Rotation 2. Most of the phosphorus added remains in the soil and contributes to increased phosphorus availability.

Further work on evaluation of accumulated residues is found in Section 6-10. Some of the models discussed there require chemical indexes of nutrient availability in soils.

6-9. Separating Nutrient Effects and Yield-Potential Effects

The best information on long-term residual effects has been obtained in the United Kingdom from field exeriments that were started in the mid 1800s. In 1970, a series of papers was published by research workers at the Rothamsted Experimental Station in England, summarizing special studies on the residual effect of applications of inorganic fertilizers and farmyard manure that had been made repeatedly over a period of years and then discontinued (see Johnston, 1970). The technique used was to superimpose small plots with different quantities of freshly applied fertilizers on the previously untreated and treated large plots. Various crops were grown.

Fig. 6-14 illustrates different extremes of the behavior observed. In Fig. 6-14A, the shapes of the response curves suggest that the difference between the responses obtained when superphosphate was freshly applied to land with and without residual phosphorus from earlier fertilization was a consequence of the residual phosphorus from the fertilizers applied many years previously. The control yield on the land with residues was equal to the yield obtained with approximately 35 kilograms of currently applied phosphorus per hectare.

In Fig. 6-14B, the shapes of the response curves indicate that the maximum yield with currently applied phosphorus was greater on previously fertilized land than on the previously unfertilized land. In other words, the previous fertilization

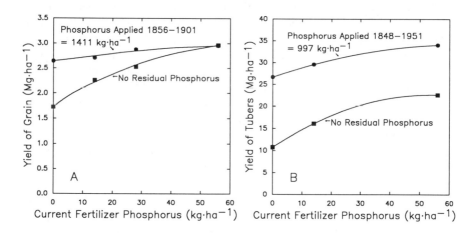

Fig. 6-14. Experiments on long-term residual effects of phosphorus in superphosphate. A. Average response of barley to freshly applied superphosphate in 1957 and 1958 on continuously cropped land to which no phosphorus had been applied since 1855 and on similarly treated land to which a total of 1411 kilograms of phosphorus as superphosphate (basic slag from 1897 to 1901) and 1568 kilograms of potassium as potassium sulfate had been applied per hectare from 1856 to 1901, with none thereafter. B. Results analogous to those in A, with potato as the test crop on another field in 1959 and 1960. The area with residues had received 997 kilograms of phosphorus as superphosphate and 3674 kilograms of potassium as potassium sulfate from 1848 to 1951 and none thereafter. In both A and B, a uniform application of fertilizer nitrogen and potassium was made to all the small plots from which the yield data were obtained. (Johnston and Warren, 1970; Johnston et al., 1970)

appears to have had some beneficial effect on the current yield that could not be duplicated by currently applied superphosphate.

Attempting to determine the nature of the beneficial effect that could not be attributed to current applications of superphosphate is the kind of problem that generally is avoided because of the improbability of obtaining definitive findings. Thinking apparently continued, however, and in 1983 a paper was published from Rothamsted by Dyke and coworkers (1983) in which response curves obtained under different conditions were shifted horizontally to estimate differences in the supply of nitrogen and vertically to estimate the effect of differences in other factors that would affect the maximum yield. The procedure was not described in detail, but the basic concept is clear from their Fig. 4. The horizontal axes of the individual response curves were shifted so that the maximum yields all corresponded to a given quantity of fertilizer nitrogen, and the vertical axes of the individual response curves were shifted so that the maximum yields were the same. The horizontal shift, for example, was used to show the magnitude of the residual effect of manure applications on nitrogen availability in terms of the effects of currently applied fertilizer nitrogen. In the example they showed, the double shift brought all the individual response curves together so they formed a single response curve to a good approximation. In general, how well a set of data could be represented by a single response curve would provide a measure of the success of the procedure and the experimental error in the individual sets of observations. The magnitude of the vertical adjustments required to equalize the maximum yields would provide an estimate of the effect of factors other than the nutrient under investigation.

Applying the double-shift technique, Johnston (1987, 1991) found evidence of an association of the magnitude of the vertical shifts with differences in organic matter content of the soils. His results were obtained in an analysis of data derived from long-term experiments in the United Kingdom in which phosphorus and potassium were in adequate supply and various quantities of nitrogen were applied. He noted that the organic matter effect appeared in data from recent years but not from earlier years. His interpretation was that the difference between the later and earlier years may be a consequence of the higher yield potential of current cultivars and the ability to protect this potential through improved control of pests and diseases. Further use of the double-shift technique for investigating the residual effects of ryegrass-clover meadows of different duration on the yields and nitrogen requirements of subsequent crops was described by Johnston et al. (1992).

Various response functions might be used to implement the double-shift technique, but the "exponential plus linear" function apparently was used in all the work described here. This function is a Mitscherlich equation (Section 1-2) modified by adding a linear term. The function requires fitting four parameters, as opposed to three in the Mitscherlich equation, but it is more realistic than the Mitscherlich equation in permitting a maximum yield with a finite application of a nutrient, followed by a decrease in yield.

The important point is that the double-shift concept provides an additional evaluation tool. It can be used not only in the study of residual effects but also in other ways for investigating differences in yield potential and their causes. Use of a horizontal shift by others has been illustrated in Figs. 4-3 and 6-8, but combining the horizontal shift with a vertical shift apparently was original with Dyke et al. (1983).

6-10. Models

A number of contributions have been made to the mathematical modeling of residual effects, almost all directed primarily at phosphorus because of the well known residual effects and the relative abundance of experimental data. The models range from the simple to the complex.

6-10.1. Half-Life Models

Several models may be traced back to research by Larsen et al. (1965) and a theoretical paper by Larsen and Probert (1968) in the United Kingdom. In a uniform experiment conducted at 24 sites, these investigators took samples of the soils and measured the L value for labile (isotopically exchangeable) soil phosphorus by growing plants following application of different quantities of superphosphate in the field. They subtracted the control L values from those obtained where superphosphate had been added, and then plotted the logarithm of the increases in L values against time, representing the plots as linear decreases. They found in this way that the time required for the increases in labile phosphorus to drop to half the initial value (the half-life) ranged in the different soils from 1.1 to 56 years, with all but one of the values equal to 6 years or less. The half-life of 56 years was found in a peat soil.

Probert and Larsen (1972) pointed out that the process of loss of availability of fertilizer phosphorus upon application to soil is more complex than a single first-order reaction, which is in effect what Larsen et al. (1965) fitted to their field experimental data. Nonetheless, in situations such as estimating residual effects, in which factors unaccounted for are admittedly at work, simplicity has its merits.

The concept of an exponential decrease in effectiveness of fertilizer phosphorus with time was adopted by Cox et al. (1981) in a model describing the overall downward trend of extractable soil phosphorus toward a limiting minimum in successive years following application of a phosphate fertilizer. These investigators assumed that the rate of change in the level of an extractable nutrient with time after the initial rapid equilibration with the soil is proportional to the difference between the extractable nutrient level and an equilibrium level:

$$\frac{dx}{dt} = -k(x - x_{equilibrium}).$$

In this equation, x is the extractable nutrient level in kilograms per hectare, t is the time in years, and k is a loss constant. The authors integrated the equation on the assumption that when $t = 0$, $x = x_{initial}$, where $x_{initial}$ is the extractable nutrient level after the initial rapid sorption, obtaining

$$x = x_{equilibrium} + (x_{initial} - x_{equilibrium}) \cdot e^{-kt}.$$

The half-life concept introduced by Larsen et al. and employed by Cox et al. is analogous to the familiar Mitscherlich plant yield equation in reverse. Instead of a yield that trends upward to a maximum value as the supply of the nutrient increases indefinitely, with one Baule unit of the nutrient increasing the yield to half of the difference between the current yield and the maximum, the effectiveness of an added nutrient trends downward to an equilibrium value at infinite time. The time required for the fertilizer value at a given moment to decrease to half its initial value is the half-life.

Probert (1985) also used the theory of an exponential decrease of effective soil phosphorus toward a limiting minimum, but he did not use experimental indexes of the effective phosphorus because the data he analyzed did not include such information. Probert introduced two other variables. One was the quantities of phosphorus removed by succeeding crops, for which he did have data. The second was the gradual release of phosphorus from a slowly available source (phosphate rock), which could account in part for a gradual increase in the effectiveness of the slowly available source relative to the standard source (superphosphate in the data he analyzed) in consecutive residual years. He had no data on the release of phosphorus from the phosphate rock included in the experiment providing the data he analyzed, but estimated it at 23% per year from the response data. Gonçalves et al. (1989) fitted a different exponential equation to data on extractable phosphorus and uptake of phosphorus by plants. Ackello-Ogutu et al. (1983) proposed a more complex econometric application of the half-life concept to soil tests.

6-10.2. Middleton Model

In contrast to the use of a Mitscherlich-type equation to represent the half-life concept described in Section 6-10.1, Middleton (1983b) employed a modified form of the Mitscherlich equation to express the increase in residual effect from repeated applications of nutrients. Middleton's equation was

$$y_{ij} = k_r A_1 [1 - R_1^{(a_i + b_i x_j)}],$$

where y_{ij} = crop yield in the i'th year with the j'th quantity of fertilizer (x_j); A_1 = the maximum yield in a "completely upgraded system" (presumably one in which the sum of the current addition and the residual effects of prior additions produces the maximum yield) under the most favorable environmental conditions; $R_1 = \exp - c_1$, which is a measure of the effectiveness of the fertilizer in increasing the yield in a completely upgraded system where the maximum yield is A_1 (c_1 here is the natural logarithmic equivalent of c in the regular Mitscherlich equation, which uses logarithms to the base 10); a_i = the residual effect of applications made in preceding years, measured in the same units as x; and b_i = a measure of the effectiveness of fertilizer applied in the i'th year. The parameter k_i adjusts the annual crop yields for differences from year to year; k_i = y_{io}/y_{no}, where y_{io} is the yield with no fertilizer in the i'th year and y_{no} is the yield with no fertilizer in the n'th year when upgrading is complete.

Middleton applied his model to the results of a 7-year experiment in New Zealand in which a primitive pasture was upgraded by annual applications of 0, 250, 500, and 1000 kilograms of superphosphate per hectare, with a blanket application of potassium chloride. Yields were measured annually, and cut herbage was returned to the soil to simulate a pasture system with the major part of the nutrients returned to the soil in the animal droppings. He used values of A_1 = 17.0 and R_1 = 0.35, presumably obtained by iterative procedures. With these numerical values, together with annual values for k_i calculated from the yield data using $n = 7$, he was able to estimate a_i and b_i by rearranging the foregoing equation and taking the natural logarithm,

$$\text{Ln}\frac{\left[1 - \dfrac{y_{ij}}{k_i A_1}\right]}{\text{Ln} R_1} = a_i + b_i x_j,$$

to obtain a linear equation. He used the data for the three quantities of superphosphate in each year to estimate a_i and b_i for the individual years. His final equation for estimating the individual yields was then

$$y_{ij} = k_i \cdot 17.0[1 - 0.35^{(a_i + b_i x_j)}],$$

where values of k_i, a_i, and b_i, as calculated for individual years, together with the known values of x_j, could be substituted to estimate the individual y_{ij} values.

Using his procedure, Middleton found that the estimated yields agreed closely with the observed yields. That the treatments had upgraded the pasture was indicated by the fact that the average adjusted yields of the fertilized plots increased for the first 5 years. The a_i values increased for the first 4 years and then decreased for unknown reasons. Because of the complexity of the situation and the numerous parameters in the model, further testing with other data seems desirable.

6-10.3. Gunnarsson Model

Gunnarson (1982) also used a modification of the Mitscherlich equation to express the increase in yields of cereal crops in Sweden as a consequence of annual applications of 0, 5, 15, and 45 kilograms of fertilizer phosphorus per hectare annually. He applied his model to data from a group of field experiments carried out for 12 years. His equation was:

$$y = 680(1 - e^{-0.05P})(1 - 0.67e^{-0.15t}),$$

where y = increase in yield of grain from phosphorus fertilization (kg·ha^{-1}·year^{-1}), P = fertilizer phosphorus added (kg·ha^{-1}·year^{-1}), and t = time after start of the experiments (years). The constants are fitted parameters that apply to the experiments studied.

The foregoing equation represents the increases in yields with increasing applications of fertilizer phosphorus and time, but it includes no implicit indication of the change of residual phosphorus. The latter was illustrated by Gunnarson in a graph that showed how the soil phosphorus extractable with the ammonium lactate-acetic acid extractant used in Sweden to provide an index of soil phosphorus availability increased with time and application of fertilizer phosphorus.

6-10.4. Jones Model

Perhaps the most ambitious model produced thus far in terms of recognizing the complexity of soil phosphorus behavior is the one proposed by Jones et al. (1984a,b) and Sharpley et al. (1984). In the first of three papers, these authors described a model involving seven phosphorus compartments: stable inorganic, active inorganic, stable organic, fresh organic, labile, fertilizer, and crop. To describe the system, 53 parameters were defined. The second paper dealt with predicting the labile, organic, and sorbed phosphorus, and the third with testing the model. The phosphorus model was incorporated in a much more comprehensive model involving other factors.

The model was shown to simulate changes in soil organic phosphorus content, changes in soil test values for phosphorus with and without fertilizer applications, responses of corn and wheat to concentrations of soil test phosphorus, and fertilizer phosphorus required to maintain the phosphorus supply of various soils at adequate levels for corn and wheat. The precision of some of the simulations was shown to decrease considerably with an extension of the time period.

6-10.5. Wolf and Janssen Model

Wolf et al. (1987), Janssen et al. (1987), Janssen and Wolf (1988), and Wolf and Janssen (1989) described the operation of what they called a summary model to estimate the recovery of fertilizer phosphorus in crops in years following the year of application. Although the operation of the model involves numerous steps, the individual calculations are simple and are based upon a relatively elementary concept of the behavior of native and added phosphorus in the soil-plant system.

In the model, the crop derives its phosphorus from the "labile pool." The labile pool equilibrates slowly with a "stable pool," so that if the labile and stable pools are at equilibrium, as is assumed to be the case in a soil that has received no fertilizer phosphorus for a long time, uptake of phosphorus by the crop would cause phosphorus to move from the stable pool into the labile pool.

The model assumes that the control soil is in a steady state as regards phosphorus. That is, weathering of soil minerals and other processes supply phosphorus to the labile and stable pools at a rate equal to the removal of phosphorus by the crop and continue to do so each year within the time period of interest.

The rates of transfer between the labile and stable pools are described as specific fractions of the sizes of the pools at the beginning of each time interval. The rate of transfer from the labile to the stable pool was estimated at 0.2 per year and the reverse rate at 0.033 per year on the basis of iterative calculations carried out to produce results that would fit field experimental data from different locations.

The model accommodates the behavior of different types of fertilizers by an adjustment for the labile and stable fractions of phosphorus they contain. For example, the authors represented the labile fraction as 1.0 and the stable fraction as 0 in ammonium phosphates, which are completely water-soluble. In superphosphates, the labile and stable fractions were taken as 0.8 and 0.2, and in phosphate rocks, the labile fraction was represented as 0.1 to 0.2 and the stable fraction as 0.9 to 0.8, depending upon the rock.

Aside from the foregoing information, which is provided by the originators of the model, the information required is the type of fertilizer, the quantity of phosphorus added, and the uptake of phosphorus from the fertilized and unfertilized soil in the year the fertilizers are applied. The authors summarized the calculations in two tables, which are readily understood and followed.

A limitation of the model is that where phosphate fertilizer has been applied recently, the soils will not be at the exhaustion level at which the removal of phosphorus by an unfertilized crop will remain constant from year to year. Prior application of phosphorus-bearing fertilizer will have built up the labile and stable pools. Thus, the rate of removal of phosphorus by the crop will decrease over a period of years as the labile and stable pools are drawn down to their equilibrium levels. The assumption in the model that in the absence of fertilization the net input to the labile and stable pools is equal to the rate of uptake of phosphorus by the crop on an annual basis then is not satisfied. Moreover, the size of the stable pool in the unfertilized soil cannot be estimated because the calculation involves the assumption that the rate of transfer from the stable to the labile pool is equal to the rate of input from weathering of soil minerals and other processes.

One of the data sets used by Janssen et al. (1987) in testing the model showed a decrease in uptake of phosphorus by the crop in succeeding years, which indicated that the assumption about the equilibrium condition made in developing the model did not apply. To deal with the problem, the authors used iterative estimates of the stable pool and the net input from sources other than the stable pool until the results presented "a reasonable fit" of the measured control yields over a period of years. Whether the difficulty in applying the model to previously fertilized soils could be circumvented by some adaptation of methods of soil analysis, such as estimating the size of the stable pool by the slow isotopic dilution of added tagged phosphorus with soil phosphorus studied by Talibudeen (1958) and Probert and Larsen (1972), would be worthy of investigation.

6-10.6. Saroa and Biswas Model

Saroa and Biswas (1989) described a system that is analogous to the technique of estimating the water content of a soil at a particular time on the basis of the initial water content, measured additions due to precipitation and irrigation, and estimated losses due to drainage and evapotranspiration. To estimate the soil phosphorus extractable by the 0.5 molar sodium bicarbonate method of Olsen et al. (1954) at any particular time, they started with the current or initial value for extractable phosphorus, added the amount of phosphorus supplied in the fertilizer, and subtracted the amounts of phosphorus removed by crops, with an adjustment for the relation between the quantity of fertilizer phosphorus and the extractable phosphorus.

According to the concept used by Saroa and Biswas,

$$x = x_0 + cF - R,$$

where x = extractable phosphorus at the end of a year or other selected time interval, x_0 = initial extractable phosphorus, F = fertilizer phosphorus added between the initial and final times, c = increase in extractable phosphorus per unit increase in F, and R = phosphorus removed by harvested crops between the initial and final times. The units of x, x_0, F, and R are kilograms per hectare.

The coefficient c depends upon the value of F. For the quadratic equation used by Saroa and Biswas to represent their empirical findings, the foregoing equation became

$$x = x_0 + b_1F + b_2F^2 - R,$$

where b_1 and b_2 are the regression coefficients. They evaluated the coefficients in each of 7 years in a field experiment at Ludhiana, Punjab, and obtained average values of $b_1 = 0.68$ and $b_2 = -0.0021$.

To estimate the extractable phosphorus at the end of 2 years, the value of x is calculated for the first year using the quantity of fertilizer phosphorus added, the quantity of phosphorus removed in the crop, and the values of the coefficients b_1 and b_2 found empirically. The value of x found for the first year is entered as the initial value x_0 for the second year. The appropriate values of F and R for the second year are inserted, and the equation is solved again for the new x, which may be introduced as x_0 for the third year, and so on.

To test the applicability of the model, Saroa and Biswas (1989) applied it to data from similar field experiments covering a 12-year period at seven other locations in India. They started with the original extractable phosphorus values, which among the various experiments ranged from a low of 8 to a high of 38 kilograms per hectare. They used the additions of fertilizer phosphorus and removals of phosphorus in harvested crops year by year to predict the extractable phosphorus in the final year.

The validation procedure used by the authors involves the assumption that the functional relationship between the quantities of added phosphorus and extractable soil phosphorus was the same at all locations after correction for the phosphorus removed by the crops. The results in Fig. 6-15 indicate that the assumption was satisfactory for the soils at five locations. The predicted and observed values followed a 1:1 relationship rather closely (correlation coefficient = 0.99). Crops other than rice were grown at all of these locations except one (Barrackpore), where rice was grown once in each 3-year crop rotation. The initial extractable phosphorus was highest at this location.

At the two low-phosphorus locations cropped to rice (Hyderabad and Pantnagar), the crops removed more phosphorus from the soil than was added in the fertilizer. As a result, the model predicted a decrease in extractable phosphorus. The data points for these soils in Fig. 6-15 are far below the 1:1 line that fits the observations from the other sites.

Presumably the chemical reduction accompanying flooding of the soils for rice temporarily increased the solubility and availability of the soil phosphorus, having a transient effect analogous to phosphate fertilization. The sodium bicarbonate extractable phosphorus measured on dried and oxidized samples of soil did not reflect this temporary increase in phosphorus availability. Special

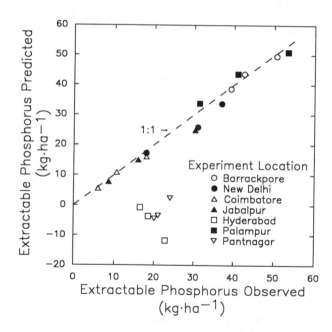

Fig. 6-15. Observed and predicted quantities of phosphorus extracted by the 0.5 molar sodium bicarbonate method from soils receiving different quantities of fertilizer phosphorus over a period of 12 years in experiments at seven locations in India. Predictions were made by use of a model that started with the initial quantity of extractable phosphorus and took into account the additions of fertilizer phosphorus and the removals of phosphorus in harvested crops year by year. (Saroa and Biswas, 1989)

techniques involving incubation of the fertilized samples of soil under flooded conditions and extraction of the soils without reoxidation probably would be needed for soils used for paddy rice (see Brandon and Mikkelsen (1979) for more information on this subject). The Saroa and Biswas model probably would give the best results if applied independently to soils used for dryland crops and soils used for paddy rice.

To adapt the Saroa and Biswas model to conditions under which the dependency of c on F varies with the soil, an adjustment for this effect could be made by determining the extractable phosphorus in samples of the soils that have been incubated with different quantities of fertilizer phosphorus for a time long enough to simulate field conditions over a season. An appropriate regression of extractable phosphorus on fertilizer phosphorus then could be developed.

6-10.7. Swartzendruber and Barber Model

The foregoing models all address the issue of residual effects of a soluble phosphate fertilizer that reacts quickly with the soil and loses availability with time thereafter. Some fertilizers do not behave in this way. Swartzendruber and Barber (1965) developed an equation for the reaction of limestone particles with soils that has been found useful in studies of the oxidation of elemental sulfur in soils (Watkinson, 1989) and also in the reaction of phosphate rock with soils (Edmeades et al., 1991). The Swartzendruber and Barber equation was based upon the proposition that the rate of disappearance of the particles of limestone is proportional to the surface area exposed by the particles.

Edmeades et al. (1991) used the equation in the form

$$\frac{m}{m_0} = (1 - kt)^3,$$

where m = residual mass of phosphate rock at time t, m_0 = mass of phosphate rock applied, and k = dissolution rate constant of the phosphate rock at a given particle-size distribution in the soil. This equation applies to the range of t values in which $t \leq 1/k$.

Fig. 6-16 illustrates the fit of the equation to the data for different phosphate rocks in a series of 19 field experiments on phosphate rock fertilization of permanent pastures in New Zealand. Fig. 6-16A shows the decline in the residual unreacted phosphate rock with time when the phosphate rocks were applied alone, and Fig. 6-16B shows the behavior in parallel treatments in which the phosphate rocks were applied with elemental sulfur. The more rapid decline of residual phosphorus in the presence of sulfur illustrates the importance of soil acidity in promoting the decomposition of phosphate rocks. From their average data, the authors calculated that if a reactive phosphate rock such as those from Chatham Rise and North Carolina in Fig. 6-16 is applied annually, the sums of the phosphorus released each year from the current and residual applications in the first 6 years would be equivalent to 30, 53, 70, 82, 91, and 96% of the total phosphorus applied per year.

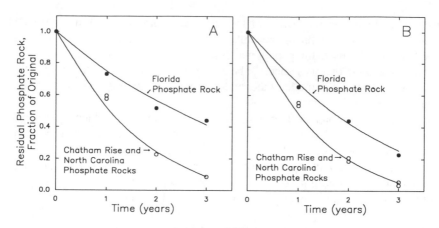

Fig. 6-16. Decline in residual quantity of phosphate rocks with time after application as top dressings to soils in experiments on permanent pastures on acid soils in New Zealand. Chatham Rise and North Carolina phosphate rocks have relatively high reactivity, and Florida phosphate rock has relatively low reactivity. A. Average results with phosphate rocks applied alone. B. Average results with phosphate rocks applied in mixtures with elemental sulfur to acidify the soil and hasten the reaction of the phosphate rocks with the soil. (Edmeades et al., 1991)

6-11. Economics of Residual Effects

The residual effects of fertilizers add another dimension to the problem of maximizing financial returns from fertilization. Some have described accumulated fertilizer residues in the soil as "money in the bank." There is much merit in this analogy; however, the "bank" pays no interest and does not permit withdrawals from the account at will. Until the nutrients are removed by a crop, they are not producing a financial return, as they could be doing if the money used to purchase them could be withdrawn and invested in some other way. Moreover, the value of the money in the bank is continuously decreasing because the value of the fertilizer residues to crops diminishes with time, even in the absence of a crop, as a consequence of loss from the soil or reaction with the soil to produce forms with decreased availability to plants.

From the standpoint of a return on the investment, it would be better if all the benefit from fertilizers could be realized in the first year instead of being spread out over several years. The return received after the first year is worth less than it would have been if received in the first year because if it had been received in the first year the money could have been invested and earning a return. Additionally, the magnitude of the return to be expected from money invested in fertilizers is uncertain as a consequence of less than precise knowledge about the most appropriate quantities to apply, as well as the unforseeable hazards of production and the unpredictability of the price the crop will bring.

These problems provide justification for discounting, which is the opposite of compound interest. The interest rate is the rate at which the value of an investment increases, and the discount rate is the rate at which it decreases.

Anticipated returns from fertilization may be discounted by multiplying the expected return by $(1 + d)^{-n}$, where d is the discount rate expressed as a decimal fraction, and n is the number of times a return would have been received from an alternative investment. Discounting may be used, not only with the first crop following the application of fertilizers, but also with succeeding crops because of residual effects.

Pesek et al. (1961) used the discounting concept in evaluating the influence of residual effects upon the optimum application of phosphate fertilizer for corn in an experiment in Iowa. They applied the principle to the returns from the first crop following fertilization as well as the returns from three succeeding crops that received no phosphate fertilizer. According to their analysis, discounting reduced to a relatively small degree the estimated quantity of fertilizer phosphorus required to yield the maximum net profit, but markedly decreased the estimated maximum net profit. They found, however, that as the residual response of the crop in each succeeding year was taken into account, the estimated quantity of fertilizer required to produce the maximum net profit increased.

The yields of the control plots in the experiment analyzed by Pesek et al. (1961) were relatively uniform in succeeding years (coefficient of variation estimated at 20%). In applying the discounting concept to residual effects of phosphorus fertilization in two experiments in Australia where the growing conditions were more variable (coefficients of variation of control plot yields estimated at 42 and 56%), Colwell (1985b) called attention to a pronounced effect of specific seasonal conditions on the adjustment of applications needed to take residual effects into account.

Gunnarson (1982) applied the discounting concept to the results of phosphorus fertilization experiments in Sweden. In his experiments, different quantities of fertilizer phosphorus were added, apparently annually. Thus, instead of decreasing with time, as in the experiment by Pesek et al. (1961), where only a single application of fertilizer was made, the residual effects increased with time. Colwell (1983) described in summary form the procedure for calculating optimal economic quantities of fertilizer where repeated applications are made and discounting is applied. Discounting was used also by Halvorson et al. (1987) in a computer program they developed for estimating the profitability of fertilizer phosphorus for wheat in the northern Great Plains region of the United States. The program takes into account the residual effect of single and repeated applications of fertilizer.

Kennedy (1986) summarized his considerations of residual effects and discounting by the general rule that the maximum net profit is obtained when fertilizer is applied in the quantity "at which the expected present value of the current crop and of the savings in future fertilizer applications obtained from the marginal unit of fertilizer equals the current price of fertilizer.'' He noted that residual effects decrease the quantity of fertilizer required to yield the maximum net profit where each crop is fertilized to yield the maximum net profit. This is in contrast to the situation considered by Pesek et al. (1961), in which residual effects increased the quantity of fertilizer yielding the maximum net profit when fertilizer was applied in the first year only.

Scobie and St-Pierre (1987) developed an economic model for phosphorus fertilization of pasture in New Zealand that estimated the most profitable application of fertilizer phosphorus each year using discounting. From the soils standpoint, their model took into account the disappearance of available phosphorus from one year to the next due to soil processes and removal in the grazing animals. They assumed that the phosphorus extracted from the soil by the sodium bicarbonate method of Olsen et al. (1954) was proportional to the available phosphorus, and used the change in extractable phosphorus from year to year in their calculations.

In a study of the residual effect of nitrogen fertilization of grassland in experiments in South Dakota and Montana, Stauber and Burt (1973) and Stauber et al. (1975) developed an estimation procedure on the basic assumption that nitrogen availability and weather conditions are the most important factors affecting hay yields in the area of concern. They used the increase in grass yield in the current year as a measure of nitrogen carry-over from previous years.

They developed two different equations for residual nitrogen, one in the first paper and the other in the second. They fitted one or the other of these equations for residual nitrogen simultaneously with a quadratic equation for response to fertilizer nitrogen in the current year. The latter equation included fertilizer nitrogen and current precipitation as independent variables.

According to their first analysis of the results of a 6-year experiment (Stauber and Burt, 1973), the average optimum total amount of additional nitrogen under the conditions investigated was 90 kilograms per hectare. Of this amount, 54% was residual and 46% would need to be applied annually.

In the later paper (Stauber et al., 1975), they described an alternative decision-making process. If the available nitrogen found in the soil had fallen below a certain level at the time fertilizer nitrogen was ordinarily applied, enough nitrogen was added to bring the total up to a specified higher level; otherwise, none was added. The optimal level at which an addition should be made and the level to which the total supply should be brought were calculated from the decision-making model. The numerical values they gave appear to be hypothetical because the basic data for available soil nitrogen were not found in the paper by Thomas and Osenbrug (1964) cited as the source of the data they analyzed.

Lanzer and Paris (1981) developed a novel but complex model in which they meshed economic principles with soil test data, the relative-yield concept, and the Cate and Nelson (1971) response model of a linear yield increase followed by a flat plateau with increasing soil test values. Their objective was to maximize the discounted net revenues in successive years, as inferred from the yield responses and the residual effects estimated by soil tests. On the basis of data from 38 field experiments with wheat and soybean in a double-cropping system in Brazil, they concluded that about 20% too much fertilizer was being recommended by the procedures used by local agronomists. Middleton (1983a) reanalyzed the same data with a different response model and obtained a figure intermediate between that of Lanzer and Paris and the local agronomists.

Other publications on the economics of residual effects are cited by the authors whose works are reviewed here. Some of the models are predictive, whereas others are suitable for analyzing the economics of data in which the residual effects have already been measured. The list of possible theoretical economic models does not yet seem to have been exhausted.

Although different theoretical ways have been proposed to maximize the economic return from fertilizers in the presence of residual effects, the concepts have not been widely adopted. The reasons may include the lack of concordance among theoretical approaches and the fact that agronomic data on residual effects are generally inadequate to complement the theoretical considerations. Most field experiments with fertilizers have a relatively large experimental error and are not well adapted to measuring residual effects that usually are considerably smaller and more difficult to measure than the first-year effects.

For the most part, the economic aspects of residual effects seem to receive little direct attention in practice. Soil tests, however, provide an index of residual effects. To the extent of their validity and the suitability of their interpretation, soil tests have economic implications in terms of residual effects because they indicate a downward adjustment of the economic quantities of fertilizers as the soil test values increase. They have the additional advantage of providing a site-specific prognosis.

As perhaps an extreme example, a study of the economics of irrigated corn in Texas (Onken et al., 1985; Stoecker and Onken, 1989) showed that, as might be expected, soil nitrate substituted for fertilizer nitrogen. The variability of soil nitrate from year to year was so great that an annual soil test for nitrate was recommended to permit proper adjustment of the application of fertilizer nitrogen. In 3 successive years, the soil nitrate nitrogen per hectare to a depth of 0.9 meter was found to be 173, 83, and 91 kilograms with no nitrogen fertilizer and 223, 122, and 101 kilograms in the respective years where 225 kilograms of fertilizer nitrogen had been applied annually. Residual effects of other nutrients would be expected to be much less variable.

Another economic aspect of residual effects that merits increasing attention in areas where the supplies of certain nutrients in the soil have been built up to relatively high levels is the possible benefit to be derived from reducing or eliminating the applications of the residual nutrient for a time. This matter was investigated by Clark et al. (1990) in New Zealand for grassland that had been fertilized with superphosphate for years. They compared the economics of continuing to apply superphosphate each year at the previous rate in several experiments versus stopping the applications. On the basis of the assumptions they employed, their analyses showed that in each instance continuing to apply the previous quantities of superphosphate was initially unprofitable, but that after 4 to 8 years the cumulative benefit of continuing to apply the fertilizer exceeded the cost. They concluded that stopping the fertilizer application was a sound strategy for producers to survive short periods of financial difficulties and low ratios of product price to fertilizer cost, but only if the ratio of product price to fertilizer cost was expected to increase.

Research with a similar objective, also in New Zealand, was reported by Mackay et al. (1990). In their work, withholding superphosphate for 6 years decreased total pasture herbage production by 30% and decreased production of the legume component of the herbage by 45%, as compared with production on plots receiving 250 kilograms of ordinary superphosphate per hectare annually, which their evidence indicated was approximately a maintenance application. According to their economic analysis, continuation of the annual maintenance applications or using part of the money otherwise invested in maintenance applications to establish a grass species shown by their experimental work to be more productive and drought resistant than the resident species would provide greater economic return than discontinuing the superphosphate applications.

An economic study of residual effects of phosphorus fertilization with a different objective was made by Sinclair et al. (1990a), also in New Zealand. They analyzed the consequences of changing from a practice of annual maintenance topdressings of superphosphate to annual maintenance topdressings of one of the reactive phosphate rocks for permanent pastures. In contrast to superphosphate, which releases most of its phosphorus to moist soils in a few days, reactive phosphate rocks release their phosphorus only over a period of a year or more in acid soils up to a period of many years in alkaline soils.

Although the price per kilogram of phosphorus is lower in phosphate rock than in superphosphate, phosphate rocks contain only a trace of sulfur. Sulfur is often deficient in New Zealand soils if superphosphate is not applied (about half of ordinary superphosphate by weight is calcium sulfate). Thus, an extra charge was made against the phosphate rock for conditions in which sulfur was needed.

Unexpectedly, the phosphate rock being used experimentally was found to contain 43 milligrams of molybdenum per kilogram, and the quantities of phosphate rock being used increased the molybdenum content of the vegetation and eliminated a molybdenum deficiency in the plants in some of the experiments (Sinclair et al., 1990b; Edmeades et al., 1991). The cost of the molybdenum fertilizer needed to bring about this effect was negligible in comparison with the cost of the phosphorus, but for the molybdenum-deficient soils the molybdenum supplied by the phosphate rock invalidated the use of crop response data for evaluating the phosphorus effects.

The phosphorus evaluations were made by determining the phosphorus released to the soil from the phosphate rock (see Perrott, 1992) and assuming that unit quantity of phosphorus released had the same value for crop production as unit quantity of citric acid-soluble phosphorus supplied by superphosphate during the same time. This assumption was found valid for the experiments on the sites with adequate supplies of molybdenum (Edmeades et al., 1991), but not for the molybdenum-deficient sites because of the response to the molybdenum supplied by the phosphate rock.

On average, the sums of the phosphorus released from the current application of phosphate rock plus the release from residual applications as a percentage of the quantity of phosphorus applied annually for years 1 through 6 were 30, 53, 70, 82, 91, and 96 (Edmeades et al., 1991). As a result of the slow release of phosphorus from the phosphate rock, the effective rate of phosphorus applications was diminished until the residual release had been built up, and this would result in some decrease in production. On the basis of the experimental results and the costs, Edmeades et al. (1991) concluded that 3 to 7 years would be required for a reactive phosphate rock to become more profitable than superphosphate if the fertilizers were applied at equal quantities of phosphorus per hectare. In the meantime, superphosphate would be the more profitable phosphorus source. If sulfur were needed, a longer time would be required for the returns from phosphate rock to surpass those from superphosphate.

Evidently there are many different economic situations where residual effects of previous applications of fertilizers are concerned. The New Zealand work on pastures is related to the concept of maintenance applications discussed in Section 4-4.5.1. Ideally the maintenance application will take into account the residual level of the nutrient in the soil and will provide the quantity of the nutrient that will lead to the practice that is of maximum economic benefit to the producer. A common denominator through which this goal may be approached is the soil test level and its relation to crop responses, as indicated in Section 4-4.5.1.

6-12. Annual Versus Intermittent Applications

Whether to make a relatively large application of fertilizer at intervals of years or to apply the same total amount in annual applications is an issue that has been addressed in a number of investigations. The discounting principle, the loss of effectiveness of the fertilizer with time, and the effects of different quantities of freshly applied fertilizer on the response curves are all involved in the outcome. The usual finding is that intermittent heavy applications are less desirable from the standpoint of both agronomic and economic returns than are smaller applications.

The most favorable results from intermittent heavy applications are obtained under circumstances in which the residual amounts remain in available form in the soil. Experiments with use of fertilizer nitrogen on grassland under conditions of little or no leaching in Saskatchewan have shown residual effects of single heavy applications on yield of forage and soil nitrate for at least 10 years (Read and Winkelman, 1982). Under similar conditions in Saskatchewan and Alberta,

Zentner et al. (1989) and Penney et al. (1990) found that despite the pronounced residual effect of heavy intermittent applications of fertilizer nitrogen, annual applications still produced the higher yields. A disadvantage not factored into the economic analysis made by Zentner et al. (1989) was the potential for toxicity of the forage to ruminants due to accumulation of nitrate when heavy applications of fertilizer nitrogen were made.

In a 3-year experiment in Niger with two phosphate rocks, with and without partial acidulation, plus ordinary superphosphate and concentrated superphosphate, Bationo et al. (1990) found that the total increase in yield of millet grain was greater in all cases when 19.5 kilograms of phosphorus were applied per hectare in the first year only than when 6.5 kilograms were applied in each of the 3 years. In the extreme case, the single heavy application produced about twice as much grain as the three lighter applications. The authors attributed their results to the fact that the soil employed was an acid siliceous sand in which the phosphorus-fixing capacity was low and the residual phosphorus availability was high. Important also was the fact that 6.5 kilograms of phosphorus per hectare initially produced yields well below those produced by 19.5 kilograms. Incidental tests of a single heavy application versus lighter annual applications suggest that the single heavy application could be superior for a few years on some calcareous soils (Smith, 1956).

Larsen et al. (1965) found that the half-life of the increase in the *L* value for phosphorus produced by fertilization of 23 mineral soils with concentrated superphosphate in the United Kingdom averaged 2 years. In an extensive experiment on residual effects by Arndt and McIntyre (1963) in Australia, the half-life of the effectiveness of superphosphate was about 1 year. Halvorson et al. (1987) used the equivalent of a half-life of 10 years for phosphorus added as concentrated superphosphate for wheat to soils in the northern Great Plains of the United States. In an experiment in Iowa, the half-life of the increase in soil test values for phosphorus following a heavy initial application of concentrated superphosphate was about 6 years (Webb et al., 1992). The relative yields of corn followed the soil test values.

The least favorable results from intermittent heavy applications are obtained when the rate of loss of availability is great. Da Silva et al. (1990) reported findings under such conditions from an investigation of cotton fertilization with concentrated superphosphate on a phosphate-fixing soil in Brazil. In a 6-year experiment, a single initial broadcast application of 318 kilograms of fertilizer phosphorus per hectare produced a greater yield than 53 kilograms banded in the same year, but in each of the succeeding 5 years the continuing 53-kilogram annual applications outyielded the single 318-kilogram application by a wide margin. Placement and residual effect were confounded in this experiment, but as a result of tillage, each annual banded application would become, in effect, a broadcast application during the remaining years.

6-13. Literature Cited

Ackello-Ogutu, Q. Paris, and W. A. Williams. 1983. A methodology for estimating residual fertilizer nutrients. *East African Agricultural and Forestry Journal* 49:14-20.

Arndt, W., and G. A. McIntyre. 1963. The initial and residual effects of superphosphate and rock phosphate for sorghum on a lateritic red earth. *Australian Journal of Agricultural Research* 14:785-795.

Barrow, N. J., and N. A. Campbell. 1972. Methods of measuring residual value of fertilizers. *Australian Journal of Experimental Agriculture and Animal Husbandry* 12:502-510.

Bationo, A., S. H. Chien, J. Henao, C. B. Christianson, and A. U. Mokwunye. 1990. Agronomic evaluation of two unacidulated and partially acidulated phosphate rocks indigenous to Niger. *Soil Science Society of America Journal* 54:1772-1777.

Berryman, C. 1971. Residual value of slurries (cattle, pig and poultry). Pp. 326-333. In *Residual Value of Applied Nutrients*. Ministry of Agriculture, Fisheries and Food, Technical Bulletin No. 20. Her Majesty's Stationery Office, London.

Bittman, S., D. A. Pulkinen, and J. Waddington. 1991. Effect of N and P fertilizer on establishment of alfalfa with a wheat companion crop. *Canadian Journal of Plant Science* 71:105 113.

Bolland, M. D. A. 1985. Residual value for wheat of phosphorus from calciphos, Duchess rock phosphate and triple superphosphate on a lateritic soil in south-western Australia. *Australian Journal of Experimental Agriculture* 25:198-208.

Bolland, M. D. A. 1992. Residual value of superphosphate measured using yields of different pasture legume species and bicarbonate-extractable soil phosphorus. *Fertilizer Research* 31:95-110.

Bolland, M. D. A., and J. W. Bowden. 1982. Long-term availability of phosphorus from calcined rock phosphate compared with superphosphate. *Australian Journal of Agricultural Research* 33:1061-1071.

Bolland, M. D. A., R. J. Gilkes, and F. F. D'Antuono. 1988. The effectiveness of rock phosphate fertilisers in Australian agriculture: a review. *Australian Journal of Experimental Agriculture* 28:655-668.

Bouldin, D. R., S. D. Klausner, and W. S. Reid. 1984. Use of nitrogen from manure. Pp. 221-245. In R. D. Hauck (Ed.), *Nitrogen in Crop Production*. American Society of Agronomy, Crop Science Society of America, and Soil Science Society of America, Madison, WI.

Brandon, D. M., and D. S. Mikkelsen. 1979. Phosphorus transformations in alternately flooded California soils: I. Cause of plant phosphorus deficiency in rice rotation crops and correctional methods. *Soil Science Society of America Journal* 43:989-994.

Cassman, K. G., S. M. Brouder, B. A. Roberts, D. C. Bryant, and T. A. Kerby. 1990. Cotton root growth in relation to localized nutrient supply, and residual value of potassium fertilizer in vermiculitic soils. Pp. 140-159. In *Advances in Fluid Fertilizer Agronomic and Application Management Technology*. 1990 Research Symposium Proceedings. Fluid Fertilizer Foundation, Scottsdale, AZ.

Cate, R. B., Jr., and L. A. Nelson. 1971. A simple statistical procedure for partitioning soil test correlation data into two classes. *Soil Science Society of America Proceedings* 35:658-660.

Clark, D. A., S. F. Ledgard, N. G. Lambert, M. B. O'Connor, and A. G. Gillingham. 1990. Long term effects of withholding phosphate application on North Island hill country: Economics. *Proceedings of the New Zealand Grassland Association* 51:29-33.

Colwell, J. D. 1983. Fertilizer requirements. Pp. 795-815. In *Soils: An Australian Viewpoint*. CSIRO, Division of Soils, Melbourne.

Colwell, J. D. 1985a. Fertilizing programs. 1. Variability in responses of successive crops to fresh and previous applications of phosphorus fertilizer, in Australia and Brazil. *Fertilizer Research* 8:21-38.

Colwell, J. D. 1985b. Fertilizing programs. 2. Optimal programs of fertilizer application. *Fertilizer Research* 8:39-47.

Cooke, G. W. 1967. The value and valuation of fertilizer residues. *Journal of the Royal Agricultural Society of England* 128:7-25.

Cope, J. T., Jr. 1981. Effects of 50 years of fertilization with phosphorus and potassium on soil test levels and yields at six locations. *Soil Science Society of America Journal* 45:342-347.

Cox, F. R., E. J. Kamprath, and R. E. McCollum. 1981. A descriptive model of soil test nutrient levels following fertilization. *Soil Science Society of America Journal* 45:529-532.

Da Silva, N. M., L. H. Carvalho, J. C. Sabino, L. G. L. Lellis, N. P. Sabino, and J. I. Kondo. 1990. Modo e época de aplicação de fosfatos na produção e outras características do algodoeiro. *Bragantia* 49:157-170.

Devine, J. R., D. Gunary, and S. Larsen. 1968. Availability of phosphate as affected by duration of fertilizer contact with soil. *Journal of Agricultural Science (Cambridge)* 71:359-364.

Duffy, M. J., and J. Webb. 1990. Crop rotation-fertility experiment. Pp. 8-12. In J. Webb (Coordinator), *Annual Progress Reports — 1989. Northern Research Center, Clarion-Webster Research Center, Kanawha, Iowa.* Iowa State University Agriculture and Home Economics Experiment Station, Cooperative Extension Service, ORC 89-14, 22. Ames.

Dyke, G. V., B. J. George, A. E. Johnston, P. R. Poulton, and A. D. Todd. 1983. The Broadbalk wheat experiment 1968-78: Yields and plant nutrients in crops grown continuously and in rotation. *Rothamsted Experimental Station Report for 1982, Part 2*:5-44. Harpenden Herts, U.K.

Edmeades, D. C., J. H. Watkinson, K. W. Perrott, A. G. Sinclair, S. F. Ledgard, S. S. S. Rajan, M. W. Brown, A. H. Roberts, B. T. Thorrold, M. B. O'Connor, M. J. S. Floate, W. H. Risk, and J. Morton. 1991. Comparing the agronomic performance of soluble and slow release phosphate fertilisers: the experimental basis for RPR recommendations. *Proceedings of the New Zealand Grassland Association* 53: In Press. (From a preprint supplied by A. G. Sinclair)

Eik, K., J. R. Webb, C. A. Black, C. M. Smith, and J. T. Pesek. 1961. Evaluation of residual effects of phosphate fertilization by laboratory and plant-response methods. *Soil Science Society of America Proceedings* 25:21-24.

Farina, M. P. W., and P. Channon. 1987. Season and phosphorus age effects on the relationship between maize yield and phosphorus soil test on a highly weathered soil. *South African Journal of Plant and Soil* 4:21-25.

Gessel, S. P., T. N. Stoats, and K. J. Turnbull. 1965. *The Growth Behaviour of Douglas-Fir With Nitrogenous Fertilizer in Western Washington.* University of Washington (Seattle), Forest Resources Bulletin 1. (Quoted by Turner, J. 1977. Effect of nitrogen availability on nitrogen recycling in a Douglas-fir stand. *Forest Science* 23:307-316.)

Ghosh, A. K., and R. J. Gilkes. 1987. The initial and residual agronomic effectiveness of some Indian, USA and Australian rock phosphates. *Fertilizer Research* 12:201-218.

Gonçalves, J. L. M., R. F. Novais, N. F. Barros, J. C. L. Neves, and A. C. Ribeiro. 1989. Cinética de transformação de fósforo-lábil em não-lábil, em solos de cerrado. *Revista Brasileira de Ciência do Solo* 13:13-24.

Gunnarson, O. 1982. Economics of long term fertility building by means of phosphate. *Fertilizers and Agriculture* No. 83:39-52.

Hall, A. D. 1917. *The Book of the Rothamsted Experiments.* Second Edition (Revised by E. J. Russell). E. P. Dutton and Company, NY.

Halvorson, A. D., E. H. Vasey, and D. L. Watt. 1987. PHOSECON: A computer economics program to evaluate phosphorus fertilization of wheat. *Applied Agricultural Research* 2:207-212.

Hannam, R. J., G. J. Judson, D. J. Reuter, L. D. McLaren, and J. D. McFarlane. 1982. Current requirements of copper for pasture and sheep production on sandy soils in the upper south-east of South Australia. *Australian Journal of Experimental Agriculture and Animal Husbandry* 22:324-330.

Hanson, R. G., J. A. Stecker, and S. R. Maledy. 1988. Effect of soybean rotation on the response of sorghum to fertilizer nitrogen. *Journal of Production Agriculture* 1:318-321.

Hergert, G. W. 1987. Status of residual nitrate-nitrogen tests in the United States of America. Pp. 73-88. In J. R. Brown (Ed.), *Soil Testing: Sampling, Correlation, Calibration, and Interpretation.* SSSA Special Publication No. 21. Soil Science Society of America, Madison, WI.

Holford, I. C. R. 1982. Effect of phosphate sorptivity on the long-term plant recovery and effectiveness of fertilizer phosphate in soils. *Plant and Soil* 64:225-236.

Holford, I. C. R., and G. J. Crocker. 1991. Residual effects of phosphate fertilizers in relation to phosphate sorptivities of 27 soils. *Fertilizer Research* 28:305-314.

Janssen, B. H., D. J. Lathwell, and J. Wolf. 1987. Modeling long-term crop response to fertilizer phosphorus. II. Comparison with field results. *Agronomy Journal* 79:452-458.

Janssen, B. H., and J. Wolf. 1988. A simple equation for calculating the residual effect of phosphorus fertilizers. *Fertilizer Research* 15:79-87.

Johnston, A. E. 1970. The value of residues from long-period manuring at Rothamsted and Woburn. I. Introduction. *Rothamsted Experimental Station, Report for 1969, Part 2*:5-6. (Four succeeding papers in the series by Johnston and coworkers are found on pages 7-90. The final paper on pages 91-112 deals with residual effects at Saxmundham.) Harpenden Herts, U.K.

Johnston, A. E. 1981. Accumulation of phosphorus in a sandy loam soil from farmyard manure (FYM) and sewage sludge. Pp. 273-290. In T. W. G. Hucker and G. Catroux (Eds.), *Phosphorus in Sewage Sludge and Animal Waste Slurries*. D. Reidel Publishing Company, Dordrecht, Holland.

Johnston, A. E. 1987. Effects of soil organic matter on yields of crops in long-term experiments at Rothamsted and Woburn. *INTECOL Bulletin* 15:9-16.

Johnston, A. E. 1991. Soil fertility and soil organic matter. Pp. 299-314. In W. S. Wilson, T. R. G. Gray, D. J. Greenslade, R. M. Harrison, and M. H. B. Hayes (Eds.). 1991. *Advances in Soil Organic Matter Research: The Impact on Agriculture and the Environment*. Royal Society of Chemistry, Special Publication No. 90. Thomas Graham House, Cambridge.

Johnston, A. E., J. McEwen, P. W. Lane, M. V. Hewitt, P. R. Poulton, and D. Yeoman. 1992. Effects of one to six year old ryegrass clover leys on soil nitrogen and the subsequent yields and fertilizer nitrogen requirements of the arable sequence winter wheat, potatoes, winter wheat, winter beans (*Vicia faba*). Manuscript submitted to the *Journal of Agricultural Science (Cambridge)*.

Johnston, A. E., and R. G. Warren. 1970. The value of residues from long-period manuring at Rothamsted and Woburn. III. The experiments made from 1957 to 1962, the soils and histories of the sites on which they were made. *Rothamsted Experimental Station, Report for 1969, Part 2*:22-38. Harpenden Herts, U.K.

Johnston, A. E., R. G. Warren, and A. Penny. 1970. The value of residues from long-period manuring at Rothamsted and Woburn. IV. The value to arable crops of residues accumulated from superphosphate. *Rothamsted Experimental Station, Report for 1969, Part 2*:39-68. Harpenden Herts, U.K.

Jones, C. A., C. V. Cole, A. N. Sharpley, and J. R. Williams. 1984a. A simplified soil and plant phosphorus model: I. Documentation. *Soil Science Society of America Journal* 48:800-805.

Jones, C. A., A. N. Sharpley, and J. R. Williams. 1984b. A simplified soil and plant phosphorus model: III. Testing. *Soil Science Society of America Journal* 48:810-813.

Jourdan, O., and P. Villemin. 1990. Field experiments for fertilizer advice — Design — Execution — Interpretation. Pp. 263-278. In *Development of K-Fertilizer Recommendations*. Proceedings of the 22nd Colloquium of the International Potash Institute held at Soligorsk/USSR. International Potash Institute, Worblaufen-Bern, Switzerland.

Kennedy, J. O. S. 1986. Rules for optimal fertilizer carryover: An alternative explanation. *Review of Marketing and Agricultural Economics* 54, No. 2:3-10.

Krantz, B. A., F. E. Broadbent, W. A. Williams, K. G. Baghott, K. H. Ingebretsen, and M. E. Stanley. 1968. Fertilize crop not crop residues. *California Agriculture* 22, No. 8:6-8.

Lanzer, E. A., and Q. Paris. 1981. A new analytical framework for the fertilization problem. *American Journal of Agricultural Economics* 63:93-103.

Larsen, S., D. Gunary, and C. D. Sutton. 1965. The rate of immobilization of applied phosphate in relation to soil properties. *Journal of Soil Science* 16:141-148.

Larsen, S., and M. E. Probert. 1968. A theoretical treatment of the maintenance of soil phosphorus status. *Phosphorus in Agriculture* No. 51:21-26.

Lawes, J. B., and J. H. Gilbert. 1884. On the continuous growth of wheat on the experimental plots at Rothamsted during the 20 years, 1864-1883, inclusive. *Journal of the Royal Agricultural Society of England* 45:391-481.

Lozano, J. M., and F. Abruña. 1981. Nitrogen rates in single and split applications and yield of flooded rice. *Journal of Agriculture of the University of Puerto Rico* 65:35-42.

Mackay, A. D., R. W. Tillman, W. J. Parker, and D. J. Barker. 1990. Effect of super-phosphate, lime, and cocksfoot on summer dry hill country pasture production. *Proceedings of the New Zealand Grassland Association* 51:131-134.

Mattingly, G. E. G., A. E. Johnston, and M. Chater. 1970. The residual value of farmyard manure and superphosphate in the Saxmundham Rotation II Experiment, 1899-1968. *Rothamsted Experimental Station, Report for 1969, Part 2*:91-112. Harpenden Herts, U.K.

Middleton, K. R. 1983a. Economic control of fertilizer in highly productive pastoral systems. I. A theoretical framework for the fertilization problem. *Fertilizer Research* 4:301-313.

Middleton, K. R. 1983b. Economic control of fertilizer in highly productive pastoral systems. III. A hypothetical model to quantify upgrading from a primitive into a highly productive state. *Fertilizer Research* 4:331-345.

Ministry of Agriculture, Fisheries and Food. 1953. *Compensation Value of Fertilizers (England and Wales)*. Statutory Instruments 1953, No. 456. Her Majesty's Stationery Office, London. (Cited by Cooke, G. W. 1966. *The Control of Soil Fertility*. Crosby Lockwood & Son, London)

Ministry of Agriculture, Fisheries and Food. 1971. *Residual Value of Applied Nutrients*. Ministry of Agriculture, Fisheries and Food, Technical Bulletin No. 20. Her Majesty's Stationery Office, London.

Mitchell, H. L., and R. F. Chandler, Jr. 1939. The nitrogen nutrition and growth of certain deciduous trees of northeastern United States. *Black Rock Forest Bulletin* No. 11. Cornwall-on-the-Hudson, NY.

Morel, C., and J. C. Fardeau. 1991. Phosphorus bioavailability of fertilizers: a predictive laboratory method for its evaluation. *Fertilizer Research* 28:1-9.

Nelson, L. B., H. R. Meldrum, and W. H. Pierre. 1946. Nitrogen fertilization of oats and its relation to other crops in the rotation. *Soil Science Society of America Proceedings* 11:417-421.

Ng Kee Kwong, K. F., and J. Deville. 1989. Timing potassium fertilizer applications to sugarcane in Mauritius. *Fertilizer Research* 20:153-158.

Olsen, S. R., C. V. Cole, F. S. Watanabe, and L. A. Dean. 1954. *Estimation of Available Phosphorus in Soils by Extraction With Sodium Bicarbonate*. U.S. Department of Agriculture, Circular No. 939. Washington.

Onken, A. B., R. L. Matheson, and D. M. Nesmith. 1985. Fertilizer nitrogen and residual nitrate-nitrogen effects on irrigated corn yield. *Soil Science Society of America Journal* 49:134-139.

Penney, D. C., S. S. Malhi, and L. Kryzanowski. 1990. Effect of rate and source of N fertilizer on yield, quality and N recovery of bromegrass grown for hay. *Fertilizer Research* 25:159-166. (*Soils and Fertilizers* 54, Abstract 16035. 1991.)

Perrott, K. 1992. Effect of exchangeable calcium on fractionation of inorganic and organic soil phosphorus. *Communications in Soil Science and Plant Analysis* 23: (In Press).

Pesek, J. T., E. O. Heady, and L. C. Dumenil. 1961. Influence of residual fertilizer effects and discounting upon optimum fertilizer rates. *Transactions of 7th International Congress of Soil Science, Madison, Wisc., U.S.A., 1960*, 3:220-227. Elsevier Publishing Company, Amsterdam.

Pommel, B. 1981a. Détermination au moyen d'un test biologique de la cinétique de libération du phosphore à partir d'une boue résiduaire. *Agronomie* 1:467-472.

Pommel, B. 1981b. A plant test for determination of phosphorus value of urban wastes. *Acta Horticulturae* 126:237-244.

Pratt, P. F., F. E. Broadbent, and J. P. Martin. 1973. Using organic wastes as nitrogen fertilizers. *California Agriculture* 27, No. 6:10-13.

Pratt, P. F., S. Davis, and R. G. Sharpless. 1976. A four-year field trial with animal manures. II. Mineralization of nitrogen. *Hilgardia* 44:113-125.

Probert, M. E. 1985. A conceptual model for initial and residual responses to phosphorus fertilizers. *Fertilizer Research* 6:131-138.

Probert, M. E., and S. Larsen. 1972. The kinetics of heterogeneous isotopic exchange. *Journal of Soil Science* 23:76-81.

Rajan, S. S. S., R. L. Fox, W. M. H. Saunders, and M. Upsdell. 1991a. Influence of pH, time and rate of application on phosphate rock dissolution and availability to pastures I. Agronomic benefits. *Fertilizer Research* 28:85-93.

Rajan, S. S. S., R. L. Fox, and W. M. H. Saunders. 1991b. Influence of pH, time and rate of application on phosphate rock dissolution and availability to pastures II. Soil chemical studies. *Fertilizer Research* 28:95-101.

Read, D. W. L., and G. E. Winkleman. 1982. Residual effects of nitrogen and phosphorus fertilizer on crested wheatgrass under semiarid conditions. *Canadian Journal of Plant Science* 62:415-425.

Russell, R. D., and T. Batey. 1971. Residual effects of potassium fertilizers in five rotation experiments. Pp. 197-219. In *Residual Value of Applied Nutrients*. Ministry of Agriculture, Fisheries and Food, Technical Bulletin No. 20. Her Majesty's Stationery House, London.

Sander, D. H., E. J. Penas, and B. Eghball. 1990. Residual effects of various phosphorus application methods on winter wheat and grain sorghum. *Soil Science Society of America Journal* 54:1473-1478.

Saroa, G. S., and C. R. Biswas. 1989. A semi-descriptive model for predicting residual-P from fertilizer P applications. *Fertilizer Research* 19:121-126.

Scobie, G. M., and N. R. St-Pierre. 1987. Economics of phosphorus fertiliser use on pastures. 2. Incorporating the residual effect. *New Zealand Journal of Experimental Agriculture* 15:445-451.

Sharpley, A. N., C. A. Jones, C. Gray, and C. V. Cole. 1984. A simplified soil and plant phosphorus model: II. Prediction of labile, organic, and sorbed phosphorus. *Soil Science Society of America Journal* 48:805-809.

Shrader, W. D., W. A. Fuller, and F. B. Cady. 1966. Estimation of a common nitrogen response function for corn (*Zea mays*) in different crop rotations. *Agronomy Journal* 58:397-401.

Silva, N. M. da, L. H. Carvalho, J. C. Sabino, L. G. L. Lellis, N. P. Sabino, and J. I. Kondo. 1990. Modo e época de aplicação de fosfatos na produção e outras características do algodoeiro. *Bragantia* 49:157-170.

Sinclair, A., M. Bowditch, K. Perrott, and J. Watkinson. 1990a. A new approach to the economic evaluation of reactive phosphate rock as a replacement for single superphosphate. *Proceedings of the New Zealand Fertiliser Manufacturers' Research Association Conference 1990*:249-252. Auckland, New Zealand. (From a preprint provided by the senior author)

Smith, C. M. 1956. *Availability of Residual Fertilizer Phosphorus and its Evaluation in Iowa Soils*. Ph.D. Thesis, Iowa State University, Ames.

Smith, J. H., C. L. Douglas, and M. J. LeBaron. 1973. Influence of straw application rates, plowing dates, and nitrogen applications on yield and chemical composition of sugarbeets. *Agronomy Journal* 65:797-800.

Smith, K. A., R. S. Howard, and I. J. Crichton. 1988. Efficiency of recovery of nitrogen fertilizer by winter barley. Pp. 73-84. In D. S. Jenkinson and K. A. Smith (Eds.), *Nitrogen Efficiency in Agricultural Soils*. Elsevier Applied Science, London.

Stauber, M. S., and O. R. Burt. 1973. Implicit estimate of residual nitrogen under fertilized range conditions in the Northern Great Plains. *Agronomy Journal* 65:897-901.

Stauber, M. S., O. R. Burt, and F. Linse. 1975. An economic evaluation of nitrogen fertilization of grasses when carry-over is significant. *American Journal of Agricultural Economics* 57:463-471.

Stoecker, A. L., and A. B. Onken. 1989. Optimal fertilizer nitrogen and residual nitrate-nitrogen levels for irrigated corn and effects of nitrogen limitations: An economic analysis. *Journal of Production Agriculture* 2:309-317.

Strong, W. M., J. Harbison, R. G. H. Nielsen, B. D. Hall, and E. K. Best. 1986. Nitrogen availability in a Darling Downs soil following cereal, oilseed and grain legume crops. 2. Effects of residual soil nitrogen and fertiliser nitrogen on subsequent wheat crops. *Australian Journal of Experimental Agriculture* 26:353-359.

Suntheim, L., W. Matzel, and B. Kuhlmann. 1987. Untersuchungen zum Verlauf der P-Alterung im Boden. *Archiv für Acker- und Pflanzenbau und Bodenkunde* 31:285-291.

Swartzendruber, D., and S. A. Barber. 1965. Dissolution of limestone particles in soil. *Soil Science* 100:287-291.

Talibudeen, O. 1958. Isotopically exchangeable phosphorus in soils III. The fractionation of soil phosphorus. *Journal of Soil Science* 9:120-129.

Thomas, J. R., and A. Osenbrug. 1964. *Interrelationships of Nitrogen, Phosphorus, and Seasonal Precipitation in the Production of Bromegrass-Crested Wheatgrass Hay*. U.S. Department of Agriculture, Agricultural Research Service, Production Research Report No. 82. Washington.

Tunney, H. 1990. A note on a balance sheet approach to estimating the phosphorus fertiliser needs of agriculture. *Irish Journal of Agricultural Research* 24:149-154.

Turner, J. 1977. Effect of nitrogen availability on nitrogen recycling in a Douglas-fir stand. *Forest Science* 23:307-316.

Ukrainetz, H., and C. A. Campbell. 1988. N and P fertilization of bromegrass in the dark brown soil zone of Saskatchewan. *Canadian Journal of Plant Science* 68:457-470.

Ukrainetz, H., C. A. Campbell, R. P. Zentner, and M. Monreal. 1988. Response of bromegrass to N, P and S fertilizer on a gray luvisolic soil in northwestern Saskatchewan. *Canadian Journal of Plant Science* 68:687-703.

Watkinson, J. H. 1989. Measurement of the oxidation rate of elemental sulfur in soil. *Australian Journal of Soil Research* 27:365-375.

Webb, J. R., A. P. Mallarino, and A. M. Blackmer. 1992. Effects of residual and annually applied phosphorus on soil test values and yields of corn and soybean. *Journal of Production Agriculture* 5:148-152.

Widjaja-Adhi, I. P. G., M. Sudjadi, and J. A. Silva. 1985. Maize response to phosphorus application at different levels of residual phosphorus in a Paleudult and a Eutrustox. *Indonesian Journal of Crop Science* 1:93-104.

Williams, E. G., and J. W. S. Reith. 1971. Residual effects of phosphate and the relative effectiveness of annual and rotational dressings. Pp. 16-33. In *Residual Value of Applied Nutrients*. Ministry of Agriculture, Fisheries and Food, Technical Bulletin No. 20. Her Majesty's Stationery Office, London.

Wolf, J., and B. H. Janssen. 1989. Calculating long-term crop response to fertilizer phosphorus. Pp. 237-242. In J. van der Heide (Ed.), *Nutrient Management for Food Crop Production in Tropical Farming Systems*. Institute for Soil Fertility and Universitas Brawijaya. Haren, The Netherlands.

Wolf, J., C. T. de Wit, B. H. Janssen, and D. J. Lathwell. 1987. Modeling long-term crop response to fertilizer phosphorus. I. The model. *Agronomy Journal* 79:445-451.

Zentner, R. P., H. Ukrainetz, and C. A. Campbell. 1989. The economics of fertilizing bromegrass in Saskatchewan. *Canadian Journal of Plant Science* 69:841-859.

CHAPTER 7

Fertilizer Placement

F ERTILIZER PLACEMENT refers to the practice of positioning fertilizers in a local area, generally near the plants, as contrasted to applying them more or less evenly (broadcast) to the entire soil. Fertilizer placement is by no means new. According to Semple (1928), localized placement of fertilizers was practiced in ancient Mediterranean agriculture.

In scientific agriculture, localized placement of fertilizers has been employed primarily to achieve improved economic efficiency in fertilizer use. For the most part, fertilizer placement practices have been guided by strictly empirical experiments with little reference to general theories. Reviews of the subject by Batchelor (1983), Dilz and Van Brakel (1985), Randall et al. (1985), Harapiak and Beaton (1986), Richards (1977), Costigan (1988), and Knittel (1988) may be consulted for general information on fertilizer placement.

The coverage of the subject in this chapter emphasizes the theoretical aspects as a guide to understanding and as a stimulus to further advances. The theoretical aspects are based upon the principles discussed in Section 7-1.

7-1. Principles

The effectiveness of fertilizer placement as a means of increasing the yield obtainable from a given quantity of fertilizer depends in part upon plant behavior and in part upon interactions of fertilizers with soils. Both are important.

7-1.1. Internal-Deficiency Effect

When plant roots are located in a uniform environment, such as a stirred nutrient solution, the rate at which they absorb nutrients depends upon three factors related to the law of mass action: the concentrations of the nutrients in solution, the internal deficiencies in the plants, and the root area available for absorption. There are other modifying factors, such as temperature, oxygen supply, and character of the absorbing surfaces. In soil, the water supply is one of the modifying factors.

The rate of nutrient intake by roots increases in response to an internal deficiency in the plants. Thus, Hoffmann (1968) found that when corn plants grown in a nutrient solution were allowed to develop a deficiency of potassium or phosphorus by depriving them of one of these nutrients for a time, returning the potassium or phosphorus to the solution at the original concentration resulted in rates of intake exceeding those in control plants that had been supplied continuously with potassium and phosphorus. As indicated by the delivery of phosphorus in the xylem exudate from plants that had been decapitated at the time of return of phosphorus to the solution, rates of phosphorus uptake during the first day were as much as seven times greater than those in plants supplied with phosphorus continuously.

The increased energy gradient that results from withholding the nutrient enhances the rate of absorption when the nutrient is resupplied. There has been some question regarding the location of the energy gradient because the rates of uptake and translocation to the tops may be increased by a period of nutrient deprivation even though the concentration of the nutrient in the absorbing root may not have changed to a detectable degree. Drew and Saker (1984) suggested that the gradient responsible for the increased uptake rate exists where the nutrients enter the xylem tissue leading to the tops. The tops do show reduced concentrations of nutrients when the root medium is deprived of nutrients.

By the same token, if a fertilizer supplying a deficient nutrient is added broadcast to a soil, the rate of uptake of the nutrient increases with its concentration in solution in contact with the roots up to a maximum rate that depends upon the conditions. And if the same fertilizer is added at the same concentration to only a portion of the soil volume in a localized placement, the initial rate of uptake from the roots in the fertilized zone will be the same, per unit of root surface area, as the initial rate of uptake where the fertilizer is applied broadcast. But because of the smaller absorbing area, a longer time will be required to eliminate the deficit in the plants. Thus, the uptake will continue at an increased rate for a longer time where the fertilizer is placed in only a portion of the soil volume, and the amount of the nutrient absorbed will be greater than the amount absorbed from an equal volume of soil receiving the broadcast fertilizer.

7-1.2. Root-Growth Effect

The internal-deficiency behavior that causes the rate of absorption of a nutrient per unit of root surface area from a zone of placed fertilizer to exceed that from broadcast fertilizer at the same concentration shows up within a period of a few hours. Over longer periods of time, the short-term internal-deficiency effect is supplemented by a root-growth effect that further increases the efficiency of absorption from the zone of placed fertilizer. That is, when only part of the root system of a plant is exposed to a higher external concentration of a deficient nutrient, that portion of the root system makes more growth, with the result that the area of absorbing surface per unit volume of soil is greater in the portion of the soil containing placed fertilizer than in soil receiving the same concentration of broadcast fertilizer.

As an example of the root-growth effect, Edwards and Barber (1976) found that when they placed one of four seminal roots of 5-day-old corn seedlings in a nutrient solution containing mineral nitrogen (ammonium, nitrate, or half ammonium and half nitrate) and the other three roots in a comparable solution without mineral nitrogen, the total length of the one plus-nitrogen root and its branches at the end of 18 days was 30 meters per plant, whereas the total length of the three roots and their branches in the minus-nitrogen solution amounted to only 12 meters per plant. The rate of nitrogen uptake per meter of root at the end of 18 days was three times as great by the 30 meters of roots developed from the single root placed in the nitrogen-containing solution as from the 54 meters of roots developed where all four roots had been in this solution.

The effect of localized application of fertilizer on root growth is not necessarily limited to the zones containing the fertilizer. De Miranda et al. (1989) divided the roots of individual sorghum plants between equal volumes of soil in a growth chamber experiment in which only one of the volumes received phosphate fertilizer. Phosphorus was translocated from plus-phosphorus roots to minus-phosphorus roots. The minus-phosphorus roots did not produce as much dry weight or length as the plus-phosphorus roots, but they still responded greatly to the phosphorus supplied to the plus-phosphorus roots.

7-1.3. Field Experimental Evidence of Combined Internal-Deficiency and Root-Growth Effects

Field experimental evidence illustrating the combined effects of internal deficiency and root growth is shown in Fig. 7-1. These data were obtained in an experiment in which equal quantities of concentrated superphosphate were either drilled with the seed of durum wheat or broadcast and worked into the surface of the calcareous soil employed. Measurements made at the time of tillering showed that fertilization had increased the concentrations of both extractable phosphorus and roots in the surface 5 centimeters of soil. The tendency in the control plots for an increased concentration of roots in the soil zone where the seeds had been planted was much enhanced by fertilization, especially with drilled fertilizer. Photographic evidence of the root proliferation effect has been published by De Wit (1953), Cooke (1954), and others.

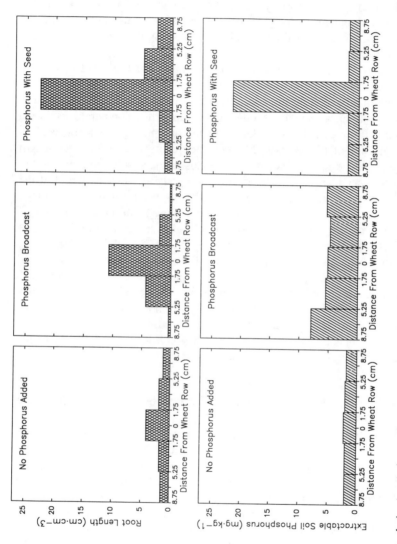

Fig. 7-1. Length of wheat roots (upper three graphs) and extractable soil phosphorus (lower three graphs) in segments of the surface 5 centimeters of calcareous soil from a cross section perpendicular to a row of drilled wheat in Syria. Rows were spaced 17.5 centimeters apart. The graphs on the left are for the control soil; those in the center are for the soil receiving 52.5 kilograms of fertilizer phosphorus per hectare, broadcast and worked in just before planting; and those on the right are for the soil receiving 52.5 kilograms of fertilizer phosphorus drilled with the seed. The fertilizer was concentrated superphosphate. (Matar and Brown, 1989)

7-1.4. Incomplete Compensation

The adjustments in uptake rate per unit area of root surface and in total area of root surface that are made by roots in zones of placed fertilizer do not necessarily compensate completely for a deficiency of the nutrient around the remaining roots. This principle may be illustrated by the data for uptake per plant in the experiment by Edwards and Barber (1976) discussed in Section 7-1.2. Per plant, the rate of nitrogen uptake by the 18-day-old corn plants was 65% as great where only one of the four roots had been in the nitrogen-containing solution as where all four roots had been in that solution.

7-1.5. Interaction With the Soil

Fertilizers interact with soil (see Sections 1-3.2, 4-3.6, and 4-4.3), and this generally affects the availability of the nutrient or nutrients added. The consequences in terms of availability of a nutrient to plants as the method of application changes from a mixture with the total volume of soil (broadcasting) to mixtures with decreasing proportions of the soil volume (various placements) depend upon the circumstances. If interaction with the soil decreases the concentration of the nutrient in solution and its availability to plants (the usual case), placement of a given amount of a soluble form of a nutrient in a progressively smaller portion of the soil can increase its concentration and, up to a limit, its availability.

From the standpoint of interaction with the soil, the net effect of fertilizer placement on nutrient availability is a result of a balance between factors with opposing effects. If the concentration of the nutrient in soil receiving broadcast fertilizer is below that needed by the plants to produce the maximum yield, the increase in the concentration of the nutrient when the same quantity of a soluble fertilizer is applied to only a portion of the soil can be a positive effect of placement. The rate of absorption of nutrients by roots increases with the concentration of the nutrients in solution up to a limit, above which the rate of uptake remains constant or decreases. Acting to decrease the availability is the smaller proportion of the root system capable of absorbing the nutrient from the soil containing the placed fertilizer. The greater the loss of availability due to reaction with the soil, the smaller is the proportion of the soil volume with which a given amount of fertilizer must be mixed for maximum uptake (see Kovar and Barber, 1989).

Fig. 7-2 illustrates the effect of placement on concentration in an experiment in which equal quantities of phosphorus as superphosphate granules of different size were applied to a given soil. As a consequence of decreases in availability due to reaction with the soil, the relative benefit from placement for the major fertilizer nutrients is generally phosphorus > potassium > nitrogen. See a paper by Kovar and Barber (1990) for a discussion of the behavior of phosphorus and potassium in different terms.

With phosphorus and potassium, the major interaction effects are due to inorganic chemical reactions. With nitrogen, principal emphasis has been placed upon immobilization by microorganisms decomposing plant residues low in nitrogen. See, for example, Malhi et al. (1989).

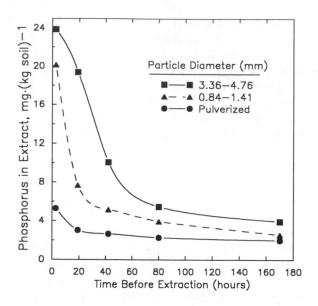

Fig. 7-2. Phosphorus extracted from a soil by shaking the soil for 1 minute with 15.4 parts of water by weight after the moist soil had been incubated for different lengths of time with superphosphate of different granule sizes, all supplying 39.6 milligrams of phosphorus per kilogram of soil. (Lawton and Cook, 1955)

On the other hand, if interaction with the soil increases the concentration of a nutrient in solution, fertilizer placement may decrease the availability of the nutrient to plants. Under these circumstances, the total increase in the amount of the nutrient in solution and the increase in availability will be greatest where the nutrient is mixed with the total volume of soil. Such behavior is probably uncommon, and it has been generally overlooked in relation to placement, but it may occur with nitrogen. Under some circumstances, additions of fertilizer nitrogen may enhance the rate of mineralization of soil organic nitrogen. Fig. 1-11 is an example.

7-1.6. Differential Effects of Small and Large Quantities of Fertilizer
Where soluble placed and broadcast fertilizers are applied in equal quantities per hectare, the yields obtained from small quantities of placed fertilizer generally exceed those from broadcast fertilizer. The reverse generally is found with heavy applications. This principle is illustrated schematically in Fig. 7-5. The higher yields with small quantities of placed fertilizer than with equal quantities of broadcast fertilizer may be explained on the basis of the principles discussed in Sections 7-1.1, 7-1.2, and 7-1.5.

Several principles may contribute to the explanation for lower yields from large quantities of placed than broadcast fertilizer at equal rates of fertilizer per hectare. First, there is an upper limit to the rate of uptake per unit area of root surface. When the concentration of the nutrient around the root is great enough to evoke the maximum uptake rate, further increases in concentration resulting from adding a greater quantity of a highly soluble fertilizer to the same volume of soil will not increase the rate of uptake.

Second, with fertilizers of low solubility, the concentration of the nutrient in solution will not be increased with additions of fertilizer per unit volume of soil that exceed the addition necessary to produce a chemically saturated solution. This effect is of special importance with phosphate fertilizers. The general rule with phosphate fertilizers is that placing a limited quantity of fertilizer in a concentrated band gives favorable results with fertilizers of high solubility and unfavorable results with fertilizers of low solubility.

Third, with fertilizers of high solubility, high concentrations of salts and related factors, including induced toxicities and extremes of pH, may damage the roots and prevent their growth in portions of concentrated bands of fertilizer. Ishizuka et al. (1965) published photographs illustrating this effect for nitrogen fertilizer.

7-2. The De Wit Theory

The theory developed by De Wit (1953) relates the nutrient uptake to be expected from localized placement of the nutrient in a fertilizer to the uptake when the fertilizer is applied broadcast under the special conditions explained in the following paragraphs. Although most fertilizer placement experiments are not conducted according to the required conditions, the theory provides a framework for understanding and for making inferences from most experiments, and to date this has been its principal value.

7-2.1. Basis of Comparison

According to the De Wit theory, the yields obtainable with a wide range of fertilizer placement patterns can be estimated if the relationships among yield, quantity of nutrient absorbed from the fertilizer, and quantity of fertilizer added are known for fertilizer applied broadcast. The comparison among methods is made on the basis of equal quantities of fertilizer per unit volume of soil that actually receives fertilizer.

The basis of comparison is illustrated by Fig. 7-3, in which two application patterns are compared under otherwise equal conditions. The upper sketch shows a common broadcast pattern, in which the plant rows are X_b centimeters apart. The lower sketch shows a placement pattern, in which the fertilizer is applied in bands X_r centimeters in width and parallel to the crop rows. If the vertical distribution of the fertilizer is identical in the two methods, the relative quantities of fertilizer applied per hectare are the same as the relative surface areas under which the fertilizer is located. Thus, if the quantity of fertilizer applied broadcast is M_b kilograms per hectare, the quantity of fertilizer placed in bands along the row is $(X_r/X_b)M_b$ kilograms per hectare. Or, in the case not shown in which

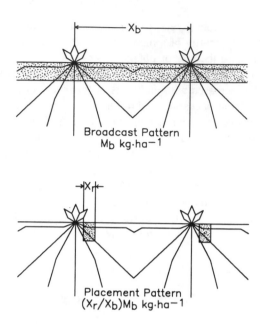

Fig. 7-3. Schematic representation of the basis for comparing the efficiency of broadcast (top) and row (bottom) applications of fertilizer according to the De Wit theory. The shaded areas represent the soil zones that receive fertilizer with equal vertical distribution and equal concentration per unit volume of fertilized soil. X_b is the distance between crop rows, and X_r is the width of the fertilizer band placed along the row. (De Wit, 1953)

identical bands of fertilizer are placed on each side of each crop row, the amount of placed fertilizer is $(2X_r/X_b)M_b$ kilograms per hectare.

Because the lateral diffusion of fertilizer salts in soil is slight, and because movement of salts with movement of soil water is mostly in a vertical direction, it can be assumed that the placed fertilizer remains confined beneath the area of soil to which it is applied. If such is the case, the extent to which the fertilizer is affected by reaction with the soil, by leaching, or by any other soil process will be the same with placed fertilizer as with broadcast fertilizer. Thus, although the vertical distribution and the chemical condition of the fertilizer may vary with time, the initial availability to plants of the placed fertilizer beneath the fertilized area X_r should be identical with that beneath an equal area receiving broadcast fertilizer. This constancy of behavior of the soil-fertilizer system with broadcast and placed fertilizer applied in equal concentrations and with equal vertical distribution constitutes the particular advantage of the basis of comparison employed by De Wit.

7-2.2. The Compensation Function

Under the hypothetical circumstances in which plant roots have become distributed uniformly throughout the fertilized portion of the soil without having absorbed any of the nutrient supplied by the fertilizer, after which the roots are allowed to absorb the nutrient from the fertilizer in the normal manner, the ratio of the resulting initial rates of absorption of the nutrient from placed and broadcast fertilizer may be designated as $\hat{U}_r/\hat{U}_b$. Under otherwise equal conditions, the ratio $\hat{U}_r/\hat{U}_b$ evidently will be a function of X_r/X_b. The ratio $\hat{U}_r/\hat{U}_b$ will be zero where $X_r/X_b = 0$ and unity where $X_r/X_b = 1$. The question then is what is the relationship in the interval between these two limits?[1]

De Wit found no experimental data that would permit direct evaluation of the functional relationship between $\hat{U}_r/\hat{U}_b$ and X_r/X_b. He noted, however, that under certain conditions the ratio of the total uptake of the nutrient from placed fertilizer to the total uptake of the nutrient from broadcast fertilizer, U_r/U_b, will give a close approximation of the ratio $\hat{U}_r/\hat{U}_b$. Two types of data available in the literature are suitable for the purpose. First, there are experiments in which plant roots were divided between two nutrient solutions — one a complete solution and the other lacking one nutrient. In this work, the concentrations were kept approximately constant by frequent renewal of the nutrient solutions. Second, there are experiments in which the concentration was allowed to change with time, but in which the proportion of the added nutrient recovered in the plants was so small that the concentration of placed fertilizer can be assumed to be the same as the concentration of broadcast fertilizer.

De Wit assembled the data available in the literature that fell in the two categories just described. After recalculating the experimental results, he plotted the resulting values of U_r/U_b against the corresponding values of X_r/X_b, as shown in Fig. 7-4A.

He was able to fit the data by the exponential equation

$$\frac{U_r}{U_b} = \left(\frac{X_r}{X_b}\right)^{0.44}.$$

De Wit called this equation the *compensation function* because it represents the extent to which a decrease in the proportion of the soil to which the fertilizer is applied is compensated by an increase in rate of absorption of the fertilizer from the fertilized portion of the soil. The solid central line in Fig. 7-4A is the compensation function, and the upper and lower broken lines are the confidence limits calculated by De Wit for the compensation function.

---

[1]According to the statement about initial rates of absorption in Section 7-1.1, $\hat{U}_r/\hat{U}_b$ will be equal to X_r/X_b. The conflict between that statement and De Wit's approach in developing the compensation function may be explained as a consequence of a difference in interpretation of "initial." In Section 7-1.1, initial is interpreted as immediate, that is, the rate of uptake before there has been time for the movement of the absorbed nutrient into other parts of the plants to influence the rate of absorption. As indicated in Section 7-1.2, only a few hours may be needed for the deficiency of a nutrient in the total plant to cause the rate of movement of the absorbed nutrient from the roots in the placed zone to exceed the rate of movement from an equal volume of soil containing broadcast fertilizer, and thus to cause the placement effect described by De Wit, in which $\hat{U}_r/\hat{U}_b$ generally will exceed X_r/X_b.

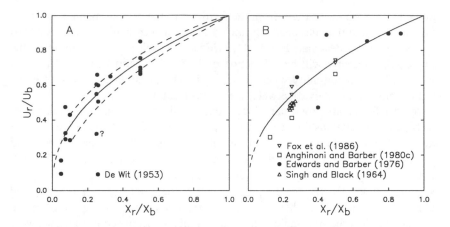

Fig. 7-4. Plots of U_r/U_b against X_r/X_b. The data in *A* on the left were assembled by De Wit (1953) from sources in the literature. The solid line with a tentative extension to small values is the best-fitting curve found by De Wit with the empirical equation $U_r/U_b = (X_r/X_b)^{0.44}$. The broken lines on either side are the confidence intervals calculated by De Wit. The curve calculated by De Wit is repeated in *B* on the right, but the data are from more recent sources.

The data in Fig. 7-4A were derived from field experiments, soil-culture experiments, and nutrient-solution experiments; from experiments with rice, corn, oat, bean, swede, potato, millet, and buckwheat; and from experiments with nitrogen, phosphorus, potassium, and iron. They were derived from experiments by independent researchers in The Netherlands, Puerto Rico, the Soviet Union, Sweden, the United Kingdom, and the United States. Because of the generally acceptable fit, one may conclude that, as a first approximation, the relationship between the ratios U_r/U_b and X_r/X_b is independent of the experimental conditions as long as the latter are the same for placed fertilizer as for broadcast fertilizer.

Although it is clear enough theoretically that $U_r/U_b = 0$ when $X_r/X_b = 0$ and that $U_r/U_b = 1$ when $X_r/X_b = 1$, the location of the function relating U_r/U_b and X_r/X_b between the two fixed points had to be determined empirically. De Wit (1953) considered the empirical equation he used to be of questionable validity in the region near $U_r/U_b = 0$ and $X_r/X_b = 0$. His doubt is reflected by his use of a broken line in this region.

Since the publication of the assembly of data shown in Fig. 7-4A, additional research has been done that satisfies the experimental conditions set forth by De Wit. These data, shown in Fig. 7-4B, are in general agreement with the original data upon which the compensation function was based. Especially in the upper reaches of the function, the agreement is satisfactory. For small values of X_r/X_b, however, the compensation function developed by De Wit (1953) still is not well defined. If eventually the empirical relationship of U_r/U_b to X_r/X_b is changed as a result of new and better experimental data, this would not invalidate De Wit's theoretical considerations because his theory is not based upon a specific form for the compensation function. But the calculations probably would be more complex.

7-2.3. Calculations Based on the Compensation Function

If use of the compensation function is restricted for the present to the specific conditions for which it was derived (that is, the same concentration of placed fertilizer as of broadcast fertilizer), the quantity of the nutrient absorbed from fertilizer placed under the soil area X_r can be estimated from the quantity absorbed from the fertilizer placed under the soil area X_b. Thus,

$$U_r = U_b\left(\frac{X_r}{X_b}\right)^{0.44}.$$

If U_b and X_b are held constant, variation of X_r within the range from 0 to X_b will cause U_r to describe a curve similar to that of the compensation function shown in Fig. 7-4. An entire curve can be calculated with the aid of the compensation function when one knows the quantity of the nutrient absorbed from only one quantity of broadcast fertilizer.

The compensation function furnishes much more information if the quantity of the nutrient absorbed from different quantities of broadcast fertilizer is known. If, for example, the curve for nutrient uptake from the broadcast fertilizer versus quantity of broadcast fertilizer is that for $X_r/X_b = 1$ in Fig. 7-5, the effect of various placement patterns may be estimated. Comparing a given quantity of

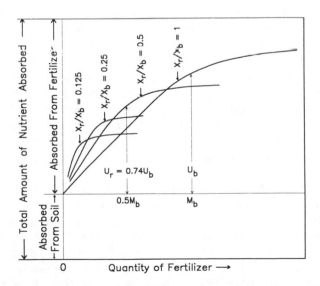

Fig. 7-5. Nutrient absorption by a crop from broadcast fertilizer ($X_r/X_b = 1$) and from placed fertilizer with $X_r/X_b = 0.125$, 0.25, and 0.5, as estimated from the compensation function. (De Wit, 1953)

placed fertilizer (M_r) with the quantity of broadcast fertilizer (M_b) for which U_r = U_b then provides a measure of the efficiency of placement relative to broadcasting for the selected values of M_r and X_r/X_b.

The procedure for estimating the effect of a given placement pattern is illustrated by the calculation of the point denoted by the intersection of the vertical line with the curve for X_r/X_b = 0.5 from the corresponding point on the curve for X_r/X_b = 1: $U_r = (U_b)(X_r/X_b)^{0.44} = (U_b)(0.5)^{0.44} = 0.74U_b$. The value 0.74 is known as the *compensation factor*. It is a constant value applicable to all values of U_b where X_r/X_b = 0.5. Because the quantity of fertilizer applied per unit volume of fertilized soil is constant, the quantity applied where X_r/X_b = 0.5 is half that applied where X_r/X_b = 1. The calculated value for U_r thus is plotted at half the value for X_r/X_b = 1, that is, at $0.5M_b$.

Fig. 7-5 shows the results of such calculations for X_r/X_b = 0.5, 0.25, and 0.125, the compensation factors 0.55 and 0.41 being used for the latter two values of X_r/X_b. An experimental verification of the shapes of the calculated curves in Fig. 7-5 was published by Fox et al. (1986).

When making calculations based on the compensation function, one must bear in mind continually the fact that the values of U_r and U_b refer to the quantities of the nutrient absorbed from the fertilizer and not to the total amount of the nutrient absorbed from fertilizer and soil. The quantity absorbed from the control soil must be subtracted from total absorption before applying the compensation function.

Fig. 7-5 shows that with small applications of fertilizer, the quantity of the nutrient absorbed from placed fertilizer exceeds that absorbed from an equal quantity of broadcast fertilizer, and that the quantity absorbed increases with decreasing values of X_r/X_b. The opposite situation prevails with heavy applications; here, the quantity of the nutrient absorbed from equal quantities of fertilizer decreases with decreasing width of the fertilizer band. Thus, the relative availability coefficients or relative efficiencies of placed and broadcast fertilizer vary with the quantity of fertilizer applied even though equal quantities of placed and broadcast fertilizer are applied per unit volume of fertilized soil.

The important point that the relative efficiency of placed and broadcast fertilizer varies with the quantity applied has not been universally recognized. Often it is not obvious because adequate response curves are not obtained.

7-2.4. Applying the Compensation Function to Crop Yields

From the practical standpoint, the criterion by which different placement methods are compared is usually crop yield, and not the quantity of the nutrient absorbed from the fertilizer. The relationship between crop yield and quantity of the nutrient absorbed from the fertilizer thus must be known before the compensation function can be used to compare yields with placed and broadcast fertilizer.

In most instances, the relationship between yield and quantity of the nutrient absorbed is independent of the placement method. Under such circumstances,

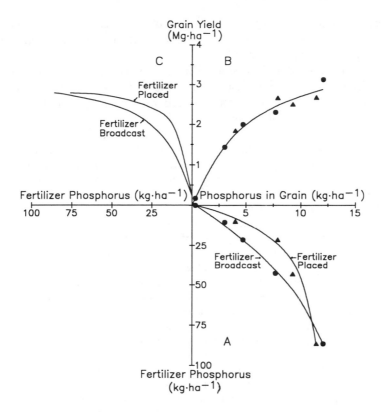

Fig. 7-6. Results of a field experiment conducted by Prummel in The Netherlands in which superphosphate was applied broadcast ($X_r/X_b = 1$) and placed ($X_r/X_b = 0.26$). A. Yield of oat grain versus quantity of fertilizer phosphorus added. B. Yield of oat grain versus yield of phosphorus in the grain. C. Yield of oat grain versus quantity of fertilizer phosphorus added. The yields for placed fertilizer in *C* have been calculated by the procedure described in the text. (De Wit, 1953)

if U_b and yield are known as a function of quantity of fertilizer applied broadcast, the compensation function may be used to estimate the yields that would be expected with different quantities of fertilizer applied in various placement patterns.

The results of a field experiment on fertilization of oat with superphosphate provide an example. See Fig. 7-6, which summarizes the pertinent data in the form in which they were represented by De Wit. In accordance with the compensation function, the plots in the fourth quadrant (*A*) show that two different curves are needed to represent the yield of phosphorus in the grain versus the quantity of fertilizer phosphorus added, one for broadcast superphosphate and one for superphosphate placed at $X_r/X_b = 0.26$. In the plot of yield of grain versus yield of phosphorus in the grain shown in the first quadrant (*B*), however, the points for broadcast and placed fertilizer can be represented by a single line.

This behavior implies that a given amount of phosphorus absorbed by the oat plants was of equal value in producing grain, whether it was absorbed from placed or broadcast superphosphate.

The following procedure is used to estimate from the yield-of-nutrient versus nutrient-added curve for broadcast fertilizer the yield-of-crop versus nutrient-added curve for placed fertilizer at any desired value of X_r/X_b. The data in Fig. 7-6 are used as a numerical example.

1. For a particular quantity of the nutrient applied as broadcast fertilizer (M_b), determine the corresponding uptake of the nutrient from the fertilizer (U_b). If an application of 25 kilograms of phosphorus per hectare is chosen as the quantity of phosphorus added broadcast, Fig. 7-6A indicates that $U_b = 5.4 - 0.3 = 5.1$ kilograms of phosphorus per hectare.

2. Calculate the quantity U_r from U_b and the desired value of X_r by means of the compensation function, $U_r = U_b(X_r/X_b)^{0.44}$. In this case, $U_r = 5.1(0.26)^{0.44} = 2.8$ kilograms per hectare. The total amount of phosphorus absorbed where the fertilizer is placed will be $2.8 + 0.3 = 3.1$ kilograms per hectare.

3. From the plot showing crop yield versus total amount of the nutrient absorbed (Fig. 7-6B), determine the yield to which the calculated value of total uptake with placed fertilizer corresponds. In this instance, the quantity 3.1 kilograms of phosphorus in the grain corresponds to a grain yield of 1.5 megagrams per hectare.

4. Plot the resulting yield value with crop yield as ordinate and quantity of the nutrient added in the placed fertilizer as the abscissa (see Fig. 7-6C), using as the quantity M_r of nutrient in the placed fertilizer the value $M_r = (X_r/X_b)M_b$. In this instance, $M_r = (0.26)(25) = 6.5$ kilograms of phosphorus per hectare.

5. Repeat steps (1) through (4) with different quantities of broadcast fertilizer until enough points have been obtained to permit drawing a response curve for the placed fertilizer.

6. Plot in the same graph the response curve for broadcast fertilizer, using as abscissa the quantity of fertilizer actually employed.

In general, the maximum crop yield obtainable with placed fertilizer approaches that obtainable with broadcast fertilizer as the ratio X_r/X_b approaches unity. If the curve for yield of nutrient versus nutrient added in placed fertilizer intersects the corresponding curve for broadcast fertilizer in the region of luxury consumption, placed fertilizer may produce a crop yield as great as the maximum obtainable with a much larger amount of broadcast fertilizer. De Wit called this behavior Case IA. The upper left graph in Fig. 7-7 illustrates this behavior for $X_r/X_b = 0.33$.

On the other hand, where the yield-of-nutrient versus nutrient-applied curve for placed fertilizer intersects that for broadcast fertilizer below the region of luxury consumption, the maximum yield is lower with placed than with broadcast fertilizer. For additions of fertilizer in excess of that corresponding to the point of intersection, placement produces a lower yield of nutrient than does broadcasting. This situation is illustrated for $X_r/X_b = 0.33$ in the upper right graph in Fig. 7-7. De Wit called this situation Case IB.

Cases IA and IB have in common the fact that the crop yield versus nutrient uptake curve is independent of the placement method. They differ in the point of intersection of the curves for yield of nutrient versus nutrient added. This difference, however, does not interfere with the use of the compensation function to estimate the uptake of the nutrient with different placements.

7-2.5. Secondary Effects

As was pointed out in Chapter 1 in connection with factors that affect the shape of response curves, the effects of fertilizers are not necessarily confined to supplying the nutrients and other substances they contain. Fertilizers may have secondary effects in the soil, which in turn influence plant behavior. The increase in availability of phosphorus in neutral or alkaline soil as a result of applying acid-forming nitrogen fertilizer is an example of a secondary effect that is important under certain circumstances.

Secondary effects of fertilization may have a significant influence on the comparative value of different fertilizer placements. In some instances, the compensation function may not apply to the secondary effects as such. In general, the greatest secondary effects per unit volume of fertilized soil and per unit quantity of fertilizer are likely to be obtained with the first increments of fertilizer. Thus, if a sufficiently higher concentration of placed than of broadcast fertilizer is employed to produce equal uptakes of the nutrient supplied by the fertilizer, the greater secondary effects probably will be associated with the broadcast fertilizer.

The lower left segment of Fig. 7-7 illustrates the behavior with negative secondary effects (De Wit's Case II) where $X_r/X_b = 0.33$. Here the yield of dry matter associated with a given yield of nutrient and the maximum yield are greater for placed fertilizer than broadcast fertilizer. The lower right segment of Fig. 7-7 illustrates the behavior with positive secondary effects (De Wit's Case III) where $X_r/X_b = 0.33$. Here the yield of dry matter associated with a given yield of nutrient and the maximum yield are greater with broadcast fertilizer than placed fertilizer.

With all the cases illustrated in Fig. 7-7, the response to placed fertilizer relative to the response to broadcast fertilizer will depend upon the shape of the response curve for broadcast fertilizer and upon X_r/X_b. As may be inferred from Fig. 7-5, for example, the maximum yield obtainable with placed fertilizer will be below that for broadcast fertilizer for small values of X_r/X_b even if fertilizer application has the negative secondary effects illustrated in the lower left segment of Fig. 7-7.

De Wit noted that inferences about the existence of secondary effects may be made by observing whether the lower end of the yield-of-dry-matter versus yield-of-nutrient curve points toward zero yield of nutrient (no secondary effects), a positive yield of nutrient (positive secondary effects), or a negative yield of nutrient (negative secondary effects). The experimental data, however, are not always good enough and the nutrient deficiency is not always great enough to

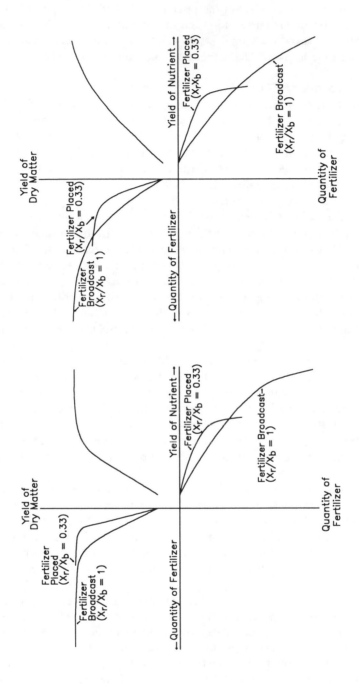

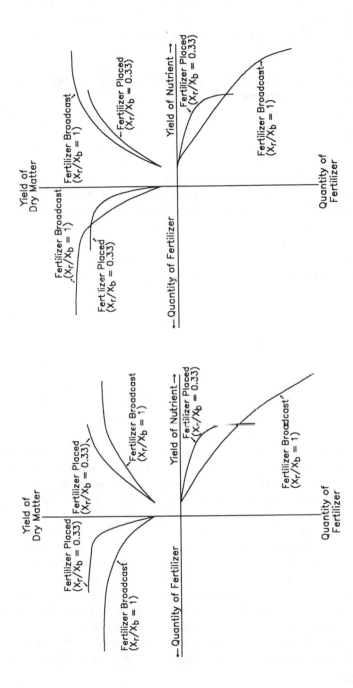

Fig. 7-7. Generalized representation of four different classes of results obtained in comparisons of placed and broadcast fertilizer according to the De Wit theory. See the text for explanation. (De Wit, 1953)

permit an unequivocal decision. Where differences in crop-yield versus yield-of-nutrient curves do occur, of course, application of the compensation function to estimate the yields obtainable with placed fertilizer would produce biased results.

7-2.6. Modification of the Compensation Function for High Nutrient Recovery Percentages

The compensation function was derived for the condition that the concentration of the fertilizer in the fertilized portion of the soil is the same with broadcast and placed fertilizer. For some of the data used in evaluating the compensation function, the assumption was made that because only a small fraction of the fertilizer was absorbed by the plants, the total quantity of the nutrient absorbed could be substituted for the rate of absorption without serious error.

If a substantial proportion of the nutrient added in the fertilizer is recovered by the crop, the concentration of the placed fertilizer falls below that of the broadcast fertilizer. Under such circumstances, the compensation function would overestimate the value of fertilizer placement. To take into account the effect of utilization of the nutrient, De Wit derived the following modification of the compensation function:

$$R_r = \frac{100(X_b/X_r)^{0.56} R_b}{100 + [(X_b/X_r)^{0.56} - 1]R_b} ,$$

where R_r is the percentage recovery of the nutrient from the placed fertilizer and R_b is the percentage recovery of the nutrient from the broadcast fertilizer. The percentage recovery R is defined by $R = 100 (U/M)$, where U is the increase in uptake of the nutrient resulting from fertilization and M is the quantity of the nutrient applied.

Table 7-1 shows for different percentage recoveries of a nutrient from broadcast fertilizer the percentage recoveries that would be calculated by use of the compensation function and the equation in the preceding paragraph. The data in the table indicate that where the recovery of the nutrient from the broadcast fertilizer is not in excess of about 10 or 15%, the compensation function can be applied to the total absorption from broadcast fertilizer without serious error. This conclusion applies only to instances in which X_r/X_b is 0.25 or greater. The error would be greater for smaller values of X_r/X_b. Field experiments with phosphorus-bearing fertilizers generally would have percentage recoveries of the added phosphorus low enough to use the compensation function without important error. For situations in which the percentage recovery of the added nutrient is too great to permit direct use of the compensation function to estimate U_r, a closer approximation can be obtained by using the equation for high recovery percentages. Experiments with nitrogen- and potassium-bearing fertilizers often need such corrections.

Table 7-1. Estimated Recoveries of a Nutrient From Fertilizer Placed With X_r/X_b = 0.25 Based on the Compensation Function and an Equation for High Recoveries, and the Estimated Additional Amount of Placed Fertilizer Required to Maintain the Concentration of Placed Fertilizer Equal to That of 100 Units of Broadcast Fertilizer

Recovery of Nutrient From Broadcast Fertilizer (%)	Recovery of Nutrient From Placed Fertilizer Estimated by Different Equations		Additional Units of Placed Fertilizer Needed
	Compensation Function (%)	Equation for High Recoveries (%)	
0	0	0	0
5	10.9	10.3	1.5
10	21.7	19.5	2.9
15	32.6	27.7	4.4
20	43.5	35.2	5.9
40	87.0	59.2	11.8
60	130	76.6	17.6
100	217	100	29.4

The units of the values in the last column of Table 7-1 are the same as those used for the fertilizer added broadcast (M_b) and have been calculated for the circumstances in which M_b − 100, X_r/X_b = 0.25, and M_r = 25. Calculations were made by De Wit's equation

$$M_{rt} - \left(\frac{X_r}{X_b}\right)M_b = \left[\frac{M_b R_b}{100}\right]\left[\left(\frac{X_r}{X_b}\right)^{0.44} - \left(\frac{X_r}{X_b}\right)\right] \quad ,$$

where M_{rt} is the total amount of the nutrient that theoretically would have to be placed initially and added gradually during absorption to maintain the same concentration of fertilizer in the placed zone as in an equivalent soil zone containing broadcast fertilizer. The other symbols have the same meanings as before.

7-2.7. Time of Availability

The hypothetical circumstances De Wit described in developing the compensation function are those in which the roots of the plants are fully extended to occupy the space between the rows without having absorbed any of the nutrient from the fertilizer. Then the roots are allowed to start absorption. These circumstances are never duplicated in practice. The ratio of the volumes of fertilized soil occupied by roots changes with time during early growth and reaches the constant value X_r/X_b only when the roots extend throughout both the placed and broadcast fertilized zones.

De Wit made an analysis of the available data in an attempt to determine the extent to which variations in time of availability affect the use of the compensation function. He concluded that where placed and broadcast fertilizer are applied at the same time, the failure of the initial ratios of volumes of fertilized soil occupied by roots to be X_r/X_b is usually not of importance. The final yield versus nutrient absorption curve is usually unaffected. By way of explanation, he gave the following reasons: (1) In actual practice, placed fertilizer usually is located near the seed, so that it will be positionally available soon after germination. (2) The major portion of each nutrient present in the mature plant is absorbed after the time when the roots are present under the entire soil surface. (3) Plants whose roots do not immediately contact the fertilizer generally catch up with plants having fertilizer in the soil around the seed, and give the same final results. The results of a more recent experiment illustrating this point are shown in Fig. 7-8.

De Wit concluded that failure of the roots to encounter the fertilizer in the first few weeks of growth has a significant effect on the final yield versus nutrient-absorption curve only when: (1) the soil is so deficient in the nutrient that the final yield without fertilizer is less than about 50% of that obtainable with an optimum supply of the nutrient, or (2) the plant is damaged by unfavorable conditions, such as seedling diseases or frosts, during the period before it contacts the fertilizer. The plants without fertilizer usually are damaged more severely than those with fertilizer. This subject will be discussed further in Section 7-7.

7-2.8. Wide Row Spacing

If crop rows are spaced so far apart that the roots of plants in adjacent rows do not meet in the soil between the rows, X_b will overestimate the relative area

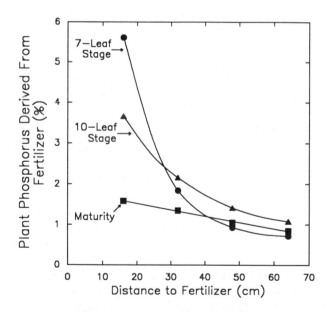

Fig. 7-8. Percentage of the phosphorus in corn plants derived from phosphorus-32-tagged ammonium polyphosphate applied at a depth of 10 centimeters at different lateral distances from the seed. Data are average values from field experiments with 5.6 and 11.2 kilograms of fertilizer phosphorus per hectare in 2 years in Nebraska. (Eghball and Sander, 1989)

from which the plants are absorbing broadcast fertilizer. Consequently, application of the compensation function will underestimate the value of banded fertilizer relative to broadcast fertilizer. A simple modification of the compensation function may be made to correct for this discrepancy. If the width of the area without roots is x, the comparative sections occupied by roots will be X_r and $(X_b - x)$. To take these effective dimensions into account, the compensation function may be written

$$\frac{U_r}{U_b} = \left(\frac{X_r}{X_b - x}\right)^{0.44}.$$

According to an analysis by Greenwood and Draycott (1988), the failure of crop roots to extend to the full distance between the rows is a characteristic of vegetable crops but not of the cereal field crops. These authors investigated the behavior of 12 different vegetable crops in experiments with various quantities of fertilizer nitrogen. In contrast to the cereal crops, for which the percentage recovery of fertilizer nitrogen applied broadcast remains approximately constant until the supply of nitrogen exceeds the demand by the crop, the percentage recovery by vegetable crops showed a linear decrease with an increase in quantity of fertilizer nitrogen. The estimated percentage recovery associated with zero application of fertilizer increased with the dry weight of the above-ground portion of the crop plus the storage roots to approximately 4 megagrams per hectare, above which it remained approximately constant.

The authors concluded that a likely explanation for the observed behavior of vegetable crops was that the ability of the roots to penetrate the soil between crop rows and to extract nitrate from it increased linearly with the plant dry weight up to 4 megagrams per hectare and that above this weight the roots had extended throughout the entire volume of soil and were able to extract virtually all the nitrate from it to the depth of rooting. Although these findings illustrate a general principle, the distance between crop rows would of course influence the yield of dry matter at which the roots of plants in adjacent rows would meet.

The De Wit theory was not developed for crops grown in solid stands, such as grasses, but placement is an important consideration with these crops also. A complicating factor in solid stands is the competition among roots of adjacent plants that reduces their lateral spread and requires that bands of fertilizer be more closely spaced than would be necessary if the plants were in rows. With grassland, additional complications are the need for repeated applications and the inability to mix broadcasted nitrogen fertilizers with the soil to prevent the significant losses of ammonia that can occur by volatilization from surface applications if the nitrogen is in the form of urea or ammonium compounds. See a paper by Malhi and Ukrainetz (1990) for experimental data and further discussion.

7-2.9. Thickness of Fertilizer Band

In most fertilizer placement experiments, the vertical distribution of placed and broadcast fertilizer is not identical, as required in the De Wit model. Usually the placed fertilizer is confined to a smaller part of the vertical soil cross section than is the broadcast fertilizer.

Small differences in vertical distribution may be unimportant if evaporation and rainfall cause upward and downward movement of water through the zones containing the fertilizer. The initial differences in thickness of the placed and broadcast fertilizer zones then will tend to disappear.

De Wit suggested a modification of the compensation function for the conditions in which the initial differences are large and not eliminated by water movement. If t_r and t_b are the vertical dimensions of the placed and broadcast fertilizer, respectively, the compensation function may be written

$$\frac{U_r}{U_b} = \left(\frac{X_r t_r}{X_b t_b}\right)^{0.44}.$$

This modified version of the compensation function may be applied like the original if the effectiveness of the fertilizer in unit volume of fertilized soil is independent of the vertical dimensions, a condition that is not satisfied in many circumstances.

Placed fertilizer usually is located some distance beneath the surface. Part of the broadcast fertilizer usually occurs in the surface layer. If plant roots are excluded from the surface layer by repeated cultivation or by dryness, a portion of the broadcast fertilizer will be used inefficiently, while the placed fertilizer may be unaffected. If the surface ineffective layer extends into the zone of placed fertilizer, both placed and broadcast fertilizer will be affected, but probably to different degrees.

The idea implied in the foregoing formulation may be used in a different sense in experiments in which placement methods involving different thicknesses of the fertilized zone are compared, and in which U_r, U_b, X_r, and X_b are known. Under these circumstances, the compensation function may be used to find an apparent t_r. The modified equation is

$$Apparent\ t_r = \left[\frac{\dfrac{U_r}{U_b}}{\left(\dfrac{X_r}{X_b t_b}\right)^{0.44}}\right]^{2.27},$$

which may be compared with the actual value. If the apparent and actual values are the same, the placement method under consideration may be said to have the effect that would be expected if the vertical distribution were the same as that of the broadcast fertilizer. If the apparent t_r is larger than the actual t_r, the placement is more effective than would be predicted from the broadcast fertilizer; and if the apparent t_r is smaller than the actual t_r, the placement is less effective than would be predicted from the broadcast fertilizer. The depth factor will be considered further in Section 7-10.

7-3. The Steenbjerg Theory

Steenbjerg (1957) called attention to the consequences of sigmoid yield response curves in terms of fertilizer placement. He noted that the quantity of fertilizer that produces the maximum increase in crop yield per unit of fertilizer (that is, the most efficient application) with a sigmoid response curve will be the one corresponding to the point at which a straight line drawn through the intersection of the yield response curve with the Y axis (zero application of the nutrient) is tangent to the yield curve. He noted further that if a smaller quantity is to be applied with the same placement, the most efficient use will be attained if the quantity available is applied to a portion of the total area at the most efficient rate, and if the remaining area is left unfertilized.

The first part of Steenbjerg's theory, described in the preceding paragraph, is a logical conclusion from the shape of a sigmoid response curve. Although the shape of the curves will affect the relative efficiencies of placed and broadcast fertilizer that emerge from application of the De Wit theory, the shape of the response curve is not involved in the De Wit theory; the response curves are accepted as they are.

As the second part of his theory, Steenbjerg stated that the concept of the most efficient application cannot be employed directly on a small scale, as in row placement, because the efficiency may be increased considerably by the competition among neighboring plants. He reported the results of an experiment verifying the existence of competitive effects between roots of plants in the fertilized soil and in adjacent unfertilized soil. Small-scale effects involving interception of fertilized soil by individual plant roots are discussed further in Section 7-7.

7-4. The Barber Theory

S. A. Barber and associates developed a theory of fertilizer placement from a different perspective. Their contribution was to provide a way to estimate the relative uptakes of a nutrient from soil where a soluble fertilizer has been added to different proportions of the total volume of soil. Most of the work has been summarized in a book by Barber (1984).

The theory is based upon the following concepts: (1) Nutrient ions are absorbed by roots from the soil solution. (2) The rate of uptake of a nutrient per unit area of root increases to a limiting value with an increase in the external concentration. (3) The rate of uptake of a nutrient ion by a root system under uniform conditions is proportional to the surface area exposed by the roots. (4) Ions are delivered from the soil to a plant root by (a) diffusion and (b) mass flow in the soil solution as the root absorbs water. (5) The rate of delivery of an ion to the root surface in response to the decrease in nutrient concentration resulting from absorption depends upon the rate of diffusion of the ion through the soil.

The rate of uptake of low concentrations of an ion from solution by unit area of root surface is usually represented by the Michaelis-Menten equation, which was developed originally to describe the rate of enzymatic reactions:

$$V = V_{max} \frac{C_1}{K_m + C_1} \quad ,$$

where V is the velocity of the reaction, V_{max} is the maximum velocity of the reaction as the concentration increases, C_1 is the concentration of the ion in solution, and K_m is the concentration at which the reaction velocity is half the maximum rate. Because it is not certain that uptake of ions by plant roots is an enzymatic process, Barber (1984) changed the notation. He also modified the equation by subtracting a value for the rate at which the ion returns from the root to the solution to take account of the fact that there is a minimum concentration below which the plant will not remove the ion from solution. He wrote the equation as

$$I_n = \frac{I_{max} C_1}{K_m + C_1} - E$$

or

$$I_n = \frac{I_{max}(C_1 - C_{min})}{K_m + C_1 - C_{min}} \quad ,$$

where I_n is the net rate of ion intake, I_{max} is the maximum net rate of ion intake, E is the rate of ion efflux, and C_{min} is the minimum concentration of the ion in solution.

Claassen and Barber (1974) found that the modified version of the Michaelis-Menten equation expressed the experimentally observed rate of potassium intake by roots in a nutrient solution to a very close approximation in short-term tests. Claassen and Barber (1976) modified the equation further for use with intact root systems to take into account the fact that the root system grows up to a certain stage in the life of the plant, which means that older roots are located in soil from which some of the ion has been absorbed, and that new roots extend into new volumes of soil. In testing the model, they made measurements of potassium uptake, root fresh weight, and root length at different times on young corn plants growing in several soils. Additional root-related factors not taken into account in the test were possible effects of root hairs, mycorrhizas, competition between roots, and changes in numerical values of the variables in the equations as the plants age. All of these factors were studied in subsequent work, although the mycorrhizal work is unpublished.

The results of the mathematical derivation given by Claassen and Barber (1976) were summarized by Claassen and Barber (1977) in the following two equations. The first equation expresses root growth as an exponential function of time:

$$L_2 = L_1 e^{k(t_2 - t_1)},$$

where L is the root length, t is the time, the subscripts 1 and 2 refer to the beginning and end of the time interval under consideration, and k is a constant representing the relative rate of root growth.

The second equation gives the average net nutrient influx over an interval of time:

$$\bar{I}_n = \frac{U_2 - U_1}{t_2 - t_1} \frac{\ln(L_2/L_1)}{L_2 - L_1} ,$$

where $\bar{I}_n$ is the average net nutrient influx, U is the amount of the nutrient in the plant, t is the time, L is the root length, and the subscripts 1 and 2 refer to the beginning and end of the time interval under consideration.

To complete the picture, the foregoing calculations from the plant side are combined with calculations from the soil side that provide estimates of the concentration of the nutrient at the surface of the root at different times and locations. Without movement of the nutrient from the soil to the root, the uptake of only an infinitesimal amount of the nutrient would be required to deplete the solution at the root surface and decrease the uptake rate to zero.

The rate of diffusion of an ion through the soil depends in part upon the rate of unimpeded diffusion of the ion through unit cross section of soil solution, the proportion of the soil volume occupied by soil solution, the viscosity of water adjacent to the particle surfaces, the electrical field around the soil particles, and the magnitude of the extra distance through which the ion must diffuse to get from one point to another because of the presence of soil particles. The viscosity, electrical field, and distance effects are expressed as an impedance or tortuosity factor.

Additionally, the rate of diffusion of an ion through the soil to a plant root depends in part upon the concentration gradient between the soil and the root surface. This gradient varies directly with the difference in concentration between the ion in solution in the bulk soil and the concentration at the root surface, and it varies inversely with the distance between the root surface and the location in the soil beyond which the concentration remains constant.

An important modifying factor is the capability of the soil solid phase to release the ion to the soil solution. The greater is this capability, the greater will be the concentration in the soil close to the root surface, and the greater will be the rate of uptake up to the concentration at which the rate of uptake reaches a maximum.

The soil factors were expressed in the following equation by Barber (1984):

$$D_e = D_1 \theta f_1 \frac{dC_1}{dC_s} \ ,$$

where D_e is the effective diffusion coefficient of the ion in the soil; D_1 is the diffusion coefficient of the ion in water; θ is the proportion of the soil volume occupied by water; f_1 is the impedance or tortuosity factor that takes into account the effects of the soil solids on the viscosity of the water, the distribution of the ions in solution, and the length of the diffusion path; C_1 is the concentration of the diffusible ion per unit volume of displaced soil solution; C_s is the concentration of the diffusible ion per unit volume of soil (the sum of the concentrations of the ion in the solution and the ion on the solid phase that equilibrates readily with the ion in solution); and dC_s/dC_1 is the soil buffer power for the ion in question. (Barber has defined the terms a little differently from Nye and Tinker (1977), who were cited as the source.)

Values for diffusion coefficients of ions in water are available in chemistry handbooks. Kovar and Barber (1988) calculated the impedance factor from the relation $f_1 = 1.58\theta - 0.17$ from Barraclough and Tinker (1981), where θ is the fraction of the soil volume occupied by water at the field capacity. For soils with >75% sand and $\theta < 0.15$, they used the constant value $f_1 = 0.25\theta$.

C_1 is measured on the displaced soil solution. For potassium, Claassen and Barber (1976) considered C_s to include both the soil solution and exchangeable forms. For phosphorus, Kovar and Barber (1988) designated the phosphorus released from the soil in 24 hours to an anion exchange resin as C_s.

The effective diffusion coefficient D_e is not evaluated directly. Rather, it is calculated by substituting appropriate values in the equation for D_e given in a preceding paragraph. Values of D_e for the various soil conditions of interest then are combined with the measurements of root length and nutrient intake rates to obtain integrated estimates of nutrient uptake. To the transport by diffusion is added the transport by mass movement, that is, the passive movement of the ion to the root in water that is brought to the root surface by the potential gradient resulting from water uptake. For potassium, nitrate, and phosphate, the ions most often deficient, transport by mass movement is less important than that by diffusion. For some ions, such as calcium, that are commonly present in soil solutions in concentrations exceeding those needed by plants, mass movement may be more important than diffusion. In fact, the concentration gradients may be the reverse of those for deficient ions, so that diffusion takes place away from the roots.

As an example of the application of the theory to fertilizer placement, Anghinoni and Barber (1980a) used the numerical values for various soil and root properties to estimate the phosphorus uptake by young corn plants from two soils that previously had been equilibrated with different quantities of fertilizer phosphorus supplied as monobasic calcium phosphate in different fractions of the soil volume. Their results showed a correlation of r = 0.85 between calculated and observed values of phosphorus uptake by the plants (see Fig. 7-9A). Fig. 7-9B shows the results of the placement calculations for one soil. As the quantity of fertilizer phosphorus added increased, the calculated maximum phosphorus uptake by the plants increased and occurred where the fertilizer was mixed with an increasing proportion of the total soil volume. This finding is consistent with Fig. 7-5 from De Wit.

Two opposing factors that affect phosphorus availability to plants are varied concurrently when a given quantity of fertilizer phosphorus is mixed with increasing proportions of the soil. On the one hand, the proportion of the soil from which plant roots can absorb fertilizer phosphorus at an enhanced concentration is increased, and this tends to increase the uptake. But on the other hand, the rate of fertilizer phosphorus uptake per unit of fertilized soil or per unit length of root is decreased because the concentration of phosphorus in solution is decreased. For the experimental work in Fig. 7-9A and B, the concentrations of phosphorus in the soil solution where the fertilizer was mixed with the smallest proportion of the soil were close to those that would cause the maximum rate of phosphorus uptake by the roots in the fertilized soil (Barber, 1984, page 381).

The foregoing description of the complex and interacting processes taken into account in the Barber theory implies that important measurements must be made while the crop is growing. Making needed measurements at that time would amount to a ''post-mortem'' as far as decisions about fertilizer placement for the current season are concerned. According to a private communication from S. A. Barber, he and associates have adapted the theoretical model in the following way to predict the proportion of a particular soil planted to corn that should be fertilized for greatest uptake of a given quantity of fertilizer phosphorus, for example.

Experimentally, one determines the relationship between the quantity x of fertilizer phosphorus applied per unit volume of soil and C_1. The equation $C_1 = ax^c + d$ is used for this purpose, where a, c, and d are fitted constants. The relationship between x and C_s is also determined. The equation $C_s = g + hx$, where g and h are fitted constants, is used for this purpose. [See a paper by Kovar and Barber (1988) for information about these equations and their use.]

Next, the relationship between the ratio y of the length of roots in the fertilized soil to the total root length and the ratio v of the volume of fertilized soil to the total volume of soil is calculated by the equation $y = v^{0.7}$. [See, for example, papers by Anghinoni and Barber (1980b, 1988), where the relationship has been evaluated experimentally.]

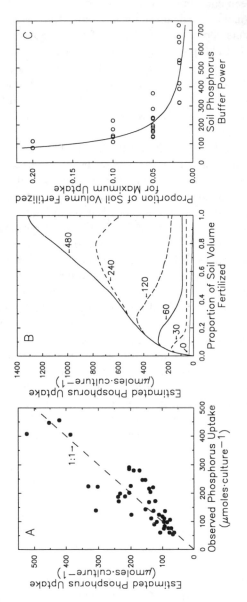

Fig. 7-9. Results of experiments on placement of phosphorus as monobasic calcium phosphate for corn plants grown 18 days in soil cultures: A. Theoretically calculated phosphorus uptake versus observed phosphorus uptake by corn plants on Raub and Wellston soils. B. Theoretically calculated phosphorus uptake versus fraction of the volume of the Raub soil treated with different quantities of fertilizer phosphorus. The lines in the graph correspond to the indicated quantities of fertilizer phosphorus in milligrams per kilogram of fertilized soil. C. Theoretically calculated proportions of the volume of various soils that should receive soluble fertilizer phosphorus at 100 milligrams per kilogram of soil to achieve the maximum phosphorus uptake by corn plants versus the phosphorus buffer power values for the soils. (Anghinoni and Barber, 1980a; Kovar and Barber, 1987)

Then the mechanistic model is used to estimate the uptake of phosphorus from the fertilized and unfertilized portions of soil, with all variables kept constant except the root growth rate k and the soil supply of fertilizer phosphorus. C_1 and C_s and consequently the buffer power of the soil for phosphorus (dC_s/dC_1) change as the plant absorbs phosphorus from the soil. The effective diffusion coefficient D_e also changes with uptake of phosphorus. These changes and their effect on phosphorus uptake are simulated by a computer program that estimates the total phosphorus uptake. The uptake from the fertilized and unfertilized portions of soil then are added to obtain the total uptake. Calculations are made for enough different fractional volumes of fertilized soil to permit generation of a curve for phosphorus uptake against volume of soil fertilized that will show the fractional volume corresponding to maximum uptake (see Fig. 7-9B for examples). A plot of yield of crop against yield of nutrient is used to convert the data to a crop-yield basis (Borkert and Barber, 1985).

Kovar and Barber (1987) found a better correlation between the volume fertilized for maximum uptake and phosphorus buffer power than between the volume and the extractable phosphorus (see Fig. 7-9C). In studies of 33 soils from different parts of the United States and Canada, Kovar and Barber (1987) found that the optimum proportion of the soil volume with which to mix the fertilizer ranged from as little as 1.7% in some soils to as much as 20% in others, where 24 kilograms of fertilizer phosphorus were supplied per hectare. Different results would be obtained with different quantities of fertilizer phosphorus, as may be inferred from Fig. 7-9B, which shows the results of calculations for additions of different quantities of fertilizer phosphorus to a given soil.

7-5. Nonuniformity of Broadcast Applications

Sections 7-2 and 7-4 deal with the purposeful horizontal distribution of fertilizer in a nonuniform manner to achieve an increase in efficiency of fertilizer use. Such nonuniformity is generally on a small scale relative to the distribution of roots of individual plants, so that, ideally, the roots of all plants contact the placed fertilizer and benefit equally from it.

At the opposite extreme are differences in quantity of fertilizer per unit area among areas located so far apart that the effects of the differences on the crop are independent. Between the extremes are uneven distributions over smaller distances such that the crop integrates the effects to some degree. This intermediate situation is relevant to the broadcast application of fertilizers by machine, where the objective is uniform application, but where the goal of uniformity is only approached in practice. Practical distribution patterns are always nonuniform to some degree, and the patterns depend upon various machine factors as well as the fertilizer and the conditions under which the fertilizer is applied.

Crowe (1985) published graphic representations of a number of measured distributions. The distribution of solid broadcast fertilizer is measured experimentally by driving the fertilizer spreader past an assembly of small boxes or

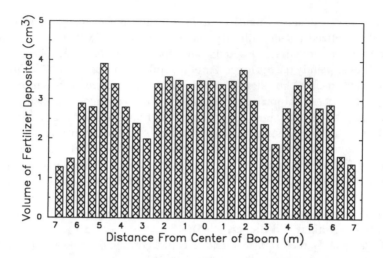

Fig. 7-10. Horizontal distribution of broadcast fertilizer by half-meter strips located different distances from the center of the traverse of a pneumatic distributor in which the fertilizer was metered into tubes and carried through the tubes to the outlets located along the boom. (Crowe, 1985)

trays arranged at right angles to the direction of movement of the spreader. The boxes are open at the top to catch the fertilizer particles as the spreader is driven by. The distribution then is represented as a plot of the weights or volumes of fertilizer in the boxes against the location of the boxes relative to the center of the traverse of the spreader.

Fig. 7-10 shows the horizontal distribution of fertilizer from a pneumatic spreader. In this instance, the coefficient of variation of the volumes of fertilizer measured in the individual segments is 39%, an unacceptably high value according to standards followed by machine manufacturers. If successive traverses of the spreader producing the results shown in Fig. 7-10 were overlapped 1 meter, the quantities of fertilizer collected in the outer two half-meter segments would be approximately doubled, and this would reduce the coefficient of variation of the distribution to 32%. According to Richards (1985), a coefficient of variation exceeding 17% is classed as poor for indoor testing under ideal conditions, and a coefficient of variation up to 20% is considered acceptable under field conditions, where the distribution may be affected by uneven ground and wind.

Machines that throw the fertilizer from one or two spinning disks located at the back of the spreader produce distributions that are relatively heavy near the center and that gradually drop to zero at the outer reaches. When successive traverses of a disk-type spreader are properly lapped, the resulting overall distribution can be considerably more uniform than the one illustrated in Fig. 7-10. If improperly lapped, or if a side wind is strong, the distribution can be less uniform than the one in Fig. 7-10.

The limiting relationships between crop yields and variability of the quantities of fertilizer applied per unit area may be illustrated by the effects of the variability under the hypothetical circumstances of crop responses that conform to the linear response and plateau function. The effect of variability in quantities applied upon the mean yield relative to the yield obtained with a uniform application of the mean quantity will depend upon where the yields fall on the response function.

If the mean quantity of fertilizer corresponds to the intersection of the linear response and plateau segments of the function, variability of quantities of fertilizer above and below the mean will have the maximum detrimental effect on the mean yield. All quantities below the mean will produce less than the maximum yield, and none of the quantities above the mean will produce higher yields.

If the mean application is far enough above the intersection so that all yields fall on the plateau, variability of quantities of fertilizer applied will have no effect on the yield. And if the mean application is far enough below the intersection so that all yields are on the linear response segment, the yields will vary, but the mean yield will be unaffected because increases and decreases in the quantity of fertilizer will produce equal positive and negative effects on the yields.

With response curves that are concave downward, yield losses from uneven application relative to the yield from uniform application of the mean quantity occur on all portions of the ascending branch of the curves because the increases in yield from quantities of fertilizer in excess of the mean are smaller than the decreases in yield from quantities below the mean. The magnitude of the losses, however, will depend upon the portion of the response curve to which the uneven applications relate.

As an example, consider the following equation for the response of corn to nitrogen fertilization in 54 experiments in North Carolina (Section 2-5):

$$y = 1.64 + 0.0437N - 0.000118N^2,$$

where y is the corn grain yield in megagrams per hectare and N is the fertilizer nitrogen applied in kilograms per hectare.

According to this equation, the yield is 5.45 megagrams per hectare with 140 kilograms of fertilizer nitrogen. If this is the target application — for example, the quantity for the maximum net profit — what would be the effects of a situation in which a third of the land receives this quantity, a third receives 100 kilograms, and the remaining third receives 180 kilograms? The calculated yields are 4.83 megagrams with 100 kilograms and 5.68 megagrams with 180 kilograms. The average yield then would be 5.32 megagrams per hectare, a loss of 0.13 megagram compared with a uniform application of the target quantity of fertilizer nitrogen, 140 kilograms per hectare.

The problem of estimating yield effects of nonuniform distribution has been approached by combining an equation for yield response with an equation for the distribution of the fertilizer. Jensen and Pesek (1962) used the quadratic equation to express the yield response and a periodic cosine function to describe the fertilizer distribution. This function represents the quantity of fertilizer versus the lateral distance as a smooth curve with alternating uniform maxima and minima.

Heymann et al. (1984) tested the value of various statistical measures of unevenness of fertilizer distribution using data for 12 different distributions and determining how well the statistical measures were correlated with the decrease in mean yield of nitrogen-fertilized winter wheat due to unevenness of distribution. They found that the coefficient of variation was the best of the measures tested. For expressing the nonuniformity of distribution, the coefficient of variation has the advantage that no particular mathematical form of the distribution pattern is implied. The coefficient of variation V is the ratio of the standard deviation of a series of measurements to their mean. As applied to the fertilizer distribution problem,

$$ V = \frac{\sqrt{\dfrac{\Sigma (X_i - X)^2}{n - 1}}}{X} \, , $$

where X_i is the quantity of fertilizer in the i'th box, X is the mean quantity of fertilizer in all the boxes, and n is the number of boxes (assuming that there is only one box at each distance from the center of the traverse of the fertilizer spreader). The coefficient of variation is shown here as a proportion. It is often expressed as a percentage by multiplying the foregoing expression by 100. Ndiaye and Yost (1989) preferred to express the statistical parameter as a coefficient of uniformity $(1 - V)$ instead of a coefficient of variation.

Richards (1985) gave equations for estimating the average yield where the fertilizer is spread unevenly for situations in which the response is represented by the Mitscherlich equation and the quadratic equation. He used the Mitscherlich equation in the form

$$ y = b_0 - b_1 e^{-b_2 x}, $$

which may be derived from the exponential version of the Mitscherlich equation in Section 1-2 (footnote 2, equation 7) by first dividing x_1 (the total supply of the nutrient) into its component parts x (the part added in the fertilizer) and b (the part supplied by the soil), whereupon the equation becomes

$$ y = A[1 - (e^{-cx})(e^{-cb})]. $$

Then letting $A = b_0$, $Ae^{-cb} = b_1$, and $c = b_2$, the equation is transformed to the form used by Richards. Combining the Mitscherlich equation with a function involving the coefficient of variation (derived from unpublished reports by J.

Dickens and R. D. Hobson of the Levington Research Station in the United Kingdom), Richards gave the mean yield y_m where the fertilizer is spread unevenly at the mean rate x_m as

$$y_m = b_0 - b_1(1 + b_2 V^2 x_m)^{-(1/V^2)}.$$

In this equation, the coefficient of variation V is expressed as a proportion.

Richards represented the quadratic equation as

$$y = b_0 + b_1 x - b_2 x^2,$$

where x is the quantity of fertilizer and b_0, b_1, and b_2 again are parameters, but are different from those for the Mitscherlich equation. Combining the quadratic equation with the Dickens and Hobson function involving the coefficient of variation, he obtained the following equation for the mean yield y_m where the fertilizer is spread uniformly:

$$y_m = b_0 + b_1 x_m - b_2(1 + V^2) x_m^2$$

In this equation, x_m is again the mean rate of application of the fertilizer, and the coefficient of variation V is expressed as a proportion. Values of y_m for observed or assumed coefficients of variation then may be calculated and compared with the values of y calculated from the corresponding response equations, which give the estimated yields with the same quantity of fertilizer spread uniformly.

Richards (1985) summarized the results of various field experiments with nitrogen fertilizers for six different crops in the United Kingdom and used the economic optimum quantities of fertilizer as the target quantities. Taking the yields with the economic optimum quantities of fertilizer spread uniformly ($V = 0$) as 100, the mean yields with the fertilizer spread nonuniformly with $V = 0.35$ or 35% were 98.6 with the Mitscherlich equation and 96.2 with the quadratic equation. The corresponding yields of the controls without nitrogen fertilizer were 59.1 and 59.6. Because of the comparative shapes of the two types of response curves, the estimated effect of nonuniform distribution of the fertilizer was greater with the quadratic equation than with the Mitscherlich equation. The important point to be derived from the study of the actual experiments, however, is that the yield effects were small, even with what is considered highly variable distribution according to the standards of machinery manufacturers.

Operation of machinery in practice, where the maintenance may not be the best, parts may be bent or worn, and successive traverses may not be closely coordinated, is another matter. The visually evident effects of uneven distribution sometimes seen, especially on lawns where the fertilization is done by nonprofessionals, may correspond to differences greater than those calculated by Richards for a coefficient of variation of 0.35 or 35%.

Applications of fertilizer by aircraft may be relatively uniform under favorable conditions, but Chiao and Gillingham (1989) noted that coefficients of variation up to 95% in aerial applications of phosphate fertilizers to pastures are not unusual in New Zealand. In New Zealand, aerial fertilization of pastures is common because much of the pasture land is too steep for land-based equipment. Application technology that reduces the coefficient of variation commands a premium. From their theoretical analysis, Chiao and Gillingham concluded that uniform spreading would be worth the most to farmers applying nitrogen fertilizers that have a low residual effect.

Uniform spreading would be valuable also to farmers applying phosphate fertilizers to new, phosphorus-deficient land. After the soil phosphorus supply has been built up by the residual effects of repeated applications, the value of technology that results in uniform spreading is much reduced. The reason is that the variability in phosphorus supply resulting from the first few uneven applications will have been reduced to a considerable extent by subsequent applications. The differences in yield between areas receiving quantities of fertilizer below and above the target average thus will be much smaller than they would have been originally when the soil was strongly deficient in phosphorus.

The coefficient of variation as such makes no distinction between uneven distributions of fertilizer on (a) a scale of distances great enough so that the effects on the crop would be independent and detrimental and (b) a scale small enough so that some of the effects would be interactive and beneficial as placements. The machine distribution illustrated in Fig. 7-10 suggests that a minor portion of the variation may have been on a small enough scale to be considered fertilizer placement for a crop with widely spaced rows. Whether any useful modifications of the coefficient of variation measure can be developed to accommodate small scale effects that might be beneficial remains to be determined.

7-6. The Burns Theory

Nitrogen-bearing fertilizers normally are not positioned in the soil below the depth of plowing. The Burns theory is considered in connection with fertilizer placement, however, because it deals with the effectiveness of nitrate at different distances below the soil surface in relation to the vertical distribution of crop roots. The theory is of special relevance to current efforts to adjust applications of fertilizer nitrogen to reflect differences in crop need according to the supply of mineral nitrogen in the soil.

The theoretical question addressed by Burns (1980a) was that of estimating the minimum depth at which soil nitrate will be below the effective depth of rooting of a specific crop. The theory can be used as a basis for deciding (a) how much to adjust the usual spring application of fertilizer nitrogen to compensate for decreases or increases in nitrate present in the soil in the spring as a consequence of deviations of winter rainfall (and nitrate leaching) from the average and (b) how much additional fertilizer nitrogen to apply during the current season to compensate for losses of spring-applied nitrogen by leaching that occurs after fertilization. The theory can also be combined with measurements of the depth distribution of mineral nitrogen in the soil to estimate the supply of mineral nitrogen still available for uptake by the crop.

The Burns theory is based upon two general premises, both supported by evidence in the scientific literature. The first is that most of a crop's requirement for nitrate can be met even if only a small portion of its root system is exposed to nitrate.

This first premise is supported by field experimental evidence from various sources. Fig. 7-11 shows the results of an experiment by Ogus and Fox (1970) illustrating the efficiency of a small portion of the root system in absorbing nitrate. In this experiment, small volumes of a calcium nitrate solution were added to the soil at closely spaced horizontal intervals at different depths in an established bromegrass sod in eastern Nebraska.

In measurements made 24 days after fertilization, the apparent recovery of fertilizer nitrogen in the above-ground plant material was 46% where the fertilizer was applied as a top dressing, and the recovery dropped rapidly with increasing depth of application. In measurements made 78 days after fertilization, the percentage recovery of the fertilizer nitrogen applied at the 30 and 60 centimeter depths had increased so that it was approximately the same as the recovery from the top dressing (Fig. 7-11A) despite the small proportion of the root system in the lower layers (Fig. 7-11B).

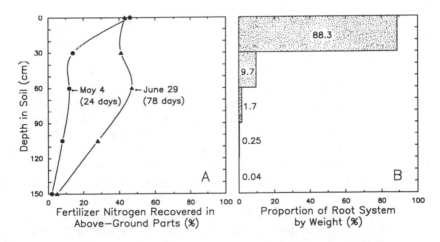

Fig. 7-11. A. Recovery of nitrogen by established bromegrass after different lengths of time from calcium nitrate placed at different depths in soil in eastern Nebraska. Before the experiment was started, grass residues from the previous season were removed by burning, and the area was irrigated to adjust the water content of the soil to field capacity to a depth of 150 centimeters. The calcium nitrate was applied in small volumes of solution at different depths on April 11, when the grass was just starting to grow. The seasonal distribution of rainfall was such that substantial movement of nitrate in the soil was unlikely. B. Vertical distribution of roots by weight in the soil. (Ogus and Fox, 1970)

To obtain indexes of the efficiency of the roots in absorbing fertilizer nitrogen, Ogus and Fox (1970) divided the fertilizer nitrogen uptake by the plants from the individual layers of soil as a percentage of the total by the weight of roots in the corresponding layers as a percentage of the total. They found that the efficiency indexes ranged from 0.4 in the surface 30 centimeters of soil to 125 in the 120 to 150 centimeter layer.

Data from experiments on winter wheat receiving fertilizer nitrogen at different depths up to 150 centimeters were reported by Daigger and Sander (1976). These experiments, designed like the experiment in Fig. 7-11, but without measurements of root distribution, showed that at the boot stage the yields of dry matter and nitrogen in the above-ground plant material decreased somewhat with increasing depth of fertilizer placement. At harvest, the yields of dry matter and nitrogen were essentially uninfluenced by depth of fertilizer placement. The experiments were conducted in western Nebraska under dryland conditions, but the soils were moist to the field capacity to a depth of 180 centimeters at the time the ammonium nitrate fertilizer was applied in April. The results of 25 field experiments in one year and 17 in another in Germany indicated that the mineral nitrogen (ammonium plus nitrate) was available to winter wheat plants to the depth of sampling, which was 100 centimeters (Wehrmann and Scharpf, 1979). On the basis of these and other findings, Burns (1980a) estimated that if the soil conditions are satisfactory, a crop's requirement for nitrate can be met by as little as 10% or less of the total root system.

The second premise underlying the Burns theory is that plant roots can take up nitrate from soils at the maximum rate even though the concentration is relatively low. Experiments indicate that the rate of uptake is not diminished until the concentration of nitrate nitrogen in solution at the root surface is 0.3 milligram per liter or less.

In soil, the solution is essentially stationary, and the solution represents only a fraction of the total soil on either a volume or weight basis. Thus, as plants absorb nitrate, a concentration gradient develops around the roots, and the average nitrate concentration in the soil that corresponds to the maximum rate of nitrate uptake by the roots exceeds the value for a stirred nutrient solution.

Using equations developed by Baldwin et al. (1973) for representing solute uptake by plant roots in soil, Burns (1980a) calculated that an average nitrate-nitrogen concentration of 12 milligrams per kilogram of dry soil would be enough to maintain the uptake of nitrate by one root at the maximum rate if the root had a radius of 0.01 centimeter, if it was the only root in a soil cross sectional area of 10 square centimeters, if only 20% of the soil volume was occupied by water, and if the root was not absorbing water. With increases in root radius and number, with more water in the soil, and with mass movement of nitrate to the root as a result of absorption of water by the root, the minimum nitrate concentration needed to maintain absorption at the maximum rate would be lessened.

If a soil has a bulk density of 1.5, the average concentration of 12 milligrams of nitrate nitrogen per kilogram that supports nitrate uptake at the maximum rate under the conditions described in the preceding paragraph would correspond to only 3.6 kilograms of nitrate nitrogen in a 20 centimeter layer of soil or 18 kilograms in a 1 meter layer per hectare. A rapidly growing crop would absorb this amount of nitrate from soil in only a few days.

In a 4-year experiment on a nitrogen-fertilized silt loam soil in Germany, Strebel and Duynisveld (1989) found that the movement of nitrogen to the crop roots in the absorbed water amounted to 15 to 33% of the total uptake and was confined mainly to the topsoil layer. From 67 to 85% of the nitrogen uptake was by diffusion. Most of the nitrate was removed from the soil by the crops.

Because the density of roots in the soil decreases with depth, the nitrate-absorbing capability of the root system also decreases with depth. In analyzing the problem of leaching of fertilizer-derived nitrate, Burns (1980a) estimated initially the minimum quantity of roots needed to absorb the amount of nitrate required by the crop to avoid growth limitation. As the first step, he determined the maximum rate of nitrate absorption per unit length of root. Using Barley's (1970) values of 0.1 centimeter as the root radius and 10^{-11} mole per square centimeter of root surface per second as the maximum rate of nitrate uptake, Burns (1980a) calculated that the maximum uptake rate per centimeter of root length would be 7.606 micrograms of nitrate nitrogen per day.

As the second step, he determined the total length of roots absorbing nitrate at the maximum rate that would be required to supply a crop with the amount of nitrate it needs to maintain growth at the maximum rate. For an uptake rate of 5.56 kilograms of nitrogen per hectare per day (55.6 micrograms per square centimeter of soil area per day) as a typical nitrogen demand for a number of crops over much of the linear phase of growth, Burns calculated that the total length of roots needed to absorb the required nitrogen would be 55.6/7.606 = 7.3 centimeters of root per square centimeter of soil surface. He called this length of root the critical length L_c. L_c is thus the minimum number of centimeters of roots of diameter 0.1 centimeter under 1 square centimeter of soil surface required to absorb the needed nitrate.

Fig. 7-12A shows schematically the vertical distribution of roots in soil. The two horizontal lines mark the depths below which the critical length and half the critical length of roots occur.

Fig. 7-12B includes the horizontal lines for root lengths at the same depths as in Fig. 7-12A. Fig. 7-12B shows Burns's concept of the availability of fertilizer and residual nitrate in a soil in which roots are established as in Fig. 7-12A. Initially the nitrate content in the upper part of the soil is adequate to meet the current needs of the crop. The nitrate then is gradually leached downward. At first, the downward movement of nitrate by leaching is not important. The concentration of roots in the upper part of the soil is adequate to absorb nitrate fast enough to avoid growth limitation even though some of the roots are in soil containing very little nitrate. When leaching (combined with the action of roots in absorbing virtually all of the small remaining quantity of nitrate) has proceeded far enough to move the upper boundary of the nitrate-containing zone to the depth below which the critical length L_c of roots occurs, the total length of roots

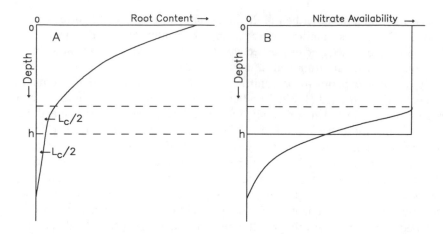

Fig. 7-12. Schematic representation of *A* the distribution of crop roots with soil depth and *B* the availability of fertilizer-derived and residual nitrate to the crop with soil depth. The critical length of roots L_c is the minimum length required to absorb nitrate at a rate sufficient to permit the crop to grow at the maximum rate, and *h* is the effective rooting depth. In *A*, the horizontal broken lines are located so that half the critical length L_c occurs between the lines and half occurs below the lower line. (Burns, 1980a)

will be just sufficient to absorb nitrate at a rate that avoids growth limitation. If the nitrate is leached to the depth below which the length of roots is $L_c/2$, the roots below will be just sufficient to absorb nitrate at half the rate required to avoid growth limitation. And when the nitrate has been leached to the lower reaches of root extension, the rate of absorption of the fertilizer and residual nitrate will be zero.

Burns (1980a) noted that the changes in nitrate availability under the conditions described could be expressed approximately by a "step function," in which the fertilizer and residual nitrate is assumed to be completely available for uptake by the plant roots down to the depth *h*, corresponding to the root concentration $L_c/2$, and completely unavailable for uptake below that depth. Burns designated *h* as the effective rooting depth.

Examination of data from other workers on root length and distribution for various crops at different stages of growth showed that the effective rooting depth increased with the total weight of the above-ground portion of the crop (*w* metric tons per hectare) and with the population density (*p* plants per hectare), and that it decreased with increasing radius (*a* centimeters) of the roots. Of the models he tested, Burns (1980a) found that the following regression equation gave the best fit:

$$h = 6.082w + 1.52 \times 10^{-5}p + (1.81 \times 10^{-3}/a^2) - 2.1.$$

Values of *w* and *p* can be measured readily. For root crops, he included the dry weight of the tap root in *w*. For *a*, he used values from a review by Brewster and Tinker (1972) for cereals and unpublished values from a colleague for vegetable crops. The value of h is determined mostly by w. Values of p and a have only relatively small effects.

Use of the foregoing equation for noncereals would require direct measurements of a or finding suitable values in the literature. Values of a are important mostly because crops with large values of a tend to be shallow rooted. Fig. 7-13 shows a plot of effective rooting depths as estimated from the foregoing equation against measured effective rooting depths for 12 crops.

Burns (1980b) investigated the utility of the method by analyzing the results of field experiments in which different quantities of ammonium nitrate had been incorporated in the surface 3 centimeters of soil just prior to planting each of several crops in different years. In each experiment, the yields were expressed as a percentage of the maximum. The method was tested by determining the extent to which the scatter of points was reduced by adjusting the quantities of nitrogen applied as ammonium nitrate for the estimated loss of nitrate nitrogen by leaching below the effective rooting depth. Fig. 7-14 shows the results obtained in 14 experiments with lettuce conducted in different years in the United Kingdom. The smaller scatter of points observed in the plot of the adjusted data on the right shows that the relative yields were more closely related to the estimated values of unleached nitrate than to the quantities of fertilizer nitrogen applied.

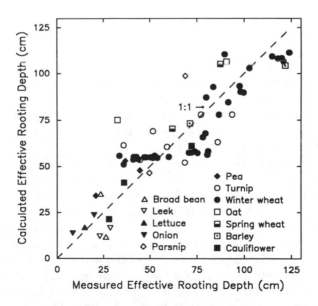

Fig. 7-13. Effective rooting depths of various crops as estimated from measurements of root distribution with depth in the soil, and as calculated using the empirical regression equation given in the accompanying text. The diagonal line indicates equal values of the calculated and measured effective rooting depths. (Burns, 1980a)

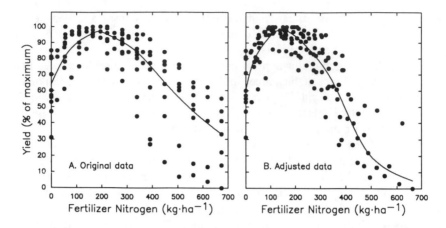

Fig. 7-14. Yield of lettuce as a percentage of the maximum obtainable by applying fertilizer nitrogen in 14 experiments in different years in the United Kingdom. The original data are shown in *A*. In *B*, the quantities of fertilizer nitrogen added have been adjusted for the estimated amounts of nitrate leached below the effective rooting depth. (Burns, 1980b)

No actual measurements of nitrate loss were made in obtaining Fig. 7-14. The loss of nitrate nitrogen L in kilograms per hectare was estimated from the equation

$$L = A\left(\frac{P}{P + \theta_m}\right)^h,$$

where h is the effective rooting depth in centimeters, A is the kilograms of nitrate nitrogen per hectare from the most recent application of fertilizer, P is the accumulated drainage in centimeters below the effective rooting depth, and θ_m is the field capacity in cubic centimeters of water per cubic centimeter of soil. Use of h as the exponent implies that the fertilizer-derived nitrate is at or near the soil surface before leaching begins.

In the work done by Burns, A was taken as the sum of the nitrate nitrogen added in the ammonium nitrate fertilizer and the nitrate nitrogen produced by nitrification of the ammonium. All of the ammonium was assumed to have been converted to nitrate unless most of the drainage occurred within 6 weeks of fertilizer application, in which case only half of the ammonium was considered to have been nitrified. This approximation was based upon experimental data indicating that the average half life of ammonium was about 3 weeks.

Drainage was estimated from the difference between rainfall and evapotranspiration in weekly increments up to and including the time of the last significant rainfall. The procedure was described in detail by Burns (1980b). Use of both rainfall and evapotranspiration in the estimation procedure improves the precision of drainage estimates because nitrate can move upward in soil in response to evapotranspiration as well as downward in response to rainfall.

Fig. 7-14 indicates that the series of approximations based upon readily available information was useful in reducing the variability of responses to fertilization due to differences in loss of nitrate by leaching from one experiment to another. The variance from the response equations based upon adjusted values of nitrate nitrogen was two thirds as great as that based upon the original unadjusted values.

The nitrate-leaching equation used by Burns involves the assumption that the nitrate in each succeeding hypothetical soil layer of infinitesimal thickness equilibrates with the water moving downward through the soil before the water moves on to the next layer (see Burns, 1974). This assumption would seem especially appropriate in areas where the intensity of rainfall is relatively low. For example, Barraclough (1989a,b) reported that in 777 hours of rainfall over a 3-year period at Jealotts Hill in the United Kingdom, only 9% of the total precipitation was received at intensities exceeding 3 millimeters per hour. The assumption would be less realistic with high rainfall intensities, where water and solutes may move downward quickly through large pores before complete equilibration with the contents of small pores has occurred (Priebe and Blackmer, 1989). Burns (1975), however, summarized experimental data in the literature from various parts of the world in which the displacement of nitrate or chloride by rainfall or irrigation in various soils had been measured, and he found generally good agreement between the observed values and the values calculated from his equation. Results obtained in experimental work on irrigation of a swelling and cracking silty clay loam were not well estimated by the equation.

An independent test of the Burns theory was made more recently by Haumann and Du Preez (1989) in a field experiment in South Africa, in which nine different initial vertical distributions of nitrate were established in a soil of sand to sandy loam texture. After the soil had been irrigated with various amounts of water to induce nitrate movement, the correlation between the measured contents of nitrate in individual 10-centimeter layers to a depth of 120 centimeters and the contents calculated from the Burns theory was $r = 0.85$. The correspondence between observations and theory was improved when the depth increments were widened. The correlation was $r = 0.94$ for the surface 60 centimeters of soil and $r = 0.95$ for the surface 120 centimeters. The rapid movement of water in the soil due to irrigation in the experiment by Haumann and Du Preez (1989) would account in part for the improved correspondence between theory and observation as the magnitude of the depth increments was increased.

The nitrate-leaching model used by Burns is relatively simple. A number of theoretical models of downward movement of nitrate have been suggested. The greater is the effort to model the detailed effects of precipitation rate, soil pore-size distribution, and pore continuity [see Barraclough (1989a,b) for a discussion of the influence of these factors], the more precise the estimates can become. At the same time, the suitability of the models for advisory purposes generally diminishes because of the increase in information specific to the situation in question that must be available before the models can be used.

More recently, a *Nitrogen Calculator* based upon the Burns procedure has been developed for practical use by farm advisors and growers to estimate the quantity of fertilizer nitrogen needed to replace the nitrate lost by leaching.[2] The *Nitrogen Calculator* also contains directions for adjusting the spring application of fertilizer nitrogen to take into account the negative or positive deviations of residual soil nitrate from the average that result from negative or positive deviations of the winter precipitation (with consequent changes in nitrate leaching) from the average.

The foregoing discussion of the work by Burns emphasizing the nitrate-absorbing capability of small quantities of roots in the subsoil has not called attention to a possible consequence of the much greater nitrate-absorbing capability of the relatively large quantities of roots in the upper portion of the soil. As may be inferred from Fig. 7-11, when a crop is grown on a soil with a high initial content of nitrate from the surface downward with little vertical transfer of nitrate in the soil during growth, the subsoil nitrate will be spared, and most of the nitrate used by the crop will be derived from the upper part of the soil. The residual nitrate in the subsoil then may be lost by leaching before the next crop is grown. Such circumstances may arise especially in vegetable crop production, where the quantities of fertilizer nitrogen added often exceed the crop needs.

7-7. Root-Fertilizer Contact and the Costigan Theory

De Wit's (1953) analysis of the effect of the time required for crop roots to reach the fertilizer was reviewed in Section 7-2.7. Here the question is considered again, with emphasis upon the conditions De Wit classed as exceptions.

Placement of fertilizer nutrients close to the seed is of special value where the soil temperature is low during the early stages of plant growth. Two special terms, *starter fertilizer* and *pop-up fertilizer*, have developed to describe the small quantities of plant nutrients applied close to the seed to promote the early growth of plants. The importance of soil temperature, inferred from field observations, has been confirmed by special tests in which plants were grown concurrently in containers with different soil temperatures but with the same air

---

[2]The calculator includes tables on the water-holding capacity of soils of different textures, critical rooting depths of different crops at harvest, and weekly evaporation data, together with directions for calculating the drainage that occurs after spring fertilizer application. An inner disk calibrated with a range of values for the water-holding capacity of the soil is rotated on an outer disk calibrated with a range of values for drainage, so that the appropriate values for water-holding capacity and drainage are lined up. Printed on one side of a window in the inner disk is a scale giving values of the critical rooting depth. Through the window, one reads on the outer disk the value for the percentage of the nitrate leached that corresponds to the critical rooting depth for the crop in question. The percentage loss of nitrate by leaching then is divided by 100 and multiplied by the quantity of fertilizer nitrogen applied to estimate the quantity of fertilizer needed as a top dressing to replace the nitrate lost by leaching. The *Nitrogen Calculator* is available from The Scientific Liaison Officer (N-Calculator), Horticultural Research International, Wellesbourne, Warwick CV35 9EF, United Kingdom.

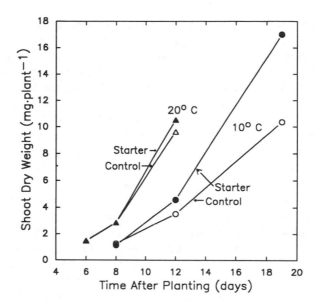

Fig. 7-15. Growth of lettuce in soil cultures maintained at 10 and 20° C with a constant aerial temperature of 15° C, with and without a starter fertilizer containing nitrogen, phosphorus, and potassium placed 1 centimeter below the seeds. (Costigan, 1986)

temperature. Fig. 7-15 from an experiment by Costigan (1986) is an example. Note the much greater effect of the starter fertilizer at a soil temperature of 10° C than at 20° C.

Concordant observations have been reported by others. Knoll et al. (1964) found that with one soil the quantity of phosphorus present in the tops of corn plants grown in soil at 15° C was more than twice as great where the phosphate fertilizer was placed in a layer close to the corn seeds as where it was mixed with the entire volume of soil. At a soil temperature of 25° C, which would favor increased root growth, the phosphorus content of the plants was about the same with both placements.

In field experimental work in North Dakota, Alessi and Power (1980) found that a band application of 15 kilograms of phosphorus as concentrated superphosphate per hectare increased the yield of spring wheat even when 160 kilograms had been broadcast and worked into the soil. In fact, the benefit from the band application was greater where 160 kilograms of fertilizer phosphorus had been broadcast than where none was applied.

A second factor in the root-fertilizer contact issue is the effect one fertilizer component may have on the uptake of another. This matter has been investigated mostly for the effect of nitrogen on phosphorus, with emphasis upon banded applications in which concentrations are relatively high. Fig. 7-16 illustrates the findings. In this experiment with barley, the uptake of phosphorus-32 from tagged fertilizer was enhanced by the presence of some nitrogen as urea with the phosphorus (supplied as concentrated superphosphate) in bands located 30 centimeters apart. The favorable effect of nitrogen on phosphorus uptake has been discussed previously. Larger quantities of urea decreased the activity of phosphorus-32 in the plants. The decrease in phosphorus-32 activity with the higher levels of urea presumably is related to the greater mobility of the nitrogen than of the phosphorus. The banded phosphorus soon is surrounded by a zone containing a high concentration of fertilizer-derived ammonium with an initially high pH value. This unfavorable environment for root growth inhibits root penetration into the inner zone containing the fertilizer phosphorus. With time, part of the ammonium is oxidized to nitrate, the soil pH decreases, and the inhibitory effect diminishes.

Beneficial results can be obtained with small quantities of fertilizer placed in direct contact with the seed, but germination of the seed and final yield of the crop can be affected unfavorably when the concentration of fertilizer salts near the seed is excessive. See, for example, data by Bullen et al. (1983) and Fig. 7-17. Crops differ considerably in their sensitivity to fertilizer placed with the seed. Ukrainetz (1975) and Ukrainetz et al. (1975) observed that sunflower, rapeseed, and flax are more sensitive than fababean to phosphate fertilizer placed with the seed. Fig. 7-17 shows the results of a direct experimental comparison between side banding and placement of concentrated superphosphate with the seed of sunflower and fababean. Concentrated fertilizer bands are often placed 2.5 to 5 centimeters to the side of the seed and 2.5 centimeters below seed level.

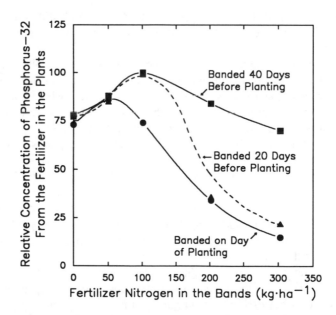

Fig. 7-16. Relative concentrations of phosphorus-32 in barley plants from concentrated superphosphate banded with different quantities of urea different lengths of time before planting in Alberta. The bands were spaced 30 centimeters apart. (Harapiak and Penney, 1984)

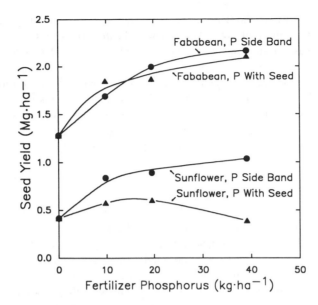

Fig. 7-17. Yield of fababean and sunflower with different quantities of phosphorus as concentrated superphosphate applied in a side band or with the seed in experiments in western Canada. (Ukrainetz, 1975; Ukrainetz et al., 1975)

A special case illustrating the sometimes critical location of banded fertilizer was reported by Cooke and Widdowson (1953) for alfalfa, a crop with a tap root. These investigators obtained no improvement in the early growth of alfalfa from fertilizer containing phosphorus and potassium that was placed in bands 5 centimeters to the side of the seed. Eight weeks after planting, the plants were found to have long straight tap roots and few lateral roots, none of which were detected near the fertilizer band. The placement effect with alfalfa is quantified in Fig. 7-18. Similar but considerably less striking results were reported by Wagner and Hulburt (1954) for a planting of tall fescue and ladino clover and by Bullen et al. (1983) for soybean.

A related observation of interest is the nonuniform lateral distribution of roots of some crops that apparently is related to the earth's magnetic field. Pittman (1964) found that the roots of wheat have a predominantly north-south orientation, and that the roots of certain varieties of wheat could be changed to a predominantly east-west orientation when the plants were placed in a strong magnetic field due to an electromagnet with its long axis in an east-west direction. The direction of root growth of fall rye did not seem to be influenced by the earth's magnetic field, but it was changed to a generally east-west orientation parallel to a strong induced magnetic field in that direction.

Woolley and Pittman (1966) found by selective tagging of superphosphate with phosphorus-32 that although the superphosphate was uniformly distributed laterally in soil around individual wheat seeds, the seedlings at 14 days had taken up two to eight times more radioactive phosphorus when the tag was located north or south of the plant than when it was located to the east or west. At 28 days, there was no significant difference in phosphorus-32 uptake among locations.

The observations on root distribution of wheat appear to fall within the conditions De Wit concluded would cause no significant problems in application of the compensation function. Small grain crops usually are drilled in close-spaced rows, with the fertilizer drilled beneath the seed or sometimes with the seed. The nonuniform lateral distribution of roots thus probably would have no significant effect with small grains.

Lateral root distribution might be of more significance with sugarbeet. Cooke (1954) observed an increased density of the "feeder" roots of sugarbeet in the vicinity of fertilizer bands. The feeder roots are fine roots that grow out of the creases in the fleshy storage root of a sugarbeet plant to a depth of about 5 centimeters below the soil surface, according to Schreiber (1958). Schreiber noted that the vertical creases and the feeder roots are oriented in an east-west direction in beets grown in the Red River Valley in Manitoba.

In field experiments at two locations, Schreiber (1958) observed that the increase in yield of sugarbeet resulting from a band application of ammonium phosphate fertilizer about 5 centimeters deep and 4 centimeters to the side of the seed was greater when the beets were planted in east-west rows than when

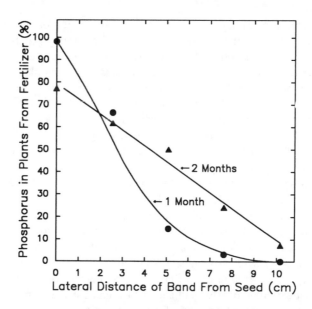

Fig. 7-18. Percentage of phosphorus in alfalfa plants derived from phosphorus-32-tagged superphosphate placed 3.5 centimeters below the seed and 4.5 centimeters below the surface of the soil or at the same depth but at different lateral distances to one side of the seed. Data are averages of results obtained with two soils and two different quantities of fertilizer in soil cultures in a greenhouse. (Tesar et al., 1954)

they were planted in north-south rows. He explained the differential response on the basis that close cultivation tends to cut off the feeder roots when the beets are planted in north-south rows, but not in east-west rows. Loss of the surface roots would tend to promote the growth of deeper roots in the north-south rows. The mostly uninjured feeder roots in east-west rows presumably could benefit from the side-banded fertilizer by growth into the fertilized soil after the first cultivation.

The distribution of fertilizer in a concentrated band can be continuous, but this is not true for fertilizer that is broadcast and mixed uniformly with the soil. Especially with modern granulated phosphate fertilizers, the diameters of the zones of soil into which the fertilizer phosphorus diffuses are much smaller than the distances between the spheres of fertilizer-affected soil in practical broadcast applications.

Costigan (1987) developed a model to elucidate the consequences of the discontinuity of zones of fertilizer-affected soil in terms of interception by roots of the spheres of soil into which the fertilizer phosphorus has moved from the individual granules. According to his model, the soluble phosphorus in each fertilizer granule moves outward into a spherical volume of soil with radius *r* within a few days. Thereafter, the movement is very slow because of the reduction in phosphorus concentration in solution that has resulted from reactions of the phosphorus with the soil.

The phosphorus supplied by a fertilizer granule is considered to be available to a root if the root passes through the sphere of soil influenced by the granule. Although a primary root might encounter only the fringe of fertilizer-affected soil surrounding a granule site, the higher phosphorus concentration could stimulate the production of secondary roots that would bring the root system into close contact with the total volume of fertilizer-affected soil. Thus, as a root grows through the soil, it can be considered to exploit the soluble phosphorus derived from fertilizer granules as much as *r* centimeters from the root surface. In terms of the locations of the original fertilizer granules, the total volume of soil within which a root will intercept spheres of fertilizer-affected soil is thus $\pi r^2 l$, where *l* is the length of the root in centimeters. If the fertilizer granules are uniformly distributed in the soil, and if primary root growth occurs at random relative to the granules, $v_p/\pi r^2 l$ centimeters of root will be required on the average to intercept the first zone of fertilizer-affected soil, where v_p is the volume of soil containing one fertilizer granule. In general, the length of root *l* in centimeters required to intercept the first zone of fertilizer-affected soil will be

$$l = \frac{v}{n \pi r^2} \quad,$$

where *v* = total volume of soil in cubic centimeters in which the fertilizer is uniformly distributed and *n* = total number of fertilizer granules.

Costigan made an illustrative calculation for an application of 131 kilograms of phosphorus per hectare as concentrated superphosphate in granules, each containing 5.4 milligrams of phosphorus. Granules were assumed to be distributed uniformly in the soil to a depth of 10 centimeters, and on the basis of experimental data from the literature, each granule was assumed to produce a sphere of phosphorus-enriched soil with a radius of 1 centimeter. Thus, v would be 10^9 cubic centimeters of soil per hectare, n would be 2.43 x 10^7 fertilizer granules per hectare, v_p would be 41.2 cubic centimeters of soil per granule, and l would be 13.1 centimeters of root.

The length of time required to produce the average length of root needed to contact the first sphere of fertilizer-affected soil will depend upon many factors, the net effect of which may be estimated by the rate of growth of the aboveground portion of the plants. For lettuce, Costigan found that a good fit to experimental data on field-grown lettuce was provided by the empirical relation

$$l = 0.1e^{\{-7.32 + 16.05/[1 - 0.078 \ln (0.001w)]\}},$$

where l is the root length in centimeters and w is the dry weight of the shoot in milligrams.

Lettuce has small seeds and hence little stored reserve. Costigan estimated that with the 131-kilogram application of fertilizer phosphorus described in a preceding paragraph, 21 days would elapse before a root from an average plant would reach the phosphorus from a fertilizer granule. The reserve of phosphorus in the seed (about 8 micrograms) is enough to supply the seedling for the first 8 days. Thus, between the 8th day and the 21st day, the average plant would have to obtain its phosphorus from unfertilized soil. He noted that the period of 13 days during which the plant would not have access to the fertilizer is equal to about 17% of the postemergence growing period of the crop. A reduced growth rate during this period, therefore, could well reduce the final yield. A presumably similar situation was described by Silcock and Smith (1982) for the tropical grass *Anthephora pubescens*. Phosphate fertilizer had to be located within 5 centimeters of the seed to have any significant effect on plant growth within the first 30 days.

Fig. 7-19A illustrates the importance of early contact with the fertilizer in a soil culture experiment on phosphorus fertilization of lettuce. Note the greater weight of the plants receiving a starter fertilizer (monobasic ammonium phosphate) placed 1 centimeter below the seed. The soil was watered from below, which should have carried some of the fertilizer phosphorus and nitrogen to the seed. Fig. 7-19B shows estimates of the average time for roots of different crop plants to reach the first sphere of fertilizer-affected soil. Lettuce, with the smallest seeds, requires the longest time, and wheat, with the largest seeds, requires the shortest time.

Consistent with the considerations of De Wit (1953) and Costigan (1987), Eghball and Sander (1987) noted that as an average of nine field experiments

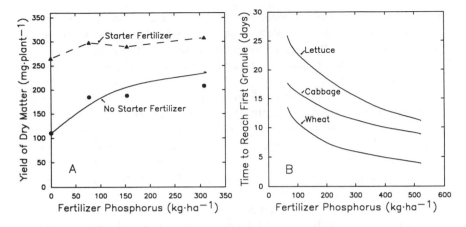

Fig. 7-19. Importance of the time required for seedlings to reach the phosphate fertilizer in broadcast applications. A. Dry weight of lettuce plants in soil cultures with different quantities of fertilizer phosphorus broadcast, with and without a starter fertilizer consisting of monobasic ammonium phosphate. The points are experimental observations, and the solid ourvod lino for yields without starter fertilizer represents values calculated from the Costigan theory. B. Estimated time for the first root of lettuce, cabbage, and wheat to reach the phosphate fertilizer in broadcast applications of concentrated superphosphate consisting of granules with diameters of about 2 millimeters, each granule containing 5.4 milligrams of phosphorus. (Costigan, 1984, 1987)

on wheat in Nebraska, applying 11 kilograms of fertilizer phosphorus per hectare with the seed in rows 30 centimeters apart produced slightly higher yields than applying the fertilizer by knifing into bands the same distance apart. The yields were closer together at 22 kilograms of fertilizer phosphorus per hectare and were identical at 33 kilograms. The knifed fertilizer was applied before planting, and the locations of the knifed bands did not coincide with the plant rows. Eghball and Sander's measurements indicated that the liquid fertilizer employed would be deposited in both the seed row and knifed applications in droplets 2.9, 0.8, and 0 centimeters apart with the applications of 11, 22, and 33 kilograms of phosphorus per hectare.

7-8. Root-Fertilizer Contact and the Greenwood Theory

Greenwood and coworkers, located at Horticultural Research International in the United Kingdom, have understandably emphasized vegetable crops in their research. In several papers they have commented on the difference in the utilization of soil nitrogen by crops of the grass family on the one hand and vegetable crops on the other.

As increasing quantities of fertilizer nitrogen are added for crops of the grass family grown on nitrogen-deficient soils, the characteristic behavior is an approximately linear increase in yield of nitrogen in the crop, with approximately constant percentage recovery of the fertilizer nitrogen in the crop until the maximum crop yield is attained. With further additions of fertilizer nitrogen, the uptake of nitrogen by the crop continues to increase, but at a much lower rate than before. At the same time, the percentage recovery of the added nitrogen in the crop drops sharply.

As increasing quantities of fertilizer nitrogen are added to vegetable crops grown on nitrogen-deficient soils, the characteristic behavior is an approximately linear or decreasing-returns type of curve for yield of nitrogen in the crops, with an approximately linearly decreasing percentage recovery of the added nitrogen. The large quantities of fertilizer nitrogen needed to produce maximum physical yields and maximum economic yields are used inefficiently. Much residual nitrate may remain in the soil at harvest.

Greenwood et al. (1989) explained the difference in performance of the two classes of crops on the basis of the relative extensiveness of the root systems. Crops of the grass family grown in the United Kingdom, including pasture grasses and the small-grain winter cereals, are close spaced. The root systems are extensive at the time the nitrogen fertilizer is applied. The authors quoted a figure of 8 kilometers of roots per square meter of soil surface area for winter wheat. Roots thus are closely spaced beneath the entire soil surface and are well positioned to absorb fertilizer nitrogen applied broadcast on the surface.

With vegetable crops, on the other hand, the nitrogen fertilizer is applied before planting or transplanting. The crops are planted in widely spaced rows, and for some time much of the mineral nitrogen in the soil is not accessible to the crops because of their limited root systems. For some crops, some of the nitrogen may not be accessible even at harvest. The authors quoted measured lengths of 1.8, 2.4, 5.7, 11.9, and 15.0 kilometers of roots per square meter at the time of harvest for onion, lettuce, parsnip, cauliflower, and turnip, respectively. Initially the root length would be zero or close to it.

Thus, according to the root-extent theory of Greenwood et al. (1989), if the roots of a vegetable crop are able to extract nitrate from half the volume of soil from which grass crops can extract it, twice as much nitrate must be available in the volume of soil accessible to the crop for a given uptake. The implication of the theory for fertilizer placement is that for row crops with limited root systems, the usual broadcast applications of fertilizer are inefficient when the crop is small, and the inefficiency may persist until harvest. The efficiency of nitrogen fertilization could be increased and the residues of nitrate subject to loss could be decreased by adjusting the fertilization practice to the crop capabilities. Localized placement of a small quantity of fertilizer nitrogen in a band below or near the row could meet the needs of the crop during much of the season. The quantity of fertilizer nitrogen applied broadcast could be reduced, or a delayed application might be made between the rows after the root extent has increased enough to absorb the nitrogen. The authors referred to an experiment at their location in which the maximum yield of onion, obtained by applying

80 kilograms of fertilizer nitrogen broadcast per hectare, was obtained also by a localized application of only 14 kilograms, with no nitrogen applied broadcast.

7-9. Nutrient Translocation in Plants and the De Wit Theory

In Section 7-2.1, the width of the fertilizer band was given as $(X_t/X_b)M_b$ where a band of fertilizer is located on one side of the row and as $2(X_t/X_b)M_b$ where bands of equal width are located on each side of the row. The implication was that the side of the row on which the fertilizer is located is immaterial, and that having a band on each side of the row has an effect equivalent to that of doubling the width of the band on one side of the row and having no band on the other side.

The De Wit theory and the experimental observations in Section 7-7 had to do with herbaceous crops. The location of the fertilizer relative to the plant may have more significant effects on woody plants. Whereas the nutrients absorbed by the roots of herbaceous crops seem to enter a common pool from which the nutrients in the above-ground portions are derived, a characteristic of at least some woody plants seems to be that the nutrients absorbed from roots on one side of the plant feed into vascular tissue that transmits the nutrients to the above-ground portions on that side of the plant without much crossing over.

For plants with limited vascular crossover between roots and tops, placing a fertilizer band on only one side of the plant may be an inappropriate way to supply fertilizer. Chandler (1957, p. 166) noted that fertilizer applied to the soil on one side of a fruit tree will cause improved growth on that side only. Gough (1984) found that when he split the root system of highbush blueberry plants and supplied half of the roots each week with a solution containing nitrogen, phosphorus, and potassium, the side of the plants receiving the nutrients grew and produced fruit, whereas the side receiving no nutrients made little growth, had chlorotic foliage, and showed some leaf reddening indicative of nitrogen deficiency. Most of the flower buds on the unfertilized side died, and the few that did open did not set fruit. This behavior supports the traditional recommendation to fertilize blueberry plants by spreading the fertilizer completely around the plants or along each side of the row.

7-10. The Depth Factor

Burns (1980a,b) considered the nitrate present in the soil above the effective depth of rooting to be equally available to plants. He meant by this that if the initial concentration is not excessive, plants can absorb nitrate almost quantitatively to the effective depth of rooting.

The nitrate-availability concept Burns used is appropriate for the conditions under which he worked — a relatively humid environment, in which distributed precipitation would keep the upper portion of the soil relatively moist, and nitrate derived from surface applications would be leached down into the soil. The concept would not be appropriate if the upper portion of the soil were dry. For example, Garwood and Williams (1967) found that where the surface soil was dry but the subsoil was moist in an experiment in the United Kingdom, established perennial ryegrass took up more fertilizer nitrogen that had been injected at a depth of 76 centimeters than it did from a surface application. Analogous findings were made by Herron et al. (1968) for corn in eastern Nebraska. (Limited nitrate absorption from the surface soil on account of dryness, however, would not interfere with use of the Burns model, because the model has to do with leaching of nitrate below the effective depth of rooting, and leaching would not be occurring if the surface soil were dry.)

The situation is more complex where nutrients undergo interactions with soil, and it is completely different in nitrogen fertilization of flooded rice. The depth factor has been investigated extensively in nitrogen fertilization of flooded rice because of the major effect it has on the efficiency of the nitrogen. Where rice receives nitrogen fertilizer in Southeast Asia, the general farm practice has been to broadcast the fertilizer by hand directly into the flood water 2 to 3 weeks after transplanting (Cochran et al., 1982). Urea, the commonly used source of nitrogen, hydrolyzes rapidly at the interface of the soil and floodwater, and the floodwater becomes a dilute solution of ammonium bicarbonate. The pH of the floodwater rises, some of the ammonium dissociates to hydrogen ions and ammonia, and some of the ammonia escapes to the atmosphere. The bicarbonate ions react with the hydrogen ions to form water and carbon dioxide (Vlek and Craswell, 1981). If ammonium is oxidized to nitrate in the flood water, the nitrogen may be lost by denitrification when the nitrate enters the soil (Zhu et al., 1989). Similarly, if ammonium in the zone of oxidized soil surrounding rice roots is oxidized to nitrate, the nitrogen may be lost by denitrification as the nitrate diffuses into the adjacent reduced soil (Reddy et al., 1989).

Many field experiments have been conducted, but the experimental work illustrated here was done in a growth chamber under conditions where precise measurements could be made of the loss of ammonia. Fig. 7-20 shows the effects of depth of placement of the urea on rice grain yield, recovery of fertilizer nitrogen in the grain, and loss of ammonia to the atmosphere. The value of placing the urea below the soil surface is evident.

Placing phosphate fertilizer below the soil surface may be beneficial for quite a different reason. Unlike nitrate, which readily moves downward in the soil with percolating water and may be lost by penetration below the effective depth of rooting, phosphate fertilizers undergo reactions with soil that greatly reduce the solubility of the phosphate and its movement with water added as precipitation or irrigation. Williams (1971) found that when he placed superphosphate on the surface of moist soils at 100 centimeters of water tension, the retention of the fertilizer phosphorus in the surface 6 millimeters of soil ranged from about 50 to 95% in the soils tested and increased with the product of the bulk density and

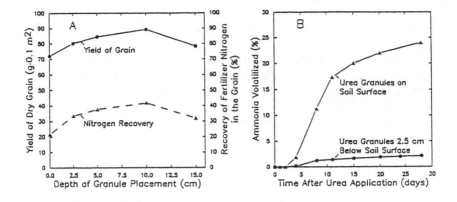

Fig. 7-20. The depth factor in urea placement for rice in soil cultures under 5 centimeters of water. The rice was fertilized with 1-gram "supergranules" of urea. A. Yield of rice grain and percentage recovery of fertilizer nitrogen in the grain with urea granules at different depths in the soil. B. Percentage of urea nitrogen lost to the atmosphere as ammonia with urea granules on the soil surface and 2.5 centimeters below the surface. (Eriksen et al., 1985)

the phosphorus sorption capacity. The soils used in this experiment were all acid.

The calcium carbonate in calcareous soils reacts rapidly with the monobasic calcium phosphate in ordinary superphosphate or concentrated superphosphate. The visible effervescence that occurs indicates dissolution of the calcium carbonate and evolution of carbon dioxide, which is accompanied by precipitation of dibasic calcium phosphate. Williams (1971) found that soils containing calcium carbonate were very effective in confining phosphorus added as superphosphate to locations immediately adjacent to the fertilizer particles.

Observations in western New South Wales in Australia and elsewhere had indicated that although the legume barrel medic is sensitive to phosphorus deficiency, its response to surface applications of superphosphate depended upon the rainfall. Fig. 7-21 shows the results of an experiment to test the hypothesis that the poor response in dry years was at least in part a result of relatively low availability of the surface-applied fertilizer. This experiment, conducted during a period of low rainfall, shows that the response increased considerably when the superphosphate was placed at depths of 5 or 10 centimeters, verifying the hypothesis about low availability of surface-applied superphosphate. The yields shown are averages of the harvests at the end of the first and second seasons after application of the fertilizer. Measurements made on the soil a year after fertilization showed that the extractable phosphorus in the 0 to 2 and 2 to 4 centimeter layers of soil had been increased by the surface application, but it was doubtful that the fertilizer phosphorus had penetrated further. Jarvis and Bolland (1990) found similarly that superphosphate placed at depths of 5 to 11 or 13 centimeters produced greater increases in yield of wheat and lupine than placement at 3 centimeters under dry conditions (30 to 38 centimeters of annual rainfall) in Western Australia. Randall and Hoeft (1988) reviewed the results of some experiments in the United States.

The relation of dryness to the depth factor in phosphorus availability was investigated under more controlled conditions by Simpson and Pinkerton (1989) by growing plants in soil cultures to which monobasic calcium phosphate and water were added differentially to surface and subsurface layers. In the experiment shown in Fig. 7-22, the convergence of the two curves and the increases in yield with increasing additions of water to the surface of the soil suggest the importance of the upper 40 millimeters of soil in meeting the crop needs, perhaps because of a preponderance of roots of the subterranean clover test crop in this portion of the soil. Similar results, but with lower yields, were obtained where the soil in the 40 to 180 millimeter layer had a lower water content (approximately two-thirds of field capacity). The failure of the yield-of-phosphorus curves to converge like the yield-of-dry-matter curves (see the original paper) suggests that the phosphorus supplied by the surface 40 millimeters of soil was almost adequate and that factors other than phosphorus in that portion of the soil were limiting the yields with the greater water supply.

The depth factor in fertilizer placement has received much attention as a consequence of no-tillage or minimum-tillage practices in which the plow, disc, and harrow, commonly employed to incorporate broadcast fertilizer into the soil,

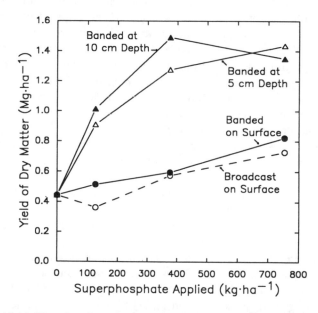

Fig. 7-21. Yield of barrel medic pasture in the first two seasons after fertilization with different quantities of superphosphate broadcast on the surface and banded on the surface and at different depths. The experiment was conducted under relatively dry conditions in western New South Wales. (Scott, 1973)

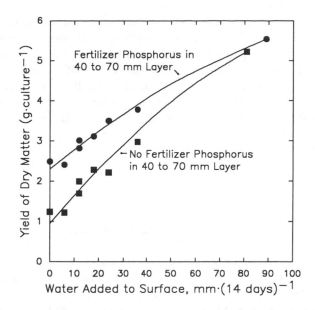

Fig. 7-22. Yield of subterranean clover versus water added to the surface of soil cultures with and without an addition of concentrated superphosphate equivalent to 50 kilograms of phosphorus per hectare of surface area to the 40 to 70 millimeter layer of soil. All cultures received a topdressing of concentrated superphosphate equivalent to 15 kilograms of phosphorus per hectare of surface area. The water content of the 40 to 180 millimeter layer of soil was maintained at approximately the field capacity. (Simpson and Pinkerton, 1989)

are not used. Fertilizer placement in reduced tillage systems was discussed at length by Randall ct al. (1985).

For broadcasting, surface applications are the most convenient in reduced-tillage systems, and this is probably the way most of the fertilizer is applied. The De Wit (1953) theory discussed in Section 7-2 applies to surface applications of fertilizer as well as to applications that are incorporated into the soil. Fig. 7-21 supplies evidence for the superiority of a banded surface application of superphosphate even under dry conditions where surface placement was inefficient relative to deeper placements.

The downward movement of phosphorus, potassium, and micronutrients applied to the surface of the soil is limited, and this reduces their effectiveness under the circumstances of soil dryness just discussed. To increase fertilizer effectiveness without at the same time losing the benefits of no tillage or limited tillage, equipment has been developed to "knife" the fertilizer into the soil in concentrated bands with little disturbance of the soil.

Microbiological immobilization may be of importance in inhibiting the movement of surface-applied fertilizer nitrogen down into the soil where it can be absorbed by plant roots. For example, at the time of harvest in a field experiment on barley in Alberta, Hartman and Nyborg (1989) found that the nitrogen from nitrogen-15-tagged urea that remained immobilized in the soil (the total amount of fertilizer-derived nitrogen in forms other than nitrate and ammonium) as an average of six water regimes was 55% where the fertilizer was broadcast and worked in to a depth of 2 centimeters compared with 40% where the fertilizer was applied in a band 2.5 centimeters wide located 10 centimeters below the soil surface and 6 centimeters below the seed. Inorganic chemical interactions of ammonium nitrogen with the soil that result in reduced availability are not of much concern. On the other hand, surface applications do have the disadvantage that they may be subject to loss to the atmosphere. Anhydrous ammonia is the extreme case. Little of the nitrogen from this source would be retained by the soil if it were applied to the surface.

7-11. The Geraldson System

Geraldson (1970) devised a system that he called *precision nutrient gradients* for fertilizing field crops under a mulch of sheet plastic. Developed for sandy soils in Florida, the system involves growing the crop in the center of a flat bed, the top of which is raised 25 to 30 centimeters above the soil between beds. The total fertilizer application per hectare includes 400 kilograms of nitrogen, 390 kilograms of potassium, 560 kilograms of superphosphate, and 22 kilograms of fritted micronutrients. The superphosphate, micronutrients, and 5 to 10% of the nitrogen and potassium are mixed with the soil in the bed, and the remaining amounts of nitrogen and potassium are placed in two bands on the surface of the soil underneath the plastic mulch near the outer edges of the bed. An ample supply of water is provided by controlling the water table at a depth of 41 to 46 centimeters.

The nutrients mixed with the soil are adequate to permit root growth throughout the bed. The concentrations of nitrogen and potassium directly below the bands are excessive and inhibit root growth, but the concentrations decrease with increasing depth. The most prolific development of roots occurs in a relatively small volume of soil beneath the bands at the junction of the gradients of water supply and nutrients that are most favorable.

Geraldson developed the system primarily for high-value crops for which fertilizer costs are of secondary concern. In seven of nine consecutive seasons in which the system was tested, tomato yields ranged from 103 to 116 megagrams per hectare. In one season in which the crop was damaged by a hurricane, the yield was 85 megagrams, and in one season in which the crop was damaged by stem canker, the yield was 87 megagrams. Yields of commercial crops with exposed root environments during these same seasons ranged from failure to 56 megagrams per hectare. The Geraldson system has now been adopted by growers in Florida, and statewide yields per hectare have doubled (Geraldson, 1990).

7-12. Foliar Fertilization and Other Application Methods

A variety of methods of application other than broadcasting and banding are used. These include spraying the nutrients on the foliage, applying the nutrients in the irrigation water, injecting the nutrients into the plants, coating the seed with a binder containing the nutrients, soaking the seed in a solution of a micronutrient, and using the seed from plants grown on soil high in a critical micronutrient. Applying the nutrients in the irrigation water is probably the most common of these methods where irrigation is practiced, and foliar application is probably the most widely used worldwide.

Plants may also absorb nitrogen and sulfur from certain gaseous forms in the atmosphere, with beneficial effects if the soil supplies of these elements are deficient, or with detrimental effects if the concentrations in the atmosphere are excessive. Gaseous sources of these elements may provide a significant proportion of the nutrient requirement of plants under some circumstances, but in practice the gases are by-products of activities other than intentional fertilization. Probably only carbon dioxide, which normally is absorbed through the foliage, is used intentionally to increase the concentration in the atmosphere for purposes of fertilization.

Each method of application has its advantages and disadvantages. Foliar application is often relatively convenient because small quantities of nutrients can be applied in conjunction with pesticide sprays more easily than they can be applied alone or with fertilizers. The repeated applications sometimes needed, however, can be troublesome.

Applying nutrients via a sprinkler irrigation system amounts to a combination of foliar and broadcast application. Applying nutrients via furrow irrigation may be considered a form of band application for nutrients that react with soil solids. Applying nutrients by trickle irrigation is a placement method in which the distribution of nitrate parallels the distribution of water, and nutrients that react with the soil are localized to some degree where the water enters the soil. With some crops, irrigation water can be used as an effective conveyor of nutrients at times when making a delayed application by ground-based machinery would create undue damage.

Foliar application has proved to be a complex subject, with promise of benefits as yet unfulfilled. By avoiding some of the losses in availability with soil applications, applying nutrients directly to the foliage presents a potential for increasing the efficiency of nutrient use. Bakker et al. (1984) in The Netherlands observed approximately equal yields of lettuce from broadcast application of ammonium nitrate before planting and application via sprinkler irrigation, but higher nitrate concentrations in the lettuce with the sprinkler irrigation method. In an 8-year experiment with citrus in California, Embleton et al. (1986) found that foliar fertilization with urea decreased the loss of nitrate in the drainage water compared with applications of calcium nitrate to the soil. Both of these investigations indicate greater efficiency of applications to the foliage than to the soil. In England, however, Poulton et al. (1990) found that although the recovery of nitrogen-15 in the crop was about the same from tagged urea applied as a spray to wheat or as a topdressing of a mixture of tagged ammonium nitrate and urea to the soil, the total recovery in soil and crop was greater with the soil application. They suggested that some of the urea nitrogen applied as a spray was lost by volatilization as ammonia.

The most important problem encountered in foliar fertilization has been damage to the foliage and sometimes the fruit. Other problems have included poor penetration, poor translocation, expense associated with the need for repeated applications, and erratic results.

The first published account of foliar application of a nutrient is generally attributed to Gris (1844) in France, who reported that applying soluble iron compounds cured chlorosis. Micronutrients, including iron, are more adaptable to foliar fertilization than are macronutrients because of the damage to the foliage that so often accompanies the application of relatively large quantities of soluble macronutrients in sprays.

Micronutrients are like macronutrients in that excesses produce unfavorable effects. The quantities of micronutrients that constitute excesses, however, are relatively small. Thus, where a deficiency of a micronutrient is significant, a large increase in yield from spraying the plants with a solution supplying a small quantity of the deficient nutrient followed by toxicity with larger applications is the norm. Fig. 7-23 is an example for zinc.

Ferrous sulfate, traditionally the most commonly used source of iron in plant nutrition, reacts quickly with soil. Relatively large quantities are needed for soil applications, and localized applications are more effective than equal quantities applied broadcast because the reversion to insoluble forms takes place less rapidly at the higher concentrations. Ferrous sulfate has often been used in foliar applications, and probably was the form used by Gris (1844). Iron sprays may need to be repeated several times during a single season because of the low mobility of iron in plants.

A major advance in treating deficiencies of iron and certain other heavy metal cations was made with the development of soluble chelates that bind the cations strongly. Some iron chelates can be applied effectively to either the foliage or the soil. In contrast to the inefficiency of soil applications of inorganic iron sources, Reed et al. (1988) found that the most effective of all the foliar and soil applications of iron compounds they tested in increasing the chlorophyll content of chlorotic peach and grape foliage in Texas was a soil application of the iron chelating agent EDDHA (ethylenediamine di-10-hydroxyphenylacetic acid) containing 6% iron. The quantity applied to the soil, however, was greater than that applied to the foliage as a spray.

Mortvedt (1986) reviewed the methods and compounds used for correcting iron deficiency chlorosis, and Raese et al. (1986) reported the results of exper-

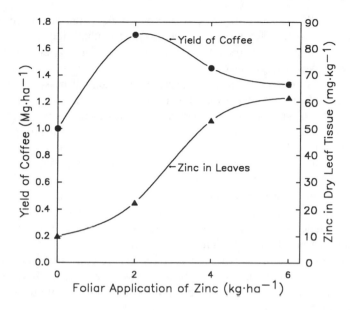

Fig. 7-23. Yield of coffee beans and zinc concentration in the leaves with different quantities of zinc applied to the foliage in Brazil. (Malavolta, 1986)

iments on different compounds and techniques for correcting iron deficiency in apple and pear trees. The latter investigators concluded that injecting the trunks with 1% ferrous sulfate solution gave better and longer lasting control of chlorosis than did foliar or soil applications. Wallace and Wallace (1986) preferred injections of ferric ammonium citrate over ferrous sulfate. Zinc chelate was effective, but zinc sulfate was not.

Molybdenum is generally applied to the soil. In contrast to iron, molybdenum is mobile in plants. In New Zealand, a single spray of sodium molybdate supplying 1 kilogram of molybdenum per hectare and applied to cauliflower seedlings before transplanting has been found more efficient than applications to the soil (Wilson and Scheffer, 1974).

Soaking the seeds of certain leguminous crops in a sodium molybdate solution before planting has been found far more efficient than applying the molybdenum to the soil (Sherrell, 1984). Scheffer and Wilson (1987) found that with cauliflower, which requires relatively large amounts of molybdenum, applying sodium molybdate in a seed coating of bentonite clay was effective.

Coating seeds with phosphate compounds has been studied (see, for example, Scott and Blair, 1988a,b). Small amounts of soluble phosphates are beneficial to the early growth, but larger amounts delay seedling emergence and reduce stands. Seed coating has not been developed as a way to supply substantial amounts of soluble phosphorus compounds.

In an early study of foliar application of a macronutrient, Hiltner (1911) published a photograph of mustard plants grown in quartz sand supplied with nutrients other than potassium. On one side of a divider, the plants received no potassium and made essentially no growth. On the other side of the divider, plants whose leaves were painted from time to time with a 2% potassium sulfate solution grew well and produced fruit.

In practice, nitrogen has been the macronutrient applied most commonly to the foliage, and urea almost invariably has been the source of choice. Foliar application of nitrogen as urea to pineapple and sugarcane has been recognized as a useful practice for many years, but the reason for the capability of these plants to tolerate relatively heavy foliar applications of urea is not understood. Perhaps it is related to the rapid hydrolysis of urea to ammonium in the plants, noted by Humbert (1968) for sugarcane. Both urea and ammonium are toxic (Krogmeier et al., 1989), but urea may pose the greater problem.

Urea applied to plant leaves is absorbed rapidly. Klein and Weinbaum (1985) found that uptake of urea by tree leaves amounted to 50% with olive and 80% with almond within the first 24 hours after application.

Spraying pineapple plants with urea replaced the previous method of applying dry ammonium sulfate by hand to the basal leaves. The latter method depended upon subsequent rainfall to dissolve the fertilizer and wash it into the soil at the base of the plant (Cannon, 1960a).

Concentrations as high as 27% and applications of as much as 65 kilograms of urea nitrogen per hectare have been used experimentally on pineapple in Queensland, with spray volume just sufficient to wet the leaves with little or no runoff. A concentration of 10% was favored for preflowering growth (Mitchell and Nicholson, 1965). Several foliar applications of fertilizer nitrogen are needed for pineapple, which requires up to 26 months from planting to harvest. Foliar fertilization with urea requires more applications than soil or basal leaf treatments because the foliage may be damaged by heavy applications, particularly if the urea contains much biuret. Sanford et al. (1954) found that urea containing 3% or more biuret produced leaf-tip dieback when applied as a spray containing 4.2% of the urea fertilizer.

In some experiments, the principal advantage of supplying nitrogen to pineapple in foliar sprays has been a saving of labor. The foliar application also acts more rapidly than the basal leaf applications, and does not have the soil acidifying effect of the ammonium sulfate treatment used formerly (Cannon, 1960a). In Queensland, urea sprays and the basal leaf treatment with ammonium sulfate were found equally effective per kilogram of nitrogen (Cannon, 1960b).

In research in South Africa, Du Plessis and Koen (1983) found that seven foliar applications of nitrogen as urea or ammonium sulfate produced higher pineapple yields than an equal number of soil applications of ammonium sulfate supplying equivalent quantities of nitrogen. Fifteen monthly sprays with urea supplying a total of 400 kilograms of nitrogen per hectare produced a yield of 90 megagrams per hectare, whereas eight soil applications of ammonium sulfate at bimonthly intervals (also supplying a total of 400 kilograms of nitrogen) produced 61 megagrams per hectare. Injury with excessive individual doses of urea nitrogen in foliar applications was indicated by a yield of 56 megagrams where the 400 kilograms of urea nitrogen were split into six applications.

Humbert (1968) described the development of urea applied in a spray from aircraft as an important source of nitrogen for sugarcane in Hawaii. The cane responds more quickly to foliar applications than to soil applications. He noted that up to 90 kilograms of nitrogen as urea could be applied per hectare without burning the leaves if the plants were well watered. If the cane was low in water, however, the urea damaged the tissues.

Although urea was effective as a foliar application, Humbert reported that the spray was eventually replaced in Hawaii by a dry granular product applied by plane in a mixture with granular potassium chloride because the urea often resulted in potassium deficiency in the plants, and the solubility of potassium chloride in water was too low to supply the amount of potassium needed in the same spray with the urea. Singh et al. (1983) reported that in Uttar Pradesh, spraying sugarcane with 75 kilograms of nitrogen as urea per hectare produced higher yields and greater profits than an equal or double amount applied to the soil. Damage to the plants from spraying appeared to be a factor in their experiments because the yield was 60 megagrams of cane per hectare where a urea solution containing 5% nitrogen equivalent was sprayed eight times during June and August, but only 50 megagrams where the urea solution contained 10% nitrogen equivalent and was sprayed four times during June and August.

In a special condition in Australia, Hodgson and MacLeod (1987) found that foliar application of urea to cotton was useful in ameliorating the damage to the crop associated with irrigating the clay soil on which the crop was produced. Upon irrigation, the soil became waterlogged, and the cotton showed symptoms of nitrogen deficiency, presumably because of loss of soil nitrate by denitrification. Spraying the plants with a solution supplying up to 20 kilograms of urea nitrogen per hectare before irrigation increased the lint yield.

The usual problem encountered when an attempt is made to spray crops with quantities of the macronutrients sufficient to meet their nutrient needs is that the foliage is damaged. Upadhyay et al. (1988) reported what is considered to be a general observation where an attempt is made to supply crops with phosphorus by foliar application. In a field experiment in which soybean was supplied with 10, 20, and 30 kilograms of phosphorus per hectare as dibasic ammonium phosphate or ordinary superphosphate, they obtained the highest yield of beans where all the fertilizer was applied to the soil, an intermediate yield where half was applied to the soil and half was applied in two sprays as 2 or 3% solutions, and the lowest yield where all the phosphorus was applied in the two sprays. The sprays damaged the leaves.

Barel and Black (1979a) found that certain condensed phosphates and compounds containing P-N bonds or P-N-P linkages could be applied at 2.5 to 3 times the quantity of phosphorus that could be applied as orthophosphate without causing leaf damage. The yields of corn and soybean in a field experiment were higher with two of these compounds (tripolyphosphate and tetrapolyphosphate) than with orthophosphate (Barel and Black, 1979b).

Sine (1986) reported that "triazone," a proprietary compound containing three nitrogen and three carbon atoms in a heterocyclic configuration, could be applied to foliage in relatively large quantities without causing damage. The compound reportedly had given favorable results in experiments and was being marketed in a liquid formulation containing 65% of its nitrogen as triazone, 28% as urea, and the remainder as other organic compounds.

Sinclair and De Wit (1975) made a contribution to the understanding of the behavior of different crop species in a theoretical paper in which they classified certain species as "self destructive" on the basis of the rate at which they required nitrogen to form proteins during seed development. According to their analysis, the requirement of seeds of these species for nitrogen during seed development was so great that they depleted the nitrogen contained in the vegetative portions of the plants, thus impairing the physiological activity of the vegetative tissues, hastening the senescence and loss of leaves, shortening the period for seed development, and limiting seed yield. The crops Sinclair and De Wit (1975) classed as self-destructive included soybean, lentil, pea, mung bean, cowpea, pigeon pea, lima bean, and chick pea. Their yield was presumed to be especially sensitive to the rate of supply of nitrogen from the soil during seed development. The remaining species they studied were said to be "not generally limited by the potential rate of nitrogen supply from the soil but rather the total amount of available nitrogen." These crops included wheat, corn, rice, sorghum, rye, and oat. In a later paper (Sinclair and De Wit, 1976), they paid particular attention to the soybean crop.

Hanway (1976) independently described the behavior of soybean during seed filling. In addition to the points made by Sinclair and De Wit (1975), Hanway added the observations that as seed filling progresses, the roots stop growing, nutrient absorption by the roots slows and stops, nitrogen fixation in the nodules stops, and the nodules die and slough off.

Garcia and Hanway (1976) added one more concept, namely, that phosphorus, potassium, and sulfur also are depleted from the leaves during seed filling, and that additional supplies of these nutrients during seed filling might be useful in delaying senescence and in increasing the yield of beans. In several field experiments conducted in Iowa over a 2-year period to test their hypothesis by making late-season foliar applications of nutrients, they obtained a number of favorable results and found that the increases in yield were greater where the sprays included nitrogen, phosphorus, potassium, and sulfur than where the mixture was incomplete.

The findings reported by Garcia and Hanway sparked a great deal of interest, and many experiments on foliar fertilization of soybean soon were conducted by investigators using fertilizer materials supplied by the National Fertilizer Development Center at Muscle Shoals, Alabama. The results of 214 comparisons reported by Gray and Akin (1984), however, showed that foliar fertilization produced an average yield decrease of 5%. Harder et al. (1982) reported a temporary decrease in photosynthesis and a decrease in final grain yield of corn plants receiving foliar applications of a solution supplying nitrogen, phosphorus, potassium, and sulfur.

In a survey of the literature, Neumann (1982) concluded that although applying foliar fertilizer late in the season sometimes increases seed yield, it ''does not inhibit leaf senescence and that such senescence is in any case not a primary factor in limiting seed yields.'' More recent results have been reported by Alexander (1986), Ashmead et al. (1986), and others. In somewhat contradictory work, Batten and Wardlaw (1987) found that applying monobasic ammonium phosphate to the flag leaf of phosphorus-deficient wheat did delay the senescence, but it did not result in a significant increase in yield of grain. Direct evidence of the connection of leaf senescence with fruiting has been provided by various investigators cited in the review by Neuman (1982), who found that removing the reproductive structures of annual plants, including soybean, inhibits the onset of leaf senescence.

To date, the situation remains unresolved. More research is needed to provide understanding of the principles and to provide an improved basis for prediction of effects.

A number of papers have described increases in yield from applying seaweed extracts as a foliar spray. Temple and Bomke (1989) conducted field experiments in British Columbia in which garden bean yield increases from foliar sprays containing a seaweed concentrate averaged 24%. Organic substances seem to be responsible for the beneficial effect, as suggested by (a) the small quantities of mineral nutrients supplied in the sprays relative to the quantities in the plants (Temple and Bomke, 1989), (b) the loss of the beneficial effect when the extract was ashed (Finnie and Van Staden, 1985), and (c) the observation that the sprays were effective for rye and cabbage only on fertilized soils (Kotze and Joubert, 1980). Plant hormones are suspected to be primarily responsible for the benefits. The seaweed concentrates contain hormones, and a hormonal extract of seaweed produced increases in yield, although they were somewhat smaller than those produced by the unfractionated extract (Temple and Bomke, 1989).

7-13. Application of Manure

''Placement'' is the norm where urine and feces are deposited by grazing animals. According to Lantinga et al. (1987), a dairy cow that spends 20 hours a day on pasture will urinate and defecate 10 times during those 20 hours.

A single urination affects plant growth on an area of about 0.68 square meter, which is often more than twice the area actually wetted. The average quantity of nitrogen applied to an area affected by a urination is the equivalent of about 500 kilograms per hectare, which exceeds the total annual application of fertilizer nitrogen recommended in the Netherlands by 100 kilograms per hectare. Although the effect is beneficial, it is unevenly distributed.

When commercial fertilizer is applied to benefit the areas that have not been fertilized by the urine, it is impractical to miss the urine spots, and indeed they often are difficult to see except when the grass is growing and is not being grazed. Thus, application of fertilizer nitrogen at the recommended rate would overfertilize the urine spots and would result in inefficiency of both the fertilizer nitrogen and the urine nitrogen. Lantinga et al. (1987) estimated that if fertilizer nitrogen were being applied at 250 kilograms per hectare, the increase in yield of herbage resulting from the urine spots would be too small to be detectable experimentally.

The area covered by the dung from a single defecation is about 0.05 square meter, and the total area on which grass growth is affected is about 0.25 square meter. The inefficiency of the localized application of feces is analogous to, but more pronounced than, that of the urine spots because grazing animals tend to avoid the herbage around dung spots, whereas they graze the herbage on urine spots. From the standpoint of nutrient efficiency, it would be preferable if the dung and urine from grazing animals could be distributed uniformly.

Where animal manures produced in concentrated areas are applied mechanically to other areas in agricultural practice, placement is a factor of major importance in determining the efficiency for crop production. This is because much of the nitrogen in animal manures is often present in the ammoniacal form and is lost by volatilization when the manure is broadcast on the surface.

As an average of experiments in The Netherlands, the apparent recoveries of applied nitrogen in grass were 23% where manure slurry was broadcast on the surface in the spring and 49% where the slurry was injected into slits in the soil (Van der Meer et al., 1987). In Finland similarly, Kemppianen (1986) found that nitrogen recovery by the vegetation was greater where slurry was injected than where it was applied on the surface. The well known inefficiency of surface applications would not be so important, but for the fact that surface applications are the most convenient and sometimes, as in the winter, may be the only practical way that manure can be applied.

Grassland is convenient for applying manure at times when growing crops make cropland inaccessible and when the soil is soft due to high water content. In some instances, little cropland is available.

The net effect of applying animal manures, especially to grassland, is a combination of positive effects from the nutrients and organic matter and negative effects of various kinds. The negative effects are generally of increasing importance as the quantities of manure applied increase. The negative effects have been given special attention in recent work in Europe with the objective of increasing the net positive effect where heavy applications of manure are made.

Top dressing grassland with animal manures may damage the vegetation as a result of the soluble and perhaps volatile constituents and the reduced light due to the coatings of solids. Prins and Snijders (1987) reported that the damage was greater from farmyard manure than from slurry.

Küntzel et al. (1987) and Prins and Snijders (1987) found that the damage was reduced by diluting slurry with water and by applying water after the slurry.

Küntzel et al. (1987) found that applying the slurry directly to the soil surface (which could be done only under experimental conditions) did not cause visible damage to the vegetation or a reduction in yield.

Injection causes some damage even in the absence of slurry. A small strip of grass along the slit may die if injection is done under dry conditions in the summer, and less desirable species may invade. As a result of these effects, widely spaced injection slits would be desirable. On the other hand, heavier applications would be needed with wide spacing, and damage could result from the high concentration of slurry. Moreover, the uneven distribution of nutrients would result in uneven fertilization of the grass. Prins and Snijders (1987) reported that even the 50-centimeter spacing commonly adopted for practical reasons in The Netherlands was too wide to produce an even distribution of nitrogen and phosphorus in the grass between slits.

Despite the damage that may occur from injecting slurry into permanent grassland, the consensus seems to be that injection is preferable to surface application, not only to increase the efficiency of slurry use in production of grass, but also to reduce odors. Luten et al. (1983) described an applicator consisting of a disc coulter, a duckfoot sweep 18 centimeters wide, and a press wheel that can be used to inject 25 to 100 megagrams of slurry per hectare per application. The injectors are 50 centimeters apart, and the slurry is pumped into the soil through a flexible pipe located behind each sweep. They found that yields were higher with injection than with broadcasting on the surface.

7-14. Literature Cited

Alessi, J., and J. F. Power. 1980. Effects of banded and residual fertilizer phosphorus on dryland spring wheat yield in the Northern Plains. *Soil Science Society of America Journal* 44:792-796.

Alexander, A. (Ed.). 1986. *Foliar Fertilization*. Martinus Nijhoff Publishers, Dordrecht, The Netherlands.

Anghinoni, I., and S. A. Barber. 1980a. Predicting the most efficient phosphorus placement for corn. *Soil Science Society of America Journal* 44:1016-1020.

Anghinoni, I., and S. A. Barber. 1980b. Phosphorus application rate and distribution in the soil and phosphorus uptake by corn. *Soil Science Society of America Journal* 44:1041-1044.

Anghinoni, I., and S. A. Barber. 1980c. Phosphorus influx and growth characteristics of corn roots as influenced by phosphorus supply. *Agronomy Journal* 72:685-688.

Anghinoni, I., and S. A. Barber. 1988. Corn root growth and nitrogen uptake as affected by ammonium placement. *Agronomy Journal* 80:799-802.

Ashmead, H. D., H. H. Ashmead, G. W. Miller, and H.-H. Hsu (Eds.). 1986. *Foliar Feeding of Plants With Amino Acid Chelates*. Noyes Publications, Park Ridge, NJ.

Bakker, M. J., J. H. G. Slangen, and W. Glas. 1984. Comparative investigation into the effect of fertigation and of broadcast fertilization on the yield and nitrate content of lettuce (*Lactuca sativa* L.). *Netherlands Journal of Agricultural Science* 32:330-333.

Baldwin, J. P., P. H. Nye, and P. B. H. Tinker. 1973. Uptake of solutes by multiple root systems from soil. III. A model for calculating the solute uptake by a randomly dispersed root system developing in a finite volume of soil. *Plant and Soil* 38:621-635.

Barber, S. A. 1984. *Soil Nutrient Bioavailability: A Mechanistic Approach*. John Wiley & Sons, NY.

Barel, D., and C. A. Black. 1979a. Foliar application of P. I. Screening of various inorganic and organic P compounds. *Agronomy Journal* 71:15-21.

Barel, D., and C. A. Black. 1979b. Foliar application of P. II. Yield responses of corn and soybeans sprayed with various condensed phosphates and P-N compounds in greenhouse and field experiments. *Agronomy Journal* 71:21-24.

Barley, K. P. 1970. The configuration of the root system in relation to nutrient uptake. *Advances in Agronomy* 22:159-201.

Barraclough, D. 1989a. A usable mechanistic model of nitrate leaching. I. The model. *Journal of Soil Science* 40:543-554.

Barraclough, D. 1989b. A usable mechanistic model of nitrate leaching. II. Application. *Journal of Soil Science* 40:555-562.

Barraclough, P. B., and P. B. Tinker. 1981. The determination of ionic diffusion coefficients in field soils. I. Diffusion coefficients in sieved soils in relation to water content and bulk density. *Journal of Soil Science* 32:225-236.

Batchelor, J. T. (Ed.). 1983. *Fertilizer Placement*. Fertilizer Issues of the 80's 1, No. 1. (Cited in *Soils and Fertilizers* 47, Abstract 4662. 1984.)

Batten, G. D., and I. F Wardlaw. 1987. Senescence of the flag leaf and grain yield following late foliar and root applications of phosphate on plants of differing phosphorus status. *Journal of Plant Nutrition* 10:735-748.

Borkert, C. M., and S. A. Barber. 1985. Predicting the most efficient phosphorus placement for soybeans. *Soil Science Society of America Journal* 49:901-904.

Brewster, J. L., and P. B. H. Tinker. 1972. Nutrient flow rates into roots. *Soils and Fertilizers* 35:355-359.

Bullen, C. W., R. J. Soper, and L. D. Bailey. 1983. Phosphorus nutrition of soybeans as affected by placement of fertilizer phosphorus. *Canadian Journal of Soil Science* 63:199-210.

Burns, I. G. 1974. A model for predicting the redistribution of salts applied to fallow soils after excess rainfall or evaporation. *Journal of Soil Science* 25:165-178.

Burns, I. G. 1975. An equation to predict the leaching of surface-applied nitrate. *Journal of Agricultural Science (Cambridge)* 85:443-454.

Burns, I. G. 1980a. Influence of the spatial distribution of nitrate on the uptake of N by plants: A review and a model for rooting depth. *Journal of Soil Science* 31:155-173.

Burns, I. G. 1980b. A simple model for predicting the effects of leaching of fertilizer nitrate during the growing season on the nitrogen fertilizer need of crops. *Journal of Soil Science* 31:175-185.

Burns, I. G. 1980c. A simple model for predicting the effects of winter leaching of residual nitrate on the nitrogen fertilizer need of spring crops. *Journal of Soil Science* 31:187-202.

Cannon, R. C. 1960a. Save on costs of fertilizer for pineapples. *Queensland Agricultural Journal* 86:473-475.

Cannon, R. C. 1960b. Pineapple research; What's in store for the grower? *Queensland Agricultural Journal* 86:635-642.

Chandler, W. H. 1957. *Deciduous Orchards.* Third Edition. Lea & Febiger, Philadelphia.

Chiao, Y.-S., and A. Gillingham. 1989. The value of stabilizing fertilizer under carry-over conditions. *American Journal of Agricultural Economics* 71:352-362.

Claassen, N., and S. A. Barber. 1974. A method for characterizing the relation between nutrient concentration and flux into roots of intact plants. *Plant Physiology* 54:564-568.

Claassen, N., and S. A. Barber. 1976. Simulation model for nutrient uptake from soil by a growing plant root system. *Agronomy Journal* 68:961-964. (The senior author's name was misspelled as Caassen in this article.)

Claassen, N., and S. A. Barber. 1977. Potassium influx characteristics of corn roots and interaction with N, P, Ca, and Mg influx. *Agronomy Journal* 69:860-864.

Cochran, B. J., I. R. Camacho, L. C. Kiamco, and G. C. Salazar. 1982. Development of deep placement fertilizer implements. *American Society of Agricultural Engineers, Paper* No. 82-5534.

Cooke, G. W. 1954. Recent advances in fertilizer placement. II. Fertilizer placement in England. *Journal of the Science of Food and Agriculture* 5:429-440.

Cooke, G. W., and F. V. Widdowson. 1953. Methods of applying fertilizers for herbage crops. *Journal of Agricultural Science (Cambridge)* 43:358-364.

Costigan, P. A. 1984. The effects of placing small amounts of phosphate fertilizer close to the seed on growth and nutrient concentrations of lettuce. *Plant and Soil* 79:191-201.

Costigan, P. A. 1986. The effects of soil temperature on the response of lettuce seedlings to starter fertilizer. *Plant and Soil* 93:183-193.

Costigan, P. A. 1987. A model to describe the pattern of availability of broadcast phosphorus fertilizer during the growth of a crop. *Plant and Soil* 101:281-285.

Crowe, J. M. 1985. The on farm application of fertilisers. Pp. 31-49. In D. A. Bull and J. M. Crowe. *Fertiliser spreading mechanisms and their performance in practice.* The Fertiliser Society (London), Proceedings No. 241.

Daigger, L. A., and D. H. Sander. 1976. Nitrogen availability to wheat as affected by depth of nitrogen placement. *Agronomy Journal* 68:524-526.

De Miranda, J. C. C., P. J. Harris, and A. Wild. 1989. Effects of soil and plant phosphorus concentrations on vesicular-arbuscular mycorrhiza in sorghum plants. *New Phytologist* 112:405-410.

De Wit, C. T. 1953. *A Physical Theory on Placement of Fertilizers.* Verslagen van Landbouwkundige Onderzoekingen No. 59.4. Staatsdrukkerij en Uitgeverijbedrijf, Wageningen, The Netherlands.

Dilz, K., and G. D. van Brakel. 1985. Effects of uneven fertiliser spreading — A literature review. Pp. 1-19. In K. Dilz, G. D. van Brakel, and I. R. Richards. *Effects of uneven fertiliser spreading on crop yield and quality.* The Fertiliser Society (London), Proceedings No. 240.

Drew, M. C., and L. R. Saker. 1984. Uptake and long-distance transport of phosphate, potassium and chloride in relation to internal ion concentrations in barley: evidence of non-allosteric regulation. *Planta* 160:500-507.

Du Plessis, S. F., and T. J. Koen. 1983. Bepaling van die bemestingsbehoeftes van pynappels. *Republic of South Africa, Department of Agriculture and Fisheries, Technical Communication* No. 180:65-71.

Edwards, J. H., and S. A. Barber. 1976. Nitrogen flux into corn roots as influenced by shoot requirement. *Agronomy Journal* 68:471-473.

Eghball, B., and D. H. Sander. 1987. Phosphorus fertilizer solution distribution in the band as affected by application variables. *Soil Science Society of America Journal* 51:1350-1354.

Eghball, B., and D. H. Sander. 1989. Distance and distribution effects of phosphorus fertilizer on corn. *Soil Science Society of America Journal* 53:282-287.

Embleton, T. W., M. Matsumura, L. H. Stolzy, D. A. Devitt, W. W. Jones, R. El-Motaium, and L. L. Summers. 1986. Citrus nitrogen fertilizer management, ground-water pollution, soil salinity and nitrogen balance. *Applied Agricultural Research* 1:57-64.

Eriksen, A. B., M. Kjeldby, and S. Nilsen. 1985. The effect of intermittent flooding on the growth and yield of wetland rice and nitrogen-loss mechanism with surface applied and deep placed urea. *Plant and Soil* 84:387-401.

Finnie, J. F., and J. van Staden. 1985. Effect of seaweed concentrate and applied hormones on *in vitro* cultured tomato roots. *Journal of Plant Physiology* 120:215-222.

Fox, R. L., W M. H. Saunders, and S. S. S. Rajan. 1986. Phosphorus nutrition of pasture species: Phosphorus requirement and root saturation values. *Soil Science Society of America Journal* 50:142-148.

Garcia, R., and J. J. Hanway. 1976. Foliar fertilization of soybeans during the seed-filling period. *Agronomy Journal* 68:653-657.

Garwood, E. A., and T. E. Williams. 1967. Growth, water use and nutrient uptake from the subsoil by grass swards. *Journal of Agricultural Science (Cambridge)* 69:125-130.

Geraldson, C. M. 1970. Precision nutrient gradients — a component for optimal production. *Communications in Soil Science and Plant Analysis* 1:317-331.

Geraldson, C. M. 1990. Conceptual evaluation of intensive production systems for tomatoes. Pp. 539-544. In M. L. van Beusichem (Ed.), *Plant Nutrition — Physiology and Applications*. Kluwer Academic Publishers, Dordrecht, The Netherlands.

Gough, R. E. 1984. Split-root fertilizer application to highbush blueberry plants. *HortScience* 19:415-416.

Gray, R. C., and G. W. Akin. 1984. Foliar fertilization. Pp. 579-584. In R. D. Hauck (Ed.), *Nitrogen in Crop Production*. American Society of Agronomy, Crop Science Society of America, and Soil Science Society of America, Madison, WI.

Greenwood, D. J., and A. Draycott. 1988. Recovery of fertilizer-N by diverse vegetable crops: Processes and models. Pp. 46-61. In D. S. Jenkinson and K. A. Smith (Eds.), *Nitrogen Efficiency in Agricultural Soils*. Elsevier Applied Science, London.

Greenwood, D. J., K. Kubo, I. G. Burns, and A. Draycott. 1989. Apparent recovery of fertilizer N by vegetable crops. *Soil Science and Plant Nutrition* 35:367-381.

Gris, E. 1844. Nouvelle expériences sur l'action des composés ferrugineux solubles, appliqués à la végétation, et spécialement au traitement de la chlorose et de la débilité des plantes. *Comptes Rendus Hebdomadaires des Séances de l'Académie des Sciences* 19:1118-1119.

Hanway, J. J. 1976. Interrelated development and biochemical processes in the growth of soybean plants. Pp. 5-15. In L. D. Hill (Ed.), *World Soybean Research*. Proceedings of the World Soybean Research Conference, University of Illinois, Urbana. Interstate Printers, Danville, IL.

Harapiak, J. T., and J. D. Beaton. 1986. Review. Phosphorus fertilizer considerations for maximum yields in the Great Plains. *Journal of Fertilizer Issues* 3:113-123.

Harapiak, J. T., and D. C. Penney. 1984. *Efficient Use of Nitrogen — the N & P Interaction.* Paper presented at the 20th Annual Alberta Soil Science Workshop, February 21-22, 1984. Edmonton, Alberta.

Harder, H. J., R. E. Carlson, and R. H. Shaw. 1982. Leaf photosynthetic response to foliar fertilizer applied to corn plants during grain fill. *Agronomy Journal* 74:759-761.

Hartman, M. D., and M. Nyborg. 1989. Effect of early growing season moisture stress on barley utilization of broadcast-incorporated and deep-banded urea. *Canadian Journal of Soil Science* 69:381-389.

Haumann, E. J., and C. C. du Preez. 1989. Voorspelling van nitraatbeweging in 'n fynsandleemgrond gedurende die braakperiode met die Burnsmodel. *Suid-Afrikaanse Tydskrif vir Plant en Grond* 6:203-209.

Herron, G. M., G. L. Terman, A. F. Dreier, and R. A. Olson. 1968. Residual nitrate nitrogen in fertilized deep loess-derived soils. *Agronomy Journal* 60:477-482.

Heymann, W., H. J. Jäschke, and K. Kämpfe. 1984. Vergleichende Bewertung von Masszahlen zur Beurteilung der Streugenauigkeit von Mineraldüngerstreuern. *Archiv für Acker- und Pflanzenbau und Bodenkunde* 28:287-294.

Hiltner, L. 1911. Über die Ernährung der Pflanzen mit mineralischen Stoffen durch die Blätter. *Praktische Blätter für Pflanzenbau und Pflanzenschutz* 10, Heft 1:6-7.

Hodgson, A. S., and D. A. MacLeod. 1987. Effects of foliar applied nitrogen fertilizer on cotton waterlogged in a cracking grey clay. *Australian Journal of Agricultural Research* 38:681-688.

Hoffmann, W. E. 1968. Mineralstofftransport in die Xylemgefässe der Wurzel in Abhängigkeit von ihrem K- und P-Status. *Landwirtschaftliche Forschung* 21:203-212.

Humbert, R. P. 1968. *The Growing of Sugar Cane.* Elsevier Publishing Co., Amsterdam.

Ishizuka, Y., M. Hayashi, S. Ogata, and I. Harada. 1965. Studies on the placement of fertilizer for upland crops. (Part 4) The diffusion of nitrogen fertilizers and the growth of crop root. *Journal of the Science of Soil & Manure* 36:289-296.

Jarvis, R. J., and M. D. A. Bolland. 1990. Placing superphosphate at different depths in the soil changes its effectiveness for wheat and lupin production. *Fertilizer Research* 22:97-107.

Jensen, D., and J. Pesek. 1962. Inefficiency of fertilizer use resulting from nonuniform spatial distribution: I. Theory. *Soil Science Society of America Proceedings* 26:170-173.

Kemppianen, E. 1986. Effect of cattle slurry injection on the quantity and quality of barley and grass yield. Pp. 64-72. In A. D. Kofoed, J. H. Williams, and P. L'Hermite (Eds.), *Efficient Land Use of Sludge and Manure.* Elsevier Applied Science Publishers, London.

Klein, I., and S. A. Weinbaum. 1985. Foliar application of urea to almond and olive: Leaf retention and kinetics of uptake. *Journal of Plant Nutrition* 8:117-129.

Knittel, H. 1988. *Placement of Solid Fertilisers in Agricultural Crops: A Review.* The Fertiliser Society (London), Proceedings No. 273.

Knoll, H. A., N. C. Brady, and D. J. Lathwell. 1964. Effect of soil temperature and phosphorus fertilization on the growth and phosphorus content of corn. *Agronomy Journal* 56:145-147.

Kotze, W. A. G., and J. Joubert. 1980. Influence of foliar spraying with seaweed products on the growth and mineral nutrition of rye and cabbage. *Elsenberg Journal* 4:17-20. (Cited by Temple and Bomke, 1989).

Kovar, J. L., and S. A. Barber. 1988. Phosphorus supply characteristics of 33 soils as influenced by seven rates of phosphorus addition. *Soil Science Society of America Journal* 52:160-165.

Kovar, J. L., and S. A. Barber. 1989. Reasons for differences among soils in placement of phosphorus for maximum predicted uptake. *Soil Science Society of America Journal* 53:1733-1736.

Kovar, J. L., and S. A. Barber. 1990. Potassium supply characteristics of thirty-three soils as influenced by seven rates of potassium. *Soil Science Society of America Journal* 54:1356-1361.

Krogmeier, M. J., G. W. McCarty, and J. M. Bremner. 1989. Phytotoxicity of foliar-applied urea. *Proceedings of the National Academy of Sciences* 86:8189-8191.

Küntzel, U., R. Krause, and C. Jonuscheit. 1987. Scorching of *Lolium perenne* caused by cattle slurry. Pp. 333-335. In H. G. Van der Meer, R. J. Unwin, T. A. Van Dijk, and G. C. Ennik (Eds.), *Animal Manure on Grassland and Fodder Crops. Fertilizer or Waste?* Martinus Nijhoff Publishers, Dordrecht, The Netherlands.

Lantinga, E. A., J. A. Keuning, J. Groenwold, and P. J. A. G. Deenen. 1987. Distribution of excreted nitrogen by grazing cattle and its effects on sward quality, herbage production and utilization. Pp. 103-117. In H. G. van der Meer, R. J. Unwin, T. A. van Dijk, and G. C. Ennik (Eds.), *Animal Manure on Grassland and Fodder Crops. Fertilizer or Waste?* Martinus Nijhoff Publishers, Dordrecht, The Netherlands.

Lawton, K., and R. L. Cook. 1955. Interaction between particle size and water solubility of phosphorus in mixed fertilizers as factors affecting plant availability. *Farm Chemicals* 118:44-46.

Luten, W., J. H. Geurink, and J. J. Woldring. 1983. Yield response and nitrate accumulation of herbage by injection of cattle slurry in grassland. Pp. 185-191. In *British Grassland Society, Occasional Symposium* No. 14. Blackwell Scientific Publications, Osney Mead, Oxford, U.K.

Malavolta, E. 1986. Foliar fertilization in Brazil — present and perspectives. Pp. 170-182. In A. Alexander (Ed.), *Foliar Fertilization*. Martinus Nijhoff Publishers, Dordrecht, The Netherlands.

Malhi, S. S., M. Nyborg, and E. D. Solberg. 1989. Recovery of ^{15}N-labelled urea as influenced by straw addition and method of placement. *Canadian Journal of Soil Science* 69:543-550.

Malhi, S. S., and D. Ukrainetz. 1990. Effect of band spacing of urea on dry matter and crude protein yield of bromegrass. *Fertilizer Research* 21:185-187.

Matar, A. E., and S. C. Brown. 1989. Effect of rate and method of phosphate placement on productivity of durum wheat in a Mediterranean climate. *Fertilizer Research* 20:83-88.

Meer, H. G. van der, R. B. Thompson, P. J. M. Snijders, and J. H. Geurink. 1987. Utilization of nitrogen from injected and surface-spread cattle slurry applied to grassland. Pp. 47-71. In H. G. van der Meer, R. J. Unwin, T. A. van Dijk, and G. C. Ennik (Eds.), *Animal Manure on Grassland and Fodder Crops. Fertilizer or Waste?* Martinus Nijhoff Publishers, Dordrecht, The Netherlands.

Miranda, J. C. C. de, P. J. Harris, and A. Wild. 1989. Effects of soil and plant phosphorus concentrations on vesicular-arbuscular mycorrhiza in sorghum plants. *New Phytologist* 112:405-410.

Mitchell, A. R., and M. E. Nicholson. 1965. Pineapple growth and yield as influenced by urea spray schedules and potassium levels at three plant spacings. *Queensland Journal of Agricultural and Animal Sciences* 22:409-417.

Mortvedt, J. J. 1986. Iron sources and management practices for correcting iron chlorosis problems. *Journal of Plant Nutrition* 9:961-974.

Ndiaye, J. P., and R. S. Yost. 1989. Influence of fertilizer application nonuniformity on crop response. *Soil Science Society of America Journal* 53:1872-1878.

Neumann, P. M. 1982. Late-season foliar fertilization with macronutrients — Is there a theoretical basis for increased seed yields? *Journal of Plant Nutrition* 5:1209-1215.

Nye, P. H., and P. B. Tinker. 1977. *Solute Movement in the Soil-Root System*. Blackwell Scientific Publications, Oxford.

Ogus, L., and R. L. Fox. 1970. Nitrogen recovery from a soil profile by *Bromus inermis*. *Agronomy Journal* 62:69-71.

Pittman, U. J. 1964. Magnetism and plant growth. II. Effect on root growth of cereals. *Canadian Journal of Plant Science* 44:283-287.

Plessis, S. F. du, and T. J. Koen. 1983. Bepaling van die bemestingsbehoeftes van pynappels. *Republic of South Africa, Department of Agriculture and Fisheries, Technical Communication* No. 180:65-71.

Poulton, P. R., L. V. Vaidyanathan, D. S. Powlson, and D. S. Jenkinson. 1990. Evaluation of the benefit of substituting foliar urea for soil-applied nitrogen for winter wheat. *Aspects of Applied Biology* 25:301-308.

Priebe, D. L., and A. M. Blackmer. 1989. Preferential movement of oxygen-18-labeled water and nitrogen-15-labeled urea through macropores in a Nicollet soil. *Journal of Environmental Quality* 18:66-72.

Prins, W. H., and P. J. M. Snijders. 1987. Negative effects of animal manure on grassland due to surface spreading and injection. Pp. 119-135. In H. G. van der Meer, R. J. Unwin, T. A. van Dijk, and G. C. Ennik (Eds.), *Animal Manure on Grassland and Fodder Crops. Fertilizer or Waste?* Martinus Nijhoff Publishers, Dordrecht, The Netherlands.

Raese, J. T., C. L. Parish, and D. C. Staiff. 1986. Nutrition of apple and pear trees with foliar sprays, trunk injections or soil applications of iron compounds. *Journal of Plant Nutrition* 9:987-999.

Randall, G. W., and R. G. Hoeft. 1988. Placement methods for improved efficiency of P and K fertilizers: A review. *Journal of Production Agriculture* 1:70-79.

Randall, G. W., K. L. Wells, and J. J. Hanway. 1985. Modern techniques in fertilizer application. Pp. 521-560. In O. P. Engelstad (Ed.), *Fertilizer Technology and Use*. Third Edition. Soil Science Society of America, Madison, WI.

Reddy, K. R., W. H. Patrick, Jr., and C. W. Lindau. 1989. Nitrification-denitrification at the plant root-sediment interface in wetlands. *Limnology and Oceanography* 34:1002-1013.

Reed, D. W., C. G. Lyons, Jr., and G. R. McEachern. 1988. Field evaluation of inorganic and chelated iron fertilizers as foliar sprays and soil application. *Journal of Plant Nutrition* 11:1369-1378.

Richards, G. E. (Ed.). 1977. *Band Application of Phosphatic Fertilizers*. Olin Corporation, Little Rock, AR.

Richards, I. R. 1985. Effects of inaccurate fertiliser spreading on crop yield and quality in the UK. Pp. 21-51. In K. Dilz, G. D. van Brakel, and I. R. Richards. *Effects of uneven fertiliser spreading on crop yield and quality*. The Fertiliser Society (London), Proceedings No. 240.

Sanford, W. G., D. P. Gowing, H. Y. Young, and R. W. Leeper. 1954. Toxicity to pineapple plants of biuret found in urea fertilizers from different sources. *Science* 120:349-350.

Scheffer, J. J. C., and G. J. Wilson. 1987. Cauliflower: molybdenum application using pelleted seed and foliar sprays. *New Zealand Journal of Experimental Agriculture* 15:485-490.

Schreiber, K. 1958. Note on an unusual tropism of feeder roots in sugar beets and its possible effect on fertilizer response. *Canadian Journal of Plant Science* 38:124-126.

Scott, B. J. 1973. The response of barrel medic pasture to topdressed and placed superphosphate in central western New South Wales. *Australian Journal of Experimental Agriculture and Animal Husbandry* 13:705-710.

Scott, J. M., and G. J. Blair. 1988a. Phosphorus seed coatings for pasture species. I. Effect of source and rate of phosphorus on emergence and early growth of phalaris (*Phalaris aquatica* L.) and lucerne (*Medicago sativa* L.). *Australian Journal of Agricultural Research* 38:437-445.

Scott, J. M., and G. J. Blair. 1988b. Phosphorus seed coatings for pasture species. II. Comparison of effectiveness of phosphorus applied as seed coatings, drilled or broadcast, in promoting early growth of phalaris (*Phalaris aquatica* L.) and lucerne (*Medicato sativa* L.). *Australian Journal of Agricultural Research* 39:447-455.

Semple, E. C. 1928. Ancient Mediterranean agriculture. Part II. Manuring and seed selection. *Agricultural History* 2:129-156.

Sherrell, C. G. 1984. Effect of molybdenum concentration in the seed on the response of pasture legumes to molybdenum. *New Zealand Journal of Agricultural Research* 27:417-423.

Silcock, R. G., and F. T. Smith. 1982. Seed coating and localized application of phosphate for improved seedling growth of grasses on acid, sandy red earths. *Australian Journal of Agricultural Research* 33:785-802.

Simpson, J. R., and A. Pinkerton. 1989. Fluctuations in soil moisture, and plant uptake of surface applied phosphate. *Fertilizer Research* 20:101-108.

Sinclair, T. R., and C. T. de Wit. 1975. Photosynthate and nitrogen requirements for seed production by various crops. *Science* 189:565-567.

Sinclair, T. R., and C. T. de Wit. 1976. Analysis of the carbon and nitrogen limitations to soybean yield. *Agronomy Journal* 68:319-324.

Sine, C. 1986. Patented fertilizer is here. *Farm Chemicals* 149, No. 8:16-17, 20.

Singh, R. G., U. Lal, and Irfanuddin. 1983. Foliar feeding of nitrogen to sugarcane in Tarai tract of Uttar Pradesh. *The Indian Sugar Crops Journal* 9, No. 4:5-8. (*Soils and Fertilizers* 48, Abstract 2872. 1985.)

Singh, R. M., and C. A. Black. 1964. Test of the De Wit compensation function for estimating the value of different fertilizer placements. *Agronomy Journal* 56:572-574.

Steenbjerg, F. 1957. A theory on the placement of fertilizers. *Den Kongelige Veterinaer-og Landbohjskole Åarskrift* 1957:1-30.

Strebel, O., and W. H. M. Duynisveld. 1989. Nitrogen supply to cereals and sugar beet by mass flow and diffusion on a silty loam soil. *Zeitschrift für Pflanzenernährung und Bodenkunde* 152:135-141.

Temple, W. D., and A. A. Bomke. 1989. Effects of kelp (*Macrocystis integrifolia* and *Echlonia maxima*) foliar applications on bean crop growth. *Plant and Soil* 117:85-92.

Tesar, M. B., K. Lawton, and B. Kawin. 1954. Comparison of band seeding and other methods of seeding legumes. *Agronomy Journal* 46:189-194.

Ukrainetz, H. 1975. Fertilizer placement for fababeans, sunflowers, rapeseed and flax. Pp. 149-161. In *Extending Crop Rotations in Saskatchewan.* Proceedings of the Soil Fertility Workshop. University of Saskatchewan, Extension Division, Publication No. 268. Saskatoon.

Ukrainetz, H., R. J. Soper, and M. Nyborg. 1975. Plant nutrient requirements of oilseed and pulse crops. Pp. 325-374. In J. T. Harapiak (Ed.), *Oilseed and Pulse Crops in Western Canada — A Symposium.* Western Co-operative Fertilizers Limited, Calgary, Alberta.

Upadhyay, A. P., M. R. Deshmukh, R. P. Rajput, and S. C. Deshmukh. 1988. Effect of sources, levels and methods of phosphorus application on plant productivity and yield of soybean. *Indian Journal of Agronomy* 33:14-18.

Van der Meer, H. G., R. B. Thompson, P. J. M. Snijders, and J. H. Geurink. 1987. Utilization of nitrogen from injected and surface-spread cattle slurry applied to grassland. Pp. 47-71. In H. G. van der Meer, R. J. Unwin, T. A. van Dijk, and G. C. Ennik (Eds.), *Animal Manure on Grassland and Fodder Crops. Fertilizer or Waste?* Martinus Nijhoff Publishers, Dordrecht, The Netherlands.

Vlek, P. L. G., and E. T. Craswell. 1981. Ammonia volatilization from flooded soils. *Fertilizer Research* 2:227-245.

Wagner, R. E., and W. C. Hulburt. 1954. Better forage stands. *National Fertilizer Review* 29:1, 13-16. (Cited by Duell, R. W. 1974. Fertilizing forage for establishment. Pp. 67-93. In D. A. Mays (Ed.), *Forage Fertilization.* American Society of Agronomy, Crop Science Society of America, Soil Science Society of America, Madison, WI.)

Wallace, G. A., and A. Wallace. 1986. Correction of iron deficiency in trees by injection with ferric ammonium citrate solution. *Journal of Plant Nutrition* 9:981-986.

Wehrmann, J., and H. C. Scharpf. 1979. Der Mineralstickstoffgehalt des Bodens als Masstab für den Stickstoffdüngerbedarf (N_{min}-Methode). *Plant and Soil* 52:109-126.

Williams, C. H. 1971. Reaction of surface-applied superphosphate with soil. II. Movement of the phosphorus and sulphur into the soil. *Australian Journal of Soil Research* 9:95-106.

Wilson, G. J., and J. J. C. Scheffer. 1974. Molybdenum deficiency in cauliflowers. *New Zealand Journal of Agriculture* 129, No. 6:29.

Wit, C. T. de. 1953. *A Physical Theory on Placement of Fertilizers.* Verslagen van Landbouwkundige Onderzoekingen No. 59.4. Staatsdrukkerij en Uitgeverijbedrijf, Wageningen, The Netherlands.

Woolley, D. G., and U. J. Pittman. 1966. P^{32} detection of geomagnetotropism in winter wheat roots. *Agronomy Journal* 58:561-562.

Zhu, Z. L., G. X. Cai, J. R. Simpson, S. L. Zhang, D. L. Chen, A. V. Jackson, and J. R. Freney. 1989. Processes of nitrogen loss from fertilizers applied to flooded rice fields on a calcareous soil in north-central China. *Fertilizer Research* 18:101-115.

CHAPTER **8**

Soil Testing and Lime Requirement

I N CONTRAST TO FERTILIZERS, the beneficial effects of which usually can be associated with the nutrients they supply, the beneficial effects of liming materials are not so clearly and easily accounted for. The varying and generally undetermined causes of the beneficial effects of liming are responsible in part for the fact that liming materials usually are termed soil amendments instead of fertilizers.

Liming materials conventionally are added to acid soils to neutralize the acidity, but acidity generally is not the cause of poor growth of plants on acid soils. If the calcium and magnesium supplied in liming materials eliminate deficiencies of one or both of these elements as plant nutrients, the materials used may be said to be acting as true fertilizers. In most instances, however, the beneficial influences of liming materials on crops are determined mostly by the indirect consequences of the reduction in soil acidity associated with their use. These indirect effects may include changes in availability of toxins and nutrients, the consequences of changes in activity of certain microorganisms, and physical effects. Determining the causes of the observed effects is a research problem.

The term *liming* is a holdover from earlier days when equipment capable of crushing limestone economically in quantities for agricultural use was not available, and the limestone was "burned" in a kiln to produce lime (calcium oxide). Adding water produced calcium hydroxide, also called slaked lime and hydrated lime. Exposing calcium hydroxide to the air permitted it to take up carbon dioxide and again become calcium carbonate. These products were relatively soft materials and could be spread on soils for soil improvement.

Marl, an unconsolidated form of calcium carbonate, was probably the first liming material used in agriculture. According to Semple (1928), soils were limed to improve crop growth in ancient Greek and Roman agriculture long before the dawn of science, and marl was the preferred form. The organic matter and clay generally mixed with the calcium carbonate in marl no doubt supplemented the effect of the calcium carbonate in some instances. Today, the liming material used most commonly is limestone from consolidated sources that must be ground or crushed, and the term *liming* is a misnomer.

Before scientific methods were available, the custom of liming for soil improvement was necessarily modified by trial and error to accommodate the observed needs of the soils and the crops grown. The development of crude chemical methods to distinguish soils that were strongly acid and could be expected to respond to liming from those that were alkaline or only slightly acid was a major advance.

The effects of soil acidity on plants have now been investigated at length. Understanding of the nature of soil acidity has advanced far beyond the stage of merely testing to determine whether acidity is present. Various methods have been developed to provide quantitative measurements of soil acidity and limestone needs. But the methods measure different things, give different answers, and are not of equal value. For example, estimates of the average quantities of calcium carbonate required to neutralize 22 acid soils of Alaska to pH 6.5 by six methods ranged from 5.2 to 18.2 megagrams per hectare (Loynachan, 1981). The discrepancies probably may be explained in part by the fact that the methods were used on soils that differed from the ones on which the original calibrations were made. But that is one of the problems. The goals of science have not yet been achieved.

This chapter treats the scientific aspects of liming. Emphasis is placed upon determining the appropriate amounts for different conditions. Recent related publications include monographs edited by Adams (1984), Robson (1989), Lewis (1989), Sposito (1989), and Wright et al. (1991) and review articles by Rowell (1988), McCray and Sumner (1990), and Van Lierop (1990).

8-1. The Nature of Soil Acidity

Soils generally are considered acid if they have a pH value below 7. The dependence of the pH value of a given soil upon the method of measurement, however, emphasizes the arbitrary nature of the definition. Moreover, even though a soil may be mixed to promote uniformity, the pH value of the portion adjacent to plant roots may differ substantially from the pH value measured on bulk soil containing no plants. The pH effects due to the presence of plants may differ from one species and cultivar to another and also from one soil to another.

The pH value of dilute solutions that contain acids or alkalies is a measure of the hydrogen ion activity. The pH of soils usually is measured on the solution phase after addition of more water or a solution of an electrolyte. The presence of solids complicates the matter, particularly since most of the acid character of soils, that is, the propensity to neutralize an alkali, resides in the solids.

A natural inference from the early evidence on soil acidity was that the hydrogen ions (now often called protons) that affected the electrodes or indicators used to measure soil pH were derived from soil acids. The acids were inferred to be weakly dissociated because the intensity of acidity indicated by the pH measurements was only a negligible fraction of the total acidity that could be measured by titration with an alkali. Five classes of substances are now understood to be possible sources of titratable acidity in the soil solids: layer silicates, organic matter, hydrous oxides, allophane, and various charged species of aluminum and iron. The first four substances are macro constituents of soils that are responsible for the cation exchange properties. Members of the fifth class interact with the cation exchange positions. The sulfide present in a few soils may be important as a potential source of titratable acidity, as will be mentioned in Section 8-3.1.

8-1.1. Macro Constituents of Soils

In most soils of temperate regions, the titratable acidity resides principally in the organic matter and layer lattice silicate minerals in the clay fraction. In layer lattice minerals, the principal source of the negative charges is considered to be isomorphous substitutions of cations in the lattice layers while the minerals are being formed. These leave the lattice layers with an excess negative charge, which is balanced by other cations outside of the layers. Some and perhaps all of these balancing cations may be exchangeable with cations present in the soil solution or added in a neutral unbuffered salt solution. In organic matter, carboxyl (COOH) groups are believed to be of greatest importance. Carboxyl groups are weakly dissociated. Other even less readily dissociated organic acid groups also may make some contribution.

Helling et al. (1964) measured the cation exchange capacity of 60 soils from Wisconsin at six pH values ranging from 2.5 to 8 and used multiple linear regression to express the dependence of cation exchange capacity at each pH value on the contents of clay and organic carbon in the soils. Because at all pH values the intercepts were close to zero, the authors used the regression coefficients as estimates of cation exchange capacity per kilogram of clay and organic carbon. They found in this way that the exchange capacity increased from 38 centimoles of charge per kilogram of clay at pH 2.5 to 64 centimoles at pH 8, but increased from 62 centimoles of charge per kilogram of organic carbon at pH 2.5 to 366 centimoles at pH 8. Because this behavior is of such fundamental

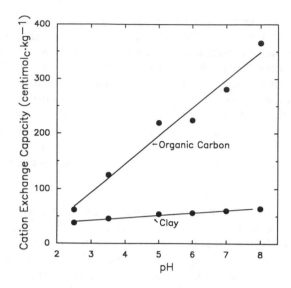

Fig. 8-1. Cation exchange capacity of the organic carbon and clay in 60 soils of Wisconsin versus the pH value at which the exchange capacity was measured. (Helling et al., 1964)

importance to the theory of soil acidity and the chemical lime requirement of soils, the data are shown graphically in Fig. 8-1.

Probably in all soils some fraction of the cation exchange positions relates directly to adsorption by hydrous oxides of aluminum and iron that occur predominantly as coatings on particle surfaces. If the pH is above a critical value, these substances are negatively charged and adsorb cations, and if the pH is below this value the substances are positively charged and adsorb anions. In soils in which the cation exchange capacity is due mostly to layer lattice clay minerals and organic matter, the contribution of the hydrous oxides is undetectable except by special experiments because their effect is small compared with the changes in cation exchange capacity with pH that are related to the clay minerals and organic matter. In soils in which the oxides are very prominent and the organic matter and layer lattice clay minerals are present in only small quantities, the adsorption by the hydrous oxides may be great enough to give the soils a positive charge and to cause them to adsorb anions, including nitrate and chloride, under strongly acid conditions. See the review papers by Thomas and Hargrove (1984) and Barrow (1985) for more information on this subject.

In some soils, especially those derived from volcanic ash, large amounts of allophane (amorphous or poorly crystalline aluminosilicate gels) are present. Experiments by Birrell and Gradwell (1956) show that allophane adsorbs cations and anions about like synthetic hydrous aluminum oxide. In contrast to layer lattice clays, the retention of cations by allophane depends strongly upon the concentration of cations in the equilibrium solution. Soils whose clay fraction is dominated by allophane thus tend to lose exchangeable bases readily by leaching (Okajima, 1977). Insufficient research has been done, however, to provide an adequate understanding of the contribution of allophane to soil acidity and lime requirement.

8-1.2. Aluminum

Aluminum plays a key role in soil acidity and lime requirement, and a long time was required to develop an understanding of this fact. An important milestone in the development of this understanding is illustrated in part by Fig. 8-2, which shows the results of an experiment in which a special technique was used to assure that the cation exchange positions in a sample of bentonite clay were occupied by hydrogen or hydronium ions. To replace the original exchangeable cations with hydrogen, a dilute suspension of the clay in water was passed through a column of a hydrogen-saturated cation-exchange resin. The cation exchange capacity of the resin column was many times that of the clay. As the clay moved past the stationary resin particles, the exchangeable cations of the clay were continuously replaced by hydrogen ions from the resin. At the moment the clay suspension emerged from the bottom of the column, most of the exchangeable cations should have been hydrogen.

As shown in Fig. 8-2, the titration curve for the clay as found immediately after emergence from the column of the hydrogen-saturated exchange resin showed a definite strong-acid character, that is, a low initial pH and consumption of considerable alkali without much rise in pH value, followed by a rapid increase in pH. When the "hydrogen-saturated" clay from the resin column was heated before titration, the titration curves looked more like those of a weak acid. When

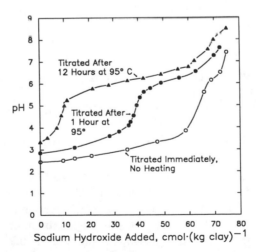

Fig. 8-2. pH values of a 1.48% aqueous suspension of bentonite clay that had been passed through a column of hydrogen-saturated cation exchange resin to remove exchangeable bases and then treated with different quantities of sodium hydroxide immediately or after heating for 1 or 12 hours at 95° C. (Davis et al., 1962)

separate samples of each of the suspensions were treated with a solution of barium chloride after 0, 1, and 12 hours of heating at 95° C, the amounts of aluminum that appeared in solution were equivalent to 8, 34, and 59 centimoles of charge per kilogram of clay. This evidence indicates that the clay was mostly a hydrogen clay as it emerged from the resin column, but that it was unstable and decomposed spontaneously to release aluminum, which then occupied some of the cation exchange positions and affected the initial pH value and the titration curve.

The rate of decomposition is lower at ordinary temperatures than in the experiment shown in Fig. 8-2, but it is still high relative to the rates of soil acidification under field conditions. For example, Coleman and Craig (1961) found that in Utah bentonite clay having hydrogen saturation of 95% or more as originally prepared; half of the exchangeable hydrogen was replaced by aluminum in 18 hours at 30° C. For a temperature of 0° C, the half-times were estimated at 11 days for the Utah bentonite, 33 days for Volclay bentonite, and 77 days for kaolinite.

A solution of aluminum chloride is acid. The aluminum ion exists as a coordination complex surrounded by six molecules of water. The hydrated aluminum ion is an acid in the general sense in that it contains removable protons (hydrogen ions). In this acid, the protons are removed from the water molecules surrounding the aluminum. These are not water molecules in the ordinary sense because the oxygen of each one is bonded to aluminum with an average charge of 0.5. Each hydrogen then is bonded to oxygen with an average charge of 0.75, leaving an average positive charge of 0.25 for each hydrogen or a total of $6 \times 2 \times 0.25 = 3$ per hydrated aluminum ion.

When an alkali is added to neutralize the hydrated aluminum ions, the steps in the hydrolysis reaction by which protons are released to form hydronium ions in water may be represented as follows:

$$[Al(H_2O)_6]^{3+} + H_2O = [Al(H_2O)_5OH]^{2+} + H_3O^+$$

$$[Al(H_2O)_5OH]^{2+} + H_2O = [Al(H_2O)_4(OH)_2]^+ + H_3O^+$$

$$[Al(H_2O)_4(OH)_2]^+ + H_2O = [Al(H_2O)_3(OH)_3] + H_3O^+.$$

Note that in each step one hydronium ion is released (and is neutralized by the added alkali), and the charge on the residual hydrated aluminum ion is reduced by one. Although the sum of the three reactions representing the formation of aluminum hydroxide with the release of three hydronium ions may be verified by the equivalence of the alkali consumed in titration and the aluminum precipitated as aluminum hydroxide, the second and third equations are hypothetical in that the process leading to final formation of aluminum hydroxide involves poorly defined interactions among individual ions. As the reaction proceeds, the individual units bond together in polymers, and precipitation of the polymers occurs before the final aluminum hydroxide stage is reached.

If neutralization of the hydrated aluminum ion has proceeded to the aluminum hydroxide stage, the final product is crystalline aluminum hydroxide [gibbsite, $Al(OH)_3$]. If the neutralization has proceeded only part way, the final products are crystalline gibbsite and the hydrated aluminum ion. Attaining equilibrium, however, is a very slow process. In soils, the hydroxyaluminum ions and polymers are adsorbed by organic matter, on surfaces of inorganic particles, and in interlayer positions in layer silicate clays. In experiments in the absence of soils, "humic acid" from soils is adsorbed on hydroxy aluminum precipitates and perturbs the crystallization that otherwise would occur (Singer and Huang, 1990). Gibbsite is a normal constituent of soils, but the adsorption of the hydroxyaluminum ions and polymers stabilizes them, so that gibbsite does not form as readily as it does in solutions. Walker et al. (1990) found that under the various experimental conditions they used, the activity of aluminum in solution after equilibration with organic soil horizons was always less than that corresponding to a saturated solution of gibbsite, which indicates that the combinations of the aluminum species with organic matter were more stable than the combination of aluminum with hydroxyl in crystalline gibbsite.

An important point where soil acidity and liming are concerned is that the partially neutralized aluminum acid ions and polymers are positively charged and are bonded strongly to cation exchange positions. Probably most of these sites then no longer function as exchange positions. Jackson (1963) discussed at length the evidence for the stability of the hydroxyaluminum polymers in interlayers in clay minerals, pointing out that aluminum released in soils during chemical weathering tends to accumulate in the form of hydroxyaluminum interlayers in preference to gibbsite.

Laboratory experiments have shown that addition of partially neutralized aluminum acid ions and polymers to clays and soils can reduce the cation exchange capacity. And other experiments have shown that special treatments, such as heating soil at 100° C with neutral sodium citrate solution (an aluminum complexing reagent), can remove aluminum interlayers and increase the cation exchange capacity.

The portion of soil acidity that is attributable to hydroxyaluminum species is still due to hydrogen, but it is not exchangeable hydrogen in the sense understood at one time, that is, hydrogen ions attached directly in exchange positions. Rather, the hydroxyaluminum ions occupy the exchange positions, and the titratable hydrogen is a part of the hydroxyaluminum complexes. The number of titratable hydrogens in the hydroxyaluminum complexes is theoretically equal to the number of exchange positions occupied by the hydroxyaluminum complexes, but the ratio of titratable hydrogens to aluminum varies.

8-1.3. Iron
A solution of ferric chloride is acid, like aluminum chloride. The ferric iron exists as the central atom in a coordination complex with six water molecules and hydrolyzes to produce hydroxyferric ions, which are analogous to hydroxyaluminum ions. Experiments indicate that the behavior of iron in relation to soil

acidity and lime requirement parallels that of aluminum, but that iron species are less effective than aluminum species in blocking cation exchange positions and more effective than aluminum species in adsorbing anions at low pH values (Thomas and Hargrove, 1984).

8-2. Forms of Soil Acidity

8-2.1. Exchange Acidity

The term *exchange acidity* was used many years ago by Kappen (1929) in his book on soil acidity to represent the titratable acidity removed from strongly acid soils by a neutral, unbuffered salt solution. It is still used with the same meaning.

To begin with an experiment from a classical investigation, Veitch (1904) shook samples of various soils with a solution of sodium chloride, filtered off the soil, and titrated the clear filtrates with standard alkali. One of the soils tested was from the District of Columbia; it grew only the "acid-loving" weed sorrel. Titration of 100 milliliters of the sodium chloride filtrate from this soil to the phenolphthalein endpoint required 1.14 millimoles of hydroxyl ion in standard alkali. The precipitate that developed during titration contained aluminum equivalent to 1.15 millimoles of charge, indicating that all the exchange acidity titrated was derived from the aluminum in the extract.

Extending the work by Veitch and relating it further to the cation exchange properties of soils, Lin and Coleman (1960) leached samples of four acid subsoils with an excess of molar potassium chloride solution and titrated the extract with standard alkali to determine the exchange acidity. They determined also the aluminum, calcium, and magnesium extracted, and found that when they attributed three positive charges to aluminum and two each to calcium and magnesium, the sums of the charges due to these three cations were approximately equivalent to the charges due to the calcium retained in exchangeable form when the samples were treated subsequently with an excess of calcium chloride. The implication, therefore, was that each aluminum ion extracted had occupied three cation exchange positions in the soil solids, that the aluminum was the acid that had consumed the alkali, and that only insignificant amounts of hydrogen or hydronium ions were occupying exchange positions. Gillman and Sumpter (1985) verified these observations in analyses of 212 samples of soils from Queensland. Bache (1970a), however, obtained evidence with two acid soils from the United Kingdom that the charge on some of the aluminum displaced by 0.5 molar barium chloride was less than three per atom.

8-2.2. pH-Dependent Acidity

In the work by Lin and Coleman (1960), the aluminum, calcium, and magnesium removed from strongly acid soils by exchange with a neutral, unbuffered salt solution, such as molar potassium chloride or sodium chloride, may be defined as the total exchangeable cations (except for quantitatively unimportant amounts of potassium, sodium, ammonium, and manganese). The calcium retained by the soils in exchangeable form after leaching the soils with a solution of calcium chloride may be defined as the effective cation exchange capacity or the cation exchange capacity at the soil pH.

The cation exchange capacity of soils generally increases considerably with the pH at which the measurement is made, as was indicated by the work by Helling et al. (1964) described previously. Raising the pH of the salt solution used for displacing the exchangeable cations of acid soils does not remove extra quantities of exchangeable bases. These remain essentially the same as the quantities displaced from acid soils at their original pH values. The extra exchange capacity that appears when the pH of the displacing solution is increased has been termed *pH-dependent acidity* because the amount that becomes effective depends upon the pH of the solution with which the soil is equilibrated. *pH-dependent acidity* appears to be a better term than *hydrolytic acidity* used by Kappen (1929) because some of Kappen's hydrolytic acidity appears to represent dissociation of weak acid groups in organic matter (Hargrove and Thomas, 1982).

The principal sources of pH-dependent acidity in soils are believed to be hydroxyaluminum species that block exchange positions in the layer silicate clays and organic matter plus some weak acid groups in the organic matter. As the pH is increased, hydroxyl ions disappear, but the aluminum species remain mostly in the soil and do not appear in the solution. If they did appear in solution, the quantities would be expected to exceed those calculated from the increase in exchange capacity by attributing three exchange positions to each aluminum because the hydroxyaluminum species are already partly neutralized by hydroxyls as they occur in the soil. Stoichiometry is assumed to apply, but verification is a problem.

In acid soils containing adsorbed sulfate, the sulfate is an additional generally minor source of pH-dependent acidity. As the pH is increased by addition of alkali, hydroxyl ions are consumed by replacement of sulfate (Mehlich, 1964).

The soluble acids (nitric, sulfuric, phosphoric, and carbonic) created in soils by microbial decomposition of organic matter contribute to soil acidification. These acids may reduce the pH value in water by as much as one unit in soils with low exchange capacity and low soluble salts (Lehr, 1950). Soils containing iron sulfide may develop extremely low pH values as a result of microbial oxidation of the sulfide to sulfuric acid (Van Beers, 1962). The oxidation of ammoniacal nitrogen derived from fertilizers to nitric acid is the most important source of soil acidity in modern agriculture. But the amounts of acidity present at any one time in the form of soluble acids from all sources range from small to negligible relative to the total titratable acidity. Most of the acidity developed by soluble acids is found by analysis as titratable acidity associated with the soil solids, and most of the titratable acidity is pH-dependent acidity because of the extensive reaction of the soluble acids with the soil. Carbonic acid of course decomposes and releases carbon dioxide to the atmosphere.

8-3. Chemical Lime Requirement

Applying increasing quantities of a liming material to an acid soil will produce a crop response curve, from which inferences may be made about the quantity required to produce the maximum net profit or to achieve some other objective. Although the response of a crop to liming may result from a combination of factors other than the active ingredient or ingredients in the liming material, the problems of curve fitting and finding the optimum application are similar to those discussed for fertilizers in Chapter 2 (Hall, 1983).

The economic interpretation is related to the effects, not the causes. Thus, whether an application of limestone to a particular soil for a particular crop is acting as a calcium and magnesium fertilizer or whether its benefit results from one or more other conditions is not of direct economic concern; however, it may be of indirect economic concern if the response to liming can be produced more economically in some other way, as for example, by applying less than a kilogram of molybdenum per hectare.

From the historical standpoint, the concept of lime requirement evolved primarily as a property of the soil, with the response of the crop being accorded at most equal importance. The concept probably developed as it did partly because of the difficulties encountered in finding by chemical methods the quantities needed for different soils, but mostly because of the great differences among soils. Differences tenfold or greater may exist in the quantities of a given material required to lime two soils with the same initial pH to a given higher pH value.

This section on chemical lime requirement will emphasize the theoretical aspects of measuring the quantity aspect of soil acidity — what might be called liming the soil instead of liming the crop. Adjusting the quantity of a liming material to the soil is an essential part of liming the crop.

The concepts of exchange acidity and pH-dependent acidity discussed in Section 8-2 are important for the theoretical understanding they provide about the nature of soil acidity and its relationship to the cation exchange properties of soils, which occupy such an important position in soil science. As a strongly acid soil is titrated, the exchange acidity reacts first; this reaction is complete at approximately pH 5.5. The classification of soil acidity into exchange and pH-dependent fractions, however, does not figure in the usual process of determining how much limestone or other liming material should be applied to a particular soil. In most cases, the total titratable acidity to the chosen pH is the criterion employed.

The task of devising a rapid and accurate soil test to determine the titratable acidity of soils has not proved to be simple, and room for improvement still exists. The primary problem is reaching an equilibrium. Numerous methods and variations of methods exist, and some of them give widely different results.

8-3.1. Total Titratable Acidity

Determining the total titratable acidity in soils has proved to be out of the question because the reaction between soils and alkali does not reach an endpoint at any pH value within the natural range of soil conditions. The pH-dependent acidity continues to increase with addition of alkali under alkaline as well as acid conditions. Under strongly alkaline conditions, part of the organic matter dissolves, and decomposition reactions occur in the mineral fraction.

Because some form of calcium carbonate is almost always used for liming acid soils and because so many soils contain calcium carbonate as a natural, sparingly soluble constituent, Bradfield and Allison (1933) argued that as good a natural basis as any for defining the total titratable acidity in soils would be to consider that the total titratable acidity is zero in soils at equilibrium with free calcium carbonate. They were looking at soil acidity as the difference between the cation exchange capacity and the exchangeable bases, and they proposed that "a soil saturated with bases is one which has reached equilibrium with a surplus of $CaCO_3$ at the partial pressure of CO_2 existing in the atmosphere and at a temperature of 25° C."

Procedurewise, they made the determination by adding a known excess of calcium hydroxide solution to a quantity of the acid soil, followed by aeration of the suspension with carbon dioxide and then air. Time was allowed for equilibration at each stage, and the calcium bicarbonate produced during aeration was changed to calcium carbonate by boiling the suspension of soil in water under vacuum. To obtain the calcium that had been taken up by the soil, they subtracted the carbonate present in the soil from the carbonate equivalent of the calcium hydroxide added. The pH of the soil after this equilibration procedure was 8.3 to 8.4.

Bradfield and Allison found that they obtained about the same value for calcium uptake by two acid clays and an acid soil when they used their regular procedure with calcium hydroxide followed by carbonation with carbon dioxide and aeration as when they used calcium carbonate followed by carbonation with carbon dioxide and aeration. With the two acid clays, they obtained essentially the same answers as before when they omitted the carbonation stage and used only air. Thus, the endpoint appeared to be the same, whether it was approached from the alkaline side or the acid side, indicating that the authors were dealing with an equilibrium condition. They also tried omitting the carbonation step with the two acid clays treated with calcium carbonate, and in that case they used a longer reaction time (14 days compared with 16 hours for the other treatments), but the calcium uptake again was essentially the same.

Shaw (1953) used the Bradfield and Allison (1933) procedure on several soils and found that the values he obtained were somewhat higher when the soils were shaken continuously during the reaction period than when the suspension was allowed to stand during the reaction period, as in the regular Bradfield and Allison procedure. These results indicate that what appeared to be an equilibrium condition in the work of Bradfield and Allison did not represent a final equilibrium.

In related work with five soils, Shaw (1953) found that the calcium uptakes by the Bradfield and Allison procedure modified by shaking the samples during equilibration (14.6 centimoles charge per kilogram on average) were essentially the same as those obtained by the residual carbonate method applied to samples of the same soils that had been mixed with very finely ground calcite, treated eight times with 30 milliliters of water per 10 grams of soil, and evaporated to dryness on a steam plate at about 95° C after each addition of water (14.3 centimoles charge per kilogram on average). The eight evaporations could be completed in two days. The measurement of residual carbonate by a rapid steam distillation method (Shaw and MacIntire, 1943) avoided the problem encountered earlier by Wheeler et al. (1900) in which the apparent decomposition of soil organic matter interfered with their attempt to determine the calcium carbonate decomposition by direct measurement of the carbon dioxide evolved when a suspension of soil in water was boiled with an excess of calcium carbonate.

Shaw (1953) also compared the values he obtained for calcium carbonate decomposition by the rapid heating method with values obtained when the same soils were incubated with calcium carbonate for a year at 30° C. The values at 30° C were invariably lower than those obtained by the rapid method at the high temperature despite the fact that during the long incubation at 30° C some calcium carbonate would have been decomposed by the nitric acid produced by nitrification in the surface soil samples. These results indicate that at ordinary temperatures a year was not long enough to reach the final equilibrium between calcium carbonate and acid soils where an excess of calcium carbonate was present.

Shaw (1953) found in a test with four soils that the apparent content of total titratable acidity averaged 12.1 centimoles of charge per kilogram of soil by the modified Bradfield and Allison method, 12.0 centimoles by the Shaw rapid high-temperature method, and 9.7 centimoles by the method of Mehlich (1948), which involved the use of a barium chloride solution buffered at pH 8.1 with triethanolamine to extract the titratable acidity. The average titratable acidity found in a different group of four soils by the rapid high-temperature method using eight evaporations at 95° C was 35.4 centimoles of charge per kilogram of soil, and the value was increased only to 35.9 centimoles with 16 evaporations. These data indicate that the modified Bradfield and Allison method and the Shaw rapid high temperature method give results that are close to the equilibrium value.

The Mehlich method was not advanced as a measure of the total titratable acidity of soils to pH 8.1, which is the approximate pH value of soils in equilibrium with calcium carbonate, but rather as a measure of all of the permanent-charge and "a large percentage" of the pH-dependent cation exchange positions (Mehlich, 1961). From the present standpoint, however, the values obtained by this method are significant because the Mehlich method sometimes is used as a secondary standard of comparison with rapid laboratory methods for estimating the lime requirement. Shoemaker et al. (1961) obtained evidence that the barium chloride-triethanolamine solution did not react with some of the acidity due to aluminum that reacted with calcium carbonate during a prolonged incubation.

A different source of acidity is found in certain soils in swampy areas along seacoasts where decomposition of organic matter by microorganisms has reduced sulfate in the seawater to sulfide and ferric iron to the ferrous form, with resulting formation of iron sulfide. If these soils are drained, the sulfide oxidizes and forms sulfuric acid. In some soils, uneconomically large amounts of calcium carbonate may be required to neutralize the acid produced. A similar problem exists in much land reclaimed from coal mining operations because coal is another product of swampy conditions that may be found along seacoasts where there is an ample supply of sulfate in the waters. The usual lime requirement methods are not applicable to these soils because the methods respond to existing acidity but not to the potential acidity that develops subsequently when the sulfide oxidizes. Special methods are needed to recognize the potential contribution of the sulfide (O'Shay et al., 1990).

8-3.2. Partial Titration With Calcium Carbonate

Knowledge of the quantities of calcium carbonate required to neutralize the titratable acidity of soils to the "calcium carbonate endpoint" is of theoretical interest, but practical interest is in estimating the much smaller quantities of calcium carbonate needed for efficient crop production. A soil pH value of 6.5 for alfalfa seems to be about the highest value recommended for any crop, although under some conditions, alfalfa has been found to respond to higher pH values.

Where the objective is liming soils to the "biological lime requirement" (see Section 8-4), calcium carbonate is still viewed as the standard titrating agent because the usual liming material is limestone or dolomitic (magnesium-bearing) limestone. In most instances, pH values are used for relating crop responses to measurements of titratable acidity.

Where calcium carbonate has been used as a standard titrating agent against which to evaluate the results obtained by other methods that are more rapid and convenient, the usual procedure has been to add pure, finely divided calcium carbonate to soil in different quantities, to moisten the soil and allow the inter-action to take place at room temperature over a time considered long enough for equilibrium to be attained, and then to measure the pH value. A plot of soil pH versus quantity of calcium carbonate added is used as a "standard curve" for judging the performance of more rapid laboratory methods. Such standard curves are not always prepared because, as usually done, they require a long time. Shoemaker et al. (1961) and McLean et al. (1961) used an incubation lasting 17 months, which is longer than most. In the absence of a standard curve generated for the soils under investigation, methods may be compared without direct reference to primary data obtained with calcium carbonate additions. In lime requirement work, the most common substitute for calcium carbonate in preparing a titration curve is calcium hydroxide.

As may be inferred from the preceding section, the residual carbonate method is appropriate for determining whether the smaller quantities of calcium carbonate used for preparing a practical standard curve have reacted with the soil. This valuable measurement has rarely been made in recent years.

Using a method for carbonate in soils published by Shaw and MacIntire (1943), Shaw (1952) found that when he incubated acid soils for 4 weeks at 30° C with increasing quantities of calcite passing a sieve with 0.044 millimeter openings, there was no residual carbonate with the smaller additions; that is, the calcium carbonate added was completely decomposed. At final pH values approaching neutrality and above, however, some residual carbonate remained. None of the pH values exceeded 8, even when considerable calcium carbonate remained. See Fig. 8-3 for the results obtained with one soil.

After a hiatus of nearly four decades following publication of Shaw's (1953) work on reaction of calcium carbonate with acid soils at elevated temperatures (see Section 8-3.1), Barrow and Cox (1990) advanced the state of knowledge by another step. Taking the rate of reaction at 10° C as unity, they found that the relative rates of reaction were 1.8 at 15°, 5.7 at 25° and 177 at 60°. Because the rate at 60° was 31 times that at 25°, reaction for 1 day at 60° would be approximately equivalent to 1 month at 25°. In their experiments, they found a 1:1 relationship between the pH values of eight soils (measured in 1:5 soil:0.01 molar calcium chloride suspensions) incubated with different quantities of calcium carbonate for 90 days at 25° C and the pH values of the same soils incubated for 1 or 3 days at 60° C. The scatter of points was much reduced where the incubation at 25° C was done in the presence of chloroform to inhibit microbial activity.

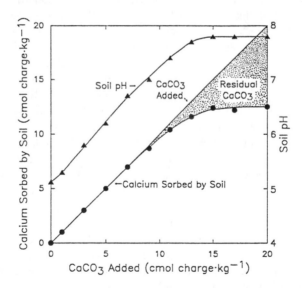

Fig. 8-3. Quantity of calcium sorbed by Hartsells sandy loam from different quantities of calcium carbonate during incubation at 30° C, together with the residual calcium carbonate and the pH of the soil paste at the end of 4 weeks. The cation exchange capacity of the soil at pH 7 was approximately 10 centimoles of charge per kilogram of soil, and the calcium carbonate was added as calcite passing a sieve with 0.044 millimeter openings. (Shaw, 1952)

For preparing a standard titration curve with calcium carbonate, the incubation at 60° C used by Barrow and Cox (1990) would have the advantage of a small time requirement and avoidance of the disturbance of pH values created by development of soluble salts (see Section 8-3.3). On the other hand, the authors noted that heating soils to 60° C for 3 days caused some release of ammonium. In their experiments, the ammonium release ranged from about 0.1 to 4 centimoles per kilogram of soil, the amount depending upon the soil and the pH. Although a release of ammonium at the lower end of the range could be neglected for lime requirement purposes, a release at the higher end of the range would interfere with measurement of the calcium carbonate effect. In soils incubated for 7 days at 60° C, Thompson (1950) had observed nitrogen mineralization in amounts up to 16 centimols per kilogram. How much this effect could be reduced by addition of biocides is not known. Barrow and Cox (1990) added chloroform in some of their incubations, but apparently not in the 3-day incubations at 60° C in which the ammonium release was measured. If some way can be devised to deal with the ammonium release, the high-temperature modification has promise for use in preparing standard titration curves with calcium carbonate and perhaps also for a quick-test method to replace the current buffer methods for lime requirement.

8-3.3. Soluble Electrolytes and pH Measurements

The seeming digression represented by this section on electrolytes and pH measurements is important at this point because evaluations of chemical and biological lime requirements generally are based upon pH measurements. The effects of electrolytes on pH values can add considerable uncertainty to these evaluations.

pH values of soils may be read to 0.01 unit or less on standard electronic pH meters. The precision, however, is deceiving because the meter readings vary with the method of measurement, and average values may differ from day to day, week to week, and month to month even if no soil treatment is applied and even if the method of measurement remains the same.

Soluble electrolytes in soils equilibrate with the exchangeable cations held by the soil solids. As the concentration of soluble electrolytes increases, more hydrogen ions are brought into the bulk solution, where they affect the pH meter, and the measured pH decreases. If the electrolytes added to the soil are acids or hydroxides and not neutral salts, they add their own effects to those of the hydrogen ions derived from the soil solids.

In acid soils that would be tested for lime requirement, microbial transformations of nitrogen are the principal cause of pH changes that occur in the absence of soil treatments. Haynes and Swift (1989) found that the pH values remained remarkably constant when moist soils were incubated at 25° C under an atmosphere of chloroform to inhibit microbial activity. But when the same soils were incubated without the chloroform, the mineral nitrogen changed markedly, and so did the pH values.

Fig. 8-4 illustrates changes in pH and mineral nitrogen found by Haynes and Swift (1989) when four soils of New Zealand were incubated in a moist condition

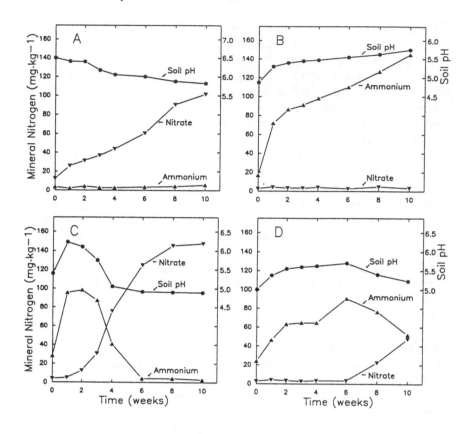

Fig. 8-4. Accumulation of ammonium and nitrate, and the associated changes in soil pH during incubation of samples of four soils of New Zealand in the laboratory at 25° C. (Haynes and Swift, 1989)

at 25° C. An increase in soil pH (measured in a 1:2.5 soil:water suspension) during incubation was associated with the accumulation of ammonium (ammonium hydroxide), and a decrease in pH was associated with the accumulation of nitrate (nitric acid).

Some of the pH effects shown in Fig. 8-4 are as great as those that might be expected from applications of calcium carbonate equivalent to 1 megagram to the surface 20 centimeters of some soils per hectare. But again, appearances are deceiving. Soluble electrolytes in the usual concentrations can have a pronounced effect on soil pH values, but they have very little effect on the titratable acidity. Calcium carbonate reduces the titratable acidity, but it has very little direct effect on the soluble electrolytes. Calcium carbonate is only sparingly soluble. The products of interaction of calcium carbonate with acid soil are exchangeable calcium, carbonic acid (which breaks down to water and carbon dioxide), and hydroxyaluminum polymers (which become attached to the soil solids). No soluble anion residue is left in the soil solution.

Although the reaction of calcium carbonate with acid soil does not generate soluble electrolytes directly, it does so indirectly. Addition of calcium carbonate generally enhances the rate of nitrate production. As an extreme example, Chew et al. (1976) found that during the first year after they limed a Malaysian peat soil, the mineral nitrogen produced in the soil increased from 1.4 grams per culture at pH 3.4 to 3.9 grams at pH 6.8. These values are the sums of nitrogen in the above-ground portion of Napier grass grown on the soil and the residual mineral nitrogen in the soil. If the organic nitrogen mineralized is assumed to have been nitrified to nitrate, the enhancement of nitric acid production in the soil due to liming would have been equivalent to 8.9 grams of calcium carbonate per culture. See a paper by Curtin and Smillie (1983) for data on accumulation of nitrate and change in pH values of the soil solutions displaced from limed and unlimed soil. The effect of liming on nitrogen mineralization diminishes with time, but it lasts longer than the time required for acid soil to react with finely divided calcium carbonate for calibration purposes. The pH effects of the electrolytes produced as a result of adding calcium carbonate are superimposed on the effects of the electrolytes present initially and those produced during incubation of the control samples that receive no calcium carbonate.

The effects of electrolytes on soil pH values may be said to impose three general restrictions where lime requirements are concerned. First, the calibrations of lime requirement methods need to be made in such a way that the pH values used as a criterion of changes in titratable acidity and as an indicator of the target value to which acid soils are to be neutralized are not confounded with electrolyte effects. Second, the measurements of soil pH values need to be made in such a way that they reflect the pH conditions to which plants respond. And third, the pH data on soils treated with calcium carbonate for calibration must be coordinated with the pH values obtained on soils to which the calibration data are to be applied.

8-3.3.1. pH Effects on Calibration

In preparing calibrations for chemical lime requirement based upon the pH values obtained with different additions of calcium carbonate to acid soils, the possible effects of soluble electrolytes have not always been recognized in the past. The changes in pH due to electrolytes and calcium carbonate have sometimes been measured in the aggregate and have been inferred to be a response to changes in titratable acidity due to addition of calcium carbonate.

The existence of electrolyte effects on pH values used in preparing calibrations for the chemical lime requirement may be inferred from observations in the literature. Thus, Shaw (1952) determined the pH values of acid soils after incubation of the soils with quantities of finely ground calcite approximately equivalent to the cation exchange capacities of the soils at pH 7. The pH values, measured on soils that had been brought to a paste consistency by addition of water, were at a maximum after the soils had been incubated for a week. After that time, the pH values of surface soils decreased, but the pH of subsoils did not. When he determined the pH values after leaching the soils with water, the values obtained were higher than those without leaching, and they decreased little or none with increasing time of incubation. These findings indicate that the drop in pH values with time was not a consequence of continued equilibration of the calcium from the decomposed calcium carbonate with the soils. The results are consistent, however, with the theory of pH effects of soluble electrolytes.

McLean et al. (1962) cited work by one of the junior authors in which the pH values of soil samples incubated with calcium carbonate in the laboratory had decreased as much as 1 pH unit after the maximum values had been reached. pH decreases were observed with unlimed samples as well as with limed samples. Although an accumulation of soluble electrolytes was not mentioned as a possible cause of the pH drop at that time, McLean (1982) later referred to an effect of accumulation of electrolytes and the need to remove them before measuring the pH values or to take them into account in some other way.

The effects of electrolytes on the pH values of soils that have been incubated with different quantities of calcium carbonate for calibration purposes may be dealt with in several ways. One is to remove the electrolytes before measuring the pH values. pH values of leached soils are higher and more stable than are pH values of soils with their normally varying complement of electrolytes. For example, Hester and Shelton (1933) found in soil samples taken from a field of a sandy loam soil in Virginia in July that the average pH in water was 4.67. When they determined the pH again after leaching the same samples with water, the pH was 5.44. Using the same method on samples taken from the same area in December, they found an average pH of 5.14 for the unleached samples and 5.48 for the leached samples. Thus, the leached samples had similar pH values, but the original samples did not. A second method, described by Loynachan (1981), is to grow plants on the variously limed soils during incubation and to add enough water to the cultures to cause some leaching. This technique would assure that the electrolyte concentration would remain low, but it is not as extreme as leaching the soils before measuring the pH values. A third way to deal with the soluble electrolyte effect is to swamp it by substituting a suitable salt solution for the water commonly added when making a pH measurement.

An experimental illustration of the soluble electrolyte problem and the first and third methods for avoiding it was published by Pierre et al. (1970). Their work involved the electrolyte effects associated with the formation of nitric acid from microbiological oxidation of the ammonium in ammonium nitrate added to soil and formation of calcium nitrate when the nitric acid was neutralized by calcium carbonate.

Theoretically, microbial oxidation of one mole of ammonium nitrate in soil should result in two moles of nitric acid. The two moles of nitric acid should react with one mole of calcium carbonate to produce one mole of calcium nitrate, and the net change in titratable acidity or lime requirement due to the overall reaction should be zero. Thus, if the method for measuring the soil pH shows a pH change as a result of the overall reaction, the indications are that the method is responding to an effect of the added calcium nitrate.

Pierre et al. (1970) combined this concept with measurements of pH and conductivity in such a way as to demonstrate the importance of the effect of soluble electrolytes. They treated an acid soil with ammonium nitrate alone or with ammonium nitrate plus enough calcium carbonate to neutralize half or all of the nitric acid produced during nitrification, and then measured the resulting pH values in 1:1 soil:water suspensions and in 1:2 soil:0.01 molar calcium chloride suspensions, both with and without prior leaching with a 65:35 mixture of methanol and water.

Figs. 8-5A and B show that in the experiment by Pierre et al. (1970), the pH values measured by all methods decreased as a result of addition of nitric acid from nitrification of the ammonium nitrate, and they increased as a result of neutralization of the acidity by calcium carbonate. The critical point, however, is the difference between the final and initial pH values.

Figs. 8-5A and B show that as a result of the overall reaction, the pH measured in water decreased where the soil was not leached, but that the pH values measured by all the other methods were but little changed as a result of the reaction. Thus, the commonly used method of measuring the pH on a suspension of unleached soil in water gave the theoretically erroneous answer that the acidity produced by microbiological oxidation of the ammonium nitrate exceeded the alkalinity of a chemically equivalent quantity of calcium carbonate and had acidified the soil. In other words, the method responded to a combination of a calcium carbonate effect and a soluble electrolyte effect. The other three methods gave results in close agreement with theory; that is, they provided information on the calcium carbonate effect that was influenced little or none by a soluble electrolyte effect.

Figs. 8-5C and D provide supporting evidence because the electrical conductivity of the filtrates from the pH measurements provides an index of the electrolyte concentration of the solutions. The conductivity of the filtered 1:1 soil:water suspensions in Fig. 8-5C increased from addition of the ammonium nitrate where the soil was not leached, but not where the soil had been leached before measuring the pH value. Because of the calcium chloride, the conductivities of all the filtrates in Fig. 8-5D were higher than those from the corresponding soil:water suspensions, but the differences in conductivity or electrolyte concentration contributed by the soil had only a small effect on the pH values (Fig. 8-5B).

8-3.3.2. pH Measurements on Soils

The second restriction is that pH measurements on soils to which a lime requirement calibration is to be applied should be made in such a way that they will reflect the pH values to which plants respond in soils. The evidence indicates that plants in soils respond to the composition of the soil solution. When meas-

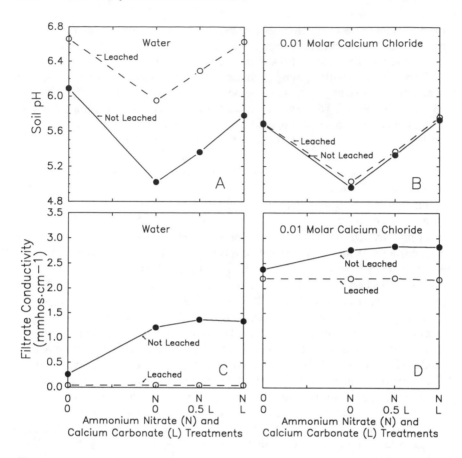

Fig. 8-5. pH and conductivity values for a silty clay loam soil after incubation for 40 days in moist condition with no treatment, with 150 milligrams of nitrogen (N) as ammonium nitrate per kilogram of soil, and with calcium carbonate (L) equivalent to 0, 0.5, and 1.0 times the acidity theoretically produced by the ammonium nitrate upon nitrification. Measurements were made on samples with and without prior leaching with a 65:35 mixture of methanol and water. The soil had been diluted with an equal weight of quartz sand. A. pH in 1:1 soil:water suspension. B. pH in 1:2 soil:0.01 molar calcium chloride suspension. C. Conductivities of the filtrates from *A*. D. Conductivities of the filtrates from *B*. (Pierre et al., 1970)

uring soil pH to determine how plants respond to the pH and to determine how much calcium carbonate is needed to increase the pH of an acid soil to the desired level, it is therefore reasonable to measure the pH of the soil solution.

Measuring the pH of the soil solution in most soils is not difficult, but it is inconvenient and time consuming because the solution must be extracted before the measurement can be made. Probably largely for this reason, practical measurements of soil pH with the usual glass electrode pH meters are invariably made on soil suspensions prepared by adding water or a salt solution. The 0.01 molar calcium chloride solution used by Pierre et al. (1970) for making some of the measurements in Fig. 8-5 was intended by its originators to yield pH values that approximate the pH of soil solutions.

The suggested measurement of soil pH on a suspension of soil in 0.01 molar calcium chloride solution is an outcome of research by Teräsvuori (1930). Schofield and Taylor (1955) extended Teräsvuori's work for soils in which the chief exchangeable cations are calcium and magnesium, showing that the ratio of the hydrogen ion activity to the square root of the sum of the activities of calcium and magnesium in equilibrium solutions was essentially constant for each of five different soils, although the pH varied. Their data for one soil are shown in Fig. 8-6, where the values for calcium plus magnesium shown on the X axis are the concentrations found by analyzing the filtrates of suspensions of 15 or 30 grams of soil in 100 milliliters of calcium chloride solutions of different concentrations used for pH measurements. The activities on the Y axes are represented in logarithmic form as pH and [pH $-$ 0.5 p(Ca + Mg)]. The value of 0.5 p(Ca + Mg) for 0.01 molar calcium chloride is 1.14. Thus, for the soil illustrated in Fig. 8-6, the pH in 0.01 molar calcium chloride would be [pH $-$ 0.5 p(Ca + Mg)] + 1.14 = 3.33 + 1.14 = 4.47. The value 3.33 is the mean of all the [pH $-$ 0.5 p(Ca + Mg)] values shown in the figure.

When water is added to soils for measurement of the pH value, the soluble electrolytes are diluted, some interactions of the soluble ions with the ions adsorbed by the soils occur, and the pH values increase. If the soil solution were

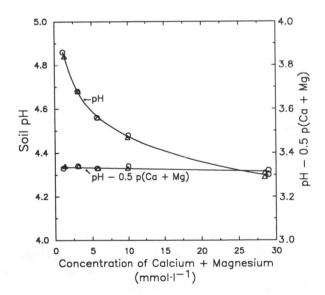

Fig. 8-6. Soil pH and [pH $-$ 0.5 p(Ca + Mg)] versus concentration of calcium and magnesium in solution for a soil equilibrated with different concentrations of calcium chloride solution in the ratio of 15 grams (triangles) and 30 grams (circles) of soil per 100 milliliters of solution. (Schofield and Taylor, 1955)

diluted by a solution having the same composition, however, the pH values should be unaffected by dilution. As a practical way to estimate the pH of the soil solution, Schofield and Taylor selected a 0.01 molar calcium chloride solution to dilute the soil solids. They considered 0.01 molar calcium chloride to be an electrolytic approximation of a "normal" soil solution in a nonsaline soil in which the principal exchangeable cations are calcium and magnesium and in which the soil solution has about the same concentration as 0.01 molar calcium chloride. (The principal ions in the soil solution of acid soils are normally calcium and nitrate.) If such a soil solution is diluted by 0.01 molar calcium chloride, the pH should remain about the same with different degrees of dilution. Fig. 8-6 shows that the pH was independent of the ratio of soil to 0.01 molar calcium chloride solution, but the range tested was from only 15 to 30 grams of soil per 100 milliliters of solution.

Bache (1970b) tested a wider range from 10 to 100 grams of soil per 100 milliliters of solution and found that the pH values in 0.001 to 0.02 molar calcium chloride solutions were essentially constant for soils for which the pH values measured in 0.01 molar calcium chloride were 6.5 and 5.2, but not for a soil with pH 4.0. The former two soils had calcium and magnesium as the main exchangeable cations, but aluminum was the dominant exchangeable cation in the strongly acid soil at pH 4.0.

Indications are that a 0.01 molar solution is a good choice for estimating the pH of the soil solution in the soils for which it was designed originally. For soils with a more dilute soil solution, a more dilute calcium chloride solution is better. See papers by White (1969), Bache (1970b), Bruce et al. (1988), and Aitken and Moody (1991) for more information.

8-3.3.3. Coordinating Calibration Data With Soil pH Measurements

At this time, the 0.01 molar calcium chloride method appears to be as good as any for the dual purpose of making pH measurements on soils treated with different quantities of calcium carbonate for calibration and of making pH measurements on soils for which the soil pH is needed as a part of the lime requirement procedure. Although 0.01 molar calcium chloride is more concentrated than the soil solution of some soils, the bias for such soils does not appear to be of major importance for lime requirement purposes because if money is to be invested in liming, money probably would be invested in other inputs that would increase the concentration of the soil solution.

An improvement on the 0.01 molar calcium chloride system could be made for calibrations with calcium carbonate by leaching all incubated samples with enough 0.01 molar calcium chloride solution to remove most of the original solutes from the samples before measuring the pH. This modification of the procedure would assure the elimination of biases that otherwise might result from the presence of enough soluble electrolytes in some of the samples to have a significant effect on the pH values.

The modification could be advantageous for samples in general, but of course more effort would be involved. As an example of the magnitude of the effects that might be observed, calculations based upon Aitken and Moody's (1991) regression equations representing measurements on 90 samples of soils from Queensland indicate that if the soil solution pH was 5.00, the average pH values measured on 1:5 soil:solution suspensions would be 5.46 in water, 4.87 in 0.002 molar calcium chloride, and 4.65 in 0.01 molar calcium chloride.

In the widely used buffer method for estimating lime requirements (see Section 8-3.5), the 0.01 molar calcium chloride leaching should be advantageous in measuring the pH of the incubated samples in the calibration process. Differences in soluble electrolytes among samples of soils to be tested, however, should be of no significant concern because the pH value measured is that of a soil-buffer suspension in the presence of large additions of electrolytes in the buffer solution.

Incubations of soils for chemical lime requirement determinations are designed to neutralize the titratable soil acidity to the equilibrium pH with the calcium carbonate or calcium hydroxide added. The acidity produced by nitrification during incubation of soils with calcium carbonate or calcium hydroxide, however, becomes a part of the titratable acidity over time, and its effect is reflected in pH measurements made on leached soils with added water or 0.01 molar calcium chloride. How much of this acidity is included in the chemical lime requirement varies with the length and conditions of incubation.

Although acidity due to nitrification is almost invariably incorporated to some degree in calibrations for chemical lime requirement, the influence of this source of acidity is ignored. It does not influence the measurements of chemical lime requirement made by rapid laboratory methods except for the possible effect of soluble electrolytes already present. Ideally, a separate measurement should be made of the acidity due to nitrification, and this component should be added to the original titratable acidity in the soil. If the calibration curve for chemical lime requirement is obtained by incubating the soils with calcium carbonate for more than a few days, it would be desirable to conduct the incubation under an atmosphere of chloroform to inhibit microbial activity so that the lime requirement components could be more clearly separated. Where acid-forming nitrogen fertilizers are to be used, the acidity generated by these could logically and reasonably be added to the acidity produced from the soil organic matter. In many instances the acidity from the fertilizers would exceed that developed from the soil organic matter. These considerations regarding acidity derived from nitrification introduce an open-ended time factor into the chemical lime requirement that increases the significance of research to determine the maintenance requirement.

8-3.4. Partial Titration With Soluble Alkalies

In titrating a soluble acid of unknown concentration with a standard alkali, the usual procedure is to add the standard solution dropwise while the solution being titrated is swirled or stirred to promote rapid mixing. The reaction is instantaneous, once the acid and alkali are in contact.

Similarly, when a potassium chloride or calcium chloride extract of soil containing exchangeable aluminum is titrated with a soluble alkali, the reaction takes place rapidly. But when the soil solids are present, the reaction is very sluggish. The pH of the suspension increases immediately upon addition of alkali, and then it begins a long, slow drift downward. For example, Dunn (1943) found that with three acid soils the average initial pH value was 10.0 after addition of a solution of calcium hydroxide equivalent to 14 centimoles of charge per kilogram of soil. After an hour, the pH had dropped to 7.0. After 12 hours, it was 6.7, and after 180 hours it was 6.5. In this experiment, the pH was measured in a 1:10 soil:0.25 molar calcium chloride suspension, and uniform distribution of ions in the solution external to the stable aggregates was assured by shaking the suspensions continuously on a mechanical shaker. In the control samples the pH dropped from 4.6 to 4.4 during the first hour, but it remained stable at that value to 180 hours. Chloroform was added to inhibit microbial activity.

Barrow and Cox (1990) found that when samples of acid soils were incubated for 42 days at 60° C with different additions of alkali, the pH decreased linearly with the logarithm of the time. Converting this behavior to the rate of change of pH with time at 25° C, which was only 3.2% as fast, would indicate that the soil pH would still be decreasing after more than 3 years.

The slow downward drift of pH values after the initial increase with addition of alkali is due in part to the slowness of neutralization of hydroxyaluminum species in blocked exchange positions in organic matter and in interlayer locations in clays. Diffusion within the soil solids is involved, and this takes place slowly, especially as equilibrium is approached and the pH gradient becomes very low. For example, between pH 5.5 and 5.6, the difference in molar activity of the hydroxyl ion, which is bringing about the neutralization of the hydroxyaluminum ions, is only 0.8×10^{-9}.

Because of the slow approach to equilibrium, two different procedures have been used when titrating acid soils with soluble alkalies in the laboratory. One has been to add relatively large increments of alkali and to allow the soil to stand for a period before measuring the pH so that the rate of decrease of the pH values will have become almost imperceptible. Presumably the high initial pH values will neutralize some hydroxyaluminum complexes to a greater degree than would have occurred, had the soil been reacting with calcium carbonate, and some back reaction will take place as the pH decreases. McLean et al. (1961) found that when they used a 3-day reaction period between standard calcium hydroxide solution and acid soils and found by interpolation the quantity of calcium hydroxide required to raise the pH to 7.2, the correlation between the calcium hydroxide requirement and the quantity of calcium carbonate required to raise the pH to 6.8 in a 17-month incubation at room temperature was $r = 0.99$, and the average values were almost the same. The higher pH value used in the calcium hydroxide titration apparently compensated for the differences in electrolyte effects and reaction with the soil.

A second procedure has been to conduct the titration in a solution of a neutral unbuffered salt, such as calcium chloride or potassium chloride. The cation of the salt solution exchanges with the so-called exchange acidity and brings it out

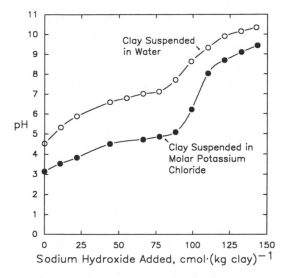

Fig. 8-7. pH of 1:100 suspensions of bentonite clay in water and molar potassium chloride solution with addition of different quantities of sodium hydroxide. The original clay had been leached with 0.1 molar hydrochloric acid to remove exchangeable bases. (Harward and Coleman, 1954)

into the solution, where it decreases the pH. But most importantly, the cation of the salt solution also seems to enhance the release of hydroxyaluminum ions from blocked exchange positions to the solution, where they will react with the hydroxyl ions supplied in very low concentrations by the alkali.

Fig. 8-7 illustrates the effect of a neutral, unbuffered salt on the titration of soil acidity. Note that with zero addition of sodium hydroxide, the pH value of the suspension was about 1.5 units lower when the clay was suspended in molar potassium chloride than in water. And note further that as sodium hydroxide was added, more acidity was titrated in reaching a given pH value when the clay was suspended in molar potassium chloride solution than in water. With time, the location of both of the curves on the graph would have changed, and the curve for the clay suspended in water, particularly at the higher pH values, would have moved closer to the curve obtained with the potassium chloride solution. The effect of the concentration of neutral salts was investigated in early work on electrometric measurements of soil pH (see, for example, Crowther, 1925), but it took some time before this knowledge was interpreted and made use of in lime requirement methods.

Depending upon the soils employed and the details of the procedures used, the values obtained by titrating soils with calcium hydroxide conceivably could provide a better estimate of the chemical calcium carbonate requirement of soils for calibration purposes than the values obtained by incubating the soils with calcium carbonate, but this cannot be determined without additional information.

Reported correlation coefficients close to unity between results by the two methods are reassuring. Nevertheless, regression equations with slopes differing substantially from unity and the sometimes considerable differences in values that may be found with some soils despite the high correlations indicate that methods development has not yet reached the stage at which the potential for improvements is negligible.

Partial titrations of soils with soluble alkalies have not found favor as practical soil tests for the chemical lime requirement of soils. Tracing a pH versus quantity of alkali curve by adding small quantities of alkali to a single sample of soil and waiting long enough after each addition for the pH value to stabilize before adding the next increment of alkali would take weeks, and the reaction with the soil still would not have reached an equilibrium. Tracing the curve by adding different quantities of alkali to separate samples so that all solutions would be equilibrating with the soil simultaneously would divide the time required by the number of samples used, but it still would be relatively slow and troublesome because of the number of samples required. This method has been used for research purposes. The usual procedure is to allow the soil suspensions to stand with occasional shaking (or to shake them continuously on a mechanical shaker) for several days before the pH values are measured.

Increasing the temperature greatly increases the rate of reaction, as indicated in the third paragraph in this section. Whether a high-temperature titration method can be developed that will be competitive with current buffer methods remains to be seen.

8-3.5. Partial Titration With Buffer Solutions

The now popular buffer methods for the chemical lime requirement of soils are an adaptation of the method of partial titration with soluble alkali. The buffer solution replaces the alkali.

Schofield (1933) devised a forerunner of current methods that illustrates the principle of the use of the buffer concept for estimating the chemical lime requirement. To estimate the amount of soil acidity to pH 7.1, he equilibrated a sample of soil with a solution of *p*-nitrophenol (a weakly dissociated acid) that had been two-thirds neutralized by calcium hydroxide and had an initial pH value somewhat above 7.1. To a volume of 40 milliliters of this solution, he added a known quantity of soil intended to produce a final pH of approximately 7.1. At this pH, *p*-nitrophenol is half neutralized and has its maximum buffering capacity. Hence, at this pH the precise quantity of soil used is not critical. Schofield noted that a 25% change in the calcium content of the buffer solution at pH 7.1 would alter the pH by only 0.2 unit. After equilibrating the soil with the buffer solution, Schofield filtered out the soil and titrated a known volume of the filtrate and an equal volume of the initial buffer solution with standard acid to a low pH value where the buffering effect of the *p*-nitrophenol was negligible. The difference between the titer of the original solution and the solution after equilibration with the soil was a measure of the soil acidity the buffer solution had neutralized.

Schofield's (1933) method involves overnight equilibration, filtration, and titration to evaluate the soil acidity that has been neutralized by the buffer. By current standards, the total time requirement is long, and the titration requires more operator time than desirable.

As now used, buffer methods have two important advantages in testing soils for chemical lime requirement: they are fast, and to a degree they titrate the soil acidity automatically, thus reducing the operator time required. In presently used buffer methods, the equilibration of the soil with the buffer solution is brief, and the filtration step is omitted. A measurement of pH on the soil-buffer suspension is substituted for the titration.

The ratio of soil acidity to buffer capacity of the solution is much greater than that in the Schofield method, so that the pH will drop far enough in the presence of soil to permit the pH measurements to yield an acceptably precise index of the amount of acidity that has been neutralized. The magnitude of the pH change associated with a given amount of soil acidity may be adjusted by varying the concentration of the buffer solution. The pH change increases with increasing dilution. This increased sensitivity is advantageous from the standpoint of measuring the change in pH, although not necessarily from the standpoint of estimating the chemical lime requirement, as is made evident in subsequent paragraphs.

The buffer solutions used in current buffer pH methods differ from the single-component buffer used by Schofield (1933). Although the initial solutions are usually neutral or slightly alkaline, the buffering properties are imparted by mixtures of buffers. Yuan (1974) published a graph showing titration curves of tris or tris(hydroxymethyl)-aminomethane, imidazol, chromic acid, and pyridine, having maximum buffering capacities at pH 8.1, 7.5, 6.5, and 5.2, respectively. Other substances used in buffer pH methods for estimating the chemical lime requirement of soils include triethanolamine, boric acid, p-nitrophenol, maleic acid, and acetic acid. Mixtures make possible an extension of both the pH range of buffering and the range in which the pH change is proportional to the quantity of acidity contributed by the soil. In most instances, a neutral unbuffered salt is included as one of the components of the buffer solution to increase the efficiency with which the soil acidity is neutralized.

The original pH of the buffer solution, and hence the pH of the soil-buffer suspension, may be adjusted by choosing different weak acids or bases, by adding a soluble salt, and by adding strongly dissociated acid or alkali. With a given buffer, the higher the initial pH, the greater is the amount of soil acidity neutralized.

The speed and economy of current buffer pH methods come at a price. The objective of the measurements, as with Schofield's (1933) neutralization by a buffer to pH 7.1 and the direct titration of an acid soil with calcium carbonate, is still to estimate the quantity of calcium carbonate required to reach a target pH value. But buffer pH methods involve more empiricism.

In Schofield's method, the target pH is achieved directly to a good approximation (although later work has shown that the quantity of calcium carbonate required to attain a given pH value exceeds the quantity found by buffer solutions). In the direct titration of acid soil by calcium carbonate, the quantity required to reach a target pH is obtained by interpolating between pH values measured when different quantities of added calcium carbonate have reacted with the soil and are presumed to have decomposed completely. This is a straightforward process.

With the buffer pH method, the process is more complex. Only part of the alkali supplied by the buffer solution is consumed in reaction with the soil. With an increase in titratable acidity in the soil, the consumption of buffer alkali increases, the soil-buffer suspension pH decreases, and the proportion of the titratable acidity neutralized by the buffer decreases. The soil-buffer suspension pH is not the target pH. It provides only an indirect indication of the soil titratable acidity, and the values must be calibrated in some way to estimate the quantity of calcium carbonate required to reach the target pH.

The usual procedure for dealing with the unknowns in the buffer pH method is to conduct a calibration experiment in which the chemical lime requirement of each of a group of acid soils selected to cover a wide range of titratable acidity is determined by a standard method, such as partial titration with calcium carbonate. The soil-buffer suspension pH is measured on the same soils. The regression of the quantities of calcium carbonate required to reach the target pH value by the standard method on the soil-buffer suspension pH values then is calculated to calibrate the soil-buffer suspension pH values in terms of chemical lime requirement. For practical application to samples of soils submitted by producers, a table of values prepared from the regression equation may be used to find the chemical lime requirement associated with a given soil-buffer suspension pH. This procedure involves the assumption that all soils follow a single relationship.

Fig. 8-8 provides a visual picture of the operation of the buffer method based upon measurements of the soil-buffer suspension pH. Part *A* shows the titration curves obtained with three soils. These soils were selected to represent the range of buffer capacities encountered in the group of 14 soils of Ohio used in the experimental work. As inferred from data published by the authors (Shoemaker et al., 1961; McLean et al., 1961), the linear regression of the quantities q of titratable acidity in the group of soils to the target pH of 6.8 (as estimated by the calcium carbonate required to raise the soil pH to that value) on the pH values of the soil-buffer suspensions (pH_b) was

$$q = 80.4 - 11.6 \, pH_b,$$

where the units of q are centimoles of charge per kilogram of soil.

The quantities of titratable acidity in Soils 1, 10, and 15 to pH 6.8 were calculated by substituting the buffer pH values in the equation, and the values

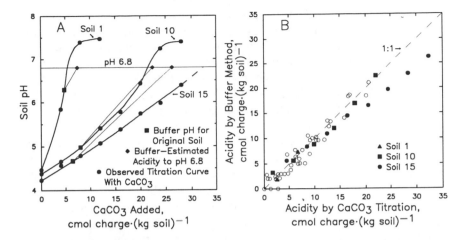

Fig. 8-8. Behavior of a buffer method for estimating the chemical lime requirement of soils. The soil-buffer suspension pH was measured on 5 grams of soil plus 5 milliliters of water and 10 milliliters of a buffer solution adjusted to pH 7.5 with sodium hydroxide and containing 1.8 grams of paranitrophenol, 2.5 grams of triethanolamine, 3.0 grams of potassium chromate, 2.0 grams of calcium acetate, and 53.1 grams of calcium chloride dihydrate per liter. A. Plot of the pH values of three soils after incubation for 17 months with different quantities of calcium carbonate. The light lines connect the soil-buffer suspension pH for each soil on the titration curve with the lime requirement to pH 6.8 as estimated by the buffer method. B. Plot of acidity in 14 soils to pH 6.8 as estimated by the buffer method against the acidity as estimated to the same pH value by adding different quantities of calcium carbonate and incubating the mixtures for 17 months. This graph includes the data from all soils with all additions of calcium carbonate resulting in pH values below 6.8. The data for the three soils featured in A are shown with solid symbols. (Shoemaker et al., 1961; McLean et al., 1961)

obtained are shown as the diamond-shaped symbols on the horizontal line representing pH 6.8. The soil-buffer suspension pH values for the original unlimed soils have been spotted on the titration curves (solid square symbols), and straight lines have been drawn between the buffer pH values and the estimated acidity to pH 6.8 calculated from the regression relationship. Ideally, the diamond shaped symbols for soil acidity to pH 6.8 should have been on the titration curves for the respective soils. One reason for the divergence is the fact that the titration curves are only approximately linear in the range of interest. The greatest difference was found with Soil 15, which was said to contain 9% organic matter. In later work, McLean (1975) used a different calibration for organic soils than for mineral soils. The differences in shapes of the titration curves indicate differences in the character of the acidity being titrated in different soils, and hence different relationships between the soil-buffer suspension pH and the quantities of calcium carbonate needed to attain pH 6.8.

McLean and coworkers used linear relationships between the acidity measured by the buffer method and a standard method for their calibrations. According to Van Lierop (1990), the calibration relationship has been found to be linear by some investigators and curvilinear by others. These findings indicate that the calibration tables prepared by the authors of some buffer methods should be regarded with skepticism when they are applied to different soils. In some instances, a calibration may be quite transferable, as found by Aitken et al. (1990), but a recheck to determine the calibration for the specific conditions under which a method is to be used is always desirable.

Fig. 8-8B shows the data published by McLean et al. (1961) on all 14 soils used in their investigation, with measurements of soil-buffer suspension pH at each level of calcium carbonate that led to a pH value below 6.8. For emphasis, the data for the three soils featured in Fig. 8-8A have been shown with solid symbols. Soil 15 with the high content of organic matter shows the greatest deviation. Despite the differences in curvature of the titration curves for the various soils, the over-all relationship looks remarkably good, especially if a different calibration is considered applicable to the soil with a high content of organic matter. Note particularly the results for Soil 10, in which the deviations from a 1:1 relationship in Part B mirror the deviations from the straight line in Part A.

Looking more closely at Fig. 8-8B, one may note a number of high percentage deviations from the 1:1 relationship where the acidity was relatively low. In an effort to develop a buffer method that would give good precision for acid soils with low cation exchange capacities and low amounts of acidity, while retaining the capability to give good precision for soils with greater exchange capacities and more titratable acidity, Yuan (1974) used a buffer differing from the one described by Shoemaker et al. (1961) that was used in obtaining the data in Fig. 8-8. Yuan also worked out what he called a *double buffer* method that provided a way to make a correction for the differences in shapes of the titration curves for different soils.

In the double buffer method, a given soil is equilibrated with two buffer solutions that differ in initial pH, and the pH value of each of the soil-buffer suspensions is determined after a period for equilibration. In effect, a straight line is drawn between the two "equilibrium" pH values on a plot of soil-buffer suspension pH versus quantity of acidity measured by the two buffers. The buffer acidity to the target pH then is estimated from the point of intersection of the straight line with the line representing the target pH. In Yuan's (1974) procedure, the total buffer acidity to the target pH was taken as the lime requirement without reference to a calibration, but he found the values were generally in good agreement with those obtained by titrating the soils with calcium hydroxide and allowing a week for equilibration. Having the target pH close to the pH obtained with the buffer solution yielding the higher soil-buffer suspension pH is important in correcting for the difference in shape of the buffer curves among soils. The double buffer method provides a way to estimate the relationship between the acidity measured by the buffer pH and titration methods on an individual-soil basis as opposed to using an overall relationship that represents the best fit for all the soils investigated experimentally. Yuan (1974) found that his method gave good precision for the sandy soils with which he was concerned in Florida.

Yuan (1974) showed graphically the geometric considerations involved in estimating the chemical lime requirement by the double buffer method. He used an illustration in which the straight line between the points was extrapolated to reach the target pH. McLean et al. (1977) described the geometric construction differently and used an illustration in which the buffer acidity to the target pH was found by interpolation.

In his double buffer work, Yuan (1974) used buffer solutions with initial pH values of 7.0 and 6.0. In a follow-up on Yuan's work, McLean et al. (1977, 1978) adjusted their own buffer to pH 7.0 and 6.0 as well as the original value of 7.5, and tested the modifications by the double buffer method. They used as standard values the quantities of calcium hydroxide required to neutralize 54 acid soils to pH 7.2 in 3-day titrations as an estimate of the quantities of calcium carbonate required to neutralize the soils to pH 6.5 after 20 months, and employed a linear regression equation to adjust the acidity values found by the double buffer method to the acidity found by titration with calcium hydroxide. In the soils used by McLean et al., the quantity of acidity estimated by the double buffer method was less than that estimated by titration with calcium hydroxide, and they found the same to be true when they used Yuan's method on their soils. They found that the various modifications of the double buffer method they tried gave results that were more highly correlated with the standard values than were the results obtained by the single buffer method. Moreover, the precision was improved for soils with relatively small amounts of acidity. Others, for example Aitken et al. (1990), have noted that the double buffer method offers what they consider only a slight improvement over the single buffer method.

The double buffer method using two solutions involves somewhat more than twice as much work as the regular single buffer method. McLean et al. (1978) accordingly suggested a modification in which they first used the single buffer method and, after measuring the soil-buffer suspension pH, added to the suspension an amount of dilute hydrochloric acid equivalent to that required to reduce to pH 6.0 the quantity of buffer solution added originally. After a period for equilibration, the soil-buffer suspension pH was measured again. The results obtained with this modification were less satisfactory than those with two separate buffers, but they were better than those with only one buffer.

See a review by Van Lierop (1990) for further discussion of chemical lime requirement methods. He gives information on each of a number of methods and detailed procedures for a few.

8-3.6. pH Model

Edmeades et al. (1985a) proposed a scheme for estimating the response of permanent pastures to liming in New Zealand. The scheme was based upon the results of many field experiments in which the percentage responses to liming with 1.25, 2.5, and 5 megagrams of limestone per hectare were expressed as linear functions of soil pH (measured on a 1:2.5 soil:water suspension over the duration of the experiment and omitting the first year). For 65 experiments with

1.25 megagrams of limestone per hectare, the percentage response (y) was given by the equation $y = 51 - 8.6$ pH. For 97 experiments with 2.5 megagrams per hectare, the equation was $y = 64 - 10.8$ pH, and for 78 experiments with 5 megagrams per hectare, the equation was $y = 79 - 13.4$ pH. Where only soil pH values are measured, this approach does not make allowance for differences in lime requirement associated with differences in cation exchange capacity of the soils. It is a good way, however, to verify the critical pH above which liming is not needed. In the three series of experiments, the critical value ranged from pH 5.90 to 5.93.

8-3.7. pH and Organic Matter Models

Beginning with the work of Pratt and Bair (1962), a number of investigators noted the importance of organic matter in determining the pH-dependent cation exchange capacity of soils and the lime requirement between a given initial pH and a higher target pH. Magdoff and Bartlett (1985) carried the concept one step further by plotting the pH (measured on a 1:2 soil:0.01 molar calcium chloride suspension) of a number of soils of Vermont against the number of charge units of calcium carbonate added per gram of soil organic matter after spotting the control pH value for each soil on the relatively complete titration curve for one soil. The many experimental points could be represented to a good approximation by a single curve with an apparent maximum of about pH 7.3. The relationship was almost linear between pH 4.5 and 6.5.

Applying the information about organic matter to a group of 23 field experiments on liming of New Zealand permanent pastures, Edmeades et al. (1985c) found that they could estimate the field lime requirement about as well by a linear regression equation having initial soil pH and organic carbon percentage as independent variables as by titration with calcium hydroxide or by two buffer methods they tested. The double buffer method of McLean et al. (1977, 1978) was a little better.

Bailey et al. (1989a,b,c) translated the findings relative to the significance of soil organic matter into a new method for estimating the chemical lime requirement of soils. Their method relates the shapes of the titration curves obtained with calcium carbonate to an indirect measure of the organic matter content of soils, making allowance for differences in titration curves in a different way than the double-buffer method discussed in Section 8-3.5.

They used the following equation for the titration curves:

$$pH_L = A + bQ^L,$$

where pH_L = soil pH with the quantity L of calcium carbonate, A is the maximum pH value obtainable by adding calcium carbonate, Q is a parameter that determines the curvature of the titration curve, b is a parameter related to the intercept of the titration curve on the pH_L or Y axis (note that when $L = 0$, $Q^L = 1$ and $pH_L = A + b$), and the pH is measured in a suspension of 10 milliliters of soil in 25 milliliters of water. This equation is equivalent to the transformed Mitscherlich equation derived in Section 4-2.5.3.

Taking the natural logarithm of the first equation and solving for L, they obtained

$$L = \frac{\ln[(pH_L - A)/b]}{\ln Q} \ .$$

This equation provides a means of estimating the quantity of calcium carbonate needed to reach a specified target pH_L when A, b, and Q are known.

The first step in simplifying the equation for use was to obtain a constant value for the maximum pH (A). From their investigation of 35 varied soils ranging from sands to peats from Northern Ireland, they found an average A value of 7.9. Substituting 7.9 for A in the equations for the individual soils, they calculated new best-fitting values for the other two parameters for each soil, which they designated as b_s and Q_s.

Examining next the parameter b_s, they noted that when $A = 7.9$ and $L = 0$, the foregoing equation becomes

$$b_s = pH_0 - 7.9,$$

where pH_0 is the soil pH value without calcium carbonate. Values of b_s estimated by this equation were almost equal to the values obtained with the best-fitting values of A for individual soils, and so they substituted $pH_0 - 7.9$ for b in the second equation to obtain

$$L = \frac{\ln[(pH_L - 7.9)/(pH_0 - 7.9)]}{\ln Q_s} \ .$$

Turning to Q_s, the curvature parameter, they assigned the values 7.89 and 3.5 to pH_L and pH_0 to represent the pH range of acid and limed soils, which gave the expression

$$\frac{-1}{\ln Q_s} = \frac{L}{6.086} \ .$$

Q_s thus is represented as a function of L. Because L is a function of the soil constituents that contribute to the titratable acidity, Q_s also is a function of these factors. Bailey et al. (1989a) found that $-1/\ln Q_s$ was closely related to the content of organic matter in the 35 soils they investigated. The ratio of $-1/\ln Q_s$ to organic matter, however, showed some variation with the initial soil pH. Taking this variation into account, and substituting the bulk density of the air dried and sieved (2 millimeters) samples (*SD*) for the organic matter content, they arrived at the following empirical equation for estimating Q_s:

$$Q_s = \frac{1}{\exp[\exp(0.27 + 3.648SD - 2.729pH_0 + 0.3040pH_0^2)]} \ .$$

This equation accounted for 98% of the variance of Q_s in the 20 surface soils in their investigation and 97% of the variance of Q_s in the 15 subsurface soils. Using the estimated value of Q_s, the equation for the chemical lime requirement is thus

$$L = \frac{\ln[(pH_L - 7.9)/(pH_0 - 7.9)]}{\ln Q_s} \ ,$$

as given previously. This equation provides an estimate of the chemical lime requirement to any target pH up to the maximum value of 7.9 on the basis of only the initial pH and the bulk density of the air dried and sieved soil used for analysis.

Bailey et al. (1989b) tested their method using data published in New Zealand (Edmeades et al., 1985b) from 23 field experiments on liming and associated laboratory estimates of chemical lime requirements based upon other methods. The correlation data indicated that the method just described was a little better than the best of the methods tested by Edmeades et al., namely, the double buffer method of McLean et al. (1977, 1978).

8-3.8. Field Versus Laboratory

The criterion for judging the accuracy of laboratory methods for estimating the chemical lime requirements of soils is the magnitude of the correlations of the measurements by the method being tested with the measurements on the same soils made by another method taken as a standard. As indicated in preceding sections, pH values are the most commonly used indicator of chemical response to liming. The preferred standard method is incubating soil with pure, finely divided calcium carbonate for an extended time such that the investigator considers an equilibrium has been obtained.

For soils with moderate to high cation exchange capacities in which calcium and magnesium are the dominant exchangeable cations, the pH could be employed as a measure of attainment of equilibrium if the electrolytes were first removed and the 0.01 molar calcium chloride method were used in measuring the pH values. As the process is normally carried out, however, pH values are not an adequate criterion of equilibrium because they do not become constant. Disappearance of calcium carbonate from the soil might provide an adequate criterion of equilibrium, but this approach has not been adequately tested. Further research on the details of the standard method using calcium carbonate should lead to improvements, including a shortening of the time required, which would increase the extent to which it is used.

If measurements obtained by a laboratory method are highly correlated with standard values, and particularly if the numerical values are almost the same as those obtained with the standard method, one may have confidence that the laboratory method is providing the kind of measurement desired. The most that can be expected from a laboratory method, however, is that it will provide a good estimate of the equilibrium titratable acidity in soils to selected target pH values.

The next step is projecting the laboratory measurements to the field. Here a new set of problems is encountered. In the laboratory testing of a particular soil, a certain ratio is established as the desired or recommended ratio between calcium carbonate and the titratable acidity. To mimic in the field what has been done in the laboratory, the same pure, finely divided calcium carbonate should be thoroughly mixed with the field soil in the same ratio to titratable acidity. When the titratable acidity per unit volume of soil is known, the ratio aspect can be accommodated by mixing the calcium carbonate with the soil in the field to the appropriate depth.

The lower temperatures in the field will reduce the reaction rates, and electrolytes will not have the effects on pH values that they have had in most standard titrations with calcium carbonate because of leaching and uptake of ions by plants. Moreover, it is not feasible to mix the calcium carbonate with the soil in the field as thoroughly as it is mixed in the laboratory studies to obtain the standard titration curve. And furthermore, it is impractical to use the pure, finely divided material employed for standard purposes in the laboratory. The practical products used in the field contain different proportions of impurities, including water. The practical products have different particle size distributions and are invariably coarser than the standard products; they also differ in chemical reactivity. Because of their differing physical and chemical properties they react at different rates with acid soil, and the reaction rates are invariably slower than those of the standard products. Fig. 8-9 shows the results of a field experiment in which a limestone described as "finely ground" was applied to acid soil under a crop rotation such that continual mixing would occur as the soil was tilled. The soil pH (presumably measured in an aqueous suspension) continued to rise for 5 years, and measured residual carbonate had disappeared from the soil by the sixth year after application.

The usual net effect of the differences in conditions between field and laboratory is that the quantities of liming materials needed to attain a target pH value are greater in the field than in the laboratory. Large differences between field and laboratory are seen in Fig. 8-10. In this instance, the laboratory estimates of lime requirement were made by equilibrating 10-gram samples of each of the six soils with different calcium carbonate equivalents as calcium hydroxide in 50 milliliters of solution, and the pH values after 3 days were measured on the resulting 1:5 soil:water suspensions. Corrections were made for the bulk density

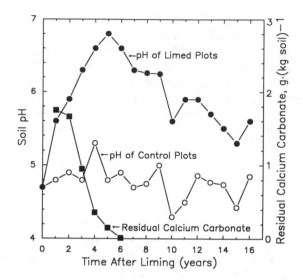

Fig. 8-9. pH values of soil in field plots and residual calcium carbonate from an application of 12.5 megagrams of finely ground limestone per hectare. A rotation of annual crops was followed during the 16 years of the experiment, which was conducted in the United Kingdom. (Walker, 1952)

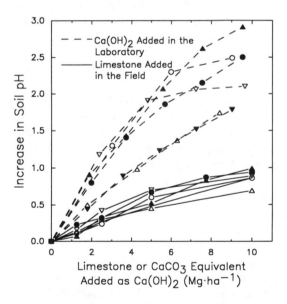

Fig. 8-10. Increases in soil pH values obtained with different quantities of limestone applied as a topdressing to six permanent pastures in New Zealand and with different calcium carbonate equivalents applied as pure calcium hydroxide to samples of the same soils in the laboratory. (Edmeades et al., 1985c)

of the soils, so that the data could be expressed as lime requirement for the surface 7.5 centimeters of soil. The smaller pH increases in the field than in the laboratory in this instance are attributable in part to plotting the total quantities of limestone applied instead of the quantities of calcium carbonate (the authors reported that New Zealand limestones typically contain from 60 to 90% calcium carbonate and that the purity of the limestone applied to two of the six sites was approximately 90%) and to incomplete decomposition of the limestones in the field. Because the limestones were applied in the field as a top dressing on permanent pastures, the rate of reaction with the soil would have been reduced, and the immediate surface of the soil would have been neutralized to the greatest degree. Small losses by leaching and uptake of calcium by plants and the pastured animals would also have affected the field results. The pH measurements shown in Fig. 8-10 for the field experiments are averages of generally biennial measurements made on the surface 7.5 centimeters of soil for 4 to 5 years following liming, with omission of the first year. There was no evidence that the soil pH values declined within the first 5 years after liming. The continued reaction of the coarser particles with the soil and the gradual penetration of the neutralizing effect of the surface applications presumably compensated for the losses during this period.

8-3.9. Limestone Quality Factors

Three limestone quality factors are generally recognized. These are purity, particle size, and chemical reactivity.

8-3.9.1. Purity

Purity of limestones for agricultural liming purposes usually is expressed as the calcium carbonate equivalent. A sample is treated with an excess of standard acid, and the residual acid is back-titrated with standard alkali. The calcium carbonate equivalent is calculated from the acid consumed. According to this convention, a unit of acid consumed by the magnesium carbonate present in many limestones along with the calcium carbonate is considered to have the same liming value as a unit of acid consumed by calcium carbonate.

Magnesium analyses often are made, especially for research purposes. The magnesium supplied by limestones is important in soils deficient in this element. Magnesium also affects the rate at which limestones decompose in acid soil.

8-3.9.2. Particle Size

Particle size distribution is important because it affects the rate at which limestone reacts with acid soil. Calcium carbonate is only slightly soluble in neutral or alkaline soil, and it becomes effective only when it has been decomposed. Thus, Allen and Hossner (1991) noted that the residual calcium carbonate equivalent from topdressings of agricultural limestone to soils of East Texas was essentially zero at pH values up to 5.5 (measured in 1:1 soil:water suspensions), but increased rapidly at pH values above 5.5. Most of the residual limestone was present in the surface 2 centimeters of soil, and it contained a higher proportion of coarse particles and a lower proportion of fine particles than did the original agricultural limestone.

Fig. 8-11 shows the decomposition of limestone particles of different sizes when added to an acid soil in quantities to supply an amount of calcium carbonate equivalent to three-fourths of the lime requirement to pH 7.0 by Schofield's (1933) *p*-nitrophenol method. The highest soil pH obtained in the experiment was 6.7. Note that the rate of decomposition increased as the particle size of the limestone decreased. Per unit mass of limestone, the area of interface with acid soil, and hence the rate of reaction with the soil, increase as the particle size decreases.

The dominant hypothesis employed in mathematical descriptions of the particle-size effect has been that the rate of reduction of the diameter of the limestone particles due to reaction with acid soils is the same for all particle sizes. This hypothesis or model is equivalent to stating that the rate of reaction with the soil is proportional to the surface area of the particles. It applies strictly to a flat surface, and has been found to describe to a good approximation the rates of decomposition of granules of 0.42 millimeter and larger sizes.[1]

Warfvinge and Sverdrup (1989) developed a more physicochemical model of limestone reaction with acid soil, which they found would apply to a very close approximation to the data in Fig. 8-11. To fit the data for all particle sizes, as opposed to only the coarser particles, they found it necessary to introduce a "mass transfer coefficient," which involves the assumption that the overall rate of decomposition of the limestone (phrased differently in their words) is determined by the rate of transport of substances between the solution at the limestone surfaces and the sites in the soil where the reactions between the substances in solution and the soil occur — that is, to the outer reaches of the shell of substantially neutralized soil around the limestone particles. According to the Warfvinge and Svedrup model, the greater rate of dissolution found per unit surface area for fine particles than coarse particles is accounted for by a smaller distance of transport as the particles become smaller because the thickness of the shell of soil around a limestone particle required to contain a given mass of soil per unit area of limestone particle surface decreases as the particle size decreases. Nye and Ameloko (1987) also developed a model for the reaction between

---

[1]Although particle size fractions of limestone usually are referred to in terms of numbers of meshes per unit length in the sieves used for fractionation, the convention followed here is to express the particle sizes in terms of the sizes of the openings in the sieves. Various systems of sieves are in use, and a given mesh number does not correspond to the same size opening in the different systems. Cadle (1955, pp. 182-183) gives a table showing the Tyler, U.S. Standard and American Society for Testing and Materials, British Standard, Institute of Mining and Metallurgy (London), and German DIN sieve series. In the U.S. Standard and American Society for Testing and Materials Sieve Series, the sieves used most commonly for particle size analyses of ground limestones have openings of 4.76, 2.38, 2.00, 0.84, 0.42, 0.250, 0.177, and 0.149 millimeters and correspond to mesh numbers 4, 8, 10, 20, 40, 60, 80, and 100, respectively.

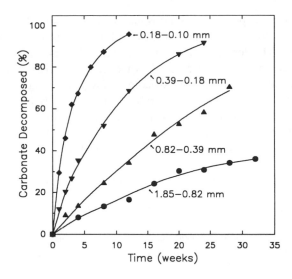

Fig. 8-11. Decomposition of limestone particles in different size ranges in an acid soil. The data shown are averages for seven limestones, and are based upon measurements of residual carbonate. The measurements shown in the graph are the apertures in the sieves used to collect the different particle size fractions. The geometric mean diameters of the size fractions from coarsest to finest were 1.23, 0.56, 0.26, and 0.13 millimeters. (Elphick, 1955a)

limestone and soil in which the rate-limiting process is the rate of transfer of acids and bases between the limestone particle surfaces and the acid soil.

As may be inferred from Fig. 8-11, the effect of particle size on the decomposition of limestone in acid soil decreases with time. Coarse particles make a poor showing initially, but in time their performance relative to fine particles improves.

Many field experiments have been conducted to evaluate the effect of particle size or fineness of limestone on crop yields. The experiments have been of two types. In one type, the limestones have been sieved, and the particle size fractions have been applied separately. In the second type, limestones differing in fineness of grinding have been used, and the comparisons have been made on the basis of the proportion of the material passing through sieves with openings of various sizes.

The experiments were conducted in earlier years without the benefit of modern biological assay concepts, and their design and heterogeneity make them difficult to summarize. Barber (1984) made an extensive analysis of the data, and his findings are summarized in Table 8-1.

The data in the table may be used as a guide to estimate the effectiveness of limestones ground to different degrees of fineness. As noted by Barber, particles larger than 0.84 millimeter are relatively ineffective, and particles finer than 0.25 millimeter are highly effective.

The importance of particle size has been widely recognized by the establishment of legal standards. For example, according to a private communication from David M. Wheeler of the Ruakura Agricultural Centre in Hamilton, New Zealand, the law in that country specifies that less than 5% of an agricultural limestone may have a particle size exceeding 2 millimeters and more than 50% must have a particle size less than 0.5 millimeter. In Iowa (Anonymous, 1990), the effective calcium carbonate equivalent of agricultural limestone is determined by the product of the chemical calcium carbonate equivalent in percent and a "fineness factor," which is to be determined as follows:

a. Multiply the percent of the total material passing the number four sieve by one-tenth.
b. Multiply the percent of the total material passing the number eight sieve by three-tenths.
c. Multiply the percent of the total material passing the number sixty sieve by six-tenths. Add the results obtained from paragraphs "a", "b", and "c" of this subsection to obtain the fineness factor.

The sieve number designations refer to the U.S. Standard and American Society for Testing and Materials sieve series. See footnote 1.

8-3.9.3. Reactivity

Carbonate rocks used for agricultural liming generally contain both calcite [$CaCO_3$] and dolomite [$CaMg(CO_3)_2$]. If the content of dolomite in the sum of calcite

Table 8-1.   Value of Limestones Differing in Fineness as Summarized From Field Experiments at Different Locations With Various Test Crops[a] (Barber, 1984)

Proportion of Limestone Passing a 0.25-Millimeter Sieve (%)	Number of Experiments	Relative Increase in Crop Yield		Calculated From Limestone Fraction Data[b]
		Direct Measurements		
		Mean	Range	
<20	2	50	30–69	58
20–30	11	77	53–100	71
30–40	5	84	62–100	75
40–50	9	92	62–100	83
50–60	4	100	100	89
60–80	5	94	84–100	94
>80	10	97	76–100	100

[a] Where the experiments were continued for many years, only the data from the first 6 or 8 years were used.
[b] The calculated relative yield increases are the summation of the products of (a) the quantity of each particle size fraction in an average ground limestone having the appropriate percentage passing through a 0.25-millimeter sieve and (b) the relative increase in yield produced by the size fraction in nine field experiments with individual particle size fractions in which the relative increases in yield were as follows: 33 (0–50) for particles of 4.76 to 2.00 millimeters, 54 (11–96) for particles of 2.00 to 0.84 millimeters, 84 (34–100) for particles of 0.84 to 0.42 millimeter, 99 (61–100) for particles of 0.42 to 0.177 or 0.42 to 0.149 millimeter, and 100 (78–100) for particles finer than 0.177 or 0.149 millimeter. The numbers in parentheses represent the ranges in the experimental values.

and dolomite is from 10 to 50% (corresponding to 13 to 65 grams of magnesium per kilogram of calcite plus dolomite), the limestone is said to be dolomitic. Carbonate rocks with higher proportions of dolomite are called calcitic dolomite and dolomite.

Dolomitic limestones are often used for agricultural liming, but they are less reactive than calcitic limestones. In a summary of available data on the rates of decomposition of fractions of different limestones of equal particle size, Barber (1984) found that, on average, dolomitic limestone disappeared half as rapidly as calcitic limestone. Thus, if the rate of decomposition of limestone particles in soil is proportional to the surface area of the particles, a dolomitic limestone would need to be ground enough more finely to have about twice the surface area of a calcitic limestone to provide a reaction rate equal to that of calcitic limestone. Alternatively, if the particle size distributions were the same, twice as much dolomitic limestone would be needed. The difference in decomposition between the two types of limestone would decrease with increasing time of contact with the soil.

Barber (1984) found that dolomitic and calcitic limestone had been compared for crop production in nine field experiments. In six experiments, the calcitic form was much superior, and in three the dolomitic form gave the greater yield increase. The results of comparisons of dolomitic and calcitic limestones for crop production cannot be interpreted unequivocally in terms of reactivity because of possible differences in effects due to the differences in calcium and magnesium they would supply. As noted by Barber, "If dolomitic limestone is superior, it is probably because of a response to Mg. If calcitic limestone is superior, it is probably because it reacted with the soil more rapidly and reduced the acid soil yield limiting factors."

Agricultural experience has indicated also that hard and soft limestones differ in rates of decomposition in acid soils. In an experimental comparison, Doak (1941) applied a hard limestone and a soft limestone over a 3-year period as a topdressing to a permanent pasture in New Zealand. He found that the hard limestone reacted much less rapidly with the soil than did the soft limestone, but by the end of 2 years after the last application, the residual carbonate from both limestones was essentially zero, even when the products added had been screened so that separates finer than 0.25 millimeter had been removed (1% of the soft limestone was retained on a sieve with 2 millimeter openings, but all of the hard limestone passed this sieve). The pH values recorded for the surface 5 centimeters of the soil in this experiment ranged from 5.0 to 6.7. Anderson (1966) and Bailey et al. (1989c) noted similarly that the rate of reaction with acid soil was greater with soft limestone than with hard limestone.

In an investigation of the rates of decomposition of 25 calcitic limestones representing the important types available in the South Island of New Zealand, Elphick (1955b) found that when 0.823 to 0.388 millimeter separates of the various limestones were incubated with an acid soil, the mean rate of diameter reduction ranged from 3.8×10^{-3} millimeter per week in a very hard limestone with porosity of 4% to 12.0×10^{-3} millimeter per week in a very soft chalk with porosity of 46%. The rates of diameter reduction were closely related (r = 0.92) to what he called the "equivalent gross volume (V)." He defined V as the gross volume in milliliters of the 0.853 to 0.422 millimeter separate that contained 1 gram of calcium carbonate equivalent:

$$V = \frac{v}{0.01mc} \, ,$$

where v = volume of the separate in milliliters, m = mass of the separate in grams, and c = calcium carbonate equivalent of the separate in grams per kilogram.

He expressed the decomposition of the limestone in the soil by the equation

$$R = \left(\frac{D - tkV}{D}\right)^3 ,$$

where R = fraction of the original mass residual after time t, D = initial mean particle diameter, and k = decomposition constant whose value is determined by the soil conditions. In this equation, kV replaced Δ, which Elphick had used previously as the mean weekly rate of diameter reduction of a limestone particle due to reaction with the soil. He found that the correlation between the observed and calculated decompositions of various ground limestones and limestone separates ranging from 1.849 to 0.823 millimeter to 0.178 to 0.099 millimeter was r = 0.99.

This study was not intended to account for the effect on decomposition resulting from the increasing volume of soil per unit of surface area within a given distance of the limestone surface as the particle size decreases. Although the correlation between calculated and observed decomposition was r = 0.99, the data show that the observed decomposition exceeded the calculated decomposition for six of the seven limestones studied at the 0.178 to 0.099 millimeter size.

Elphick (1955b) made two points about the foregoing equation that deserve special comment. One is that the parameter V is proportional to Δ and inversely proportional to c, the calcium carbonate equivalent. Therefore, the product of Δ and calcium carbonate equivalent is constant, which means that the rate at which the diameter of the limestone particle is reduced by contact with the soil is inversely proportional to the density of the calcium carbonate equivalent.

In other words, under given soil conditions, the calcium carbonate in the limestone decomposes at the same rate per unit area at all carbonate surfaces. Thus, the more porous and impure the limestone particles, the lower is the calcium carbonate equivalent per unit volume, and the greater is the rate at which their diameter is reduced by interaction with the soil. Elphick's analysis thus replaces the concepts of reactivity and hardness as factors affecting the rate of decomposition with the concept that the rate of reaction of calcium carbonate with the soil is constant under given conditions, but that the rate of erosion of the particle surfaces varies inversely with the quantity of calcium carbonate per unit volume. A similar condition would be supposed to prevail with dolomitic limestones except that the results would be complicated by the differences in rate of decomposition associated with the dolomite content.

Elphick's second point was to emphasize the fact that the progress of the decomposition depends upon the ability of the soil surrounding the limestone particles to remove the calcium from the limestone surface and, although he did not mention it, to supply the hydrogen ions that cause the decomposition. As the thickness of the layer of more or less neutralized soil increases around a given particle, the diffusion path lengthens, the gradients decrease, and the rate of decomposition diminishes. Nye and Ameloko (1986) published a theoretical treatment of the diffusion process.

The significance of the balance between diffusion of the neutralizing effect into the soil and loss of the calcium supplied by a calcium carbonate particle in an acid soil may be more readily apparent from some measurements made by DeTurk (1938) in soil under a naturally occurring glacial limestone boulder imbedded in acid soil in Illinois. The boulder apparently had been in place for the centuries during which the soil was developing and becoming acid. DeTurk found that the distance of effective (but not complete) neutralization was 0.8 centimeter and that the greatest distance at which a measurable decrease in soil acidity occurred was 1.7 centimeters.

As is true for particle sizes, the differences in decomposition of limestones in soil discussed here under the heading of "reactivity" diminish with time and disappear when decomposition is complete. The importance of the differences can range from considerable if a highly sensitive crop is to be grown on an initially strongly acid soil immediately after liming, to negligible if the crop is tolerant of soil acidity and if a year or more has elapsed since liming. In his second paper, Elphick (1955b) included a table showing the results of calculations of the decreasing differences between limestones and between two different limestone separates for times up to 24 weeks.

Some time evidently is required for limestone applications to reach their maximum effectiveness. Moreover, the rates of loss of limestone and the exchangeable calcium and magnesium that result when limestone decomposes in soil are modest. Reflecting this understanding from the scientific standpoint, practical applications normally are made between crops at convenient times, unlike most fertilizers, which are immediately effective and generally are applied just before planting or as a topdressing or sidedressing while the crop is growing to avoid losses in effectiveness.

As is true for fertilizers, it would theoretically be advantageous from the economic standpoint if there were no residual effect of limestone and if all the value could be realized in the first crop. The facts, however, are at variance with economics where limestone is concerned. Limestone has a marked residual effect. An adequate limestone application may last for a decade or longer. Simán (1989) reported that in six factorial experiments on liming and fertilization in Sweden, an application of 6 megagrams of calcium oxide equivalent as calcium hydroxide per hectare was still increasing crop yields more than 50 years later.

Because of the behavior of limestone in soils and the way it is usually applied, extreme fineness and rapid reaction with the soil are not of first importance. Particles that pass through a screen with openings of 0.84 millimeter (20-mesh size in the U.S. system) or smaller are generally adequate. Of greater concern is the fact that despite reaction of the limestone with the surface soil, many years may be required before the downward movement of the neutralizing effect has eliminated the aluminum toxicity and calcium deficiency in subsoils that are originally strongly acid.

8-3.10. Integrated Factors

This section describes the theory and testing of two methods that involve more than the chemical lime requirement of the soil. One method was developed by Bailey et al. (1989c) to adjust the quantity of specific liming materials for the aggregate effect of purity, fineness, and reactivity. The evaluation of liming materials was combined with an evaluation of the soil chemical lime requirement to obtain an overall evaluation of the requirement of a specific soil for a specific liming material to reach any selected target pH from 5.0 to 7.1. Their experimental data were obtained on soils and limestones from Northern Ireland.

The first step in the theory of the limestone evaluation was to represent the fractional increase F in the calcium carbonate addition required to compensate for the difference in effective calcium oxide content between a given commercially available liming material and reagent grade calcium carbonate by the equation

$$F = \frac{C - L}{L} ,$$

where C = equivalent quantity of calcium carbonate in the liming material required to raise the pH of a given soil to the target value pH_L (pH values measured on a suspension of 10 milliliters of soil in 25 milliliters of water), and L = quantity of reagent grade calcium carbonate required to increase the soil pH to the same target value. When reagent grade calcium carbonate is incubated with soil, $C = L$, and $F = 0$. When an ordinary liming material is incubated with soil, C exceeds L, and F has a positive value. F is a decimal fraction of L.

C is given by

$$C = \frac{M \cdot NV}{56} ,$$

where M and NV are the quantity and neutralizing value of the liming material. The denominator is the neutralizing value (calcium oxide equivalent) of reagent grade calcium carbonate. To use the more common expression for calcium carbonate equivalent, the denominator would be 100 (the molecular weight of calcium carbonate), and values of NV for the liming material would be expressed relative to a value of 100 for pure calcium carbonate. The numerical values of C would be the same with either convention. The equation for C may be substituted in the equation for F, and the resulting equation may be solved for M to obtain

$$M = \frac{56L(1 + F)}{NV} \ .$$

When a specific liming material is incubated with a standard soil under specific conditions, the value of F obtained experimentally may be designated as f. The relationship between F for a test soil and f for the standard soil may be represented by

$$F - f(1 + Z),$$

where Z is the fractional amount by which f must be adjusted to make the product $f(1 + Z)$ equal to F. The parameter Z thus is given by

$$Z = \frac{F - f}{f} \ .$$

In separate incubation experiments, the authors found that f was related to the percentage content of certain size fractions and the percentage content of magnesium in the limestones; and Z was related to the initial pH of the soils and the organic matter content (as proxied by the bulk density of the disturbed soil samples used for analysis, in accordance with the findings by Bailey et al. (1989a) described in Section 8-3.7). The relationships were expressed empirically by regression equations.

To test the model, two other experiments were conducted. In one, 21 widely differing liming materials were incubated with a single soil. Values of F calculated from the original equation $F = (C - L)/L$ were substantially proportional to the values of F derived from the equation $F = f(1 + Z)$, and the proportion of the variance accounted for was $r^2 = 0.74$. In a final experiment, five liming materials of different types were incubated with nine soils ranging from sands to peats. Single quantities of the various liming materials were used, but the quantities per kilogram of soil were varied from 3 to 45 grams, depending upon the properties of the limestones and the soils, so as to obtain a range of target pH values from 5.0 to 7.1. Values of F were calculated from the relationship $F = f(1 + Z)$, using for f and Z the values obtained by substituting the appropriate

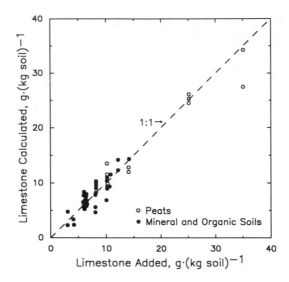

Fig. 8-12. Quantities of five different limestones added to nine different soils of Northern Ireland, and the quantities of the limestones calculated to produce the pH values actually obtained. (Bailey et al., 1989c)

numerical data in the empirical equations that had been developed to estimate these parameters from the properties of liming materials and soils, respectively. The pH values were measured after an incubation period of 8 months in this experiment, as in each of the others.

When all the theoretical and empirical relationships were put together using the data obtained in the last experiment, the quantities of liming materials actually added were found to be directly proportional to the quantities predicted to produce the target pH values, and the proportion of the variance accounted for by the predicted values was $r^2 = 0.94$ (see Fig. 8-12). In evaluating these results, which were obtained entirely in the laboratory, the authors noted that the 8-month incubation period they had used would relate to the time when the rate of reaction of the liming materials with acid soil would have become very slow. Because the rate of reaction is slower in the field than in the laboratory, they suggested that their data might correspond to a field incubation of about a year, which they considered most relevant to agricultural liming practice.

Although an 8-month incubation period was used in testing the method on various soils, no incubation would be required in use of the method on samples submitted by producers for testing. For such samples, the information needed would be the bulk density of the dried and sieved sample of soil (a convenient substitute for the organic matter content) and the initial pH, together with the neutralizing value, magnesium content, and particle size data on the limestone to be used. These items of information would be entered into the equations worked out in the calibration process, and the overall lime requirement would be calculated from the equations.

Hochman et al. (1989) devised an integrated model for liming subterranean clover pastures in New South Wales. Their model takes into account (a) soil properties that affect the chemical lime requirement, the residual effect, and exchangeable aluminum, (b) the pasture productivity, (c) potential livestock production and the stocking rate required to use the additional pasture production, and (d) the economics of liming. Only the soil aspects will be considered here.

The soil data for the model were derived from a group of 16 field experiments on several kinds of soils with initial pH values ranging from 3.9 to 4.8. To find the pH buffering capacities of the soils, different quantities of a limestone with 95% passing a 0.15 millimeter sieve and a calcium carbonate equivalent of 98% were incorporated to a depth of 10 centimeters, and the plots were sampled about a year after liming for measurement of soil properties.

The equation

$$\Delta pH = B_s(\sqrt{L + 1} - 1),$$

where B_s = pH buffering capacity of soil s (megagrams of limestone per hectare per unit pH increase due to liming the surface 10 centimeters) and L = megagrams of limestone per hectare, accounted for 93% of the variance of the ΔpH values due to liming in the 16 experiments. ΔpH$/B_s$ was directly proportional to $(L + 1)^{1/2} - 1$. Values of B_s for the various soils ranged from 0.34 to 1.55, which means that 1 megagram of limestone per hectare increased the soil pH by 0.14 to 0.64 unit. pH values were measured in a 1:5 soil:0.01 molar calcium chloride suspension.

To develop a way to estimate values of B_s for other similar soils from soil properties that could be measured more easily, the first step was to relate the total exchangeable cations (TEC) displaced from the 16 individual soils by 0.125 molar barium chloride to the pH values of the soils. (The units for TEC are centimoles of charge per kilogram of soil.) The regression equation obtained from the data on the samples from the limed and unlimed plots of the 16 soils was

$$TEC = SL_s(pH - 2.043) - 0.957,$$

where SL_s is the soil-specific variable that determines the slope of the line relating ΔpH$/B_s$ to $(L + 1)^{1/2} - 1$. The subscript s refers to a specific soil. The proportion of the variance accounted for by this equation was $r^2 = 0.96$. Solving the equation for SL_s,

$$SL_s = \frac{TEC + 0.957}{pH - 2.043} \quad .$$

A plot of $SL_s(pH - 2.043) - 0.957$ against TEC for the various soils, limed and unlimed, was linear to a good approximation.

The regression of B_s on SL_s was somewhat curvilinear, but accounted for 88% of the variance of B_s. The equation was

$$B_s = \frac{34.24}{SL_s + 9.5} - 1.741.$$

The procedure to calculate the quantity of limestone needed to produce a particular target pH from the foregoing information is (a) to calculate SL_s by solving the equation giving SL_s as a function of TEC and soil pH, (b) to substitute the value of SL_s in the equation giving B_s as a function of SL_s, (c) to substitute the resulting value of B_s and the required value of ΔpH in the equation giving ΔpH as a function of B_s and L, and (d) to solve the equation for L.

Hochman et al. (1989) proceeded further to estimate the percentage saturation of the effective cation exchange capacity with aluminum, which is an indicator of lost yield potential. When the soil pH is increased by liming, the effective cation exchange capacity (total exchangeable cations or TEC) increases, and the exchangeable aluminum decreases.

As the first step, functions of exchangeable aluminum and soil acidity were found that were proportional to each other. The resulting equation was

$$SLAl_s = \frac{Al - 1.0273}{a_{H+} - 3.7 \cdot 10^{-6}},$$

where Al = exchangeable aluminum in centimoles of charge per kilogram of soil associated with the hydrogen-ion activity a_{H+}, and $SLAl_s$ is the slope of the regression for soil s. (In the equation given by Hochman et al., the Al in the numerator was multiplied by the coefficient 90, which converted their original measured quantities of exchangeable aluminum in micrograms per gram to milligram equivalents per 100 grams. The latter units are equivalent to centimoles of charge per kilogram.) The equation accounted for 94% of the variance of the data.

Solving the foregoing equation for Al,

$$Al = SLAl_s(a_{H+} - 3.7 \times 10^{-6}) + 1.0273.$$

The percentage saturation of the effective cation exchange capacity by aluminum then is given by

$$\frac{100Al}{TEC} = \frac{100[SLAl_s(a_{H+} - 3.7 \times 10^{-6}) + 1.0273]}{SL_s(pH - 2.043) - 0.957}.$$

In this equation, the hydrogen-ion activity and the pH both refer to the limed soil, that is, the target pH. To simplify the equation, the symbol a_{H+} thus may be replaced by 10^{-pH}.

In a subsequent paper, Hochman et al. (1992) applied their model to two other groups of experiments on liming — one from a different area in New South Wales and the other from New Zealand. In this work, they separated the relationship between total exchangeable cations and pH into relationships between (a) exchangeable aluminum and pH and (b) total exchangeable bases and pH. They were able to improve the precision of the latter relationship by including a measure of soil organic carbon as a second independent variable.

To estimate the residual effect of liming, Hochman et al. (1989) modified an equation used earlier by Bromfield et al. (1983) describing soil acidification with time when subterranean clover was grown:

$$pH = 4.90 - 2.42(1 - e^{-0.0075t}),$$

where pH is the soil pH in the year t of cropping to subterranean clover. (The equation given here is the one shown by Hochman et al., which differs slightly from the one given by Bromfield et al.)

Hochman et al. modified the equation of Bromfield et al. in several ways. They generalized the coefficient 0.0075 found for t by multiplying it by $B_s/1.3$, where 1.3 was the buffering index of the soils studied by Bromfield et al., and B_s is the buffering index appropriate for soil s. They also substituted pH_v for 4.90, the mean initial pH value for the soils studied by Bromfield et al., where pH_v is a value appropriate for soil s in virgin condition. In the study by Bromfield et al., the soils had been in subterranean clover continuously since clearing, and no limestone had been applied. To take into account the fact that these circumstances would not apply in general, Hochman et al. added a value of t_i, which could be calculated from

$$t_i = -\ln\left(\frac{(pH_i - 2.42)/(pH_v - 2.42)}{0.0058B_s}\right),$$

where the subscript i applies to the current year. The overall equation for the soil pH in the year t (pH_t) counted forward from the current year (t_i) was given as

$$pH_t = 2.42 + (pH_v - 2.42)e^{-0.0058(t + t_i)B_s} .$$

Subtracting pH_t from pH_i would give a value for ΔpH, which could be substituted back in the original equation, $\Delta pH = B_s[(L + 1)^{1/2} - 1]$, to calculate the value of L equivalent to the pH change. Hochman et al. followed the alternative route of substituting the value of pH_t in a modification of an equation used by Gasser (1973) to calculate the loss of calcium carbonate and then substituting the calcium carbonate loss in the equation for ΔpH to calculate ΔpH.

Except for the work by Hochman et al. (1989) just described, the primary concern of the research discussed in preceding paragraphs has been the capacity aspect of the chemical lime requirement of soils, that is, the quantity of calcium carbonate or other liming material required to yield a given target pH when the carbonate has disappeared. The quantity needed per year for maintenance also varies widely with the conditions, but relatively little attention has been given to maintenance requirements. Gasser (1973) reviewed the data on this subject and suggested an equation that was referred to by Hochman et al. (1989) in the work just cited.

Bolton (1977) did further work on maintenance applications, making measurements of soil calcium extracted by neutral molar ammonium acetate solution from samples of soils from liming experiments in the United Kingdom. The liming materials used in these experiments were "soft cretaceous limestones (chalk)," the residues of which would have been dissolved by the ammonium acetate, leading to data that would represent the sum of exchangeable and carbonate forms of calcium. Bolton's findings in Fig. 8-13 show a marked increase in annual rate of loss with an increase in soil pH value measured in 1:2.5 soil:water suspension. In terms of maintenance applications, the higher the target pH, the greater would be the quantity needed per year. Bolton did not have experimental data on losses by leaching from the experiments he analyzed, but he pointed out that calcium would be the main cation lost and that it would be accompanied by anions. He made estimates of the loss of calcium by leaching from other published data, and his calculations, summarized by one of the curves in Fig. 8-13, fitted the data well. An important point he made was that bicarbonate would be the principal anion at pH 8. pH values close to 8 were measured in

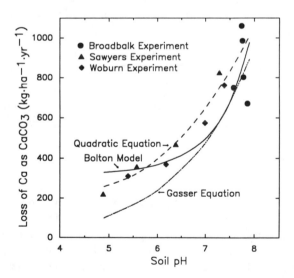

Fig. 8-13. Annual losses of calcium expressed as calcium carbonate from the surface 23 centimeters of soil in three experiments in the United Kingdom, as represented by the experimental data (points), a fitted quadratic equation, an equation by Gasser (1973), and a model by Bolton. (Bolton, 1977)

the soil from the Broadbalk wheat experiment, which contains free calcium carbonate.

From evidence obtained in long-term crop rotation experiments in Sweden, Simán (1989) reported that an average of 225 kilograms of calcium carbonate per hectare per year would be required to maintain 70% base saturation in the topsoil (apparently where base saturation was considered to be obtained at a soil pH value of 7.0 in water). To maintain 100% base saturation would require 645 kilograms of calcium carbonate per hectare annually. In Victoria, Ridley et al. (1990) reported annual soil acidification corresponding to 39 kilograms of calcium carbonate per hectare under subterranean clover pastures.

Goulding et al. (1989) noted that calcium loss should be a function of soil type, crop, rainfall, soil pH, and quantity and form of fertilizer nitrogen. On the basis of data from several liming experiments, they developed a series of linear equations to represent the change in pH with time as a function of the quantity of calcium carbonate added during a period of 12 years. The information provided was limited, but it appeared that the linear equations were used independently, with each equation representing a different condition.

8-4. Biological Lime Requirement

Although a major part of this chapter has been devoted to the chemical lime requirement of soils, liming materials are applied to soils to benefit the crops, not the soils. Ideally, the lime requirement should be a function of the crop or the cropping system and the economics of the particular situation as well as the conditions in the soil to which the limestone is to be applied.

The *biological lime requirement* may be defined as the quantity of calcium carbonate that produces a selected biological response. The biological lime requirement is usually researched on the basis of a sample of soil or the surface layer of soil, but it may also be viewed on a profile basis, as described in Section 8-4.7.

Much less attention has been given to adjusting quantities of liming materials to biological responses than has been true for fertilizers. Several reasons might be given. The great difference in chemical lime requirement among soils with similar pH values would be one. The slowness of reaction and the long residual effect of the usual applications of ground limestone would be another. Applications normally are made only at intervals of several years, and during the interval the soil pH probably will be increasing for at least a year, and then will be decreasing, with the consequence that a range will exist in the degree to which the chemical lime requirement of the soil is satisfied and in the pH values that might be measured. Another probable reason is the fact that the responses of a given crop to liming cannot be correlated broadly with any direct causal factor. pH values are the most common measure of soil acidity, and they are almost invariably listed when differences in crop response to liming are reported, but hydrogen ion activity as such is generally a minor factor in crop response. Other factors that change with the pH value are responsible for the major effects within the usual pH range of soils. What these factors are is generally not obvious, and they usually are not known because of the expense and time required to obtain the information.

8-4.1. Biological Factors

Differences in soil acidity affect plants through a complex of factors, of which toxicities, deficiencies, and diseases are of principal importance if the crop is a single species. Where several species are grown together, as in pastures and natural conditions, interspecies competition must be added as another important factor. Diseases and competition will not be considered here. See Schmitthenner and Canaday (1983) and Robson and Abbott (1989) for information on diseases.

The most important toxicity is that of aluminum. Aluminum toxicity can occur on almost any mineral soil at pH values of 5 and below, measured in soil:water suspension, but measurements of exchangeable aluminum, the percentage of aluminum in the exchangeable cations, and the concentration of aluminum in the soil solution do not provide consistently suitable criteria for judging the existence of aluminum toxicity for a given crop species. The principal reason appears to be the occurrence of the solution-phase aluminum in a variety of species differing in toxicity. Although much work has been done on this subject in recent years, the estimates of species are derived from computer programs based upon thermodynamic equilibria that have not been precisely evaluated. Dobbs et al. (1989) showed graphically the distribution of certain inorganic aluminum species with pH as estimated from four different sets of stability constants. Although the distributions did not agree, Al^{3+} was always indicated as the major species below about pH 4.7 to 5.4 (the exact pH depending upon the set of constants used), and it represented more than 90% of the total aluminum in solution at pH 4. Sometimes the thermodynamic data are supplemented with chemical studies that measure rates of transformation of unknown species to forms that react with certain color-producing reagents. Available analytical methods are not capable of determining specific metal forms.

The findings suggest that the most toxic species is Al^{3+}, that the other toxic species are for the most part monomeric, and that polymeric species and aluminum complexed with sulfate ($AlSO_4^+$), phosphate, fluoride, or organic matter have little or no toxicity (Pavan et al., 1982; Hue et al., 1986; Bell and Edwards, 1986; Kinraide and Parker, 1987; Tanaka et al., 1987; Parker et al., 1989a,b; Wright et al., 1989). Manganese toxicity also can be important.

Of the nutrient deficiencies found in strongly acid soils, calcium, magnesium, molybdenum, and phosphorus seem to be most important. Nitrogen can be important under some circumstances.

A paper by Adams and Moore (1983) describes procedures they used to determine whether aluminum toxicity and calcium deficiency are important causes of poor root growth. They grew cotton seedlings in special glass-front boxes containing subsoil horizons of a number of acid soils. The boxes were inclined so that the roots would grow against the glass, where their rate of growth and appearance could be observed. The criteria used were as follows: (1) If root growth is enhanced by addition of calcium sulfate and calcium hydroxide but decreased by magnesium oxide, calcium deficiency is indicated. (2) If root growth is enhanced by calcium hydroxide and magnesium oxide but is decreased or unaffected by calcium sulfate, aluminum toxicity is indicated. (3) If root growth is enhanced more by calcium hydroxide than by magnesium oxide and calcium sulfate, a combination of calcium deficiency and aluminum toxicity is indicated. The diagnoses derived from these criteria were coupled with observations of the visual appearance of the roots. A deficiency of calcium for cotton roots is indicated by straight, thin primary roots with brown tips, and toxicity of aluminum is indicated by primary roots that are swollen, stubby, and gnarled.

Liming acid soils to higher pH values decreases the availability of some nutrients and may induce deficiencies. Manganese, zinc, boron, copper, and sometimes phosphorus show reduced availability at high pH values. Whether or not deficiencies appear depends in part upon the initial supply, the pH attained, and the crop. In some instances, root diseases are more severe at higher soil pH values. All these factors may reduce crop yields when soils are limed beyond a critical value. The reduction in crop yields (so-called overliming injury) that may be observed in some instances, especially in crops grown on highly weathered, initially strongly acid soils, now appears to be principally a matter of induced nutrient deficiencies. Aluminum toxicity may be involved in some soils.

A situation studied by Friesen et al. (1980a,b) illustrates an induced nutrient deficiency. See Fig. 8-14. They found that the yields of corn plants 21 days after emergence on cultures of a strongly weathered and acid Nigerian soil (pH 4.1 in a 1:1 soil:water suspension) increased somewhat with liming to pH 6.5, but were essentially unaffected by adding zinc. With a heavy application of fertilizer phosphorus (300 milligrams per kilogram), the yields were far greater and increased with liming to pH 4.8, but decreased with further liming to pH 6.5. When zinc was added at 5 or 10 milligrams per kilogram of soil, the yield with phosphorus did not decrease above pH 4.8, but remained about the same, indicating that the added zinc had eliminated a deficiency induced when the soil was limed to pH values above 4.8 in the presence of fertilizer phosphorus. See a paper by Gupta et al. (1989) for data on changes in micronutrient soil test values and micronutrient concentrations in the grain of wheat grown on strongly acid soils in India with and without liming.

In solutions without complexing ions other than hydroxyl, concentrations of aluminum are very low from pH 5 to 7.5. Aluminum toxicity to plants in soils generally is dismissed as being no problem in this range except occasionally a little above pH 5. There is evidence, however, that in some instances the overliming injury may actually be an induced toxicity of aluminum in the range of pH values generally considered safe. In a greenhouse experiment with corn grown on several strongly weathered soils, Farina et al. (1980a, 1982) found that the

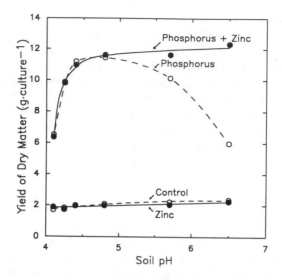

Fig. 8-14.   Yield of corn plants after growth for 21 days on cultures of a strongly weathered soil of Nigeria in a factorial experiment with different quantities of calcium carbonate, phosphoric acid, and zinc ethylenediaminetetraacetate. (Friesen et al., 1980a,b)

yield reduction with liming was associated with an increase in aluminum concentration in the plant tissue. The appearance of the roots was similar to that of roots subjected to toxic concentrations of aluminum. Consistent with the occurrence of the most marked yield depressions in soils with the higher organic matter content, their guess was that the aluminum was being brought into solution as an aluminum chelate and that some of this chelate was absorbed by the plants. Although some evidence exists for solubilization of organic matter and aluminum as a result of liming (Farina et al., 1980a; Hargrove, 1986), more critical experimental work is needed.

8-4.2. pH as a Criterion

Summaries of information on the optimum pH values for different crops may be found in several publications, including monographs edited by Adams (1984) and Robson (1989), a review article by Rowell (1988), and a bulletin by Spurway (1941) devoted entirely to this subject. The basic deficiency of soil pH as a criterion for biologically optimum applications of calcium carbonate is that when soil pH and crop responses change with additions of calcium carbonate, the cause of the differences in crop response is usually not the pH but one or more other factors that are only associated with the pH. Crops respond differently to the individual soil factors, and the soil factors appear in a variety of combinations.

The inevitable consequence of the complexity of the relationships between soil pH values and crop performance is occurrence of the maximum yield of a given crop at a pH value that may differ from one set of circumstances to another. Carefully conducted field experiments are needed to determine what the optimum pH values are and to determine the economics of liming for individual crops or for the crop rotation that may be followed on particular soils.

Extensive research on the crop aspect was done many years ago in Ohio, from which some of the data were obtained for the examples given in following paragraphs. The Ohio research was a large experiment at a single location. The resources apparently have not been available to conduct the needed field experimental work under the many conditions where liming is a profitable practice.

Examples of the variability of pH optima are given in Fig. 8-15. Fig. 8-15A shows yields of alfalfa at different soil pH values in field experiments in Virginia and Ohio. The data from the Virginia experiment indicate that the optimum pH was about 6.0 or above. Dolomitic limestone and calcitic limestone produced approximately equal yields at equal pH values. The data from the Ohio experiment indicate an optimum pH of about 6.8 where phosphate fertilizer was used and an optimum pH of 7.5 or above where no phosphate was used. To judge from the shapes of the response curves, liming increased the availability of soil phosphorus.

Fig. 8-15B shows yields of corn in rotation with alfalfa and timothy in the Ohio experiment. Here the indications are that the optimum soil pH was 6.8 where corn was in a 3-year rotation with timothy and 7.5 or above where it was in a 3-year rotation with alfalfa. No fertilizer nitrogen was used in this experiment. Because alfalfa is a nitrogen-fixing legume and timothy is a grass, after which corn generally is deficient in nitrogen, the results in Fig. 8-15B suggest that the higher optimum pH for corn in rotation with alfalfa was mostly a response to the extra nitrogen left in the soil by the alfalfa. The alfalfa yielded 30% more at pH 7.5 than at pH 6.8. If an adequate supply of fertilizer nitrogen had been added, the optimum pH for corn probably would have been relatively broad and different from both of those shown.

Fig. 8-15C illustrates the differences in response to soil pH values among three wheat cultivars grown in a field experiment in Brazil. The cultivars had been selected on the basis of their differential tolerance to soil acidity.

A striking broadening or shift in the optimum pH has been observed when legumes that are relatively insensitive to soil acidity are fertilized with molybdenum on soils deficient in molybdenum. Fig. 8-15D shows that the yield of soybean in a field experiment in Georgia was about the same over the pH range from 5.6 to 6.4 where the crop was fertilized with 299 grams of molybdenum as molybdic acid per hectare, but that where no molybdenum was added the yield rose through the same pH range as a result of additions of calcium hydroxide or dolomite. In this instance, the principal effect of liming appears to have been that of increasing the molybdenum availability, and fertilization with molybdenum would be an economical alternative to liming. Edmeades (1981) reviewed investigations in New Zealand in which the need for liming was greatly reduced by additions of as little as 70 grams of molybdenum per hectare. In this situation, molybdenum fertilization presented an especially welcome alternative because for some of the hilly terrain used as permanent pastureland, limestone could not be spread by ground equipment and had to be applied by air.

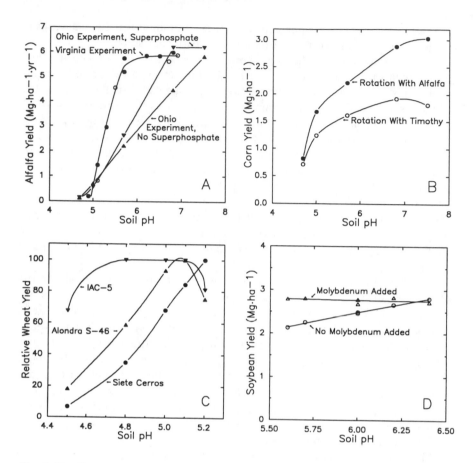

Fig. 8-15. Response of crops to different soil pH values. (pH measurements in the Virginia experiment in *A* were measured in 1:1 soil:water suspensions. pH measurements in the other experiments apparently were made in soil:water suspensions.) A. Annual yields of alfalfa in two experiments. In the Virginia experiment, the experimental points represent pH values established by different additions of dolomitic limestone (solid points) and calcitic limestones (open points). The yields are average values for a 3-year stand. (Moschler et al., 1960) In the Ohio experiment, the yields are average values for 1-year hay crops in various 3-year rotations of corn, small grain, and alfalfa. The pH value of 4.7 was produced by additions of sulfur and aluminum sulfate. The pH values of 5.7, 6.8, and 7.5 were produced by addition of finely ground high-calcium limestone. (Ohio Agricultural Experiment Station, 1938) B. Yields of corn grain in 3-year rotations of (a) corn, small grain, and alfalfa and (b) corn, small grain, and timothy. The pH values were established as for alfalfa in *A*. (Ohio Agricultural Experiment Station, 1938) C. Relative yields of three wheat cultivars on an acid soil with a high degree of aluminum saturation of the effective cation exchange capacity in Brazil. The different pH values were produced by additions of dolomitic limestone. (Bataglia et al., 1985) D. Yield of soybean grain on a loam soil in Georgia with various quantities of calcium hydroxide alone and with a mean application of 299 grams of molybdenum as molybdic acid per hectare. One pair of points at pH 6.0 was obtained with an application of dolomite. (Parker and Harris, 1962)

The experiments in Fig. 8-15 employed the technique of adding increasing quantities of limestone at a given location to find the quantity corresponding to the maximum crop yield. A suitably designed experiment of this kind would provide information from which inferences could be made about the range of quantities of limestone within which the maximum yield or other selected biological response would be expected to lie with specified probability. The same information could be expressed in terms of a range of pH values or other factors related to crop yield. The value for the range found in this way would be a function of the experimental error in the experiment of concern. Additional experiments would be needed to find the range for the particular crop on the particular soils in question. Each experiment would yield one value for the estimated optimum.

The response curves cited in Fig. 8-15 of course were affected by the usual experimental errors. Nonetheless, trends are evident. In the experiment in Fig. 8-15C there was considerable similarity in the curves in the 3 years of data given in the original paper. Similarly, Singh et al. (1987) found considerable similarity among years in the corn yield versus soil pH curves in experiments at each of three locations in Nigeria.

Burmester et al. (1981, 1988) used a different approach. They conducted field experiments on soybean at many locations over a period of years in northern Alabama in which the treatments were simply the control and limestone. When the yields of the control plots as a percentage of the yields of the limed plots in the individual experiments are plotted against the pH values of the control plots (measured in a 1:1 soil:water suspension), the critical pH value as found by the method of Cate and Nelson (1965) is between 5.4 and 5.5.

Mahler and McDole (1987) conducted field experiments in northern Idaho to determine the effects of different quantities of calcium carbonate and sulfur on the yields of various crops. Pooling the relative yield data from the experiments by crops and cultivars, they used the linear response and plateau model (see Section 1-5.3) to represent the relative yield data as a function of soil pH. The soil pH values at which the two lines intersected were considered to be the critical values.

Edmeades et al. (1984, 1985a) calculated the regression of percentage response to a given application of limestone on the pH of the soil on the control plots in various experiments as an alternative way of estimating the critical soil pH above which crop response to liming is not expected. They used this way of analyzing the results of their experiments on liming permanent pastureland in New Zealand. See Section 8-3.6.

The techniques used in analyzing the experimental data obtained by Burmester et al. (1981, 1988), Mahler and McDole (1987), and Edmeades et al. (1984, 1985a) all involve collating the results of many experiments. The critical pH values obtained thus have broader significance than a critical pH value found in a single experiment.

Bandopadhyay and Goswami (1982) used a still different approach. In a greenhouse experiment with three crops on four acid soils of India, they added six to nine different quantities of powdered calcium carbonate corresponding to fractions or multiples of the chemical lime requirements as found by various methods. This approach served the dual purpose of providing a response curve having an estimated optimum soil pH with a comparison of the methods. Where methods are to be compared, the same purpose would be served by obtaining a response curve in the regular way and then finding by interpolation the pH values corresponding to the quantities of calcium carbonate inferred from the various lime requirement methods.

8-4.3. Aluminum as a Criterion

In theory, numerical values of measurements of any of the plant growth factors affected by liming could be plotted against the quantity of calcium carbonate added to determine the quantity required to produce the desired response. Because of the great importance of aluminum toxicity in strongly acid soils, aluminum is logically the first specific plant growth factor that comes to mind as a possible alternative to soil pH as a criterion for the biological lime requirement.

The possible value of aluminum as a criterion for the biological lime requirement is suggested by different kinds of evidence. One is the existence of a marked difference among crops in sensitivity to aluminum at a given pH value. For example, Hetherington et al. (1988) found that in culture solutions maintained at pH 4.2, the relative sums of the activities of monomeric species of aluminum in micromoles per liter associated with a 10% reduction in relative root length were 1.0 for navybean, 1.8 for corn, 2.0 for soybean, and 16 for sugarcane (mean of three cultivars).

A second form of evidence suggesting the value of aluminum as a criterion for the biological lime requirement is the existence of response curves indicating maximum yields of some crops at low pH values that seem to be related to the chemical behavior of aluminum in soils (Evans and Kamprath, 1970; Reeve and Sumner, 1970; Mehlich, 1976; Farina et al., 1980b). See also Shuman (1990) for evidence on the organic matter effect.

The importance of aluminum toxicity was behind Kamprath's (1970, 1984) proposal to use elimination of aluminum toxicity instead of pH as a criterion for determining when the biological lime requirement has been satisfied. For highly weathered and leached soils, he recommended adding a quantity of calcium carbonate chemically equivalent to 1.5 times the content of exchangeable aluminum for general crops or 2 times the content of exchangeable aluminum for very sensitive crops. According to Kamprath (1984), these additions will reduce the aluminum saturation of the exchange capacity to practically zero and will supply adequate amounts of calcium or calcium plus magnesium. (The excess of calcium carbonate over the exchangeable aluminum was needed because some originally nonexchangeable aluminum appears in exchangeable form when the exchangeable aluminum is removed.) Mehlich (1976) proposed a buffer method to obtain the exchange acidity (which is essentially equal to the exchangeable aluminum), and he used this method to estimate the lime requirement of soils of North Carolina.

Fig. 8-16 illustrates the relation between aluminum behavior and the biological lime requirement in one soil. This figure shows the results of an experiment on application of finely ground dolomitic limestone (100% through a sieve with 0.15 millimeter openings) to a strongly acid but not highly weathered soil in India. As indicated by the thickening and poor branching of roots of the test crop of wheat in the unlimed soil as compared with the roots in the limed soil, the crop was suffering from aluminum toxicity. In this experiment, applying a quantity of limestone equivalent to the exchangeable aluminum neutralized 94% of the exchangeable aluminum. The maximum observed yield at pH 5.4 (measured in a 1:2 soil:0.01 molar calcium chloride suspension) was obtained by liming with twice the content of exchangeable aluminum. The authors suggested that the relatively flat curve for aluminum concentration in the plants resulted from an increase in root growth with increasing pH and uptake of aluminum from the subsoil, which was not limed.

The evidence now available indicates that the relationship between exchangeable aluminum and the amount of calcium carbonate required to raise the soil pH in water to 5.5 to precipitate the aluminum or to reduce the extractable aluminum to zero may differ considerably among soils. Researchers in Australia

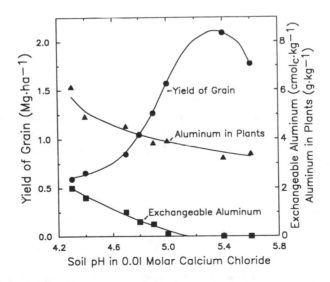

Fig. 8-16. Yield of wheat grain, aluminum content of the plants, and exchangeable aluminum in the soil at different pH values (measured in 1:2 soil:0.01 molar calcium chloride suspension) as affected by application of different quantities of finely ground dolomitic limestone to a strongly acid soil in a field experiment in India. From left to right, the data points were obtained with applications of 0, 0.4, 0.8, 1.2, 1.6, 2.0, 4.0, and 6.0 megagrams of limestone per hectare. (Patiram et al., 1990)

(Gillman and Sumpter, 1986; Aitken et al., 1990) have found that the relationship on the basis of chemical equivalence is approximately 1:1 in some soils, but that in others it may take considerably more calcium carbonate than even the factor of 2 used by Kamprath for very sensitive crops.

The percentage of the effective cation exchange capacity occupied by aluminum is superior to the exchangeable aluminum as a criterion of aluminum toxicity. As a criterion for the biological lime requirement, however, the two measurements should be of equal value if the biological lime requirement is defined as the quantity of calcium carbonate that corresponds to zero exchangeable aluminum or zero percentage of the effective cation exchange capacity occupied by aluminum.

8-4.4. pH Versus Aluminum

Experimentally, an important problem in deciding whether pH or some measure of soil aluminum is the better criterion for determining when the biological lime requirement has been met is that within individual soils the various forms and ionic species of aluminum are correlated with the pH values. Findings published by Abruña et al. (1983) illustrate the kind of relationships that may be expected within a soil. They added different quantities of limestone to a clay subsoil in a field experiment in Puerto Rico and found that correlations of sorghum yield with the various independent variables they measured were $r = 0.89, 0.88, 0.86, 0.87, 0.56$, and 0.90 with soil pH (measured in a 1:5 soil:water suspension), percentage aluminum saturation of the effective cation exchange capacity, ratio of exchangeable aluminum to exchangeable bases, exchangeable calcium plus magnesium, calcium percentage in the leaves, and phosphorus percentage in the leaves. All the correlations given were significant at the 1% level, and all except for the calcium percentage in the leaves were about the same.

Similarly, Friesen et al. (1982) found in a 5-year field experiment in Nigeria that as criteria for estimating the relative yield of corn on a strongly weathered soil that had received different quantities of calcium hydroxide at the beginning of the experiment, there was little to choose between the pH measured in a 1:1 soil:water suspension and the percentage saturation of the effective cation exchange capacity with aluminum. The proportion of the variance of relative yields accounted for by the two criteria was almost the same. In this case, however, the pH would have been the more illuminating measurement because the relative yield decreased at pH values above about 5.5, where the highest yield was observed. The aluminum data did not show this effect because the exchangeable aluminum was essentially zero throughout the pH range in which the decrease in yield occurred.

The correlations among aluminum species and pH still exist in a group of soils, but they are lower, and as a consequence it may be possible to find a distinction between the usefulness of the different criteria. To verify satisfactorily for a given crop or cultivar that aluminum is a better criterion for the biological lime requirement than pH or vice versa, the test crop needs to be grown on a variety of acid soils, each of which has been limed with a wide enough range of quantities of calcium carbonate to include the quantity that will correspond to the biological response selected to tell when the biological lime requirement has been satisfied. For each soil, the quantity of calcium carbonate associated with the selected biological response is found by interpolation in the response curves obtained when the responses are plotted against the quantities of calcium carbonate.

Reeve and Sumner (1970) used the approach just described in experimental work with samples of eight strongly weathered soils of South Africa with exchangeable aluminum contents ranging from 0 to 2.7 centimoles of charge per kilogram and pH values (measured in 0.002 molar calcium chloride) ranging from 4.4 to 5.0. Their data show that the correlations between the biological lime requirement (the quantity of calcium carbonate corresponding to the maximum yield of trudan) and the soil measurements were $r = 0.99$ with the quantity of calcium carbonate required to reduce the exchangeable aluminum to 0.2 centimol of charge per kilogram of soil and $r = 0.61$ with the quantities of calcium carbonate required to increase the soil pH to 6.0 or 6.5 according to the buffer method of Shoemaker et al. (1961). Thus, the test indicated that aluminum was the better criterion for the biological lime requirement.

In experimental work with samples of 154 soils from Portugal, Coutinho (1990) defined the biological lime requirement as the quantity of calcium carbonate associated with 90% of the maximum growth of the roots of seedlings of a wheat cultivar that was relatively sensitive to aluminum. His statistical calculations indicated that the biological lime requirements of the soils were related more closely to the quantities of calcium carbonate required to raise the soil pH to 5.5 (measured in a 1:5 soil:0.01 molar calcium chloride suspension) ($r^2 = 0.90$) than to the exchangeable aluminum ($r^2 = 0.77$). Thus, in this instance, the soil pH would be the preferred criterion for the biological lime requirement. A pH value of 5.5 in 0.01 molar calcium chloride would be equivalent to a pH value of about 6 in water, which is high enough to eliminate aluminum toxicity.[2]

Applying the general approach just described is complicated by the fact that various forms and ionic activities of aluminum will be associated with 90% of the maximum root growth or other biological measurement in each soil and by the fact that the numerical values obtained for soil pH depend upon the procedure employed in measurement. A selection must be made. Ideally, the methods selected should be the ones that yield the highest correlations with the biological

---

[2]Although Coutinho's (1990) data showed that the percentage saturation of the effective cation exchange capacity of the unlimed soil samples with aluminum was more highly correlated with relative root growth than was the pH, these measurements are not biological lime requirements. The same may be said about the findings by Farina et al. (1980b) in a greenhouse experiment and by Farina and Channon (1991) in field experiments in Natal, in which relative yields were employed as the biological response. They did not give data for the biological lime requirements.

lime requirement. Measurements on the soil solution probably will be most suitable for both aluminum and pH.

Further complications in experimentation and interpretation are presented by the criterion selected to determine when the biological lime requirement has been met, by the behavior of the various nutrients whose availabilities vary with the quantity of calcium carbonate added, and by the interaction of mycorrhizas with aluminum. In most instances, the relative root growth in the early seedling stages employed is probably indifferent to differences in supplies of nutrients other than calcium, magnesium, and perhaps phosphorus because of the supplies in the seed. If the criterion for the desired biological condition in soils were the final yield of the plants, the biological lime requirement would be affected by the changes in availability of nutrients other than calcium or calcium plus magnesium that have been produced by liming. Making a uniform application of these nutrients to all plots or all cultures would tend to eliminate this aspect of the biological lime requirement and would result in values that are more similar to those derived from measurements of relative growth of the roots of seedlings.

As regards the mycorrhizal factor, Cumming and Weinstein (1990) found that when pitch pine seedlings were inoculated with mycorrhizal fungi, they showed very little sensitivity to concentrations of aluminum that severely inhibited the nonmycorrhizal controls. The importance of this effect in other species has not been determined, but because effective inoculation requires more time than that required for the seedling root-growth test, different responses could be expected from older plants than from seedling root growth in species that behave like pitch pine.

8-4.5. Aluminum Tolerance

The importance of aluminum as an inhibitor of plant growth on strongly acid soils receives attention from another perspective, that of identifying and breeding cultivars that are tolerant to aluminum toxicity (Foy, 1988; Scott and Fisher, 1989). Cultivars of a given crop may differ considerably in tolerance to aluminum. For example, Hetherington et al. (1986) found that the sum of the micromoles of monomeric species of aluminum per liter of solution associated with a 10% reduction in root length ranged from 5 to 26 in a group of nine sugarcane cultivars.

Improved tolerance to aluminum toxicity and other unfavorable factors of soil acidity may provide an economic substitute for liming, especially if the other crops in the rotation are sufficiently tolerant and if the cost of liming is great enough. Tolerance to aluminum, however, is not a biological substitute for liming. Aluminum-tolerant cultivars usually respond to liming if grown on strongly acid soils because the tolerance is not complete, and liming has favorable effects on plant growth factors other than aluminum. Programs aiming to breed aluminum-tolerant cultivars of important crops are active at several locations throughout the world. See Section 7 in a symposium edited by Wright et al.

(1991) for a group of papers dealing with the genetics and breeding of acid-tolerant plants. Foy (1983) discussed the polemics related to the plant breeding issue.

8-4.6. Integrated Factors as Criteria

Aluminum toxicity and calcium deficiency, the most important unfavorable biological factors in acid soils for most crop plants, interact to an important degree. High calcium alleviates aluminum toxicity, and high aluminum aggravates calcium deficiency (Noble et al., 1988; Wolfe and Joslin, 1989; Alva and Edwards, 1990). The same appears to be true for aluminum and magnesium (Grimme, 1982; Edmeades et al., 1991; Keltjens and Dijkstra, 1991). Addition of calcium as the sulfate introduces another complication in the form of chemical interaction between aluminum and sulfate to form a relatively nontoxic complex (Noble et al., 1988).

Konishi (1990) reported that tea plants, which grow best in strongly acid soils, responded in reverse fashion. Tea plants were benefited by adding aluminum (11 milligrams per liter) to culture solutions maintained at pH 4.5. The benefit was greatest (an eightfold increase in yield) where the concentration of calcium was lowest, and it decreased to an approximate doubling of the yield with added calcium.

Calcium and pH also interact. Arnon and Johnson (1942) observed that the concentration of calcium required to obtain the maximum yield of lettuce in solution cultures increased as the pH was lowered from 6 to 5 and 4. Ewing and Robson (1990) found that the growth of three species of *Medicago* in culture solutions at pH 5.5 and 6.5 was unaffected by the calcium concentration, but that nodulation was poorer at pH 5.5 than at 6.5. At both pH values, nodulation increased with the calcium concentration except for one species that formed no nodules at pH 5.5 at any calcium concentration tested.

The work on the interaction of aluminum, calcium, sulfate, magnesium, and pH is a first step toward integrating all the toxic and nutrient availability factors involved in crop response to liming into a comprehensive criterion for determining the biological lime requirement. Because of the complexity of the subject and the currently limited capabilities for evaluating the critical soil factors, a long time probably will be required to make much progress toward a useful criterion for biological lime requirement based upon an integration of the various soil factors.

8-4.7. Profile Lime Requirement

The chemical and biological lime requirements of soils generally are researched and expressed in terms of the plowed layer or, for permanent pastures, a surface layer of lesser arbitrary thickness. These conventions seem adequate if the subsoil is no more than mildly acid.

Where subsoils are moderately or strongly acid, the acid condition may limit the growth of sensitive crops even if the surface layer is adequately limed. The fact that the biological lime requirement of the profile cannot be determined readily because of the difficulty of liming the subsoil usually means that the adequacy of surface soil liming for meeting the biological lime requirement of the soil profile can only be surmised.

8-4.7.1. Effects of Subsoil Acidity

The tops of plants are less sensitive to subsoil acidity than are the roots because the roots normally are concentrated in the surface soil, whether or not the subsoil is acid. Inhibition of root growth in acid subsoils is of secondary significance from the standpoint of nutrient uptake because moderately or strongly acid sub-soils generally are relatively poor sources of several nutrients, and the surface soil generally will supply portions of the nutrients that otherwise might have been absorbed from the subsoil. The issue is more critical with water. The capacity of surface soils to hold water to carry plants during drought periods is limited. Consequently, acidity that inhibits root growth in the subsoil can have an important indirect effect on plant growth by decreasing water availability.

In an experiment on a strongly acid sandy clay loam soil in Georgia, Sumner et al. (1985) found that alfalfa removed 4.3 centimeters of water from a subsoil (15 to 107 centimeter depth) during a 9-day period without rain when the subsoil had been mixed and limed to about pH 6.5 measured in water. Water removal was 1.5 centimeters from a comparable mixed but unlimed subsoil and was 2.7 centimeters from an untreated control. The poor crop growth generally observed where subsoil has been mixed but not limed has been attributed to aluminum brought up from the lower portion of the subsoil in the mixing process.

In Alabama, Doss and Lund (1975) acidified a soil to different degrees by use of ammonium sulfate as a nitrogen fertilizer. Then they limed the surface 15 centimeters to pH 6.0 to 6.5 for growth of cotton. Cotton yields and use of subsoil water were both decreased by subsoil acidification. Supplying extra water by irrigation overcame the detrimental effect of subsoil acidity on cotton yields, but it did not increase the growth of roots in the subsoil.

If plant growth tests are made on strongly acid subsoils independently of the surface soils, aluminum toxicity and calcium deficiency often may be identified as causes of poor root growth for sensitive crops (Adams and Moore, 1983; Bruce et al., 1988). Where plants are growing under natural conditions in the field, however, these nutritional problems usually are masked by surface soil effects and are not evident from visual or chemical examination of the above-ground portions of the plants.

In Western Australia, Carr et al. (1991) found that subsoil pH showed little relation to the yield of wheat. Their best test for the impact of subsoil acidity was the ratio of the concentration of aluminum to the concentration of sodium in a 0.005 molar potassium chloride extract of the 15 to 25 centimeter layer of soil. This test accounted for 52% of the variance in wheat grain yield in 53 field experiments conducted within a period of 3 years.

8-4.7.2. Crop Differences

Pinkerton and Simpson (1986) investigated the profile lime requirement for different crops in a greenhouse experiment on a soil from New South Wales that had been acidified by 55 years of continuous subterranean clover production. The soil was sampled by 10 centimeter layers and was reconstituted for the plant growth experiment (a) without liming, (b) with addition of calcium carbonate to the surface 10 centimeters to produce pH 6.5 (measured in a 1:5 soil:water suspension), and (c) with addition of calcium carbonate to the surface 10 centimeters to pH 6.5 as in (b) and with additions to the individual 10 centimeter layers from 10 to 80 centimeters to produce pH 6.0. The results in Fig. 8-17 show that liming the surface 10 centimeters of soil adequately improved the soil profile for growth of canarygrass, but not for alfalfa, wheat, and rape. The latter three species benefited from liming the subsoil in addition to the surface soil.

In field work in Alabama on differences among species in response to subsoil acidity, F. Adams et al. (1967) found that yields of cotton were reduced by about a third when subsoil acidity restricted root growth mostly to the plowed layer, in which the pH values were in the range of 6.0 to 6.5. Although peanut is sensitive to calcium deficiency in the surface few centimeters of soil where the nuts are produced, Adams and Pearson (1970) found that neither the yield of

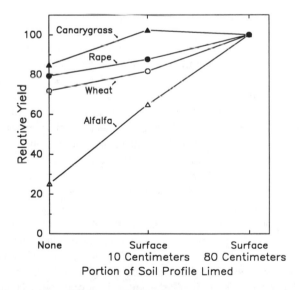

Fig. 8-17. Relative yields of four crops in a greenhouse experiment on an acid soil when different portions of the soil profile were limed as indicated. The surface 10 centimeter layer of soil was limed to pH 6.5. The 10 to 80 centimeter portion was limed to pH 6.0 by individual 10 centimeter layers, which were placed in the same relative locations as in the field. The soil profile had been acidified by 55 years of subterranean clover production without liming in New South Wales. (Pinkerton and Simpson, 1986)

nuts nor the root growth in the acid subsoil was adversely affected by the subsoil acidity at one of the same sites used by F. Adams et al. (1967) in their field experiments on cotton. Adams and Pearson quoted unpublished evidence indicating that peanut is far more tolerant to aluminum than is cotton.

8-4.7.3. Amelioration of Subsoil Acidity

Over a period of years, normal treatment of acid surface soils with limestone and fertilizers improves underlying acid subsoils to some degree as an environment for root growth (Hardy et al., 1990). Once a problem is recognized, however, operators generally prefer some overt action to improve the situation quickly if a feasible remedy is available.

For experimental purposes, soil profiles can be excavated, mixed with limestone by individual layers, and reconstituted for plant growth tests. Unlimed plots can be similarly excavated, mixed by individual layers, and reconstituted to provide appropriate controls. More practical experiments can be conducted by using deep tillage equipment that will mix limestone with soil profiles to a depth of a meter or more. Although practical in the sense that the conditions can be produced with the aid of commercial equipment, the cost of the operations may be greater than can be justified by the results. Moreover, the mixing process brings to the surface some subsoil material that generally is unfavorable to plant growth.

Other equipment has been developed to inject limestone into soils below the plowed layer without mixing all the soil. For example, Anderson and Hendrick (1983) described equipment for injecting a slurry of crushed limestone in narrow slits to a depth of 18 to 36 centimeters, and Kirchoff et al. (1991) described equipment for mixing liming materials with soil in slots 15 centimeters wide and up to 1 meter deep.

In an investigation of the effect of the uneven distribution of limestone in subsoil layers that would result from the usual mechanical applications, Pearson et al. (1973) measured the root and top growth of cotton on a simulated soil profile consisting of a 20-centimeter layer of a fertilized and limed sandy loam soil at the surface with a 173-centimeter layer of a uniformly mixed acid B horizon beneath. They limed the subsoil to various degrees and in layers with alternating layers of unlimed subsoil. Fig. 8-18 summarizes part of their findings.

A significant point about Fig. 8-18 is that the plants did not seem to respond to the differences in distribution of the limed zones. Rather, the response was related to the average pH of the subsoil. In Fig. 8-18, the pH value shown on the X axis is the average value for the subsoil as a whole, whether the calcium carbonate had been mixed with all the subsoil or only with layers that were separated by layers of unlimed subsoil. In the treatments represented by pH 4.9 and 5.5, the subsoil had been limed in layers. In the treatments represented by pH 5.1 and 6.0, the soil had been limed uniformly. pH 4.6 was the unlimed control. Supporting the absence of a significant layering effect in Fig. 8-18, Pearson et al. (1973) observed no tendency for the roots of cotton plants to

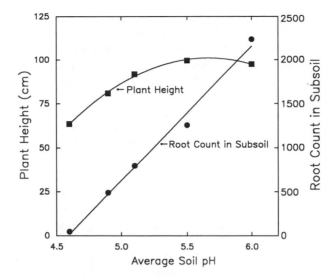

Fig. 8-18. Response of cotton tops and roots to subsoil pH on a disturbed soil profile. The upper 20 centimeters was a fertilized and limed sandy loam. The subsoil layer of 173 centimeters had been thoroughly mixed to promote uniformity and was limed uniformly or in layers, as described in the text. (Pearson et al., 1973)

follow the zone of neutralized soil when a limed layer was placed at an angle with the horizontal. Thus, the geotropic tendency of the roots seemed to be stronger than any change in chemotropic tendency that may have resulted from liming. For crops with roots that behave in this way, applying limestone to acid subsoils in vertical slots probably would be relatively ineffective for improving the growth of roots unless the slots were close together. There would be little lateral movement of the neutralizing effect or the roots.

In contrast to the findings by Pearson et al. (1973), Kauffman and Gardner (1978) observed a marked chemotropic effect of subsoil liming on the growth of wheat roots. Their paper included photographs showing that where vertical segments of an acid soil (pH 4.30 in a 1:2 soil:water suspension) were limed to pH 5.2 or above with a mixture of calcium and magnesium carbonates, almost all the roots were in the limed segments. Little root growth occurred in the intervening unlimed segments, and the roots in these segments showed aluminum toxicity symptoms. Maximum yields were obtained where only 30% of the soil volume was adequately limed. Selective growth of roots in vertical segments of limed soil in the field was observed by Anderson and Hendrick (1983) with alfalfa, barley, and wheat and by Kirchoff et al. (1991) for grape. The evidence thus indicates that mechanical incorporation of limestone in segments of acid subsoil can be useful in improving root penetration.

Making use of the knowledge that had been developed on the residual acidity and basicity of fertilizers, Pearson and Abruña (1961) found that it was possible to speed the neutralization of acid subsoil by surface-applied limestone if the limestone was applied with a nitrogen fertilizer that would produce nitric acid upon nitrification. The calcium nitrate produced by reaction of the nitric acid with the limestone would leach down into the subsoil, where preferential absorption of nitrate by the plant roots would leave a basic residue to react with some of the acidity.

Fertilizer nitrogen applied as calcium nitrate also would neutralize the subsoil. In this case, the calcium would be derived from the fertilizer instead of the limestone. Nitrate in excess of plant requirements, added either directly or indirectly, through acid-forming nitrogen fertilizers that would form nitric acid and would react with limestone in the surface soil to form calcium nitrate, would not have the neutralizing effect resulting from preferential absorption of nitrate by plants. If leached into the subsoil, however, the calcium nitrate would displace a little of the exchangeable aluminum and, after leaching, could raise the subsoil pH to a small degree.

Subsoil acidity normally develops slowly, but in intensive cropping systems involving heavy applications of acid-forming fertilizer nitrogen it can develop in only a few years. In such systems, subsoil acidification can be reduced or prevented by adding limestone to the surface soil and taking advantage of the Pearson and Abruña (1961) principle to carry the neutralizing effect to the subsoil. Fig. 8-19 shows some of the results of an experiment in Georgia in which the

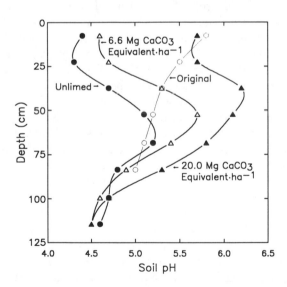

Fig. 8-19. pH values of a profile of a sandy loam soil in Georgia showing the changes during 6.4 years under Coastal bermudagrass receiving an average annual application of 801 kilograms of fertilizer nitrogen as ammonium nitrate per hectare with and without an initial application of two quantities of calcitic limestone (91% through a sieve with 2 millimeter openings and 33% through a sieve with 0.18 millimeter openings) to the plowed layer. (W. E. Adams et al., 1967)

average annual application of nitrogen as ammonium nitrate was 801 kilograms per hectare. The final pH values indicate that applications of limestone supplying either 6.6 or 20.0 megagrams of calcium carbonate equivalent per hectare to the surface soil had prevented acidification of the subsoil, but that only the heavier application had prevented acidification of the surface soil. The remarkable downward movement of the neutralizing effect in the period of only 6.4 years provides an indication of the value of the Pearson and Abruña principle in practice. Other data not reproduced in the figure showed that without liming the amount of exchangeable calcium in the soil was reduced by 90% in the surface 30 centimeters and 26% in the surface 91 centimeters.

Adams and Pearson (1969) carried the concept of indirect subsoil neutralization one step further by using calcium gluconate as a source of calcium. Like calcium nitrate, this substance would leach downward in the soil, but plant action would not be required to develop the neutralizing effect. Microbial decomposition of the gluconate would make the full amount of calcium effective in neutralization. In a field experiment on an acid loamy sand soil, they found that calcium gluconate was far more effective in raising the soil pH and exchangeable bases to a depth of 45 centimeters than was dolomitic limestone applied at three times the rate of the calcium gluconate.

Calcium acetate would have an effect similar to that of calcium gluconate. Both substances are too expensive for practical use, but they could be valuable for research purposes. The use of calcium gluconate or calcium acetate for liming selected layers in undisturbed soil profiles by the technique used by Ogus and Fox (1970) for injecting fertilizer nitrogen (See Section 7-6) is a possibility for use on very small plots.

Another substance that is less effective than calcium gluconate or calcium acetate, but which has more practical potential, is calcium sulfate or gypsum. Calcium sulfate is far less soluble than calcium gluconate or calcium acetate, but it is more soluble than calcium carbonate under neutral conditions, and it is also relatively economical. In some areas, large quantities of unwanted calcium sulfate have accumulated as a byproduct of the addition of sulfuric acid to phosphate rock to produce phosphoric acid. The byproduct, called phosphogypsum, contains a little phosphorus along with some calcium fluoride. The potentialities of calcium sulfate for ameliorating acid subsoils were not appreciated for many years. Recently, however, a number of papers have been published in which the value of surface applications of calcium sulfate has been investigated.

The general principles of the process are illustrated in Fig. 8-20, taken from an early experiment by Reeve and Sumner (1972). The graphs on the left show that calcium hydroxide mixed with a sample of surface soil eliminated the exchangeable aluminum and increased the exchangeable calcium and the effective cation exchange capacity. Some calcium leached from the surface soil accumulated in the subsoil as exchangeable calcium, but the effective cation exchange capacity was almost unaffected, indicating that the calcium had been leached mostly as a neutral, unbuffered salt, most likely calcium nitrate. (Neutralization

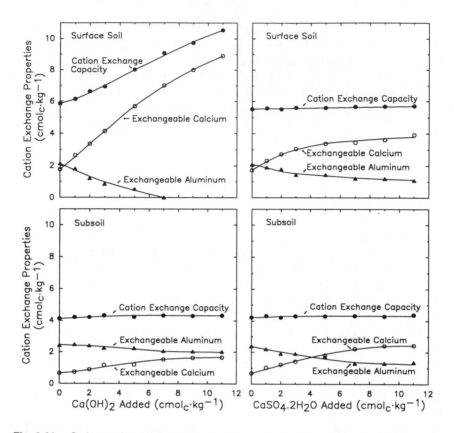

Fig. 8-20. Cation exchange properties of surface and subsoil samples of a strongly weathered soil from Natal with a high content of free oxides after addition of calcium hydroxide (left) and calcium sulfate (right) and leaching with water in a laboratory experiment. The calcium hydroxide and calcium sulfate were added only to the samples of the surface 15 centimeters of soil and were mixed with these samples. The leachates from the surface samples passed through the underlying samples of subsoil from the 15 to 30 centimeter layer. (Reeve and Sumner, 1972)

of acid soil by calcium hydroxide does not yield a soluble anion that can accompany the calcium ion in downward movement.) The exchangeable aluminum in the subsoil was reduced to only a small degree.

The graphs on the right show that calcium sulfate increased the exchangeable calcium in the surface soil, but had no significant effect on the effective cation exchange capacity. Exchangeable aluminum was decreased, but not eliminated, as it was where calcium hydroxide was used. In the subsoil, the effective cation exchange capacity was essentially unaffected. Exchangeable calcium was increased to a greater extent and exchangeable aluminum was decreased to a greater extent than where calcium hydroxide was added. Downward movement of calcium was enhanced by the presence of sulfate as a counter ion, and some of the exchangeable aluminum apparently was replaced by the calcium and leached from the sample of subsoil in the leaching column.

Calcium hydroxide alleviates aluminum toxicity by raising the soil pH, which causes aluminum to be precipitated from solution and from the exchangeable form. Additions of both calcium chloride and calcium sulfate increase the ionic strength of the solution, and this tends to counteract the toxicity of the aluminum that appears in solution by decreasing the activity coefficient of the aluminum. Calcium chloride decreases the soil pH value, but calcium sulfate may slightly decrease it in some soils and slightly increase it in others. The increase has been attributed to replacement of hydroxyl when sulfate is adsorbed by hydrous oxides of iron or aluminum.

Not appreciated initially was the fact that sulfate forms a complex ion with aluminum ($AlSO_4^+$) that appears to have little or no toxicity to plants (Tanaka et al., 1987; Kinraide and Parker, 1987). Pavan et al. (1982) found that where calcium sulfate was added to two soils in amounts equivalent to the aluminum removed by molar potassium chloride solution, about 40% of the aluminum in the soil solution was present as the sulfate complex. The stability of this ion would promote the exchange of calcium added as calcium sulfate for the aluminum in acid subsoils. Where the source of calcium sulfate is phosphogypsum, additional factors that would alleviate aluminum toxicity are the aluminum fluoride and aluminum phosphate complexes that would result from the fluoride and phosphate components. These complexes are more stable than those of sulfate with aluminum (Alva and Sumner, 1989).

Another facet of the calcium sulfate behavior is that the aluminum replaced from the exchangeable form may exceed the aluminum that appears in the leachate. O'Brien and Sumner (1988) and Courchesne and Hendershot (1990) suggested that the aluminum retained by the soil in nonexchangeable form might be alunite, a sparingly soluble potassium aluminum hydroxy sulfate $Al_3(OH)_6(SO_4)_2$.

In a greenhouse experiment with a strongly weathered acid clay soil from Brazil, Chaves et al. (1988) found that a surface application of calcium sulfate was far more effective than a surface application of calcium carbonate in increasing the calcium concentration in the drainage water and in increasing the root growth and yield of coffee seedlings. Many more roots developed in the 30 to 60 centimeter depth of soil receiving a surface application of calcium sulfate than in soil in which calcium carbonate was mixed with the surface 30 centimeters.

In a field experiment in Georgia, Sumner and Carter (1988) found that adding calcium sulfate to the topsoil improved an acid subsoil enough so that alfalfa roots grew into it far better than before. The calcium sulfate treatment was less effective than liming the soil profile, but it was much more economical because the soil did not have to be disturbed to put the soluble calcium in contact with the subsoil. See Bouton et al. (1986) for data on the changes in exchangeable plus soluble calcium in the soil profile during 3 years following a surface application of gypsum.

Van der Watt et al. (1991) found that a soluble form of calcium complexed with an organic fraction derived from coal was far more effective in raising subsoil pH than calcium ethylenediaminetetraacetate and was several times more effective than calcium sulfate in transporting calcium to the subsoil in a laboratory leaching experiment. This product would be considerably more economical than calcium acetate or gluconate if transportation costs were equal.

Bouldin (1979) summarized research related to subsoil acidity in Brazil and Puerto Rico. He noted that the limitations imposed by subsoil acidity could be reduced by (a) selecting for initial development the areas in which the subsoil acidity has the smallest detrimental effect, (b) by incorporating limestone as deeply as economically feasible and by waiting until the surface-applied limestone has ameliorated the subsoil enough to permit root growth, and (c) by selecting crop species and cultivars that are most tolerant to the unfavorable conditions.

In some circumstances, the use of tolerant crop species and cultivars may greatly extend the time during which crops may be grown profitably on soils with limiting subsoil acidity. In others, the benefit may be short lived in the absence of appropriate additions of plant nutrients. Bouldin (1979) noted that several hundred million hectares of soils in the tropics are not now used extensively for growing crops because of limitations of soil fertility and acidity, especially subsoil acidity. The supplies of available nutrients are low, and the primary minerals have been depleted by weathering, so that even if a crop or cultivar were found that would yield reasonably well on the native supplies of nutrients, these supplies would be removed in only a few crops.

8-5. Literature Cited

Abruña, F., J. Rodríguez, and S. Silva. 1983. Crop response to soil acidity factors in Ultisols and Oxisols in Puerto Rico. VI. Grain sorghum. *Journal of Agriculture of the University of Puerto Rico* 67:28-38.

Adams, F. (Ed.). 1984. *Soil Acidity and Liming.* Second Edition. Agronomy Number 12. American Society of Agronomy, Crop Science Society of America, and Soil Science Society of America, Madison, WI.

Adams, F., and B. L. Moore. 1983. Chemical factors affecting root growth in subsoil horizons of Coastal Plain soils. *Soil Science Society of America Journal* 47:99-102.

Adams, F., and R. W. Pearson. 1969. Neutralizing soil acidity under bermudagrass sod. *Soil Science Society of America Proceedings* 33:737-742.

Adams, F., and R. W. Pearson. 1970. Differential response of cotton and peanuts to subsoil acidity. *Agronomy Journal* 62:9-12.

Adams, F., R. W. Pearson, and B. D. Doss. 1967. Relative effects of acid subsoils on cotton yields in field experiments and on cotton roots in growth-chamber experiments. *Agronomy Journal* 59:453-456.

Adams, W. E., A. W. White, Jr., and R. N. Dawson. 1967. Influence of lime sources and rates on 'Coastal' bermudagrass production, soil profile reaction, exchangeable Ca and Mg. *Agronomy Journal* 59:147-149.

Aitken, R. L., and P. W. Moody. 1991. Interrelations between soil pH measurements in various electrolytes and soil solution pH in acidic soils. *Australian Journal of Soil Research* 29:483-491.

Aitken, R. L., P. W. Moody, and P. G. McKinley. 1990. Lime requirement of acidic Queensland soils. II. Comparison of laboratory methods for predicting lime requirement. *Australian Journal of Soil Research* 28:703-715.

Allen, E. R., and L. R. Hossner. 1991. Factors affecting the accumulation of surface-applied agricultural limestone in permanent pastures. *Soil Science* 151:240-248.

Alva, A. K., and D. G. Edwards. 1990. Response of lupin cultivars to concentration of calcium and activity of aluminum in dilute nutrient solutions. *Journal of Plant Nutrition* 13:57-76.

Alva, A. K., and M. E. Sumner. 1989. Alleviation of aluminum toxicity to soybeans by phosphogypsum or calcium sulfate in dilute nutrient solutions. *Soil Science* 147:278-285.

Anderson, C. A. 1966. Short term effects of several liming materials on soil pH, calcium, and magnesium in Lakeland fine sand. *Soil and Crop Science Society of Florida, Proceedings* 26:94-105.

Anderson, D. L., and J. G. Hendrick. 1983. Subsoil lime injector. *Soil Science Society of America Journal* 47:337-339.

Anonymous. 1990. Agricultural lime. Chapter 201. In *Code of Iowa 1991*. Legislative Service Bureau, General Assembly of Iowa, Des Moines.

Arnon, D. I., and C. M. Johnson. 1942. Influence of hydrogen ion concentration on the growth of higher plants under controlled conditions. *Plant Physiology* 17:525-539.

Bache, B. W. 1970a. Barium isotope method for measuring cation-exchange capacity of soils and clays. *Journal of the Science of Food and Agriculture* 21:169-171.

Bache, B. W. 1970b. Determination of pH, lime potential, and aluminium hydroxide potential of acid soils. *Journal of Soil Science* 21:28-37.

Bailey, J. S., R. J. Stevens, and D. J. Kilpatrick. 1989a. A rapid method for predicting the lime requirement of acidic temperate soils with widely varying organic matter contents. I. Development of the lime requirement model. *Journal of Soil Science* 40:807-820.

Bailey, J. S., R. J. Stevens, and D. J. Kilpatrick. 1989b. A rapid method for predicting the lime requirement of acidic temperate soils with widely varying organic matter contents. II. Testing the lime requirement model. *Journal of Soil Science* 40:821-829.

Bailey, J. S., R. J. Stevens, and D. J. Kilpatrick. 1989c. A rapid method for predicting the lime requirement of acidic temperate soils with widely varying organic matter contents. III. A lime-quality correction model. *Journal of Soil Science* 40:831-847.

Bandopadhyay, B. K., and N. N. Goswami. 1982. Evaluation of lime requirement methods for some acid soils of India. *Journal of the Indian Society of Soil Science* 30:296-302.

Barber, S. A. 1984. Liming materials and practices. Pp. 171-209. In F. Adams (Ed.), *Soil Acidity and Liming*. Second Edition. Agronomy Number 12. American Society of Agronomy, Crop Science Society of America, and Soil Science Society of America, Madison, WI.

Barrow, N. J. 1985. Reaction of anions and cations with variable-charge soils. *Advances in Agronomy* 38:183-230.

Barrow, N. J., and V. C. Cox. 1990. A quick and simple method for determining the titration curve and estimating the lime requirement of soil. *Australian Journal of Soil Research* 28:685-694.

Bataglia, O. C., C. E. de O. Camargo, O. F. de Oliveira, V. Nagai, and V. J. Ramos. 1985. Resposta à calagem de três cultivares de trigo com tolerância diferencial ao alumínio. *Revista Brasileíra de Ciência do Solo* 9:139-147.

Beers, W. F. J. van. 1962. *Acid Sulphate Soils*. International Institute of Land Reclamation and Improvement, Bulletin 3.

Bell, L. C., and D. G. Edwards. 1986. The role of aluminium in acid soil infertility. Pp. 201-223. In M. Latham (Ed.), *Soil Management Under Humid Conditions in Asia and the Pacific (Asialand)*. IBSRAM, Proceedings No. 5. International Board for Soil Research and Management, Bangkok, Thailand.

Birrell, K. S., and M. Gradwell. 1956. Ion-exchange phenomena in some soils containing amorphous mineral constituents. *Journal of Soil Science* 7:130-147.

Bolton, J. 1977. Changes in soil pH and exchangeable calcium in two liming experiments on contrasting soils over 12 years. *Journal of Agricultural Science (Cambridge)* 89:81-86.

Bouldin, D. R. (Ed.). 1979. *The Influence of Subsoil Acidity on Crop Yield Potential*. Cornell International Agriculture, Bulletin 34. Ithaca, NY.

Bouton, J. H., M. E. Sumner, J. E. Hammel, and H. Shahandeh. 1986. Yield of an alfalfa germplasm selected for acid soil tolerance when grown in soils with modified subsoils. *Crop Science* 26:334-336.

Bradfield, R., and W. B. Allison. 1933. Criteria of base saturation in soils. Pp. 63-79. In A. A. J. de'Sigmond and F. Zucker (Eds.), *Transactions of the Second Commission and Alkali-Subcommission of the International Society of Soil Science*, Volume A. Kjöbenhavn, Danmark.

Bromfield, S. M., R. W. Cumming, D. J. David, and C. H. Williams. 1983. Change in soil pH, manganese and aluminium under subterranean clover pasture. *Australian Journal of Experimental Agriculture and Animal Husbandry* 23:181-191.

Bruce, R. C., L. A. Warrell, D. G. Edwards, and L. C. Bell. 1988. Effects of aluminium and calcium in the soil solution of acid soils on root elongation of *Glycine max* cv. Forrest. *Australian Journal of Agricultural Research* 38:319-338.

Burmester, C. H., J. F. Adams, and J. W. Odom. 1988. Response of soybean to lime and molybdenum on Ultisols in northern Alabama. *Soil Science Society of America Journal* 52:1391-1394.

Burmester, C. H., A. Waggoner, and F. Adams. 1981. *On-Farm Lime and Fertilizer Experiments With Soybeans and Cotton in Northern Alabama, 1977-1980*. Alabama Agricultural Experiment Station, Circular 254. Auburn.

Cadle, R. D. 1955. *Particle Size Determination*. Interscience Publishers, NY.

Carr, S. J., G. S. P. Ritchie, and W. M. Porter. 1991. A soil test for aluminium toxicity in acidic subsoils of yellow earths in Western Australia. *Australian Journal of Agricultural Research* 42:875-892.

Cate, R. B. Jr., and L. A. Nelson. 1965. *A Rapid Method for Correlation of Soil Test Analyses With Plant Response Data*. International Soil Testing Series, Technical Bulletin No. 1. North Carolina State University Agricultural Experiment Station, Raleigh.

Chaves, J. C. D., M. A. Pavan, and M. Miyazawa. 1988. Reduço da acidez subsuperficial em coluna de solo. *Pesquisa Agropecuária Brasileira* 23:469-476.

Chew, W. Y., C. N. Williams, K. T. Joseph, and K. Ramli. 1976. Studies on the availability to plants of soil nitrogen in Malaysian tropical oligotrophic peat. I — Effect of liming and pH. *Tropical Agriculture* 53:69-78.

Coleman, N. T., and D. Craig. 1961. The spontaneous alteration of hydrogen clay. *Soil Science* 91:14-18.

Courchesne, F., and W. H. Hendershot. 1990. The role of basic aluminum sulfate minerals in controlling sulfate retention in the mineral horizons of two Spodosols. *Soil Science* 150:571-578.

Coutinho, J. F. 1990. Exchangeable aluminium and root growth of wheat (*Triticum aestivum*) as criteria of lime requirement in acid soils of northeast Portugal. Pp. 435-441. In M. L. van Beusichem (Ed.), *Plant Nutrition — Physiology and Applications*. Proceedings of the Eleventh International Plant Nutrition Colloquium, 30 July-4 August 1989, Wageningen, The Netherlands. Kluwer Academic Publishers, Dordrecht, The Netherlands.

Crowther, E. M. 1925. Studies on soil reaction. III. The determination of the hydrogen ion concentration of soil suspensions by means of the hydrogen electrode. *Journal of Agricultural Science (Cambridge)* 15:201-221.

Cumming, J. R., and L. H. Weinstein. 1990. Aluminum-mycorrhizal interactions in the physiology of pitch pine seedlings. *Plant and Soil* 125:7-18.

Curtin, D., and G. W. Smillie. 1983. Soil solution composition as affected by liming and incubation. *Soil Science Society of America Journal* 47:701-707.

Davis, L. E., R. Turner, and L. D. Whittig. 1962. Some studies of the autotransformation of H-bentonite to Al-bentonite. *Soil Science Society of America Proceedings* 26:441-443.

DeTurk, E. E. 1938. *Properties of Illinois Soils Which are Related to Their Need for Limestone and Factors Controlling the Effectiveness of Limestone*. Illinois State Geological Survey, Circular No. 23:191-203. Urbana.

Doak, B. W. 1941. Effect of fineness of grinding and hardness of limestones on their efficacy as pasture top-dressing. I. Analytical data. *New Zealand Journal of Science and Technology* 23A:24A-30A.

Dobbs, A. J., P. French, A. M. Gunn, D. T. E. Hunt, and D. A. Winnard. 1989. Aluminum speciation and toxicity in upland waters. Pp. 209-228. In T. E. Lewis (Ed.), *Environmental Chemistry and Toxicology of Aluminum*. Lewis Publishers, Chelsea, MI.

Doss, B. D., and Z. F. Lund. 1975. Subsoil pH effects on growth and yield of cotton. *Agronomy Journal* 67:193-196.

Dunn, L. E. 1943. Lime-requirement determination of soils by means of titration curves. *Soil Science* 56:341-351.

Edmeades, D. C. 1981. Lime research by the Ministry of Agriculture and Fisheries. *New Zealand Agricultural Science* 15:192-199.

Edmeades, D. C., R. M. Pringle, G. P. Mansell, P. W. Shannon, J. Ritchie, and K. M. Stewart. 1985a. Effects of lime on pasture production on soils in the North Island of New Zealand 5. Description of a lime recommendation scheme. *New Zealand Journal of Experimental Agriculture* 13:47-58.

Edmeades, D. C., R. M. Pringle, P. W. Shannon, and G. P. Mansell. 1984. Effects of lime on pasture production on soils in the North Island of New Zealand 4. Predicting lime responses. *New Zealand Journal of Agricultural Research* 27:371-382.

Edmeades, D. C., D. M. Wheeler, F. P. C. Blamey, and R. A. Christie. 1991. Calcium and magnesium amelioration of aluminium toxicity in Al-sensitive and Al-tolerant wheat. Pp. 755-761. In R. J. Wright, V. C. Baligar, and R. P. Murrmann (Eds.), *Plant-Soil Interactions at Low pH*. Kluwer Academic Publishers, Dordrecht, The Netherlands.

Edmeades, D. C., D. M. Wheeler, and O. E. Clinton. 1985b. The chemical composition and ionic strength of soil solutions from New Zealand topsoils. *Australian Journal of Soil Research* 23:151-165.

Edmeades, D. C., D. M. Wheeler, and J. E. Waller. 1985c. Comparison of methods for determining lime requirements of New Zealand soils. *New Zealand Journal of Agricultural Research* 28:93-100.

Elphick, B. L. 1955a. Studies in use of agricultural limestone II. Solubility of limestone in acid soil as influenced by particle size. *New Zealand Journal of Science and Technology* 37A:156-173.

Elphick, B. L. 1955b. Studies in use of agricultural limestone III. Solubility of limestone in acid soil as influenced by type of stone and particle size. *New Zealand Journal of Science and Technology* 37A:241-258.

Evans, C. E., and E. J. Kamprath. 1970. Lime response as related to percent Al saturation, solution Al, and organic matter content. *Soil Science Society of America Proceedings* 34:893-896.

Ewing, M. A., and A. D. Robson. 1990. The effect of solution pH and external calcium concentration on the early growth and nodulation of several annual *Medicago* species. *Australian Journal of Agricultural Research* 41:933-939.

Farina, M. P. W., and P. Channon. 1991. A field comparison of lime requirement indices for maize. Pp. 465-473. In R. J. Wright, V. C. Baligar, and R. P. Murrmann (Eds.), *Plant-Soil Interactions at Low pH*. Kluwer Academic Publishers, Dordrecht, The Netherlands.

Farina, M. P. W., P. Channon, and M. E. Sumner. 1982. Lime induced yield depressions in maize (*Zea mays* L.) on highly weathered soils. Pp. 162-167. In A. Scaife (Ed.), *Plant Nutrition 1982*. Proceedings of the Ninth International Plant Nutrition Colloquium, Warwick University, England, August 22-27, 1982. Volume 1. Commonwealth Agricultural Bureaux, Slough, U.K.

Farina, M. P. W., M. E. Sumner, C. O. Plank, and W. S. Letzsch. 1980a. Aluminum toxicity in corn at near neutral soil pH levels. *Journal of Plant Nutrition* 2:683-697.

Farina, M. P. W., M. E. Sumner, C. O. Plank, and W. S. Letzsch. 1980b. Exchangeable aluminum and pH as indicators of lime requirement for corn. *Soil Science Society of America Journal* 44:1036-1041.

Foy, C. D. 1983. Plant adaptation to mineral stress in problem soils. *Iowa State Journal of Research* 57:339-354.

Foy, C. D. 1988. Plant adaptation to acid, aluminum-toxic soils. *Communications in Soil Science and Plant Analysis* 19:959-987.

Friesen, D. K., A. S. R. Juo, and M. H. Miller. 1980a. Liming and lime-phosphorus-zinc interactions in two Nigerian Ultisols: I. Interactions in the soil. *Soil Science Society of America Journal* 44:1221-1226.

Friesen, D. K., A. S. R. Juo, and M. H. Miller. 1982. Residual value of lime and leaching of calcium in a kaolinitic ultisol in the high rainfall tropics. *Soil Science Society of America Journal* 46:1184-1189.

Friesen, D. K., M. H. Miller, and A. S. R. Juo. 1980b. Liming and lime-phosphorus-zinc interactions in two Nigerian Ultisols: II. Effects on maize root and shoot growth. *Soil Science Society of America Journal* 44:1227-1232.

Gasser, J. K. R. 1973. An assessment of the importance of some factors causing loss of lime from agricultural soils. *Experimental Husbandry* 25:86-95. (Cited by Hochman et al., 1989.)

Gillman, G. P., and E. A. Sumpter. 1985. KCl-extractable aluminium in highly weathered soils. Is it exchangeable? *Communications in Soil Science and Plant Analysis* 16:561-568.

Gillman, G. P., and E. A. Sumpter. 1986. Surface charge characteristics and lime requirements of soils derived from basaltic, granitic, and metamorphic rocks in high-rainfall tropical Queensland. *Australian Journal of Soil Research* 24:173-192.

Goulding, K. W. T., S. P. McGrath, and A. E. Johnston. 1989. Predicting the lime requirement of soils under permanent grassland and arable crops. *Soil Use and Management* 5:54-58.

Grimme, H. 1982. The effect of Al on Mg uptake and yield of oats. Pp. 198-203. In A. Scaife (Ed.), *Plant Nutrition 1982*. Proceedings of the Ninth International Plant Nutrition Colloquium. Volume 1. Commonwealth Agricultural Bureaux, Slough, U.K.

Gupta, R. K., R. D. Singh, R. N. Rai, and R. N. Prasad. 1989. Effect of lime and micronutrients on yield and uptake of micronutrients by wheat on acid soil. *Journal of the Indian Society of Soil Science* 37:70-73.

Hall, H. H. 1983. Economic evaluation of crop response to lime. *American Journal of Agricultural Economics* 65:811-817.

Hardy, D. R., C. D. Raper, Jr., and G. S. Miner. 1990. Chemical restrictions of roots in Ultisol subsoils lessened by long-term management. *Soil Science Society of America Proceedings* 54:1657-1660.

Hargrove, W. L. 1986. The solubility of aluminum-organic matter and its implication in plant uptake of aluminum. *Soil Science* 142:179-181.

Hargrove, W. L., and G. W. Thomas. 1982. Titration properties of Al-organic matter. *Soil Science* 134:216-225.

Harward, M. E., and N. T. Coleman. 1954. Some properties of H- and Al-clays and exchange resins. *Soil Science* 78:181-188.

Haynes, R. J., and R. S. Swift. 1989. Effect of rewetting air-dried soils on pH and accumulation of mineral nitrogen. *Journal of Soil Science* 40:341-347.

Helling, C. S., G. Chesters, and R. B. Corey. 1964. Contribution of organic matter and clay to soil cation-exchange capacity as affected by the pH of the saturating solution. *Soil Science Society of America Proceedings* 28:517-520.

Hester, J. B., and F. A. Shelton. 1933. Seasonal variation of pH in field soils as a factor in making lime recommendations. *Journal of the American Society of Agronomy* 25:299-300.

Hetherington, S. J., C. J. Asher, and F. P. C. Blamey. 1986. Tolerance of sugarcane to Al in soil and solution culture. *Proceedings of the Australian Society of Sugar Cane Technologists, 1986 Conference*:63-68. [Cited by Hetherington et al. (1988).]

Hetherington, S. J., C. J. Asher, and F. P. C. Blamey. 1988. Comparative tolerance of sugarcane, navybean, soybean and maize to aluminium toxicity. *Australian Journal of Agricultural Research* 39:171-176.

Hochman, Z., D. C. Edmeades, and E. White. 1992. Changes in effective cation exchange capacity and exchangeable aluminium with soil pH in lime-amended field soils. *Australian Journal of Soil Research* 43: In Press.

Hochman, Z., D. L. Godyn, and B. J. Scott. 1989. The integration of data on lime use by modelling. Pp. 265-301. In A. D. Robson (Ed.), *Soil Acidity and Plant Growth*. Academic Press, Sydney.

Hue, N. V., G. R. Craddock, and F. Adams. 1986. Effect of organic acids on aluminum toxicity in subsoils. *Soil Science Society of America Journal* 50:28-34.

Jackson, M. L. 1963. Aluminum bonding in soils: a unifying principle in soil science. *Soil Science Society of America Proceedings* 27:1-10.

Kamprath, E. J. 1970. Exchangeable aluminum as a criterion for liming leached mineral soils. *Soil Science Society of America Proceedings* 34:252-254.

Kamprath, E. J. 1984. Crop response to lime on soils in the tropics. Pp. 349-368. In F. Adams (Ed.), *Soil Acidity and Liming*. Second Edition. Agronomy 12. American Society of Agronomy, Crop Science Society of America, and Soil Science Society of America, Madison, WI.

Kappen, H. 1929. *Die Bodenazidität*. Julius Springer, Berlin.

Kauffman, M. D., and E. H. Gardner. 1978. Segmental liming of soil and its effect on the growth of wheat. *Agronomy Journal* 70:331-336.

Keltjens, W. G., and W. J. Dijkstra. 1991. The role of magnesium and calcium in alleviating aluminium toxicity in wheat plants. Pp. 763-768. In R. J. Wright, V. C. Baligar, and R. P. Murrmann (Eds.), *Plant-Soil Interactions at Low pH*. Kluwer Academic Publishers, Dordrecht, The Netherlands.

Kinraide, T. B., and D. R. Parker. 1987. Non-phytotoxicity of the aluminum sulfate ion, $AlSO_4^+$. *Physiologia Plantarum* 71:207-212.

Kirchoff, G., J. Blackwell, and R. E. Smart. 1991. Growth of vineyard roots into segmentally ameliorated acidic subsoils. Pp. 447-452. In R. J. Wright, V. C. Baligar, and R. P. Murrmann (Eds.), *Plant-Soil Interactions at Low pH*. Kluwer Academic Publishers, Dordrecht, The Netherlands.

Konishi, S. 1990. Stimulatory effects of aluminium on tea plant growth. *Transactions 14th International Congress of Soil Science* IV:164-169.

Lehr, J. J. 1950. Seasonal variations in the pH value of the soil, as influenced by nitrification. *Transactions of the Fourth International Congress of Soil Science* 2:155-157.

Lewis, T. E. (Ed.). 1989. *Environmental Chemistry and Toxicology of Aluminum*. Lewis Publishers, Chelsea, MI.

Licrop, W. van. 1990. Soil pH and lime requirement determination. Pp. 73-126. In R. L. Westerman (Ed.), *Soil Testing and Plant Analysis*. Third Edition. Soil Science Society of America, Madison, WI.

Lin, C., and N. T. Coleman. 1960. The measurement of exchangeable aluminum in soils and clays. *Soil Science Society of America Proceedings* 24:444-446.

Loynachan, T. E. 1981. Lime requirement methods for cold-region soils. *Soil Science Society of America Journal* 45:75-80.

Magdoff, F. R., and R. J. Bartlett. 1985. Soil pH buffering revisited. *Soil Science Society of America Journal* 49:145-148.

Mahler, R. L., and R. E. McDole. 1987. Effect of soil pH on crop yield in northern Idaho. *Agronomy Journal* 79:751-755.

McCray, J. M., and M. E. Sumner. 1990. Assessing and modifying Ca and Al levels in acid subsoils. *Advances in Soil Science* 14:45-75.

McLean, E. O. 1975. Recommended pH and lime requirement tests. Pp. 6-9. In *Recommended Chemical Soil Test Procedures for the North Central Region*. North Dakota Agricultural Experiment Station, Bulletin 499 (North Central Regional Publication No. 221). Fargo.

McLean, E. O. 1982. Soil pH and lime requirement. Pp. 199-224. In A. L. Page, R. H. Miller, and D. R. Keeney (Eds.), *Methods of Soil Analysis. Part 2. Chemical and Microbiological Properties.* Second Edition. Agronomy Number 9 (Part 2). American Society of Agronomy and Soil Science Society of America, Madison, WI.

McLean, E. O., D. J. Eckert, G. Y. Reddy, and J. F. Trierweiler. 1978. An improved SMP soil lime requirement method incorporating double-buffer and quick-test features. *Soil Science Society of America Journal* 42:311-316.

McLean, E. O., W. R. Hourigan, H. E. Shoemaker, and D. R. Bhumbla. 1964. Aluminum in soils: V. Form of aluminum as a cause of soil acidity and a complication in its measurement. *Soil Science* 97:119-126.

McLean, E. O., H. E. Shoemaker, and W. R. Hourigan. 1961. Some effects of aluminum on lime requirement tests of soils. *Transactions of 7th International Congress of Soil Science* II:142-151.

McLean, E. O., J. F. Trierweiler, and D. J. Eckert. 1977. Improved SMP buffer method for determining lime requirements of acid soils. *Communications in Soil Science and Plant Analysis* 8:667-675.

Mehlich, A. 1948. Determination of cation- and anion-exchange properties of soils. *Soil Science* 66:429-445.

Mehlich, A. 1961. Charge characterization of soils. *Transactions of 7th International Congress of Soil Science* II:292-302.

Mehlich, A. 1964. Influence of sorbed hydroxyl and sulfate on neutralization of soil acidity. *Soil Science Society of America Proceedings* 28:492-496.

Mehlich, A. 1976. New buffer pH method for rapid estimation of exchangeable acidity and lime requirement of soils. *Communications in Soil Science and Plant Analysis* 7:637-652.

Moschler, W. W., G. D. Jones, and G. W. Thomas. 1960. Lime and soil acidity effects on alfalfa growth in a red-yellow podzolic soil. *Soil Science Society of America Proceedings* 24:507-509.

Noble, A. D., M. V. Fey, and M. E. Sumner. 1988. Calcium-aluminum balance and the growth of soybean roots in nutrient solutions. *Soil Science Society of America Journal* 52:1651-1656.

Nye, P. H., and A. Ameloko. 1986. A comparison of measured and theoretical soil acidity diffusion coefficients over a wide range of pH. *Journal of Soil Science* 37:191-196.

Nye, P. H., and A. Y. Ameloko. 1987. Predicting the rate of dissolution of lime in soil. *Journal of Soil Science* 38:641-649.

O'Brien, L. A., and M. E. Sumner. 1988. Effects of phosphogypsum on leachate and soil chemical composition. *Communications in Soil Science and Plant Analysis* 19:1319-1329.

Ogus, L., and R. L. Fox. 1970. Nitrogen recovery from a soil profile by *Bromus inermis*. *Agronomy Journal* 62:69-71.

Ohio Agricultural Experiment Station. 1938. *Handbook of Experiments in Agronomy.* Ohio Agricultural Experiment Station, Special Circular No. 53. Wooster.

Okajima, H. 1977. Reclamation and improvement of soils derived from volcanic ash. Pp. 64-84. In Y. Ishizuka and C. A. Black (Eds.), *Soils Derived From Volcanic Ash in Japan.* Centro Internacional de Mejoramiento de Maiz y Trigo, Mexico 6, D.F. Mexico.

O'Shay, T., L. R. Hossner, and J. B. Dixon. 1990. A modified hydrogen peroxide method for determination of potential acidity in pyritic overburden. *Journal of Environmental Quality* 19:778-782.

Parker, D. R., T. B. Kinraide, and L. W. Zelazny. 1989a. On the phytotoxicity of polynuclear hydroxy-aluminum complexes. *Soil Science Society of America Journal* 53:789-796.

Parker, D. R., L. W. Zelazny, and T. B. Kinraide. 1989b. Chemical speciation and plant toxicity of aqueous aluminum. Pp. 117-145. In T. E. Lewis (Ed.), *Environmental Chemistry and Toxicology of Aluminum*. Lewis Publishers, Chelsea, MI.

Parker, M. B., and H. B. Harris. 1962. Soybean response to molybdenum and lime and the relationship between yield and chemical composition. *Agronomy Journal* 54:480-483.

Patiram, R. N. Rai, and R. N. Prasad. 1990. Effect of liming on aluminium and yield of wheat in acidic soils. *Journal of the Indian Society of Soil Science* 38:719-722.

Pavan, M. A., F. T. Bingham, and P. F. Pratt. 1982. Toxicity of aluminum to coffee in Ultisols and Oxisols amended with $CaCO_3$, $MgCO_3$, and $CaSO_4 \cdot 2H_2O$. *Soil Science Society of America Journal* 46:1201-1207.

Pearson, R. W., and F. Abruña. 1961. Nitrogen speeds lime movement in subsoil. *Crops and Soils* 13, No. 9:23.

Pearson, R. W., J. Childs, and Z. F. Lund. 1973. Uniformity of limestone mixing in acid subsoil as a factor in cotton root penetration. *Soil Science Society of America Proceedings* 37:727-732.

Pierre, W. H., J. Meisinger, and J. R. Birchett. 1970. Cation-anion balance in crops as a factor in determining the effect of nitrogen fertilizers on soil acidity. *Agronomy Journal* 62:106-112.

Pinkerton, A., and J. R. Simpson. 1986. Responses of some crop plants to correction of subsoil acidity. *Australian Journal of Experimental Agriculture* 26:107-113.

Pratt, P. F., and F. L. Bair. 1962. Cation-exchange properties of some acid soils of California. *Hilgardia* 33:689-706.

Reeve, N. G., and M. E. Sumner. 1970. Lime requirements of Natal Oxisols based on exchangeable aluminum. *Soil Science Society of America Proceedings* 34:595-598.

Reeve, N. G., and M. E. Sumner. 1972. Amelioration of subsoil acidity in Natal Oxisols by leaching of surface-applied amendments. *Agrochemophysica* 4:1-6.

Ridley, A. M., K. R. Helyar, and W. J. Slattery. 1990. Soil acidification under subterranean clover (*Trifolium subterraneum* L.) pastures in north-eastern Victoria. *Australian Journal of Experimental Agriculture* 30:195-201.

Robson, A. D. (Ed.). 1989. *Soil Acidity and Plant Growth*. Academic Press, San Diego.

Robson, A. D., and L. K. Abbott. 1989. The effect of soil acidity on microbial activity in soils. Pp. 139-165. In A. D. Robson (Ed.), *Soil Acidity and Plant Growth*. Academic Press, San Diego.

Rowell, D. L. 1988. Soil acidity and alkalinity. Pp. 844-898. In A. Wild (Ed.), *Russell's Soil Conditions and Plant Growth*. Eleventh Edition. Longman Scientific and Technical, Essex, England.

Schmitthenner, A. F., and C. H. Canaday. 1983. Role of chemical factors in development of *Phytophthora* diseases. Pp. 189-196. In D. C. Erwin, S. Bartnicki-Garcia, and P. H. Tsao (Eds.), *Phytophthora Its Biology, Taxonomy, Ecology and Pathology*. American Phytopathological Society, St. Paul.

Schofield, R. K. 1933. Rapid methods of examining soils. II. The use of *p*-nitrophenol for assessing lime status. *Journal of Agricultural Science* 23:252-254.

Schofield, R. K., and A. W. Taylor. 1955. The measurement of soil pH. *Soil Science Society of America Proceedings* 19:164-167.

Scott, B. J., and J. A. Fisher. 1989. Selection of genotypes tolerant of aluminium and manganese. Pp. 167-203. In A. D. Robson (Ed.), *Soil Acidity and Plant Growth.* Academic Press, San Diego.

Semple, E. C. 1928. Ancient Mediterranean agriculture Part II. Manuring and seed selection. *Agricultural History* 2:129-156.

Shaw, W. M. 1952. Interrelationship between calcium sorption, exchangeable hydrogen and pH values of certain soils and subsoils. *Journal of the Association of Official Agricultural Chemists* 35:597-621.

Shaw, W. M. 1953. Reaction of calcium carbonate with soils and determination of their calcium sorption capacities. *Journal of the Association of Official Agricultural Chemists* 36:421-444.

Shaw, W. M., and W. H. MacIntire. 1943. Steam distillation — A new procedure for the determination of carbonate CO_2. *Journal of the Association of Official Agricultural Chemists* 26:357-396.

Shoemaker, H. E., E. O. McLean, and P. F. Pratt. 1961. Buffer methods for determining lime requirement of soils with appreciable amounts of extractable aluminum. *Soil Science Society of America Proceedings* 25:274-277.

Shuman, L. M. 1990. Comparison of exchangeable Al, extractable Al, and Al in soil fractions. *Canadian Journal of Soil Science* 70:263-275.

Simán, G. 1989. Effect of lime on soil and crop in Swedish field experiments. *Agrokémia és Talajtan* 38:765-774.

Singer, A., and P. M. Huang. 1990. Effects of humic acid on the crystallization of aluminum hydroxides. *Clays and Clay Minerals* 38:47-52.

Singh, L., U. R. Pal, and Y. Arora. 1987. Direct and residual effect of liming on yield and nutrient uptake of maize (*Zea mays* L.) in moderately acid soils in the savanna zone of Nigeria. *Fertilizer Research* 12:11-20.

Sposito, G. (Ed.). 1989. *The Environmental Chemistry of Aluminum.* CRC Press, Boca Raton, FL.

Spurway, C. H. 1941. *Soil Reaction (pH) Preferences of Plants.* Michigan Agricultural Experiment Station, Special Bulletin 306. East Lansing.

Sumner, M. E., J. H. Bouton, J. Hammel, and H. Shahandeh. 1985. Enhanced alfalfa production through profile modification in the southeastern U.S. Pp. 240-243. In *Forages: The Keystone of Agriculture.* Proceedings of the 1985 Forage and Grassland Conference. American Forage and Grassland Council, Lexington, Kentucky.

Sumner, M. E., and E. Carter. 1988. Amelioration of subsoil acidity. *Communications in Soil Science and Plant Analysis* 19:1309-1318.

Tanaka, A., T. Tadano, K. Yamamoto, and N. Kanamura. 1987. Comparison of toxicity to plants among Al^{3+}, $AlSO_4{}^+$, and Al-F complex ions. *Soil Science and Plant Nutrition* 33:43-55.

Teräsvuori, A. 1930. Über die Bodenazidität mit besonderer Berücksichtigung des Elektrolytgehaltes der Bodenaufschlämmungen. *Valtion Maatalouskoetoiminnan, Julkaisuja* N:o 29.

Thomas, G. W., and W. L. Hargrove. 1984. The chemistry of soil acidity. Pp. 3-56. In F. Adams (Ed.), *Soil Acidity and Liming.* Second Edition. Agronomy Number 12. American Society of Agronomy, Crop Science Society of America, and Soil Science Society of America, Madison, WI.

Thompson, L. M. 1950. *The Mineralization of Organic Phosphorus, Nitrogen and Carbon in Virgin and Cultivated Soils*. Ph.D. Thesis, Iowa State College, Ames.

Van Beers, W. F. J. 1962. *Acid Sulphate Soils*. International Institute of Land Reclamation and Improvement, Bulletin 3.

Van der Watt, H. v. H., R. O. Barnard, I. J. Cronje, J. Dekker, G. J. B. Croft, and M. M. van der Walt. 1991. Amelioration of subsoil acidity by application of a coal-derived calcium fulvate to the soil surface. *Nature* 350:146-148.

Van Lierop, W. 1990. Soil pH and lime requirement determination. Pp. 73-126. In R. L. Westerman (Ed.), *Soil Testing and Plant Analysis*. Third Edition. Soil Science Society of America, Madison, WI.

Veitch, F. P. 1904. Comparison of methods for the estimation of soil acidity. *Journal of the American Chemical Society* 26:637-662.

Walker, T. W. 1952. The estimation of the lime requirements of soils. Soil and crop yield data from the Harper Adams and other liming experiments. *Journal of Soil Science* 3:261-276.

Walker, W. J., C. S. Cronan, and P. R. Bloom. 1990. Aluminum solubility in organic soil horizons from northern and southern forested watersheds. *Soil Science Society of America Journal* 54:369-374.

Warfvinge, P., and H. Sverdrup. 1989. Modeling limestone dissolution in soils. *Soil Science Society of America Journal* 53:44-51.

Watt, H. v. H. van der, R. O. Barnard, I. J. Cronje, J. Dekker, G. J. B. Croft, and M. M. van der Walt. 1991. Amelioration of subsoil acidity by application of a coal-derived calcium fulvate to the soil surface. *Nature* 350:146-148.

Wheeler, H. J., B. L. Hartwell, and C. L. Sargent. 1900. Chemical methods for ascertaining the lime requirement of soils. *Journal of the American Chemical Society* 22:153-157.

White, R. E. 1969. On the measurement of soil pH. *Journal of the Australian Institute of Agricultural Science* 35:3-14.

Wolfe, M. H., and J. D. Joslin. 1989. Honeylocust (*Gleditsia triancanthos* L.) root response to aluminium and calcium. *Plant and Soil* 119:181-185.

Wright, R. J., V. C. Baligar, and R. P. Murrmann (Eds.). 1991. *Plant-Soil Interactions at Low pH*. Kluwer Academic Publishers, Dordrecht, The Netherlands.

Wright, R. J., V. C. Baligar, K. D. Ritchey, and S. F. Wright. 1989. Influence of soil solution aluminum on root elongation of wheat seedlings. *Plant and Soil* 113:294-298.

Yuan, T. L. 1974. A double buffer method for the determination of lime requirement of acid soils. *Soil Science Society of America Proceedings* 38:437-440.

Appendix

Common and Scientific Names of Plants Referred to in the Text

Alfalfa, *Medicago sativa*
Apple, *Malus* spp.
Barley, *Hordeum vulgare*
Barrel medic, *Medicago truncatula*
Bean, *Phaseolus vulgaris*
Broad bean, *Vicia faba*
Bromegrass, *Bromus inermis*
Browntop, *Agrostis tenuis*
Buckwheat, *Fagopyrum esculentum*
Cabbage, *Brassica oleracea*
Canarygrass, *Phalaris aquatica*
Carrot, *Daucus carota*
Cassava, *Manihot utilissima*
Cauliflower, *Brassica oleracea*
Chick pea, *Cicer arietinum*
Chinese cabbage, *Brassica oleracea*
Citrus, *Citrus* spp.
Cluster bean, *Cyamopsis tetragonoloba*
Coastal bermudagrass, *Cynodon dactylon*
Coffee, *Coffea arabica*
Corn, *Zea mays*
Cotton, *Gossypium hirsutum*
Cowpea, *Vigna unguiculata*
Crested wheatgrass, *Agropyron cristatum*
Dallis grass, *Paspalum dilatatum*
Durum wheat, *Triticum turgidum*
Eucalyptus, *Eucalyptus* spp.
Fababean, *Vicia faba*
Finger millet, *Eleusine coracana*
Flax, *Linum usitatissimum*
Fodder beet, *Beta vulgaris*
Fraser fir, *Abies fraseri*
French bean, *Phaseolus vulgaris*
Garden beet, *Beta vulgaris*
Grain sorghum, *Sorghum bicolor*
Grape, *Vitis vinifera*
Grapefruit, *Citrus paradisi*
Hemp, *Cannabis sativa*
Highbush blueberry, *Vaccinium corymbosum*
Honeylocust, *Gleditsia triancanthos*
Indian mustard, *Brassica juncea*
Italian ryegrass, *Lolium multiflorum*

Jack pine, *Pinus banksiana*
Kidney bean, *Phaseolus vulgaris*
Kiwifruit, *Actinidia deliciosa*
Ladino clover, *Trifolium repens*
Leek, *Allium ampeloprasum*
Lentil, *Lens culinaris*
Lettuce, *Lactuca sativa*
Lima bean, *Phaseolus limensis*
Lodgepole pine, *Pinus contorta*
Lupine, *Lupinus* sp.
Medic, *Medicago truncatula*
Millet, *Pennisetum americanum*
Monterey pine, *Pinus radiata*
Mung bean, *Vigna radiata*
Mustard, *Brassica hirta, Brassica juncea*
Napier grass, *Pennisetum purpureum*
Navy bean, *Phaseolus vulgaris*
Oat, *Avena sativa*
Oil palm, *Elais guineensis*
Oilseed rape, *Brassica napus*
Onion, *Allium cepa*
Orange, *Citrus sinensis*
Panic grass, *Panicum* spp.
Parsnip, *Pastinaca sativa*
Pea, *Pisum sativum*
Peach, *Prunus persica*
Peanut, *Arachis hypogaea*
Pear, *Pyrus communis*
Perennial ryegrass, *Lolium perenne*
Pigeon pea, *Cajanus cajan*
Pineapple, *Ananas comosus*
Pitch pine, *Pinus rigida*
Poplar, *Populus* spp.
Potato, *Solanum tuberosum*
Proso millet, *Panicum miliaceum*
Pumpkin, *Cucurbita* spp.
Radish, *Raphanus sativus*
Rape, *Brassica napus*
Rapeseed, *Brassica campestris*
Red clover, *Trifolium pratense*
Red oak, *Quercus coccinea*
Rice, *Oryza sativa*
Rubber tree, *Hevea* spp.

Rye, *Secale cereale*
Ryegrass, *Lolium rigidum, Lolium multiflorum*
Scots pine, *Pinus sylvestris*
Setaria, *Setaria sphacelata*
Siratro, *Macroptilium atropurpureum*
Slender serradella, *Ornithopus pinnatus*
Sorghum, *Sorghum vulgare*
Sorghum-sudangrass hybrid, *Sorghum sudanense*
Sorrel, *Rumex acetosella*
Soybean, *Glycine hispida*
Spinach, *Spinacea oleracea*
Squash, *Cucurbita spp.*
Subterranean clover, *Trifolium subterraneum*
Sudangrass, *Sorghum vulgare*
Sugarbeet, *Beta vulgaris*
Sugarcane, *Saccharum officinarum*
Sunflower, *Helianthus annuus*

Swede, *Brassica napus*
Sweet orange, *Citrus sinensis*
Sweetclover, *Melilotus officinalis*
Tall fescue, *Festuca arundinacea*
Timothy, *Phleum pratense*
Tobacco, *Nicotiana tabacum*
Tomato, *Lycopersicon esculentum*
Triticale, *X Triticosecale* spp.
Trudan, *Sorghum sudanense*
Tufted hairgrass, *Deschampsia caespitosa*
Turnip, *Brassica rapa*
Valencia orange, *Citrus sinensis*
Vetch, *Vicia benghalensis, Vicia sativa*
Weeping lovegrass, *Eragrostis curvula*
Wheat, *Triticum aestivum*
White clover, *Trifolium repens*
Wild oat, *Avena fatua*
Winter bean, *Vicia faba*
Yellow mustard, *Brassica hirta*

Index